# Physical Principles and Techniques of Protein Chemistry

PART C

## MOLECULAR BIOLOGY

An International Series of Monographs and Textbooks

Editors: BERNARD HORECKER, NATHAN O. KAPLAN, JULIUS MARMUR, AND HAROLD A. SCHERAGA

A complete list of titles in this series appears at the end of this volume.

# PHYSICAL PRINCIPLES AND TECHNIQUES OF PROTEIN CHEMISTRY

PART C

Edited by SYDNEY J. LEACH
SCHOOL OF BIOCHEMISTRY
UNIVERSITY OF MELBOURNE
PARKVILLE, VICTORIA, AUSTRALIA

1973

ACADEMIC PRESS   New York and London
A Subsidiary of Harcourt Brace Jovanovich, Publishers

Copyright © 1973, by Academic Press, Inc.
ALL RIGHTS RESERVED.
NO PART OF THIS PUBLICATION MAY BE REPRODUCED OR
TRANSMITTED IN ANY FORM OR BY ANY MEANS, ELECTRONIC
OR MECHANICAL, INCLUDING PHOTOCOPY, RECORDING, OR ANY
INFORMATION STORAGE AND RETRIEVAL SYSTEM, WITHOUT
PERMISSION IN WRITING FROM THE PUBLISHER.

ACADEMIC PRESS, INC.
111 Fifth Avenue, New York, New York 10003

*United Kingdom Edition published by*
ACADEMIC PRESS, INC. (LONDON) LTD.
24/28 Oval Road, London NW1

**Library of Congress Cataloging in Publication Data**

Leach, Sydney J
 Physical principles and techniques of protein chemistry.

 (Molecular biology: an international series of monographs and textbooks)
 Includes bibliographical references.
 1. Proteins.  I. Title.  II. Series.
[DNLM:  1.  Chemistry, Physical.  2.  Proteins.
QU 55 L434p]
QD431.L4   547'.75   68–23488
ISBN 0–12–440103–1

PRINTED IN THE UNITED STATES OF AMERICA

# Contents

| | |
|---|---|
| LIST OF CONTRIBUTORS | ix |
| PREFACE | xi |
| CONTENTS OF OTHER PARTS | xiii |

## 17. Density and Volume Change Measurements
### D. W. Kupke

| | |
|---|---|
| Glossary of Symbols | 1 |
| I. Introduction | 2 |
| II. Definitions and Basic Considerations | 3 |
| III. Direct Volume Change | 22 |
| IV. Density | 32 |
| Appendix A: Derivation of the Partial Specific Volume from the Density | 62 |
| Appendix B: Splitting-Up of the Parameter: Exclusion of Diffusible Components | 66 |
| References | 72 |

## 18. Osmotic Pressure
### M. J. Kelly and D. W. Kupke

| | |
|---|---|
| Glossary of Symbols | 77 |
| I. Introduction and Phenomenon | 78 |
| II. Elementary Theory | 87 |
| III. Applications | 102 |
| IV. Techniques | 121 |
| References | 137 |

## 19. Small-Angle X-ray Scattering
### Ingrid Pilz

| | |
|---|---|
| Glossary of Symbols | 141 |
| I. Introduction | 142 |

II. Remarks on the Theory of Particulate Scattering . . . . . 143
III. Remarks on the Experimental Technique . . . . . . . 168
IV. Selected Examples of Applications . . . . . . . . 188
V. Prospects . . . . . . . . . . . . . . . 238
References . . . . . . . . . . . . . . . 239

## 20. Pulsed Nuclear Magnetic Resonance

*W. J. O'Sullivan, K. H. Marsden, and J. S. Leigh, Jr.*

Glossary of Symbols . . . . . . . . . . . . . 246
I. Introduction . . . . . . . . . . . . . . 247
II. Scope of the Technique . . . . . . . . . . . 247
III. Theory of Relaxation . . . . . . . . . . . . 253
IV. Experimental Methods . . . . . . . . . . . 267
V. Experimental Design . . . . . . . . . . . . 273
VI. Individual Enzymes . . . . . . . . . . . . 286
VII. Absorption of Water by Diamagnetic Molecules . . . . . 294
References . . . . . . . . . . . . . . . 298

## 21. The Use of Least Squares in Data Analysis

*R. D. B. Fraser and E. Suzuki*

Glossary of Symbols . . . . . . . . . . . . . 301
I. Introduction . . . . . . . . . . . . . . 302
II. Outline of Theory . . . . . . . . . . . . 304
III. The Iteration Process . . . . . . . . . . . . 309
IV. Statistical Aspects . . . . . . . . . . . . 316
V. Model Functions . . . . . . . . . . . . . 321
VI. Computational Procedure . . . . . . . . . . 331
VII. Applications . . . . . . . . . . . . . . 352
References . . . . . . . . . . . . . . . 353

## 22. Optical Rotatory Dispersion and the Main Chain Conformation of Proteins

*Kazutomo Imahori and N. A. Nicola*

Glossary of Symbols . . . . . . . . . . . . . 358
I. Introduction . . . . . . . . . . . . . . 360
II. The Basic Relations for Optically Active Molecules . . . . 362
III. Visible Rotatory Dispersion . . . . . . . . . . 371
IV. Peptide Cotton Effects . . . . . . . . . . . . 394
V. Cotton Effects Due to Side-Chain Chromophores . . . . . 412

|        |                                                      |     |
|--------|------------------------------------------------------|-----|
| VI.    | Other Secondary Structures                           | 421 |
| VII.   | Conformational Transitions                           | 427 |
| VIII.  | Experimental Considerations                          | 430 |
| IX.    | The Relative Advantages of ORD and CD                | 436 |
|        | Appendix: Table I                                    | 439 |
|        | References                                           | 440 |

## 23. Circular Dichroism

*Duane W. Sears and Sherman Beychok*

|        |                                                              |     |
|--------|--------------------------------------------------------------|-----|
|        | Glossary                                                     | 446 |
| I.     | Introduction                                                 | 447 |
| II.    | Theory of Optical Activity and Its Applications              | 449 |
| III.   | Secondary Structure of Proteins                              | 533 |
| IV.    | Side-Chain Optical Activity in Model Compounds and Proteins  | 541 |
| V.     | Selected Proteins                                            | 554 |
| VI.    | Concluding Remarks                                           | 583 |
|        | Appendix: Electromagnetic Units                              | 583 |
|        | References                                                   | 585 |

AUTHOR INDEX . . . . . . . . . . . . . . 595

SUBJECT INDEX . . . . . . . . . . . . . 609

# List of Contributors

Numbers in parentheses indicate the pages on which the authors' contributions begin.

SHERMAN BEYCHOK, Department of Biological Sciences, Columbia University, New York, New York (445)

R. D. B. FRASER, Division of Protein Chemistry, CSIRO, Parkville, Victoria, Australia (301)

KAZUTOMO IMAHORI, Department of Agricultural Chemistry, University of Tokyo, Tokyo, Japan (357)

M. J. KELLY,* Department of Biochemistry, School of Medicine, University of Virginia, Charlottesville, Virginia (77)

D. W. KUPKE, Department of Biochemistry, School of Medicine, University of Virginia, Charlottesville, Virginia (1, 77)

J. S. LEIGH, Jr., Johnson Foundation, University of Pennsylvania, Philadelphia, Pennsylvania (245)

K. H. MARSDEN, Department of Physics, University of New South Wales, Kensington, North South Wales, Australia (245)

N. A. NICOLA, School of Biochemistry, University of Melbourne, Parkville, Victoria, Australia (357)

W. J. O'SULLIVAN, Department of Medicine, University of Sydney, North South Wales, Australia (245)

*Present address: Department of Chemistry, Pomona College, Claremont, California.

INGRID PILZ, Institut für Physikalische Chemie, der Univrsität Graz, Universitätsplatz, Graz, Austria (141)

DUANE W. SEARS, Department of Biological Sciences, Columbia University, New York, New York (445)

E. SUZUKI, Division of Protein Chemistry, CSIRO, Parkville, Victoria, Australia (301)

# Preface

Our present understanding of the way in which the effects of intermolecular interactions are transmitted between ligands and proteins and from protein to protein is, to say the least, imperfect. Yet these interactions and their molecular effects are central to our understanding of the many facets of protein function. Certainly there has been a shift from the picture of a rigid protein structure toward one incorporating Linderstrøm–Lang's concept of motility. This shift in thinking owes much to the application of more searching physical methods to protein systems and is reflected in much of the material in the volumes of this treatise. Each technique described within these three volumes will continue to play a major part in filling in the serious deficiencies in our picture of the molecular dynamics of protein action during biological function.

This volume follows the pattern set by Parts A and B of the treatise, the stress varying between the theory, the experimental execution, and the problems of interpreting data according to the current stage of development of each physical technique and its probable familiarity to potential users.

I wish to thank the authors for their patient collaboration and the staff of Academic Press for their continued help at each stage of production.

Sydney J. Leach

# Contents of Other Parts

## PART A

Electron Microscopy of Globular Proteins
   *Elizabeth M. Slayter*

X-Ray Methods
   *R. D. B. Fraser and T. P. MacRae*

Ultraviolet Absorption
   *John W. Donovan*

Fluorescence of Proteins
   *Raymond F. Chen, Harold Edelhoch, and Robert F. Steiner*

Perturbation and Flow Techniques
   *B. H. Havsteen*

Dielectric Properties of Proteins I. Dielectric Relaxation
   *Shiro Takashima*

Dielectric Properties of Proteins II. Electric Birefringence and Dichroism
   *Yoshiro Yoshioka and Hiroshi Watanabe*

Electrophoresis
   *John R. Cann*

Analytical Gel Filtration
   *D. J. Winzor*

Author Index–Subject Index

## PART B

Ultracentrifugal Analysis
   *J. H. Coates*

Viscosity
   *J. H. Bradbury*

Light Scattering
  *Serge N. Timasheff and Robert Townend*

Infrared Methods
  *R. D. B. Fraser and E. Suzuki*

Nuclear Magnetic Resonance Spectroscopy
  *J. C. Metcalfe*

Binding of Protons and Other Ions
  *Frank R. N. Gurd*

Differential Thermal Analysis
  *Hirokazu Morita*

Author Index–Subject Index

# 17 □ Density and Volume Change Measurements

## D. W. KUPKE

| | |
|---|---|
| Glossary of Symbols | 1 |
| I. Introduction | 2 |
| II. Definitions and Basic Considerations | 3 |
|    A. The Volume Property | 3 |
|    B. The Density | 6 |
|    C. The Partial Specific Volume | 8 |
|    D. The Apparent Specific Volume | 10 |
|    E. The Isopotential Specific Volume | 12 |
|    F. Density and Composition | 16 |
| III. Direct Volume Change | 22 |
|    A. Dilatometry | 22 |
|    B. Applications | 25 |
| IV. Density | 32 |
|    A. Methods | 32 |
|    B. Applications | 45 |
|    Appendix A. Derivation of the Partial Specific Volume from the Density | 62 |
|    Appendix B. Splitting-Up of the Parameter $\xi$: Exclusion of Diffusible Components | 66 |
|    References | 72 |

## Glossary of Symbols

| | |
|---|---|
| Single primes | refer to the solvent medium, usually containing diffusible components only; note that $v'$ and $\phi'$ are exceptions |
| Double primes | refer to the undialyzed (isomolal) protein solution |
| c | grams per milliliter of a component |
| $f_j$ | weight fraction of a species $j$ in component $i$ relative to the total component |
| g | grams of a component |
| $g$ | acceleration of gravity |
| $i$ | any component in the solution |
| $j$ | any nonprotein component (usually membrane-diffusible) except where $(j \neq i)$ is specified |

| | |
|---|---|
| $j_e$ | that amount of component $j$ which is in excess relative to dialyzate composition |
| m | molality |
| ml | milliliter |
| n | number of moles |
| $n$ | refractive index |
| v | specific volume (milliliters per gram) of a pure component |
| $\bar{v}$ | partial specific volume (milliliters per gram) |
| v' | isopotential specific volume (milliliters per gram) |
| z, −z | net charge (positive and negative, respectively) |
| $z$ | axial distance from solenoid core |
| E | internal energy |
| G | Gibbs free energy |
| H | enthalpy |
| $H$ | magnetic field intensity |
| $I$ | electric current (amperes) |
| M | molecular weight |
| $M$ | magnetic moment |
| N | total number of components in a macrophase |
| $N$ | number of molecules of a species |
| P | pressure |
| R | gas constant |
| S | entropy |
| T | temperature, degrees Kelvin |
| V | total volume in milliliters |
| $\bar{V}$ | partial molal volume |
| W | weight fraction of a component |
| TYMV | turnip yellow mosaic virus |
| $\alpha$ | polarizability of a species |
| $\delta_j$ | net grams of component $j$ (per gram protein) transported across a semipermeable membrane during dialysis |
| $\mu$ | chemical potential |
| $\mu l$ | microliter |
| $\xi_j$ | grams of component $j$ (per gram protein) included in the definition of the protein component at osmotic equilibrium |
| $\xi^*_j$ | grams of diffusible component $j$ (per gram protein) which excludes any other diffusible component $j$ relative to the composition of the diffusible components in the dialyzate |
| $\rho$ | density (grams per milliliter) |
| $\tau$ | oscillation time; inverse of resonance frequency |
| $\phi$ | apparent specific volume |
| $\phi'$ | isopotential apparent specific volume |
| $\Omega$ | milliliter fraction of protein solution having properties like dialyzate |
| $\Gamma_j$ | moles of component $j$ (per mole protein) included in the definition of the protein component at osmotic equilibrium |

## I. Introduction

Until recently, the potential usefulness of density measurements for the study of proteins in solution was not widely appreciated and the elementary relationships which form a basis for more general application were not often expressed. The quest for values of the partial specific volumes of proteins has provided the principal impetus for determining the density of protein solutions accurately. The application of density to other purposes, as a routine measurement, is quite new. The reason for this limited view on density applications is self-evident; density measurements of sufficient accuracy by the classical methods were time consuming, awkward to perform, severely limited with respect to variation of the common variables, and, with generally available procedures, often required excessive amounts of a purified protein. Because of these difficulties, even the partial specific volume was frequently borrowed from a literature value on a similar preparation, or was approximated by additive procedures if the amino acid composition was known. It is well known, however, that differences in the material presented as the protein component or differences in the nature of the solvent medium may result in significantly different values of the partial specific volume. With newer procedures, it is now possible to apply density measurements much more conveniently to the problems raised by the study of proteins in multicomponent solutions and to follow the change in density as a function of temperature, pressure, composition, and time. We are, in fact, at the threshold of being able to define protein solutions routinely in terms of the partial volumes of all the components present in a system. With a knowledge of the density $\rho$ of a protein solution, the volume V may be obtained, since by definition the density is mass per unit volume. The difference in volume $\Delta V$, such as the volume change on mixing a protein solution with a particular reactant, is sometimes the parameter of principal interest. This change can be measured with an accuracy comparable to that of careful density determinations by the use of dilatometers. Heretofore, this technique was more convenient than the direct measurement of densities. A substantial amount of investigation, therefore, has been carried out on protein solutions by measurements of volume changes, and the dilatometer method continues to be a useful approach and to receive developmental effort.

## II. Definitions and Basic Considerations

### A. THE VOLUME PROPERTY

The total volume V of a liquid mixture in terms of grams g of each component $i$ in a milieu of N components at a specified temperature and pressure is

$$V = \sum_{i=1}^{N} \bar{v}_i g_i \tag{1}$$

where V is in milliliters and $\bar{v}_i$ is the partial specific volume of the $i^{th}$ component. Components are definable substances which can be added quantitatively so that the composition of a mixture in terms of independently added components can always be known. The quantity $\bar{v}$ is defined as the change in total volume per unit mass upon adding an infinitesimal amount of component $i$ at constant temperature T, pressure P, and masses in grams $g_j$ of all other components $j$. Thus

$$\bar{v}_i = \left(\frac{\partial V}{\partial g_i}\right)_{T,P,g_j} \qquad (j \neq i) \qquad (2)$$

Alternatively, one might imagine the finite change in volume after adding a measurable quantity of component $i$ to a very large sea of the solution so that the concentrations of the various components remain essentially unchanged. Partial specific volumes, like other partial specific or partial molal quantities, appear as partial derivative coefficients in the total differential of the extensive (i.e., mass-dependent) property being considered. Thus, for the volume property of a solution, which is a function only of the masses of the components present at a fixed temperature and pressure, that is $V = f(g_1, g_2, \ldots)_{T,P}$, partial differentiation gives

$$dV = \left(\frac{\partial V}{\partial g_1}\right)_{T,P,g_{(N-1)}} dg_1 + \left(\frac{\partial V}{\partial g_2}\right)_{T,P,g_{(N-2)}} dg_2 + \cdots$$

$$= \sum_{i=1}^{N} \left(\frac{\partial V}{\partial g_i}\right)_{T,P,g_j} dg_i \qquad (j \neq i) \qquad (3)$$

where subscript numerals designate the components. (In this chapter, water is component 1, protein is component 2, and component 3 is usually a membrane-diffusible solute in the solvent medium for the protein, such as urea, salts, and sugars.) Integration then yields Eq. (1) provided that the fractional mass of each component is maintained constant over each successive step dV (cf. Tanford, 1961; Klotz, 1964). Accordingly, the partial specific (or molal) volumes remain constant as long as the composition does not vary, regardless of the total mass (and volume) of the system. The partial molal volume $\bar{V}$ is defined analogously by substituting the number of moles n for grams of mass g. If temperature and/or pressure are not held constant, the partial volumes may vary even at constant composition, because the volume is also a function of these "intensive" (i.e., mass-independent) variables. Unlike the partial quantities of the other commonly employed extensive properties of solutions, partial volumes may be obtained by direct measurement and need not be evaluated relative to an assigned value at some arbitrarily chosen reference state. The corresponding total differential of the Gibbs free energy G in terms of moles of the components, at constant temperature and pressure, is

$$dG = \sum_{i=1}^{N} \left(\frac{\partial G}{\partial n_i}\right)_{T,P,n_j} dn_i \qquad (j \neq i) \qquad (4)$$

where the coefficients within parentheses are called partial molal free energies or chemical potentials and are denoted usually as $\bar{G}$ or $\mu$. The free energy G, like entropy S and enthalpy H, is a derived extensive property rather than a directly measured one such as volume, so that total values for the free energy are not ordinarily obtainable. Hence, difference values between the coefficient $\mu_i$ and that at some chosen reference state $\mu_i^0$ are used in practice. Since no such limitation applies to the volume property, the total volume can be expressed in terms of the experimentally available partial volumes of the individual components. Thus, changes in the partial volumes may be correlated with chemical events when the composition of a solution is altered or when reactants are mixed together. As any chemist knows, however, the change in volume of a solution during a reaction at constant temperature and pressure is usually trivial and it is often neglected in the thermodynamic balance sheet. That is, the difference in the total Gibbs free energy G between two states of an open system (T and P are constant) is largely a result of changes in the internal energy E and in the entropy function S as expressed in the well-known relation

$$\Delta G = \Delta E + P\Delta V - T\Delta S \qquad (5)$$

where $\Delta E$, for many purposes, is virtually equivalent to the enthalpy change $\Delta H$ ($=\Delta E + P\Delta V$). If an accurate account can be taken of the volumes in protein-mediated processes, a kind of information may be obtained which is unique. The change in volume $\Delta V$ need not necessarily parallel the changes seen by other probes which reflect more closely the total free energy change. In $\Delta V$ we are dealing with a difference in the volumes as a result of changes in the partial volume of each component when chemical bonds are broken and formed and/or where secondary forces are altered during the approach to equilibrium. The latter forces are often important in the changes which occur in the conformation and solvent–solute interactions of proteins. A variety of reactions have been studied which exhibit measurable volume changes without affecting the primary structure of proteins. Thus, changes in the partial volume of one or more (perhaps all) of the added components relative to some initial state must have taken place in accordance with Eq. (1) if the total masses are constant or properly accounted for. Hence, if partial volume measurements are feasible, the overall $\Delta V$ for a process can be separated into quantities related to each component. Usually, only the partial specific volume, or its change $\Delta \bar{v}$, of the protein component alone is measured. Clearly, however, a knowledge of $\Delta \bar{v}$ of the other components in the mixture may be very instructive also. The science of relating partial volume changes to chemical events is still in its infancy with respect to the chemistry of proteins and other macromolecules. The acquisition of detailed background information on the changes in total volume of simpler systems has been initiated as an approach to a sound interpretation of the volume property in protein reactions. This same approach applied to changes in the partial volume of the individual components should contribute much to our understanding of the structural changes taking place during transconformations and biologically important interactions of proteins.

## B. THE DENSITY

The volume of a solution is related to its density $\rho$ according to the definition of density at a given temperature and pressure

$$\rho = \frac{g_1 + g_2 + \cdots}{V} = \frac{\sum_{i=1}^{N} g_i}{V} \qquad (6)$$

where, as before, g is the number of grams of the component designated by a subscript and milliliters are chosen as the units of volume.[1] The density of a solution, therefore, is the sum of the masses in grams of all the components in the mixture divided by the number of milliliters in the total volume. For ease in handling the relationships concerned with density and volume in this chapter, concentrations will be expressed

---

[1] When cubic centimeters ($cm^3$) are designated instead of milliliters (ml), it should be kept in mind that the difference in these volume units, while small, may be important for some purposes. $1.000027 \ cm^3 = 1$ ml, where the milliliter is defined as the volume of 1 gm of pure water at 3.98°C in the absence of air at standard gravity (International Bureau of Weights and Measures, 1910; but see Wagenbreth and Blanke, 1971). Thus, the density of pure water at 20°C according to Plato (1900) is 0.998234 gm/ml or 0.998207 $gm/cm^3$. On the other hand, the value listed for pure water at 20°C in the "Handbook of Chemistry and Physics" (1962) (computed from the relative values by Thiesen et al., 1900) is given in the dimensions of grams per cubic centimeter, i.e., 0.998203 $gm/cm^3$; this value corresponds to 0.998230 gm/ml using the volume conversion factor. Whereas the difference in the experimental values, reduced to the same dimensions, is usually quite small, the difference of about 3 parts in $10^5$ obtained by intermingling the volume units becomes quite important when calibrating a procedure with known standards.

primarily in units of the "c" scale, where $c_i = g_i$ per milliliter. These dimensions are consistent with those in the definitions of density and specific volume. Thus, an alternative definition of density at a particular temperature and pressure is

$$\rho = \sum_{i=1}^{N} c_i \tag{7}$$

The volume expression Eq. (1) in these units of concentration becomes

$$\sum_{i=1}^{N} \bar{v}_i c_i = 1 \tag{8}$$

which simply says that, for any liquid, the sum of the partial specific volumes of all the included components, each multiplied by its respective concentration in grams per milliliter, must equal unity. It will be useful to keep in mind also the more detailed definition of density

$$\rho = \frac{\sum_{i=1}^{N} g_i}{\sum_{i=1}^{N} \bar{v}_i g_i} \tag{9}$$

For example, one may determine the concentration of a pair of nonvolatile components, such as protein and salt, in an aqueous mixture by a combination of dry weight and density measurements. A mass of the three-component solution to be dried is weighed on the balance to give $\Sigma g_i$. The density determined on this solution is $\rho = (\Sigma g_i)/V$. Hence, a value for the total volume is given by the ratio $(\Sigma g_i)/\rho$. From the drying procedure we obtain a value for the evaporated mass of water $g_1$ and a value $(g_2 + g_3)$ for the residue comprising the combined masses of protein and salt. Recalling Eqs. (1) and (9), it is evident that two simultaneous equations exist for calculating $g_2$ and $g_3$ in the dried residue if the partial specific volumes are available for the components in the pre-dried solution. Thus

$$\sum_{i=1}^{N} g_i - g_1 = g_2 + g_3$$
$$\left(\sum_{i=1}^{N} g_i\right) \bigg/ \rho = \bar{v}_1 g_1 + \bar{v}_2 g_2 + \bar{v}_3 g_3 \tag{10}$$

A similar set of equations may be written if component 3 is completely volatile. With dilute protein solutions it is found that the partial specific volumes of the nonprotein components are usually the same, within experimental error, as those listed for the solvent medium itself. [It should be recognized here that the determination of the dry weight of the two-component solvent medium, usually carried out concurrently with that of the protein solution, can be related to the density of the solvent medium. This density, in fact, gives the composition of the solvent (Section II,F) and serves as a check on whether the drying and heating protocol is adequate for quantitative results; for example, the protocol may cause partial volatilization of component 3 or changes in its mass through oxidations.]

C. THE PARTIAL SPECIFIC VOLUME

The evaluation of the partial specific volume $\bar{v}$ will now be considered. According to the definition for this volume quantity, Eq. (2), we wish to measure the change in volume of a solution upon addition of an infinitesimal amount of the component in question. This can be done most conveniently by measuring the densities of a series of solutions in which only the mass of the designated component is varied. Either the actual masses or the mass ratios of all the other components must be held constant in preparing the solutions of the series. This is a straightforward exercise on the analytical balance with components which can be added in their pure states. Proteins, on the other hand, cannot be conveniently added in the anhydrous form to a solvent. Since the solvent medium generally contains water, a water solution of the isoionic protein or of the charged protein ion having a known amount of counterions is added. The composition of the protein–water solution is first determined by dry weight analysis. If the solvent medium is not pure water, its composition must be adjusted on the balance, e.g., by adding nonwater solvent components, in order to compensate for the amount of water added with the protein. An alternative procedure is to add the nonwater components included in the solvent medium to the protein–water solution until the mass ratios, or molalities, of all nonprotein components in the protein solution exactly match those of the solvent. Dilutions of this stock protein solution are then made on the balance with aliquots of the solvent medium in order to prepare a series of solutions in which only the mass of the protein component is varied. The density values obtained for the series, including the density $\rho'$ of the solvent medium itself, are plotted *versus* the concentration $c_i$ of the component being varied (in this case, the protein). A line which best fits the data is then prepared with the aid of a desk computer or a standard program at a computer center (see also Chapter 24). Ordinarily, a rectilinear fit is observed with protein solutions, in which case a simple calculator suffices. If concentrations are expressed in grams per milliliter, differentiation of the solution density in terms of $\rho'$ with respect to $c_i$ gives for any component $i$

$$\left(\frac{\partial \rho}{\partial c_i}\right)^0_{g_j} = (1 - \bar{v}_i^0 \rho') \qquad (j \neq i) \tag{11a}$$

where superscript zero refers to vanishing concentration of the $i^{\text{th}}$ component and $g_j$ denotes the fact that the masses of all other components are held constant. [The differentiation leads to an expression with additional terms (see Appendix A). However, these terms drop out as $c_i \to 0$.] The concentrations in terms of the "c scale" are obtained from the weight fraction W of the component ($W_i = g_i/\Sigma g_j$) and the density of the solution, since by definition $c_i = \rho W_i$. Hence, the alternative expression in terms of W is

$$\left(\frac{\partial \ln \rho}{\partial W_i}\right)^0_{g_j} = (1 - \bar{v}_i^0 \rho') \qquad (j \neq i) \tag{11b}$$

The partial specific volume of the protein component is usually taken as the value at vanishing $c_2$. Transposition of Eq. (11a) and evaluation of the limiting slope $(\partial \rho/\partial c_2)_m^0$ yields

$$\bar{v}_2^0 = \frac{1}{\rho'}\left[1 - \left(\frac{\partial \rho}{\partial c_2}\right)^0_m\right] \tag{12}$$

where subscript m is substituted for $g_j$ to express constant molality of the other components, as is the practice when considering the protein in multicomponent systems. It should be noted that the partial specific volume may be similarly defined at a finite concentration of any component $i$ by including a specified mass of that component in the initial solution; in effect, the solvent medium of density $\rho'$ already contains the component for which $\bar{v}_i$ is desired at concentration $c_i$. An alternative expression for the partial specific volume at any value of $c_i$, but not employing $\rho'$, is derived in Appendix A.

In dealing with partial specific volumes of proteins it must be kept in mind that the values of $\bar{v}_2$ apply to the protein component as defined. If the partial specific volume of an isoionic protein is desired in a medium which would alter the net charge z on the protein (e.g., a solvent containing a buffer pair, acid or alkali, which is not at the isoionic pH), the value of $(\partial \rho/\partial c_2)_m^0$ obtained by adding, in effect, the dry isoionic material still yields the thermodynamically defined value of $\bar{v}_2^0$. Unless the protein is isoionic at the pH of the solvent, the pH of unbuffered solutions may vary with $c_2$ in the density series. It may be useful for some purposes to determine a value of $\bar{v}_2$ for an electrically neutral protein component in which the protein moiety possesses a net charge. The protein component may then be defined to include a stoichiometric number of small ions of opposite charge in addition to the protein ion. When this component is added to an unbuffered solvent medium, the change in pH with concentration of component 2 may be attenuated by the presence of sufficient salt having an ion in common with the small ion of the protein component.

In general, it has been found that $\bar{v}_2$ changes somewhat with pH (e.g., Adair and Adair, 1947; Charlwood, 1957), because of the electrostrictive effect of ion–water dipole interactions. Hence, a plot of density versus $c_2$ in nonisoionic, unbuffered solutions may give rise to observable curvature; i.e., $(\partial\rho/\partial c_2)_m$ is not constant and $\bar{v}_2{}^0$ must be evaluated carefully from the data obtained at finite values of $c_2$. The complication of superimposed density changes arising from a small change in buffer composition when protein is added might bring curvature even if the pH varies only a small amount. The species of a buffer pair often have highly different partial specific volumes when added independently; it may be presumed that significant volume changes can occur as a result of internal shifts in the ionic composition, even though the protein is maintained at a fixed charge as $c_2$ is varied.

## D. The Apparent Specific Volume

In most cases, the density plotted as a function of concentration in grams per milliliter of the protein component yields a straight line throughout the range of the rather low protein concentrations usually employed (<5%). The apparent invariance of $(\partial\rho/\partial c_2)_m$ in buffer solutions has been found to hold to unusually high concentrations for some globular proteins in the few cases studied (e.g., Adair and Adair, 1947). Accordingly, the apparent specific volume $\phi_2$ of proteins is often measured in practice. The value of $\phi$ for a component is derived from the slope of a straight line drawn between the density of the solvent medium and that of a solution at some finite concentration of the component added. Thus, for the protein component

$$\left(\frac{\rho - \rho'}{c_2}\right)_m = (1 - \phi_2\rho') \qquad (13)$$

since the density of the protein solution $\rho$ in these terms is defined as $\rho = \rho' - \rho'\phi_2 c_2 + c_2$ (see Appendix A). When Eq. (13) is transposed to solve for $\phi_2$, the resulting relation takes a form analogous to that for the differential expression of Eq. (12), and

$$\phi_2 = \frac{1}{\rho'}\left[1 - \left(\frac{\rho - \rho'}{c_2}\right)_m\right] \qquad (14)$$

Hence, if the density varies in direct proportion to $c_2$, the partial specific volume is constant and $\phi_2 = \bar{v}_2^0 = \bar{v}_2$ at any $c_2$; if $(\partial \rho/\partial c_2)_m$ is not constant, then only as $c_2 \to 0$ will $\phi_2 \to \bar{v}_2^0$.

Since there has been confusion about the difference between the apparent specific volume and the partial specific volume, some additional remarks may be helpful to the student. The apparent volume of a dissolved solute, by definition, is the difference in volume between that of the solution and the solvent medium containing none of the solute in question (at the same temperature and pressure). Hence, the value of $\phi$ of the added solute is simply this difference in the volumes divided by the number of grams of the solute. (If the volume difference is divided by the number of moles of the solute per kilogram of water, the apparent molal volume is obtained.) The definition implies that the solvent medium behaves ideally with respect to its volume contribution. The solvent, of course, cannot be assigned a definite three-dimensional domain in the solution containing the solute; nonetheless, the approach of arbitrarily assigning to the solvent a volume contribution which is identical to its volume in the absence of the solute has practical utility. With the definition of $\phi$ in terms of these volumes, $\phi$ is easily related to the partial specific volume *via* Eqs. (15–17). Thus, for any solute $i$

$$\phi_i = \frac{V - V'}{g_i} = \frac{V - \sum_{j \neq i}^{N} \bar{v}'_j g'_j}{g_i} \tag{15}$$

where $V' = \Sigma \bar{v}'_j g'_j$ is the volume of the solvent medium in which $\Sigma \bar{v}'_j g'_j / \Sigma g'_j$ represents the specific volume of the combined components $j$ of the pure solvent. The total volume V is then defined in these terms by rearranging and by multiplying through by $g_i$ so that

$$V = \phi_i g_i + \sum_{j \neq i}^{N} \bar{v}'_j g_j' \tag{16}$$

Differentiating this equation with respect to $g_i$, noting that $\Sigma \bar{v}'_j g'_j$ is arbitrarily treated as a constant independent of $g_i$, gives finally

$$\bar{v}_i = \phi_i + g_i \left(\frac{\partial \phi_i}{\partial g_i}\right)_{g_j} \quad (j \neq i) \tag{17}$$

From Eq. (17) it is clear that at vanishing protein concentration ($i = 2$), the partial specific volume and the apparent specific volume of the protein becomes identical. The value of $\bar{v}_i$ at any concentration of component $i$ may be determined, as noted before, by taking the limiting slope to a fixed value of $c_i$. In this case, the value of $\phi_i$ should also relate to the solvent medium containing an amount of component $i$ rather than to the pure solvent. Although data are lacking in the case of protein solutes, it is not inconceivable that the apparent specific volume, which is related only to the pure solvent, may vary in a bizarre fashion with the true partial specific volume determined at each concentration; in fact, such $\phi_2$ may vary over a concentration range in which $\bar{v}_2$ is constant. This is likely if the protein interacts with or is titrated by a minor component in the solvent medium; after the titration is complete a different value of $\bar{v}_2$ may be obtained which is essentially constant upon further addition of titrant. Similarly, if a volume change accompanies the self-association of a protein as a function of $c_2$, we expect that such $\phi_2$ will continue to vary at concentrations where the association is complete and $\bar{v}_2$ is constant.

In passing, it should be noted that a value of $\phi$ for the solvent medium is not ordinarily a useful quantity in practice. In general, the volume or density of a dry solute is not known accurately; adding solvent to the solute therefore in order to obtain the apparent specific volume of the solvent medium is often impractical. Moreover, the objective is to determine the apparent specific volume of the solute of interest, usually at low concentrations. It is not useful, then, to reverse the procedure and arbitrarily assign a volume contribution to the solute, because in all likelihood the apparent specific volume of the solvent medium will be little or no different from that of its specific volume at the solvent–solute proportions usually studied.

## E. The Isopotential Specific Volume

Along with the growing emphasis in the study of multicomponent systems, another specific volume quantity $v'$ has become of interest; the apparent quantity $\phi'$ is usually applied in practice.[2] These parameters refer to a nondiffusible component, such as the protein, in equilibrium dialysis experiments, and subscript designations of components, therefore, are often omitted. When a nondiffusible component is at osmotic equilibrium in a medium containing two or more diffusible components, the distribution or mass ratios of the diffusible components on the two sides

---

[2] The primed designation for the specific volume $v$ and for the apparent specific volume $\phi$ is a deviation from other usage in this chapter, wherein primes are used to designate the solvent or dialyzate phase. Because these quantities $v'$ and $\phi'$ are still recognized generally by the presence of the primes, a change in the designations at this time would probably contribute to the confusion which has surrounded this subject.

of the membrane are not necessarily the same. This condition can arise from a number of factors as a result of the presence of a nondiffusible protein on one side of the membrane or of different protein concentrations on both sides. Donnan effects, preferential binding, and/or differential accessibility of the diffusible components to certain regions in the protein solution are some of the factors which have been considered. Although the compositions, or molalities, of the diffusible components on the two sides may be unequal, the chemical potentials $\mu$ of each component at equilibrium are equal in the two phases. In plotting the densities of a series of solutions varying in $c_2$ which have been equilibrated against a common dialyzate (of unlimited volume), the derivatives $(\partial \rho/\partial c_2)_\mu$ are obtained rather than $(\partial \rho/\partial c_2)_m$; the subscript $\mu$ denotes that each of the diffusible components has the same chemical potential in every phase. The values of $(\partial \rho/\partial c_2)_\mu$ may be quite different from the values $(\partial \rho/\partial c_2)_m$ at the same concentrations of protein. The latter derivatives, it may be recalled, are the tangents obtained by maintaining constant the mass ratios of the nonprotein components as $c_2$ is varied for density measurements without dialyzing. It is evident that if the molalities of the diffusible components change with $c_2$, a value of $\bar{v}_2$ or $\phi_2$ cannot be obtained by Eqs. (12) or (14), respectively, because these parameters are defined on the basis of constant composition of the nonprotein components. In order to determine $\phi_2$ at each $c_2$, information is required on the changing mass ratios of the diffusible components in each equilibrated protein solution of different $c_2$ against a common dialyzate. This information must then be related at each $c_2$ to the density of a different solvent medium, one which is isomolal in each case with the diffusible components of the respective protein solutions. Thus, a series of values for $\rho'$ along the ordinate at $c_2 = 0$ would be manifested in the conventional $\rho$ vs. $c_2$ plots, and these values would approach the density of the dialyzate as $c_2 \to 0$. (Since the isomolal solvent media for each $c_2$ ordinarily would change only slightly, the various values of $\phi_2$ would probably be identical.)

The slopes at vanishing protein concentration, $(\partial \rho/\partial c_2)_m{}^0$ and $(\partial \rho/\partial c_2)_\mu{}^0$, for the isomolal and isopotential density series, respectively, will not merge but will remain distinct if the presence of protein gives rise to a redistribution of the diffusible components at membrane equilibrium. In practice $(\partial \rho/\partial c_2)_\mu$ tends to be constant over the usual dilute range employed for $c_2$ as is the case with $(\partial \rho/\partial c_2)_m$. Thus, the apparent value $\phi'$, obtained from the density difference between that at a finite value of $c_2$ and that of the dialyzate, is often determined instead of v' (or $v^{0'}$ at $c_2 = 0$). $v^{0'}$ for a nondiffusible component is defined by a relation similar to Eq. (12) for $\bar{v}_2{}^0$ and

$$v^{0\prime} = \frac{1}{\rho'}\left[1 - \left(\frac{\partial\rho}{\partial c_2}\right)^0_\mu\right] \tag{18}$$

wherein the important difference in subscript, $\mu$ for m, is emphasized. The corresponding relation for $\phi'$ is

$$\phi' = \frac{1}{\rho'}\left[1 - \left(\frac{\rho - \rho'}{c_2}\right)_\mu\right] \tag{19}$$

Since the masses of the diffusible components are not held constant under isopotential conditions but may vary unpredictably as the concentration of a nondiffusible component is varied, v' cannot be a partial specific quantity. A suggested name for this kind of specific volume might be the *isopotential specific volume*. The corresponding quantity $\phi'$ is then an *apparent isopotential specific volume*. These two specific volumes, however, are not easily related as are $\bar{v}_2$ and $\phi_2$ by Eq. (17) (cf. Casassa and Eisenberg, 1964).

Until rather recently, the quantity v' was seldom recognized by investigators (the writer included) as being distinct from $\bar{v}_2$. Often the evaluation of $\bar{v}_2{}^0$ of a protein was carried out on a dilution series in $c_2$ by adding dialyzate to a dialyzed protein solution. This procedure, in effect, yields values for $v^{0\prime}$ rather than $\bar{v}_2{}^0$ (Casassa and Eisenberg, 1961). In a solvent consisting of essentially one component there is, of course, no difference between v' and $\bar{v}_2$ if Donnan effects are absent. As the molality of a second diffusible component is increased, the difference in these two specific volumes may be substantial (Reisler and Eisenberg, 1969). v' is a useful quantity when dealing with multicomponent systems, particularly in ultracentrifugation (Casassa and Eisenberg, 1961; Coates, 1970) and low-angle X-ray scattering (Eisenberg and Cohen, 1968; Pilz, this volume).

It may be instructive now to relate the isopotential specific volume to the true partial specific volume of a redefined protein component; in addition, this operation leads to a parameter $\xi$ which is of more general interest and one which can be evaluated by the density method (described in Section IV,B,3). We begin by employing the conceptual device of redefining the protein component to include the amount of a diffusible component which is in excess relative to the composition of dialyzate (Güntelberg and Linderstrøm-Lang, 1949; Casassa and Eisenberg, 1960, 1961). The composition of the diffusible components not included in the redefined protein component is then equivalent to the composition of the dialyzate. By this formalism, if anhydrous protein is added or removed from a system at osmotic equilibrium, a corresponding amount of the excess component is likewise added or taken away, respectively. One might take away instead, the appropriate amount of the deficient component when adding the anhydrous protein. Thus, variation of the concentration of this redefined protein component (usually designated with an asterisk) does not involve a change in the mass ratios of the diffusible components on the two sides of the membrane. (This is true provided that the number of moles of excess, or of deficient, diffusible component per mole of the anhydrous protein remains constant over the range in $c_2{}^*$ under investigation. Ordinarily, the change in osmotic pressure with $c_2{}^*$ is $\ll 1$ atm, yielding a trivial effect if any, and the concentration of a second diffusible component is sufficient to override any substantial change in the total concentration of this component when component $2^*$ is varied.) Obviously, the value of $c_2{}^*$, within the constraint imposed, is related to $c_2$ by a constant factor (positive or negative) so that

$$c_2{}^* = c_2 + \xi_j c_2 \qquad (j \neq 2) \tag{20}$$

where $\xi_j$ is the grams of excess (or deficient) diffusible component $j$ per gram of the anhydrous protein. For example, if component 3 is the excess diffusible component, the amount $\xi_3$ must accompany a gram of the anhydrous protein being added or removed from the solution at osmotic equilibrium. Component 3, however, may be deficient relative to the dialyzate composition, in which case $\xi_3$ is negative; i.e., $\xi_3$ is the amount of component 3 which must be removed from the solution when a gram of the protein is added. In this event, it may be more convenient to visualize $c_2{}^*$ as the sum, $c_2 + \xi_1 c_2$, since water is then the excess diffusible component. If the protein is an ion with net charge z and component 3 is a salt having an ion in common with the counter ion, $\xi_3$ may be defined with respect to the electrically neutral protein–salt component, i.e., $Na_z(protein)^{-z}$ or $(protein)^z Cl_z$, as component 2. Hence, $\xi_3 g_2$ becomes the amount of the salt, component 3, which must be added to or taken away from (along with $g_2$ grams of the electrically neutral protein–salt component) a solution at osmotic equilibrium, in order to maintain the same distribution of water and salt on the two sides of the membrane (Casassa and Eisenberg, 1961).

The quantity $\xi_j$ is a thermodynamically defined parameter which is readily expressed in terms of the c scale of concentration for evaluation by density measurements (Section IV,B,3). For $\xi_3$ we simply subtract the number of grams of component 3 relative to its proportion with the number of grams of water per milliliter of dialyzate from the corresponding mass ratio in 1 ml of the equilibrated protein solution. This difference in mass of component 3 relative to water on the two sides of the membrane is reduced to that per gram of the dry protein relative to water. Thus

$$\xi_3 = \left(\frac{c_3}{c_1} - \frac{c'_3}{c'_1}\right) \bigg/ \left(\frac{c_2}{c_1}\right) = \left[c_3 - c'_3\left(\frac{c_1}{c'_1}\right)\right] \bigg/ c_2 \tag{21}$$

The definition of $\xi_1$ is identical except that all subscripts 1 and 3 are interchanged. Hence, $\xi_1$ and $\xi_3$ are easily interconverted; by setting the respective equations for $\xi_1$ and $\xi_3$ equal to $c_2$ in Eq. (21), it is seen that

$$\xi_1 = -\xi_3 \frac{c'_1}{c'_3} \tag{22}$$

Finally, it is evident that by varying $c_2^*$ (the concentration defined to include the excess or deficient amount of a diffusible component), a true partial specific volume $\bar{v}_2^*$ is possible, because the molalities of the diffusible components not included in the definition of component $2^*$ remain constant. Hence, at vanishing $c_2^*$, the quantity $\bar{v}_2^{0*}$ is given by Eq. (12) by merely substituting $c_2^*$ for $c_2$. The corresponding apparent quantity $\phi_2^*$, at a specified concentration $c_2^*$, is then defined analogously to that for $\phi_2$ as in Eq. (14), and

$$\phi^* = \frac{1}{\rho'}\left[1 - \left(\frac{\rho - \rho'}{c_2^*}\right)_m\right] \tag{23}$$

Since the observed density difference between the dialyzed solution and dialyzate is the same regardless of how we define the nondiffusible component, we may equate the definitions of $\phi'$, Eq. (19), and $\phi^*$, Eq. (23), by factoring out $(\rho - \rho')$. Recalling Eq. (20) for $c_2^*$ in terms of $c_2$ and $\xi_j$, we find

$$\phi' = \phi^* - \xi_j(1/\rho' - \phi^*) \tag{24}$$

The relationship between the conventional partial specific volume $\bar{v}_2$ determined under isomolal conditions, and the isopotential specific volume $v'$, in equilibrium dialysis, via the interaction parameter $\xi$ will be treated in the discussion on preferential interaction [Eqs. (49) and (50); Section IV,B,3].

F. DENSITY AND COMPOSITION

It is sometimes important to ascertain whether there have been small changes in composition or the proportionate masses of the nonprotein components in both the solvent medium and protein solution after an equilibration procedure (e.g., equilibrium dialysis or column experiments). As pointed out before, water and component 3 may redistribute obtaining partial or apparent specific volumes. Density determinations and weighings are usually done with high accuracy compared with the evaluation of $c_2$, which is the principal source of error in all the applications requiring this quantity. Direct dry weight analysis of protein in a water solution is the best means, generally available, of determining the absolute mass of the protein component. This cannot be done with some systems for a variety of reasons; moreover, the procedure is very time consuming and requires meticulous care in order to achieve high precision. Since many evaluations of the protein concentration are often required in any study of these substances, more convenient indices of the concentration are employed of necessity. For protein assays in two-component solvents, the method chosen must be insensitive to the concentration of component 3. These indirect methods, ultimately, are based on a unit of the dry mass of the protein component. Thus, the error in two analytical methods must be contended with when using an indirect procedure, such as a light-absorption technique. Therefore, an uncertainty in $c_2$ of $\pm 1\%$ obtained by indirect assays can seldom be sustained, even though the precision of the indirect method is shown to be $\pm 1\%$. For example, the average value of a concentration coefficient, such as an extinction coefficient, for a given protein may be significantly different between laboratories which use the same assay method and exercise similar precautions (see Section IV,B,3 for a procedure involving the density). In relating an indirect procedure to a dry mass of a protein, there is unfortunately no unanimity on the best procedure to follow for evaluating the dry mass. An assessment of some of the variables has recently been reported by Hunter (1966). Different values for the "constant weight" were observed depending on the drying protocol used; the differences in weight, while slight, were significant. This work is also useful as a guide for developing greater precision in the dry weight analysis of proteins. Routine precisions which approach 1 to 2 parts in $10^3$ are obtainable by careful control, such as by the extrapolation of weights *versus* time to an anhydrous atmosphere. In the absence of a universally accepted dry weight protocol, the principal point to be made here is that a dry weight definition be established locally which gives consistent results. In addition, the history and properties of a particular protein component should be defined as specifically as possible. In this way, the internal agreement on the work in a given laboratory can be quite good. In those studies where parameters are sought which involve differences between related systems, such as for the evaluation of $\xi$ (Section IV,B,3), the effect of error in $c_2$ can be reduced substantially by performing the measurements on the related systems concurrently using the same preparation of the protein.

## III. Direct Volume Change

The classical means of determining the change in volume $\Delta V$ for a process is by direct measurement of the volume difference. The procedure is known as dilatometry. $\Delta V$ derived from density measurements is discussed in Section IV,B. The volume change can be obtained indirectly from the effect of pressure on the equilibrium constant of a reaction

$$\Delta V = -RT\left(\frac{d \ln K}{dP}\right) \tag{30}$$

where K is the equilibrium constant and P is the pressure. The equilibrium is usually observed by optical means (e.g., spectral changes) in a pressure bomb. The application of pressure change to the study of macromolecules is discussed by Johnson et al. (1954), and recent advances in technique, applicable to protein systems, are described by Gill and Glogovsky (1964, 1965), Rainford et al. (1965), Brandts et al. (1970), and Kliman (1969). $\Delta V$ may be derived also from changes in the refractive index utilizing the Lorentz–Lorenz equation

$$\left(\frac{n^2 - 1}{n^2 + 2}\right) = \frac{4}{3}\pi \left(\frac{\sum_{i=1}^{N} \alpha_i N_i}{V}\right) \tag{31}$$

where $n$ is the refractive index and $\alpha$ is the polarizability of the $i^{th}$ component consisting of $N$ molecules. If the polarizability and number of molecules of each component remain constant or can be evaluated accurately for a process, $\Delta V$ may be calculated (cf. Kauzmann, 1959). A recent application of this method, including comparison with direct volume change measurements, is described by Noguchi and Yang (1963). The present discussion is restricted to direct volume change measurements by dilatometry.

### A. Dilatometry

The direct measurement of the change in volume $\Delta V$ on mixing proteins with other substances in solution goes back several decades (e.g., Weber and Nachmansohn, 1929; Weber, 1930). The basic equipment for such measurements are capillary dilatometers. This method was greatly refined and vigorously applied at the Carlsberg Laboratorium and the technique has undergone little change since then (Linderstrøm-Lang and Lanz, 1938; Linderstrøm-Lang, 1940, 1952). The dilatometer consists of two glass parts, an inverted V-shaped reaction vessel, and an accurately calibrated microcapillary. The two solutions (or a dry material and a solution) to be mixed are introduced carefully into separate limbs of the reaction vessel; total volumes of 4–20 ml are usually employed. The space above the two phases is then filled with an immiscible, nonreactive liquid, such as purified kerosene or heptane. The capillary tube is connected to the top of the reaction vessel by a carefully fitted, standard, tapered joint. After capillary height adjustments are made [these are based on past experience (Kauzmann, 1958)], the dilatometer is placed in a constant-temperature bath for thermal equilibration. When the height of the meniscus in the capillary reaches a steady state and its reading is taken, the dilatometer is gently rocked to mix the phases. After thermal equilibrium is restored, the final height, or rate of change of height, of the meniscus is observed relative to the calibration markings on the capillary; these heights can be estimated to ±0.01 $\mu$l (a cathetometer is sometimes used for this purpose). The mixing is carried out cautiously to prevent breaks in the column of capillary liquid. The formation of bubbles from dissolved air also causes such breaks and the reactants and the inert phase are therefore degassed by evacuation prior to use in the dilatometer. Since leaks are a severe problem in dilatometry, various procedures have been used to make tight joints without causing irreversible bonding (Johansen, 1948; Johansen and Thygesen, 1948; Katz and Ferris, 1966; Gerber and Noguchi, 1967; Krivacic and Rupley, 1968). In Fig. 1 are shown photographs of the two parts of the dilatometer. These also show the Teflon sheaths bonded to the tapered joint of the capillary, like those used by Katz and Ferris (1966) to minimize leaking, and the fusing together of the two parts after prolonged use. The serrated sheaths (left capillary) recently developed by S. Katz (private communication) were found to be more successful in general than the cylindrical ones; a leak rate of less than 0.01 $\mu$l/hr is observed. The capillaries, about 25 cm long, are calibrated to 10 $\mu$l, full scale, with graduations at 0.05 $\mu$l intervals. Calibration is achieved by measuring the length of a short column of mercury which is moved along the bore of the capillary. The volume is calculated by weighing the mass of mercury corresponding to a specified length of the scale at a given temperature. Volume changes are reproducible to a few hundredths of a microliter if the temperature is controlled to ±0.001°C; a change of 0.001°C corresponds to a change of about 0.05 $\mu$l in the dilatometer reading when heptane is the manometric liquid. In terms of volume changes observed in protein reactions, the precision in $\Delta V$ is about 20 ml/mole if approximately 1 $\mu$mole of protein is present in the reaction mixture of the dilatometer (Kauzmann, 1958; Katz and Ferris, 1966). Further details on procedures for various experimental purposes are

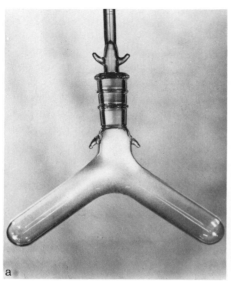 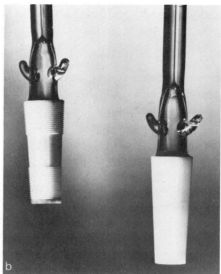

Fig. 1. The Dilatometer. (a) Glass reaction vessel to which a capillary column is joined (see text). (b) Capillaries showing two types of Teflon-bonded sheaths on the standard taper male joint. (Courtesy Dr. S. Katz, 1970.)

found in the papers by Linderstrøm-Lang and Jacobsen (1941), Kauzmann (1958), Kauzmann et al. (1962), Rasper and Kauzmann (1962), Katz and Ferris (1966), Gerber and Noguchi (1967), and Krivacic and Rupley (1968).

The measurement of volume change by dilatometry affords a direct observation of the change in an extensive property during a chemical process. A particular advantage of the dilatometer method is its relative insensitivity to impurities. If the impurity does not react in the system, it serves only as a diluent, having no effect on $\Delta V$; if the impurity does react, it must be present in substantial amounts to contribute to the volume change significantly. In density measurements, on the other hand, an impurity of 1 part in $10^3$ can lead to relatively large errors. For example, if the specific volume of the impurity is about 0.4 ml/gm lower than the specific volume of the aqueous solution, an error in the density of about $4 \times 10^{-4}$ gm/ml is encountered. The difference in specific volumes with inorganic impurities is usually greater than 0.6 ml/gm so that errors exceeding $10^{-3}$ gm/ml may be suffered if such impurities are present to the extent of 0.1% and are not compensated for with adequate controls. Obviously, if the specific volumes are similar, the effect of impurities on the density becomes negligible. Furthermore, since dilatometers contain 10 ml or so of solution, a difference of a few microliters in the reproducibility of the volumes added to the reaction vessel will usually have no measurable effect on ΔV. Where strictly comparable companion experiments are performed, as in measuring differences in partial volumes, greater uniformity in the delivery of identical samples to the reaction vessels is required (Katz and Ferris, 1966; Katz, 1968). Finally, the rate of change of the volume during a reaction can be followed easily in slow reactions (Linderstrøm-Lang and Lanz, 1938; Linderstrøm-Lang and Jacobsen, 1941; Gerber and Noguchi, 1967). With suitable control dilatometers, it is possible to extrapolate to zero time more accurately by correcting for slight volume changes accompanying the mixing process.

B. APPLICATIONS

### 1. *Model Systems*

What do volume changes mean in terms of chemical events? The fact that ΔV is an obligatory quantity, however small, in the thermodynamic expression for the total Gibbs free energy change, Eq. (5), is of no direct help. Attempts to relate ΔV to known chemical changes are at the basis of most of the studies on volume change in protein chemistry. The earlier studies on following the changes in volume accompanying the hydrolysis of peptides, enzymic splitting of proteins, and denaturation reactions gave answers for the overall ΔV, characteristic of the processes. The number of distinct chemical events was often rather large and precluded any precise correlations being drawn between volume changes and the individual events. For example, the splitting of a peptide bond is accompanied by the dissociation of water, by hydroxyl group attachment to the carbonyl stump, and by protonation of the new primary amine; in addition, we surmise that a substantial restructuring of solvent water takes place especially around the newly created charged groups. Ion–water dipole interaction, in this sense, is defined as electrostriction and gives rise to a volume contraction, the effect being smaller for zwitterions than for greater charge separations (Cohn and Edsall, 1943; Kauzmann *et al.*, 1962). Obviously, the quest for volume change information on simple systems is to be encouraged. This fundamental approach was initiated by Kauzmann and his associates in studying the ionization of proteins. To interpret the volume change accompanying the titration of proteins with hydrogen and hydroxyl ions, many model acids and bases and their derivatives were examined under different conditions (Kauzmann *et al.*, 1962); an extensive survey of volume change data compiled from the literature was also provided for comparison. The data were assessed relative to theoretical considerations and to changes in other thermodynamic properties. This paper is a highly informative reference source, reaching beyond the bounds of a report on experimental results and associative interpretations. By dilatometry also, the volume change for the formation of water from hydrogen and hydroxyl ions was reevaluated (Bodanszky and Kauzmann, 1962) in order to interpret more accurately the results from the model compound studies; previously accepted values, based on other procedures at relatively high concentrations of alkali, were found to be in error by about 5%. A particularly interesting observation with proteins is pertinent for illustration purposes. It was noted that the volume change accompanying the protonation of carboxyl groups of various proteins coincided essentially with that obtained with simple carboxylic acids (about $+11$ ml/mole $H^+$); on the other hand, removal of protons from basic groups on the proteins produced volume increases which were substantially lower (by about one-third) than that observed in the analogous reactions with simple amines (Rasper and Kauzmann, 1962). Evidently, the environments of these groups on the proteins are distinguishable from that of the model compounds, even though the free energies and enthalpies of dissociation of the groups on proteins appeared not to be affected. A recent study of this unusual feature was carried out on a lysine-rich histone (Krausz and Kauzmann, 1970). In this protein, the proportion of basic groups to the total residues with charge potential far outweighs the proportion found in proteins generally. The volume change on titrating with alkali in this case (22.3 ml/mole $H^+$) was in rather striking agreement with that for model amines (e.g., lysine gives 22 ml/mole between pH 10 and 11). Except for minor electrostatic effects in both lysine and lysyl residues on the protein, the volume change is similar to that for simple organic amines (about 25 ml/mole). Thus, the normalized value of $\Delta V$ with the histone suggests that the abnormal volume changes observed in the alkaline region for more standard proteins results from the presence of many other groups with charges (carboxyl, imidazole, and most probably phenolic; cf., Krausz, 1970); possibly also, the effect of changes at an alkaline pH in the tertiary structure, which is not very intricate in histone, contributes to the suppression of a normal volume change.

By emulating the model systems approach, attempts have been made to relate $\Delta V$ to conformation changes in proteins. These studies were stimulated by viewing the partial molar volume property of proteins in solution in terms of volume increments as proposed by Kauzmann (1959; also Rasper and Kauzmann, 1962). In this concept the observed property is dissected into (a) the major contribution, a constitutive volume envelope deduced from the radii and bond lengths of the constituent atoms in the primary structure, and (b) the minor contributions arising from conformational arrangements of the polypeptide and the variety of possible responses of the solvent molecules to the macromolecule. As examples of model systems, Noguchi and Yang (1963) measured the volume changes accompanying the titration of the homopolypeptide poly-L-glutamate, which changes from a helix to a random coil conformation; Bradbury et al. (1965) have studied the conformational transition of the benzylester derivative of this polypeptide as a function of temperature. In the latter experiments, a change in $\Delta V$ after subtraction of the expansion of the solvent (dichloroacetic acid–dichloroethane, 3:1 v/v) was observed over approximately the same temperature range (27°–30°C) as that for the helix to random-coil transition determined by optical rotation. The change in volume, however, disagreed in both sign and magnitude with that deduced pycnometrically for the change in apparent specific volume attending the transition. Barring experimental uncertainties, the results suggest that the partial volumes of the solvent components changed also. In the former study, Noguchi and Yang observed a substantially larger $\Delta V$ with the refractometric method than that by dilatometry; the discrepancy, they suggest, is a result of changes in the polarizability of some of the components as the charge on the polypeptide is altered by the titration, Eq. (31). They also point out various possible complications which must be dealt with on titrating model polymers: overlap of electrostatic fields between ionic groups which change with pH, hydrogen bonding between polymer residues and between residues and water during helix to random-coil transitions, the change in pK of the dissociable groups before and after the conformational change, and electrostatic effects from increasing the concentration of counterions during the titration.

## 2. The Parameter $\Delta \phi_2$

The change in the apparent specific volume $\Delta \phi_2$ has been explored to test whether known or assumed chemical events are explicable in terms of this parameter; such changes have been measured when proteins undergo an independently observable interaction or when they are transported from one medium to another. Stevens and Lauffer (1965), for example, have measured $\Delta \phi_2$ for the polymerization of tobacco mosaic virus protein and conclude that the positive value found during this self-association is in substantial agreement with other evidence supporting a mechanism for the release of bound water from the subunits. The change in $\phi_2$ with denaturation of serum albumin, using a highly precise dilatometric protocol, has been studied as a function of urea and guanidinium chloride concentration (Katz and Ferris, 1966; Katz, 1968). In these experiments, $\Delta \phi_2$ was determined upon transporting the isoionic albumin from a water solution to one containing the denaturant. The procedure for extracting the quantity $\Delta\phi_2$ from dilatometric results will now be described.

Since we wish to isolate the change in $\phi_2$ from other factors contributing to the volume change, we must know the changes caused by the mixing of water with components other than the protein, as well as the change from the dilution suffered by the protein when it is mixed with the companion solution in the other limb of the dilatometer. In addition to the isoionic protein and water, one additional component will be used, such as a denaturing agent, which is called component 3. We will consider three processes (I, II, and III) in three separate dilatometers. In one dilatometer, a water solution of the protein at a sufficiently high concentration is placed in one limb and pure water is put into the other. To define the volumes before and after mixing [see Eq. (1)], the subscripts 1, 2, and 3 of the masses g and partial specific volumes $\bar{v}$ denote the components water, protein, and denaturant, respectively. Sub-subscript numerals refer to the other components which are present in the solution in which the component with a particular value of g and $\bar{v}$ exists. It will be recognized that a given value of g for a component can be applied both before and after the mixing process, whereas the value of $\bar{v}$ for the component may change as a result of the mixing. Superscripts B and A denote the quantities before and after the mixing, respectively. Thus, for process I in the first dilatometer, the volumes $V_I^B$ and $V_I^A$ are defined as

$$V_I^B = [g_{1_2}\bar{v}_{1_2}^B + g_{2_1}\bar{v}_{2_1}^B] + [g_1\bar{v}_1^B]$$
$$V_I^A = g_{1_2}\bar{v}_{1_2}^A + g_{2_1}\bar{v}_{2_1}^A + g_1\bar{v}_{1_2}^A$$

where the bracketed terms designate the volumes occupied by the separated phases before mixing; in passing, it may be noted that $\bar{v}_1^B$ is identical with the specific volume of pure water $v_1$. Subtracting the total volume of the two separated phases from the volume after mixing gives

$$\Delta V_I = g_{1_2}(\bar{v}_{1_2}^A - \bar{v}_{1_2}^B) + g_{2_1}(\bar{v}_{2_1}^A - \bar{v}_{2_1}^B) + g_1(\bar{v}_{1_2}^A - \bar{v}_1^B)$$

In a second dilatometer (process II), we wish to learn the change in volume on mixing water with component 3. In one limb is placed pure water having the same volume as that which was present with protein in process I; that is, the protein contribution to this volume has been subtracted. The second limb contains known amounts of water and of component 3 in the same volume as that used for the pure water of process I. These volumes before and after mixing are stated analogously as in process I, and their difference $\Delta V_{II}$ is

$$\Delta V_{II} = g_{1_3}(\bar{v}_{1_3}^A - \bar{v}_{1_3}^B) + g_{3_1}(\bar{v}_{3_1}^A - \bar{v}_{3_1}^B) + g_1(\bar{v}_{1_3}^A - \bar{v}_1^B)$$

Finally, in process III, we mix (water plus protein) with (water plus component 3). The volume and protein concentration of the former mixture is like that in process I, and the volume and denaturant concentration of the latter mixture is the same as in process II. The total volumes before and after mixing are written

$$V_{III}^B = [g_{1_2}\bar{v}_{1_2}^B + g_{2_1}\bar{v}_{2_1}^B] + [g_{1_3}\bar{v}_{1_3}^B + g_{3_1}\bar{v}_{3_1}^B]$$
$$V_{III}^A = g_{1_2}\bar{v}_{1_{23}}^A + g_{2_1}\bar{v}_{2_{13}}^A + g_{1_3}\bar{v}_{1_{32}}^A + g_{3_1}\bar{v}_{3_{12}}^A$$

Accordingly, the difference in volume for process III is

$$\Delta V_{III} = g_{1_2}(\bar{v}_{1_{23}}^A - \bar{v}_{1_2}^B) + g_{2_1}(\bar{v}_{2_{13}}^A - \bar{v}_{2_1}^B)$$
$$+ g_{1_3}(\bar{v}_{1_{32}}^A - \bar{v}_{1_3}^B) + g_{3_1}(\bar{v}_{3_{12}}^A - \bar{v}_{3_1}^B)$$

If we assume that the dilution of protein causes no change in $\bar{v}_2$, whether in a water or water plus component 3 medium—which is usually the result found experimentally—we may substitute $\phi_2$ for $\bar{v}_2$ in the second term of the expression for $\Delta V_{III}$. We then simply equate the other three terms to the difference in volume $\Delta V_{II}$ produced by mixing water with component 3. This is equivalent to the statement that the specific volume of water and that of water plus component 3 are the true contributions to the total volume when protein is added to such solutions in accordance with Eqs. (15) and (16) for the definition of $\phi$. Hence, the change in volume $\Delta V_{III}$ becomes

$$\Delta V_{III} = \Delta V_{II} - g_{2_1}(\phi_{2_{13}} - \phi_{2_1}) \tag{32}$$

As support for these substitutions, the result of the experiment with the first dilatometer (process I) should show no volume change ($\Delta V_I = 0$) within experimental error. We might also have conducted a fourth experiment, with volumes identical to those in processes I and III, where the two limbs contained the same mass ratio of water and component 3 (i.e., isomolal), but with one limb containing also the same mass of protein as that used in the first and third experiments. In this fourth experiment, the value of $\Delta V$ should again be zero. The quantity $\Delta V_{II}$ may be applied directly from the second experiment to the definition of $\Delta V_{III}$, provided that the amount of pure water in the one limb corresponds precisely to that in the limb containing protein plus water in processes I and III. (Since $g_2$ is known for a given weight of the stock protein–water solution, the mass of pure water to be added in process II is determined from the mass of protein–water in the first limb of process I.) If density–composition data are available for the changes produced by mixing water with component 3 at the desired temperature, $\Delta V_{II}$ can be calculated. By curve fitting and subsequent calculation of the quantities $\bar{v}'_1$ and $\bar{v}'_3$ from such data (Section II,F), and by substituting the appropriate values of $\bar{v}'_i$ into the relations for process II, the value of $\Delta V_{II}$ is predicted. With the quantities $\Delta V_{III}$, $\Delta V_{II}$, and $g_2$ (or $c_2$) known, Eq. (32) gives the value for $\Delta \phi_2$. Obviously, if $\phi_{2_1}$ is known for the protein in water, then $\phi_{2_{13}}$, the value in the presence of component 3, may be calculated.

### 3. *Solid-Phase Studies*

A special feature of the dilatometer method is its amenability to experiments on proteins in the solid state. A recent protocol for measuring the volume change accompanying the dissolution of crystalline protein upon adding an aqueous phase has been described (Katz and Ferris, 1966; Katz, 1968). In this way, the change in the apparent specific volume $\Delta \phi_2$ of albumin was determined by adding water or an aqueous denaturant solution. Essentially the same result was obtained by this procedure as when the protein was in the dissolved state initially (see Section III,B,2). The volume change during crystallization of a protein has also been measured successfully (Krivacic and Rupley, 1968). Through these experiments, it was possible to show that horse methemoglobin underwent no substantial change in volume during crystallization, as if the conformation of the protein and the solvent properties were unaffected by the process. In these studies a solution of the protein in an aqueous salt medium was placed in one limb of the dilatometer, and a stronger concentration of the salt in water, at identical pH, was put into the other limb. The final salt strength on mixing was sufficient to cause crystallization of the hemoglobin. The volume changes with time were followed until most of the protein had crystallized (15 to 20 hr). In other experiments the salt was added directly to a solution of the protein, then deaerated, and placed in the dilatometer before crystallization commenced. By the latter procedure, almost all of the protein could be converted to the crystalline state while inside the dilatometer. Virtually the same results were obtained by either protocol. Leakage at the dilatometer joints during these relatively long-term experiments gave rise to a number of failures; hence, a greater than normal number of samples for replicate experiments may be needed at the outset.

By contrast, the bacterial protein, flagellin, which polymerizes to flagella, thus mimicking crystallization, is accompanied by a substantial volume increase (about 157 ml/mole flagellin; Gerber and Noguchi, 1967). The polymerization process appears to be a monomer addition to pre-existing short filaments or seed polymer, reminiscent of glycogen formation. Neither the seeds nor the monomer polymerize in the absence of the other under the conditions employed. Polymerization does not continue, however, until a separate phase is formed as in crystal growth;

instead the solutions become increasingly opalescent. The experimental system is ideally simple. One limb of the dilatometer initially contains the monomer and the other limb contains seed polymer in the identical solvent medium. On mixing the phases, polymerization proceeded rapidly to completion (about 30 min), depending on the temperature. Since a change in conformation of the flagellin was held to be unlikely, or at most small, during polymerization below 28°C, the observed volume change was believed to result primarily from a change in the solvent structure. An interaction of several hydrophobic residues was suggested on the basis of activation parameters and the high content of nonpolar residues. Above 28°C a much larger maximum $\Delta V$ was obtained, which suggested that the flagellin monomer at these temperatures was in equilibrium with an inactive monomer of different conformation. The additional increase in the maximum volume was ascribed to a negative conformational volume increment to the partial volume of the monomer when it changed to the inactive form.

### 4. *Kinetic Experiments*

The rate of change of volume during a process can often be followed quite easily by carefully adjusting the experimental system. A cathetometer accurate to 0.001 cm or, simply, capillary calibration lines have been employed to measure the vertical displacement of the meniscus in the capillary after the reactants have been mixed. Applications to the kinetics of the splitting of peptide bonds and of alterations in the noncovalent structures of proteins have been set out in detail by Linderstrøm-Lang and his associates (cf. Linderstrøm-Lang, 1962). The experiments on the hydrolysis of proteins led to the hypothesis of a dual effect of proteases, that of changing the conformation of the substrate polypeptides prior to the actual splitting of the peptide bonds; large contractions in volume were observed with substrate proteins, having intricate tertiary structures, as a prelude to the hydrolytic reaction (Linderstrøm-Lang and Jacobsen, 1941). On the other hand, the rate of hydrolysis of RNA by ribonuclease showed a marked increase in volume prior to the splitting of the phosphate bonds (Chantrenne *et al.*, 1947). This study also called attention to the fact that the subsequent decrease in volume per phosphate bond hydrolyzed was much larger than that known for simple phosphate esters. [It was later established that a transient, 2′,3′-cyclic phosphodiester intermediate appears in the overall reaction mechanism; in contrast the hydrolysis of DNA exhibits a rectilinear volume decrease with time, the slope of which is characteristic of that for the hydrolysis of simple phosphate esters as shown by Vandendriessche (1953).] More recent examples of studies on the time course of volume changes are those on the splitting of ATP by myosin (Noguchi et al., 1964) and on the transformation of G-actin to the F-form (Ikkai et al., 1966).

The kinetics of polymerization of flagellin to flagella filaments have been studied in some detail by dilatometry (Gerber and Noguchi, 1967). A constant limiting value of $\Delta V$ was observed at all temperatures below 28°C. Above this temperature, however, the maximum $\Delta V$ was twice this amount despite the fact that no difference in the amount or type of polymerization was evident. Below 28°C the time curve for the volume change paralleled that for the specific viscosity which is used as an index for the degree of polymerization. From the kinetic data at 25°C the activation parameters were calculated, suggesting that the structure of the activated, monomer–seed polymer complex was similar to that of the components before the activation. Above 28°C, the decreasing rate of polymerization and higher maximum $\Delta V$ were ascribed to a conformational transition between the active and inactive flagellin monomer. This study is perhaps an ideal example of the use of dilatometry for kinetic purposes since the reaction for polymer growth is initiated simply by the introduction of some seed polymer as nucleating agent without altering the solvent medium. [The effect of diluting the flagellin upon addition of the seeding solution was negligible; i.e., $(\partial \phi_2 / \partial c_2)_m$ is usually zero for proteins.]

## IV. Density

### A. Methods

The densities of protein solutions should be known to about $10^{-5}$ gm/ml to be useful for most of the purposes described in this chapter. To achieve such accuracy, the sensitivity of the measurements should approach a level of $10^{-6}$ gm/ml. As a rule of thumb, three significant figures in the value of $\phi_2$ require five significant figures in the densities of the solution and solvent (viz., 1% of added protein increases the density of water by about 0.003 gm/ml). With aqueous systems it is important to keep in mind also that the density of water changes by about 2 to $3 \times 10^{-6}$ gm/ml per 0.01°C over the temperature range (20°–30°C) used for most measurements. The means of measuring the densities of liquids are not designed as absolute methods, because of the inherent difficulty of establishing and maintaining exactly known volumes of containers. By international convention, however, the milliliter is a defined quantity on the basis of a weight of pure water (see footnote 1). Thus, it is far more convenient to relate volumes to accurate weights of water or to weights of standard solutions based on water. In some methods, the volume of a small object immersed in sample liquids is required, but such volumes are either cancelled out by the procedure or are obtained by calibration in liquids of known density. The methods discussed below are those currently in use for the study of protein solutions, and all are capable of returning values accurate to $10^{-5}$ gm/ml, or better, relative to values for the reference liquids. A subsection is also given to emergent methods which may become useful for measuring densities of protein solutions or the partial specific volumes of proteins.

## 1. *Pycnometry*

The classical method for measuring densities of liquids is by direct weighing of a container (pycnometer) filled reproducibly at a given temperature with liquids of known and unknown density. This method remains the simplest and least expensive procedure in terms of readily available equipment but may be the most expensive in terms of time and amounts of protein required.

The pycnometer for our purpose is a glass vessel, usually containing between 10 and 25 ml of fluid when filled, and which can be conveniently weighed without evaporation. The handling of the vessel must at all times be performed in ways to minimize any tiny gain or loss of its mass (e.g., not touching the vessel directly with the hands). The filling of the pycnometer reproducibly is an art which must be mastered by trial and error for the style of equipment chosen; the formation of bubbles is a frequent source of error. Often, the temperature of the liquid is kept lower than that of the thermostated bath so that the excess liquid from expansion will flow out of a capillary hole. Another procedure is to fill while both solution and pycnometer are equilibrated in a constant temperature bath. Usually the balance chamber is maintained at a temperature somewhat lower than that of the bath to prevent any loss of liquid by expansion during the weighing, and to minimize condensation of water vapor from the surrounding air. The temperature equilibration system may be a special container of very clean water set within an ordinary thermostated water bath. Most of the water adhering to the pycnometer is carefully wiped off with lint-free absorbing material and the remainder is allowed to evaporate by equilibrating in the environment of the balance chamber.

Details on methodology and diagrams of various types of pycnometers are given in a thorough treatment by Bauer and Lewin (1959). The student is encouraged to refer to this comprehensive source and also to an accompanying article on weighing by Corwin (1959) before commencing work with pycnometers. A useful supplementary source work oriented to protein solutions is found in a discussion of the partial specific volume by Schachman (1957). When examining details on pycnometric procedures, it should be noted whether they apply to the equal-arm type of balance or to the single-arm substitution balances now extant. A summary of the principles of different types of balances is discussed in a chapter on the measurement of mass by Macurdy (1967).

If the liquid volume in a pycnometer can be maintained exactly constant, although undefined, it is evident that the density of an unknown liquid $\rho_u$ can be determined relative to the density assigned to a reference liquid $\rho_r$. The ratio of the true weights of unknown to known liquids having identical volumes, i.e., the specific gravity, is related to the densities by

$$\rho_u = \frac{g_u}{g_r} \rho_r \tag{33}$$

For studies of aqueous systems, the reference liquid is simply degassed pure water. Both the weights of the protein solution and the corresponding solvent medium are measured and related to the weight of water; in some cases the density of the solvent medium at the temperature of interest is already known from density–composition tables, and this can be compared with the observed density as a test of the procedure. The true weights in Eq. (33) are derived from the apparent weights observed at the balance by applying those corrections which are necessary to achieve the level of accuracy desired (cf. Corwin, 1959; Macurdy, 1967). The principal correction after the weights of the balance are calibrated, is to account for the effect of buoyancy in air of the materials weighed relative to the buoyant force on the calibrated counterbalancing weights. Even though the amount of air displaced is the same for reference and unknown solutions (with the pycnometers ordinarily used in protein chemistry), it is obvious that the ratio of the weights in Eq. (33) will be different after adding the same air correction to each because the masses are unequal. Ordinarily, this correction is trivial because the densities of aqueous solutions are usually no more than a few percent higher than that of water, the reference liquid. For many purposes, an average correction of 1.2 mg/ml for the weight of air is applied; tables for air buoyancy corrections on this basis, relative to the type of counterbalancing weights employed, are in general use. The accuracy of the weighings, however, cannot be stated to better than a few parts in $10^5$ in uncontrolled weighing rooms because the density of air fluctuates over 10% annually in many localities. For highest accuracy, the actual density of the air at the time of each weighing should be used. [This is calculated from readings taken for temperature, barometric pressure, and relative humidity in the balance room (cf., e.g., Bauer and Lewin, 1959).]

For the purpose of focussing on the level of precision often demanded of this technique, we will assume that approximately 100 mg of a purified protein is available for determining a value of $\phi_2$. This will allow 10 ml of a 1% protein solution to be prepared. As noted before, the density of an aqueous solution is increased by about 0.003 gm/ml (the density increment) for each 1% increase in the concentration of a protein. (Note: salts, buffer ions, and many other solutes in relatively large amounts in the solvent medium also increase the density substantially. This has the effect of reducing the density increment contributed by the protein, because the latter is displacing a high-density instead of a low-density solvent.) If the pycnometer contains exactly 10 ml when it is equilibrated at 20°C, the solvent (pure water) should weigh 9.98234 gm. If the above density increment (0.003 gm/ml) applies exactly, the equilibrated protein solution in the same pycnometer will weigh only 30 mg more than the water. Thus, an overall error of 0.1 mg resulting from the various manipulations (cleaning, filling, drying, and wiping the pycnometer, weighing, temperature regulation, etc.) gives rise to an error of 1 part in 300 for the density increment. Simply touching the pycnometer with the hands or an adhering speck of dirt from the thermostat bath can easily cause this much error. An overall precision of 0.1 mg for this kind of operation is not easy to achieve. The effect of this small error in the value of $\phi_2$ for this case is easily calculated. If the experimental weight of the protein solution is 0.1 mg less than the true value, a value of 10.01224 gm will be recorded instead of 10.01234 gm. Hence, the density of the protein solution $\rho$ is

$$\rho = \left(\frac{10.01224}{9.98234}\right)\rho_{H_2O} = 1.001224$$

where $\rho_{(H_2O)}$ at this temperature corrected to vacuum is taken as 0.998234 gm/ml from the extensive tables of Plato (1900) on the density of sucrose–water mixtures. Applying Eq. (14) and noting that $(\rho - \rho') = 1.001224 - 0.998234 = 0.002990$ gm/ml and that $c_2 = 0.01000$ gm/ml, we have for $\phi_2$

$$\phi_2 = \frac{1}{0.998234} - \frac{0.002990}{0.998234 \times 10^{-2}} = 0.7022 \text{ ml/gm}$$

On the other hand, the correct value of $\phi_2$ is $0.701_2$ ml/gm, i.e., 0.001 ml/gm lower, which is a quite acceptable error for most purposes. By diluting the protein solution with the diluent (water) on the balance to obtain an accurate weight fraction of the known stock solution, other values of the density may be determined in like manner. In this way, one may prepare a plot of $\rho$ vs. $c_2$ from which a least-squares fit is made to obtain the limiting slope $(\partial \rho / \partial c_2)_m^0$. Ordinarily, a higher concentration (2 to 5%) of the protein is used, if available, for the stock solution so that good precision may be obtained with the diluted samples. With a sufficient number of accurately determined points, the possibility of curvature may be explored if a value of $\bar{v}_2^0$ is required rather than $\phi_2$.

Volumes of solution which are much larger than in the example above can lead to only a marginal increase in the precision. The loss in balance sensitivity, the uncertainties associated with a larger glass surface, and the greater probability of changes in volume with size, generally, are but some of the causes for a falloff in precision. A volume of 30 ml has been considered an upper limit for good precision with the usual laboratory conditions and equipment. Although smaller pycnometers have been used successfully (e.g., volumes of 2 ml or less), it should be obvious from the foregoing example with a volume of 10 ml, that more elaborate precautions and skill in weighing are required to achieve precisions an order of magnitude better than 0.1 mg.

The pycnometer method, as ordinarily applied, is not suitable for those studies requiring many accurate density values or measurements which must be carried out quickly. Also, studies on the variation of pressure and temperature would pose severe difficulties. Nonetheless, this classical method is very accurate and will remain preeminent until precision instruments which can measure densities rapidly on small volumes of solution become generally available.

## 2. *Density-Gradient Columns*

Density-gradient columns received extensive development and application at the Carlsberg Laboratorium (Linderstrøm-Lang and Lanz, 1938). Details of the method are found also in the review by Bauer and Lewin (1959) and by Schachman (1957). The principle is that of floating equilibrium, in which an object free to move in a medium of varying density stabilizes at a level where object and medium are isodense. In this method, the medium is a vertical column of a nonpolar liquid mixture, in which the density increases almost linearly from top to bottom, and the object is a tiny, immiscible droplet of the sample solution. A series of these droplets will align themselves when at rest, at a common height in the column if the gradient is uniform and remains undisturbed. For samples in aqueous solution, the density gradient is prepared with a pair of solvents, such as purified kerosene and bromobenzene. The mixing of these solvents to form a fairly stable, linear gradient at a particular temperature is an art. Usually columns about 20 cm in height are used for convenience; to achieve density differences accurate to 1 part in $10^5$, the density difference from top to bottom of such columns should not be greater than about $10^{-2}$ gm/ml. The columns are then saturated with water vapor so that sample droplets will not lose water when equilibrating in the nonpolar mixture. Droplets of less than a microliter in volume are introduced, without catching air bubbles, via specially prepared micropipets. Samples of crystalline proteins have also been used (McMeekin *et al.*, 1954). The final positions of the droplets are measured with a cathetometer and these heights are then related to the average height of different standard droplets which are used for calibrating the gradient. Overall precisions of better than $1 \times 10^{-5}$ gm/ml in density have been achieved with this method. It has been used for a variety of purposes in protein chemistry other than for partial volume determinations, e.g., experiments on hydrogen–deuterium exchange, on enzyme mechanisms via isotopic intermediates, and on the kinetics of enzyme-mediated reactions (cf. Linderstrøm-Lang, selected papers 1962).

Despite the low equipment costs, the high precision, and the unique advantage of using negligible amounts of sample, the method has yielded to other techniques. The expertise required to prepare and maintain the gradient columns and to perform the measurements with meticulous care poses a stumbling block. In addition, the method must be considered slow by current standards. Each droplet requires 15 min to reach equilibrium position and several droplets of a given sample are applied to obtain an average height; also, calibration droplets of different density, to bracket the unknowns, must accompany each measurement. Further, several hours are required for a given column to recover from the disturbance to the gradient after a measurement is concluded. Several columns in a standard-sized water bath, however, can be accommodated. Although somewhat versatile, the method does not lend itself easily to studies as a function of temperature or pressure nor to those in which drastic differences in density of the samples are encountered. Theoretical questions have been raised which may be difficult to answer in a given case. Is the droplet truly immiscible (except for water exchange between the phases)? Are electrostatic charge effects at the interfaces negligible? Does the protein or other solute within the droplet redistribute as a function of composition of the gradient medium? In general, the self-consistency of the data and comparisons with other data, where available, support the conclusion that such effects usually are trivial.

### 3. *Isopycnic Temperature Method*

This method, known historically as the *flotation temperature method* (cf. Bauer and Lewin, 1959), depends on the precise variation of the temperature of a solution until its density is identical to that of a calibrated glass diver. The temperature of flotation, or isopycnic temperature, where diver and solution have the same density, can be approached with good accuracy by noting the rate of rise or fall of the diver as the temperature is varied. In practice, the isopycnic temperature is usually proportional to the concentration of a solute which is used to increase the density of a medium (Richards and Shipley, 1914; Randall and Longtin, 1939). The application of this principle to the measurement of the density of dilute protein solutions was developed by Hunter (1966, 1967). The apparatus, shown in Fig. 2, is simple and inexpensive

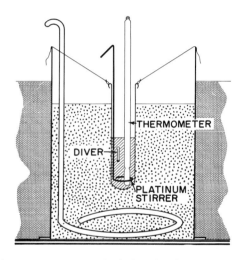

Fig. 2. Isopycnic temperature method for density. A glass diver, a platinum stirring rod, and about 3 ml of the solution are placed in the test tube (1.5 × 12.5 cm) which is suspended in the large cylinder (containing about 1 liter of water) within a thermostat bath. The test tube is adjusted for height and vertical alignment by three thin Nylon lines looped through fused hooks around the top of the tube; the lines are then drawn taut under a rubber band around the top of the large cylinder. Graduated marks on the large cylinder are used to judge the rate of movement of the diver in order to estimate temperature adjustments; the temperature is too low if the diver rises and too high if the diver sinks. The temperature of the bath is adjusted rapidly until it is close to the isopycnic temperature. The system is then allowed to equilibrate for 20 min, after which about five observations of the diver's behavior are made at increments of 0.01°C in the region of the isopycnic temperature. A density measurement usually can be completed in less than 45 minutes.

The divers are made from melting point capillaries. Lengths of 1 cm are heat sealed to contain more glass at one end so that the diver will stand vertically in the solution. [From Hunter (1966); reproduced by permission from the author and the American Chemical Society.]

to assemble if a good variable temperature control system is available. The technique can yield densities of high precision; the uncertainty currently realized for the isopycnic temperature is 0.01°C when this temperature is approached slowly from either direction. In aqueous solution, an interval of 0.01°C corresponds to an uncertainty in the density of about $2 \times 10^{-6}$ gm/ml. A given diver is calibrated with solutions of known density, such as with aqueous KCl mixtures. Density–temperature profiles at a given composition of a calibrating solution must be prepared from existing data or must be determined. The density of KCl–water solutions exhibits a rectilinear relationship with temperature for calibrating the glass divers; a difference of about 0.0002 gm/ml in density between 20°C and 35°C is observed with dilute KCl solutions. Since the densities of a series of protein solutions varying in $c_2$ are determined at somewhat different temperatures, only the apparent specific volume $\phi_2$ can be obtained with a given diver. If $\phi_2$ is not concentration dependent at any of the isopycnic temperatures, the value of each $\phi_2$ is equivalent to $\bar{v}_2$ at the specified temperature (Section II,D). With this method, the change in $\phi_2$ per degree of temperature ($d\phi_2/dT$) may be obtained directly from the data on a dilution series in $c_2$. Precisions in $\phi_2$ at a given temperature have been attained which are better than $\pm 0.0002$ ml/gm. In summary, the technique is an inexpensive and reasonably rapid one for determining the apparent specific volume of proteins on relatively small amounts of solution ($<3$ ml). In addition, the variation of $\phi_2$ with temperature can be obtained by using slightly different concentrations of a protein.

### 4. *Magnetic Densimetry*

A rapid method based on the magnetic balancing principle currently offers the most versatility while requiring only 0.3 ml or less, of solution per measurement. Lamb and Lee (1913) first applied the principle of suspending magnetically a ferromagnetic float or buoy in solutions to obtain their density. The magnetic force, measured as electrical current to a solenoid, balances any difference in the opposing effects of gravity and buoyancy on the buoy. The technique was modified and adapted to various uses with great accuracy but it remained a rather laborious one and required relatively large volumes of solution for good precision (cf. Geffcken *et al.*, 1933; MacInnes *et al.*, 1951; Dayhoff *et al.*, 1952). The introduction of a highly stable magnetic support system for the buoy by Beams and collaborators (Beam *et al.*, 1955; Beams and Clarke, 1962; Clarke *et al.*, 1963; Ulrich *et al.*, 1964) enabled measurements to be made rapidly on very small amounts of sample. A schematic diagram of a laboratory model used for experiments on proteins is shown in

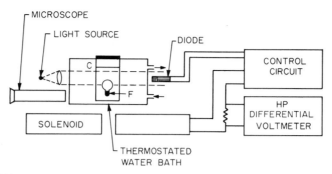

Fig. 3. Diagrammatic sketch of the components of the magnetic densimeter with optical sensing. The float or buoy (F) in the thermostated cell (C), containing 0.2 to 0.3 ml of liquid, is pulled down to a chosen height in the microscopic field by the air-core solenoid. The position of the buoy is maintained by automatic current adjustment to the solenoid via the amount of light reaching the photodiode receptor in circuit with the solid-state servo system. A precision resistor in series with the solenoid allows the voltage drop across the resistor to be measured with a differential voltmeter; the voltage drop is directly proportional to the solenoid current. A change of 0.1 mV corresponds to approximately $2 \times 10^{-6}$ gm/ml difference in the density of the liquid, depending on the value of the denominator of Eq. (35). A sensitivity of $\pm 0.01$ mV may be achieved with close temperature control. [From Senter (1969); reproduced by permission from the author and the American Physical Society.]

Fig. 3 (Senter, 1969). A tiny ferromagnetic cylinder, made magnetically soft by annealing in hydrogen at 1120°C and jacketed with glass or Kel-F, is held at a precise height in the solution by an electronic servomechanism to an air-core solenoid. As the buoy tends to sink or float, a change of impedance in a remote sensing coil below the solution, or a change in the light striking a tiny photodiode (Fig. 3), signals an adjustment of the current to the solenoid, in order to prevent any up or down movement. The magnetic force required to suspend the buoy is a function of the height (along a vertical distance $z$ from the solenoid) and of the buoyant force exerted by the solution. The magnetic force generated by the current is a product of the moment $M$ and the field gradient $(\mathrm{d}H/\mathrm{d}z)$. Thus, if the buoy is held at precisely the same height in all measurements, the current to the solenoid reduces to a function only of the density of the solution. If the buoy is more dense than the solution, the magnetic force required to balance a difference in the opposing forces of gravity and buoyancy is

$$M\left(\frac{\mathrm{d}H}{\mathrm{d}z}\right) = V_\mathrm{b} g(\rho_\mathrm{b} - \rho_\mathrm{s}) \qquad (34)$$

where the subscripts b and s refer to buoy and solution, respectively, and $g$ is the acceleration of gravity. To retain positive force, the densities in brackets are reversed if the buoy floats; in this case, the buoy must be submerged to a position $z$ by a magnetic force emanating below the cell, as shown in the figure. The solenoid current $I$ is proportional to the moment within very narrow limits for high permeability materials, and it is strictly proportional to the field gradient. Hence, the magnetic force is essentially proportional to the square of the current, which in turn is proportional to the difference in density between buoy and solution; the proportionality constant includes $g$ and the fixed volume of the buoy, which need not be measured. Since the density of the buoy is constant at a given temperature and pressure, varying $\rho_s$ with the current, using reference liquids, yields a calibration expression for use with unknown solutions which is independent of $\rho_b$. In practice, a series of standard solutions is prepared on the analytical balance for calibrating a given buoy at the desired temperature. The absolute densities of sucrose solutions by Plato (1900) are the most extensive tables available at 20°C. However, the densities of a variety of other solutes in water at various temperatures may be found in the International Critical Tables (Beattie *et al.*, 1928) and in the Landolt-Börnstein Tabellen (Bein and Langbein, 1905). These data can be utilized to give densities at any concentration by curve fitting, as pointed out in Section II,F. [In this connection, it is very timely to press for a new table of absolute densities using a solute, such as CsCl, which can be maintained ultrapure and be dissolved in water to cover a very broad range of densities (viz., 1.0 to 1.8 gm/ml) without becoming viscous.] The densities of the calibrating solutions as a function of the current to the solenoid are then fitted to a curve so that the density at any value of the current is available. In the writer's experience, the relationship between density and the observed voltage (measured across a precision resistor) is so closely quadratic that a simple, rectilinear, least-squares fitting of the square of the voltage *versus* density invariably correlates to within 2 parts in $10^6$ for a given buoy. This operation is easily handled with a desk computer or calculator. The densities of unknown solutions are then calculated from the square of the observed voltage and the slope and intercept values of the rectilinear calibration equation. Correlation coefficients ($r$) with a similar or slightly lower goodness of fit ($r \cong 0.99999$) are routinely obtained assuming a straight-line relationship for $\rho$ *vs.* $c_2$ when the protein concentration is varied by direct weighing, over a concentration range of several percent. Thus, a high level of confidence may be given the values of $(\partial\rho/\partial c_2)_\mu^0$ and $(\partial\rho/\partial c_2)_m^0$ from which to calculate $v^{0\prime}$ and $\bar{v}_2^0$, respectively (i.e., confidences relative to the densities assigned to the reference solutions and to the concentration of protein assigned to the stock solution).

Instruments with solid-state control designed for routine use in biochemical experiments have been described (Senter, 1969; Goodrich et al., 1969). In the Senter model, a small solenoid is placed below the solution to permit greater access to the sample area, and optical sensing of the buoy position instead of electrical impedance is utilized so that no particular skill is needed for operating the controls. Special circuitry allows the buoy to be maintained at a preset height automatically. With this feature, the current required to hold the buoy may be displayed on a time-chart recorder when the density changes as the result of a reaction. A total time of about 5 min between measurements on successive samples can be sustained; this includes approximately 2 min for temperature equilibration. The smallness of the cell requires little circulating fluid for thermostating; the temperature can therefore be changed readily via a small capacity temperature controller. Since glass expands only about 10% as much as water at the usual experimental temperatures, the temperature control with glass-jacketed buoys should be maintained within 0.005°C for $1 \times 10^{-6}$ gm/ml precision ($\Delta\rho/0.01°C$ for water is about $2 \times 10^{-6}$ gm/ml at 20°C). The bulk expansion of Kel-F with temperature is much closer to that of water so that similar precision can be achieved with temperature fluctuations of about 0.025°C. Glass-jacketed buoys, however, require calibration less frequently. For best sensitivity, the density of a given buoy should be within about 0.03 gm/ml of that of the solutions to be measured. A series of buoys of different density are therefore needed in order to make accurate measurements over a wide range of solution densities; usually, however, one buoy can cover a difference in protein concentration of nearly 0.10 gm/ml in a given solvent. The sensitivity $(dI/d\rho)$ is given by the relation

$$\frac{dI}{d\rho} = -\frac{I}{2(\rho_b - \rho_s)} \tag{35}$$

Instrumentation on magnetic densimetry is under continuing development. Efforts are being made to eliminate the need for multiple buoys and to cancel the magnetic moment so that the current is strictly proportional to the field gradient only (Beams, 1969). In addition, with remote drive coils around the cell, it has been found possible to rotate the buoy slowly and steadily while in support, and thus to obtain the viscosity simultaneously with the density (Beams and Hodgins, 1971; Kupke et al., 1972). The magnetic balancing principle is also suitable for carrying out density measurements as a function of pressure. With the cell enclosed in a pressure bomb, compressibilities of solutions and apparent specific volumes of proteins have been reported over a range of pressures corresponding to those developed in centrifugal fields (Fahey et al., 1969).

## 5. Other Methods

Two additional density methods have recently been proposed for the study of protein solutions. These have received as yet only preliminary application from investigators; attention, however, is drawn to the developmental aspects. One (Cahn Instrument Co., Paramount, California) is an adaptation of the Kohlrausch hydrostatic weighing method which is based on Archimedes principle. A sinker attached to an analytical balance is suspended by a fine wire in the liquid to be measured. The density of the liquid is the loss in weight of the sinker when immersed in the liquid divided by the volume of the sinker. The method has been developed to handle small volumes of solution ($<2$ ml) by utilizing an automatic electrobalance of high sensitivity and small glass sinkers (about 1 cm$^3$). The density cell is so constructed and located that evaporation and the effect of different operating temperatures on the balance are considered negligible. The surface tension effect from the fine wire, while greatly minimized, must still be reckoned with; the contact angle between wire and liquid is a function of the composition of the solution——including impurities. The sinkers are calibrated for volume by immersing them in liquids of known density, preferably liquids having compositions as close as possible to that of the samples. Special precautions are required to minimize vibrations which affect the measurement of the contact angle for the surface tension correction.

A densimeter based on a measuring principle different from those of the classical procedures has now been described in some detail (Kratky et al., 1969; cf. Stabinger et al., 1967). In brief, the device is a small pycnometer ($<1$ ml) in which masses are compared by applying a vibrational force. The density of a medium is related to the change in resonance frequency of a laterally vibrating tube made of a special glass. The total mass of the tube and that of the enclosed liquid (or gas) is inversely proportional to the square of the resonance frequency; the oscillation period $\tau$ is the reciprocal of this frequency. By determining a calibration constant A for the instrument, using air and water as reference media, the difference in density between unknown liquids and water is related to the difference in the square of the oscillation periods. Thus

$$(\tau_u^2 - \tau_{H_2O}^2) = A(\rho_u - \rho_{H_2O}) \tag{36}$$

where the subscript u refers to the unknown material. A quartz timer-oscillator system measures accurately the time for the completion of a relatively small number of oscillations (e.g., $2 \times 10^4$); in this way it is possible to obtain values of $\tau$ in less than a minute ($\tau$ is given in units of $10^{-5}$ sec, and the numerical value of $\tau$ for aqueous solutions is on the order of $5 \times 10^6$ units). A seven-digit numerical display acts in concert with the precision timer. Results with a standard NaCl–water solution relative to pure water indicate that a difference of about 1.5 units in $\tau$ corresponds to $1 \times 10^{-6}$ gm/cm$^3$ in $\Delta\rho$. It appears that standards closer in density to that of the samples may be employed to increase the sensitivity. A measuring cycle of <30 min is indicated, which includes insertion of the sample, temperature equilibration (about 15 min), rinsing, drying, and a measurement in air. The apparent specific volume of a number of proteins and other polymers were reported. By the use of stronger materials, it was found possible to measure density differences under pressures up to 10 atm and at temperatures up to 230°C; the accuracy, however, was limited to about $10^{-4}$ gm/cm$^3$ because of the more rugged design.

As in pycnometry, the volume of the liquids being compared must be appropriately cancelled. Apparently, an overfilling procedure is currently being employed so that the masses in the measurement region are at constant volume. With containers which are about 1 ml in volume, the precision of filling the sensitive region should be better than 0.01 $\mu$l in order that the density data be generally useful in the study of proteins. Also, the overall effect of the mechanical agitation on protein solutions has yet to be evaluated. Nonetheless, the convenience, speed, and versatility which are indicated suggest that this approach will be useful for many purposes.

Finally, brief mention is made of two recent methods, not employing density measurements directly, by which the partial specific volume of proteins have been determined. Edelstein and Schachman (1967) have described a procedure with the analytical ultracentrifuge in which $\bar{v}_2$ at very low concentrations may be obtained by a differential technique using waters of different densities ($D_2O$, $D_2{}^{18}O$). The absolute concentration of the protein need not be known. The technique makes use of the fact that the conventional plot of the logarithm of a concentration parameter *versus* the square of the radial distance at equilibrium yields a slope proportional to the molecular weight and the density derivative $(\partial\rho/\partial c_2)_\mu{}^0$. Hence, if the same monodisperse protein is measured in ordinary water and in water of greater density, the difference in the slopes from the sedimentation data reflects the difference in the density derivatives; the latter are related to the partial specific volume, provided that the solvent is a single component [in multicomponent systems, $\phi'$ can be estimated if preferential interaction is assumed to be constant (Thomas and Edelstein, 1971)]. By combining the two equilibrium equations, the value of $\bar{v}_2$ is extracted after adjusting for deuteration of the protein. Since the method depends on the density increment ($\Delta\rho/c_2$) decreasing as the density of the solvent increases, greater accuracy is achieved with $D_2^{18}O$ than with the less dense $D_2O$ (the latter component, however, is readily available).

The other method for determining partial specific volumes of proteins is to measure the buoyancy on a dialysis bag of the protein solution while it is suspended in the solvent medium (Jaenicke and Lauffer, 1969). If the solvent is water, or one which behaves essentially as a one-component solvent, the net buoyant force on the bag reflects the partial specific volume of the protein. The change in weight of the bag membrane relative to one containing no protein is measured with a very sensitive calibrated quartz spring (Stevens and Lauffer, 1965). The overall procedure was designed to give a measure of the degree of hydration of proteins in multicomponent systems. By assuming $\bar{v}_2$ to be constant while a quasi-inert, diffusible component, such as glycerol, is added to the solvent, any change in the predicted buoyancy is assigned to a defined hydration value for the protein. (The buoyancy due to $\bar{v}_2$, if constant, decreases predictably as the density of the solvent is increased by the added glycerol.) The method must be considered slow with respect to determining $\bar{v}_2$ (48-hr equilibration time plus control experiments); with additional effort, however, the opportunity is available for securing a measure for the hydration (which requires density values for the solvents). By the direct density method, the extraction of hydration values (also under special conditions) similarly requires equilibrations and multiple measurements as the solvent composition is varied (Appendix B).

## B. Applications

### 1. *Volume Change*

The change in volume $\Delta V$ for a process may be derived from accurate density measurements if a suitable protocol can be devised for the system under study. Dilatometry (Section III) is the predominant method which has been heretofore used for this purpose. Excellent agreement has been obtained, however, between density and dilatometry measurements on certain enzymic reactions (Linderstrøm-Lang and Lanz, 1938). In principle, the change in volume on mixing reactants follows from the detailed definition of density, Eq. (9). The density of both sets of reactants is measured, giving values denoted as $\rho_a$ and $\rho_b$; weighed amounts of each set $\left(\sum_{i=1}^{N} g_i\right)_a$ and $\left(\sum_{i=1}^{N} g_i\right)_b$ are then mixed and the density of the mixture $\rho_{(a+b)}$, is determined at the same temperature as that for the individual sets of reactants. The volume before mixing $V^B$ is given by

$$V^B = \frac{\left(\sum_{i=1}^{N} g_i\right)_a}{\rho_a} + \frac{\left(\sum_{i=1}^{N} g_i\right)_b}{\rho_b} \qquad (37a)$$

and the volume after mixing $V^A$ is

$$V^A = \frac{\left(\sum_{i=1}^{N} g_i\right)_a + \left(\sum_{i=1}^{N} g_i\right)_b}{\rho_{(a+b)}} \qquad (37b)$$

The difference $V^A - V^B$ gives the volume change $\Delta V$ for the reaction. The density method for this purpose has not been used widely for proteins because of the inconvenience and because of the relatively large amounts of solution required by conventional pycnometric procedures. The experiments with density by Linderstrøm-Lang and Lanz, cited above, were carried out with gradient tubes, which required only microamounts of protein. Obviously, if densities can be measured conveniently and accurately on small amounts of protein, the measurement of $\Delta V$ by the density method becomes expedient. Also, a wider latitude of experimental systems than that just described in Eqs. (37a) and (37b) can be studied, depending on the versatility of the particular density technique (e.g., variation of temperature and pressure).

## 2. Partial Volumes

Not infrequently, the question is raised: "Which of the values reported for the partial specific volume of protein X is the correct one?" We may first inquire: "Why is there a variety of significantly different values of $\bar{v}_2$ (or $\phi_2$, usually) reported for protein X?" Values of $\bar{v}_2$ differing by more than 5% are found in the literature for some of the best characterized proteins. [We eliminate here those values reported as $\bar{v}_2$ for which, in fact, $\phi'$ was determined (Section II,E); without additional information, it is not possible to relate these quantities.] The answer to this second question cannot be simply that most, if not all, of the investigations on protein X were bungled. Almost invariably, any published work on the partial specific volume of a protein was done carefully. We need not suspect, moreover, that the various authors did not properly evaluate the stated precision of their work. Other factors must also be considered, such as (a) the definition of the protein component actually used, (b) the nature of the solvent medium, and to a lesser extent (c) differences in temperature of the various determinations. The latter factor we may virtually discount except for possible cases in which the protein undergoes a transition over the temperature range covering the reported experiments. The data of Hunter (1966) and others suggests that a difference of 10°C affects the value of $\bar{v}_2$ by less than 1%. The effect of different solutes in the aqueous medium has not been well documented. However, except for significant changes in $\bar{v}_2$ with H⁺ concentration (Charlwood, 1957; Kay, 1960), the presence in the solvent of moderate amounts of salt and buffer species, as commonly employed, has not been shown to cause gross changes in $\bar{v}_2$. Solvents which severely affect the structure of proteins by causing hydrolysis, covalent additions, or complete unfolding are of course, excluded, in these comparisons. Hence, we are left with variations in the definition of protein X as a major source of discrepancy among the reported values of $\bar{v}_2$. The term *definition* refers here to the composition of the material which is called *the protein component*. Values of $\bar{v}$ for different preparations of a given protein may vary significantly. For example, a preparation which was not made isoionic despite prolonged dialysis may include significant amounts of small ions which themselves have partial specific volumes in water about half that of the polypeptide material (viz., 0.35 versus 0.70 ml/gm). Applying the relation

$$\bar{v}_i = \sum_{j=1}^{N} \bar{v}_j f_j \tag{38}$$

for a mixed component $i$ having N species $j$, a 5% w/w impurity of counterion ($f_j = 0.05$) with $\bar{v}_j = 0.35$ ml/gm yields $\bar{v}_i = 0.683$ ml/gm rather than 0.70 ml/gm for the pure isoionic polypeptide. This much impurity may correspond to only about 1 mole of small ions per $10^3$ gm of polypeptide. The discrepancy (about 2½%) is about six times greater than that which is expected from an error of 1% in evaluating the protein concentration, which usually is the major error of the operation. Nucleotides, and many other organic substances which may accompany protein preparations extracted from tissues, also have partial specific volumes which are substantially lower than that of polypeptides; lipids, on the other hand, tend to raise the average value of $\bar{v}_2$. Evidence in support of the hypothesis—that variable preparations have led to variable values of $\bar{v}_2$—is the fact that ultraviolet extinction coefficients of some refined proteins in water differ as much as 5% between laboratories; this is quite outside a reasonable margin for experimental error. Unquestionably, methods of purification have advanced much in the recent past and more uniform preparations of many proteins are now available. The preparation of deionized samples via mixed-bed ion-exchange resins has become more commonplace, and it is becoming accepted practice to begin quantitative studies on a protein which was first made isoionic, if at all possible. From these considerations we might hazard an answer to the question posed at the outset. The investigator who wishes to know the correct value of $\bar{v}_2$ for protein X very likely has in mind to borrow the value to use in his experiments. Probably, the most appropriate value of $\bar{v}_2$ for protein X is the one determined on the same preparation and solvent medium which are to be used in his experiments.

The application of specific volumes to molecular weight determinations by both velocity and equilibrium sedimentation is well known. In a solvent consisting of essentially one component (e.g., water), the partial specific volume of the protein appears in the buoyancy term $(1 - \bar{v}_2\rho')$ of the sedimentation equations as derived originally (Svedberg and Pedersen, 1940). Since the values of $\bar{v}$ for proteins fall in the range 0.70 to 0.75 ml/gm, it is evident that any error in $\bar{v}_2$ is magnified about 3-fold in the buoyancy term, which is inversely proportional to the molecular weight. However, no solvent medium used for sedimentation studies on proteins is truly a single component. The presence of salts to overcome the primary charge effect and/or the introduction of buffers to maintain pH may give rise to preferential interaction (see Section IV,B,3). In equilibrium sedimentation, for example, we deal with a host of phases which are in osmotic equilibrium with one another along the radial path of the solution. Hence, the isopotential specific volume $v'$, or usually $\phi'$ (Section II,E), instead of $\bar{v}_2$ is inserted in the buoyancy term (Casassa and Eisenberg, 1961). More conveniently, the derivative $(\partial\rho/\partial c_2)_\mu^0$ at vanishing protein concentration (Section II,E) is substituted for the buoyancy term in multicomponent systems (cf. Reisler and Eisenberg, 1969). The derivative has recently been applied also to low-angle X-ray scattering measurements in multicomponent solutions (Eisenberg and Cohen, 1968). Many of the aqueous solvents used in sedimentation studies, however, contain only small amounts of added solutes ($<1\%$), and may be considered as essentially one-component solvents. Hence, the use of the classical partial specific volume leads to no great error from preferential interaction if Donnan effects are minimized. Accordingly, it is still common practice to measure only the density of the solvent or dialyzate and to borrow a value for $\bar{v}_2$. Nonetheless, it is incumbent on the investigator who is striving for more than an approximate molecular weight value to carry out density determinations on both the equili-

$$\delta_1 = (c_1 - c''_1)/c_2$$
$$\delta_3 = (c_3 - c''_3)/c_2 \tag{42}$$

Since $\phi_2$ is the quantity most easily measured at finite values of $c_2$ rather than the quantity $\bar{v}_2$, the relations leading to and including Eqs. (41) may be substituted with like relations containing $\bar{v}'_1$ for $\bar{v}_1$ and $\bar{v}'_3$ for $\bar{v}_3$ (see Section II,D for apparent quantities). This simplifies the experimental procedure, because $\bar{v}'_1$ and $\bar{v}'_3$ may be taken directly from density–composition relations for the two-component solvent media. As a rule, of course, there is no appreciable difference between the values $\phi_2$ and $\bar{v}_2$, the difference being a second-order variation over a fairly broad range in protein concentration. Nonetheless, $\phi_2$ is not necessarily equal to $\bar{v}_2$ in all experiments; therefore, the apparent specific volume $\phi_2$ and the corresponding quantities $\bar{v}'_1$ and $\bar{v}'_3$ will be substituted in the ensuing discussion. With this in mind, it will be noted that Eqs. (42) can easily be cast in terms of the dialyzate quantities for components 1 and 3 instead of $c''_1$ and $c''_3$. That is, the volume $V'$ of the solvent medium (the dialyzate) and that of the isomolal protein solution $V$, containing identical amounts of components 1 and 3, are related according to Eqs. (15) and (16) by $V' = V(1 - \phi_2 c_2)$. Hence, the term $(1 - \phi_2 c_2)$ represents the fraction of a milliliter of the solvent containing the same masses of the diffusible components as are in 1 ml of the preequilibrated (isomolal) protein solution, and

$$c''_1 = c'_1(1 - \phi_2 c_2)$$
$$c''_3 = c'_3(1 - \phi_2 c_2) \tag{43}$$

which are identical to Eqs. (28) except that double primes are employed to distinguish the preequilibrated solution from the equilibrated one.[5] Since the dialyzate is unlimited in amount (for instance, because it is exchanged with fresh solvent, occasionally, during the dialysis), the concentrations $c'_1(1 - \phi_2 c_2)$ and $c'_3(1 - \phi_2 c_2)$ before or after dialysis correspond precisely to the concentrations of these diffusible components in the protein solution *prior* to the dialysis. Equations (42) then become

---

[5] When apparent quantities are used, the thermodynamic volume contribution of the diffusible components, $\sum_{j \neq 2} \bar{v}_j c''_j$, in 1 ml of the isomolal protein solution is not necessarily identical with the apparent volume contribution, $\sum_{j \neq 2} \bar{v}'_j c''_j$; i.e., $c'_j(1 - \bar{v}_2 c_2) \neq c'_j(1 - \phi_2 c_2)$. Hence, the values of $c''_1$ and $c''_3$ as derived in Eqs. (42) can be different, in theory, from the corresponding values seen when Eqs. (42) are derived with the use of $\phi_2$ as was done in Eqs. (43).

$$\delta_1 = [c_1 - c'_1(1 - \phi_2 c_2)]/c_2$$
$$\delta_3 = [c_3 - c'_3(1 - \phi_2 c_2)]/c_2 \quad (44)$$

We may define the interaction parameter $\delta$ as the difference in mass of a given diffusible component per gram of protein between "apparent" equivalent volumes of the diffusible components in the equilibrated solvent and protein solutions. If preferential interaction is not zero, i.e., if $(\rho - \rho'')/c_2 \neq 0$, one subscripted value of $\delta$ will be positive and the other subscripted value will be necessarily negative. With this definition of $\delta$ it becomes clear that the experimental protocol does not depend on maintaining either a constant volume of the protein solution or a fixed concentration of the protein during the equilibration. The composition of the dialyzate, determined by a density measurement, predetermines the concentration of the diffusible components in the isomolal, predialyzed protein solution at that value of $c_2$ which is found in the postdialyzed protein solution. Hence, the method for calculating $\delta_1$ and $\delta_3$ from the density measurements rests only on an accurate analysis of the protein concentration after equilibrium is achieved. A reasonably accurate value of $\phi_2$ in the particular two-component solvent chosen must also be known.

The net masses of the diffusible components transported across the membrane per unit of protein during the equilibration are related to the conventional thermodynamic interaction parameter $\xi$ by examining the definition of $\xi$, Eq. (21). At equilibrium, the concentrations $c_1$ and $c_3$ in the protein solution are $(c''_1 + \Delta c_1)$ and $(c''_3 + \Delta c_3)$, respectively. This follows from the foregoing discussion on the mass exchange, $\Delta g_j = (g_j - g''_j)_{j=1,3}$ when $\Delta V = 0$. For $\xi_3$ we rewrite Eq. (21) and

$$\xi_3 = \left[\left(\frac{c''_3 + \Delta c_3}{c''_1 + \Delta c_1}\right) - \left(\frac{c''_3}{c''_1}\right)\right] \bigg/ \left(\frac{c_2}{c''_1 + \Delta c_1}\right) = \left(\Delta c_3 - \Delta c_1 \frac{c''_3}{c''_1}\right) \bigg/ c_2$$
$$= -\left(\Delta c_1 \frac{\bar{v}'_1}{\bar{v}'_3} + \Delta c_1 \frac{c''_3}{c''_1}\right) \bigg/ c_2 = \left(\Delta c_3 + \Delta c_3 \frac{\bar{v}'_3 c''_3}{\bar{v}'_1 c''_1}\right) \bigg/ c_2$$

Substituting $\delta_3$ for $\Delta c_3/c_2$, we find $\xi_3$ in terms of $\delta_3$ to be

$$\xi_3 = \delta_3 \left(1 + \frac{\bar{v}'_3 c''_3}{\bar{v}'_1 c''_1}\right) \quad (45)$$

The corresponding relation for $\xi_1$ in terms of $\delta_1$ follows identically except that all subscripts 1 and 3 are interchanged. [For convenience, the ratio of the dialyzate concentrations $(c'_3/c'_1)$ or $(c'_1/c'_3)$ may be substituted for the corresponding ratios of these components in the undialyzed, isomolal protein solution without loss of generality.[6]] According to Eq. (45), $\xi_j$ ($j = 1, 3$) includes not only the gain or loss in mass of the subscripted diffusible component per gram of dry protein during equilibration (i.e., $\delta_j$), but also the mass of this component which was originally together with the amount of the other diffusible component transported in the opposite direction; the word *originally* refers to the condition before protein was added to the pure solvent. It must be emphasized that positive values of the parameter $\xi$, as well as of $\delta$, do not imply that a preferred component is bound or interacts with protein while the ejected component does not interact. In more general terms, both diffusible components could exclude each other from different microregions relative to the uniform composition of the pure solvent (the dialyzate) as a result of the presence of protein. $\xi$ and $\delta$ represent only a net amount of preference, which may be zero, depending on the composition of the solvent. In thermodynamic terms, the activity of each diffusible component in various microregions of the predialyzed protein solution, these regions having properties different from that of the pure solvent, is such that the total activity of each component may balance with that of the respective activities in the pure solvent. At other compositions of solvent, an imbalance in the total activities of each component on the two sides might obtain, so that a net transport of diffusible components at constant volume of the protein solution, i.e., $\Delta g_1 = -\Delta g_3 (\bar{v}'_3/\bar{v}'_1)$, is observed upon dialysis.

In an actual experiment, the values of $\delta$ and $\xi$ are determined at a given value of $c_2$ by dialyzing several samples of the protein solution in dialysis bags against the two-component solvent at the desired temperature. For more rapid equilibration, the composition of the diffusible components in the bags should be nearly like that of the solvent, and good stirring, without evaporation, should be provided. [See also Gordon and Warren (1968) for other equilibration procedures including a reverse technique in which solvent, in effect, is dialyzed against the protein solution.] Periodic removal and analysis of one of the bags establishes when equilibrium is attained. If relatively small amounts of solvent are used, repeated changes may be necessary if one wishes to achieve a dialyzate which has a molality equal to that of the original solvent. This is checked by a density measurement on the dialyzate (Section II,F). At equilibrium, the concentration of protein is measured

---

[6] Substitution of $(c'_3/c'_1)$ for $(c''_3/c''_1)$ in Eq. (45) results in a further simplification of this equation. The term within parentheses may then be written as $(\bar{v}'_1 c'_1 + \bar{v}'_3 c'_3)/\bar{v}'_1 c'_1$. Since the numerator is unity [Eq. (8)], the quantity $\xi_3$ in Eq. (45) reduces to $\delta_3/(\bar{v}'_1 c'_1)$; correspondingly, $\xi_1 = \delta_1/(\bar{v}'_3 c'_3)$. The reduction is made possible by the employment of apparent volumes rather than partial volumes.

on replicate samples, e.g., by ultraviolet absorption, where the extinction coefficient has been determined accurately on a dry weight basis. Usually, dilution of the protein solution is required for light-absorption measurements. This is most accurately done by weighing an aliquot of the solution into a weighed amount of dialyzate. Since the densities in each set of components are known from the density measurements after equilibration, the volume dilution is known almost exactly by

$$\text{Volume dilution} = \frac{\left(\sum_{i=1}^{N} g_i\right)/\rho + \left(\sum_{j=1,3}^{N} g'_j\right)/\rho'}{\left(\sum_{i=1}^{N} g_i\right)/\rho} \quad (46)$$

where $\Sigma g_i$ is the number of grams of the equilibrated protein solution. The error is trivial if $\bar{v}_2$ changes by a small amount when the protein solution is diluted. In some cases, dilution with dialyzate may be considered too inaccurate, because of substantial light absorption by component 3. Another diluent, such as water or other aqueous medium, may be employed to dilute equal aliquots of protein solution and dialyzate, the latter acting as a blank. In general, the amount of diluent should be large by comparison with the solution or dialyzate in order to minimize error as a result of slightly unequal amounts of component 3 in the weighed samples of solution and dialyzate. Precise volume dilutions are necessary for best accuracy when values for the protein concentration are desired on a weight per volume basis.

With a firm value for $c_2$ and the measured values of $\rho$ and $\rho'$ in hand, along with a value of $\phi_2$ predetermined for the protein in the solvent chosen [Eq. (14)], the value of $\rho''$, the density of the corresponding isomolal pre-dialyzed solution, is calculated by

$$\rho'' = \rho'(1 - \phi_2 c_2) + c_2 \quad (47)$$

since $\rho' = c'_1 + c'_3$ and $\rho'' = c''_1 + c_2 + c''_3 = (c'_1 + c'_3)(1 - \phi_2 c_2) + c_2$ [cf. Eqs. (43)]. The values of $\delta_1$ and $\delta_3$ are then obtained with Eqs. (41) by substituting the values of $\bar{v}'_1$ and $\bar{v}'_3$, obtained from density–composition curves on the pure solvent, in place of the values of $\bar{v}_1$ and $\bar{v}_3$, respectively. [This is hardly a substitution at relatively low concentrations of protein, because it is very difficult to determine with certainty any slight differences in these values for the diffusible components between those in the protein solution and those in the pure solvent. If, in the determination of $\bar{v}_2$ (Section II,C) no apparent curvature is observed in the plot of density versus $c_2$ for the isomolal series, it may be presumed that $\bar{v}'_1 = \bar{v}_1$ and $\bar{v}'_3 = \bar{v}_3$, since $\bar{v}_2 = \phi_2$. Conceivably, differences in this property for both of the diffusible components might balance each other and still show $\bar{v}_2 = \phi_2$; however, this seems very unlikely.] Following the calculation of $\delta_1$ and $\delta_3$, evaluation of $\xi_1$ and $\xi_3$ is then carried out with the use of Eq. (45).

$\xi$ may also be calculated from a determination of the components by dry weight analysis on the protein solution and dialyzate. If component 3 is either completely nonvolatile or volatile, Eqs. (10) may be applied to determine the excess or deficiency of component 3 at equilibrium. Again, $\bar{v}'_1$ and $\bar{v}'_3$ are substituted for the corresponding quantities in the protein solution since $\phi_2$ rather than $\bar{v}_2$ is usually the experimental quantity measured. (If the definition of component 2 includes small ions to balance a net change on the protein, both $\phi_2$ and $g_2$ must correspond with that definition in either the dry weight or density methods.) Ordinarily, the tedium of and the painstaking care required in the determination of $\xi$ by dry weight analysis militates against the use of this procedure if values of $\xi$ are to be determined routinely; it is important, however, to check the efficacy of a chosen protocol for the density technique occasionally, and the dry weight method, if applicable, is the most accessible one in many laboratories.

It is evident that greater confidence is gained in the values of $\xi$ obtained by the density method if a dilution series in $c_2$ is performed concurrently on both the isomolal and the equilibrated (isopotential) systems. In this way, also, any curvature in the plots of density *versus* protein concentration can be evaluated, and values for $(\partial \rho / \partial c_2)_m^0$ and $(\partial \rho / \partial c_2)_\mu^0$ at vanishing protein concentration can be assessed with greater accuracy. For the limiting case, the difference between the isopotential and isomolal slopes is

$$\left(\frac{\partial \rho}{\partial c_2}\right)^0_\mu - \left(\frac{\partial \rho}{\partial c_2}\right)^0_m = \xi_1^0(1 - \bar{v}'_1 \rho') = \xi_3^0(1 - \bar{v}'_3 \rho') \tag{48}$$

This follows from the development of Eqs. (41) for $\delta$ at a given value of $c_2$ such that as $c_2 \to 0$, the values $c_1$ and $c''_1$ approach $c'_1$; $c_3$ and $c''_3$ approach $c'_3$; and the values of $\bar{v}_1$ and $\bar{v}_3$ become $\bar{v}'_1$ and $\bar{v}'_3$ respectively. Hence, the two slopes at $c_2 = 0$ are, respectively

$$\left(\frac{\partial \rho}{\partial c_2}\right)^0_\mu = \left(\frac{\rho - \rho'}{c_2}\right)^{c_2 \to 0} = \left(\frac{c_1 + c_2 + c_3}{c_2}\right)^{c_2 \to 0} - \left(\frac{c'_1 + c'_3}{c_2}\right)^{c_2 \to 0}$$

and

$$\left(\frac{\partial \rho}{\partial c_2}\right)^0_m = \left(\frac{\rho'' - \rho'}{c_2}\right)^{c_2 \to 0} = \left(\frac{c''_1 + c_2 + c''_3}{c_2}\right)^{c_2 \to 0} - \left(\frac{c'_1 + c'_3}{c_2}\right)^{c_2 \to 0}$$

The difference between these slopes is

$$\left(\frac{c_1 - c''_1 + c_3 - c''_3}{c_2}\right)^{c_2 \to 0} = (\delta_1 + \delta_3)^0$$
$$= [\delta_1(1 - \bar{v}'_1/\bar{v}'_3)]^0 = [\delta_3(1 - \bar{v}'_3/\bar{v}'_1)]^0$$

Thus, the difference in the limiting slopes of Eq. (48) is related to $\delta^0$ analogously as in the derivation of Eqs. (41) for $\delta$. In practice, one simply subtracts these slopes and divides by the term $(1 - \bar{v}'_j\rho')_{j=1,3}$ to obtain $\xi_j^0$.

At this point we digress to note that the difference in these limiting slopes, Eq. (48), is related to the difference in $v^{0'}$ and $\bar{v}_2^0$, respectively. Thus, if the quantity $\xi_j$ is known, $\bar{v}_2^0$ may be calculated from a determination of $v^{0'}$, or vice versa, by

$$(1 - v^{0'}\rho') - (1 - \bar{v}_2^0\rho') = \xi_j^0(1 - \bar{v}'_j\rho') \qquad (j = 1, 3) \qquad (49)$$

which follows by substituting from Eqs. (18) and (12) into Eq. (48). In terms of the corresponding apparent quantities $\phi'$ and $\phi_2$, rearrangement gives

$$\xi_j = \left(\frac{\phi_2 - \phi'}{1/\rho' - \bar{v}'_j}\right) \qquad (j = 1, 3) \qquad (50)$$

which is slightly more general but otherwise the same as that derived by Casassa and Eisenberg (1961; their equation 7, rearranged) wherein $\xi_3$ was measured independently and $\phi'$ was obtained by density measurements.

An example of the results which can be expected from the procedure just cited, using Eq. (48), is seen in Fig. 4. Turnip yellow mosaic virus (M $\sim 5.5 \times 10^6$ Daltons) was studied in a mixture of 5% w/w sucrose in water. About half of the stock TYMV solution in water was mixed with approximately the appropriate proportion of dry sucrose and was dialyzed to equilibrium. A dilution series with dialyzate as the diluent was made up by weight with the equilibrated nucleoprotein solution. The protein concentration was determined on the dialyzed solution spectrophotometrically, as outlined above, and the value of $c_2$ at each dilution with dialyzate was calculated from the volume dilution, Eq. (46). Densities were measured with the magnetic densimeter on the dialyzate and on all of the diluted nucleoprotein solutions, yielding the data shown in the lower curve of density *versus* $c_2$ in Fig. 4. The other half of the stock TYMV–water solution was made isomolal with the sucrose–water dialyzate of the isopotential experiment by adding the appropriate amount of anhydrous sucrose on the analytical balance. The amount of water in the stock solution was calculated by difference from the percent dry weight. Weight dilutions of the isomolal TYMV–water–sucrose solution were made with the pure solvent (which had the same density as

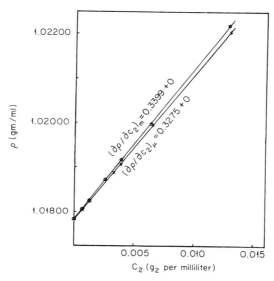

Fig. 4. Preferential interaction. Density $\rho$ vs. TYMV concentration $c_2$ for dialyzed (isopotential) series (crosses) and for undialyzed (isomolal) series (circles) in 5% w/w sucrose at 20°C (see text). Correlation coefficients $r$ for linear least-squares fitting of the data yield $r_\mu = 0.99991$ for the isopotential series, and $r_m = 0.99995$ for the isomolal series. [D. W. Kupke, unpublished experiments.]

that of the final dialyzate). About 1 ml or more of each diluted sample was prepared for both of the dilution series so that density measurements could be made in triplicate. The protein concentration and densities of the isomolal dilution series were measured as before; these results are shown in the upper curve of Fig. 4. Both slopes gave a correlation coefficient $r$ (i.e., goodness of fit) of better than 0.9999 to a rectilinear, least-squares fit of the data. Values of $r$ an order of magnitude better than this have been obtained with this procedure. Obviously, the correlation coefficients would have been poorer if the protein concentration had been measured on each diluted sample. The purpose here is to demonstrate that density measurements and the work at the balance can be routinely carried out with the precision necessary to yield slopes from which $\xi$ can be obtained with some confidence. The effect of bias in estimating the protein concentration is largely cancelled when the data for both slopes are determined concurrently from the same preparations of solvent and macromolecule. The uncertainty in $c_2$ is the only serious source of error which may vitiate the entire effort to derive a useful value of $\xi$. From the values of the slopes in Fig. 4, $\xi_1$ by Eq. (48) is $+0.63_6$ $g_1/g_2$ and $\xi_3 = -0.03_3$ $g_3/g_2$. Evidently, at this composition of solvent, water is preferred over sucrose. The high value of $\xi_1$ compared to that usually found for proteins in solvents containing inert-type solutes ($\sim +0.3$ $g_1/g_2$) suggests that water, but not sucrose, can permeate the sizeable interior space calculated for this virus from its crystallographic dimensions, molecular weight, and apparent specific volume.

A further treatment of the parameter $\xi$ is proposed in Appendix B. The interested student may wish to explore the possibility of splitting up this parameter in such a way that it may be used to estimate both the degree of hydration and the degree of interaction with component 3 by the protein. The procedure is outlined in terms of the density method and it is shown that in certain cases reasonable values may be expected for the net amounts of mutual exclusion between the two diffusible components.

### 4. *Other Applications*

The purpose of these remarks is to call attention to other pertinent areas of study which are now feasible with improved density techniques. These subjects have been alluded to in the discussion of the techniques (Section IV,A), but applications in depth are awaited. Measurements of changes in volume and partial volumes of components as a function of temperature form a significant area of study which hardly needs to be emphasized. Structural transitions and self-association equilibria of proteins are but two of the obviously important subjects for investigation. Selecting the magnetic densimeter method (which is the technique most familiar to the writer), it is clear that the change in density with temperature may be followed easily and precisely through the use of a variable temperature controller.[7] The magnetically suspended buoy is maintained at the preset height in the solution by the integrated circuit (Senter, 1969) as the density changes. The change in current to the solenoid, which is a function of the density, is then monitored, with a strip-chart recorder or a printout voltmeter. Hence, $(d\phi_2/dT)$ may be obtained by measuring $(d\rho'/dT)$ subsequently on the solvent medium as carried out by Hunter (1966), using the isopycnic diver method. With additional effort, the function $(d\bar{v}_i/dT)$ for all components of a solution may be obtained. The latter would appear to be particularly important at temperatures at which a transition in the conformation or at which self-association of the protein is encountered; the change in $\phi_2$, which assumes that the volume contribution of the solvent is fixed at each temperature, may be less informative than $(d\bar{v}_2/dT)$ if $\bar{v}_2$ is concentration dependent in a transition or self-association range. Finally,

---

[7] Unpublished experiments on densities at the critical state. D. V. Ulrich, Bridgewater College, Bridgewater, Virginia (1968).

it should be stressed that the change in the interaction parameter $\xi$ with temperature, is a misty domain which may provide some fundamental surprises in view of our rudimentary understanding of water structure in the presence of proteins.

We have seen also that the variation of density with pressure is now becoming a more feasible undertaking with protein solutions. The acquisition of accurate isothermal compressibility data on solutions containing proteins is a timely objective. Experiments carried out in centrifugal fields will rely increasingly upon a knowledge of the density change with pressure; the important effect of small changes in $\phi_2$ with pressure changes in these centrifugal fields on the equilibria of self-associating proteins has been amply discussed (Kegeles et al., 1967; TenEyck and Kauzmann, 1967; Josephs and Harrington, 1967). The practicability of measuring the density under applied pressure with the magnetic technique has been reported (Fahey et al., 1969); it is anticipated that refinements will lead to the evaluation of very small changes in $\phi_2$ which may have been obscured by the relatively primitive equipment employed. The study of the behavior of protein solutions under pressure, generally, is fast becoming an exciting field, and the assessment of $\Delta V$ and of changes in partial volumes is a prominent feature of these experiments. The significance of studying denaturation as a function of pressure is illustrated by the findings and discussions concerning the role of hydrophobic bonds in proteins (Kliman, 1969; Brandts et al., 1970).

The usefulness of density measurements for studying the kinetics of certain reactions was demonstrated years ago (e.g., Linderstrøm-Lang, 1962). With newer techniques (e.g., magnetic densimetry[8]), it should be possible to utilize density in a much more direct way and so to expand application to a greater variety of kinetic problems. Also, where the density can be monitored automatically, it becomes convenient to determine whether a protein solution is at equilibrium before beginning other studies upon it. For example, slow reactions, involving viscosity or other changes, may usually be observed by simply leaving the solution in a recording densimeter for a few hours. As noted before, the breaking and reforming of bonds in any reaction will probably not have exactly opposing volume changes and the net change may therefore be monitored.

Finally, we must not dismiss the possibility of employing density

[8] Preliminary experiments in the author's laboratory show that changes in density can be monitored continuously with an apparent precision of about $10^{-6}$ gm/ml. This is facilitated with the use of a servomechanism which holds the suspended buoy at a fixed height in the reacting solution.

measurements in relaxation phenomena. There seem to be no current technical obstacles to such an approach. Since relaxation studies appear not to have been reported upon, one might suggest that measurements of the change in density with time on pure water alone, after a temperature or pressure jump, may be an exciting adventure.[9]

## Appendix A

### Derivation of the Partial Specific Volume from the Density

METHOD I

We wish to obtain an expression by which the partial specific volume of a component, as defined in Eq. (2), may be calculated from density measurements and the concentrations of the component. The approach used for this method utilizes the solvent density $\rho'$ and introduces the apparent quantity $\phi$ in place of $\bar{v}$, initially. This kind of approach is more easily envisioned and the result seems to be more practical when dealing with proteins than is the equation derived for $\bar{v}$ at any concentration of the component (method II). We define a fixed density $\rho'$ of a solution to which the component of interest (here, component 2) is to be added. (This initial solution is usually the solvent medium. However, for more general application a fixed amount of protein may already be present. For the present, we shall take the more simple course and consider a solvent medium devoid of protein.) At constant temperature and pressure, the addition of $g_2$ grams of anhydrous protein to a given volume $V'$ of the solvent changes the density from $\rho'$ to $\rho$ such that

$$\rho = \frac{\sum_{j \neq 2}^{N} g'_j + g_2}{V} = \frac{\rho' V' + g_2}{V' + \phi_2 g_2} \tag{51}$$

where the new volume $V$ is related to $V'$ and $\phi_2$ by Eqs. (15) and (16). We express the added mass of component 2 in terms of its concentration in grams per milliliter, $c_2$, by

$$c_2 = W_2 \rho = \left(\frac{g_2}{\rho' V' + g_2}\right) \rho$$

Solving for $g_2$ we obtain

[9] From a discussion with Prof. H. F. Frank (1968).

$$g_2 = \rho'V'\left(\frac{c_2}{\rho - c_2}\right)$$

Substituting this expression into Eq. (51), thus eliminating V', the density of the solution in terms of $\rho'$, $c_2$ and $\phi_2$ becomes

$$\rho = c_2(1 - \phi_2\rho') + \rho'$$
$$\rho = \rho'(1 - \phi_2 c_2) + c_2 \quad (52)$$

This expression for the density, upon adding a finite mass of component 2 to the solvent, is also obtained by starting with Eq. (7), the definition of density in terms of the "c" scale. In this case the concentrations $c_j$ of all components $j$ ($j \neq 2$) in the protein solution are cast in terms of their concentrations in the solvent $c'_j$. Since $V' = V - \phi_2 g_2$ [Eq. (15)], it follows that the masses of the components $j$ in 1 ml of the protein solution of density $\rho$ must be identical to the masses of these components in $(1 - \phi_2 c_2)$ milliliters of the solvent of density $\rho'$ as seen in Eqs. (43). The term $(1 - \phi_2 c_2)$, by definition, is the volume contribution of the components $j$ in 1 ml of the final solution. Hence, by substituting via Eqs. (43) into Eq. (7)

$$\rho = (1 - \phi_2 c_2) \sum_{j \neq 2}^{N} c'_j + c_2$$

Since $\Sigma c'_j = \rho'$, the right-hand side becomes identical with that of Eqs. (52). As indicated previously, any mass of component 2 already present in the solvent medium of density $\rho'$ may be treated as one of the components $j$ without loss of generality.

We now obtain the variation in density as component 2 is being added in a continuous manner. Differentiation of Eqs. (52) with respect to $c_2$ gives

$$\left(\frac{\partial \rho}{\partial c_2}\right)_{g_j} = (1 - \phi_2 \rho') + \frac{\partial \rho'}{\partial c_2}(1 - \phi_2 c_2) - c_2 \rho' \frac{\partial \phi_2}{\partial c_2} \quad (j \neq 2)$$

The second term on the right-hand side is eliminated because $\rho'$ is a fixed quantity. As the concentration of the added component 2 approaches zero, the third term also drops out, resulting in the expression

$$\left(\frac{\partial \rho}{\partial c_2}\right)_{g_j}^0 = (1 - \phi_2^0 \rho') \quad (j \neq 2)$$

We have seen however, by Eq. (17) that $\phi_2 = \bar{v}_2$ at vanishing concentration of component 2. Hence,

$$\left(\frac{\partial \rho}{\partial c_2}\right)_{g_j}^0 = (1 - \bar{v}_2^0 \rho') \quad (j \neq 2) \quad (53)$$

which is the same as Eq. (11a). More generally, the superscript zero refers to the initial state rather than the vanishing concentration of the component being varied. Thus, if protein is present in the initial solution at a concentration $c_2^0$ and of density $\rho'$, the quantity $\phi_2$ must refer to this solution rather than to the pure solvent containing no protein.

The differential expression for the variation of density with respect to the weight fraction $W_2$ of component 2 by this method is seen most easily by substituting $\rho W_2$ for $c_2$ in Eqs. (52). Differentiating and dividing through by $\rho$ gives

$$\left(\frac{1}{\rho}\frac{\partial \rho}{\partial W_2}\right)_{g_j} = \frac{W_2}{\rho}\frac{\partial \rho}{\partial W_2}(1 - \phi_2 \rho') + (1 - \phi_2 \rho')$$
$$+ \frac{\partial \rho'}{\partial W_2}\left(\frac{1}{\rho} - \phi_2 W_2\right) - W_2 \rho'\frac{\partial \phi_2}{\partial W_2} \qquad (j \neq 2)$$

In the limit, as $W_2 \to 0$ (or the zeroeth addition of component 2 at a previously fixed level of this component), and $\phi_2 \to \bar{v}_2$, we see that

$$\left(\frac{\partial \ln \rho}{\partial W_2}\right)_{g_j}^0 = (1 - \bar{v}_2^0 \rho') \qquad (j \neq 2) \qquad (54)$$

which is identical with Eq. (11b).

## Method II

To obtain $\bar{v}_2$ at any value of $c_2$ by the density method, a more general differentiation of the density with respect to $c_2$ is outlined. We again begin with Eq. (7) for the density in terms of the concentrations of all components in the c scale. Thus

$$\rho = c_2 + \sum_{j \neq 2}^{N} c_j$$

where $c_j$ refers to the number of grams per milliliter of the nonprotein components $j$. Differentiation with respect to $c_2$ and dropping notations above and below the summation sign gives

$$\left(\frac{\partial \rho}{\partial c_2}\right)_{g_j} = 1 + \frac{\partial}{\partial c_2}\sum c_j = 1 + \sum \frac{\partial c_j}{\partial c_2}$$

Since $c_2 = g_2/V$

$$\left(\frac{\partial \rho}{\partial c_2}\right)_{g_j} = 1 + \sum \frac{\partial c_j}{\partial (g_2/V)}$$

$$= 1 + \sum \frac{\partial c_j}{\partial g_2} \cdot \frac{\partial g_2}{\partial (g_2/V)}$$

$$= 1 + \sum \frac{(\partial c_j/\partial g_2)}{\partial (g_2/V)/\partial g_2}$$

$$= 1 + \sum \frac{(\partial c_j/\partial g_2)}{\left(V - g_2 \frac{\partial V}{\partial g_2}\right)/V^2}$$

From Eq. (2) for the definition of $\bar{v}_2$

$$\left(\frac{\partial \rho}{\partial c_2}\right)_{g_j} = 1 + \sum \frac{(\partial c_j/\partial g_2)V}{1 - \frac{g_2}{V}\bar{v}_2}$$

$$= 1 + \sum \frac{(\partial c_j/\partial g_2)V}{1 - c_2\bar{v}_2} = 1 + \frac{V}{1 - c_2\bar{v}_2} \sum \frac{\partial c_j}{\partial g_2}$$

Since $c_j = g_j/V$, the last summation becomes

$$\sum \frac{\partial c_j}{\partial g_2} = \sum \frac{\partial (g_j/V)}{\partial g_2} = -\sum \frac{g_j}{V^2} \cdot \frac{\partial V}{\partial g_2} = -\sum \frac{g_j \bar{v}_2}{V^2}$$

Substituting into the total expression gives

$$\left(\frac{\partial \rho}{\partial c_2}\right)_{g_j} = 1 - \frac{V}{1 - c_2\bar{v}_2} \sum \frac{g_j \bar{v}_2}{V^2} = 1 - \frac{\bar{v}_2}{1 - c_2\bar{v}_2} \sum c_j$$

By Eq. (7) $\Sigma c_j = \rho - c_2$; hence

$$\left(\frac{\partial \rho}{\partial c_2}\right)_{g_j} = 1 - \frac{\bar{v}_2(\rho - c_2)}{1 - c_2\bar{v}_2} = \left(\frac{1 - \bar{v}_2\rho}{1 - \bar{v}_2 c_2}\right) \tag{55}$$

Equation (55) is identical with the one stated by Casassa and Eisenberg (1964; their Eq. 4.22). Thus, if the data are sufficiently precise where the density is a curved function of $c_2$, values of $\bar{v}_2$ at any $c_2$ may be calculated from the tangents of a fitted curve by rearranging Eq. (55). Obviously, as $c_2 \to 0$, this equation reduces to Eq. (53), since $\rho$ becomes $\rho'$.

In practice, it may be more expeditious with proteins to begin with a known stock solution of the protein in the chosen solvent medium, and to add the isomolal solvent to aliquots of this stock solution on the analytical balance. In this way, the protein concentration may be lowered by known increments, until the desired value of $c_2^0$ is attained.

Density determinations on the series of mixtures then yield the curve which intercepts the ordinate at $c_2^0$. The limiting slope at $c_2 = c_2^0$ is determined and the value of $\bar{v}_2$ at this concentration is obtained by Eq. (53) or (54) where $\rho'$ is the density at $c_2^0$. The dilutions may be continued, of course, to give density values at concentrations lower than $c_2^0$. The fitted curve then permits the tangent at point $c_2^0$ to be evaluated rather than a limiting slope; this may have some advantage in accuracy, because the changes in slope on either side of the point at $c_2^0$ are given weight in narrowing the choices for the tangent at $c_2^0$. With this extra-effort protocol, Eq. (55) would be used to calculate $\bar{v}_2$.

Most often, however, the value of $\bar{v}_2$ at $c_2 = 0$ is determined. Hence, Eq. (53) or (54) is the relevant relation ordinarily. The density of the protein-free solvent medium $\rho'$ is the most accurate measurement of the entire procedure and serves to anchor the curve with respect to all other values of the density at finite values of $c_2$.

## Appendix B

### Splitting-Up of the Parameter $\xi$: Exclusion of Diffusible Components

The parameter $\xi_j$ represents the derivative in the variation of the number of grams of diffusible component $j$ on varying the number of grams of protein at constant chemical potential; i.e., $\xi_j = (\partial g_j/\partial g_2)_{\mu_j} = M_j/M_2 (\partial m_j/\partial m_2)_{\mu_j}$, where $j = 1, 3$ for our purposes. It is understood that temperature is constant and that small pressure differences may be safely neglected. The corresponding parameter on a mole per mole basis $\Gamma_j$ is equated with the derivative $(\partial m_j/\partial m_2)_{\mu_j}$. Usually, the plots of $g_j$ or $m_j$ vs. $g_2$ or $m_2$, respectively, appear rectilinear and the derivatives are constant over the usual range of protein concentration studied. On the other hand, these derivatives are often quite different on varying the concentration of components $j$; i.e., $\xi_1$ and $\xi_3$ may change remarkably with solvent composition. It can easily be shown by a simulation procedure with the density method[10] that if the two diffusible components

---

[10] The simulation procedure consists of calculating a value for the density of the equilibrated protein solution $\rho$ by assigning values to $\xi_1^*$, $\xi_3^*$. Thus, $\rho = \rho'\Omega + \xi_1^* c_2 + \xi_3^* c_2 + c_2$ [Eq. (56)]. $\rho'$ and the values of $\bar{v}'_1$ and $\bar{v}'_3$ used in the calculation of $\Omega$ [Eq. (57)] are obtained from the density versus composition curve for the two-component solvent employed. The use of these partial volumes accounts for virtually all of the nonideality in the volume of the protein solution as the solvent composition is changed. The value chosen for $\phi_2$ is usually held constant because normally $\bar{v}_2$ changes slightly, if at all, with solvent composition; hence, substitution of $\bar{v}'_1$ and $\bar{v}'_3$ for the true experimental quantities, $\bar{v}_1$ and $\bar{v}_3$, respectively, involves only a trivial

exclude one another—relative to dialyzate composition—from some of the volume elements in the protein solution, $\xi_1$ and $\xi_3$ vary hyperbolically with solvent composition, even though the amounts of the exclusions remain fixed. Where one component, e.g., water, excludes component 3, and the latter component does not exclude water at all (relative to dialyzate composition), $\xi_1$ remains constant and $\xi_3$ varies hyperbolically from zero at $c'_3 = 0$ to $-\infty$ at $c'_3 = \rho'$ (i.e., from $W'_3 = 0$ to $W'_3 = 1$). (The density method cannot be used, however, at very low concentrations of component 3 in preferential interaction studies, because slight uncertainties in the values of $c_2$ give rise to enormous errors in $\xi_i$.) In such simulations, it is assumed as a first approximation that the partial specific volumes of the diffusible components in the volume elements where exclusions occur are the same as those in the dialyzate.

Whether the excluded region which is partly or completely inaccessible to one of the diffusible components because of a higher concentration of the other one represents hydration, binding, or another interaction with protein or simply steric exclusion (cf. Schachman and Lauffer, 1949) is not at issue. The basis for the preceding remarks on excluded volumes is the statement that there are regions, such as microphases, in the equilibrated protein solution in which the activity coefficients of a diffusible component are greater or less than that in the dialyzate; the total activity at equilibrium is the same, of course, on each side of the membrane. For the purpose of this method, we assume that a difference in the activity coefficient between microphases is proportional to the difference in the concentration of the component. We might conceive the exclusion by component 1 as referring to those masses of water which occupy various volume elements in the equilibrated protein solution for more of the time than they do in any like volume element of the dialyzate. If the time of occupation in one volume element is long compared to the time occupied in another, the average density of the first element is lower than that of the second (when $\bar{v}_1 > \bar{v}_3$, as is usually the case). The sum of the masses in each element divided by the total volume of such elements having densities lower than that of the dialyzate, yields an average density from which the excess number of grams of

---

volume change from that of real systems. The density of the preequilibrated protein solution $\rho''$ is calculated by Eq. (47) from the known value of $\rho'$. With $\rho$ and $\rho''$ calculated at a chosen value of $c_2$, $\delta_1$ and $\delta_3$ are calculated by Eqs. (41) and $\xi_1$ and $\xi_3$ by Eq. (45). The values of these preferential interaction parameters obtained by this simulation procedure agree precisely with those calculated by use of Eq. (62) for $\delta$, and Eq. (64) or Eq. (65) for $\xi$; the latter expressions are independent of volumes and densities and clearly show the relationship between $\xi$ and the pair of linked parameters $\xi_1^*, \xi_3^*$ into which $\xi$ was split up.

water, relative to an equal volume of dialyzate, may be calculated. A similar consideration applies to the volume elements with densities greater than that of the dialyzate—in order that the excess number of grams of component 3 may be evaluated. Thus, by dividing the apparent volume of the diffusible components in 1 ml of equilibrated protein solution $(1 - \phi_2 c_2)$ into regions of higher and lower average density than that of the dialyzate, the excluding masses per milliliter of solution for both water and component 3 are obtained. For analytical purposes, it will be convenient to assign the total excess (i.e., excluding) mass of each diffusible component a volume contribution as if it existed as a compact mass of the pure material. Thus, in 1 ml of the equilibrated protein solution there will be a volume contribution for the anhydrous protein component $\phi_2 c_2$ and a volume contribution for each of the excluding masses of the diffusible components $\bar{v}_j \xi_j c_2$. We are left then with a mass of diffusible components with a composition like that of dialyzate. The volume of this mass per milliliter of protein solution is designated as $\Omega$ (milliliter per milliliter). We might call $\Omega$ the volume of bulk solvent phase in 1 ml of the equilibrated protein solution with properties like those of the dialyzate; the mass of this phase per milliliter of the solution, therefore, is $\rho'\Omega$. Of course, there may be no real phase in the protein solution with a composition like that of the dialyzate; however, in dilute protein solutions we presume this to be so.

It is evident from the preceding discussion that the thermodynamically defined parameter $\xi$ is being decomposed into two other linked parameters. We define, therefore, the excess (in grams) of component 1 per gram of the protein in a milliliter of equilibrated solution as $\xi_1^*$, and any excess (in grams) of component 3 as $\xi_3^*$. We may then write the density $\rho$ of the equilibrated protein solution as the sum of the masses per milliliter of: protein $c_2$, of the water which excludes component 3 $(\xi_1^* c_2)$, of the mass of component 3 which excludes water $(\xi_3^* c_2)$, and of the bulk solvent phase with a composition like that of the dialyzate $\rho'\Omega$. Hence

$$\rho = c_2 + \xi_1^* c_2 + \xi_3^* c_2 + \rho'\Omega \tag{56}$$

The volume fraction $\Omega$ as described above is defined in terms of $\xi_1^*$ and $\xi_3^*$ by

$$\Omega = 1 - \phi_2 c_2 - \bar{v}'_1 \xi_1^* c_2 - \bar{v}'_3 \xi_3^* c_2 \tag{57}$$

Subtracting from $\rho$ the density of the pre-equilibrated protein solution $\rho''$, which is isomolal with respect to the diffusible components [Eq. (47)], we obtain

$$\left(\frac{\rho - \rho''}{c_2}\right) = \xi_1^*(1 - \bar{v}'_1 \rho') + \xi_3^*(1 - \bar{v}'_3 \rho') \tag{58}$$

Recall that we have substituted $\phi_2$ for $\bar{v}_2$ in expressing densities and volumes so that the values $\bar{v}'_1$ and $\bar{v}'_3$ appear in Eq. (58) for practical purposes. These partial volumes are approximations, as mentioned before, for the average of the true partial specific volumes $\bar{v}_{je}$, of the excess (in grams) of water and of component 3 in microphases having densities different from that of the dialyzate. The quantities $\bar{v}_{1e}$ and $\bar{v}_{3e}$ are unknown. It would be only slightly more correct to apply the values $\bar{v}_1$ and $\bar{v}_3$ for the macro phase on the protein side of the membrane, if they were experimentally distinguishable from $\bar{v}'_1$ and $\bar{v}'_3$, respectively. For dilute protein solutions, the determination of $\bar{v}_j$ is usually not required.

We may now recast Eq. (58) into a form which is more useful for attempting the evaluation of $\xi_1^*$ and $\xi_3^*$ by the density method. Returning to the parameter $\delta$, which is the quantity directly observed in equilibrium dialysis by the density method, we express $\delta_1$ or $\delta_3$ in terms of $\xi_1^*$ and $\xi_3^*$ via Eqs. (41) and (58). Carrying through for $\delta_1$, we see that

$$\delta_1 = \xi_1^* \left( \frac{1 - \bar{v}'_1 \rho'}{1 - \bar{v}'_1/\bar{v}'_3} \right) + \xi_3^* \left( \frac{1 - \bar{v}'_3 \rho'}{1 - \bar{v}'_1/\bar{v}'_3} \right) \tag{59}$$

The corresponding expression for $\delta_3$ is identical except that the denominator in both terms in parentheses is replaced by $(1 - \bar{v}'_3/\bar{v}'_1)$. Applying the definitions $\Sigma \bar{v}'_j c'_j = 1$ and $\rho' = \Sigma c'_j$, Eqs. (1) and (7), reduces the last expression to

$$\begin{aligned} \delta_1 &= \bar{v}'_3(\xi_1^* c'_3 - \xi_3^* c'_1) \\ \delta_3 &= -\bar{v}'_1(\xi_1^* c'_3 - \xi_3^* c'_1) \end{aligned} \tag{60}$$

Since $\bar{v}'_1$ and $\bar{v}'_3$ are slowly varying functions of the dialyzate composition, the above pair of expressions may be combined after transferring these partial volumes across the equal signs to give

$$(\delta_1/\bar{v}'_3) = -(\delta_3/\bar{v}'_1) = \xi_1^* c'_3 - \xi_3^* c'_1 \tag{61}$$

By shifting to the weight fractions scale (W) for dialyzate composition, we cast the variable quantities $c'_1$ and $c'_3$ in terms of a variation only in component 3. Using the definitions $W'_3 = c'_3/\rho'$ and $(1 - W'_3) = c'_1/\rho'$, we arrive at a linear expression for the excluded masses in terms of the experimental quantities, and

$$\left( \frac{\delta_1}{\bar{v}'_3 \rho'} \right) = -\left( \frac{\delta_3}{\bar{v}'_1 \rho'} \right) = (\xi_1^* + \xi_3^*) W'_3 - \xi_3^* \tag{62}$$

It is immediately apparent that if both $\xi_1^*$ and $\xi_3^*$ are constant ($\geq 0$) when the combined experimental parameter $(\delta_1/\bar{v}'_3 \rho')$ is plotted over a range of solvent composition ($W'_3$), the intercept at $W'_3 = 0$ gives $-\xi_3^*$, and the intercept at $W'_3 = 1$ yields $\xi_1^*$. Hence, if two essentially rectilinear segments are obtained, such as before and after a transition reaction, it may be possible to determine the two pairs of values $\xi_1^*$, $\xi_3^*$ from the intercepts in order to deduce the overall change in exclusions. With an equation of this form, any curvature cannot be interpreted precisely. The slope at any value of $W'_3$ is

$$\frac{\partial(\delta_1/\bar{v}'_3\rho')}{\partial W'_3} = \xi_1^* + \xi_3^* + W'_3 \left(\frac{\partial \xi_1^*}{\partial W'_3}\right) + (W'_3 - 1)\left(\frac{\partial \xi_3^*}{\partial W'_3}\right) \quad (63)$$

Where $\xi_1^*$ and/or $\xi_3^*$ vary over a range in dialyzate composition so that one or both of the last two terms are not zero, the intercept method is not applicable. Applying the same approximations for $\bar{v}_{je}$ as those in the derivation of Eq. (58), the experimental parameter $(\delta_1/\bar{v}'_3\rho')$ may be replaced by $\xi_1 W'_3$ in Eq. (62), since $\delta_1 = \xi_1 \bar{v}'_3 c'_3$ (see footnote 6). Hence, an expression for exclusions in terms of the parameter $\xi$ may also be written, where for $\xi_1$

$$\xi_1 = -\xi_3^* \left(\frac{1}{W'_3}\right) + (\xi_1^* + \xi_3^*) \quad (64)$$

which is less easy to visualize. Equation (64) may be expressed in terms of $\xi_3$ by using Eq. (22) and then converting to the weight fraction scale.

Although unique values for $\xi_1^*$ and $\xi_3^*$ may not be obtained by density alone over a given range in $W'_3$, it is evident that $\xi_1^*$ can be calculated if a value for $\xi_3^*$ is available from an independent method. Ifft and Vinograd (1966) have measured chloride ion binding by the pH change method of Scatchard and Black (1949) in systems containing albumin at high concentrations of CsCl for density-gradient sedimentation experiments. The density-gradient procedure gave a value equivalent to $\xi_1$ of $+0.20$ $g_1/g_2$. By counting each mole of chloride bound to the protein as a mole of the total component 3 (CsCl) bound, a value comparable to that defined here for $\xi_1^*$ was obtained which was about $2\frac{1}{2}$ times greater than for $\xi_1$.[11] Generally, $\xi_1$ will be positive at high concentrations of component 3 as may be expected from Eq. (64), since proteins invariably appear to be hydrated to some extent (cf. Fisher, 1965). Occasionally, an assumed value for the hydration of simple proteins (e.g., $+0.3$ $g_1/g_2$) is introduced and a value equivalent to $\xi_3^*$ is then calculated. The various methods employed to ascertain the hydration

[11] Virtually exact correspondence with their value of $\xi_1$ was obtained by the writer by applying Eq. (59) for $\delta_1$ to the values found for hydration and bound CsCl. $\xi_1$ was then calculated by Eq. (45). The values of $\bar{v}'_1$ and $\bar{v}'_3$ were obtained from density versus composition curves of aqueous CsCl (at the buoyant density and temperature used) prepared by the fitting procedure of Godschalk (1968). Equation (64), of course, provided a simpler correspondence, which gave essentially the same value for $\xi_1$ as did the density simulation method.

of proteins, however, (depending somewhat on the definition of hydration used) have not as yet provided reasonably unequivocal values. On the other hand, values for $\xi_1$ in the presence of quasi-inert solutes, such as the component 3, may provide a more uniform basis for discussions on protein hydration. If component 3 appears not to interact with protein over a range in dialyzate composition, values for $\xi_1$ vs. $c'_3$ or $W'_3$ will appear invariant so that $\xi_1 = \xi_1^*$. Ribonuclease and TYMV in the presence of varying concentrations of sucrose ($c'_3 = 0.05$ to $0.44$ gm/ml) have exhibited such behavior.[12] Through the use of other inert solutes for component 3, it may be possible to assess the fraction of hydration as the result of the steric exclusion principle ascribed to Kauzmann (Schachman and Lauffer, 1949). The difference in effective molecular radii between water and component 3 implies that a statistical layer of the smaller molecule (usually water) must exist at the surface of the macromolecule from steric considerations alone.

It is especially important to note that different values of $\xi_1$ or $\xi_3$ obtained with different solvent compositions cannot be interpreted to mean that the protein has gained or lost a certain amount of water or of component 3 as was sometimes thought. Simulations are easily constructed with Eq. (62) or Eq. (64), which show clearly that by introducing reasonable (positive) values for both $\xi_1^*$ and $\xi_3^*$, which are held constant over a difference in $W'_3$, values of $\Delta\xi_1$ or $\Delta\xi_3$ are not zero. For example, in proceeding from a dialyzate with a higher concentration of component 3 to one where $W'_3$ is lower, the value of $\Delta\xi_1$ may be highly negative, while $\Delta\xi_3$ is positive. This may be seen more easily by recasting Eq. (64) in terms of the concentrations in grams per milliliter where

$$\xi_1 = \xi_1^* - \xi_3^* \left(\frac{c'_1}{c'_3}\right) \tag{65}$$

As water becomes more concentrated, even a small value for $\xi_3^*$ will cause $\xi_1$ to become negative.

### Acknowledgments

A substantial portion of the material relating to density could not be drawn from the literature or utilized from it without modification. Hence, some of the discussion reflects the considerations only of the author as derived from unpublished experiments and simulations on density problems. This aspect of the work was supported by a grant (GB-8284) from the U. S. National Science Foundation.

It is a pleasure to express my gratitude to Prof. W. Godschalk, who cheerfully and frequently gave aid in depth in the preparation of this manuscript, particularly

---

[12] D. W. Kupke and J. P. Senter, unpublished experiments.

by humbling my inept derivations and exposing imprecise formulations. I was privileged also to have the suggestions of Drs. Eric E. Brumbaugh, Margaret Hunter, Sam Katz, and Michael J. Kelley. J. W. Beams, of course, was the inspiration.

## References

Adair, G. S., and Adair, M. E. (1947). *Proc. Roy. Soc. Ser. A* **190**, 341.
Bates, F. J. (1942). *Nat. Bur. Stand. (U. S.), Circ.* **440**, 626.
Bates, F. J., Phelps, F. P., and Snyder, C. F. (1927). *In* "International Critical Tables" (E. W. Washburn, ed.), Vol. II, p 342. McGraw-Hill, New York.
Bauer, N., and Lewin, S. Z. (1959). *In* "Physical Methods of Organic Chemistry" (A. Weissberger, ed.), 3rd ed., Vol. 1, Part I, Chapter IV, p. 131. Wiley (Interscience) New York. (An updated chapter is in preparation.)
Beams, J. W. (1969). *Rev. Sci. Instrum.* **40**, 167.
Beams, J. W., and Clarke, A. M. (1962). *Rev. Sci. Instrum.* **33**, 750.
Beams, J. W., and Hodgins, M. G. (1971). *Rev. Sci. Instrum.* **42**, 1455.
Beams, J. W., Hulbert, C. W., Lotz, W. E., Jr., and Montague, R., Jr. (1955). *Rev. Sci. Instrum.* **26**, 1181.
Beattie, J. A., Brooks, B. T., Gillespie, L. J., Scatchard, G., Schumb, W. C., and Tefft, R. F. (1928). *In* "International Critical Tables" (E. W. Washburn, ed.), Vol. III. p. 51 *ff*. McGraw-Hill, New York.
Bein, W., and Langbein, G. (1905). *In* "Landolt-Börnstein Physikalisch-Chemische Tabellen" (R. Börnstein and W. Meyerhoffer, eds.), p. 315, Springer-Verlag, Berlin and New York.
Bodanszky, A., and Kauzmann, W. (1962). *J. Phys. Chem.* **66**, 177.
Bradbury, J. H., Fenn, M. D., and Gosney, I. (1965). *J. Mol. Biol.* **11**, 137.
Brandts, J. F., Oliveira, R. J., and Westort, C. (1970). *Biochemistry* **9**, 1038.
Casassa, E. F., and Eisenberg, H. (1960). *J. Phys. Chem.* **64**, 753.
Casassa, E. F., and Eisenberg, H. (1961). *J. Phys. Chem.* **65**, 427.
Casassa, E. F., and Eisenberg, H. (1964). *Advan. Protein Chem.* **19**, 287.
Chantrenne, H. K., Linderstrøm-Lang, K. U., and Vandendriessche, L. (1947). *Nature (London)* **159**, 877.
Charlwood, P. A. (1957). *J. Amer. Chem. Soc.* **79**, 776.
Clarke, A. M., Kupke, D. W., and Beams, J. W. (1963). *J. Phys. Chem.* **67**, 929.
Coates, J. H. (1970). *In* "Physical Principles and Techniques of Protein Chemistry" (S. J. Leach, ed.), Part B, p. 1. Academic Press, New York.
Cohn, E. J., and Edsall, J. T. (1943). "Proteins, Amino Acids and Peptides," p. 157. Van Nostrand-Reinhold, Princeton, New Jersey.
Corwin, A. H. (1959). *In* "Physical Methods of Organic Chemistry" (A. Weissberger, ed.), 3rd ed., Vol. 1, Part I, Chapter III, p. 71. Wiley (Interscience), New York.
Dayhoff, M. O., Perlmann, G. E., and MacInnes, D. A. (1952). *J. Amer. Chem. Soc.* **74**, 2515.
DeRosier, D. J., and Haselkorn, R. (1966). *J. Mol. Biol.* **19**, 52.
Edelstein, S. J., and Schachman, H. K. (1967). *J. Biol. Chem.* **242**, 306.
Eisenberg, H., and Cohen, G. (1968). *J. Mol. Biol.* **37**, 355.
Fahey, P. F., Kupke, D. W., and Beams, J. W. (1969). *Proc. Nat. Acad. Sci. U. S.* **63**, 548.
Fisher, H. F. (1965). *Biochim. Biophys. Acta* **109**, 544.
Geffcken, W., Beckmann, C., and Kruis, A. (1933). *Z. Phys. Chem., Abt. B* **20**, 398.

Gerber, B. R., and Noguchi, H. (1967). *J. Mol. Biol.* **26**, 197.
Gill, S. J., and Glogovsky, R. L. (1964). *Rev. Sci. Instrum.* **35**, 1281.
Gill, S. J., and Glogovsky, R. L. (1965). *J. Phys. Chem.* **69**, 1515.
Godschalk, W. (1968). "VBARTAB," a program in Algol. Computer Sciences Center, University of Virginia, Charlottesville.
Goodrich, R., Swinehart, D. F., Kelly, M. J., and Reithel, F. J. (1969). *Anal. Biochem.* **28**, 25.
Gordon, J. A., and Warren, J. R. (1968). *J. Biol. Chem.* **243**, 5663.
Gucker, F. T., Jr., Gage, F. W., and Moser, C. E. (1938). *J. Amer. Chem. Soc.* **60**, 2583.
Güntelberg, A. V., and Linderstrøm-Lang, K. U. (1949). *C. R. Trav. Lab. Carlsberg, Ser. Chim.* **27**, 1.
"Handbook of Chemistry and Physics." (1962). (C. D. Hodgman, ed.), 44th ed., p. 2197. Chem. Rubber Publ. Co., Cleveland, Ohio.
Hunter, M. J. (1966). *J. Phys. Chem.* **70**, 3285.
Hunter, M. J. (1967). *J. Phys. Chem.* **71**, 3717.
Ifft, J. B., and Vinograd, J. (1966). *J. Phys. Chem.* **70**, 2814.
Ikkai, T., Ooi, T., and Noguchi, H. (1966). *Science* **152**, 1756.
Jaenicke, R., and Lauffer, M. A. (1969). *Biochemistry* **8**, 3077.
Johansen, G. (1948). *C. R. Trav. Lab. Carlsberg, Ser. Chim.* **26**, 399.
Johansen, G., and Thygesen, J. E. (1948). *C. R. Trav. Lab. Carlsberg, Ser. Chim.* **26**, 369.
Johnson, F. H., Eyring, H., and Polissar, M. J. (1954). "The Kinetic Basis of Molecular Biology," Chapter 9. Wiley, New York.
Josephs, R., and Harrington, W. F. (1967). *Proc. Nat. Acad. Sci. U. S.* **58**, 1587.
Kaper, J. M. (1968). *In* "Molecular Basis of Virology" (H. Fraenkel-Conrat, ed.), Chapter I, p. 1. Van Nostrand-Reinhold, Princeton, New Jersey.
Kaper, J. M., and Litjens, E. C. (1966). *Biochemistry* **5**, 1612.
Katz, S. (1968). *Biochim. Biophys. Acta* **154**, 468.
Katz, S., and Ferris, T. G. (1966). *Biochemistry* **5**, 3246.
Kauzmann, W. (1958). *Biochim. Biophys. Acta* **28**, 87.
Kauzmann, W. (1959). *Advan. Protein Chem.* **14**, 1.
Kauzmann, W., Bodanszky, A., and Rasper, J. (1962). *J. Amer. Chem. Soc.* **84**, 1777.
Kawahara, K., and Tanford, C. (1966). *J. Biol. Chem.* **241**, 3228.
Kay, C. M. (1960). *Biochim. Biophys. Acta* **38**, 420.
Kegeles, G., Rhodes, L., and Bethune, J. L. (1967). *Proc. Nat. Acad. Sci. U. S.* **58**, 45.
Kliman, H. L. (1969). Ph.D. Dissertation, Princeton University, Princeton, New Jersey.
Klotz, I. M. (1964). "Chemical Thermodynamics: Basic Theory and Methods," rev. ed., Chapter 13. Benjamin, New York.
Kratky, O., Leopold, H., and Stabinger, H. (1969). *Z. Angew. Phys.* **27**, 273.
Krausz, L. M. (1970). *J. Amer. Chem. Soc.* **92**, 3168.
Krausz, L. M., and Kauzmann, W. (1970). *Arch. Biochem. Biophys.* **139**, 80.
Krivacic, J., and Rupley, J. A. (1968). *J. Mol. Biol.* **35**, 483.
Kupke, D. W. (1966). *Fed. Proc., Fed. Amer. Soc. Exp. Biol.* **25**, 990.
Kupke, D. W., Hodgins, M. G., and Beams, J. W. (1972). *Proc. Nat. Acad. Sci. U. S.* **69**, 2258.
Lamb, A. B., and Lee, R. E. (1913). *J. Amer. Chem. Soc.* **35**, 1666.

Lewis, G. N., and Randall, M. (1961). "Thermodynamics" (revised by K. S. Pitzer, and L. Brewer), p. 207. McGraw-Hill, New York.
Linderstrøm-Lang, K. U. (1940). In "Methoden der Enzymforschung" (K. Myrbäck and E. Bauman, eds.), p. 970. Leipzig.
Linderstrøm-Lang, K. U. (1952). "Lane Medical Lectures," Vol. 6. Stanford Univ. Press, Stanford, California.
Linderstrøm-Lang, K. U. (1962). In "Selected Papers of Kaj Linderstrøm-Lang" (H. Holter, H. Neurath, and M. Ottesen, eds.). Academic Press, New York.
Linderstrøm-Lang, K. U., and Jacobsen, C. F. (1941). C. R. Trav. Lab. Carlsberg, Ser. Chim. 24, 1.
Linderstrøm-Lang, K. U., and Lanz, H. (1938). C. R. Trav. Lab. Carlsberg, Ser. Chim. 21, 315. [Reprinted in Mikrochim. Acta 3, 210 (1938) and in "Selected Papers of Kaj Linderstrøm-Lang" p. 145. Academic Press, New York, 1962.]
MacInnes, D. A., Dayhoff, M. O., and Ray, D. R. (1951). Rev. Sci. Instrum. 22, 642.
McKeekin, T. L., Groves, M. L., and Hipp, N. J. (1954). J. Amer. Chem. Soc. 76, 407.
Macurdy, L. B. (1967). Treatise Anal. Chem. 7, Part I, p. 4247 (I. M. Kolthoff, P. J. Elving, and E. B. Sandell, eds.), Wiley, (Interscience) New York.
Noguchi, H., and Yang, J. T. (1963). Biopolymers 1, 359.
Noguchi, H., Kasarda, D., and Rainford, P. (1964). J. Amer. Chem. Soc. 86, 2077.
Pilz, I. (1972). In "Physical Principles and Techniques of Protein Chemistry" (S. J. Leach, ed.), Part C, p. 141 this volume. Academic Press, New York.
Plato, F. (1900). Kaiserl. Normal-Eichungs-Komm., Wiss. Abh. 2, 153 (quoted in Bates et al., 1927, and Bates, 1942).
Rainford, P., Noguchi, H., and Morales, M. (1965). Biochemistry 4, 1958.
Randall, M., and Longtin, B. (1939). Ind. Eng. Chem., Anal. Ed. 11, 44.
Rasper, J., and Kauzmann, W. (1962). J. Amer. Chem. Soc. 84, 1771.
Reisler, E., and Eisenberg, H. (1969). Biochemistry 8, 4572.
Richards, T. W., and Shipley, J. W. (1914). J. Amer. Chem. Soc. 36, 1.
Scatchard, G., and Black, E. S. (1949). J. Phys. Colloid Chem. 53, 88.
Schachman, H. K. (1957). In "Methods in Enzymology" (S. P. Colowick and N. O. Kaplan, eds.), Vol. 4, p. 65. Academic Press, New York.
Schachman, H. K., and Lauffer, M. A. (1949). J. Amer. Chem. Soc. 71, 536.
Scheel, K. (1905). In "Landolt-Börnstein Physikalisch-Chemische Tabellen" (R. Börnstein and W. Meyerhoffer, eds.), p. 37, Springer-Verlag, Berlin and New York.
Senter, J. P. (1969). Rev. Sci. Instrum. 40, 334.
Stabinger, H., Leopold, H., and Kratky, O. (1967). Monatsch. Chem. 98, 436.
Stevens, C. L., and Lauffer, M. A. (1965). Biochemistry 4, 31.
Stott, V., and Bigg, P. H. (1928). In "International Critical Tables" (E. W. Washburn ed.), Vol. III, p. 24 ff. McGraw-Hill, New York.
Svedberg, T., and Pedersen, K. O. (1940). "The Ultracentrifuge." Oxford Univ. Press (Clarendon), London and New York.
Tanford, C. (1961). "Physical Chemistry of Macromolecules." Chapter 4. Wiley, New York.
TenEyck, L. F., and Kauzmann, W. (1967). Proc. Nat. Acad. Sci. U. S. 58, 888.
Thiesen, M., Scheel, K., and Diesselhorst, H. (1900). Wiss. Abh. d. Physikalish. Technischen Reichsanstalt 3, 68 (Quoted in Scheel, 1905; cf. also Stott and Bigg, 1928).
Thomas, J. O., and Edelstein, S. J. (1971). Biochemistry 10, 477.

Ulrich, D. V., Kupke, D. W., and Beams, J. W. (1964). *Proc. Nat. Acad. Sci. U. S.* **52,** 349.
Vandendriessche, L. (1953). *Acta Chem. Scand.* **7,** 699.
Wagenbreth, H., and Blanke, M. (1971). *Phys. Tech. Bundestagen Mitteilungen* **6,** 412.
Weber, H. H. (1930). *Biochem. Z.* **218,** 1.
Weber, H. H., and Nachmansohn, D. (1929). *Biochem. Z.* **204,** 215.

# 18 □ Osmotic Pressure

## M. J. KELLY and D. W. KUPKE

| | |
|---|---|
| Glossary of Symbols | 77 |
| I. Introduction and Phenomenon | 78 |
|   A. The Phenomenon | 80 |
| II. Elementary Theory | 87 |
| III. Applications | 102 |
|   A. Molecular Weight | 102 |
|   B. Dissociation and Subunits | 106 |
|   C. Mixtures, Association Equilibria, and Hybrids | 108 |
|   D. Other Applications | 116 |
| IV. Techniques | 121 |
|   A. Methods of Measurement | 122 |
|   B. Membranes | 131 |
|   C. Treatment of Data | 135 |
| References | 137 |

## Glossary of Symbols

| | |
|---|---|
| primes | denote the side of the membrane containing only diffusible components |
| a, b, c | first, second, and third virial coefficients, respectively [Eq. (17)] |
| $d$ | diameter of a rodlike macromolecule |
| f | weight fraction of a nondiffusible species relative to the total nondiffusible component(s) |
| $f_i$ | activity coefficient of component $i$ on the mole fraction scale |
| $g_i$ | number of grams of component $i$ |
| $g$ | acceleration of gravity |
| $h$ | corrected height of liquid column in terms of pressure above atmospheric |
| $i$ | any component or species in the solution |
| $j$ | any component or species where $j \neq i$ |
| $m_i$ | molality of a component or species $i$ |
| $m$ | mass |
| $n_i, n_j$ | number of moles of a component $i$ or $j$ |
| $n$ | refractive index |
| r | radius |
| $\bar{v}_i$ | partial specific volume of component $i$ |
| $v_e$ | excluded volume between closest approach of two centers of mass |
| $w_i$ | grams of component $i$ per kilogram of water |

| | |
|---|---|
| $x_i$ | mole fraction of protein component or protein species $i$ with respect to the total number of moles of protein |
| $y_i$ | activity coefficient of component or species $i$ in mass per unit of volume concentration scales |
| $z_i$ | magnitude of the charge on diffusible ions |
| A | $RT/V_m^0 M_n$ |
| B | variable parameter of the second virial coefficient, Eq. (18) |
| $C_i$ | concentration of component $i$ in grams per liter |
| D | dimer species of a nondiffusible component (usually as a subscript) |
| G | Gibbs free energy |
| K | association constant |
| L | length of a rodlike molecule |
| M | monomer species of a nondiffusible component (usually subscript) |
| $M_i$ | molecular weight of component or species $i$ |
| $\bar{M}_n$ | number-average molecular weight |
| $\bar{M}_w$ | weight-average molecular weight |
| N | number of components or species included within a summation |
| $N_\lambda$ | number of fringes of wavelength $\lambda$ |
| $\mathcal{N}$ | number of molecules in a mole |
| P | pressure |
| R | molar gas constant |
| T | temperature (Kelvin scale) |
| V | total volume |
| $\bar{V}_i$ | partial molal volume of component $i$ |
| $V_m^0$ | volume of solution containing 1 kg of water at vanishing concentration of nondiffusible components |
| $W_i$ | weight fraction of component $i$ relative to all components of a solution |
| $X_i$ | mole fraction of component $i$ relative to all components in a solution |
| $Z_2$ | net charge on nondiffusible ampholyte |
| $\alpha_i$ | relative activity of component $i$ |
| $\beta_i$ | $\Sigma \nu_i \ln \alpha_i$ |
| $\beta_{ij}$ | $\dfrac{\Sigma \nu_i \partial \ln \alpha_i}{\partial m_j} = \dfrac{\partial \beta_i}{\partial m_j} = \dfrac{\partial \beta_j}{\partial m_i}$ |
| $\gamma_i$ | activity coefficient of component or species $i$ on a molal concentration scale |
| $\lambda$ | wavelength |
| $\mu_i$ | chemical potential of component or species $i$ |
| $\nu$ | number of species in a component (e.g., for NaCl, $\nu = 2$) |
| $\xi_j$ | excess or deficient number of grams of diffusible component $j$ per gram of protein relative to dialyzate composition at osmotic equilibrium |
| $\pi$ | osmotic pressure |
| $\rho$ | density |
| $\phi_i$ | apparent specific volume of component $i$ |

## I. Introduction and Phenomenon

The measurement of osmotic pressure can be employed as a routine method for the thermodynamic study of many kinds of proteins and mixtures of proteins in solution with the special advantage that great freedom is allowed in the choice of a solvent medium. The principles and apparatus are comparatively unsophisticated. A single investigator can assemble materials available at any laboratory bench and devise a measuring system which yields data as relevant and reliable as that obtained using more complicated macromolecular methods representative of the advancing fronts of technology and theoretical expertise. The principal limitation of the method is that low molalities of protein ($<10^{-4}$) cannot usually be studied with accuracy.

Quite generally, one associates osmotic pressure with the molecular weight M. The quest for unambiguous values of M has not abated and will not be resolved in the near future by the perfunctory use of a single, universally accepted technique. Sharp controversies will continue, as in the past, to hinge upon different values of M or the number-average value, $\overline{M}_n$ for seemingly the same protein component. The comparatively uncomplicated evaluation of M by osmotic pressure, even in complex solvent systems, can be used to resolve differences in a variety of cases. Less appreciated, however, is the fact that the method may be exploited for a number of purposes other than that of collecting molecular weights of proteins. These objectives have come into view by studying the departure between the idealized and the observed osmotic pressure as a function of the concentration of protein and other variables. Examination of such departures has been related to protein self-association equilibria, molecular shape, diffusible-ion imbalance on the two sides of the membrane, protein–small molecule or ion interactions and interactions between mixed macromolecular components, which sometimes are compounded by cross-association of their subunits. Hence, osmotic pressure can be a very useful method in combination with other techniques in carrying out imaginative research on proteins.

The theory and practice of the osmotic pressure method have been reviewed in detail on several occasions and the subject is discussed at length in a number of texts.[1] The purpose of this chapter is to call attention to the scope of applications and of techniques, with particular

[1] A selection of readings which is minimally sufficient to provide a suitably complete background in the theory and practice of osmotic pressure is listed as follows: Wagner and Moore (1959; also Wagner, 1949)—a comprehensive survey of technical principles as well as general theory and applications; Kupke (1960—a review on applications and techniques directed primarily to the study of proteins; Tanford (1961, Chapter 4, particularly Sections 13 and 14)—a very usable and generally complete theoretical guide for the treatment of systems containing proteins and ions. For those desiring a thorough and strict theoretical background in osmotic pressures of protein systems, the paper by Scatchard (1946) has become a classic; subsequent developments and other treatments are largely extensions and variations of this basic treatment and to an extent also, of the treatment by Güntelberg and Linderstrøm-Lang (1949). The earlier contributions, e.g., Adair (1937, and preceding papers). Donnan (1935, and preceding papers), and Marrack and Hewitt (1929), may also be useful. Other sources for specific purposes are referred to in the text.

emphasis on the more timely aspects. Examples have been chosen to illustrate the method to the student, giving little regard to recording the most significant works and names associated with this subject; references are employed primarily to direct the student to a source for further study. The elementary theory is presented for orientation in discussing the applications given; however, the readings cited in footnote 1 are to be consulted for detailed theory and for general background in depth. In conforming with the aim of this series of volumes, theoretical completeness and numerous equations pertinent to the applications described have been omitted.

As a prelude to the thermodynamic definition of osmotic pressure given in Section II, a description of the phenomenon is proposed at this point. It is hoped by this means to ease the path of the unprepared student when he is confronted with thermodynamic formalism[2]; it is also intended to dispel certain notions of osmotic pressure which have led to confusion, especially at the physiological level.

A. The Phenomenon

All isolated systems tend toward equilibrium; i.e., ultimately, a system comes to rest and remains so unless subjected to other influences. With solutions, we know from experience that this state of rest is achieved when macroscopic differences in composition can no longer be detected in the various parts with time. At temperatures above $0°K$, of course, there will be submicroscopic differences in composition occurring constantly. However, these balance out at the macroscopic level so that a real difference within the samples used for analysis is extremely improbable. For our purpose, it is sufficient to call any bounded region wherein the composition of a solution is constant a *phase*. According to the opening statement (which is a way of stating the second law of thermodynamics), if another phase is now placed in contact with the first one, spontaneous changes will take place until a new state of rest is achieved. That is, the composition again becomes uniform throughout and the two phases have become one (barring phase separation). Very frequently, we encounter systems in nature where a physical barrier intervenes between aqueous phases so that one or more of the chemical species cannot pass through, while one or more other species,

---

[2] An unusual presentation of thermodynamics, especially suitable for students with little or no background in the physical sciences, is given in a small volume by Spanner (1964). Also recommended is the clear discussion of the energy quantities in thermodynamics by Klotz (1967), available in paperback form.

such as water molecules, can do so. In this case, constant composition throughout the regions accessible to some of the species cannot be achieved. An equilibrium state is reached, nonetheless, because we know from experience that an undisturbed system cannot remain out of equilibrium indefinitely. If it could do so, we would be confronted with perpetual motion, which, on a macroscopic level, has never been demonstrated.

The approach to equilibrium, when a selective barrier separates phases, may be envisioned by considering a very real example in terms of some concepts employed in thermodynamics. In Fig. 1a, we have a glass tube in the shape of a large U which has a membrane permeable to water seated in a cross-section at the bottom. If we add water to one of the limbs, it is self-evident that in a short time the water level in both limbs will be at the same height. This state of rest may be described by stating that the ability of water to pass through the membrane from either side is now the same; before equilibrium was reached this ability was not the same in limb A and limb B; hence, a net transport of water across the membrane was observed. We may call this ability of the water in the two regions the *relative activity*, $\alpha$. At equilibrium, then, $\alpha_{H_2O}^A = \alpha_{H_2O}^B$ since net transport is no longer seen. At a given tempera-

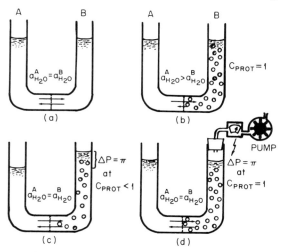

FIG. 1. Diagramatic representation of osmosis and osmotic pressure utilizing a U-tube with limbs A and B separated by a semipermeable membrane. (a) Equilibrium with water on both sides. (b) Conditions existing shortly after introducing protein to limb B. (c) Equilibrium between sides A and B following the osmosis resulting from the added protein. (d) Equilibrium between sides A and B when a counter pressure equal to the osmotic pressure is applied before any net transport of water can take place after adding the protein (Kupke, 1964).

ture and pressure, the activity of a pure component[3] is arbitrarily assigned a value of unity by convention.

If we now dilute the water by substituting some of it with molecules which do not have the ability to pass through the membrane, the system will probably not remain at rest. In Fig. 1b we have substituted such a component in limb B at a designated concentration of unity; for example, a protein, which for simplicity is uncharged and causes no hydrolysis of water into ions. Ideally, the activity of water in this limb is now less than that of the pure water in limb A simply because we have diluted the water in B with something not having the activity of water. Hence, a net transport of water, the diffusible component, ensues from A to B; this process is called *osmosis*. The spontaneous net transport of water has the effect of diluting the protein, or more importantly, of concentrating the water in limb B. If no other influences were operative, equilibrium would not be achieved until all the water had crossed the membrane, because the activity of a solvent in ideal systems is always greatest in the pure phase. Owing to the particular geometry of the system, however, we observe that a difference develops between the heights of the menisci of limbs A and B. Ultimately, the difference in pressure $\Delta P$, i.e., the difference in force per unit of area, has become sufficient to prevent further osmosis. Thus, the activity of water in side B is again equal to that in side A as shown in Fig. 1c. The extra pressure $\Delta P$, on side B, which raised the activity of water (to unity in this case) is called the *osmotic pressure* $\pi$. Values of $\pi$ are most often given in terms of the number of centimeters in height $h$ of either the solvent medium employed in the experiment or of pure water; sometimes $\pi$ is given in millimeters of mercury or in atmospheres. The resisting head of pressure may be considered in terms of the potential energy available for osmosis to continue. The potential energy in the cgs system is the pressure (dynes/cm$^2$) times the volume (cm$^3$) or the force acting over a distance (dyne-cm). The pressure in turn is the product of the height, the density $\rho$ of the selected liquid (gm/cm$^3$ for this purpose), and the acceleration of gravity, $g$. The volume V is given by the height times the appropriate unit of cross-sectional area (cm$^2$). Since $V\rho$ is the mass, $m$, the potential energy becomes immediately the familiar product $mgh$.

There are many kinds of quasi-isolated systems in the world about us, natural or artificial, in which liquid phases are in contact but cannot achieve a uniform composition by virtue of a selective barrier be-

---

[3] A component is a definable material, not necessarily a single compound, which can be added to a system independent of all other components so that the masses of the system in terms of the components can always be known.

tween them. Pressure differences arise spontaneously, because the systems are bounded and have some kind of geometry. It is difficult to imagine a real system (i.e., one not perfectly elastic) having a selective barrier (which is not a gas phase) wherein a pressure difference may not be developed spontaneously. It is to be noted that other forces acting over a distance would also serve to raise the activity of the diffusible species. As must be evident, however, other forces around us, such as gravitational, electrical, and magnetic forces, are essentially uniform over the entire domain of systems usually seen, unless purposely controlled (e.g., by centrifugation). There are many possible systems, of course, in which there is virtually no room for expansion to develop an obvious column of liquid for a countering pressure. Returning to Fig. 1, it is clear that we could have simply closed off side B at the meniscus after the protein was added so as to prevent any osmosis from occurring. In order to measure the pressure difference, however, we have applied an additional pressure via a pump sufficient to maintain the menisci at the same level. This situation is visualized in the diagram of Fig. 1d. In this case we note that the additional pressure observed by reading the meter must be greater than the value of $\pi$ in Fig. 1c, because the activity of water had not been increased by dilution from osmosis (i.e., $C_{prot}$ remains unity). Although no volume change has occurred in this instance, the potential energy $mgh$ may be calculated as before by converting the pressure units read from the meter to that for a column of the same liquid as used in Fig. 1c. Ordinarily, however, $mgh$ is not calculated in osmotic pressure experiments.

As indicated previously, the activity of a component is some function of its concentration. With respect to a solvent, such as water, the activity becomes proportional to the concentration as the solvent component approaches the pure state (Raoult's Law). The activities of dissolved solutes, on the other hand, become proportional to their concentrations as the latter approach zero (Henry's Law). There are different ways of expressing concentrations, and for this introductory section it is most convenient to use the mole fraction X. The mole fraction $X_i$ of a component $i$ in a mixture of N components is

$$X_i = \frac{n_i}{\sum_{j=1}^{N} n_j} \tag{1}$$

where n is the number of moles. Hence, if the value of $\alpha_{H_2O}$ is set at unity for pure water at a particular temperature and pressure, then both $X_{H_2O}$ and $\alpha_{H_2O}$ will approach unity as the concentration of solute approaches zero. Since the relative activity is some measurable response of the change in properties of a system as a component is added or removed, the response is not necessarily strictly proportional to the concentration of the component. The deviation from such a proportionality, or from ideal behavior, is reflected in the value of an activity coefficient. Activity coefficients are empirical proportionality constants relating activities and concentrations at a particular composition; thus

$$\alpha_i = f_i X_i$$
$$= \gamma_i m_i \quad (2)$$
$$= y_i C_i$$

(To prevent confusion, different symbols are used for the activity coefficient when different units of concentration are employed; a number of authors will use $f$ with mole fractions, $\gamma$ with molalities $m$, and $y$ with concentrations $C_i$ in terms of mass per unit of volume of the solution. Regardless of the units chosen, however, the values of the various coefficients for the solvent must approach unity as the solution becomes ideal where $X_{H_2O} \to 1$.)

In applying thermodynamics to solutions, the activity is related to the Gibbs free energy G rather than to a potential energy in mechanical terms. (At this point the student might wish to review the derivation of G, also denoted as F, in any elementary textbook of physical chemistry, or in one of the special presentations recommended in footnote 2.) The Gibbs free energy is an extensive property of a system; that is, its value depends on the size or total amount of matter at a particular temperature and pressure. Ordinarily, we do not know the absolute value of G; it is the difference in G, however, which is of importance in studying changes in chemical systems. In a closed system, one in which there can be no exchange of matter, an infinitesimal change in the total value of G is completely defined by its variation with temperature T (Kelvin scale) and pressure P. The latter two variables are intensive properties (i.e., mass-independent or qualitative properties) of a system. Hence, by varying each intensive property in turn, for a constant number of moles, n of all the components, the change in G is

$$dG = \left(\frac{\partial G}{\partial T}\right)_{P,n} dT + \left(\frac{\partial G}{\partial P}\right)_{T,n} dP = -SdT + VdP \quad (3)$$

The coefficients in parenthesis are the negative of the entropy S and the volume V of the system, respectively. In an open system, where matter can be added or subtracted, such as we have considered in Fig. 1, the change in G (being an extensive property) must depend also on any change in the number of moles of the components. Holding the temperature and pressure constant, an infinitesimal change in G for N number of components is described by

$$dG = \left(\frac{\partial G}{\partial n_1}\right)_{T,P,n_{N-1}} dn_1 + \left(\frac{\partial G}{\partial n_2}\right)_{T,P,n_{N-2}} dn_2 + \cdots$$

$$= \sum_{i=1}^{N} \left(\frac{\partial G}{\partial n_i}\right)_{T,P,n_j} dn_i \qquad (j \neq i) \quad (4)$$

where the masses of all other components $j$ are held constant, in turn, as an infinitesimal mass of a particular component $i$ is varied. The coefficient $(\partial G/\partial n_i)_{T,P,n_j}$ for a component $i$ is the partial molal free energy of the component, usually called the *chemical potential* and designated as $\mu_i$ or $\bar{G}_i$. (The chemical potential may be defined by varying other thermodynamic energy quantities with the number of moles of matter. These definitions, however, involve holding constant one or more extensive properties, such as the entropy or volume; temperature and pressure, on the other hand, are conveniently held constant in experiments on liquids.) The term *potential*, in general, refers to a potential energy per unit of mass. The chemical potential $\mu$ is thus seen to be a partial quantity which is an intensive property independent of the size of the system. As long as the proportion of the various components is not changed, the value of $\mu$ for each component remains fixed at a particular temperature and pressure, regardless of how many or how few moles of matter there are.

Chemical potentials are related exponentially to absolute activities; i.e., $\mu_i = RT \ln [\alpha]_i$, where $[\alpha]$ is the absolute activity and R is the molar gas constant. The activity of interest, $\alpha$, is a relative property which is arbitrarily equated with the actual concentration at some specified standard state. Since the chemical potential of a component may or may not vary in a strictly parallel way with the concentration, the relative activity is related to the difference between $\mu_i$ at the concentration of interest and the potential $\mu_i^0$, at the chosen standard state. Thus

$$\mu_i - \mu_i^0 = RT \ln \alpha_i = RT[\ln X_i + \ln f_i]$$
$$= RT[\ln m_i + \ln \gamma_i] \quad (5)$$
$$= RT[\ln C_i + \ln y_i]$$

where the actual concentrations and activity coefficients are taken from Eqs. (2). If we take pure water to be the standard state, it follows from Eqs. (5) that the relative activity of pure water on side A of Fig. 1 must be unity, the arbitrary assignment which we gave it earlier. It follows also that the chemical potentials of a diffusible component must be equal on both sides of the membrane at equilibrium because the activities are equal. Hence, in Figs. 1(a), 1(c), and 1(d) we have

$$\mu_{H_2O}^A = \mu_{H_2O}^B \qquad (6)$$

This condition must apply equally for the potentials of all other diffusible components, regardless of how many, which have access to both sides of the system. This is in accord with the second law of thermodynamics; a more specialized statement of it is that a component will move spontaneously from a region of higher potential to one where its potential is lower. Since the activity we are interested in is manifested by the relative ability of a component to pass through a membrane, we are not concerned at this time with the activities and potentials of those components which do not have this ability.

From this preliminary discussion, it is evident that we may view the phenomenon in terms of the solvent without inquiring at this point into the nature and behavior of the molecules which cannot cross the membrane. It should be clearly recognized that an osmotic pressure can only refer to the total, connected system and that there is no such phenomenon associated with one phase when separated from the other.[4] In summary, we see that the chemical potential of a component must eventually become the same in all parts of the system to which it has access—and that this potential is a function only of its concentration at constant temperature and pressure. Where it is not possible for the composition of a mixture to become uniform throughout, because of the presence of nondiffusible components, the spontaneous movement of diffusible components is nullified by a counterpressure arising as a result of the boundaries on the system. This pressure difference presents an additional potential on one side of the semipermeable membrane so that the total potential of each diffusible component is identical with that on the other side. In the elementary theory which follows, we shall equate the pressure difference to this extra potential, and then see how the latter

---

[4] Seemingly, it should not be necessary to emphasize this point. Writers have frequently exposed this and other misconceptions of the osmotic pressure. The notion remains widespread, however, arising perhaps from didactic expressions and popular writings, that a protein or salt solution itself has an osmotic pressure—as if the solute behaved like a gas within the liquid volume according to the gas law, $PV = nRT$. It is this conception which appears to cause the greatest hardship to students when relating osmotic pressure to more complicated systems in biology. If the actual pressure were measured at a given depth in a flask containing a protein solution, it would be difficult to detect any difference in the pressure from that at the same depth in another flask of the corresponding solvent medium containing no protein; any trivial pressure difference would be a function of the slight difference in the densities.

relates to the number of moles and, therefore, to the molecular weight of the nondiffusible component.

## II. Elementary Theory

We wish first to find how the observed osmotic pressure $\pi$ is related to the molecular weight of a macromolecular component which cannot diffuse through a semipermeable membrane. Second, the departure from strict proportionality between the observed pressure difference and amount of nondiffusible component added to the system is to be examined; that is, how may we interpret the amount of pressure which is in excess of, or is deficient from, the amount predicted on the basis of such proportionality? It is the manner in which this departure varies which is the important challenge to the method and which presents opportunity for creative thought and timely applications in the quest for understanding how proteins behave in solution.

Components are denoted by numerals, principally as subscripts, where 1 is water, 2 is a nondiffusible component such as a protein, and 3 is a second diffusible component, such as a salt. Higher even numbers refer to additional nondiffusible components and higher odd numbers to additional diffusible components, respectively. Single primes are used to designate the side of the membrane containing no protein (i.e., the solvent medium containing only components 1, 3, 5, . . .) and the absence of primes denotes the phase containing both diffusible and nondiffusible components. Initially, the simplest case is considered namely, an ideal solution of an uncharged protein in pure water at neutral pH. If a semipermeable membrane separates the protein solution from the water at constant temperature and external (atmospheric) pressure $P_0$, equilibrium is achieved, as noted in Section I, when the chemical potentials of water in the two phases are the same [Eq. (6)]. The potential of water on the protein side, however, is related to the potential of the protein-diluted water at atmospheric pressure $(\mu_1)_{P_0}$ plus the gain in potential from increasing the pressure on this phase from $P_0$ to $P_0 + \pi$ (the additional pressure $\pi$ being just sufficient to prevent any net transfer of water from side A to side B, as in Fig. 1d). In order to evaluate this gain in chemical potential of water on the protein side, we must know how the potential changes with pressure. Thus, at constant temperature and constant number of moles of the components n, the value of $(\partial \mu_1/\partial P)_{T,n}$ is required for each change in the pressure dP. Since $\mu_1 = (\partial G/\partial n_1)_{T,P,n_2}$ and recalling that $(\partial G/\partial P)_{T,n} = V$, the total volume [Eq. (3)], it follows from the cross-differentiation rule that

$$\frac{\partial \mu_1}{\partial P} = \frac{\partial}{\partial P}\left(\frac{\partial G}{\partial n_1}\right) = \frac{\partial}{\partial n_1}\left(\frac{\partial G}{\partial P}\right) = \frac{\partial V}{\partial n_1} = \bar{V}_1 \qquad (7)$$

where subscripts denoting constancy of variables are dropped for convenience. The change in total volume with the number of moles of a component $i$ is called the partial molal volume $\bar{V}_i$, which is an intensive property of the phase and is independent of the size. Hence, the extra potential on the protein side required to maintain equilibrium is $(\mu_1)_{P_0}$ plus the sum of each $\bar{V}_1 dP$ in proceeding from $P_0$ to $P_0 + \pi$. The equilibrium expression becomes

$$(\mu'_1)_{P_0} = (\mu_1)_{P_0} + \int_{P_0}^{P_0+\pi} \bar{V}_1 dP \qquad (8)$$

The partial volume of water changes so little with pressure as to be negligible over any reasonable range of $\pi$ so that $\bar{V}_1$ may be placed outside the integral sign for this purpose. Integrating and transposing gives

$$\pi \bar{V}_1 = (\mu'_1)_{P_0} - (\mu_1)_{P_0} \qquad (9)$$

The potential energy for osmosis to occur spontaneously is thus the difference in the chemical potential of water on the two sides at a common pressure, and this difference is given by the product of the osmotic pressure and the partial molal volume of the water. Recalling from Section I that the chemical potential is related to the relative activity $\alpha$ of a component [Eq. (5)], the preceding equation becomes

$$\pi \bar{V}_1 = RT \ln \alpha'_1 - RT \ln \alpha_1 = -RT \ln \frac{\alpha_1}{\alpha'_1} \qquad (10)$$

The standard chemical potential, $\mu_1^0$ is a constant for both sides and has dropped out of the expression. (The term on the right is cast as a negative expression as a convenience for this special case for reasons which will be evident presently.) Since the solution is ideal (i.e., activity coefficients are unity), the mole fraction of the diffusible component, water $X_1$, equals the activity and Eq. (10) reduces to

$$\pi \bar{V}_1 \underset{(X_1 \to 1)}{=} -RT \ln X_1 \qquad (11)$$

because $\alpha'_1 = X'_1 = 1$. For this system, it is obvious that $X_1 = 1 - X_2$. The mole fraction of macromolecules is usually a small number (e.g., $X_2 < 10^{-4}$) owing to the high molecular weights compared to that of diffusible components. Moreover, the solution must be dilute for Eq. (11) to apply. Hence, $\ln(1 - X_2)$ is equivalent to $-X_2$ [i.e., $\ln(1 - X_2) = -X_2 - 1/2 X_2^2 - \ldots$] and, as the mole fraction of component 2 approaches zero $(X_2 \to 0)$, Eq. (11) becomes

$$\pi \bar{V}_1 = RTX_2 \quad (X_2 \to 0) \tag{12}$$

whereby the pressure difference is now related directly to the concentration of the nondiffusible component. Thermodynamic relationships for aqueous systems are often expressed in terms of molal concentrations or in weights of the components relative to a weight of water. At vanishing concentrations of protein, the mole fraction $X_2$ becomes equal to $n_2/n_1$. The molality $m_2$ of component 2, however, is $1000 n_2/M_1 n_1$, since $M_1 n_1$ is the number of grams of water. Further, the volume of a mole of pure water, $V_1^0$ may be substituted for $\bar{V}_1$ because volumes are additive in ideal solutions. The volume containing 1000 gm of water is then $1000 (V_1^0/M_1)$, which we may denote with the symbol $V_m^0$, where the superscript zero refers to vanishing concentration of component 2. Then, by dividing through both sides of Eq. (12) with $(M_1/1000)$ we obtain

$$\pi V_m^0 = RTm_2 \quad (m_2 \to 0) \tag{13}$$

where $V_m^0$ is given in liters. As will be more apparent presently, the quantity, $V_m^0$ is the volume of the solvent medium containing 1 kg of water even though the medium consists of several diffusible components. We introduce the molecular weight by noting that the number of grams of a component $i$ is $g_i = n_i M_i$. Denoting the number of grams of component 2 in a solvent containing 1 kg of water by $w_2$, we observe that $m_2 = w_2/M_2$. This equality may be utilized in Eq. (13) to relate the pressure difference $\pi$ to the molecular weight of the protein. Thus

$$\pi V_m^0 = RT \left( \frac{w_2}{M_2} \right) \quad (w_2 \to 0) \tag{14}$$

Equation (14) is one form of the relation known as the van't Hoff limiting law. The corresponding relation in terms of concentrations in grams per unit of volume is

$$\pi = RT \left( \frac{C_2}{M_2} \right) \quad (C_2 \to 0) \tag{15}$$

where $C_2$ is grams protein per liter of solution.[5]

---

[5] The values of $\pi$ will be nearly the same using these two equations (depending on the temperature) with $w_2$ and $C_2$ numerically the same, if low concentrations of protein in dilute salt or buffer solutions are employed. If substantial amounts of salt or denaturants are present, the values of $w_2$ and $C_2$ yielding the same value of $\pi$ will diverge because $V_m^0$ increases with addition of components 3, 5, etc. The weight

From the above considerations, it is clear that the osmotic pressure is proportional, in the limit, to the number of nondiffusible moles or molecules per unit of volume. The limiting slopes $(\partial \pi/\partial w_2)^0_{T,n_{j\neq 2}}$ or $(\partial \pi/\partial C_2)^0_{T,n_{j\neq 2}}$ given by $RT/V_m^0 M_2$, or $RT/M_2$, respectively [Eqs. (14) and (15)], are independent of the size and kind of nondiffusible species. For a mixture of nondiffusible species, the molecular weight at infinite dilution is a number-average value $\overline{M}_n$. This is seen by considering the sum of the molalities of a number of nondiffusible components at vanishing concentration whereby from Eqs. (13) and (14) it follows that

$$m_2 + m_4 + \cdots = w_2/M_2 + w_4/M_4 + \cdots$$
$$= \sum_{i=2,4}^{N} (w_i/M_i) = \pi V_m^0/RT \quad (\Sigma w_i \to 0)$$

Dividing through by $\Sigma w_i$ to give the number of moles of nondiffusible components per gram of these components, we obtain, after transposing

$$\left(\frac{\pi V_m^0}{RT \sum\limits_{i}^{N} w_i}\right)_{\Sigma w_i \to 0} = \frac{\sum\limits_{i}^{N} (w_i/M_i)}{\sum\limits_{i}^{N} w_i} = \frac{1}{\overline{M}_n} \quad (i = 2, 4, 6, \ldots) \quad (16)$$

where $\overline{M}_n$ is the number-average molecular weight at infinite dilution of a mixture of nondiffusible components or of a single component containing two or more nondiffusible species. Concentrations in weight per unit of volume may be substituted for $w_2$ in the equation.

It is to be emphasized that the van't Hoff law is only exact at vanishing concentrations of the nondiffusible component(s). Deviations from the value of the limiting slope $(\partial \pi/\partial w_2)^0_{T,n_{j\neq 2}}$ are commonly observed

---

per weight scales of concentration, such as $w_i$ and $m_i$, are independent of temperature and are more convenient for relating to chemical potentials.

By tradition, concentration units in weight per unit of volume (i.e., the "c" scale) have usually been employed in practice. Weights per unit of volume are related to weights per weight of a particular component by the density $\rho$, given in units of grams per milliliter. Thus, $C_2 = 1000\rho W_2$ where $W_2$ is the weight fraction $g_2/\sum_{i=1}^{N} g_i$, of component 2 in the solution. Also, $W_2 = w_2/(1000 + w_2 + w_3 + \cdots)$ so that $C_2 = \rho w_2/1 + \Sigma w_i/1000)$, where $i \neq 1$; thus for a multicomponent solvent medium it follows that $w_2 = C_2(1000 + w_3 + w_5 + \cdots)/(1000\rho - C_2)$. Henceforth, except where noted otherwise, we will use grams and moles per kilogram of water (w and m, respectively) rather than grams and moles per unit of volume.

in the range of concentrations employed experimentally. (A superscript zero generally denotes a reference condition, in this case, zero nondiffusible components.) These deviations are a result of a number of factors, some well identified, others not. Hence, $M_2$ or $\overline{M}_n$ must be obtained from a series of measurements at different concentrations. Usually, values of the reduced osmotic pressure, $\pi/w_2$ or $\pi/C_2$, are plotted *versus* the concentration and an extrapolation to infinite dilution of the protein is made; the value of the reduced pressure at the intercept (i.e., at $w_2 = 0$) is a more convenient representation than is a limiting slope. By analogy with gases, a virial expansion is the general way of describing the variation of the osmotic pressure with concentration of protein. Thus

$$\pi = am_2 + bm_2^2 + cm_2^3 + \cdots \qquad (17)$$

In the study of proteins, the data often appear to be represented up to relatively high concentrations by only two terms in the expansion. Thus, the second virial coefficient accounts for the observed deviation from linearity of $\pi$ vs. $w_2$ (the van't Hoff proportionality) according to the relation

$$\frac{\pi}{w_2} = A(1 + Bw_2) \qquad (18a)$$

or equivalently by

$$\frac{\pi}{m_2} = A\overline{M}_n(1 + B\overline{M}_n m_2) \qquad (18b)$$

where $A = RT/V_m^0 \overline{M}_n$ (Scatchard, 1946).[6]

Figure 2 is a representation of the types of plots which are usually encountered with protein solutions (plots of $\pi/C_2$ vs. $C_2$ are, of course, very similar). Globular-shaped proteins (i.e., compact and symmetrical conformations) which are highly pure and neither dissociate nor cause an imbalance in the number of diffusible ions on the two sides of the membrane have exhibited nearly ideal behavior in dilute solutions approximating curve (b) of zero slope. Positive departure from the $\pi$ vs. $w_2$ proportionality is usually manifested by a constant value of B in Eqs. (18), as indicated in curve (a). This kind of plot is seen with mixtures of nonaggregating proteins, with proteins which are elongated or behave as random coils, and with charged proteins which give rise to a diffusible-ion imbalance, or with combinations of these. Except for the ideal diffusible-ion imbalance (see below), positive effects may be lumped together under the term *solution nonideality*. Finally, negative deviations are

---

[6] The coefficient B is sometimes defined by $\pi = Am_2 + ABm_2^2$ as in Scatchard's later papers. By this definition the coefficient is equivalent to $B\overline{M}_n$ of Eqs. (18).

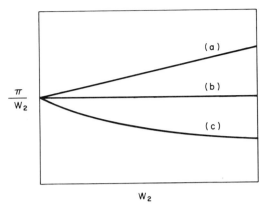

Fig. 2. Types of plots encountered experimentally upon varying the protein concentration. Curve (a), nonassociating system which exhibits nonideality and/or diffusible-ion imbalance; curve (b), ideal, nonassociating system or compensation between positive and negative deviations from the van't Hoff limiting law; curve (c), associating system where any nonideal and/or Donnan effects are overwhelmed by the attractive interactions (Kupke, 1960).

observed, superimposed on the positive effects, when proteins self-associate or aggregate into larger units (curve c). In this event, B is not constant except at concentrations where the molecules are completely dissociated or have associated into a limiting size, e.g., into dimers or trimers; an upward curvature with increasing concentration may be the experimental result after a limiting size is reached. Examples have been cited, however, in which an apparent cancellation of positive and negative effects resulted in plots simulating ideal behavior.

It is exceedingly difficult to obtain reliable data on protein systems in the absence of salt or buffer ions (e.g., Scatchard et al., 1946a). The presence of a third component, contributing ionic strength to the solvent, is generally a necessity even where the protein possesses no net charge. Because of this requirement, the investigator must also be concerned with the chemical potential of the membrane-diffusible salt and/or of buffering components. If we consider only a uni-univalent salt as component 3, for simplicity, it is a condition of equilibrium that, $\mu'_3 = \mu_3$. Unlike water, the principal component of the solvent medium, the potentials of salt and other diffusible solutes vary so slightly as a function of pressure over the ranges in $\pi$ encountered experimentally as to be trivial (cf. Tanford, 1961). Thus, if we substitute all subscripts 1 in Eq. (8) with subscript 3, the integral term in going from $P_0$ to $P_0 + \pi$ may be neglected in practice.

A far more important effect of diffusible ions on the osmotic pressure may result from there being a greater number of these ions on the protein side at equilibrium. This condition, known as the Donnan or Gibbs-Donnan effect, arises when the protein has a net charge Z (positive or negative). Although proteins have many positive and negative charges, there is an imbalance at most pH values so that the protein behaves somewhat as a polyvalent ion, called a *macroion*. To satisfy electroneutrality, the charge on the protein is compensated by diffusible ions of opposite charge so that the neutrality condition on the protein side of the membrane requires that

$$Z_2 m_2 = m_- - m_+ \qquad (19)$$

where $m_+$ and $m_-$ are the molalities of the univalent diffusible ions. On the solvent medium side, the neutrality condition is simply

$$m'_+ = m'_- \qquad (20)$$

If $Z_2$ is positive, a greater number of negative than positive diffusible ions are required for electrical neutrality on the protein side, and vice versa if $Z_2$ is negative. At equilibrium, where net transport of all diffusible components must be zero, the activities of the diffusible ions on each side must be equal, just as the activities of water must be the same on the two sides. The activities of the diffusible ions are taken as the product of the activities of the constituent ions of a salt even though the number of moles of each kind of ion on the protein side is not the same [Eq. (19)]. Assuming, as is usually done, that the mean activity coefficients $\gamma_\pm$ are essentially the same on the two sides of the membrane, we may simply use the molal concentrations. The condition at osmotic equilibrium is then

$$m'_+ m'_- = (m'_\pm)^2 = m_+ m_- \qquad (21)$$

since ordinary pressure differences have no significant effect on the activities. But if $Z_2$ is positive, the neutrality condition specifies that

$$m_- > m'_-$$
$$m_+ < m'_+$$

and the inequalities are reversed if $Z_2$ is negative. Since $m'_+ = m'_-$, it is numerically impossible for the total number of diffusible ions on the two sides to be the same, because $m'_+ + m'_-$ will always be less than $m_+ + m_-$ in ideal solutions (i.e., the least sum of any two numbers yielding a common product is given by the pair in which the numbers are identical). Because of the greater number of diffusible ions on the protein side, the activity of water is reduced by an additional amount so that a greater osmotic pressure is required to achieve equilibrium than that which would result from the presence of the protein species alone [cf.

Overbeek (1956) for detailed theory and nonideal effects when $Z_2$ is large and $m_2$ is small].

Returning to Eq. (11), in which the osmotic pressure is related to the mole fraction of water, it will be recognized that the presence of a third component, such as a diffusible salt, no longer permits the simplification, $\ln X_1 \cong -X_2$. The mole fractions, $X_1$ and $X'_1$, are reduced appropriately, to account for the number of moles of salt in each phase. If $X_1$ and $X'_1$ are expressed in terms of moles of protein, $m_2$ and salt ions, $m_+$ and $m_-$ per kilogram of water, Eq. (11) becomes

$$\pi = \frac{RT}{\bar{V}_1} \ln \left[ \frac{1 + M_1(m_2 + m_+ + m_-)/1000}{1 + M_1(m'_+ + m'_-)/1000} \right] \quad (22)$$

where $M_1$ is the molecular weight of water. The values of the second term in both numerator and denominator are usually quite small (e.g., $<0.005$, if $m_3 < 0.15$, $m_2 < 10^{-3}$, and $M_1 = 18$). Hence, the logarithms of both numerator and denominator are nearly equal to the values of the second terms in each case. Equation (22) may now be written to show the ideal relationship of $\pi$ to the molality of the protein and to the difference in ion concentrations on the two sides of the membrane. Thus

$$\pi = \frac{RT}{\bar{V}_1} \frac{M_1}{1000} (m_2 + m_+ - m'_+ + m_- - m'_-) \quad (23)$$

[Since the differences between the molalities of each mobile ion on the two sides of the membrane arising from the Donnan effect *per se* decrease with increasing ionic strengths, ultimately approaching zero, this equation is still virtually exact even at much higher concentrations of salt insofar as the Donnan equilibrium is concerned. In any case, salt concentrations of 0.15 molal are already sufficient to essentially eliminate the ideal Donnan effect of many proteins over a broad range of pH (4 to 10) in very dilute solutions (e.g., 0.1%), such as we are considering.] We wish to cast the effect of the ion imbalance resulting from the Donnan equilibrium in terms of the coefficient B, because this imbalance appears as a deviation in the van't Hoff proportionality between $\pi$ and the protein concentration. The osmotic pressure, however, may still reflect the ideal value, both with respect to the number of moles of protein and the difference in the number of moles of the diffusible ions on the two sides so that $\pi_{total} = (\pi_{prot} + \pi_{ions})_{ideal}$ at a given concentration of charged protein. For this purpose, a number of steps is required, including an expansion in a power series (cf. Scatchard *et al.*, 1944; Wagner, 1949; Edsall, 1953; Tanford, 1961). The ion pressure $\pi_{ions}$, as a result of the ideal Donnan equilibrium (where a substantial excess of salt ions over macroions exists so that higher terms are dropped), is

$$\pi_{\text{ions}} \cong \frac{RT}{V_m^0} \left( \frac{m_2^2 Z_2^2}{2\Sigma m_{\pm} z_{\pm}^2} \right) \qquad (24)$$

which becomes an approximation as $m_3 \to 0$. The summation $\Sigma m_{\pm} z_{\pm}^2$ in the denominator is twice the ionic strength of the salt. Hence, the total osmotic pressure, in terms of molalities when a uni-univalent salt is employed, becomes

$$\pi_{\text{total}} \cong \frac{RT}{V_m^0} \left[ m_2 + \frac{Z_2^2}{4m_3^0} m_2^2 \right] \qquad (25)$$

where the term $Z_2^2/4m_3$ is the coefficient B for the ideal case. By inspection, it is clear that as $m_2$ or $w_2 \to 0$, the Donnan pressure drops off rapidly so that at the intercept, where $w_2 = 0$, the value of $\pi/w_2$ is the same value as that for the uncharged protein. The positive slope (like curve (a), Fig 2) in plots where Donnan effects are severe, however, may be quite steep so that extrapolation to $w_2 = 0$ is more hazardous. As is evident from the equation, the Donnan effect may be reduced to negligible proportions by simply raising the ionic strength of the aqueous solvent medium. (The student can easily prove to himself that the ionic imbalance decreases with salt concentration by simulating a system at fixed net charge and number of moles of protein and then increasing the input of salt so that in each case an equal number of charges appear on both sides of the membrane initially. At equilibrium, where $[Na'_+][Cl'_-] = [Na'_+]^2 = [Na_+][Cl_-]$, one needs only to count the total number of moles of sodium and chloride ions on both sides after solving for the number of moles of each ion which have been transported in either direction.)

If other diffusible, ionizable components are present in the system, the same equivalence in their chemical potentials on the two sides of the membrane must also apply at equilibrium. Thus, the molalities of multivalent salts and of strong acids and bases, each taken as the total diffusible component on the protein side, are similarly equal on the two sides (with the assumption, again, that the activity coefficients of the individual ions are the same in both phases). It can be shown (e.g., Tanford, 1961) that the ratio of the activities of a diffusible ion on the two sides is very simply related to such ratio of any other ion. If a Donnan condition exists as a result of protein being present on one side, and if molal concentrations (indicated by brackets) are substituted instead of the activities, we have for a mixture of NaCl and $MgSO_4$ in water

$$\frac{[Na'_+]}{[Na_+]} = \frac{[H'_+]}{[H_+]} = \frac{[Mg'_{++}]^{1/2}}{[Mg_{++}]^{1/2}} = \frac{[Cl_-]}{[Cl'_-]} = \frac{[OH_-]}{[OH'_-]} = \frac{[SO_{4=}]^{1/2}}{[SO'_{4=}]^{1/2}} \qquad (26)$$

The common ratio relating all diffusible ion species at equilibrium is better known as the Donnan ratio. It is seen that a given ratio for any one diffusible ionic species fixes the concentration ratio for any other ion in the particular system. We note also that a Donnan effect dictates that the two sides at equilibrium be at a different pH, a commonly observed phenomenon with proteins in osmotic equilibrium.

At this point we must record the fact that the actual difference in molality of a particular ion on the two sides does not necessarily correspond to the diffusible ion imbalance. Often, proteins selectively bind ions of a salt. Such bound ions, of course, may not be counted in the Donnan equilibrium. The binding of one kind of ion, however, changes the net charge on the protein and a different Donnan effect will thus be seen than in the absence of binding. The paper by Scatchard *et al.* (1946a) provides a detailed illustration of how the osmotic pressure data differ from those calculated by Eq. (25) when chloride ion binding to the protein occurs. Moreover, because of other contributions to non-ideality, such as preferential hydration of proteins compared to their interactions with a salt component, it is not only possible, but frequently observed, that the molality of a salt component is actually lower on the protein side at equilibrium. This point will be discussed presently.

Before leaving the subject of charged proteins, it is pertinent to inquire how the protein ion may be defined in order that the protein component, as such, can be varied independently in osmotic pressure studies. The macroion component must be treated as an electrically neutral component in some consistent fashion so that all the components, viz. water, salt ions, and protein ions (and perhaps other ingredients), are accounted for when varying the concentration of component 2. Considerable latitude is available in making a choice, but convenience is important. For example, in a solvent medium containing a fixed mass ratio of water and NaCl, the continued addition of a protein component, defined as the neutral salt [protein·$Cl_{z_2}$], would increase the total number of chloride ions. In other words, the protein component might contain several times more small ions than macroions and, hence, the chemical potential of several moles of diffusible ion in combination with one mole of macroion must be considered. Clearly, a definition in which the component more nearly conforms to the macroions alone is desirable. The most often used definition is that given by Scatchard (1946) and the name *Scatchard components* has become commonplace. The protein component by this definition is the protein ion of net charge $Z_2$, plus $(Z_2/2)$ diffusible ions of opposite charge, minus $(Z_2/2)$ diffusible ions of the same sign as $Z_2$. In other words, half the charge of the macroion is balanced by an excess of oppositely charged counterions and the other half is balanced by a deficiency of the other diffusible ion. In effect, the neutral protein component is defined in such a way that the composition of the solvent medium, containing small ions in common with the counterions of the protein ion, is adjusted each time the protein component is added or taken away. For example, if the macroion has a net charge of $Z_2 = +10$, we add conceptually 10/2 moles of chloride ion and remove 10/2 moles of sodium ion when considering the addition or subtraction of one mole of the protein component from the solution. In practice, for any amount of the neutral protein–salt complex [protein·$Cl_{10}$] which is added to the solvent medium of water + NaCl, 10/2 moles of NaCl would be removed per mole of the macroion complex added. The component defined in this way has its charges balanced and at the same time represents virtually 1 mole of the macroion per mole of the protein component. [This statement cannot be considered as strict, because one assumes that the chemical potentials of the different small ions balance out in the definition. That is, the chemical potential $\mu_2$ is that of one mole of macroions plus $(Z_2/2)$ times $\mu_-$ minus $(Z_2/2)$ times $\mu_+$; however, $\mu_\pm$ of the separate ion species cannot be defined thermodynamically.] The ion concentrations, $m_+$ and $m_-$, on the protein side may now be written as follows:

$$m_+ = m_3 - Z_2 m_2/2$$
$$m_- = m_3 + Z_2 m_2/2$$

where again, $Z_2$ may be a positive or negative number. This definition of the components is the one applied in arriving at the relation for the ideal diffusible-ion pressure shown in Eq. (24) (cf. Tanford, 1961). Other definitions, which involve also the binding of salt ions to the protein, have been proposed (Güntelberg and Linderstrøm-Lang, 1949; Scatchard and Bregman, 1959; Casassa and Eisenberg, 1960).

The factors other than the Donnan equilibrium which contribute to the value of B in the second virial coefficient are less well understood in molecular terms. Hence, the quantitative contributions to solution nonideality from the various possible kinds of molecular interactions are not as predictable. In specially designed experiments, however, the value of a particular contribution may be approximated. A complete thermodynamic expression for the second virial coefficient in terms of activity coefficients is given in the comprehensive general theory of osmotic pressure by Scatchard (1946). Since we are interested in the deviation of the slope $(\partial \pi/\partial m_2)_{T,\mu}$ from its limiting value at $m_2 = 0$, the second derivative of the quadratic expression, Eq. (18b), in terms of molalities defines the constant B in curve (a) of Fig. 2; i.e., $V_m^0/RT \, (\partial^2 \pi/\partial m_2^2)_{T,\mu} = 2B\overline{M}_n$. For a system of one pure protein in a solvent medium of water plus a uni-univalent salt, as is frequently employed, the coefficient B in Eq. (18b) as derived by Scatchard is

$$B = \frac{1}{2M_2}\left[\frac{Z_2^2}{2m_3^0} + \frac{\partial \ln \gamma_2}{\partial m_2} - \frac{\left(\frac{\partial \ln \gamma_2}{\partial m_3^0}\right)^2 m_3^0}{2 + \frac{2\partial \ln \gamma_3}{\partial m_3^0}m_3^0}\right] \quad (27)$$

assuming incompressibility and that the partial molal volume of the salt is small compared with the volume of the solvent medium containing 1 kg of water. The superscript zero denotes vanishing protein concentration and the numerals 2 in the denominator of the third term within large brackets denotes that there are 2 moles of ion species per mole of the salt. The logarithm of the activity coefficient in the second and third terms is utilized because it represents the excess chemical potential which is the quantity we are concerned with in deviations from ideal behavior. That is, the potential for a component $i$, Eqs. (5), in terms of molality is $RT \ln \alpha_i = RT \ln m_i + RT \ln \gamma_i$, wherein the latter term is the amount which is greater (or less) than the amount predicted from a strict proportionality between the potential and the concentration of a component Thus, the change in the quantity $\ln \gamma_i$ (obtained by dividing through by RT) when varying the number of moles of each component in the system (per kilogram $H_2O$) independently, describes the net sum of the total deviations from ideal behavior in dilute aqueous protein solutions containing a low molecular weight salt.[7]

It may be noted by inspection that the first term within brackets in Eq. (27) is equivalent to the Donnan effect seen in Eq. (25). It is the only term which remains if all excess potentials are zero, i.e., if all activity coefficients are unity. The second term reflects the net attraction or repulsion between protein molecules themselves. In theory, it would be the only term remaining in a two-component system where the nondiffusible component cannot ionize in water. If the protein molecules repel, relative to inert, volumeless units of nondiffusible matter, the activity coefficient increases as more molecules are added to the system; hence, real nondiffusible molecules (having finite volumes) will show an increase in the amount of net repulsion as their concentration increases, even if they are chemically inert, because one molecule cannot occupy the space taken up by another at the same instant. If attractions be-

---

[7] It has become somewhat standard procedure to denote $\Sigma \nu_j \ln \gamma_j$ as $\beta_j$ ($j = 1, 2, 3, \ldots$); i.e., $\beta_j$ is the sum of the number of moles of chemical species or ions per mole of a component $j$ times the logarithm of the activity coefficient of the total component. Thus, $\beta_2 = \ln \gamma_2$, since only 1 mole of species is usually defined for the protein component; if component 3 is NaCl, $\beta_3 = 2 \ln \gamma_3$. The variation of $\beta_j$ with molality of a component $i$ or $j$ (holding other variables constant) is denoted $\beta_{ij}$ ($i, j = 1, 2, 3, \ldots$); hence, $(\partial \beta_2/\partial m_2) = \beta_{22}$, whereas $(\partial \beta_2/\partial m_3) = \beta_{23}$, etc. Note also, that $\beta_{ij} = \beta_{ji}$.

come greater with concentration than in the noninteracting case, the coefficient decreases and, therefore, the activity of water increases and this reduces the pressure difference. If the attractions are sufficient to produce stable associations as protein is added, the positive effects may be overcome so that the data appears as in curve (c) of Fig. 2. The third term combines the change in the activity coefficient of the protein with respect to the salt concentration and to the change in the coefficient of the salt with its own concentration. The change in the activity coefficient of salt is such that the total denominator in the third term is comparatively small and positive. The numerator, however, is magnified by the squared term if interaction occurs between the salt ions and the protein, such as in the binding of a few chloride ions to albumin (Scatchard et al., 1946a). Such interactions tend to decrease the osmotic pressure, because the third term is generally positive and is subtracted from the sum of the other terms contributing to B. Evaluation of the parameters in Eq. (27) for nonideality as observed by osmotic pressure measurements is illustrated in the thorough study by Scatchard and co-workers (1946a) on serum albumin in solutions of sodium chloride at different pH and ionic strengths. As must be evident, a number of other measurements were required to obtain all the values in the various terms of Eq. (27).

A particular source of nonideality, which has become of general interest in protein chemistry, is the phenomenon of apparent hydration of proteins which is often observed at osmotic equilibrium. When a substantial concentration of protein is present with relatively high concentrations of diffusible components, such as salts, sugars, and polyhydric alcohols, a significantly lower molality of such diffusible components may be found on the protein side relative to the outer, solvent phase (or dialyzate). This apparent or preferential hydration of proteins in the presence of relatively high concentrations of salt was shown long ago in the many experiments conducted by Sørensen (1917) and co-workers. The phenomenon is also seen at high salt concentrations, even where some salt is found to be bound or interacting with protein. Thus, a plot of the number of grams of water *versus* the number of grams of protein gives a positive slope (which is usually constant) at constant chemical potential of the diffusible components. [This constant slope $(\Delta g_1/\Delta g_2)_{T,\mu}$ is sometimes denoted as $\xi_1$; a similar plot with the salt or third component to give $(\Delta g_3/\Delta g_2)_{T,\mu}$, denoted as $\xi_3$, is then negative. At relatively low salt concentrations (e.g., <0.2 molal, as is commonly used to overcome the Donnan effect), a higher molality of salt may be observed on the protein side, indicating preferential salt interaction or negative hydration (i.e., $\xi_3$ is positive while $\xi_1$ is negative). This apparent discrepancy, depending on the concentration of component 3, is discussed in

Chapter 17, Appendix B.] For the purpose at hand, it is sufficient to point out that preferential hydration contributes only a small effect to the total nonideality. Generally, values of $\xi_1$, which represent, in effect, the excess number of grams of water per gram of protein relative to the proportion of water and component 3 in the outer solution, are of the order of 0.1–0.4 $g_1/g_2$. The hydration contribution to nonideality has been dealt with in thermodynamic terms by Güntelberg and Linderstrøm-Lang (1949). The effect is included in the third term of Eq. (27) and it is distinguished by rearranging this term so that the hydration factor, equivalent to $\xi_1$, appears in this term (or in terms of $\xi_3$, since $\xi_3 = -\xi_1 g'_3/g'_1$, where $g'_3/g'_1$ is the weight ratio of salt to water in the outer phase). For the simple case of a pure, isoelectric protein in relatively high concentrations of a uni-univalent salt, the apparent hydration $\xi_1$ is related to B, utilizing Eq. (27), and the $\beta$ notation (footnote 7), by

$$B = \frac{1}{2M_2}\left(\beta_{22} - \beta_{23}\xi_1 \frac{g'_3}{g'_1}\right) \qquad (28)$$

where the Donnan effect and salt–salt interactions are considered absent. The values of $\xi_1$ and $(g'_3/g'_1)$ may be determined in two-component solvent media by density measurements (see Chapter 17). The experimental results by these authors demonstrated that the contribution to the nonideal osmotic pressure by apparent hydration is only a few percent of the total nonideal pressure.[8] Lauffer (1964, 1966) has given another thermodynamic treatment for protein hydration at osmotic equilibrium with similar conclusions as to the magnitude of the nonideal contribution.

Finally, attention is called to the contribution to nonideality which arises from the presence of different kinds of proteins in a solution. If equimolal amounts of two proteins exist together in solution, will the degree of nonideality merely be the sum of each value of B which is contributed by the two proteins separately under comparable conditions? From theoretical considerations alone it is highly improbable that such additivity will be observed unless the two proteins are very much alike. Also, experimental results have shown that significantly different values

---

[8] Note: The osmotic pressure equation as shown by Güntelberg and Linderstrøm-Lang (1949) may also be cast in terms of the molecular weight of the hydrated protein, whereby the protein component is defined to include the excess amount of water. In this case, the third term of Eq. (27) drops out and the middle term then refers to the variation of the logarithm of the activity coefficient of the hydrated component 2 with its own concentration. The redefinition of the protein component to include the excess or deficient amount of water, or equivalently, the deficient or excess amount of component 3 on the protein side at osmotic equilibrium, has since become of general application and is utilized in a number of other kinds of equilibrium measurements in multicomponent systems.

of B are obtained with mixtures of nondiffusible components than that predicted from the results on the separate components. Moreover, if Donnan effects can be accounted for or suppressed and if molecular associations are absent (e.g., because electrostatic effects are swamped out with salt), an enhanced positive nonideality is conferred by a mixture of nondiffusible components. Inspection of the terms in Eq. (27) should suggest that the addition of a second protein will give rise to a new quantity in the middle term of this expression. In this term, the variation of the activity coefficient of a protein component with its own concentration is expressed (i.e., as $\partial \ln \gamma_2/\partial m_2 = \beta_{22}$; see footnote 7 for the $\beta$ notation). Hence, a second protein (component 4) requires not only the addition of the comparable value $\beta_{44}$ but also the cross-variations by which the logarithms of the activity coefficient of each protein component vary with the concentration of the other protein component, namely the cross-derivatives $\beta_{24}$ and $\beta_{42}$. The latter two derivatives are identical since $\beta_{ij} = \beta_{ji}$. The general equations for dealing with mixtures of nondiffusible components at osmotic equilibrium have been derived by Scatchard (1946). For two nondiffusible components, Eq. (27) may be expanded appropriately, as shown by Scatchard et al. (1954), to include the contributions to B when introducing the second protein. For this purpose, it is convenient to utilize the mole fraction x of each protein component with respect to the total number of moles of protein so that $x_2 = m_2/(m_2 + m_4)$ and $x_4 = m_4/(m_2 + m_4)$. (Weight fractions may also be used.) In this way all quantities relating to component 4 and cross-products drop out as $x_2 \to 1$. The value of B in these terms, employing a solvent medium containing water and a uni-univalent salt, as before, becomes

$$B = \frac{1}{2\overline{M}_n}\left[\frac{(Z_2x_2 + Z_4x_4)^2}{2m_3^0} + (\beta_{22}x_2^2 + 2\beta_{24}x_2x_4 + \beta_{44}x_4^2) - \frac{(\beta_{23}^0 x_2 + \beta_{43}^0 x_4)^2 m_3^0}{2 + \beta_{33}^0 m_3^0}\right] \quad (29)$$

where $\overline{M}_n$ replaces $M_2$ of Eq. (27). The Donnan term can be calculated if the net charges are known or it may be virtually eliminated by choosing conditions which suppress this effect. The third term within the large brackets similarly may be evaluated if the extent of ion binding is known for each protein or it may be eliminated by defining the protein components to include any bound ions. Thus, the principal difference in the value of B for a mixture over that predicted from results on the separate components resides in the middle term within the large brackets of Eq. (29). At moderate ionic strengths, the cross-derivatives, $\beta_{24} + \beta_{42} = 2\beta_{24}$, override the weighted effects of $\beta_{22}$ and $\beta_{44}$; i.e., $2\beta_{24} -$

$\beta_{22} - \beta_{44} > 0$ (Scatchard et al., 1954). Because of the latter inequality, a more positive departure from ideality, i.e., a higher value of $\pi$, is to be expected from mixtures, barring associations, than that which can be calculated without a knowledge of the cross-derivatives. If the two proteins associate, $\beta_{24}$ is negative and the total middle term of Eq. (29) may become less than zero.

The empirical parameter B, which has been used throughout this discussion to represent the departure from a strict proportionality between $\pi$ and $w_2$ or $m_2$, is that defined by Eqs. (18). Various other definitions of B are employed, which involve the inclusion or exclusion of certain constants. Also, B cannot be precisely the same when other concentration units are used. Hence, the student must examine the definitions when comparing values of B.

## III. Applications

The determination of molecular weight continues to be the principal application of the osmotic pressure method in the study of proteins. Other applications, however, while frequently involving the molecular weight, have evoked renewed interest in this method. Principal emphasis will be given here to those applications which have undergone significant development in relating to the central theme of present-day research, the study of interacting systems in the literal sense. Thus circumscribed this term connotes not only the self-association of proteins, but also the behavior of subunits *per se* and the interactions between different proteins and between proteins and other substances. This direction is quite natural, because, ultimately, we wish to know how proteins behave in the complex mixtures within compartments in and between living cells.

### A. Molecular Weight

The van't Hoff limiting law, Eqs. (14) and (15), relating the molecular weight to the osmotic pressure has been verified experimentally with proteins on countless occasions. For the student, the principal theme to be stressed here is a recognition of the unique features of the osmotic pressure method in order to aid him in the choice of a method, or combination of methods, which is best suited for the system of interest.

A very important consideration in present-day experimentation concerns the case with which definitive results can be obtained on multicomponent systems. Although the presence of protein may cause a diffusible ion imbalance and affect the activity coefficients of the various diffusible components, it is clear that as the concentration of protein approaches zero, differences in the diffusible ion concentration and in the activity coefficients on the two sides of the membrane become negligible. Moreover, the interactions between the nondiffusible molecules themselves vanish at infinite dilution, provided that all nondiffusible species attain a limiting stable size. Hence, the intercept at $(\pi/w_2)^0$ when $w_2 = 0$ gives the value $RT/V_m^0 \overline{M}_n$ (or $\pi/C_2 = RT/\overline{M}_n$ at $C_2 = 0$), regardless of the number and amount of diffusible components in the system. [See Scatchard (1946) for a discussion of the conditions which are implied in order for the van't Hoff law to have the same validity as the other laws of dilute solutions.] This distinct advantage given by the osmotic pressure method has not always been exploited when proteins have been studied in multicomponent solvent media, such as those containing large amounts of a denaturant; in some examples, the intensive efforts made with other methods have resulted in controversy, whereas a few experiments with handmade osmometers would have told the essential truth for a small fraction of the effort.

Second, for molecular weight purposes we do not require a buoyancy term as, for example, in sedimentation methods. Hence, the additional exercise of measuring very accurately the density $\rho$ and concentration of a number of protein solutions, in order to evaluate the partial specific volume or, for multicomponent systems, the derivative $10^3(\partial\rho/\partial C_2)_{\mu^0}$, is not required. More importantly, the accuracy of the molecular weight value is not threatened by the effect which small errors in the protein concentration have on this derivative (see Chapter 17). The density of the solvent medium, $\rho'$, however, is required in order to evaluate the volume $V_m^0$, because the medium is rarely pure water when proteins are studied. The density value, however, need not be more accurate than $10^{-4}$ gm/ml. This measurement is easily carried out, for example, by comparing the weight of the solvent medium with that of pure water in a constriction pipet of 1 ml or more at the temperature of the analytical balance chamber. (Usually, a sufficiently accurate density at the temperature of the osmotic experiment may be calculated by assuming the same proportionate change in density with temperature as that of water.) $V_m^0$ (in liters) is then obtained by

$$V_m^0 = \frac{1000 + w'_3 + w'_5 + \cdots}{1000\rho'} \tag{30}$$

where $\rho'$ is the density of the solvent in grams per milliliter. When concentrations are given in terms of grams protein per unit of volume, a solvent density is not required for the molecular weight obtained at infinite dilution of the protein. At finite protein concentrations, however, an effective volume contribution by the protein is included in the deviation from ideal behavior. The role of the volume of the nondiffusible component in osmotic theory has been the subject of discussion (cf. Wagner, 1949; Adair, 1961). In an ideal system and one without a Donnan effect, the second virial coefficient has been equated with the apparent specific volume $\phi_2$ of the protein (Adair, 1961). Thus, a volume-corrected concentration $C_{2(v)}$ may be substituted for $C_2$, where $C_{2(v)} = 1000 C_2 / (1000 - \phi_2 C_2)$. By this substitution, the concentration is cast in terms of a volume of solvent, where the solvent volume is arbitrarily assigned the volume it has in the absence of the protein. This manner of treating protein concentrations on the "grams per volume" scale is analogous to the formalism we have adopted in this chapter where $V_m^0$ is the volume of solvent arising from the "w" scale, which is based on the molal definition of concentration. Adams (1965b) has also employed $\phi_2$ (or its virtual equivalent, the partial specific volume) with the "grams per unit of volume" definition of concentration in his theoretical development on nonideal, self-associating protein systems. Evaluation of $\phi_2$, however, requires the same care and effort as does the determination of the derivative $10^3 (\partial \rho / \partial C_2)_\mu^0$; i.e., $\phi_2$ is obtained from the average of several very accurate density and protein concentration measurements by

$$\phi_2 = \frac{\Delta V}{g_2} = \frac{1}{\rho'} \left[ 1 - \left( \frac{\Delta \rho}{C_2} \right)_m 10^3 \right] \tag{31}$$

where V is total volume, g is in grams, and $\Delta \rho$ is the difference in density between solution and solvent. The subscript m refers to constant molality of the nonprotein components in both the solvent and solution. (For osmotic pressure experiments, it is probably not as important to distinguish between protein specific volumes at constant molality and at constant chemical potential as it is in sedimentation equilibrium experiments. In the former method, the use of a highly accurate value of $\phi_2$ when using $C_2$ instead of $w_2$ is perhaps not justified anyway, because the protein specific volume usually represents a relatively small departure from the van't Hoff proportionality compared with other effects.)

Other special points may be noted about the osmotic pressure method for the estimation of molecular weights. Impurities of low molecular weight which can pass through the membrane do not contribute to the value of $(\pi/w_2)^0$; such impurities can give rise to large error in other colligative methods. Moreover, the method is relatively insensitive to the presence of very large particles. Dust or other particulates make very little difference in the pressure, whereas these must be stringently removed when using a method which becomes more sensitive with solute size, such as in the weight-average methods. Indeed, the molecular weight of average-sized proteins may be evaluated, within limits, in solutions containing very large macromolecules, such as viruses. This kind of study also provides a clear measure of the mixed interaction if the macromolecule concentrations are varied independently. More practically, the protein of interest may contain small amounts of aggregates of itself or of contaminants of high molecular weight from the tissue source, which are usually difficult to remove entirely. Since the osmotic pressure method yields a number-average result, the effect of these additional species on the pressure may be trivial. Finally, because the osmotic pressure method does not depend on optical properties, neither refractive index, color, nor other light absorption of the solutions makes any difference to the measurements.

The range of molecular weights determinable by osmotic pressure is not unlimited. As already indicated, the method becomes less sensitive as the size of the nondiffusible species increases, because the magnitude of $\pi$ depends on the number of moles of such species rather than on the number of grams per unit of volume. Osmometry, as ordinarily practised, is not very useful for the accurate estimation of $M_2$ of large proteins; values of $M_2 > 0.5 \times 10^6$ Daltons, obtained without benefit of specialized procedures, are probably rough approximations. Special procedures have been developed for the purpose of studying macromolecules of the order of $10^6$ Daltons (cf. Kupke, 1960); one such technique is described briefly in Section IV. With commercial osmometers, precisions in the pressure difference of about ±0.02 to 0.03 cm of solvent have been reported. This represents a substantial improvement during the past decade over the usual precision of 0.1–0.2 cm obtained with simple, handmade osmometers. If we assume that $M_2 = 100{,}000$, $V_m^0 = 1$ liter at a temperature of 20°C and a solvent density of about 1 gm/ml so that $RT = 24{,}870$ liter-cm $H_2O$ per mole, an osmotic pressure of 1 cm solvent corresponds, ideally, to a concentration of $w_2 = 4.02$ $g_2$ per kilogram water, or about 4.0 mg/ml [Eq. (14)]. With a precision of ±0.03 cm $H_2O$, this concentration is about at the lower limit of usefulness when designing a concentration series with this protein. Obviously, lower concentrations of lower molecular weight proteins can be studied with similar precision (viz., 0.4 mg/ml if $M_2 = 10{,}000$). The minimum molecular weight of proteins which can be studied effectively depends on the differential permeability of the membrane with respect to the protein and the solvent components. A lower limit of 1 to $2 \times 10^4$ Daltons has often been mentioned. Membranes are available, however, by which useful data may be obtained on proteins substantially smaller than $10^4$ Daltons (Section

IV,B). The time required for attainment of equilibrium becomes substantially longer, however, as the average pore diameter of the membranes is reduced. In this case, the time to achieve equilibrium may be the critical factor when attempting to apply the method to the study of fragile proteins.

Since the molecular weight obtained at infinite dilutions of the protein is a number-average value, no information on homogeneity is given by the method. The degree of heterogeneity must be established by other methods. In combination with a weight-average procedure, such as light scattering, a ratio of the weight- and number-average molecular weights is obtained which may be a sensitive indicator of polydispersity in certain cases. More routinely, the velocity method with the ultracentrifuge is employed to assess the relative homogeneity of a preparation prior osmotic pressure measurements.

## B. Dissociation and Subunits

Of the more timely applications of the osmotic pressure method are the evaluation of the number of polypeptide chains and/or subunits in a protein and the study of these chains or subunits. Subunits, of course, may mean many things, including one or more chains; it is appropriate here to refer to a very readable review by Reithel (1963) on subunits and the semantic problems involved therein. The history of the osmotic pressure method for the purpose of determining the minimum molecular weights of proteins in dissociating solvents is long and illustrative (cf. Kupke, 1960). Probably no other method has been utilized as extensively or successfully in providing clear-cut answers where the subunits are identical or are nearly of the same size. Recent examples are the impressive array of experiments by Castellino and Barker (1968) on a number of proteins, including the controversial enzyme aldolase, and the comprehensive work by Jeffrey (1968, 1969) on the difficult study of the subunits from wool. Perusal of these sources and of the references noted therein, apart from their heuristic value, provides a current perspective of this relevant topic and illustrates the role of osmotic pressure as an anchoring method. While the sophistication of sedimentation procedures (Chapter 10 of this series) and the advent of the powerful gel filtration method (Chapter 9), will together resolve many old controversies and contribute greatly in the subunit quest, it remains incumbent upon the investigator to apply osmotic pressure techniques whenever possible to authenticate a particular conclusion drawn from these and other methods.

The study of the subunits *per se* which are obtained in dissociating solvents has also been explored by osmotic pressure. Lapanje and Tanford (1967) have considered the random-coil behavior of disulfide-reduced polypeptide chains in 6 $M$ guanidinium chloride. Since polypeptides in this solvent appear to be devoid of any noncovalent structure, their behavior compared to that of linear polymers, for which theory exists, is of importance. In order to determine the dimensions of randomly coiled polymer chains, the viscosity is usually measured. As pointed out by these authors, an independent measure of thermodynamic nonideality is more reliable for this purpose. Thus, the second virial coefficients obtained from osmotic pressure measurements are used in combination with viscosity data to yield the unperturbed dimensions. A discussion of the theoretical aspects of this work is removed from the purpose at hand, and it suffices to say that the results for the unperturbed dimensions are in reasonable agreement with the predictions by Flory and associates (Brant and Flory, 1965; Miller *et al.*, 1967). Similar values for the second virial coefficients by osmometry on two of the same proteins in this denaturing solvent were obtained by Castellino and Barker (1968); the coefficients for all seven proteins studied by these authors were in close agreement with those predicted for random coils. The power of the osmotic pressure method has been exploited in depth on the long-standing enigma surrounding the dissociation of hemoglobin (Guidotti, 1967a). This study provided evidence that the unliganded tetramer (deoxyhemoglobin) is more stable to dissociation into the assumed dimeric form than the liganded tetramers (i.e., oxy-, carbomonoxy-, and cyanmethemoglobins). The effects of ionic strength and different salt ions provided substantial evidence for structural differences between liganded and unliganded dimer subunits; these results on the different behavior of the various presumed dimer species presented a basis for an interesting discussion on the central role of the dimer in the physiological role of hemoglobin (Guidotti, 1967c).

Limited proteolysis may be considered within the category of dissociation, albeit not in the usual connotation. It is sometimes of interest to compare the properties of the native or precursor protein with that of a derived material following limited proteolysis. Discrete products from proteolytic action are often found in nature, such as in proenzyme to enzyme conversion or in the formation of fibrin monomer from fibrinogen, a reaction in the blood clotting mechanism (cf. Ottesen, 1958, for additional examples). Osmometric comparison of the unaltered protein with the derived material may indicate whether a number of nondiffusible species is formed per molecule, or whether there is a separation of fragments at all after scission of peptide bonds. In other cases, diffusible fragments are produced so that the molecular weight of the nondiffusible material is lower than that of the initial protein. Also, measurements of the change in thermodynamic nonideality of the nondiffusible species before and after limited proteolysis is an important undertaking for understanding the nature of the proteins being studied. A classic example is the osmotic comparison of ovalbumin with the product plakalbumin, which is derived from the former by the microorganism enzyme, subtilisin (Güntelberg and Linderstrøm-Lang, 1949). Although the molecular weight of plakalbumin is only about 1.3% lower than ovalbumin (a small amount of peptide is split off in the reaction), careful osmotic pressure measurements showed that the apparent molecular weight of the former molecule was lower at all concentrations employed. While the difference in molecular weights was within the experimental error of the method, statistical analysis yielded molecular weights (viz., 45,000 for ovalbumin *versus* 44,700 for plakalbumin) which were in essential agreement with data from other methods. More importantly, the second virial coefficient for plakalbumin was substantially more positive. The implications of this finding on the physical-chemical nature of the two proteins is discussed by the authors. This example demonstrates the discrimination potential of the method when it is practised carefully even with simple, handmade equipment (Section IV, Fig. 6).

## C. Mixtures, Association Equilibria, and Hybrids

Within the general context of protein–protein interactions are included the interactions between different proteins and/or their subunits, between subunits of the same species of protein, and between subunits of closely related species wherein hybrid molecules may be formed. The dissociation products or subunits, just discussed in Section III,B, were considered as though the dissociation were both complete and independent of other protein species. It is also important, however, to study the dissociable protein under conditions where the dissociation is incomplete over a range of protein concentrations. In this way the equilibrium constants for the reactions may be evaluated in order to better understand the behavior of proteins in biological systems. Similarly, we may wish to learn whether a mixture of different proteins exhibits a net repulsion or net attraction relative to the ideal case, under various conditions. That living systems contain a varying mixture of macromolecules acting in concert within and between the different subcompartments is a compelling fact which presents a tremendous challenge to the protein chemist. The deviations from the van't Hoff proportionality arising from protein–protein interactions are confined to the middle term of the Scatchard equation for B, Eq. (27). Molecular repulsions are usually reflected by a constant positive value for this term, while molecular associations yield negative and generally inconstant values as a function of protein concentration. With mixtures, the overall problem is one of sorting out the contributions of each derivative, such as $\beta_{22}$, $2\beta_{24}$, or $\beta_{44}$ in the expanded term for two nondiffusible components [Eq. (29)]. Also, the behavior of a single self-associating protein component might be a compromise of net repulsion between the oligomers and between oligomers and monomers, while the monomers themselves are mutually attractive.

### 1. *Mixtures*

For technical and theoretical orientation, the reader might review the experiments by Scatchard et al. (1954) on equimolal mixtures of serum albumin and γ-globulin as a function of salt strength and pH. The results serve to emphasize that unexpectedly large changes in interaction between different proteins can arise with minor variations in conditions; as noted in Section II, the middle term of Eq. (29) cannot be predicted from the osmotic results on the separate protein components. Another application (to a more immediately practical problem at that time) is illustrated in the osmotic experiments by Rowe (1955) on mixtures of serum and dextran. The latter material (a polymer of glucose largely in 1 → 6 linkage) was commonly introduced into patients as a plasma volume expander after acute loss of blood or serum proteins. The study on the serum protein–dextran mixtures showed clearly that the expected pressure increments from the added dextran, predicted from the concentration and number-average molecular weight of the dextran ($\overline{M_n}$ for dextran was similar to that of serum protein), were very much lower than the observed values as normal serum total colloid concentrations were approached. Moreover, there was no predictable dependence of the observed pressure increment on the average molecular weight of different dextrans. As might be expected, the osmotic pressure of a mixture could not be predicted from the osmotic data on a given dextran and on the serum protein separately (the concentration of each component at a given osmotic pressure of a mixture, however, was found empirically to be approximately predictable). Of particular interest was the observation that as the concentration of the serum protein was reduced, the pressure increment from a given amount of dextran decreased substantially. In this regard, it is instructive to note that this plasma expander material has virtually no charged groups, whereas the albumin of serum has a large net negative charge at normal pH ($Z_2 > 15$ negative charges). Recalling the discussion of the Gibbs-Donnan equilibrium (Section II), an increase in the plasma volume caused by the addition of the expander will dilute the existing albumin. Although the reduced osmotic effect from diluting the albumin should be compensated by the dextran, there is superimposed a loss in the so-called Donnan volume. Scatchard et al. (1944) estimated that the Donnan volume of plasma (largely a result of the albumin component) at the osmotic pressure and pH of normal plasma may be fully one third of the total plasma volume. (The latter volume is maintained against the hydrostatic pressure on the blood.) Dilution of the albumin then, reduces the diffusible ion osmotic pressure drastically, because, as we have seen, this additional pressure depends on the square of the molal concentration of the charged nondiffusible species [Eqs. (24) and (25)].

## 2. *Association*

It was pointed out in Section II that a departure from the van't Hoff proportionality, i.e., $\partial(\pi/w_2)/\partial w_2 \neq 0$, is also obtained if the number of moles of nondiffusible species changes with protein concentration (e.g., self-association of the protein into larger stable molecules as $w_2$ increases; curve (c), Fig. 2). If other sources of nonideality are small and Donnan effects are absent, the coefficient B is not constant, but will decrease as the concentration of the protein component is increased over the concentration range in which association constants are finite. A procedure for evaluating the association constants in self-associating protein systems by osmotic pressure was originally outlined by Güntelberg and Linderstrøm-Lang (1949) for ideal, isoelectric systems. We may consider their simplest example, that of monomer molecules $A_1$ in equilibrium with dimers $A_2$; i.e., $A_1 + A_1 \rightleftarrows A_2$ and K is the molar equilibrium constant. For this purpose, let us assume there is no Donnan effect and that $\beta_{23}$ is negligible so as to eliminate the third term in Eq. (27). The equilibrium constant for the association may be given in terms of grams (rather than in moles as is usually done) so that $K' = w_D/w_M^2$, where the subscripts **M** and **D** refer to monomer and dimer, respectively; also, $w_2 = w_M + w_D$, for the total number of grams of protein component per kilogram of water. Since the molecular weight of the dimer $M_D$ is equal to $2M_M$, the value for the number-average molecular weight $\overline{M}_n$ observed osmotically in an ideal system can be expressed using the van't Hoff law, Eq. (14), in the form

$$\frac{\pi}{w_2} \frac{V_m^0}{RT} = \frac{1}{\overline{M}_n} = \frac{1}{2M_M} \left[ 1 + \left( \frac{\sqrt{1 + 4K'w_2} - 1}{2K'w_2} \right) \right] \qquad (32)$$

For this equation, the authors have related the change in chemical potential of the protein component to the change in the mass of the monomer so that $\partial \mu_2/\partial w_2 = RT/M_M (\partial \ln w_M/\partial w_2)$. An equation, in general form, for higher association products is also outlined by these authors.

Many globular proteins do appear to behave almost ideally when at low concentrations, if Donnan effects are virtually excluded. Steiner (1954) has also taken advantage of this fact and has developed equations for obtaining the consecutive association constants for a simply associating system of the type $A_1 + A_i = A_{1+i}$, where $i = 1, 2, 3, \ldots$ in the consecutive reactions involving the smallest subunit $A_1$. If the molecular weight of the monomer $M_M$ is known, it may be combined with the value determined for the apparent number-average molecular weight (which is assumed to be equal to the true number-average molecular weight, $\overline{M}_n$) and the protein concentration to yield the number or mole fraction of the monomer, $x_M$. (That is, $x_M$ is the number of moles of monomer per total number of moles of all of the protein species.) Thus, over a given range in concentration

$$\ln x_M = \int_0^m \left(\frac{M_M}{\overline{M}_n} - 1\right) d\ln m \tag{33}$$

where m is the total molality (or molarity). The total number of moles of the protein species (as molalities here) is related to the overall association constants $K_i$, by $m = \Sigma K_i m_M{}^i = m_M + K_2 m_M{}^2 + K_3 m_M{}^3 + \ldots$. The constant $K_2$ for the monomer–dimer case is

$$K_2 = \left[\frac{d}{dm_M}\left(\frac{m - m_M}{m_M}\right)\right]_{m_M = 0} \tag{34}$$

The overall association constant $K_3$ for the consecutive association from monomer to trimer is

$$K_3 = \left[\frac{d}{dm_M}\left(\frac{m - m_M - K_2 m_M{}^2}{m_M{}^2}\right)\right]_{m_M = 0} \tag{35}$$

An example of a recent application of this method is shown in the work on a low molecular weight protease–inhibitor protein complex using high-speed membrane osmometry (Harry and Steiner, 1969). The association constants from the experimental curves ($\overline{M}_n$ versus protein concentration) as a function of ionic strength and pH were combined with titration data and evaluated with respect to electrostatic theory.

Adams (1965b) has extended the analysis of self-associating proteins to nonideal systems by employing the weight fraction f of the monomer species in place of the number or mole fraction. The apparent weight fraction $f_a$ is related to the ideal weight fraction f by $\ln f_a = \ln f + B'M_M C_2$, where $f = C_M/C_2$. $C_M$ is the concentration of monomer, $C_2$ is the total protein concentration (grams per unit of volume), $M_M$ is the molecular weight of the monomer, and $B'$ is the corresponding nonideality coefficient for this concentration scale. The latter is defined by assuming that the activity coefficient $y$ (on this scale) is described by $\ln y_i = iB'M_M C_2$. In terms of this concentration scale, the general equation is

$$\ln f_a = \int_0^{C_2} \left( \frac{M_M}{\overline{M}_{n,a}} - 1 \right) d \ln C_2 + \frac{M_M}{\overline{M}_{n,a}} - 1 \tag{36}$$

where $\overline{M}_{n,a}$ is the apparent number-average molecular weight.

By this procedure, the weight-average molecular weight $\overline{M}_w$ may be obtained from the number-average result (or the apparent weight-average value, $\overline{M}_{w,a}$, when nonideality is indicated by the data). Thus

$$\ln f_a = \int_0^{C_2} \left( \frac{M_M}{\overline{M}_{w,a}} - 1 \right) d \ln C_2 \tag{37}$$

where $M_M/\overline{M}_{w,a} = M_M/\overline{M}_{n,a} + C_2 d(M_M/\overline{M}_{n,a})/dC_2$.

The analysis for the association equilibria in final form is identical with that which was derived by Adams (1965a) for sedimentation equilibrium experiments. (The student may recall at this point that in sedimentation equilibrium each cross-sectional plane along the radial path in the solution is a phase in osmotic equilibrium with all other planes or phases in the centrifuge cell.) Equations are given for the monomer–dimer–trimer case, as well as for the monomer–dimer association, and also for the multicomponent ionizable system containing a charged protein in water plus supporting electrolyte. The detail required for the proper development of these analyses in terms of association constants is not reproduced here; the reader who is interested in the osmotic study of nonideal, associating systems is encouraged to examine the original paper. As is evident, the problem is one of separating out the nonideal solution effects which are superimposed on the change in pressure accompanying a shift in the proportion of each nondiffusible species when the protein concentration is varied. Hence, successive approximation procedures are indicated in order to deduce the association constants by applying a reasonable assumption for the activity coefficient of the total protein component.

### 3. *Hybrids*

An illustrative application which combines the methods for the study of both associating and nonassociating mixtures is the work bearing on the hybridization of hemoglobin and the reaction mechanism of hemoglobin with ligands, such as oxygen (Guidotti, 1967a,b,c). The experiments were designed to test the fundamental hypothesis that dissimilar or "unlike" dimer subunits, arising from the dissociation of unlike parent tetramers in a mixture, may combine to form stable "hybrid" tetramers. The term *unlike* may refer simply to liganded *versus* unliganded molecules or to *unlike* in the sense of differences in the polypeptide chains, e.g., normal *versus* variant hemoglobins or to hemoglobins from different animal species.[9] The existence of hybrid molecules of tetrameric hemoglobin in hemoglobin mixtures, while indicated by some techniques, has been in question. The discriminating power of the osmotic pressure method was challenged in the light of the following considerations. If tetrameric hybrids are in equilibrium with the two parent tetramers and with the dimers from each parent, the weight fraction of total dimers, $(\Sigma f_j)_h$ (weight of dimer species $j$ to weight of total hemoglobin) should be less than $(\Sigma f_j)_{nh}$ where such hybrids cannot form; the subscripts h and nh refer to hybridizable and nonhybridizable, respectively. Hence, the osmotically measured value of $\overline{M}_{n,a}$, barring unusual differences in solution nonideality, will be larger if hybrids are present than will that calculated if hybrid formation is absent. For example, a 50:50 mixture should show a lower value of $\pi/w_2$, ideally, when hybrids are formed than the mean value of $\pi/w_2$ observed for the separate hemoglobin components all at the same total protein concentration. (The molecular weights of the unlike dimers, or of the unlike tetramers, are so similar in each case as to be indistinguishable.) A qualitative argument similar to that used by Guidotti (1967b) may be employed to predict that $(\Sigma f_j)_{nh}$ should be greater than $(\Sigma f_j)_h$. In the simplest case, an equimolar mixture of two nondiffusible components of approximately the same molecular weight, each dissociating with the same ease into two similar-sized monomers, may be considered. Each monomer then has an equal probability of colliding with an alike or an unlike monomer. If there is a significant tendency for unlike monomers to combine into parent-sized hybrid dimers, each monomer will spend less time, on the average, in the monomeric state than if the only collisions which were fruitful were those between like monomers to form the respective parent molecules. Hence, by analogy, the number of hemoglobin dimers in a mixture of two unlike hemoglobins will tend to be larger if hybrids cannot form than will the number of dimers in the case where stable hybrid tetramers are

---

[9] There appears to be substantial agreement that hemoglobin dimers are composed of one $\alpha$ and one $\beta$ type polypeptide chain; these dimers are designated as $\alpha\beta$. The tetramer is designated $(\alpha\beta)_2$ or $(\alpha\beta)_2^A$ for the normal A type, and a given tetramer is held to be in equilibrium with its $\alpha\beta$-dimers (and to a lesser extent with the $\alpha$- and $\beta$-monomers). Appropriate superscripts are often used to denote differences either in the individual chains from the normal type or in the liganded state of the dimers.

created. This conclusion is given in more quantitative terms by Guidotti (1967b) where it is shown that the weight fraction of hemoglobin dimer is necessarily smaller if an association constant for hybrid formation is finite.

As pointed out in Section II, the value of the coefficient B in the absence of association is positive (cf. Fig. 2, curve (a)). Hence, the value of $\pi/w_2$ (or $\pi/C_2$) in a nonideal associating system is larger at any finite concentration of protein than it would be if the solution were otherwise completely ideal. That is, the value of $\overline{M}_n$ tends to be smaller than the observed value $\overline{M}_{n,a}$ if solution nonideality is not negligible. As noted previously, for a single protein component one may discriminate between the effects of associative processes and those of solution nonideality (Adams, 1965b). In a mixture of proteins, however, the situation is more complicated (cf. Scatchard, 1946). The work of Scatchard *et al.* (1954; cf. also, Scatchard and Pigliacampi, 1962) on mixtures of two protein components showed that, at ionic strengths sufficient to suppress electrostatic interactions and by eliminating effects of bound diffusible ions, the nonideality coefficient B may be essentially expressed by dropping the first and third terms in Eq. (27) so that for two nondiffusible components Eq. (29) becomes

$$B = \frac{1}{2\overline{M}_n} (\beta_{22}x_2^2 + 2\beta_{24}x_2x_4 + \beta_{44}x_4^2) \tag{38}$$

where, again, even-numbered subscripts refer to nondiffusible components and x is the mole fraction of the subscripted component with respect to total protein (weight fractions may be used also). As Scatchard and co-workers have shown, the deviation from ideal behavior is more positive than that predicted from considering the sum of the effects with each protein alone. Accordingly, by comparing the osmotic data on an associating system of two mixed components, which might hybridize, with a simulated system drawn from osmotic pressure data on the two components separately (as was done by Guidotti in order to determine whether hybrid formation is indicated), the additional positive nonideality which is not included in the simulated case has a special importance. That is, the simulated plot of $\pi/(w_2 + w_4)$ *vs.* $(w_2 + w_4)$ will tend to show a lower curve than a plot obtained on the mixture if the latter contains no hybrids. If hybrids are formed in the mixture, the experimental curve will tend to be lowered from this cause, as discussed above. Hence, if the experimental curve is lower than that for the simulated curve drawn from the data on the individual components, then substantial hybrid formation is strongly indicated in the equilibria. This kind of result was indeed observed with mixtures of liganded and unliganded hemoglobin, as well as with mixtures of human and horse hemoglobin. Figure 3 shows an example of the results using a mixture of oxyhemoglobin and deoxyhemoglobin in a dissociating solvent medium. As is evident, the simulated result (dashed curve) calculated from the data of the top and bottom curves is clearly above the experimental curve for the mixture. This result has fundamental implications for the mechanism of oxygenation of hemoglobin (cf. Guidotti, 1967c). An additional example from this study is shown for a mixture of serum albumin and oxyhemoglobin in Fig. 4. In this case the albumin and hemoglobin cannot form hybrids. Since the molecular weight values of the hemoglobin tetramer and the albumin are nearly the same, it seems evident that in this solvent medium the former component, when alone, is partly dissociated at all concentrations studied. The albumin, on the other hand, behaves as a single molecular species, exhibiting a linearly positive deviation from ideality. Furthermore, the experimental curve for the mixture is clearly above that for the simulated curve obtained from the data on

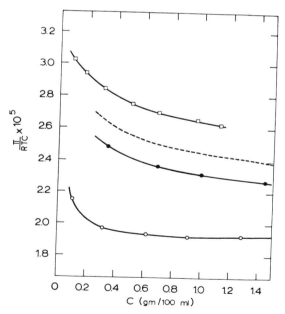

FIG. 3. Osmotic pressure data for oxyhemoglobin (□), deoxyhemoglobin (◉), and a mixture of the two (deoxyhemoglobin to oxyhemoglobin, 34:66%) (●), in 0.4 $M$ MgCl$_2$, pH 7. The dashed line is the theoretical curve for the mixture calculated from the data of the upper and lower curves assuming no cross-terms in the second virial coefficient [cf. Eq. (29)]. [Taken from Guidotti (1967b) with the permission of the author and the Journal of Biological Chemistry.]

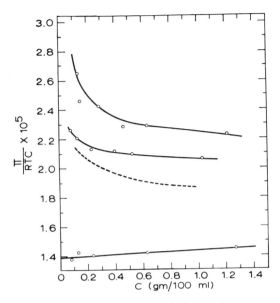

Fig. 4. Example of a mixture of an associating and a nonassociating protein. Osmotic pressure data for oxyhemoglobin (upper curve), bovine serum albumin (lower curve), and an equimolar mixture of the two proteins (middle curve), in 2 $M$ NaCl, pH 7. The dashed line is the theoretical curve for the mixture calculated from the data of the upper and lower curves, assuming no cross-terms in the second virial coefficient [cf. Eq. (29)]. [Taken from Guidotti (1967b) with the permission of the author and the Journal of Biological Chemistry.]

hemoglobin and albumin independently wherein the extra nonideality, contributed by the so-called cross-derivatives $\beta_{24} + \beta_{42} = 2\beta_{24}$ [Eq. (38)], is not included. The relevant equations describing the methods employed also include the relations for calculating the equilibrium constants for the association to hybrid species (Guidotti, 1967b). It is shown that if more than two dissociable components are mixed together, a unique solution for the weight fraction of a particular hybrid tetramer is not given. Detailed theoretical analyses of hybrids other than hemoglobin (more generally the association of mixed species of protein) have been developed by Steiner (1968) for the ideal case and by Adams et al. (1969) and Steiner (1970) for the nonideal case.

D. OTHER APPLICATIONS

This group of applications includes those which may be considered as emergent or comparatively unexploited. As a preface, the versatility of the osmotic pressure method is mentioned to alert the student to the fact that the technique can conform to the application rather than the reverse. Great freedom is given for applying one's imagination in designing a measuring system to suit an application. The humorous essay by Linderstrøm-Lang (1962) on "The Thermodynamic Activity of the Male House Fly" embodies this theme of versatility in the osmotic pressure method. The measurement of a pressure difference *per se* is not even necessary. Hill (1959) has discussed the possibility of measuring very small osmotic pressures in terms of an electric or magnetic field. Ogston and Wells (1970) have shown that very high osmotic pressures (about 1 atm) may be determined by measuring the change in dimensions of a single bead of Sephadex impaled on a needle. It is also pertinent to emphasize under this general heading that osmotic pressure determinations as a function of temperature and pressure have so far received commensurately little attention. The gain in information which awaits the application of these variables is potentially great. The few works with temperature which have been published demonstrate the usefulness of varying this intensive property (e.g., Gutfreund, 1948; Phelps and Cann, 1956; Banarjee and Lauffer, 1966). Osmotic pressure determinations as a function of pressure on self-associating protein systems are a very timely application which appears not to have been touched upon as yet. The effect of pressure on the average molecular weight and structure of proteins may be studied with a minimum of the complexity, ambiguity, and data-handling problems inherent in the methods now being used toward these ends. Another timely aspect is the determination of the density on each side of the membrane after the osmotic pressure has been evaluated. In this way, the change in the apparent molecular weight can be correlated with the preferential interaction parameter $\xi$ (see Chapter 17).

### 1. *Ion Binding*

Scatchard *et al.* (1959) have shown that Eq. (27) for the coefficient B may be utilized for the estimation of ion binding to proteins at relatively high ionic strengths. For this purpose, the protein component is defined to include the bound ion. Using the example of chloride binding in a medium of NaCl, the definition in terms of the Scatchard components described in Section II becomes

$$\mathrm{Na}_{\nu/2}[\text{protein} \cdot \mathrm{Cl}_\nu]\mathrm{Cl}_{-\nu/2} = \mathrm{Na}_{\nu/2}[\text{protein}]\mathrm{Cl}_{\nu/2}$$

where $\nu$ is the number of moles of the anion bound per mole of the protein. Thus, if $\nu = 10$ chloride ions bound, the component is composed of one molecule of the protein plus five molecules of NaCl. By this means, the third term of Eq. (27) drops out if it is assumed also that the activity coefficient of the salt is otherwise independent of protein concentration.

If the protein is initially isoionic, the bound chloride gives rise to a Donnan term in this equation so that B may be described simply by

$$B = \frac{1}{2M_2}\left(\frac{\nu^2}{2m_3} + \beta_{22}\right)$$

where $\nu$ replaces the net charge $Z_2$ in Eq. (27) and the ionic strength is equal to the molality of an uni-univalent salt. Transposing and solving for $\nu$ gives

$$\nu = \sqrt{4m_3(BM_2 - \beta_{22}/2)} \qquad (39)$$

A large value for the Donnan term has been found to contribute the major positive deviation from the van't Hoff proportionality; $\beta_{22}$ is comparatively small if self-association and electrostatic effects are negligible and if the proteins are globular in shape so that the excluded volume is not large. Since ion binding usually increases with the concentration of the salt, the value of B should be largely a function of the Donnan term at relatively high salt concentrations. The results by Scatchard et al. (1959) with a number of different anions and albumin showed that the values of $\nu$ were in approximate agreement with those obtained by other methods, depending somewhat on whether $\beta_{22}$ was assumed to be zero or was given an average value derived from other experiments.

## 2. Shape

It was observed in Section II that a source of positive deviation from ideality arises from the fact that real molecules are not points of matter having no volume and, therefore, one molecule cannot occupy simultaneously the identical space taken up by another. Accordingly, the proportion of the solution volume which is inaccessible to a nondiffusible molecule increases with the concentration of such molecules; the loss in available space per molecule increases its effective concentration or activity relative to that of ideal, volumeless nondiffusible molecules. We might think of the centers of the nondiffusible molecules as the ideal, volumeless units of matter, which are repelled from one another by the outer shells, thus giving rise to a net repulsion related to the size and shape of the shells. Using the simple example of noninteracting spheres, it is obvious that the center of one such sphere cannot approach the center of another sphere more closely than 2r, where r is the radius. Hence, the volume which is excluded to the center of an incoming sphere by the presence of a like sphere at a particular position in space is eight times the volume of these spheres (the volume of a sphere is proportional to the cube of the radius; by doubling the radius, the resultant volume is $2^3$ times greater). This kind of excluded volume is used in statistical mechanics to predict the macroscopic properties of solutions and, hence, the contribution of this volume in terms of the coefficient B.[10] The excluded volume $v_e$ of identical, noninteracting spheres of component 2 is

$$v_e = 8 \frac{\phi_2 M_2}{N} \tag{40}$$

where the molecular volume of a dissolved molecule of component 2 is defined as its apparent volume per mole $\phi_2 M_2$ divided by the number of molecules in a mole $N$. [We can make no definite statements about the actual space occupied by a molecule in solution; however, the apparent specific volume $\phi_2$ at least tells us how much the volume of a solution is increased (or changed) when a particular number of grams of the component is introduced. In the case of macromolecules, particularly the compact globular ones, we have some correlations for regarding $\phi_2$ as approximating what we might reasonably expect the space occupied by a gram of the molecules to be.] Shapes other than perfect spheres tend to increase the excluded volume. For example, long thin rods exhibit a far greater exclusion per unit of molecular volume (or mass equivalent). The excluded volume, from statistical reasonings, is proportional to the axial ratio $(L/d)$, where $L$ is the length of the rod and $d$ is the cross-sectional diameter. For stiff rods (cf. Tanford, 1961), $v_e$ becomes

$$v_e = 2 \left(\frac{\phi_2 M_2}{N}\right) \left(\frac{L}{d}\right) \tag{41}$$

The more difficult problem of calculating the exclusion by flexible chains is also outlined by Tanford (1961) as are the contributions by $v_e$ to the coefficient B for the three conformations just cited. The effect of $v_e$, of course, is manifested in the variation of the activity coefficient with concentration of component 2 (i.e., $\beta_{22}$) in Eq. (27). As is to be expected, the excluded volume of globular proteins (i.e., where the shape is somewhat spherical) contributes to B in a comparatively minor way and the positive effect appears to be linear in $w_2$ over the dilute range of concentrations usually studied. The effect on the osmotic pressure of an extra diffusible ion on the protein side, however, gives rise to a pressure difference which is comparable to that of one protein molecule (large spheres contribute an extra pressure only fractionally more than the same number of much smaller nondiffusible spheres). Rodlike proteins and de-

---

[10] The statement appearing in the review by Kupke (1960), p. 83, that "the excluded volume is equal to four times the molecular volume" is clearly wrong. The coefficient B (depending on its definition) is four times the molecular volume (or proportional to this amount) in the case of noninteracting spheres (cf. Edsall, 1953; Tanford, 1961).

natured or random chain polypeptides can be expected to cause much more positive deviations from the ideal.

Application of the osmotic pressure method to the question of molecular conformation of proteins has received only moderate attention. The work of Lapanje and Tanford (1967) and of Castellino and Barker (1968) on protein subunits denatured in guanidinium chloride has already been noted (Section III,B). A classic example for illustrating the potential of the method in the study of rodlike proteins is demonstrated in the osmotic experiments on the elongated protein myosin (Portzehl, 1950). Because of the high molecular weight, a special osmometer was required to measure unusually small pressure differences accurately, e.g., <0.1 to 1.5 cm $H_2O$ (Weber and Portzehl, 1949). Although later work on myosin suggests that these preparations consisted mainly of dimers, the results clearly showed a highly positive value of B under conditions where the Donnan term was not of significance (low protein concentration and high ionic strength). Applying statistically derived relations, such as Eq. (41), an axial ratio $(L/d)$ of >100 may be calculated from the data (cf. also, Edsall, 1953). These values, though somewhat high, are in approximate agreement with those now accepted for monomeric myosin. An historical point worth noting is that subsequent work with other methods (sedimentation and light scattering) erroneously indicated that this protein exhibited but little deviation from ideality. When these methods were further refined, confirmation was obtained for the large axial ratio of myosin as was originally indicated by osmotic pressure measurements.

### 3. *Gel Phases and Osmotic Pressure*

Is there an osmotic pressure in a sponge? Are there pressure differences between microcompartments of varying permeability within a gel? Such questions have been the subject of lively debate (cf. Ogston, 1966). Edmond *et al.* (1968) have shown that gels used for molecular sieve columns swell in accordance with osmotic theory and have demonstrated that the reversible changes in the inner volume of such gels can be used to measure the osmotic pressure (cf. Ogston and Wells, 1970). The osmotic potential and the responses as a result of variable exclusion of proteins and other solutes from gel matrices is fundamental to an understanding of the function of the many gel-like compartments within living tissues. Expositions on the role, in health and disease, of mucopolysaccharide matrices, such as hyaluronic acid, within the spaces of the joints must account for the osmotic effects arising from the exclusions and interactions of solutes in the mobile phase bathing these matrices. The studies initiated and carried forward by Ogston and associates on the application of osmotic principles represent a fresh and imaginative approach to the complex problems presented by the many gel-like phases in living organisms. The work also has important application to the emerging, powerful gel filtration method. Theoretical considerations for implementing studies of protein solutions in the presence of mucopolysaccharide complexes and on gel columns have been developed by Ogston (1962), Preston et al. (1965), and Ogston and Silpananta (1970). From the elementary theory (Section II), the systems may be viewed as a special case of mixed nondiffusible components. The gel component generally causes no substantial pressure difference, when very dilute, owing to its very large molecular weight; at higher concentrations, however, nonideal behavior gives rise to positive values of $\pi$. In the absence of Donnan effects and binding interactions, the average exclusion of a protein, such as albumin, from the region within the framework of large polysaccharides, such as hyaluronic acid, should result in a much larger value of the middle term in the Scatchard equation for B, Eq. (29), than is seen with mixtures of compact nondiffusible proteins. (Albumin appears to be the principal circulating protein bathing connective tissue spaces.) Indeed, the exclusion effect is seen to be the principal source of the extra pressure (above the ideal value) in controlled experiments with soluble proteins in the presence of gel-like particles (cf. Laurent and Ogston, 1963; Laurent, 1964; Preston et al., 1965).

## IV. Techniques

Historically, osmometers have been classified as static and dynamic types. With the static type, the pressure is measured when the system is finally at osmotic equilibrium, whereas with the dynamic type, the rate of approach to osmotic equilibrium is measured and related to $\pi$. In practice, both kinds of measurement can usually be made with either kind of osmometer and a combination of the two principles is often applied. The special advantage of the kinetic approach is the saving of time, especially if the equilibrium pressure is achieved very slowly. On the other hand, much more attention and general expertise is required for dynamic measurements to be meaningful; moreover, a number of static osmometers can be handled simultaneously by an investigator. Nonetheless, if a protein system tends to be unstable, speed in a given experiment may be of critical importance. A detailed comparison of these two approaches, including the particular problems encountered with each, has been made (Kupke, 1960). With the advent of commercially available rapid-membrane osmometers, the many classical osmometers described in the literature are becoming of less interest for general practice. Accordingly, only a brief description of the two classical types is presented below.

A. METHODS OF MEASUREMENT

1. *Static Osmometers*

The diagram of a simple and easily fabricated static osmometer is shown in Fig. 5. Static osmometers may consist only of a bag membrane tightly connected to a length of open-ended glass capillary, appropriately suspended in a thermostated solvent; the bag membrane contains the protein solution. For better overall precision, a water-immiscible organic liquid of low density (e.g., *n*-decane) is layered over the solution and fills the capillary. By this means, pressure displacements are magnified somewhat in the capillary and greater reproducibility is achieved in the position of the meniscus than that given with aqueous media. When the apparatus has been assembled and equilibrated, the change with time in the position of the meniscus may be observed (and measured) in response to the degree of disequilibrium. At equilibrium, the corrected hydrostatic pressure $h$ is the osmotic pressure $\pi$ in terms of the organic liquid of known density in the capillary. The final position of the meniscus is corrected for capillarity by introducing a reference capillary (preferably of the same tubing as used for the osmometer) into the immiscible liquid above the solvent (not shown in the figure). Capillaries of smaller internal diameter shorten the time of measurement; however, the greater effect from temperature fluctuations and the larger capillary correction may contribute additional error. A more useful way of shortening the time to reach equilibrium is to adjust the level in the capillary initially (with a syringe and long needle in the example shown) to near the equilibrium level; the adjustment can be predicted when experience with a particular protein system has been gained. Special details for accelerating measurements in static osmometry have been outlined (Kupke, 1960). Other osmometers of the static type are discussed by Adair (1949, 1961), Alexander and Johnson (1949), Wagner (1949), and Bonnar *et al.* (1958).

2. *Dynamic Osmometers*

In the kinetic approach, $\pi$ is expressed in terms of the rate of flow $dh/dt$ of the diffusible components per unit of time $t$ toward the protein solution; $h$ is the corrected height of the capillary liquid and is proportional to the total pressure P on the inner solution (sometimes a negative pressure relative to atmospheric is applied to the outer solution). Thus,

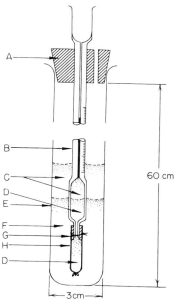

FIG. 5. Illustration of the static method. Stopper A, with small hole to permit entry and to allow in air, supports capillary B, of about 1 mm bore and having a cylindrical bulb at the lower end. Protein solution D in cellophane bag H is slipped over rubber sleeve G and the bag is tied tightly around the sleeve with two loops of nylon monofilament trout leader line. Protein solution (or solvent medium) D′ is placed in the lower half of the bulb with a syringe having an extra-long needle; the remaining space is filled with $n$-decane, C, to a desired level in the capillary. Solvent medium F fills the outer tube E to the indicated level and $n$-decane is layered over F to a level above the bulb. The interfaces between C and D′ and between C and F are matched by adjusting the capillary through the stopper. The outer tube is fixed vertically in a thermostated water bath with a window for observation and measurements.

The meniscus in B is observed by means of a cathetometer, reading to 0.01 mm. The height of the meniscus may be adjusted during the experiment by adding or removing decane with the syringe and needle. The static pressure level is corrected for capillarity by introducing another capillary of the same bore into the outer decane phase. Magnetic stirring of F is easily effected if found to be advantageous [cf. Kupke (1961)].

$P = \rho g h$, where $\rho$ is the density of the capillary liquid and $g$ is the acceleration of gravity. When the system is in temperature equilibrium and is "steady" so that the measured rate $\Delta h/\Delta t$ over a very small difference in height is essentially constant, the rate of flow is related to $h$ and $\pi$ by

$$\frac{dh}{dt} = k(h - \pi) = \frac{k}{\rho g}[P - (P_0 + \pi)] \tag{42}$$

where k is a proportionality constant dependent on a number of factors unique to the particular osmometer and $P_0$ is the atmospheric pressure in whatever units are chosen. According to Eq. (42), measurements of $\Delta h/\Delta t$ as a function of applied pressure (usually in terms of $h$) both above and below the equilibrium osmotic pressure (i.e., positive and negative rates) give the value $\pi$ at $\Delta h/\Delta t = 0$ by extrapolation. A shift in the extrapolated line toward lower pressures along the zero rate ordinate suggests permeation of the protein through the membrane or a loss in the effective protein concentration from any cause. Equation (42) may be cast as a definite integral, $d\ln(h - \pi) = k dt$ which, when integrated over the limits of three equal, consecutive time intervals, gives the equation by Phillip (1951)

$$\pi = (h_1 h_3 - h_2^2)/(h_1 + h_3 - 2h_2) \tag{43}$$

where subscripts for $h$ correspond to the heights at the three time intervals. This procedure is an alternative to the extrapolation method and may be more precise provided that suitable equal time intervals are found and the temperature is strictly controlled over the entire period covering the measurements.

The data obtained with slowly permeating proteins can still be useful if extrapolation back to zero time is carried out. Staverman (1951) has shown that the osmotic pressure in leaky membranes does not yield the thermodynamic apparent number-average molecular weight but a permeability-average molecular weight. With mixed solutes of different sizes or permeability, such as synthetic polymers, the discrepancy between the observed osmotic pressure and the true osmotic pressure may be serious. When protein leakage is slow relative to transport of the solvent components, the permeability coefficient which relates these two pressures is near unity and the time-extrapolation procedure can be used. With pure proteins which do not dissociate, the effect of slow leakage is unimportant. In lieu of extrapolating to zero time, the experiment can be concluded at some point where the pressure loss is constant with time. The difference in protein concentration at this time, relative to the initial concentration has been found to be strictly proportional to the loss in pressure between zero time and the time at the conclusion of the experiment (cf. Kupke, 1961). It must be recognized with dynamic osmometry that the redistribution of diffusible solutes may require much more time for some systems than for others. Hence, the pressure value at $\Delta h/\Delta t = 0$ obtained early in the experiment may not represent $\pi$ correctly and, therefore, the experiment cannot be safely concluded until repetitive determinations agree. In general, for either method, it is advantageous to dialyze the solution against the solvent medium whenever practicable.

A simple osmometer used for dynamic measurements is illustrated in Fig. 6 (Güntelberg and Linderstrøm-Lang, 1949). This apparatus was assembled from materials at hand and was applied to the osmotic discrimination between ovalbumin and plakalbumin discussed in Section III,A. The problem of differential ballooning of the membrane when changing the applied pressure can be minimized by the use of a planoconvex surface upon which to seat the membrane instead of a flat surface (Carter and Record, 1939).

### 3. *Ultrasensitive Osmometers*

A number of highly sensitive osmometers have been designed where pressure differences are measured in microns instead of centimeters of water. These were developed primarily for the study of very high molecular weight solutes where small pressure differences at equilibrium were expected [for example, the osmotic balance of Jullander and Svedberg (1944) and Enoksson (1951), the inverted microosmometer of Christiansen and Jensen (1953), and the special osmometer for large asymmetric proteins by Weber and Portzehl (1949)]. The interferometric osmometer (Claesson and Jacobsson, 1954), while not designed for the study of proteins, appears to hold the most promise as an ultrasensitive method for their study. The device may be useful regardless of molecular size, particularly where very low concentrations are required, such as for self-associating systems. The technique involves passing light of wavelength λ vertically through two parallel upright tubes containing solvent and solution; the liquids in the tubes are connected to a cell divided by a membrane mounted on a rigid support, such as is employed in the Fuoss-Mead osmometer (1943). Using a very simple optical system, the difference in height $h$ between solution and solvent is related to the number of interference fringes $N_\lambda$ so that

$$h = N_\lambda \left(\frac{\lambda}{\Delta n}\right) \tag{44}$$

where $\Delta n$ is the difference in refractive index between the liquids and vapor-saturated air. Since the method is designed for very dilute solutions, the value of $\Delta n$ for both solution and solvent is identical in practice (Jacobsson, 1954). The width of the tubes (2 cm) is sufficient to eliminate differential capillary effects of solution and solvent and of menisci curvature without sacrificing optical resolution (evaporation is controlled with glass plates which function also as optical compensators).

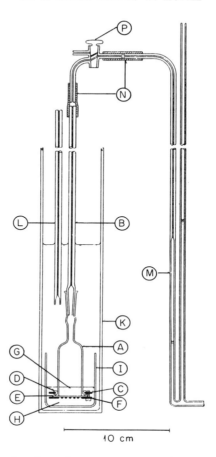

Fig. 6. Osmometer for dynamic measurements (Güntelberg and Linderstrøm-Lang, 1949). Cylindrical glass vessel A is connected with capillary B by means of a ground-glass joint. Membrane E is pressed against the glass flange, ground plane, at the lower open end of A by tightening three screws holding together the stainless steel flat ring C and the perforated plate F. The inner rim of ring C is cushioned with thin rubber tubing D. The protein solution G is situated in A, and the solvent H is in a short cylindrical vessel I. Toluene fills the remainder of A and extends into B. Both A (with B) and I are submerged in toluene in a tall glass cylinder K together with reference capillary L. The upper end of B is connected via rubber sleeves N with water manometer M, which is furnished with a pressure regulator (not shown) of the type described by Holter (1943). K and M are placed in a water thermostat, controlled to ±0.003°C, which contains a window for observation.

A desired pressure is applied to the protein solution by regulating the manometer liquid in M and manipulating stopcock P. The rate of movement of the meniscus in B is observed through a cathetometer with micrometer. The pressure at a given setting is the algebraic sum of the difference in levels at M and in B and L, corrected

$N_\lambda$ is determined with a precision of one fringe for most liquids, which corresponds to an error of 1 to 2 microns in $h$. By this means, it has been possible to measure osmotic pressures on nondiffusible solutes which are an order or more in magnitude lower in concentration than by conventional osmometers (e.g., nondiffusible solutes of $M > 10^5$ are routinely studied over a range in $C_2$ of 0.5 to 5 gm/liter). The total instrument must be mounted on a substantial supporting framework of great rigidity for the sake of the optical system and, therefore, it presents an imposing appearance for an osmometer. Although this osmometer, as originally designed, is not ideally suited for the study of proteins, with appropriate modifications there seems to be no reason preventing its application to these substances (S. Claesson, private communication).

## 4. *Electronic Osmometers*

Technological advances in the field of electronics have also had an impact on osmometry. Hansen (1950; 1952) reported the application of an electric condenser manometer for osmotic measurements by the compensation method of Sørensen (1917). In this method a pressure equivalent to the osmotic pressure is exerted on the protein side of the membrane to prevent any net flow of solvent to the solution compartment (as suggested by the diagram in Fig. 1d). These early attempts inspired a number of commercial models. Rowe (1954) reported the development of a workable pressure transducer system which used the compensation approach; this model was later refined by Rowe and Abrams (1957) for manufacture (Nash and Thompson, Chessington, Surrey, England). Adair (1961) has reviewed this osmometer in detail and Davies (1966) has described further modifications. Reiff and Yiengst (1959) successfully applied the strain-gauge transducer principle by which minute pressure changes are sensed as solvent passes through the membrane. The strain-gauge system consists of a fine wire which responds electrically to the distension of a thin, pressure-sensitive diaphragm. The change in electrical resistance in the wire is a linear function of the strain put upon it by changes in the pressure.

One of the new commercial osmometers now available is a recording instrument utilizing the basic principles of the Reiff and Yiengst osmometer (Melabs, Inc., Palo Alto, California, 94304). This osmometer is diagramed in Fig. 7. The osmotic chamber consists of two stainless steel

---

for densities to that of water; an additional correction is made if the interfaces above G and H become unmatched. The pressure is reset in the manometer, after closing P, to a value not far from the previous setting. A few minutes are allowed after reopening P, until the movement of the meniscus becomes constant before measuring the new rate.

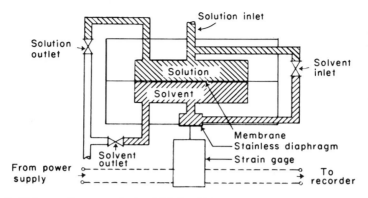

Fig. 7. Side-on diagram of the Melabs high-speed membrane osmometer. See text for details. (Courtesy of Melabs, Palo Alto, California, 94304.)

circular plates with interconnecting grooves (similar to the Fuoss-Mead design; 1943) having suitable inlets and outlets for solvent and solution. A circular flat membrane is placed between the plates, which are then clamped together along with Teflon gaskets for a tight seal. The pressure transducer consists of a stainless steel diaphragm connected to the strain gauge electronics, all of which is sealed in the solvent (lower) chamber. Temperature is controlled with a thermoelectric cooler and heater assembly to $\pm 0.02°C$ in the range 5°–20°C (the full range of temperature is 5° to $>100°C$). The calibration of the pressure transducer is carried out after the membrane is preequilibrated. A piece of calibrated glass tubing is placed over the solution inlet and is filled with solvent. With the solvent inlet valve open to allow free passage of solvent around the membrane, a measured difference in the solvent head in the calibrated tube above the solution inlet is adjusted to a full scale deflection on the recorder. Alternatively, a pressure head of solvent may be applied to the solution inlet using a calibrated manometer. A series of fixed resistors, which control the voltage to the recorder, permits signals of 1 mV to be produced for pressure differences in the ranges 0–5, 0–10, 0–50, and 0–100 cm of the solvent. (A 0–1 cm solvent scale is feasible with proper modification; this scale is especially useful for studies on self-associating systems and very large macromolecules.) After calibration, the solvent level is reduced to a level just above the upper stainless steel plate and the solvent inlet valve is then closed to allow the membrane to come to equilibrium again. The level of solution or solvent in the solution inlet just above the membrane clamp is detected by a system (not shown) which measures the change in light intensity as the meniscus of the liquid crosses the light path. The optical system converts this change in the intensity between air and liquid into an electrical signal. A particular current reading on the meter is then chosen as the reference position and the recorder is zeroed when the solvent is reequilibrated. Protein solution is introduced in 0.5 ml aliquots and the solution chamber is flushed of all solvent. Additional aliquots of the protein solution are introduced and measurements of the osmotic pressure are recorded until reproducible results, to within ±0.02 cm of solvent, are obtained. Generally, the dilution of the solution by the solvent is trivial during the period of measurements, because the solvent chamber is sealed and the distension of the diaphragm in this chamber from the pressure changes is very small. Measurement times are usually not more than 30 min, depending on the properties of the particular membrane being used.

The Mechrolab osmometer (Hewlett Packard Co., Avondale, Pennsylvania, 19311), shown in Fig. 8, operates on the compensation principle and employs an optical sensing system to measure the osmotic pressure. The design of the membrane compartment is similar to the Melabs osmometer except that a glass capillary leads from the solvent chamber and this tube is connected by polyethylene tubing to a reservoir of the solvent resting on a servo-driven elevator. A bubble is introduced into the capillary before the membrane is placed on the grid for clamping by the steel plates. The pressure difference arising from the insertion of the protein solution on the upper side of the membrane causes the bubble to move in the capillary. The bubble position is sensed by a photocell via the change in light intensity from a small light source. The change

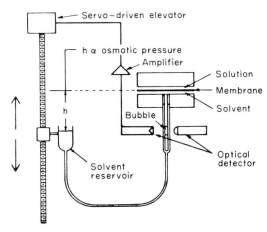

FIG. 8. Side-on diagram of the Mechrolab high-speed membrane osmometer. See text for details. (Courtesy of Hewlett-Packard Co., Avondale, Pennsylvania, 19311.)

in light intensity activates the servo-driven elevator containing a solvent-filled cup or reservoir to provide a balancing head of pressure. The osmotic pressure is read as a difference between reference points in centimeters of solvent when the system is in balance and the elevator is stationary. A recording system is also available for the model. Temperature control is provided by heater and cooler assemblies over the range 5°–65°C. Reported stability of temperature is ±0.2°C in the range 5°–25°C (Castellino and Barker, 1968; Harry and Steiner, 1969). Paglini (1968) has written a special protocol for the cleaning and operation of this instrument.

Rolfson and Coll (1964) have developed a commercial osmometer which is also based on the compensation principle (Hallikainen Instruments, Richmond, California, 94804). This instrument, shown in Fig. 9, maintains a balanced pressure head by means of a servo system connected to a float which displaces solvent into the osmotic chamber. The

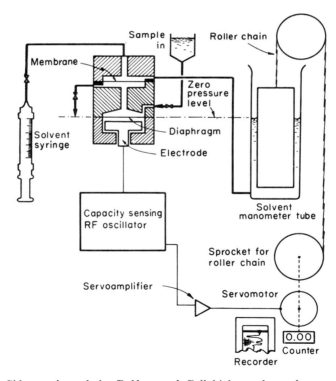

FIG. 9. Side-on view of the Rolfson and Coll high-speed membrane osmometer. See text for details. [Reprinted from *Anal. Chem.* **36**, 888 (1964). Copyright (1964) by the American Chemical Society. Reproduced by permission of the authors and copyright owner.]

pressure sensing device is a thin metal diaphragm, as in the Melabs osmometer, which is connected to a radio frequency oscillator; the latter produces an electric signal to run the servo system. Temperature control is by heating above ambient temperature, which makes the unmodified instrument more desirable for use with organic solvents at elevated temperatures than for standard work on proteins. Also, the design utilizes more solution than the previously described electronic osmometers.

The electronic osmometers offer the advantages of shorter measurement times (10 to 30 min per measurement in many cases), more standardized techniques, and precisions on the order of 0.02–0.03 cm of solvent height. Except for the Hallikainen model, the volumes of sample solution are smaller (e.g., about 0.5 ml) than for most classical models. For best results with the commercial osmometers, specially designed cellulose acetate membranes have been recommended; type B-19 by Schleicher and Scheull (Keene, New Hampshire, 03431) is used most frequently. For small proteins ($M < 20,000$) the B-20 type is preferable, although the time of measurement is increased. Dialysis tubing, which has often been used for very small proteins in the classical osmometers, has not been found to be useful in the Mechrolab instrument (Lapanje and Tanford, 1967).

The advantages cited for the commercial, high-speed, electronic osmometers represent a general advancement in the technique, especially where the work requires relatively rapid measurements to be made and where lower concentrations of protein must be used than is possible with the standard, classical osmometers. Partly for this reason, the renewed interest in the method, which is now evident, has come about. This is manifested by recent imaginative studies directed toward the solution of problems not clearly solved by other techniques. Examples of these applications using electronic osmometers are those concerned with self-associating proteins and the quaternary, tertiary, and even secondary structures of proteins (Preston et al., 1965; Banarjee and Lauffer, 1966; Lapanje and Tanford, 1967; Guidotti, 1967a,b; Castellino and Barker, 1968; Harry and Steiner, 1969; Jeffrey, 1969; Andrews and Reithel, 1970).

The vapor-pressure osmometers and the commercial membrane osmometers which have been designed for physiological experiments are not discussed here because these instruments have seen little application to the thermodynamic study of proteins.

## B. Membranes

The quest for the ideal membrane undoubtedly has stirred the imagination of any investigator who has worked with the osmotic pressure method. The behavior of membranes continues to be the most vexing materiel problem facing the method. Real membranes are neither ideally thin, rigid, nor inert structures, nor do they have uniform holes which prevent all molecules above a certain effective volume from passing through. Factors other than simple mechanical exclusion are involved in the permeation process by which we distinguish between diffusible and nondiffusible components. Currently, osmotic membranes are hundreds to thousands of times thicker than the diameters of most proteins and they appear to consist of haphazard layers of randomly oriented polymer strands; these materials exhibit gel-like properties in aqueous media and may interact with the components of the solutions. Hence, the term *membrane phase* is sometimes encountered. The nature of the processes and interactions in the membrane phase, while interesting, do not enter into the thermodynamic considerations. If the membrane is essentially semipermeable with respect to the protein of interest and does not affect the activity coefficients significantly, we may safely ignore its presence just as we ignore the vessels containing the system. At the present state of the art, we must be content with membranes which are far from ideal but with which virtual equilibrium can be achieved, albeit with discouraging slowness. Fortunately, in the study of proteins, we deal with nondiffusible components which have discrete sizes so that even when associating systems are studied the objective is merely to use membranes which do not permit appreciable passage of the smallest protein subunit. In the study of synthetic polymers, the problems with membranes become comparatively more serious, because the array of polymer sizes in a preparation is often extensive, sometimes approaching a continuum of sizes. Accordingly, reproducible results are more difficult to achieve. [A differential loss of the smaller polymer molecules would have a comparatively large effect on $\overline{M}_n$. This uncertainty cannot be easily overcome, because a sharp cutoff in the size of polymer to be passed by a particular membrane is hardly to be expected. The preparation of sufficiently similar membranes for parallel experiments is also very difficult to achieve. A further complication arises because membranes tend to change with time. Bonnar *et al.* (1958) give a detailed discussion of these problems and describe the results of cooperative testing programs on standard polymer samples.]

Although organized testing programs have not been carried out for comparing osmotic pressures on protein systems, the general agreement over the years (at $C_2 = 0$) on nonassociating proteins is quite remarkable. Thus, extreme uniformity in the preparation, pretreatment, and type of membrane is unnecessary for most purposes. The membranes most commonly used for aqueous systems are derivatives of cellulose, such as collodion, cellophane, and, more recently, cellulose acetate. The two latter types are available commercially; the cellophane is usually in the form of dialysis tubing made from regenerated cellulose. A particular shape and degree of rigidity or a close tolerance in the average pore diameter may be important for some purposes. Such membranes can be best tailored in the laboratory from collodion. References to selected papers and reviews dealing with collodion membranes have been listed (Kupke, 1960) and a special discussion on the preparation and permeabilities of collodion membranes is given by Adair (1961).

For the study of very small proteins (M < 20,000) some additional remarks on membranes may be useful. The membranes which are generally available for dialysis and osmometry, whether coming from commercial sources or made from standard collodion recipes, do not hold back small proteins sufficiently well over long-term experiments. In particular, when proteins of molecular weights about 10,000 or less are studied, special procedures are required. The Schleicher and Schuell B-20 type cellulose acetate membranes recommended for the rapid electronic osmometers when studying small proteins have been found to exhibit large variations in their permeability (Harry and Steiner, 1969); a suitable membrane must be selected by trial and error or double membranes may be tried. The cellophane and collodion membranes used with the classical osmometers are much more amenable to alteration by the investigator for specific purposes; very likely, these have not as yet had fair trial in the electronic osmometers and further efforts to adapt such membranes to these instruments are indicated. The permeability of cellophane membranes can be reduced for the study of very small proteins by dry-heating them for a few days at 90°C (Kupke, 1961) or by acetylation (Craig *et al.*, 1957; Craig and Konigsberg, 1961); under some conditions, the acetylated cellophanes were found to hold back bacitracin (M = 1428). Vink (1966) compared different derivatives of cellophane for the purpose of holding back very small solutes. The retentiveness of the modified cellophanes for various saccharides was found to be in the order; acetylated > nitrated > sulfonated > untreated. Both acetylated and nitrated membranes virtually retained raffinose (M = 504) and were mechanically superior, both in rigidity and strength, to the sulfonated and untreated cellophanes. Similarly, collodion membranes can be prepared in order to measure very small proteins osmotically. Partridge *et al.* (1955) studied a component derived from elastin (M = 5500) by using glycerol-treated collodion (Adair, 1956). Carr *et al.* (1957) have described a permeability control procedure with ethanol-treated collodion whereby solutes smaller than $M = 10^3$ may be studied.

At the other extreme of the molecular weight spectrum, different membrane problems must be considered. Equilibrium pressure differences in systems containing very large proteins (M > 500,000) are necessarily small, the entire span of pressures covering perhaps only a few millimeters of water. Hence, conditions which contribute to small uncertainties in assessing the true pressure difference must be stringently minimized. The use of wide tubes instead of capillaries in the osmometers to reduce capillary effects has already been pointed out. For the membranes, the initial requirement is to select or prepare the most porous kind compatible with just holding back the macromolecules of interest. Smaller macromolecular impurities, which tend to contribute a disproportionate increment to the pressure difference, thereby become permeable. Generally, however, the very large macromolecules should be highly asymmetric or have a flexible-chain character because their large excluded volumes allow more porous membranes to be used; a globular protein of M = 500,000 is only about two times larger in diameter than one of M = 50,000 and it is very difficult to control membranes porosities to this level of discrimination. Various procedures are available for making collodion membranes more porous (cf. Wagner, 1949). Craig and Konigsberg (1961) have described procedures using $ZnCl_2$ and mechanical stretching for increasing the porosity of cellophane. The use of porous membranes also permits a much faster attainment of equilibrium A second objective for the study of very large proteins is to minimize anomalous membrane pressures. This is sometimes denoted as *asymmetry pressure*, i.e., small positive or negative pressure differences ($< \pm 2$ mm $H_2O$) may be noted with certain membranes when solvent is placed on both sides of the membrane. Membranes, therefore, should be tested before use for this kind of work. The reasons for anomalous pressures are not altogether clear; it is insufficient, however, simply to regard all such anomalies as a failure to achieve thermodynamic equilibrium. Small pressure differences can result from relatively impermeable impurities and polymeric material released from the membrane during the experiment if the inner phase is small compared to the outer one. Also, polymer chains of the membrane which have become partially disengaged may affect the activity of the solvent differentially on the two sides. An asymmetry pressure as a function of applied pressure has also been proposed; however, the application of a correction varying with pressure based on solvent calibrations may not be valid (Phillip, 1951). Long soaking of the membranes has been found to reduce these anomalies; also, boiling in water a few minutes is known to leach out material detectable by ultraviolet absorption and dry weight analysis (bag membranes should be turned inside out or such materials may be trapped within). Anomalous pressure differences apparently of mechanical origin have been found to contribute significant error when very small osmotic pressures are being measured. In this case it is observed that the movement of the meniscus during the approach to zero pressure difference from either direction with solvent on both sides of the membrane stops short of true zero. The phenomenon has been described as a membrane inertia (*tragheit;* cf. Weber and Portzehl, 1949) which is virtually eliminated by using very porous membranes having high degrees of rigidity. The exacting experiments on myosin by Portzehl (1950) illustrates the precision which can be expected with highly asymmetric macromolecules, despite the very low osmotic pressures, when membrane and capillary anomalies are made negligible.

## C. Treatment of Data

In order to determine the molecular weight at infinite dilution from osmotic pressures at finite protein concentrations, a variety of procedures has been employed. These procedures have been reviewed (Kupke, 1960) and may be quickly surveyed because the data from nonassociating systems are simple to describe. For a more detailed statistical approach, see Bonnar et al. (1958), who have devoted a substantial section to the theory and calculation of the molecular weight from osmotic pressure data; the discussion is directed primarily to the more complex behavior exhibited by flexible-chain polymers.

As noted in Section II, the data obtained with nonassociating proteins, as single components or as mixtures, can usually be described by a straight line when $\pi/w_2$ is plotted against $w_2$, Eq. (18a). Even the data on very asymmetric proteins, at the low concentrations used, obey the quadratic form. Very little in the way of sufficiently extensive data on protein systems has been elucidated in order to apply statistical analyses meaningfully. Notice may be taken of the extensive data on albumin by Rowe and Abrams (1957) and the studies at very high concentration of this protein by Scatchard et al. (1946b) wherein the cubic term (i.e., the third viral coefficient) appeared to be significant. In the case of associating systems, however, the variation of $\pi/w_2$ is not usually so simple. It was noted in Fig. 2 (curve c) that in such experiments the data describe a curve where the deviation from a van't Hoff proportionality is negative. This is pointed up also in Figs. 3 and 4 on the studies with hemoglobin. Particular attention is directed to the observation by Guidotti (1967b) that an upward curvature followed the typical association curve when the concentration of hemoglobin was sufficient to support principally the tetrameric form of the protein (Fig.

10, lower curve); in this segment of the curve, the positive, nonideal effects become predominant as the equilibrium position shifts away from the dimer form. In general, this type of behavior may be expected upon reaching concentrations where the association reaction is essentially complete. The analysis of such curves on associating systems in which nonideal behavior varies with the proportions of the species present as a function of $w_2$ is now receiving attention; further comment, however, is not warranted at this writing. Obviously, the analysis should be simpler with osmotic data than with data obtained from transport methods and centrifugal fields encompassing a pressure gradient.

Of the four principal sources of error which have traditionally attended the osmotic pressure method, three have already been discussed, i.e., capillarity, permeation of the protein, and membrane asymmetry. Obviously, the evaluation of $w_2$, or the protein concentration, is critical if accurate molecular weights are to be obtained. A short discussion on the evaluation of the protein concentration, where special accuracy is required, is given in Chapter 17. It should be emphasized at this time that methods used for estimating protein content should reflect the concentration of the protein component as defined. If the isoionic, unsolvated component is to be varied, the assay methods should be insensitive to the presence of other diffusible components, because their composition, as

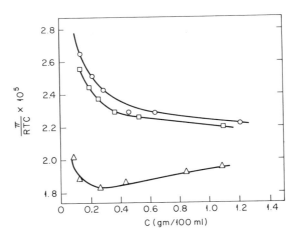

Fig. 10. Osmotic pressure data for hemoglobins in $2\,M$ NaCl, pH 7. ○, Oxyhemoglobin; (□), CO hemoglobin; (△), deoxyhemoglobin. The lower curve exhibits pronounced positive deviations from the van't Hoff proportionality for the tetramer of deoxyhemoglobin at the higher protein concentrations. [Taken from Guidotti (1967a) with permission of the author and the Journal of Biological Chemistry.]

we have noted, may vary with $w_2$ under the conditions of osmotic equilibrium. Hence, if refractive index measurements are made on the equilibrated protein solution and outer solvent, the protein component may then be defined to include the excess (or deficient) amount of a diffusible solute (component 3) as Casassa and Eisenberg (1961) have shown. In general, the true concentration of the polypeptide material will be different from that indicated by refractive index measurements on the two phases at osmotic equilibrium. With the classical osmometers, where the time of measurement is relatively long, the concentration should be determined after the experiment whenever possible. With the use of the rapid electronic osmometers, the concentration determined on a stock solution before the experiment, may be sufficiently accurate because dilution and permeation of protein are minimal; for the dilutions of a stock solution by this procedure, it is most accurate to prepare the mixtures by weight on the analytical balance. If the concentrations are to be expressed in terms of grams of protein per unit of volume, a correction to the weight dilutions, using the densities of the protein solution and solvent, can be made as noted in Chapter 17 [Eq. (46)].

## References

Adair, G. S. (1937). *Trans. Faraday Soc.* **33**, 1106.
Adair, G. S. (1949). *In* "Haemoglobin" (F. J. W. Roughton and J. C. Kendrew, eds.), p. 191. Wiley (Interscience), New York.
Adair, G. S. (1956). *Biochem. J.* **62**, 26P.
Adair, G. S. (1961). *In* "A Laboratory Manual of Analytical Methods of Protein Chemistry" (P. Alexander and R. J. Block, eds.), Vol. 3, p. 23. Pergamon, Oxford.
Adams, E. T., Jr. (1965a). *Biochemistry* **4**, 1646.
Adams, E. T., Jr. (1965b). *Biochemistry* **4**, 1655.
Adams, E. T., Jr., Pekar, A. H., Soucek, D. A., Tang, L. H., Barlow, G., and Armstrong, J. L. (1969). *Biopolymers* **7**, 5.
Alexander, A. E., and Johnson, P. (1949). "Colloid Science," Vol. I. Oxford Univ. Press (Clarendon), London and New York.
Andrews, A. T. deB., and Reithel, F. J. (1970). *Arch. Biochem. Biophys.* **141**, 538.
Banarjee, K., and Lauffer, M. A. (1966). *Biochemistry* **5**, 1957.
Bonnar, R. U., Dimbat, M., and Stross, F. H. (1958). "Number Average Molecular Weights," Wiley (Interscience), New York.
Brant, D. A., and Flory, P. J. (1965). *J. Amer. Chem. Soc.* **87**, 2791.
Carr, C. W., Anderson, D., and Miller, I. (1957). *Science* **125**, 1245.
Carter, S. R., and Record, B. R. (1939). *J. Chem. Soc., London* p. 660.
Casassa, E. F., and Eisenberg, H. (1960). *J. Phys. Chem.* **64**, 753.
Casassa, E. F., and Eisenberg, H. (1961). *J. Phys. Chem.* **65**, 427.
Castellino, F. J., and Barker, R. (1968). *Biochemistry* **7**, 2207.
Christiansen, J. A., and Jensen, C. E. (1953). *Acta Chem. Scand.* **7**, 1247.
Claesson, S., and Jacobsson, G. (1954). *Acta Chem. Scand.* **8**, 1835.
Craig, L. C., and Konigsberg, W. (1961). *J. Phys. Chem.* **65**, 166.
Craig, L. C., King, T. P., and Stracher, A. (1957). *J. Amer. Chem. Soc.* **79**, 3729.

Davies, M. (1966). *Makromol. Chem.* **90**, 108.
Donnan, F. G. (1935). *Trans. Faraday Soc.* **31**, 80.
Edmond, E., Farquhar, S., Dunstone, J. R., and Ogston, A. G. (1968). *Biochem. J.* **108**, 775.
Edsall, J. T. (1953). *In* "The Proteins" (H. Neurath and K. Bailey, eds.), 1st ed., Vol. 1, Part B, p. 549. Academic Press, New York.
Enoksson, B. (1951). *J. Polym. Sci.* **6**, 575.
Fuoss, R. M., and Mead, D. J. (1943). *J. Phys. Chem.* **47**, 59.
Guidotti, G. (1967a). *J. Biol. Chem.* **242**, 3685.
Guidotti, G. (1967b). *J. Biol. Chem.* **242**, 3694.
Guidotti, G. (1967c). *J. Biol. Chem.* **242**, 3704.
Güntelberg, A. V., and Linderstrøm-Lang, K. (1949). *C. R. Trav. Lab. Carlsberg, Ser. Chim.* **27**, 1.
Gutfreund, H. (1948). *Biochem. J.* **42**, 156.
Hansen, A. T. (1950). *Amer. J. Physiol.* **163**, 720.
Hansen, A. T. (1952). *Acta Med. Scand., Suppl.* **266**, 473.
Harry, J. B., and Steiner, R. F. (1969). *Biochemistry* **8**, 5060.
Hill, T. L. (1959). *J. Chem. Phys.* **30**, 1161.
Holter, H. (1943). *C. R. Trav. Lab. Carlsberg, Ser. Chim.* **24**, 399.
Jacobsson, G. (1954). *Acta Chem. Scand.* **8**, 1843.
Jeffrey, P. D. (1968). *Biochemistry* **7**, 3345 and 3352.
Jeffrey, P. D. (1969). *Biochemistry* **8**, 5217.
Jullander, I., and Svedberg, T. (1944). *Nature (London)* **153**, 523.
Klotz, I. M. (1967). "Energy Changes in Biochemical Reactions." Academic Press, New York.
Kupke, D. W. (1960). *Advan. Protein Chem.* **15**, 57.
Kupke, D. W. (1961). *C. R. Trav. Lab. Carlsberg, Ser. Chim.* **32**, 107.
Kupke, D. W. (1964). *In* "Serum Proteins and the Dysproteinemias" (F. W. Sunderman and F. W. Sunderman, Jr., eds.), Part 1, Chapter 4, p. 19. Lippincott, Philadelphia, Pennsylvania.
Lapanje, S., and Tanford, C. (1967). *J. Amer. Chem. Soc.* **89**, 5030.
Lauffer, M. A. (1964). *Biochemistry* **3**, 731.
Lauffer, M. A. (1966). *Biochemistry* **5**, 1952.
Laurent, T. C. (1964). *Biochem. J.* **93**, 106.
Laurent, T. C., and Ogston, A. G. (1963). *Biochem. J.* **89**, 249.
Linderstrøm-Lang, K. (1962). *In* "Selected Papers of Kaj Linderstrøm-Lang," (H. Holter, H. Neurath, and M. Ottesen, eds.), p. 573. Academic Press, New York.
Marrack, J., and Hewitt, L. F. (1929). *Biochem. J.* **23**, 1079.
Miller, W. G., Brant, D. A., and Flory, P. J. (1967). *J. Mol. Biol.* **23**, 67.
Ogston, A. G. (1962). *Arch. Biochem. Biophys., Suppl.* **1**, 39.
Ogston, A. G. (1966). *Fed. Proc., Fed. Amer. Soc. Exp. Biol.* **25**, 1112.
Ogston, A. G., and Silpananta, P. (1970). *Biochem. J.* **116**, 171.
Ogston, A. G., and Wells, J. D. (1970). *Biochem. J.* **119**, 67.
Ottesen, M. (1958). *C. R. Trav. Lab. Carlsberg, Ser. Chim.* **30**, 211.
Overbeek, J. T. (1956). *Progr. Biophys. Biophys. Chem.* **6**, 57.
Paglini, S. (1968). *Anal. Biochem.* **23**, 247.
Partridge, S. M., Davis, H. F., and Adair, G. S. (1955). *Biochem. J.* **61**, 11.
Phelps, R. A., and Cann, J. R. (1956). *Arch. Biochem. Biophys.* **61**, 51.
Phillip, H. J. (1951). *J. Polym. Sci.* **6**, 571.
Portzehl, H. (1950). *Z. Naturforsch. B* **5**, 75.

Preston, B. N., Davies, M., and Ogston, A. G. (1965). *Biochem. J.* **96**, 449.
Reiff, T. R., and Yiengst, M. (1959). *J. Lab. Clin. Med.* **53**, 291.
Reithel, F. J. (1963). *Advan. Protein Chem.* **18**, 124.
Rolfson, F. B., and Coll, H. (1964). *Anal. Chem.* **36**, 888.
Rowe, D. S. (1954). *J. Physiol. (London)* **123**, 18P.
Rowe, D. S. (1955). *Nature (London)* **175**, 554.
Rowe, D. S., and Abrams, M. E. (1957). *Biochem. J.* **67**, 431.
Scatchard, G. (1946). *J. Amer. Chem. Soc.* **68**, 2315.
Scatchard, G., and Bregman, J. (1959). *J. Amer. Chem. Soc.* **81**, 6095.
Scatchard, G., and Pigliacampi, J. (1962). *J. Amer. Chem. Soc.* **84**, 127.
Scatchard, G., Batchelder, A. C., and Brown, A. (1944). *J. Clin. Invest.* **23**, 458.
Scatchard, G., Batchelder, A. C., and Brown, A. (1946a). *J. Amer. Chem. Soc.* **68**, 2320.
Scatchard, G., Batchelder, A. C., Brown, A., and Zosa, M. (1946b). *J. Amer. Chem. Soc.* **68**, 2610.
Scatchard, G., Gee, A., and Weeks, J. (1954). *J. Phys. Chem.* **58**, 783.
Scatchard, G., Wu, Y. V., and Shen, A. L. (1959). *J. Amer. Chem. Soc.* **81**, 6104.
Sørenson, S. P. L. (1917). *C. R. Trav. Lab. Carlsberg, Ser. Chim.* **12**, 1.
Spanner, D. C. (1964). "Introduction to Thermodynamics." Academic Press, New York.
Staverman, A. J. (1951). *Rec. Trav. Chim. Pays-Bas* **70**, 344.
Steiner, R. F. (1954). *Arch. Biochem. Biophys.* **49**, 400.
Steiner, R. F. (1968). *Biochemistry* **7**, 2201.
Steiner, R. F. (1970). *Biochemistry* **9**, 4268.
Tanford, C. (1961). "Physical Chemistry of Macromolecules." Wiley, New York.
Vink, H. (1966). *J. Polym. Sci., Part A-2* **4**, 830.
Wagner, R. H. (1949). *In* "Techniques of Organic Chemistry; Physical Methods" (A. Weissberger, ed.), 2nd ed., Vol. I, Part I, p. 487. Wiley (Interscience), New York.
Wagner, R. H., and Moore, L. D. (1959). *In* "Techniques of Organic Chemistry; Physical Methods" (A. Weissberger, ed.), 3rd ed., Vol. I. Part I, p. 815. Wiley (Interscience), New York.
Weber, H. H., and Portzehl, H. (1949). *Makromol. Chem.* **3**, 132.

# 19 ☐ Small-Angle X-ray Scattering

INGRID PILZ

| | |
|---|---|
| Glossary of Symbols | 141 |
| I. Introduction | 142 |
| II. Remarks on the Theory of Particulate Scattering | 143 |
|    A. Fundamentals | 143 |
|    B. Approximation of Guinier and Radius of Gyration | 146 |
|    C. Relation between the Form of the Scattering Curve and the Shape of the Particle | 148 |
|    D. Radial Density Distribution of Particles with Spherical Symmetry | 158 |
|    E. Rodlike and Lamellar Particles | 159 |
|    F. Volume and Tail End of the Scattering Curve | 163 |
|    G. Molecular Weight and Mass per Unit Length | 165 |
|    H. Internal Solvation | 167 |
| III. Remarks on the Experimental Technique | 168 |
|    A. Special Requirements in the Investigation of Protein Solutions | 169 |
|    B. Remarks on Cameras, Monochromatization, and Recording | 170 |
|    C. Absolute Intensity | 177 |
|    D. Collimation (Smearing) Effects | 179 |
|    E. Preparation of Suitable Protein Solutions | 182 |
| IV. Selected Examples of Applications | 188 |
|    A. Comparison of the Results of Small-Angle Scattering and X-Ray Studies on Crystals | 188 |
|    B. Determination of Models Equivalent in Scattering | 197 |
|    C. Changes in Conformation on Denaturation | 209 |
|    D. Determination of the Cross-Section of Elongated Particles | 211 |
|    E. Radial Electron Density Distribution by Fourier Transformation | 214 |
|    F. Characteristics of Hollow Bodies | 226 |
|    G. Allosteric Effects and Similar Changes in Conformation | 234 |
| V. Prospects | 238 |
| References | 239 |

## Glossary of Symbols

| | |
|---|---|
| $a$ | distance between sample and recording plane |
| $A_q$ | area of cross-section of elongated particles |
| $c$ | concentration of the solute (protein) in milligrams per milliliter |

| | |
|---|---|
| $d$ | interplanar spacing $d$ in Bragg's law |
| $D$ | thickness of the sample |
| $f_a$ | anisotropy or shape factor |
| $f_s$ | swelling factor |
| $F$ | area of the counter tube slit |
| $F(h)$ | structure factor of the particle |
| $h$ | $2\pi s = 4\pi(\sin\theta)/\lambda$ |
| $i_0$ | Thompson scattering constant of a free electron |
| $I$ | scattered intensity at angle $2\theta$ of the desmeared (slit-corrected) curve |
| $I_0$ | intensity at zero angle |
| $I_q$ | intensity of the cross-section factor |
| $\tilde{I}$ | scattered intensity at angle $2\theta$ of the slit-smeared curve |
| $L$ | length of the particle |
| $M$ | molecular weight |
| $M/1\text{Å}$ | mass per unit length of an elongated particle |
| $N_L$ | Avogadro's number |
| $P_0$ | intensity of the primary (incident) beam |
| $Q$ | invariant of the desmeared (slit-corrected) scattering curve |
| $\tilde{Q}$ | invariant of the slit-smeared scattering curve |
| $r$ | distance to the center of a particle |
| $R$ | radius of gyration |
| $\tilde{R}$ | apparent radius of gyration obtained from slit-smeared scattering curves |
| $R_q$ | radius of gyration of the cross-section |
| $R_t$ | radius of gyration of the thickness |
| $s$ | $2(\sin\theta)/\lambda$ |
| $S$ | total area of interface separating the dispersed phase (solute) and the dispersant phase (solvent) |
| $S/V_1$ | specific inner surface |
| $V_1$ | volume occupied by the dispersed phase |
| $t$ | thickness of a lamellar particle |
| $\bar{v}$ | partial specific volume |
| $\bar{v}_1$ | partial specific volume of the solute (protein) |
| $V$ | volume of one (average) particle |
| $\Delta z$ | excess moles of electrons |
| $\theta$ | half scattering angle equals Bragg angle |
| $2\theta$ | scattering angle |
| $\lambda$ | wavelength of X-rays |
| $\rho$ | measured electron density in $e$ per cubic angstrom |
| $\rho_1$ | electron density of the solute |
| $\rho_2$ | electron density of the solvent |
| $\rho(r)$ | electron density at a distance $r$ from the center of the particle |

## I. Introduction

When the numerous studies which have been carried out on homogeneous proteins are reviewed, it is seen that the molecular weights of the proteins vary within very wide limits—from a few thousand to several million. Since the molecular weight of a single polypeptide chain is seldom greater than 50,000, most proteins are built up from several polypeptide chains, which are spatially connected in a definite way and form the so-called quaternary structure of proteins. We know today that this spatial structure of proteins is of decisive importance for their biological effectiveness. In the case of enzymes, for instance, splitting into individual peptide chains or subunits usually leads to a loss of enzyme activity. The absorption of oxygen by hemoglobin is also attended by a change of conformation, and allosteric effects of enzymes play an essential part in the kinetics of their reactions.

In order to understand the mechanism of action of proteins it is thus of fundamental importance to obtain as much information as possible on the size and shape of protein molecules.

It has proved possible in the past decade to improve the experimental technique of small-angle X-ray scattering to such an extent that this method can now make an important contribution to the investigation of proteins. Small-angle X-ray scattering offers above all the advantage that proteins can be investigated in solution (containing any buffer desired), that is to say, as they were in their native state. The method provides information on the size, shape, and degree of swelling of proteins and is particularly sensitive to changes in conformation. Changes in the shape of proteins caused by adding effectors or varying temperature, pH, or concentration can be easily followed with small-angle X-ray scattering.

Most suitable for small-angle X-ray investigations in solution are macromolecules, the molecular weights of which are between 10,000 and 10,000,000 and the largest dimensions of which lie between 50 Å and 1000 Å. As most proteins fall within this range the technique is clearly applicable to protein investigations.

## II. Remarks on the Theory of Particulate Scattering[1]

A. FUNDAMENTALS

### 1. General

Small-angle X-ray scattering is analogous to the well-known scattering of visible light. The only difference is that in small-angle scattering a wavelength one thousand times shorter is used (as a rule the copper $K\alpha$ line with $\lambda = 1.54$ Å). In comparison with this wavelength, protein molecules are very large. Thus, in accordance with the well-known law of reciprocity in optics (the size of the diffracting angle and that of the diffracting object are always inversely related) the diffraction effect,

---

[1] All equations given in this section are for slit-corrected scattering curves (compare Section III,D).

which in light scattering fills the entire angle from 0° to 180°, is reduced to a few degrees only in small-angle scattering.

This effect can be explained very clearly and I should like here to draw the attention of the reader to the review articles of Kratky (1960a, 1963).

When electrons lie in the path of a beam of X-rays, they are stimulated to undergo forced vibrations. Since an accelerating charge acts as a source of radiation, these electrons produce scattered secondary waves, which have the same wavelength as the incident beam (coherent scattering) and will be superimposed according to their phase. In Fig. 1 the scattering is considered in two electrons A and B of a macromolecule chosen at random (Kratky, 1963). We note that at larger scattering angles, $2\theta$, phase differences of the magnitude of one wavelength appear between the diffracted waves; that is to say, the waves scattered by all the electrons of the macromolecule will be superimposed in all phases, so

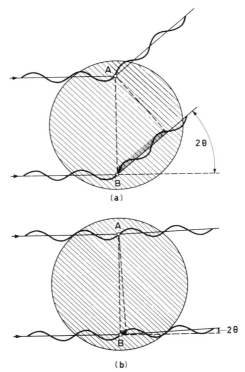

Fig. 1. Coherent scattering of a particle which is large compared with the wavelength. The phase difference of the two waves scattered at A and B is for (a) about one $\lambda$ and for (b) only a fraction of $\lambda$.

that there is practically complete cancellation (Fig. 1a). Only those waves which are scattered at very small scattering angles—of a few minutes to a few degrees—and which possess correspondingly smaller phase differences are not extinguished (Fig. 1b).

The plot of the scattered intensity as a function of the scattering angle will accordingly give a curve, the so-called scattering curve, whose maximum is at the zero angle, where all the waves are in phase, and which rapidly decreases as the scattering angle increases (Fig. 2).

Figure 1 makes clear two most essential facts about small-angle X-ray scattering.

1. The scattering curve is condensed to smaller and smaller angles, as the particles become larger, because the phase differences for a definite scattering angle $2\theta$ increase with the size of the particles when the wavelength is unchanged. This is to say, the shape of the scattering curve depicts the particle shape in reverse. Dust and other contaminants, which cause considerable disturbance in light scattering, are thus negligible in small-angle scattering, since they scatter to angles which are too small to be measurable.

2. The form of the scattering curve depends upon the shape of the particle (see Fig. 2).

In solutions of monodisperse macromolecules which are sufficiently diluted, the distances between the individual particles (for instance, protein molecules) are so great and irregular that no phase relations exist any longer between the waves scattered by the individual particles. The scattering intensities of the individual particles simply add up and the shape of the scattering curve corresponds to the scattering curve of the particles, which represents all orientations with equal probability.

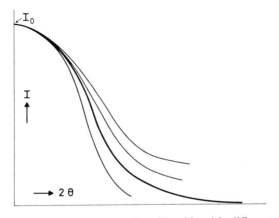

FIG. 2. Theoretical scattering curves for ellipsoids with different axial ratios.

Such dilute systems of particles allow a definitive theoretical interpretation. The expression and theory of particulate scattering was first introduced by Guinier (1937, 1939, 1943) and the theoretical principles have been greatly extended in recent decades. In the present article only this particulate scattering is discussed and all the investigations mentioned have been made on dilute solutions, so that the theory of particulate scattering can be applied to them.

Summaries of the method and application of small-angle X-ray scattering have been given by Guinier and Fourmet (1955), Kratky (1955a, 1963), and Beeman *et al.* (1957).

## 2. *Angular Nomenclature*

In the literature, besides the scattering angle $2\theta$, the following angular functions are used, and they are connected with the Bragg spacing $d$ in the following way

$$s = \frac{1}{d} = \frac{2 \sin \theta}{\lambda} \tag{1a}$$

$$h = 2\pi s = \frac{2\pi}{d} = \frac{4\pi \sin \theta}{\lambda} \tag{2a}$$

Because of the smallness of the angles, these equations may be written as

$$s = \frac{1}{d} = \frac{2\theta}{\lambda} \tag{1b}$$

$$h = 2\pi s = \frac{2\pi}{d} = \frac{4\pi\theta}{\lambda} \tag{2b}$$

### B. Approximation of Guinier and Radius of Gyration

Guinier (1937, 1939, 1943) was the first to show that the scattering curves of corpuscular particles in a dilute, monodisperse system can be approximated by a Gaussian curve (Fig. 3)

$$I = I_0 \exp(-KR^2\theta^2) \qquad K = \left(\frac{4\pi}{\lambda}\right)^2 \cdot \frac{1}{3} \tag{3a}$$

Here $I_0$ means the scattering intensity at zero angle, $I$ is the scattering intensity at angle $\theta$, $R$ is the radius of gyration, $\theta$ is the half scattering angle, and $K$ is a universal constant. In Fig. 3, $tg\alpha$ corresponds to $-KR^2$. From Eq. (3a) it is immediately seen that the width of the scattering curve is in inverse proportion to the radius of gyration. In order to determine $R$ the above equation is used in logarithmic form

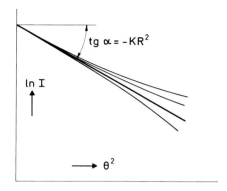

Fig. 3. Guinier plot (ln $I$ vs. $\theta^2$) of the inner portion of the scattering curves.

$$\log_e I = \log_e I_0 - KR^2\theta^2 \tag{3b}$$

In this plot (log $I$ versus $\theta^2$)—called the *Guinier plot*—the scattering curves are more or less linear over a range of angles and tend, at zero angle, toward a straight line with a slope (Fig. 3) of

$$\text{Slope} = -KR^2 \tag{3c}$$

which is directly related to the radius of gyration $R$.

The radius of gyration is the root-mean-square of the distances of the electrons from the electronic center of gravity of the representative particle. When particles with equal volumes are considered (equal numbers of electrons), the particle with a spherical shape will have the smallest possible radius of gyration. $R$ will increase with increasing anisotropy of the particle and also with an increasing hollow space inside the particle, because the distances of the electrons from the center of gravity become larger.

If the volume of the particle is ascertained by Eq. (11) (Section II,F) an anisotropy factor $f_a$ can be defined as the ratio of the experimental $R$ value to that of a sphere with equal volume (Kratky, 1963)

$$f_a = \frac{R_{\text{exp}}}{R_{\text{sphere}}}$$

The larger the value of $f_a$ the more anisotropic is the particle or the larger is the hollow space inside it.

The radius of gyration and the molecular weight are precise parameters, which can be obtained by small-angle scattering without invoking supplementary hypotheses. Table I gives the calculated radii of gyration of some bodies of regular shape.

## TABLE I
Radii of Gyration $R$ for Bodies of Certain Particle Shapes and Radii of Gyration $R_q$ for Certain Particle Cross-Sections

| Body | $R^2$ |
|---|---|
| Sphere of radius $r$ | $\dfrac{3}{5}r^2$ |
| Hollow sphere, of inner radius $r_i$ and outer radius $r_a$ | $\dfrac{3}{5} \cdot \dfrac{r_a^5 - r_i^5}{r_a^3 - r_i^3}$ |
| Ellipsoid of semiaxes $a$, $b$, and $c$ | $(a^2 + b^2 + c^2)/5$ |
| Elliptical cylinder of height $h$ and semiaxes $a$ and $b$ | $(a^2 + b^2 + h^2/3)/4$ |
| Hollow circular cylinder of height $h$ and radii $r_a$ and $r_i$ | $(r_a^2 + r_i^2)/2 + h^2/12$ |

| Section | $R_q^2$ |
|---|---|
| Circle of radius $r$ | $r^2/2$ |
| Circular ring of radii $r_a$ and $r_i$ | $(r_a^2 + r_i^2)/2$ |
| Ellipse of semiaxes $a$ and $b$ | $(a^2 + b^2)/4$ |

### C. Relation between the Form of the Scattering Curve and the Shape of the Particle

The scattering curve of a protein is determined not only by the radius of gyration but also by the shape of the protein. The relation between the shape of a particle and the form of its scattering curve is not immediately apparent, since the variety of colloid structures is very great.

It is important to realize that it is never possible to give an unequivocal structural interpretation on the basis of small-angle X-ray scattering data alone. However, it is possible to get information about the overall shape of the particle and to find a model which is equivalent in scattering (see Section IV,B), for instance, by comparing the experimental scattering curve with the theoretically calculated scattering curves for various models.

Protein molecules in solution are, of course, in any position to the incident X-ray beam. For this reason small-angle scattering of molecules in solution gives poorer resolution of the conformation of a molecule than does diffraction by crystals, where the molecule has a fixed position. In small-angle scattering similar structures can give nearly identical scattering curves.

To obtain a picture of how the scattering curves of proteins are determined by their shapes it is best to observe theoretical scattering curves of simple bodies with random orientations. By altering the axial ratios of simple triaxial bodies the influence of the shape of the scattering molecules on the form of the scattering curve can be studied. This will be illustrated in the following sections with some practical examples.

## 1. *Scattering Curves of Simple Triaxial Bodies*

*a. Calculation of Scattering Curves.* Kratky (1947) was one of the first to investigate the influence of the shape of a particle on its scattering curve; he did so by approximating particles of various shapes by assemblies of spheres. Guinier (1939), Shull and Roess (1947), Malmon (1957a), and Schmidt and Hight (1959) calculated scattering curves for ellipsoids of revolution. The scattering curves of hollow spheres go back to Leonard (1952), Oster and Riley (1951), Schmidt (1955), and more particularly Porod (1948) and Mittelbach, who calculated, with the aid of computers, scattering curves for right-angled prisms (Mittelbach and Porod, 1961a), elliptical cylinders, hollow cylinders (Mittelbach and Porod, 1961b), and ellipsoids (Mittelbach and Porod, 1962).

In calculating theoretical scattering curves, the following assumptions are usually made:

1. The solution is sufficiently dilute that interparticle interferences can be neglected.

2. Only coherent scattering is taken into account. (The proportion of incoherent scattering is of no importance at such small angles.)

3. The electron density within the dissolved macromolecules and within the solvent is assumed constant.

In order to study the influence which the shape of a particle has upon the shape of its scattering curve it is best to compare particles of identical radii of gyration $R$. That is to say, one takes as abscissa $hR$, whereby $h = (4\pi \sin \theta)/\lambda$ [see Eq. (2a)].

In this plot, the dependence of the scattering curve on the size of the particle is eliminated and the dependence of the scattering curve on the shape of the particle can be studied.

In Fig. 4a theoretical scattering curves of ellipsoids (Mittelbach, 1964) of axial ratio $1:1:c$ are presented in a log–log plot and in Fig. 4b scattering curves of right-angled prisms of axial ratio $1:1:c$; the longest axis $c$ is varied from 1 to 10. It is immediately apparent that the scattering curves of the most isometric bodies (spheres or cubes) are the steepest. An increase in anisotropy leads to a flattening of the curve. It is also observed that certain small differences exist between the curves for various triaxial bodies with identical axial ratios. Thus, the scattering curve of an ellipsoid, for instance, is always more steep than that of a right-angled prism.

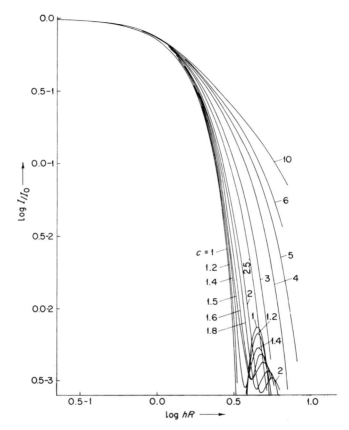

Fig. 4a. Theoretical scattering curves for ellipsoids with an axial ratio of $1:1:c$ in a log–log plot. The intensity is plotted as $I/I_0$, where $I_0$ is the intensity at zero angle; the curves are normalized to $hR$, where $h = (4\pi \sin \theta)/\lambda$ and $2\theta$ is the scattering angle (Mittelbach, 1964).

These differences, however, are not very marked. If, therefore, the analysis of a scattering curve is restricted to the first intensity decrease through two orders of magnitude, we can obtain information only on the axial ratios of models equivalent in scattering.

*b. Subsidiary Maxima of the Scattering Curves.* To get more detailed information on the shape of a molecule, the subsidiary maxima in the outside portion of the scattering curve (which is weak in intensity) must be included in the interpretation. It is, therefore, necessary to measure not only the relatively intense inner portion of the scattering curve (the main maximum) with sufficient precision but also the outer portion.

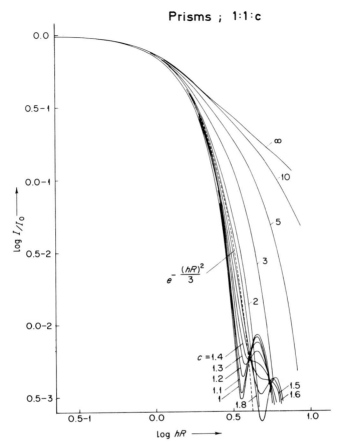

FIG. 4b. Theoretical scattering curves for right-angled prisms with the axial ratio 1:1:c in a log-log plot (Mittelbach, 1964).

This contains the subsidiary maxima which are extremely dependent upon the shape of the body. The intensity of the subsidiary maxima of triaxial bodies is less than 1/100 of the intensity of the main maximum at zero scattering angle. Extremely isometric bodies, such as spheres or cubes, have the most intense subsidiary maxima. Scattering curves of spheres are easily recognized, since they are the only type of body in which the intensity between the maxima is reduced to zero. In Fig. 5, the outer part of scattering curves of elongated ellipsoids of revolution with the axial ratio 1:1:c is presented. The intensity of the main maximum is normalized to 1 and it is clearly seen that with the sphere (1:1:1) the first subsidiary maximum is the most intense (8/1000 of

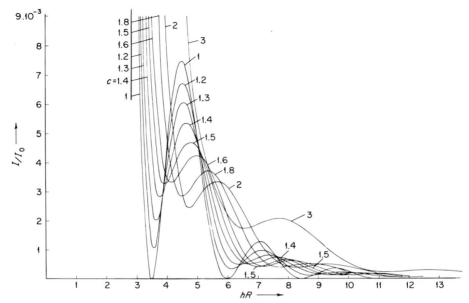

Fig. 5. Outer portions of theoretical scattering curves for ellipsoids with an axial ratio of $1:1:c$ (Mittelbach, 1964).

the main maximum) and the intensity between the maxima is reduced to zero. With increasing anisometry the height of the first subsidiary maximum decreases rapidly.

With quadratic prisms (axial ratio $1:1:c$) the subsidiary maximum quickly disappears with increasing anisometry and with $c = 1.4$ only a slight bulge remains in the position of the subsidiary maximum (Fig. 6).

Besides the height of the first subsidiary maximum, its position is also characteristic. With all isometric bodies (spheres or cubes) the position is given by the following equation

$$d = \lambda/2\theta = 1.40R \qquad (4)$$

This equation allows one to calculate the value of the radius of gyration from the position of the first maximum. The $R$ value can then be used to check the value of the radius of gyration determined from the slope of the inner portion of the curve [Eq. (3c)]. With increasing anisotropy the subsidiary maximum shifts to larger angles as is clearly seen in Fig. 5; it is located then at smaller Bragg spacings $d$ than $1.40R$ [see Eq. (1b)].

These few remarks should serve to show the importance of also in-

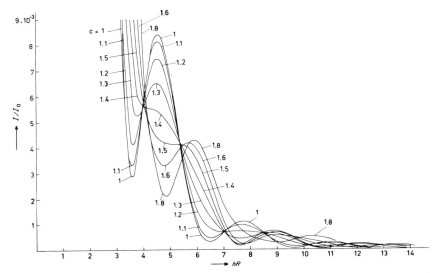

FIG. 6. Outer portions of theoretical scattering curves for right-angled prisms with an axial ratio of $1:1:c$ (Mittelbach, 1964).

vestigating the low-intensity outer part of the scattering curves and, above all, the position and height of the subsidiary maxima which may appear. A periodicity (minima and maxima) in the outer part of the scattering curve can also be caused by the substructure of a protein (see 2c).

*c. Scattering Curves of Hollow Bodies.* Theoretical scattering curves of hollow spheres and hollow cylinders differ only a little from the scattering curves of the corresponding full bodies with respect to the decrease in their main maxima. On the other hand, hollow bodies differ very markedly from full bodies in their subsidiary maxima. The height of the first subsidiary maximum is a direct measure of the size of the cavity relative to the size of the particle, and the height of this first subsidiary maximum increases rapidly as the ratio of the inner radius to the outer radius of the hollow body increases (Fig. 7). With a full sphere, for example, the height is 8/1000 of the intensity of the main maximum and increases as the hollow space increases up to about 5/100, which is the value for a hollow sphere consisting of only a thin skin.

## 2. *Shape Determination by Comparing Experimental with Theoretical Scattering Curves*

The simplest and most frequently applied procedure for obtaining the shape of proteins is the following. The experimental scattering curve

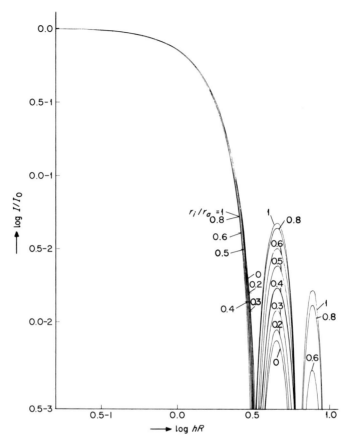

FIG. 7. Theoretical scattering curves for hollow spheres of various hollowness. The ratio of inner radius to outer radius ($r_i/r_a$) is varied from 0 (solid sphere) to 1 (sphere, consisting only of a thin skin).

is compared with the theoretical scattering curves for various models. As already mentioned, it is not, of course, possible to carry out an unequivocal determination of the shape of a macromolecule from small-angle X-ray scattering data alone. By comparing the data with theoretical curves for various models in a trial-and-error manner, many models which are not equivalent in scattering properties can be excluded and others which are equivalent in scattering properties can be selected for further examination. Only those models, however, which show clear differences in shape can be differentiated. As a rule it is not possible to differentiate between variations in folding of the protein chains which yield a similar overall shape.

In practice there are a number of ways of carrying out the comparison with theoretical scattering curves.

*a. Comparison with Theoretical Curves of Bodies with Uniform Electron Densities.* The assumption of a uniform electron density within the particle is of course a simplification, because no protein has a completely uniform electron density. In spite of this, it is possible in this way to obtain very important information on the overall shape of proteins, as will be explained later by means of practical examples.

In practice, the best approach is to determine roughly, by comparing the form of the main maximum of the experimental scattering curve with the curves of various triaxial models, the overall shape (the axial ratio) of the macromolecule.

More detailed information is obtainable if the subsidiary maxima (when these are present) are included in the comparison between experimental and theoretical curves. With particles of anisotropic shape the cross-section can often be determined too (see Section II,E,1).

When we have selected a model in this way with a scattering curve which corresponds well with the experimental curve, the overall dimensions of the macromolecule can be calculated directly from the axial ratio and the radius of gyration.

It is of course obvious that not all macromolecules and, more especially, all proteins can be sufficiently well described by a simple triaxial body. There is often no scattering curve for a simple body which corresponds satisfactorily to the experimental curve. In such cases we must try to calculate scattering curves for more complex models. In our institute, a program has been worked out which renders it possible, with the aid of a computer, to calculate scattering curves for various models built up from spheres, cylinders, or triaxial ellipsoids, which can be arranged in any manner in the space (Haager, 1970).

As the calculation of such models requires time and a considerable amount of work, it is advantageous to use experimental evidence from other physical techniques, such as electron microscopy, to select appropriate models for testing the scattering data.

*b. Comparison with Theoretical Curves of Models Calculated from the Atomic Coordinates.* In special cases, if a great deal is known about the primary and secondary structure of a protein, it may be of interest to calculate the scattering curves according to the Debye (1925) formula from the atomic coordinates of the suggested model. In this way, one may determine whether the tertiary structures suggested for the protein are equivalent in scattering or not.

With large proteins, the computer time required to calculate scattering curves from the atomic coordinates is very high. As a rule, therefore, one does not use the individual atoms as scattering centers but combines certain atoms into groups. The way in which atomic groups are used as scattering elements is described by means of a practical example in Section IV,A,4.

Since the information about the shape of the molecules obtained by small-angle scattering is usually too general to permit a detailed analysis of the structure, it is not worthwhile to calculate such detailed models, especially with larger proteins.

*c. Determination of the Shape of Proteins Which are Composed of Subunits.* Sometimes it is possible to dissociate a large protein into well-defined subunits, and if we can obtain homogeneous solutions of these, the shape of the individual subunits can be determined. Knowing the size and shape of the subunits, one can then arrange them in different ways to represent the original protein. The scattering curves of such models can then be calculated and compared with the experimental curves obtained for the whole molecule, as shown in detail by an example in Section IV,B,1.

Certain very large proteins seem to be built up of a large number of more or less identical subunits, as for instance the hemocyanin of *Helix pomatia* or the protein shells of various spherical viruses. While the inner section of the scattering curve and the first maxima or minima depend in the first instance on the overall shape of the protein, with large angles long-wave periodicities appear in the form of minima. These are independent of the overall shape of the protein and give information on the size of the identical subunits (Glatter, 1972; Pilz *et al.*, 1972). If measurements are sufficiently accurate, these minima may be used to estimate the size of the subunits and to construct models of the protein. The scattering curves of the models can then be calculated and checked against the experimental curves (see Section IV,F,1).

### 3. *Other Ways of Determining Shape*

In the procedures discussed so far, the shape of the protein was obtained by comparing experimental scattering curves with those calculated for models. There are alternative procedures for obtaining information on the shape of the proteins in solution.

*a. Analysis of Experimental Scattering Functions.* A new and interesting way of determining the shape of particles from the scattering function is described by Stuhrmann (1970b). Information on the shape of proteins can be obtained, as already mentioned, by comparing the experimental scattering curves with the theoretical curves for models with a uniform electron density. This procedure is a good approximation, but we cannot expect the scattering curves of such models to agree completely with those of proteins since no protein has a completely uniform electron density. While the deviation from uniform electron density is negligible at small angles, it causes some variation in the scattering function at large angles. Stuhrmann showed that it is possible to separate the scattering due to the shape of the molecule from the scattering due to deviations from uniform electron density. By measuring proteins in solvents of different electron densities, the whole scattering function can be split into different terms; the shape scattering can be separated from the scattering due to deviations in local electron density from the mean value and a third term containing the influence of both factors. By correlating the coefficients of the power series of the shape scattering function with the coefficients of the expansion of the shape component in the form of a series of spherical harmonics, the shape of the protein can then be determined.

With this method, relatively exact determinations of the shape of a protein are possible as long as we are not dealing with very complex shapes. The shape can be determined without assumptions concerning the distribution of electron density. However, with this procedure the scattering of the protein must be measured in at least three solvents of different electron density. This introduces experimental difficulties because proteins can as a rule be studied only in aqueous solution. The electron density difference between protein and water is not in itself very large. To produce solvents of differing electron density, glycerine or sucrose is normally added (see Section II,F) but this decreases the electron density difference still further. Such measurements therefore require an extremely high degree of accuracy.

The use of added solvents introduces a further problem in that the conformation of the protein may be altered. This possibility must be checked by the use of auxiliary physical measurements, as discussed by Donovan in Part A of this series (p. 129 ff.).

b. *Determination of the Greatest Diameter and of Other Characteristic Parameters.* Another way to obtain information without making assumptions about the distribution of electron density within the macromolecule and without the use of models is to determine "characteristic parameters."

It is possible, for instance, as Damaschun *et al.* (1968a) have shown in detail, to determine the greatest diameter of a macromolecule. This is done by calculating the correlation function (Debye and Bueche, 1949) from small-angle X-ray scattering curves. The determination can be applied to all corpuscular particles the maximum diameters of which (that is, the greatest distance between the atoms) are between 20 Å and 5000 Å. Determinations of the maximum length have been carried out, for instance, on catalase by Malmon (1957b) and on ribonuclease by Filmer and Kaesberg (1962). It is not usually possible to determine the value of the maximum length with any greater accuracy than would be obtained by matching to the largest dimension of a model equivalent in scattering; however, when the theoretical scattering curves of various bodies are not available, the above procedure is less laborious.

Besides the maximum length there are other characteristic parameters which can be obtained analytically from the experimental scattering curve and which are independent of a model.

Such characteristic parameters are, for instance, the radius of gyration, the volume (cf. Section II,F), the mean transverse length (Porod, 1951a,b, 1960), the characteristic length (Klimanek, 1964; Mittelbach and Porod, 1965), the characteristic area (Porod, 1951a,b; Damaschun and Pürschel, 1968), the surface (von Nordstrand and Hach, 1953; Porod, 1951a,b), and the mean distance between the electrons. Damaschun and Pürschel (1969) have calculated the mean electron distance for various models. A knowledge of these parameters considerably restricts the number of possible molecular shapes as all models whose characteristic constants do not correspond to the values obtained from the experimental scattering curve may be safely rejected.

## D. Radial Density Distribution of Particles with Spherical Symmetry

For particles with spherical symmetry it is possible to determine a radial density distribution from the Fourier transform of the scattering amplitudes (Guinier and Fournet, 1955).

The structure factor $F(h)$ of the particle—that is, the ratio of the amplitude scattered by the particle to the amplitude scattered by a single electron—is for particles with spherical symmetry

$$F(h) = \int_0^\infty \rho(r) \frac{\sin hr}{hr} 4\pi r^2 \, dr \tag{5a}$$

In this equation $h = 4\pi \sin \theta/\lambda$ and $2\theta$ is the scattering angle. The electron density $\rho(r)$ at a distance $r$ from the center of the particle can be determined from the Fourier transform of $h\,F(h)$, according to the following equation:

$$\rho(r) = \frac{1}{2\pi^2 r} \int_0^\infty h\,F(h) \sin hr\,dh \tag{5b}$$

$F(h)$ can be determined from the square root of the scattering intensities but some additional information is necessary to determine the sign.

Further details of this method are explained in Section IV,E by means of a practical example.

E. Rodlike and Lamellar Particles

Although most proteins have a globular shape, some are known to be elongated either in the monomeric form or in the form of rodlike aggregates.

Proteins rarely exist in a platelike or lamellar form but may show lamellar regions in their substructure. It is therefore of interest to indicate the types of scattering curves to be expected of rodlike and lamellar particles.

1. *Rodlike Particles*

It is assumed that the solution of the particles is sufficiently dilute that regularities in position and orientation among the particles are negligible.

*a. Very Elongated Particles.* The scattering curve of a particle, one dimension of which is very large in comparison to its other dimensions, can be split into two factors (Kratky and Porod, 1948; Porod, 1948): the Lorentzian factor (Porod, 1949), which is related to the length of the rodlike particles and depends upon $1/2\theta$, and the cross-sectional factor $I_q$ of a Gaussian form. That is

$$I = I_q \cdot \frac{1}{2\theta} \quad (6a)$$

Therefore, the scattering curve of an elongated macromolecule is steeper than that of a globular particle in the innermost portion, approximately obeying a $1/2\theta$ dependence. Multiplying the scattering intensity by the scattering angle $2\theta$, the Lorentzian factor of the length can be eliminated and the curve then obtained represents only the cross-sectional factor $I_q$

$$I \cdot (2\theta) = I_q \quad (6b)$$

by analogy with Eq. (3b) the equation for the cross-sectional factor is

$$\log_e(I \cdot 2\theta) = \log_e(I \cdot 2\theta)_0 - K'R_q^2\theta^2 \qquad K' = \left(\frac{4\pi}{\lambda}\right)^2 \cdot \frac{1}{2} \quad (7)$$

$R_q$ is the radius of gyration of the cross-section, which can be directly obtained from the slope of the scattering curve in the Guinier plot, by analogy with Eq. (3c).

*b. Dependence of the Cross-Section Factor on the Form of the Cross-Section.* The shape of the cross-section curve gives information about the form of the cross-section, just as the shape of the whole scattering curve of a globular particle gives information about the whole shape. In Fig. 8 curves for very long elliptical cylinders with the cross-sectional axial ratio $a:1$ are given as an example (Mittelbach, 1964). Here too,

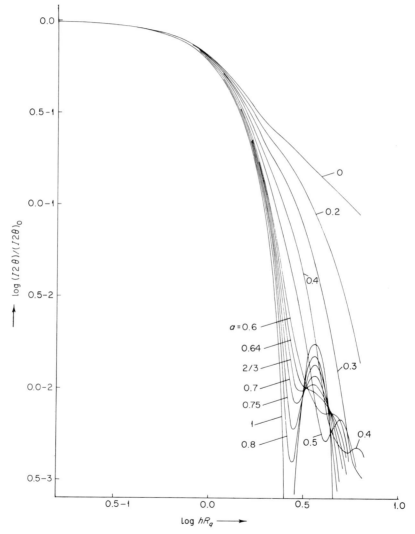

FIG. 8. Theoretical cross-section factors for elliptical cylinders in a log–log plot (axial ratio $\alpha:1:\infty$); the curves are normalized to $hR_q$, where $h = (4\pi \sin \theta)/\lambda$ and $R_q$ is the radius of gyration of the cross-section (Mittelbach, 1964).

isotropic cross-sections show the steepest curves, while increasing anisotropy leads to a flattening of the cross-section curves. From Fig. 8 it can also be observed that the cross-section curves again show subsidiary maxima. They are most pronounced for extremely isotropic cross-sec- It thus appears at obviously larger angle than the subsidiary maxima in this case is given by

$$\frac{\lambda}{2\theta} = 1.73 R_q \qquad (8a)$$

It thus appears at obviously larger angles than the subsidiary maxima of isotropic globular particles which lie at $1.4\ R$.

Cross-section curves of elongated ellipsoids show much more diffuse subsidiary maxima than do prisms. Besides this, their first side maximum is at larger angles than that of prisms; that is

$$\frac{\lambda}{2\theta} = 1.9 R_q \qquad (8b)$$

By comparing experimental cross-sectional curves with theoretical ones, information can accordingly be obtained about the form of the cross-section of elongated particles, just as in Section II,C for globular particles. In practice, only the anisotropy of the shape of the cross-section is determined, since proteins would not be expected to have regular cross-sections corresponding to, say, rectangles or ellipses.

*c. Elongated Particles.* There are not many proteins which are extremely long in comparison with their cross-sections, and thus correspond to the requirements stated above. In practice it is therefore of more general interest to see whether it is also possible to obtain information on the cross-sections of particles with a length which is only a few times larger than the other dimensions.

In Fig. 9 the cross-section curves (Mittelbach, 1964) of ellipsoids with the axial ratio $1:1:c$ are shown in the Guinier plot. These curves lead to two conclusions:

1. The cross-section curves of particles the length of which is only several times their diameter differ from the cross-section curves of very long particles (dashed line) by their low intensities at very small angles. This low intensity becomes the more pronounced, that is, the curve deviates already from the Guinier straight line at correspondingly larger angles, the smaller the length of the particle in comparison to its diameter.

This low intensity at small angles is readily understood since particles which are not very long naturally lack large intraparticular distances and thus also the corresponding scattering at small angles.

2. It is seen from Fig. 9 that the Guinier linearity of the cross-section

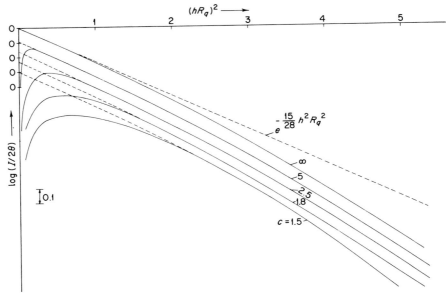

Fig. 9. Theoretical cross-section curves for ellipsoids with the axial ratio $1:1:c$ in Guinier plot (Mittelbach, 1964).

factor is already clearly apparent at a ratio of diameter to length of about 1:2. One can thus obtain information about the cross-section of a protein molecule which is only about twice as long as its diameter. This is done by correcting the low-intensity region of the cross-section factor, in the manner illustrated in Fig. 9, that is, by extrapolating the scattering curves to zero angle (dashed lines).

With these corrected cross-section curves for particles, which are not very elongated, all of the procedures discussed above for very long particles, can be carried out, in particular the determination of the radius of gyration of the cross-section $R_q$ and of its anisotropy. From $R_q$ and the anisotropy we obtain directly the dimensions of the cross-section axes. In Table I, the calculated $R_q$'s of some cross-sections are given.

The cross-section can be determined not only for rigid rods, but also for slightly bent rods.

*d. Length of the Particles.* When the length of a protein particle does not exceed 1000 Å, it is possible to determine not only the radius of gyration of the cross-section $R_q$ but also the radius of gyration $R$ of the whole particle, provided that in very long particles sufficiently small angles are measured. When $R$ and $R_q$ are known, the length $L$ of the particle can be directly calculated from

$$R^2 - R_q^2 = \frac{L^2}{12} \tag{9a}$$

for a prism and

$$R^2 - R_q^2 = \frac{c^2}{5} \tag{9b}$$

for an ellipsoid, where $c$ is the largest semiaxis. The length can also be estimated from the decrease in intensity at small angles (see Fig. 9).

### 2. *Lamellar Particles*

By a lamella is understood a particle in which two dimensions are large compared with the third dimension (the thickness). The scattering curve of a lamellar particle is composed of the Lorentzian factor of the area, which is proportional to $(2\theta)^{-2}$, and a "thickness factor" of the Gaussian type. By multiplying the scattering curve by $(2\theta)^2$ the Lorentzian factor can be eliminated, as shown for rodlike particles in Eqs. (5a) and (5b). The remaining "thickness factor" can be plotted according to the method of Guinier. From the slope of the straight line thus obtained, the radius of gyration $R_t$ of the thickness is estimated, which when multiplied by $(12)^{1/2}$ gives the thickness $t$ itself (Porod, 1948; Kratky and Porod, 1948)

$$t = R_t(12)^{1/2} \tag{10}$$

Though these relations are calculated for particles the thickness of which is very small compared with the dimensions in the plane of the lamellae, the thickness of lamellar particles can still be determined even if the length and breadth of the particle are only twice as large as the smallest dimension (the thickness). The precision of these determinations, is however, not as great as that for the determination of cross-sections of elongated particles. Thickness can be determined not only with plane lamellar particles but also with bent lamellae (Pilz *et al.*, 1970a) (see also Section IV,F,2).

### F. Volume and Tail End of the Scattering Curve

A further parameter, which can be determined from the scattering curve, is the volume $V$ of the protein molecule in solution. The appropriate equation is as follows:

$$V = \frac{\lambda^3 a^3}{4\pi} \cdot \frac{I_0}{Q} \tag{11}$$

Using the cross-section curves, the area $A_q$ of the cross-section can be determined from

$$A_q = \frac{\lambda^2 a^3}{2} \cdot \frac{(I2\theta)_0}{Q} \tag{12}$$

$I_0$, or $(I2\theta)_0$, is the scattering intensity at zero angle and can be obtained by extrapolating the Guinier straight line (see Figs. 3 and 9). The values of $\log I_0$ or $\log(I2\theta)_0$ are determined from the ordinate intercepts.

$Q$ is a parameter which depends only on the total volume of the macromolecules and is independent (Debye and Bueche, 1949) of the degree of dispersion, that is, of the volume of the individual particles. This quantity was described by Porod (1951) as "invariant." It can be obtained as follows:

$$Q = \int_0^\infty I(2\theta)^2 \, d(2\theta) \tag{13}$$

The scattering curve is multiplied by the square of the scattering angle and the area under the curve thus obtained is determined.

Since the very low-intensity tail end of the scattering curve is multiplied by the square of the scattering angle it makes a not inconsiderable contribution to the invariant $Q$. That is, the determination of the volume will be the more exact, as the intensities at the tail end of the scattering curve are determined more exactly.

As a rule, the scattering curves at large angles should follow the asymptotic $(2\theta)^{-4}$ dependence or oscillate around such a dependence. Therefore, the contribution made to the invariant by very small intensities at large angles can be satisfactorily evaluated by calculating the tail end of the curve according to a theoretical $(2\theta)^{-4}$ dependence. [If one uses the so-called "slit-smeared" curves, which are discussed in greater detail in Section III,D, the tail end conforms to a $(2\theta)^{-3}$ dependence; the scattering curve need then be multiplied only by $2\theta$ and the invariant is given by $\tilde{Q} = \int_0^\infty \tilde{I}(2\theta) \, d(2\theta)$.]

We have so far assumed that the electron density within the protein molecule can be regarded as constant to a first approximation. However, the scattering intensity at the tail end of the curve does not decrease to zero but to a final value which is often constant over a large range of angles. This additional constant term in the scattering curve arises from fluctuations in electron density within the protein molecule (Luzzati et al., 1961a). In practice, therefore, this constant term is often determined approximately and subtracted from the entire scattering curve (Kratky, 1963).

This procedure is not correct for very exact measurements since the additional term is not constant over the whole range of angles but approaches zero at small angles.

Stuhrmann and Kirste (1965) have shown that it is possible to take account of the effects caused by fluctuations in electron density, if the scattering curves of a protein are investigated in at least three different solvents of differing electron density. From these scattering curves one may determine the scattering function which the particle would have exhibited if the mean electron density within it had been completely homogeneous (Kirste and Stuhrmann, 1967).

From the tail end of the scattering curve and the invariant, a further parameter, namely, the *specific inner surface*, can be obtained. This specific inner surface is defined as the ratio of the area of the phase interface to the volume occupied by the dispersed phase (volume of the solute) (Porod, 1951a,b).

### G. Molecular Weight and Mass per Unit Length

All of the parameters so far mentioned—such as the radius of gyration, volume, and overall dimensions—can be obtained directly from the shape of the scattering curve.

For the determination of the molecular weight and the mass per unit length of rodlike particles, an additional piece of information is necessary—the so-called *absolute intensity;* that is, the scattered intensity $I_0$ must be known as a fraction of the primary intensity $P_0$. This means that not only the scattered intensities but also the intensity of the primary (incident) beam must be measured. The experimental determination of this parameter is discussed in Section III,C. $I_0$ represents the scattering intensity at zero angle and can, as already mentioned, be determined by extrapolating the Guinier straight line to zero angle (see Fig. 3).

This intensity $I_0$ depends not only on the primary intensity $P_0$ but also on the molecular weight of the protein and its concentration $c$ in solution (see also Fig. 11); $I_0$ is proportional to these quantities. In addition, the intensity increases with the contrast in electron density ($\Delta z$) between the solute (protein) and the solvent.

When the values of the intensities $I_0$ and $P_0$, the concentration $c$ (in milligrams per milliliter), and the contrast in electron density ($\Delta z$) are known, the mass of one particle in molecular weight units $M$ can be calculated directly from the following equation (Kratky et al., 1951; Kratky, 1956)

$$M = K \cdot \frac{I_0}{P_0 F} \cdot \frac{a^2 10^3}{D(\Delta z)^2 c} \qquad K = \frac{1}{i_0 N_L} = 21.0 \qquad (14a)$$

In this equation $a$, $F$, and $D$ are apparatus constants: $a$ is the distance of the sample from the registering plane in centimeters, $F$ is the area of the counter tube slit in square centimeters, and $D$ is the thickness of the capillary or cell used in centimeters; $i_0$ is the Thomson constant, i.e., the scattering of the single electron, and $N_L$ is Avogadro's number. As already mentioned, the quantity $\Delta z$ represents the contrast in electron density between the macromolecule and the solvent and is responsible for the intensity of small-angle scattering. If there is no contrast in electron density between the macromolecule and solvent then no small-angle scattering occurs. We may note that $\Delta z$ represents the excess moles of electrons in the volume occupied by the macromolecule over those which would be present in the same volume occupied by solvent, as shown schematically in Fig. 10. It is of no importance whether $\Delta z$ is positive or negative, since the square of this value $(\Delta z)^2$ is used in Eq. (14a).

The value of $\Delta z$ can be calculated from the following equation

$$\Delta z = (z_1 - \bar{v}_1 \cdot \rho_2) = \left(\frac{\Sigma O}{\Sigma A}\right)_1 - \bar{v}_1 \cdot d_2 \left(\frac{\Sigma O}{\Sigma A}\right)_2 \quad (15)$$

Here, $z_1$ is the number of moles of electrons per gram of the (protein) solute. $\Sigma O/\Sigma A$ (sum of atomic numbers/sum of atomic weights) generally has the value 0.535 for proteins, and $\bar{v}_1$ is the partial specific volume of the (protein) solute in cubic centimeters per gram. $\rho_2$ is the electron density of the solvent and is calculated from $\Sigma O/\Sigma A \cdot d_2$ in which $d_2$ is the density of the solvent.

As there is no great contrast between the electron density of proteins and the buffers which are often used as solvents, the difference between $z_1$ and $\bar{v}_1 \cdot \rho_2$ shown in Eq. (15) is small. Since the square of the difference $(\Delta z)^2$ is used in Eq. (14a), an incorrect value for $\bar{v}_1$ can lead to large errors in the calculated values for $M$ (errors of 1% in the $\bar{v}_1$ value, for instance, produce errors of 4 to 5% in the value of $M$ when proteins are investigated in the usual dilute buffers). With proteins it is therefore of great importance to know the partial specific volume as exactly as possible. For this reason, special apparatus has been developed which renders it possible to carry out exact $\bar{v}$ determinations with small amounts of substance (Stabinger et al., 1967; Kratky et al., 1969; see also Chapter 17).

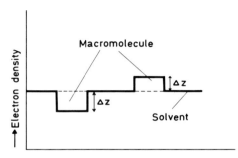

Fig. 10. Schematic illustration of electron density difference ($\Delta z$) between macromolecule and solvent.

It is also essential always to determine the partial specific volume of the protein in the solvent used in the scattering investigation, since the $\bar{v}$ value is by no means constant for a given protein. Eisenberg and Cohen (1968) have studied the change in the partial specific volume of macromolecules in various saline solutions and have found that considerable errors can arise if macromolecules are investigated in concentrated saline solutions (for instance, several moles per liter of NaCl, guanidine hydrochloride, or urea) and one merely uses a literature value for the partial specific volume. The same is also true of course for measurements with the ultracentrifuge (see Chapter 10 in Part B of this series), and Cohen and Eisenberg (1968) recommend that the designation $\phi$ be used for the experimentally determined partial specific volume of the macromolecule in the corresponding solution rather than the usual designation $\bar{v}$. This problem is discussed in greater detail by Kupke in Chapter 17 of this volume.

With elongated particles the mass (molecular weight units) per 1 Å unit length of the particle, $M/1\,\text{Å}$, can also be determined. The equation to be used is analogous to Eq. (14a) (Kratky and Porod, 1953)

$$M/1\,\text{Å} = K_1 \cdot \frac{(I2\theta)_0}{P_0 F} \cdot \frac{a^2 10^3}{D(\Delta z)^2} \qquad K_1 = \frac{2}{i_0 \lambda N_L} = 27.3 \qquad (14b)$$

$N_L$ being Avogadro's number. $(I2\theta)_0$ is obtained from the ordinate intercept of the plot of $\log(I2\theta)_0$ vs. $(2\theta)^2$ of the cross-sectional factor as shown in Fig. 9.

For lamellar particles the mass per square angstrom unit can also be calculated in an analogous manner to Eqs. (14a) and (14b).

H. INTERNAL SOLVATION

The normal solvents for proteins are aqueous and, because of the numerous hydrophilic groups which proteins possess, many water molecules are associated with their structures. These water molecules are essential structural features in proteins, whether crystalline or in solution, and it is therefore of interest to know the so-called swelling factor or degree of swelling $f_s$ for various proteins (Kratky, 1963).

In the solvation of a protein we might, in general, distinguish between two hypothetical limiting cases:

1. The intercalation of the solvent water molecules takes place in such a way that the protein molecule resembles a sponge, which can then be regarded as a body of homogeneous electron density.

2. The protein has large, coherent inclusions of solvent water; for instance, it may be a hollow sphere filled with water.

In the first case, the volume determined by means of the Porod invariant (see Section II,F) will be considerably larger than in the second case, since the invariant makes it possible to determine the volume of homogeneous, spongelike, swollen particles as a whole. Furthermore, conversion of an unswollen particle to a large hollow sphere filled with water (as assumed in the second case) would not lead to an increase in volume but only to an increase in the radius of gyration. Small-angle X-ray scattering thus makes it possible to differentiate between these two extreme kinds of solvation.

If, on the one hand, we know the volume of the swollen protein and, on the other, the volume of the unswollen protein from the molecular weight and the partial specific volume $\bar{v}_1$, the swelling factor $f_s$ corresponding to the degree of hydration of a protein is determined to a good approximation by

$$f_s = \frac{N_L V}{\bar{v}_1 M 10^{24}} \tag{16}$$

$V$ being the volume of a single protein molecule in solution. From this, the number of grams of water per gram of protein can be calculated

$$\text{grams } H_2O/\text{grams protein} = \bar{v}_1 \cdot (f_s - 1) \tag{17}$$

The swelling factor (which is perhaps a misleading term) depends upon the conformation of individual protein subunits and the interstices between them when packed together. We have found that the swelling factor is greater for large than for small proteins because of the less dense packing when many subunits are present. For example, hemocyanin *Helix pomatia*, which consists of about 400 subunits (Pilz et al., 1970c, 1972), has a swelling factor of 0.44 gram water per gram protein, while the whole myeloma IgG molecule has a value of 0.37 gm and its fragments have only 0.32 gram water per gram protein (Pilz et al., 1970d). These models are especially relevant in interpreting observations on protein denaturation.

For globular particles, Luzzati (1960) has defined an internal solvation ratio $\alpha$, which is related to Kratky's $f_s$ (also designated by $q$) as follows: $f_s = \alpha + 1$.

### III. Remarks on the Experimental Technique

There is a number of publications dealing with experimental techniques for measuring small-angle X-ray scattering (Beeman et al., 1957; Kratky, 1954, 1958; Bonse and Hart, 1967; Alexander, 1969). This chapter is therefore limited to enumerating a few techniques which are especially suitable for investigating dilute protein solutions.

## A. Special Requirements in the Investigation of Protein Solutions

Exact investigations of dilute protein solutions make great demands upon the intensity and constancy of the source of radiation for the following reasons:

1. As already mentioned, the electron density difference between protein and water (or buffer) is not very large; that is to say, protein solutions do not scatter very much.

2. In order to eliminate concentration effects (see Section III,E,1 and IV,B,3), the inner portion of the scattering curves must be investigated with relatively dilute solutions (about 1 to 5 mg/ml).

3. It is not the scattering of the protein alone which is measured but the scattering of the whole solution in the capillary or cell. The difference between the scattering curve of the solution and that of the solvent (blank scattering curve) gives the desired scattering of the protein. This, of course, demands that the scattering curves of solution and solvent be recorded under identical conditions. In Fig. 11, the scattering curves of two protein solutions (curves 1 and 2) and the blank curve (curve 3) belonging to them are shown. Using identical concentrations, the particles with a 10-fold greater molecular weight produce a 10-fold greater intensity $I_0$ (curve 1) than the smaller particles (curve 2); on the other hand, the intensity for the larger particles decreases much more rapidly.

In order to record the scattering of solution and solvent under identical conditions, the recommended procedure is to measure the scattering curves of the solutions and the solvent in the same Mark capillary. The difference between the two curves gives the scattering for the protein solute. The capillary has a wall thickness of about 0.01 mm and a diameter of about 1 mm for aqueous solutions.

Since statistical errors are considerably increased in subtracting the experimental scattering curves of solution and solvent, it is desirable—in the usual method of impulse counting—to record at least $10^5$ pulses for every point measured, corresponding to a mean statistical error of ±0.3%.

4. For exact investigations of dilute solutions it is essential to have an X-ray source of sufficiently high intensity and stability, intensity oscillations being less than 1%. High-intensity tubes with shortened focus (Kratky et al., 1960; Kratky, 1967) to permit much better utilization of the intensity are of great advantage. High intensity is

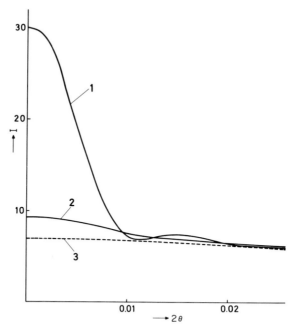

Fig. 11. Experimental scattering curves. (Curve 1) Hemocyanin in $H_2O$; concentration $c = 3$ mg/ml, molecular weight $M = 9 \times 10^6$. (Curve 2) Hemocyanin in borate buffer; $c = 3$ mg/ml, $M = 9 \times 10^5$. (Curve 3) Blank curve (Pilz et al., 1970c).

obtained from tubes with rotating anodes; these were first described by Leonard (1951) and applied by Worthington (1956) and Beeman (1967) to small-angle X-ray scattering investigations. With the very narrow entrance slits which are necessary for measurements at the smallest angles, the instability of the focal point is a serious problem.

It is also desirable to use cells which may be thermostated at low temperatures (e.g., 5°C), since the exact measurement of a scattering curve may take several hours and the protein solution may otherwise deteriorate.

## B. Remarks on Cameras, Monochromatization, and Recording

In Fig. 12, a schematic plan of small-angle scattering geometry and a camera are shown. Normally, the whole camera is covered with a vacuum hood to eliminate air-scattered X-rays, or at least the space between the sample and the counter is evacuated.

Since a high intensity is absolutely necessary in the investigation of dilute protein solutions, cameras with slit collimations are used since

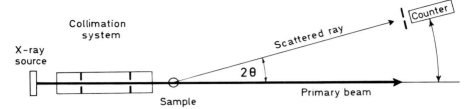

FIG. 12. Schematic geometry of a small-angle scattering camera. A collimated beam (primary beam) strikes the sample and the scattered X-rays are measured by a counter, which is moved step by step through the angular range.

they yield much higher intensities than do cameras with pinhole collimators. While the latter produce primary beams of pointlike cross-section, cameras with slit collimation produce primary beams of line-shaped cross-section.

An X-ray tube with a copper target yields an X-ray beam with the characteristic spectrum shown in Fig. 13. It consists of a broad band of continuously varying wavelengths with two characteristic lines $K\alpha$ and $K\beta$, of which the $K\alpha$ line ($\lambda = 1.54$ Å) is the more intense and is therefore used for small-angle scattering measurements. Since it is not possible to obtain a beam of pure $K\alpha$ X-rays directly from the X-ray tube, monochromatization of the primary beam or the scattered X-rays is necessary. Various ways of obtaining monochromatic scattering curves

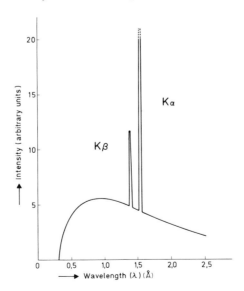

FIG. 13. Intensity curve for X-rays from a copper target operated at 40 kV.

and the various types of cameras used are described in detail by Alexander (1969). A brief description of methods of monochromatization and of some cameras which are suitable for measurements with dilute solutions, is given here.

A monochromatic scattering curve can be obtained by the following methods:

1. Bragg's reflection or total reflection.
2. Ross' filter difference method.
3. Use of a proportional counter in combination with a discriminator and elimination of the $K\beta$ line either with the aid of filters or mathematically.

### 1. *Bragg's Reflection or Total Reflection*

Monochromatization by Bragg's reflection was described in detail by Guinier (1946), who applied a bent Johansson (1933) crystal, and more recently by Jagodzinsky and Wohlleben (1960). A Johansson-type quartz monochromator crystal is also used in the small-angle camera developed for precise measurements by Luzzati and Baro (1961).

In recent years it has been found that hot-pressed pyrolitic graphite is eminently suitable as a monochromator. The use of flat graphite crystals (Sparks, 1966) overcomes the serious loss in scattered intensity suffered by the other crystals mentioned above. When pyrolitic graphite is used as a monochromator, however, care must be taken to insure that the mosaic spread of the material has the correct size.

Small-angle X-ray cameras which work on the principle of total reflection were first used by Lely and van Ryssel (1951), as well as by Ehrenberg and Franks (1952) and Franks (1955, 1958). The Franks camera produces a relatively intense beam by total reflection of the incident X-rays from two crossed glass plates.

In recent years, Damaschun (1964, 1965) and Damaschun et al. (1968b) have used the Kratky camera (see Section III,B,3) in conjunction with a totally reflecting plane mirror, making monochromatization possible without great loss of intensity. Moreover, because of the special collimation system of the camera (see Fig. 16), adjustment of the reflector (which is normally very difficult for Cu $K\alpha$ radiation on account of the small reflection angles of $\leq 4 \times 10^{-3}$ rad), is no longer necessary.

An entirely new method of obtaining an intense monochromatic primary beam was invented by Bonse and Hart (1965, 1966, 1967). They developed a diffractometer which produces a primary beam by means of multiple reflections at germanium crystals and is specially suitable for investigations at the smallest angles.

All cameras based on reflection have the advantage of a monochromatic primary beam, but most of them give rather low intensities, at least at the angles which are generally of interest for protein solutions; as already stressed, this is a disadvantage when very dilute solutions are to be investigated. For intensity measurements at very small angles (equivalent to Bragg spacings of more than, say, 1000 Å), that is, in the investigation of very large particles, the camera of Bonse and Hart has advantages.

## 2. *Filter Difference Method*

In this method (Ross, 1928), the intensity is measured by alternately interposing in the X-ray beam two filters, the absorption edges of which are just above and just below the wavelength of the characteristic line. For copper radiation, cobalt and nickel are used as balanced filters to obtain monochromatic Cu $K\alpha$ radiation. As it is only the difference between the two curves which gives the monochromatic curve, each scattering curve must be measured twice—once with a Ni and once with a Co filter; the method is therefore time consuming.

## 3. *Use of Proportional Counter and Discriminator*

The highest beam intensity in the angle range essential for proteins is afforded by any method which does not completely monochromatize the primary beam before this passes through the sample. A primary beam which is not monochromatized has the wavelength spectrum of the copper-target X-ray tube shown in Fig. 13 and, of course, the resultant scattering curve is also polychromatic. For measuring the intensity of scattered radiation, proportional counters are used, and these have a certain degree of inherent wavelength selectivity depending on their filling gas. If, therefore, a suitable counter is used, some enrichment of the characteristic wavelength is possible. If the proportional counter is used in connection with pulse-height discrimination, one can then reject all pulses due to wavelengths which are much longer or shorter than the characteristic wavelength on which the discriminator window is centered. This means that in the special case of a copper-target X-ray tube only the pulses due to the wavelength of the Cu $K\alpha$ line and its immediate neighborhood are registered.

Complete monochromatization, however, is not possible with this method, because pulse-height discrimination is largely ineffective in excluding pulses due to wavelengths differing only slightly from the characteristic wavelength. For example, the $K\beta$ line can be weakened "but not eliminated" by a suitable adjustment of the discriminator. Therefore, one may choose to weaken the $K\beta$ line to the extent of a few percent by using a 10 $\mu$ Ni filter which causes an intensity loss of about 40%, or one may register the scattering curve derived from both the $K\alpha$ and the $K\beta$ line and then eliminate the $K\beta$ line mathematically with the aid of a computer.

A program to eliminate the $K\beta$ line by calculation has been developed in our group at Graz by Zipper (1969). For the complete elimination of the $K\beta$ line it is necessary, however, to determine the real share of the $K\beta$ line exactly, and this is easily done by means of a 30 $\mu$ Ni filter, which absorbs the $K\beta$ line completely. The real share of the $K\beta$ line is certainly not given by the proportion ($\sim$87% $K\alpha$, $\sim$13% $K\beta$) radiation known to be emitted by the tube. After emission, the shorter $K\beta$ line is much less attenuated than is the $K\alpha$ line on its passage through the sample, foils, air, etc., so that the proportional intensity greatly increases and in the registered scattering curve it can reach twice the initial value (about 25%). A favorable combination of slit length correction (see Section III,D) and elimination of the $K\beta$ line was worked out recently by Hossfeld (1968).

A camera using a combination of the various methods of monochromatizing the X-ray beam was recently described by Thomas (1967). A cylindrically bent glass plate is used as a total reflector. The monochromatization of the radiation takes place partly by means of a nickel filter, partly by reflection, and partly by the use of a proportional counter with a discriminator. The authors (Conrad et al., 1969b) also use a Soller slit system in order to avoid the smearing effect which is caused by the length of the primary beam (see also Section III,D).

Two small-angle X-ray cameras which are very suitable to study dilute solutions of proteins are those of Kratky (Kratky, 1954, 1958, 1963, 1967; Kratky and Skala, 1958) and Beeman and his collaborators (Ritland et al., 1950; Chonacky and Beeman, 1969). Beeman and his collaborators collimate a slit-shaped primary beam using four-slit scattering geometry as shown in Fig. 14. The advantage of the Kratky camera (Fig. 15) is mainly that it has a special collimation system in which blocks are used instead of the usual pair of long slits, as shown schematically in a longitudinal section in Fig. 16. This system excludes parasitic scattering originating from the collimating apertures. It is essential that the beam-defining faces of the blocks $B_2$ and $B_3$ be perfectly planar and that they coincide exactly with one and the same horizontal plane, indicated in Fig. 16 by the plane trace H (dashed line). Above this plane there is no parasitic scattering from the collimating apertures. The entrance slit is formed by the two blocks $B_1$ and $B_2$, which limit the breadth of the primary beam. For the investigation of dilute protein solutions, entrance slits of 0.04 mm to 0.2 mm width are generally

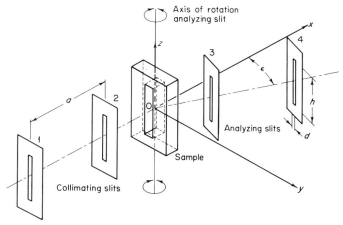

Fig. 14. Schematic diagram of a symmetrical four-slit diffractometer for small-angle scattering studies. The incident beam is collimated by slits 1 and 2. The scattered beam is defined by slits 3 and 4, which rotate about the vertical axis $z$ through the center of the sample. The scale is greatly compressed in the $x$ direction.

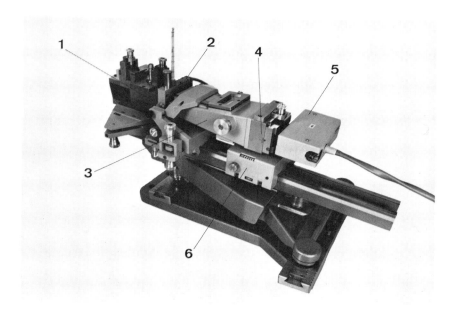

Fig. 15. View of Kratky small-angle camera (1) Collimation system; (2) sample holder; (3) vacuum tube; (4) counter slit; (5) counter; and (6) step scanning device.

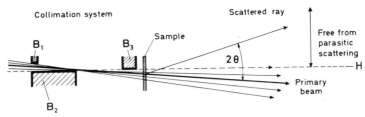

Fig. 16. Longitudinal section through the collimating system of the Kratky camera. The three blocks $B_1$, $B_2$, and $B_3$ collimate the primary beam. The beam-defining faces of $B_2$ and $B_3$ must be perfectly plane and coincide with the same horizontal plane, as indicated by the plane trace H.

used. If necessary, smaller slits, permitting resolution of Bragg spacings of several thousand angstroms, may also be used.

Both systems, which are described in detail by Beeman (1967) and Kratky (1967) provide sufficiently high intensities for the investigation of very dilute protein solutions (up to 1 mg/ml).

### 4. Resolution Required of Cameras

Since the angular limit within which the scattering occurs is inversely proportional (see Section II,A,1) to the size of the macromolecules in solution, it is necessary to carry out the measurements to smaller angles, the larger the particles to be investigated. This means that the larger the particle, the higher the required resolution.

Generally, the resolution is chosen so that the smallest angle measured corresponds to a Bragg spacing $d$ which is at least some four times larger than the largest dimension of the protein under investigation $(1/d = 2\theta/\lambda)$.

Of course, the intensity decreases rapidly if very narrow entrance slits are used in an effect to obtain high resolution. In these cases, the outer portion of the scattering curve is measured using larger slit widths and lower resolution to get enough intensity.

### 5. Measurement of Radiation

The method of pulse counting already mentioned is decidedly preferable to the photographic method owing to its greater precision. The counter is moved step by step through the angle range of interest. Since thorough investigations of dilute protein solutions usually require at least several hours of measurement, it is a great help to make use of a step scanning device which is mechanically or electronically programable (C. Kratky and Kratky, 1964; Leopold, 1965). The number and size of the steps, as well as the number of pulses to be registered at each angle, can be chosen and programed as desired. In this way, it is possible to adapt the method of collecting scattering curve data to each type of macromolecule.

C. ABSOLUTE INTENSITY

### 1. *Methods for Determining Absolute Intensity*

By *absolute intensity* is meant the ratio of the scattered intensity to that of the incident primary beam; by means of this ratio the molecular weight of the macromolecules in solution can be determined, as already described in detail in Section II,G.

The determination of the absolute intensity presents special problems which have been discussed by Kratky (1964). As the intensity of the primary (incident) beam is some $10^4$ to $10^7$ times greater than that of the scattered X-rays, the primary beam must either be attenuated in a controlled way or else its intensity must be determined indirectly.

*a. Rotator Method.* The first determinations of absolute intensity were carried out by Kratky *et al.* (1951). A defined attenuation of the primary beam using a rotating sectorial diaphragm—the so-called rotator method was developed later by the same group (Kratky and Wawra, 1963). In this method, a rotating disk containing a number of small holes is placed directly in front of the counting tube and these holes cross the primary beam when the disk is rotated. It is only through these holes, the dimensions of which must be exactly known, that single quanta of the primary beam can pass. In this way, the primary beam can be attenuated in a well-defined manner without altering its spectral composition. This procedure can also be applied with the methods described in Section III,B,3 which do not work with a fully monochromatized primary beam.

*b. Attenuation by Means of Filters.* Luzzati (1960, 1963) has developed a procedure for attenuating the primary beam with absorbing nickel filters—a method which had also been used often in photographic measurements (Cleeman and Kratky, 1960). The calibration of these filters must be carried out very accurately because very small differences in their thickness cause changes in the intensity of the attenuated primary beam. Since the absorption of X-rays is strongly dependent on the wavelength, nickel filters can change the spectral composition of the primary beam. This method is therefore acceptable only if the primary beam is strictly monochromatic; that is to say, when a monochromatic beam is produced as set out in Section III,B,1 using Bragg's or total reflection.

Recently, Damaschun and Müller (1965) have also made absolute measurements with absorption filters using a total reflector (Damaschun, 1964, 1965).

c. *Indirect Measurement.* Beeman's group (Beeman, 1967) has developed a generally applicable method which allows exact determination of absolute intensity by indirect means. This procedure, which was worked out by Katz (1958) and Shaffer (1963), consists of measuring the scattering of a gas of precisely known composition and equation of state. Since the scattering of such a gas ($C_4F_8$, $SF_6$, or $CClF_2$ have been found suitable) can be calculated theoretically, it is possible to determine the absolute intensity by comparing the scattering of the standard gas with the scattering of an unknown sample.

d. *Indirect Measurement with the Help of the Invariant.* Hermans et al. (1959) described a method for determining the absolute intensity by measuring the invariant (see Section II,F) of a precious metal sol of known composition. The scattering power of the metal sol, which serves as a standard, can be calculated if the electron density and volume fraction of both the dispersant and the dispersed phase (metal) are known. Since the absolute value of the invariant depends on the scattering power and on the intensity of the primary beam, the latter—and with it the absolute intensity—can be obtained by measuring the invariant of the metal sol.

## 2. *Use of a Standard Sample*

It seemed desirable to have a working standard sample which, once calibrated with an absolute method, would permit the rapid determination of the absolute intensity by comparing its scattering quantitatively with the scattering of the sample under investigation. We have therefore developed a sample which is suitable as a working standard (Kratky et al., 1966). The criteria used in choosing a suitable standard were as follows:

a. The sample should be convenient to handle and as far as possible have the form of a uniformly thick platelet which is always ready for use without further preparation.

b. The sample should scatter satisfactorily over a range of angles which, in small-angle X-ray investigations, is always easily measurable (Bragg spacing 100–200 Å). The scattering curve in this range should not fall too sharply, so that small errors in setting the angle do not produce appreciable errors in the determination of the absolute intensity. The scattering curve in this range should as far as possible show a linear decrease, so that the influence of slit width may be neglected (cf. Section III,D). Furthermore, the scattering curve should decrease almost to zero before the angle increases unduly; the scattering intensity will then be independent of the length of the primary beam over a wide range of angles.

c. The sample should not change its scattering intensity after repeated irradiation.

Experiments with various substances have shown that plates of synthetic material are the most suitable for this purpose. Unfortunately, synthetic materials which are resistant to irradiation, for instance, those (such as polystyrene) which contain aromatic rings, show too weak a scattering in the desired range of angles. However, after a suitable preparatory treatment (tempering and very slow cooling) polyethylene[2] met the above requirements (Pilz and Kratky, 1967; Pilz, 1969).

Using the standard Lupolen platelet and a Bragg spacing of 150 Å, the ratio of the scattered intensity to the intensity of the primary beam is then determined once and for all by means of the rotator method. The absolute intensity is thus obtained in a few minutes.

Recently, Peret and Ruland (1971) have found that glassy carbon is highly suitable for standard samples. It is very stable to irradiation and the intensity of the scattered radiation has a negligible temperature dependence.

### D. Collimation (Smearing) Effects

All theoretical scattering curves and equations so far discussed are related to a primary beam with a pointlike cross-section. It has already been pointed out, however, that a primary beam of this kind, which can be obtained only by collimating the beam with very small pinholes, has a much too low intensity for investigating solutions. In general, therefore, a slit-shaped collimated primary beam is used, the cross-section of which has the shape of a long, thin line. Both the length[3] and the width of this primary beam have an influence on the shape of the scattering curve. Scattering curves obtained with such slits are generally described as "slit smeared" or as containing a "collimation error." Two ways of dealing with slit-smeared curves have been devised. On the one hand, theoretical treatments have been developed to obtain the principal geometric and mass parameters directly from the slit-distorted

---

[2] Lupolen 1811 M, supplied by the Badische Anilin- and Sodafabrik, Ludwigshafen am Rhein, Germany.

[3] In the horizontally positioned X-ray tubes, the longer dimension of the cross-section of the primary beam is often referred to as the *height*. However, in this chapter, this dimension will be referred to as the *length* so that no assumption about the positioning of the tube need be made.

scattering curve, and on the other, formulas have been derived to correct the scattered intensities for the collimation error (that is, to "unsmear" the scattering curve). Both procedures will be discussed, though the second is the more usual.

### 1. *Interpretation of Slit-Smeared Curves*

Certain data can be obtained from slit-smeared curves, especially when the primary beam can be regarded as infinitely long, that is, when it is so long that its edges no longer cause additional smearing effects in the recorded scattering curve.

Luzzati (1958) described a procedure which makes it possible to determine the radius of gyration and molecular weight from slit-smeared scattering curves of corpuscular particles; later, Luzzati (1960) extended the procedure to slit-smeared scattering curves of rodlike particles. Burge and Draper (1967) made an exact study of the conditions under which it is possible to obtain certain data directly from slit-smeared scattering curves of rodlike and lamellar particles. Using various models for rodlike and lamellar particles the authors showed that such quantities as the radius of gyration of the cross-section and the mass per unit length of rods (or the radius of gyration of the thickness and the mass per unit area of disks) can be obtained directly from the slit-smeared curves.

A further quantity which can be obtained directly from the slit-smeared curves is the invariant (see Section II,F), as Porod has demonstrated.

A serious deficiency in using slit-smeared curves is that the theoretical scattering curves of models are calculated usually for a point-shaped primary beam and therefore no direct comparison of experimental and theoretical scattering curves is possible.

### 2. *Correction of Collimation Errors (Smearing Effects)*

Owing to the limited amount of information which can be obtained using slit-smeared curves and, in particular, to the fact that no information can be obtained about particle shape, the experimental scattering curves are usually corrected for collimation errors (that is, they are freed from smearing effects). Here, we must distinguish between the slit length (height) correction which is due to the length of the primary beam used and the slit width correction. The width smearing effect can often be neglected when a narrow primary beam is collimated, that is, when the width of the primary beam cross-section is very small.

Guinier and Fournet (1947) and Du Mond (1947) were the first to describe an exact method for eliminating length smearing when the primary beam is infinitely long. Kratky *et al.* (1951) extended this method to the use of primary beams of finite length. Kratky *et al.* (1960) also discussed slit length correction (desmearing) of scattering curves and a method for eliminating width smearing. Additional papers on the theory underlying corrections for slit collimation errors are those by Shull and Roess (1947) and Gerold (1957).

As slit correction of scattering curves is time consuming, the procedures have been programed for digital computers by Schmidt and Hight (1960), Heine and Roppert (1962), Kent and Brumberger (1964), and Schmidt (1965).

The collimation effects can be expressed mathematically by a threefold integral equation of the first kind; the main smearing effect is caused by the slit length, further collimation effects being due to the slit width and the geometry of the counter slit.

The main difficulties in collimation error corrections are solving the integral equation and the fact that the experimentally obtained curve is not smooth. The measured intensities are subject to statistical fluctuations and it is difficult to find a procedure which smoothes out these fluctuations without losing data, such as weak subsidiary maxima. One must also avoid weighting the values, and therewith the errors in measurement, to differing extents.

For solving the integral equation, an arbitrary primary beam geometry and nonequidistant measuring angle intervals must be allowed and the termination errors should be negligible. The reintegration of the solution of the integral equation should give a weighted least-squares approximation of the measured intensity values. This means that all of these values should be approximated with the same precision in relation to the measuring errors. The correction procedure should not magnify the effect of random errors in the experimental data. The most frequently used methods for solving these problems are those of Heine and Roppert (1962), Lake (1967), and Schmidt (1965) but these older methods only partly fulfill the conditions mentioned above.

Newer and more satisfactory methods are those of Hossfeld (1967, 1968), Schelten and Hossfeld (1971), and Vonk (1971). A new general iterative procedure by Glatter (1972) has proved very satisfactory and fulfills the required conditions. It allows arbitrary primary beam geometry and arbitrary measuring angle steps and gives a weighted least-squares approximation with a negligible termination error. The degree of smoothing can be adjusted by a flexible parameter. The iteration stops automatically when the required accuracy is achieved. A further advantage of the methods of Glatter (1972) and Hossfeld (1968) is that correction of the nonmonochromatic radiation (correction of Cu $K\beta$ line) can be included in the general correction procedure since the $K\beta$ line correction is also carried out using an integral equation of the convolution type.

Other desmearing programs have been published by Syneček (1960), Shehedrin and Feigin (1966), Porteus (1968), Federov (1968), and Federov et al. (1968).

### E. Preparation of Suitable Protein Solutions

#### 1. *Concentration*

It should again be borne in mind that all of the equations discussed in Section II are valid only for sufficiently dilute solutions. A solution is only sufficiently dilute if the distances between the macromolecules

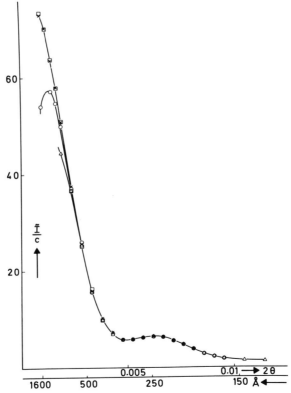

Fig. 17. Scattering curves for hemocyanin in $H_2O$ with the following concentrations $c(mg/ml)$: 2.48 ($\square$); 4.92 ($\times$); 11.33 ($\bigcirc$); and 25.4 ($\triangle$) (Pilz et al., 1970c).

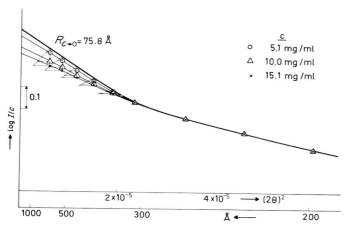

FIG. 18. Scattering curves for IgG-Immunoglobulin in 0.05 $M$ Tris buffer, pH 8, with the quoted concentrations $c$ in a Guinier plot. Extrapolation to zero concentration (thick curve) is carried out as indicated in the figure (Pilz et al., 1970d).

in solution are so irregular and so large that no phase relations (interparticle interferences) exist between the waves scattered by the single macromolecules.

The effect of such interparticle interference is usually to decrease the scattering intensity at small angles as shown in Fig. 17 for different concentrations of hemocyanin in $H_2O$. These concentration effects can be eliminated either by sufficient dilution of the solution or by investigating a series of concentrations and then extrapolating to zero concentration as schematically shown in Fig. 18. In practice, the second method is usually used, though here too it is necessary to include measurements at suitably low concentrations. It is useful to plot the values of $I/c$, that is, to relate the curves to unit concentration, since a clearer picture of the concentration effects is obtained in this way (see Figs. 17 and 18). In practice, the extrapolation to infinite dilution is carried out as follows. Radii of gyration are first obtained from the Guinier plot of the scattering curves for the various concentrations (Fig. 19). These values are then plotted against the concentration, as shown in Fig. 20 (see also Figs. 26 and 37).

The magnitude of concentration effects depends upon many other factors besides the concentration itself; for instance, it depends on the number of macromolecules in solution, their shape, and their charge, as well as on the ionic strength and character of the solvent. In extreme

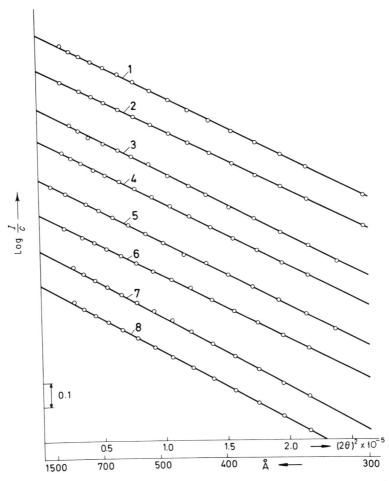

Fig. 19. Scattering curves of fatty acid synthetase in potassium phosphate buffer, pH 6.5, in a Guinier plot. The concentrations $c$ of the enzyme have the following values (mg/ml): (1), 22.2; (2), 18.3; (3), 15.4; (4), 10.1; (5), 9.84; (6), 9.45; (7), 6.90; (8), 5.6 (Pilz et al., 1970a).

cases—at high concentrations—interactions between the macromolecules can lead to regular arrangements which produce a kind of lattice factor. It is therefore impossible to evaluate concentration effects quantitatively.

As a general rule, depending on the size and nature of the proteins under investigation, dilutions down to 1 mg/ml or 5 mg/ml are necessary.

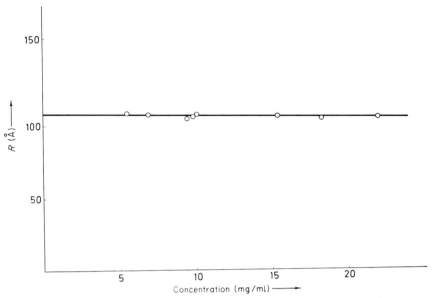

FIG. 20. Radii of gyration $R$ of fatty acid synthetase as a function of the concentration $c$ of the enzyme. The values of $R$ are obtained from the Guinier plots shown in Fig. 19 (Pilz et al., 1970a).

Dilutions below 1 mg/ml are difficult to investigate, as the scattering intensity is very low.

As concentration effects are apparent only at relatively small angles, much higher concentrations (50 to 100 mg/ml) can be used for investigating the tail end of the scattering curve where the intensity is weak.

## 2. Homogeneity of Macromolecules

To derive exact data from small-angle scattering, all the macromolecular particles in the system must be identical in size and shape. Only in truly monodisperse solutions is it possible to determine the size and shape of the solute. From polydisperse solutions only general information about the colloid system can be obtained and some average parameter estimated. Polydisperse solutions do not yield linear Guinier plots: an example of such a plot for a polydisperse solution of silk fibroin is shown in Fig. 21 (see also Section IV,A,1).

If a small percentage of the protein in solution is present in the form of very large aggregates or of very small fragments (about one order of magnitude larger or smaller than the main component), such components are often without appreciable effect on the scattering curve of the protein.

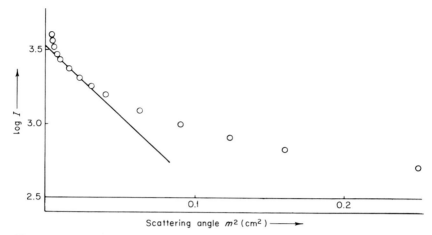

FIG. 21. Course of a Guinier plot which is typical for very polydisperse solutions (Solution of silk fibroin, Kratky et al., 1964); $m$ corresponds to the scattering angle.

Since proteins often tend to form aggregates, it is always desirable to test the monodispersity of the protein solution using other physical techniques before investigating small-angle X-ray scattering. A minimal criterion is that the solute show a single symmetrical kink on velocity sedimentation in the analytical ultracentrifuge (see Chapter 10 in Part B of this series).

With monodisperse solutions an increase in concentration usually leads to a progressive loss of intensity at small angles, as shown in Figs. 17 and 18. When a reversed dependence on concentration is observed, that is, when the values of the radii of gyration decrease rather than increase with decreasing concentration, it must be presumed that aggregates are present at the higher concentrations. These split when diluted into monomers, causing a decrease in the $R$ value. We were able to observe a reversed dependence on concentration with the Fab fragments of IgG-immunoglobulins, as shown in Fig. 22.

If there are aggregates, one must try to remove them by such methods as gel filtration using Sephadex, electrophoresis, or electrofocusing. These methods, of course, fail if there is an association–disassociation equilibrium. This is often found, especially with enzymes, and is indicated also in Fig. 22. In such cases one must try to work under conditions in which the equilibrium (which is often dependent on pH, concentration, and temperature) is far enough to one side to get an approximately monodisperse solution.

With equilibria which are dependent upon concentration, monodispersity is often impossible to achieve within the concentration range

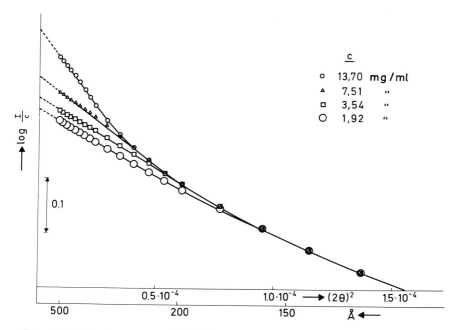

FIG. 22. Scattering curves of Fab fragments in 0.05 $M$ Tris buffer, pH 8.0. The different concentrations $c$ are given in the figure (Pilz et al., 1970d).

necessary for small-angle X-ray scattering; however, in such cases it is sometimes possible to obtain essential information on the protein, as explained in Section IV,D,1 using a practical example.

The formation of aggregates can often be prevented by adding urea, guanidine hydrochloride, and similar hydrogen bond splitting reagents. Solutions of native silk fibroin, for example, contain aggregates of the most varied size (Kratky and Pilz, 1955) (see also Fig. 21). Addition of 6 $M$ urea prevents the formation of aggregates and fairly monodisperse solutions are obtained (Kratky et al., 1964). Adding of hydrogen bond splitting reagents on the other hand changes the native conformation of the protein. Therefore the data obtained for the protein in such solutions correspond to the denatured particle.

Contamination by very large particles (large aggregates, dust, bacteria, etc.) can be removed by using sterile Millipore filters or by centrifuging. As already mentioned, particles of dust do not interfere with small-angle X-ray scattering, since they scatter to immeasurably small angles.

It is important that the exact composition of the buffer be known and that the buffer used for the blank curve be identical with that in which the protein is dissolved. An identical pH value for the buffer is by no means sufficient (as is sometimes thought), since for small-angle scattering the difference in electron densities is decisive. In every case, it is best to dialyze the protein solution against the buffer used as the blank solution.

## IV. Selected Examples of Applications

In this section, the efficiency and limits of the method will be demonstrated with selected examples. In discussing particular cases only those results which are characteristic of small-angle X-ray scattering investigations will be discussed in detail. The enumeration of all data, such as molecular weight, radius of gyration, volume, internal solvation, surface, and shape, which are usually determined with small angle scattering, would be too extensive. Furthermore, so many small-angle X-ray investigations have been carried out on proteins in recent years that it is impossible to mention them all in this chapter.

### A. Comparison of the Results of Small-Angle Scattering and X-Ray Studies on Crystals

The following examples were chosen because the structures of myoglobin (Kendrew et al., 1960), hemoglobin (Perutz, 1963, 1970), chymotrypsin (Kraut et al., 1967), and lysozyme (Blake et al., 1965) are already known from X-ray studies on crystals and a comparison of the results shows the efficiency and limitations of small-angle scattering. Moreover it is interesting to compare the well-defined conformation of protein molecules in the crystal with that observed in dilute solutions.

#### 1. *Myoglobin*

Beeman (1967) made precise measurements on dilute solutions of myoglobin (sperm whale metmyoglobin) in phosphate buffer at pH 7.3. The values of 18 Å for the radius of gyration found in earlier investigations seemed high compared with the value of 15.5 ± 0.5 calculated by Watson from the atomic coordinates and also from the electron density maps of Kendrew's crystallographic model at a resolution of 6 Å (Kendrew et al., 1960). An exact check on the monodispersity of the solution showed the presence of more than 5% dimer and some higher aggregates

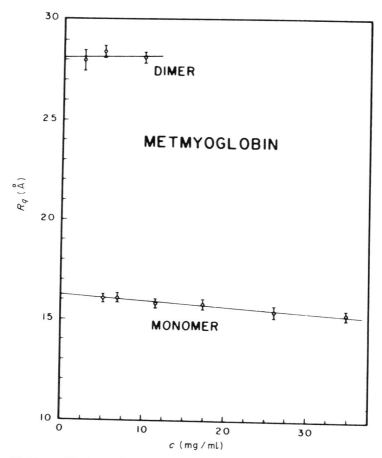

Fig. 23. The radii of gyration of metmyoglobin monomer and dimer as a function of concentration $c$ of the protein (Beeman, 1967).

in addition to the monomer. It was possible to separate these aggregates using gel filtration with Sephadex. The radii of gyration ($R_g$) of the monomer and the separated dimer are plotted in Fig. 23 as a function of the concentration. The value for the radius of gyration of the monomer, which is extrapolated to zero concentration, is now only 16 Å and thus agrees very well with that calculated from the atomic coordinates.

This example shows clearly the importance of using monodisperse solutions. Small amounts of dimer do not necessarily cause curvature in the Guinier plot (see Section III,E,2); values for the radius of gyration can, however, be appreciably increased by the presence of small amounts

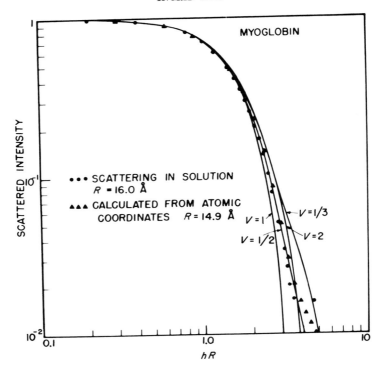

Fig. 24. Experimental and calculated scattering curves plotted against $hR$. The intensities have been made equal at zero angle. The solid curves are the theoretical scattering curves for ellipsoids of revolution of constant electron density and of axial ratio $v$ as indicated (Beeman, 1967).

of aggregates, because, in arriving at a Z-average,[4] the larger values are weighted more heavily than are small ones.

In Fig. 24 the form of the experimental scattering curve is compared with that calculated using the Debye formula and the atomic coordinates. The two curves are very similar in their inner segments; only at

[4] The average $R$-value ($\bar{R}$) found for a mixture of dissolved particles is a Z-average and given by the equation

$$\bar{R} = \frac{\sum_{i=1}^{N} c_i \cdot R_i^2}{\sum_{i=1}^{N} c_i \cdot R_i}$$

if there are $N$ components each possessing the concentration $c$ and the radius of gyration $R$.

larger angles is there a noticeable difference. This is to be expected, since the theoretical curve is calculated for the molecule *in vacuo*, while the experimental curve refers to the molecule in water and the course of the curve at large angles is dependent on the surrounding medium.

Figure 24 also shows the theoretical scattering curves for ellipsoids of revolution with various axial ratios $v$. These show that the experimental curve agrees very well with the curve for $v = \frac{1}{2}$. Since the shape of the molecule as determined by X-ray studies on crystals is best described by an oblate ellipsoid of revolution with an axial ratio of 0.55, the agreement is very satisfactory.

Kirste and Stuhrmann (1967) have carried out exact small-angle scattering studies on myoglobin. They tried to eliminate the background scattering caused by internal fluctuations in the electron density of the protein molecule by investigating the myoglobin in a series of different solvents. As already mentioned, by varying the electron density of the solvent it is possible to determine the scattering function which the myoglobin would have if it had a uniform electron density throughout its volume. The radius of gyration which the myoglobin molecule would have under these conditions was calculated by the authors to be 15.6 Å. This value agrees very well with that calculated by Watson from the electron density maps. One obtains the same value if one considers the dependence of the radius of gyration $R$ on the solvent and if the reciprocal value of the mean electron density difference between myoglobin and solvent $1/\overline{\Delta\rho}$ is extrapolated to zero. (Fig. 25). The solvents used in

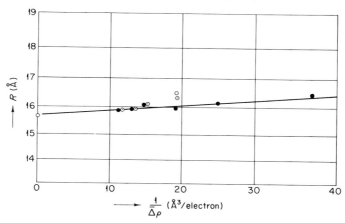

Fig. 25. Dependence of the radius of gyration of myoglobin on the solvent: glycerol–water (●); glucose–water (⊙); and sucrose–water (○). $R$ is plotted *versus* the reciprocal of the electron density of the solvent (Kirste and Stuhrmann, 1967).

these investigations were: glycerol–water mixtures (up to 90% by weight of glycerol), glucose solutions (up to 50% glucose·$H_2O$), and sucrose solutions (up to 45% sucrose).

## 2. Oxy- and Deoxyhemoglobin

Similar investigations to those on myoglobin have been carried out on oxy- and deoxyhemoglobin by Conrad et al. (1969a,b).

In Fig. 26, the apparent (slit-smeared) radii of gyration for both states of hemoglobin are plotted against concentration. Deoxyhemoglobin clearly has a larger radius of gyration than does oxyhemoglobin. The reversible change in the $R$ value, which characterizes the movement of the four chains in the hemoglobin molecule upon the addition and removal of oxygen, could be repeated as often as desired by oxygenation and deoxygenation. After slit-correction and extrapolation to zero concentration, the radii of gyration given by small-angle X-ray scattering were compared with those calculated from the electron density maps of the model of Cullis et al. (1962) (Table II).

In both cases, deoxyhemoglobin shows the larger radius of gyration. The authors found a difference of 1.5 ($\pm 0.9$) Å experimentally between the radii of gyration of the two hemoglobin states. From the models proposed by Muirhead et al. (1967) for the hemoglobin transformation, a difference in $R$ value of 0.35 may be calculated. (The absolute value of the radii of gyration calculated from the electron density depends to a

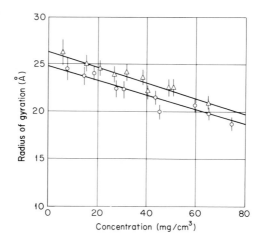

Fig. 26. Apparent radii of gyration and their extrapolation to infinite dilution for deoxy- and oxyhemoglobin. Deoxyhemoglobin ($\triangle$) and oxyhemoglobin ($\bigcirc$), showing probable error bars (Conrad et al., 1969b).

TABLE II

(a) Radii of Gyration $R$ and Volumes of Human Deoxy- and Oxyhemoglobin Calculated from Small-Angle X-Ray Scattering[a]

|  | Deoxyhemoglobin | Oxyhemoglobin | Deoxy- minus oxyhemoglobin |
|---|---|---|---|
| $R$ (Å) | 26.2 ± 1.2 | 24.7 ± 1.1 | 1.5 ± 0.9 |
| volume (Å³) | 140,000 ± 20,000 | 135,000 ± 20,000 | |

(b) Radii of Gyration Calculated from the Electron Density Maps of Horse Hemoglobin at 5.5 Å Resolution[a]

| Contour level of molecular boundary used for calculations | Radii of gyration | | Deoxy- minus oxyhemoglobin |
|---|---|---|---|
|  | Deoxyhemoglobin | Oxyhemoglobin |  |
| 0.33 (density of water) | 24.6 | 24.3 | 0.3 |
| 0.40 (average density in protein crystal) | 24.3 | 24.0 | 0.3 |
| 0.46 | 23.9 | 23.5 | 0.4 |
| 0.53 (first contour on map published by Cullis et al., 1962) | 23.6 | 23.2 | 0.4 |

[a] Conrad et al., 1969b.

certain extent on the level of electron density chosen, but the difference between the values of the radii of gyration was always 0.35 Å.)

As a matter of interest, Rupley (1968) carried out some interesting investigations by other methods to determine how closely the structure in solution resembled that in the crystal. In particular, he measured changes in protein solubility (Rupley and Gates, 1968), in proton binding, and in volume (Krivacic and Rupley, 1968) when the protein crystallized. As a result of these investigations he has concluded that in hemoglobin, for instance, there can be no essential conformational difference between the molecule in the crystal and in solution.

### 3. *Chymotrypsinogen and Chymotrypsin*

Chymotrypsin, as is well known, is secreted in the same way as many enzymes of the stomach and intestine in the form of an inactive precursor, namely chymotrypsinogen. The activation to active chymotrypsin takes place by the action of trypsin. According to whether the activation takes place slowly or rapidly and whether one or two dipeptides are split off from the chymotrypsinogen, various chymotrypsins are formed. These are described as $\alpha$-, $\beta$-, $\gamma$-, $\delta$-, and $\pi$- chymotrypsin.

It is naturally of great interest to know whether this transformation of inactive chymotrypsinogen into active chymotrypsin is accompanied by a change in conformation. Krigbaum and Godwin (1968) investigated dilute solutions of chymotrypsinogen and of $\alpha$-, $\beta$-, $\delta$-, and $\gamma$-chymotrypsin using small-angle X-ray scattering. For each sample, they obtained the radius of gyration $R$, the ratio of surface to volume $S/V$, and the axial ratio of the ellipsoid of revolution, which is equivalent in scattering. The values are summarized in Table III and it can be seen that they are very similar. This means that the activation of the enzyme is clearly not connected with any essential change in conformation. Chymotrypsin, which is largely formed by slow activation, provides exactly the same data as inactive chymotrypsinogen; $\delta$-chymotrypsin, which is formed by rapid activation, is slightly larger and somewhat more symmetrical, as indicated by the larger $R$ value and the smaller axial ratio.

It is interesting to compare these values with the results of the crystallographic studies by Kraut et al. (1967). These authors, too, find that the conformation of inactive chymotrypsinogen is very similar to that of the active chymotrypsin. On the other hand, the shape of the molecule in the crystal appears to differ from that in solution. The shape of chymotrypsinogen in the X-ray crystallographic model is best described by an ellipsoid with semiaxes of $25 \text{ Å} \times 20 \text{ Å} \times 20 \text{ Å}$, which corresponds to an axial ratio of 1.3:1 and a radius of gyration of 16 Å. According to the investigations by Krigbaum and Godwin, however, chymotrypsinogen in solution is more asymmetric; it has, as already mentioned, an axial ratio of 2:1 and a radius of gyration of 18 Å, which correspond to ellipsoid semiaxes of $32 \text{ Å} \times 16 \text{ Å} \times 16 \text{ Å}$. The authors conclude from this that there is a considerable difference between the shape of the molecule in solution and the shape in the crystalline state, as seen in Fig. 27. In this figure the experimental curve is compared with theoretical scattering curves for the two axial ratios 1.3:1 (X-ray crystallography) and 2:1 (small-angle scattering). The authors presume that, in solution, a portion of the molecule unfolds, leading to a more asymmetric shape.

TABLE III
COMPARISON OF THE SIZE AND SHAPE PARAMETERS FOR CHYMOTRYPSINOGEN AND CHYMOTRYPSINS[a]

| Substance | $R$ (Å) | $S/V$ (Å$^{-1}$) | Axial ratio |
|---|---|---|---|
| Chymotrypsinogen | 18.1 | 0.160 | 2.0:1 |
| $\alpha$-Chymotrypsin | 18.0 | 0.157 | 2.0:1 |
| $\beta$-Chymotrypsin | — | 0.155 | 2.0:1 |
| $\delta$-Chymotrypsin | 19.0 | 0.146 | 1.8:1 |
| $\gamma$-Chymotrypsin | 18.3 | 0.153 | 1.9:1 |

[a] From Krigbaum and Godwin (1968), with permission.

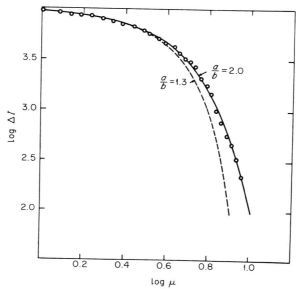

Fig. 27. Comparison of the experimental scattering curve for a 1.832% solution of chymotrypsinogen-A (○) with the theoretical scattering curve for ellipsoids with an axial ratio of 2:1 (full line) and 1.3:1 (dashed line) (Krigbaum and Godwin, 1968).

### 4. Lysozyme and α-Lactalbumin

Krigbaum and Kügler (1970) have carried out measurements on egg-white lysozyme and bovine α-lactalbumin, comparing the molecular conformation of lysozyme in solution and in the crystalline state. The overall dimensions and the radii of gyration—summarized in Table IV—show that the shape of the lysozyme molecule in solution is very similar to that determined from crystal structure studies.

The radius of gyration of the molecule in the crystalline state was calculated from the known crystallographic atomic coordinates (Blake

TABLE IV
Comparison of Size and Shape Parameters for α-Lactalbumin and Lysozyme[a]

| Substance | State | $R$ (Å) | $V$ (Å$^3$) | Overall dimensions (Å) |
|---|---|---|---|---|
| α-Lactalbumin | solution | 16.7 ± 0.4 | 26,500 | 22 × 44 × 57 |
| Lysozyme | solution | 14.3 ± 0.3 | 19,800 | 28 × 28 × 50 |
| Lysozyme | crystal | 13.8[b] | | 30 × 30 × 45 |

[a] From Krigbaum and Kügler (1970) with permission.
[b] Calculated for the molecule in a vacuum.

*et al.*, 1965) and the scattering curve was calculated from the atomic coordinates using the relation of Debye (1925). The computer time required to calculate the scattering curve using the atomic coordinates of lysozyme was rather long. For comparative purposes, the authors also performed the calculation using groups of atoms instead of atoms as scattering elements (the real scattering elements are the electrons). In one case, each amino acid residue was represented by one scattering element located at the $\alpha$-carbon atom (that is, 129 scattering points). In the second approximation, each residue was replaced by two weighted scattering elements. One of these was located at the $\alpha$-carbon and the other at an atom at the extremity of the side chain (that is, 244 points). Finally, all the known atomic positions (that is, 961 points) were used.

The results are shown in Fig. 28 and are compared with the experimental scattering curve obtained using small-angle scattering in solution. The curves differ only at their tail end. Hyman and Vaughan (1967) have demonstrated, by different calculations, that the tail end of the

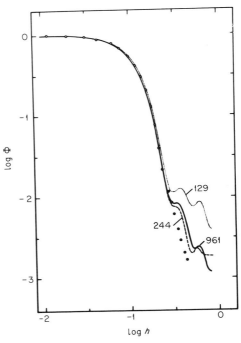

FIG. 28. Lysozyme scattering curve (O) compared with those calculated using crystallographic coordinates for the indicated number of points (scattering elements) per molecule. $\phi$ corresponds to $I/I_0$ and $h = 2\pi \sin 2\theta/\lambda$, where $2\theta$ is the scattering angle (Krigbaum and Kügler, 1970).

curve is influenced by the medium surrounding the particles. The theoretical scattering curve calculated for the molecule *in vacuo*, therefore, cannot be expected to agree completely with the experimental curve at the tail end, and as a rule it is not necessary to calculate the theoretical scattering curve from the atomic coordinates. Small-angle scattering usually gives information only about the overall shape and, in calculating theoretical scattering curves, models can be approximated by bodies of uniform electron density or atomic groups.

The data for lactalbumin and lysozyme are also presented in Table IV. The amino acid sequences of lysozyme and lactalbumin exhibit considerable homology (Blake *et al.*, 1965; Brew *et al.*, 1967) and this has led to the suggestion that they may have similar tertiary structures. As shown by their data in Table IV, Krigbaum and Kügler found, on the contrary, that the two enzymes have different molecular conformations in solution.

### B. Determination of Models Equivalent in Scattering

The following procedure is usually applied. A mean axial ratio is determined by comparing the experimental and theoretical scattering curves for simple triaxial bodies or by calculating various characteristic quantities (cf. Section II,C). The agreement between the experimental curve and the theoretical curve for a simple triaxial body is often so good that the macromolecule can be described satisfactorily by means of the model. For instance, as already shown in Section IV,A, the shape of myoglobin can be approximated by an ellipsoid of revolution of a definite axial ratio (see Fig. 24). Malmon (1957b) found that the shape of the enzyme catalase is best described by a slightly elongated body, the maximum dimension of which is $146 \pm 10$ Å and this length is about twice as great as the mean diameter.

Protein molecules, however, often have a more complex form which cannot be described by simple triaxial bodies. We shall use the example of the IgG-immunoglobulins to demonstrate that even here it is possible to choose appropriate models so as to give useful information about the shape of the protein molecule in solution.

#### 1. *IgG-Immunoglobulins*

For a long time there were difficulties in investigating $\gamma$-immunoglobulins, since only a very heterogeneous mixture of these macromolecules could be isolated from blood serum. The chemical and physical properties of the various IgG-immunoglobulins were so similar that it was almost impossible to isolate a completely homogeneous sample in sufficient quantity. This heterogeneity also interfered with small-angle X-ray scattering measurements. Values for the radii of gyration varied, according to the previous history and origin of the samples, from 65 to 78 Å and molecular weights from 130,000 to 190,000.

The one feature which we found common to all the IgG-immunoglobulins we investigated was the unusual shape of the cross-section curve, which consisted of an inner, steeper portion and an outer, flatter portion. In Fig. 29, the characteristic cross-section curves for some samples investigated in our institute are shown. (The scattering curves for the horse IgG are taken from unpublished measurements by Kratky, Kügler, and Holasek.) The flatter portion of the cross-section curves was very similar in all the samples investigated, yielding a radius of gyration of 14–15 Å,

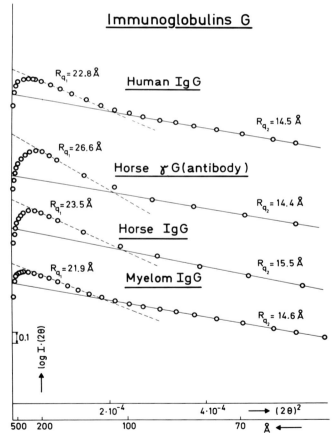

Fig. 29. Cross-section curves of different IgG-globulins in solution in Guinier plots, $R_{q_1}$ and $R_{q_2}$ are the different radii of gyration of the cross-section (Pilz, 1970).

while that of the steeper portion varied from 19 to 26 Å for different samples.

The data, such as radius of gyration, molecular weight, volume, and degree of swelling (Kratky et al., 1963), could be determined without difficulty. A model which agreed satisfactorily with the experimental data and had a scattering curve which showed the characteristic shape of the cross-section curves, could not be found for a long time, the number of possible shapes being too great. It was possible only to state that the IgG-globulin molecule was elongated; the length was determined to be 240 Å, which agreed well with earlier investigations by Kratky et al. (1955b).

An important step forward in investigating the IgG-immunoglobulins was the finding that very homogeneous IgG-immunoglobulins, the so-called myeloma globulins, could be isolated from the serum of patients with multiple myeloma. Using the homogeneous γG1 myeloma protein Eu, Edelman et al. (1969) were able to carry out the first complete amino acid sequence determination and to determine the location of the disulfide bonds which link the single chains (2H chains and 2L chains, see Fig. 30).

Small-angle scattering investigations were carried out (Pilz et al., 1970c) on the same homogeneous sample, which Edelman and Gall had placed at our disposal. The cross-section curves of this homogeneous sample showed the same characteristic shape as the curves for the heterogeneous samples, as shown in Fig. 29. The three upper curves are for

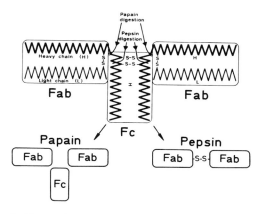

Fig. 30. Schematic illustration of the digestion of the IgG-immunoglobulin molecule into different fragments by the enzymes papain and pepsin. The two specific combining sites for antigens are expected to be located at the opposite ends of the molecule in the Fab regions (Pilz, 1970).

heterogeneous samples; the bottom one is for the homogeneous IgG-immunoglobulin (myeloma protein). It was thus clear that the unusual shape of the curve was not due to heterogeneity in the usual serum IgG-immunoglobulins.

To obtain further information about the undoubtedly complex shape of the immunoglobulin molecule, fragments of the macromolecule were investigated as well as the intact molecule. Figure 30 shows schematically the well-defined fragments which can be obtained by digestion with papain or pepsin. We therefore investigated homogeneous solutions of the Fab, Fc, and $F(ab')_2$ fragments.

First of all, it should be noted (Fig. 31) that the outer, flatter portions of the cross-section curves are identical for the Fab fragment, the $F(ab')_2$ fragment, and the whole IgG molecule. This suggests that they have one almost identical radius of gyration of the cross-section and that the cross-section of the Fab fragment is unchanged on fragmentation.

It was not immediately apparent, however, whether the steeper, inner portion of the cross-section curve of the whole molecule was due to a second larger cross-section or to a particular arrangement of the fragments.

A second conclusion may be drawn from Fig. 31. In Section II,E,1,c it was pointed out that the shorter the particle, the larger the angles at which the cross-section curves deviate from the Guinier straight line (i.e., the larger the angles at which loss of intensity occurs). Figure 31 shows clearly that the length of the intact molecule is greatest and that of the

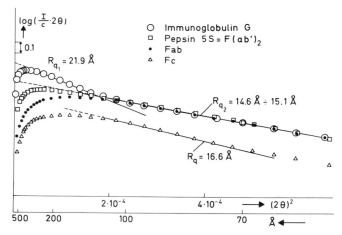

FIG. 31. Cross-section curves for IgG-immunoglobulin and its fragments $F(ab')_2$, Fab, and Fc in Guinier plots. $R_{q_1}$ and $R_{q_2}$ are the radii of gyration of the cross-section (Pilz et al., 1970d).

Fab fragments least, while the length of the F(ab')$_2$ fragment lies between the two. The Fc fragment is probably just as short as the Fab fragment, as it deviates from the Guinier straight line at approximately the same value of $2\theta$.

The data on the intact molecule and its fragments are summarized in Table V. The agreement between the sum of the molecular weights or volumes of the (2 Fab + Fc) fragments and the molecular weight for the intact molecule can be regarded as satisfactory: as already shown in Fig. 22 no exact extrapolation to infinite dilution could be made. It was also possible to describe the shape of the Fab and Fc fragments very well by ellipsoids or elliptical cylinders with the dimensions given in the table.

The next step was to arrange the three fragments (2 Fab + Fc) in different ways in order to arrive at a model with a scattering curve which agreed with the experimental curve for the molecule as a whole. The model had to fulfil the following requirements: (a) In accord with the chemical structure of the molecule (H chains running through from Fab to Fc) the Fc fragment had to be connected with both Fab fragments (see Fig. 30). (b) The radius of gyration and the volume had to correspond to the values found experimentally for the whole molecule. (c) The cross-section of the Fab fragments had to be retained, as already stated, so that the individual fragments could not be packed closely together. Furthermore, the cross-section curve for the model had to show the same two characteristic portions, differing in slope, just as in the experimental curve; this requirement seemed the most important one to us.

Using a program written by Haager, of our group (Pilz et al., 1970d),

TABLE V
$\gamma$G1-Immunoglobulin and Fragments[a]

| Sample | $R$ (Å) | $M$ | $R_q$ (Å) | $V$ (Å$^3$) | Grams H$_2$O per gram protein | Elliptical cylinder | | |
|---|---|---|---|---|---|---|---|---|
| | | | | | | $2a$ | $2b$ | $L$ (Å) |
| Fab | 32.0 | 48,000 | 15.0 | 83,000 | 0.32 | 22 | 56 | 98 |
| Fc | 33.1 | 55,000 | 16.6 | 96,500 | 0.32 | 21 | 63 | 99 |
| F(ab')$_2$ | 53.0 | 93,000 | 15.1 | 154,000 | 0.27 | | | |
| 2 Fab + Fc | | 151,000 | | 263,000 | | | | |
| $\gamma$G1-Immunoglobulin | 75.8 | 162,000 | 14.6 / 21.9 | 297,000 | 0.37 | | | |

[a] Pilz et al. (1970d).

we approximated the three fragments by ellipsoids and used these to build up various models.

Following a proposal by Noelken *et al.* (1965), according to which regions corresponding to the three fragments are linked together by extended, flexible regions of polypeptide chains, we calculated a series of star-shaped models in which the individual compact fragments did not touch each other (Fig. 32, curves 1–4). None of the curves agreed with the experimental curve (curve 5). Curve 4 represents the scattering curve for shapes obtained by varying the angle between the ellipsoids from 0° to 90° in such a way as to average the models and minimize the deviation from the experimental curve.

In Fig. 33 models are shown in which a closer contact between the individual fragments is assumed, according to the well-known electron micrographs by Valentine and Green (1967). An elongated model (curve

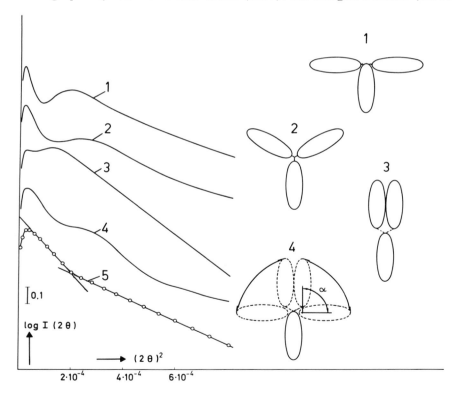

Fig. 32. Comparison of calculated cross-section curves for various models (1–4) with the experimental cross-section curve (5) of IgG-immunoglobulin (Pilz *et al.*, 1970d).

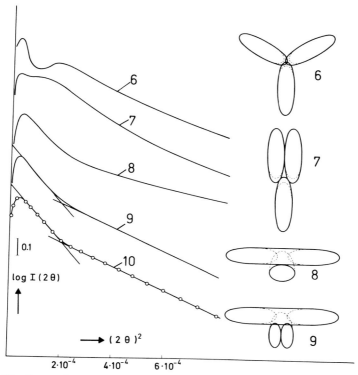

Fig. 33. Comparison of the calculated cross-section curves for various models (6-9) with the experimental cross-section curve for IgG-immunoglobulin (10). The dashed lines correspond to the ellipsoids equivalent to the individual fragments (Pilz et al., 1970d).

9) in which the Fc fragment is directly linked with the Fab fragments, leads to a scattering curve which is practically identical with the experimental curve (curve 10). The dashed lines delineate the Fab fragments; the two small ellipsoids correspond only to that part of the Fc fragment protruding from the main body of the molecule.

The inner portion of the cross-section curve represents the middle part of models 8 and 9 (Fig. 33) in which both the Fab regions and Fc regions make contributions to the cross-section; the outer, flatter portion of the cross-section curve corresponds to the cross-section of the Fab fragments (Pilz et al., 1970c).

On the basis of this model, which is "equivalent in scattering" (model 9) we have suggested that IgG-immunoglobulin in solution can be described by an elongated T-shaped body as shown schematically in Fig. 30. This T-shaped model was recently confirmed by the first crystal X-ray study (at 6 Å resolution) on an IgG-immunoglobulin (Sarma et al., 1971).

Edelman et al. (1969) have suggested that the Fab and Fc regions of the molecule are folded in a series of compact domains linked by less tightly folded stretches of peptide chain. This would agree with our finding that the intact molecule shows a higher swelling factor than its fragments (Table V). Obviously the individual compact fragments are linked together by a somewhat less tight region, which may act as a flexible hinge.

An indication that the fragments are linked by a flexible hinge was given by investigations on the F(ab')$_2$ fragment. If the two Fab fragments in this F(ab')$_2$ fragment were arranged end-to-end, as shown in Fig. 34, the F(ab')$_2$ would have a radius of gyration of 58.5 Å. Since the experimental value is only 53 Å the two fragments must be arranged at an angle to each other. An arrangement with an angle of 90° would, for instance, correspond to the experimental radius of gyration of 53 Å, although there is no necessity that this angle be fixed.

This result for the "model equivalent in scattering" leads us to expect a flexible hinge between the fragments. A hinged molecule of this kind, with a variable distance between the two antigen binding sites, would be more efficient at crosslinking antigens than would a rigid rod with a fixed geometry. A review on the structure of immunoglobulins is given by Pilz (1970).

Feinstein and Rowe (1965) and Valentine and Green (1967) also assume a flexible hinge between the fragments from electron microscopy. Feinstein and Rowe, however, observed a length of about 200 Å for the extended molecule, while Valentine and Green found a mean length of only 120 Å; the radius of gyration of such an immunoglobulin would be only 46 Å.

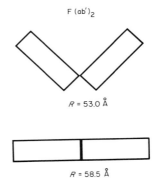

FIG. 34. Radii of gyration $R$ for different arrangements of the two Fab fragments in the F(ab')$_2$ fragment, obtained by digestion with pepsin.

## 2. α-Lactoglobulin

Witz et al. (1964) discussed various models for α-lactoglobulin based on the results of small-angle scattering. In acetate buffer of 0.1 ionic strength at room temperature, α-lactoglobulin is monomeric. The authors found very similar radii of gyration and molecular weights for both of the genetic variants, α-lactoglobulins A and B. The radius of gyration of α-lactoglobulin A was 21.6 ($\pm 0.4$) Å and its molecular weight was 36,600 ($\pm 1000$); the values for α-lactoglobulin B were 21.7 ($\pm 0.2$) Å and 36,900 ($\pm 1100$).

Green and Aschaffenburg (1959) proposed a model for α-lactoglobulin based on X-ray crystallographic studies. This model consists of two spheres, which not only touch each other but overlap somewhat. Below pH 3.7, α-lactoglobulin splits into two identical almost spherical subunits with molecular weights of 18,000, and each of these subunits represents a sphere in the model of Aschaffenburg and Green.

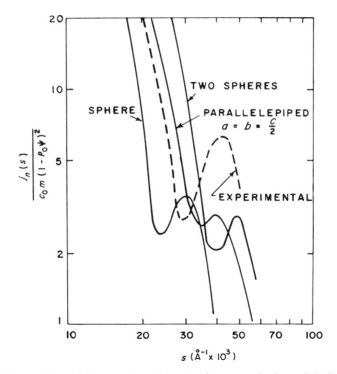

Fig. 35. Comparison of the experimental scattering curve for lactoglobulin (dashed line) with theoretical curves calculated for different models (solid lines) (Witz et al., 1964).

The radius of gyration of 21.7 Å obtained from small-angle scattering agrees with this two-sphere model. However, comparison of the tail end of the experimental scattering curve with the theoretical curve for this two-sphere model shows that it does not satisfactorily fit the data for α-lactoglobulin (Fig. 35); thus, there is no agreement in the height and position of the subsidiary maxima. In addition, the theoretical scattering curves for simple triaxial bodies, such as spheres or parallelepipeds, do not agree with the experimental curve, which indicates a more complex shape for α-lactoglobulin.

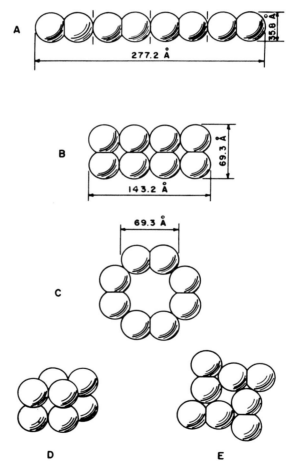

Fig. 36. Various models for α-lactoglobulin A tetramer. The monomeric units are represented in terms of the two-sphere model of Green and Aschaffenburg. The radii of gyration of the models are given in Table VI (Witz et al., 1964).

TABLE VI
CALCULATED RADII OF GYRATION ($R$) FOR VARIOUS TETRAMERIC MODELS[a]

| Model | $R$ (Å) |
|---|---|
| Sphere | 27.7 |
| A | 81.0 |
| B | 47.1 |
| C | 48.7 |
| D | 35.4 |
| E | 40.5 |
| Experimental value | 34.4 ± 0.4 |

[a] From Witz et al. (1964), with permission.

Witz et al. also investigated α-lactoglobulin A at pH 4.5 and 3°C in an acetate buffer of 0.1 ionic strength. Under these conditions a large part of the lactoglobulin is present as a tetramer. The scattering of the small percentage of monomer which is in equilibrium with the tetramer was eliminated using the known equilibrium constant and the radius of gyration and molecular weight of the monomer. In this way it was possible to arrive at a radius of gyration of 34.4 (±0.4) Å for the tetramer.

Witz et al. compared this value with the radii of gyration for various models. If one retains the simple model of Green and Aschaffenburg, which consists of two overlapping spheres, the tetramer is built up from eight spheres, each with a diameter of about 35 Å. In Fig. 36, various possible models are shown for α-lactoglobulin A, and in Table VI the calculated radii of gyration $R$ for these models are compared with the experimental value. The best agreement is obtained with a cubic arrangement of the eight spheres (model D). The two nonsymmetrical models (A and B) can be excluded for various reasons. They are shown only to demonstrate the effect of overall shape on the radius of gyration for particles of identical volume.

### 3. Ribosomes

Hill et al. (1969) have studied 70 S ribosomes and the 50 S and 30 S subunits from *Escherichia coli*. In Fig. 37, the apparent (slit-smeared) radii of gyration are plotted against the concentration. The scatter in the 70 S data seems to be due to a small amount of aggregation. After correcting the values for the smearing effect and extrapolation to infinite dilution the radii of gyration summarized in Table VII were found.

The dimensions of the ribosomes were estimated by comparing the experimental scattering curves with theoretical scattering curves for ellipsoids and cylinders. The best fit was with the models, the shapes and dimensions of which are given in Table VII.

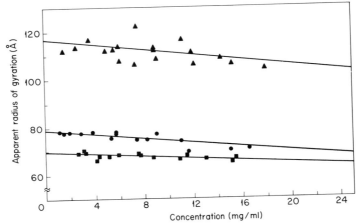

Fig. 37. An extrapolation to infinite dilution of the apparent radii of gyration (values of the slit-smeared scattering curves) of the 70 S ribosomes (▲), 50 S ribosomal subunits (●), and 30 S ribosomal subunits (■) (Hill et al., 1969).

From these results it was not obvious how to assemble the 30 S and 50 S subunits together in order to make a 70 S ribosome. Putting the two flattened ellipsoids side by side results in a particle which has too small a radius of gyration and is too symmetrical. Stacking the two ellipsoids on top of each other results in too large a radius of gyration and a ribosome which is too asymmetric.

The authors therefore suggested a model for the 70 S ribosome in which the 30 S subunit is nearly folded in half and then attached to the edge of the 50 S subunit in such a way as to form a hole or tunnel between the two particles. Such a molecule would have approximately the dimensions (135 Å × 200 Å × 400 Å) found for the 70 S ribosome.

TABLE VII
SIZE AND SHAPE OF *Escherichia coli* RIBOSOMES[a]

| Particle | $R$ (Å) | Shape | Dimensions (Å) |
|---|---|---|---|
| 70 S | 125 | elliptical cylinders 0.67:1:2 | 135 × 200 × 400 |
| 50 S | 84 | ellipsoid 0.75:1:1.8 | 130 × 170 × 310 |
| 50 S (with RNase) | 77 | ellipsoid 1:2:2 | 115 × 230 × 230 |
| 30 S | 69 | ellipsoid 1:4:4 | 55 × 220 × 220 |

[a] From Hill et al. (1969), with permission.

C. Changes in Conformation on Denaturation

As already mentioned, small-angle X-ray scattering is especially suitable for studying changes in the shape and size of macromolecules brought about, for example, by changes in temperature (Pilz et al., 1970b) or pH or by the addition of certain reagents.

### 1. Acid Denaturation of Myoglobin

Kirste et al. (1969) studied the changes in conformation of metmyoglobin caused by changes in pH. They found that myoglobin was stable in aqueous solution in the absence of salt between pH 7 and pH 4.6. Below pH 4.6, a weak association was observed; an actual change in conformation, the unfolding of the native structure, begins only at pH 3.7. Below pH 2, increased association occurs. In the presence of salt a somewhat different type of conformational change occurred.

The authors investigated the reversible acid denaturation using various methods and found that small-angle scattering was particularly suitable for observing the change in structure. The change involves two simultaneous events: molecular unfolding and molecular aggregation. Using small-angle X-ray scattering, it was possible to differentiate between aggregation (which leads to an increase in the average molecular weight and in the radius of gyration R) and the effect of unfolding (which leads only to an increase in the $R$ value while the $M$ value remains unchanged).

In Fig. 38, the change in the radius of gyration $R$ is plotted against pH for salt-free and for saline solutions of myoglobin at an ionic strength of 0.1.

### 2. Denaturation of Bovine Serum Albumin

Luzzati et al. (1961b) studied the change in conformation of bovine serum albumin on acid denaturation. At pH 5.3, the macromolecule is in a compact form, and the authors obtained a radius of gyration of 30.6 Å and a molecular weight of 81,000. When the pH is lowered to 3.6, the molecule shows completely different scattering characteristics. The radius of gyration increases considerably—from 30.6 Å at pH 5.3 to 68.5 Å at pH 3.6—while the molecular weight remains unchanged. This means that the molecule expands and unfolds extensively. From the shape of the scattering curve, the authors concluded that at pH 3.6 only a part of the molecule had unfolded, and the remaining part of the molecule still had the same compact structure as at pH 5.3. The unfolded portion of the molecule (about 35%) was considered to form a loosely folded coil surrounding the compact core of the folded molecule. This model of bovine

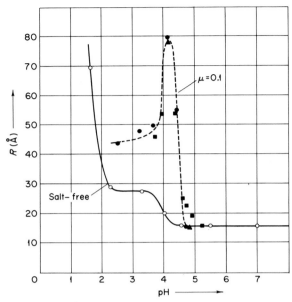

FIG. 38. Radius gyration of metmyoglobin solutions as a function of pH; solution without salt (○ and solid line); solution with salt, ionic strength 0.1 (dashed line); 0.1 N KCl (■), 0.1 N NaCl (●), 0.1 N acetate buffer (▲) (Kirste et al., 1969).

serum albumin at low pH values agreed with all the data that the authors assembled from small-angle scattering, such as radius of gyration, volume, molecular weight, surface and internal solvation.

The urea denaturation of bovine serum albumin has been studied by Echols and Anderegg (1960). The native protein was investigated using 0.2 M acetate buffer at pH 4.7 and the denaturation in 6 M urea solution. Echols and Anderegg, too, observed that during the denaturation the radius of gyration increased by 27% from 33.5 Å to 46 Å; urea caused a partial unfolding of the molecule by splitting hydrogen bonds. Attempts to renaturate the denatured bovine serum albumin by removing the urea by dialysis were unsuccessful because the denatured protein underwent an aggregation process involving the thiol group. However, prior treatment of the thiol groups of the native protein with $p$-chloromercuribenzoate prevented the aggregation process and the urea denaturation process could then be reversed on removal of the urea to yield a product which was indistinguishable from the native protein with respect to size and shape.

## D. Determination of the Cross-Section of Elongated Particles

Since small-angle scattering uses a wavelength of only 1.5 Å, one is using a very fine probe, and it is possible to obtain information not only about the molecule as a whole but also about the smaller dimensions of the molecule, for instance, the cross-section. Data on the cross-section of a particle provide a more detailed picture of its shape. With very long particles (several thousand angstroms), the length of which cannot be determined by small-angle scattering, at least the shape and mass per unit length of the cross-section can be determined. With associating proteins, the type of association may be determined from the cross-section, as will be shown with glutamate dehydrogenase.

### 1. *Glutamate Dehydrogenase*

In the range of concentrations suitable for small-angle scattering (1 mg/ml to 30 mg/ml), beef liver glutamate dehydrogenase exists as a mixture of aggregates, the size of which depends upon the concentration. In such a concentration-dependent association–dissociation equilibrium (Sund *et al.*, 1969) exact information on the subunit size of the enzyme is difficult to obtain, since the usual method—investigation of a concentration series and extrapolation to zero concentration— is difficult. Sund *et al.* (1969) investigated glutamate dehydrogenase in $1/15\,M$ phosphate buffer at pH 7.6. The concentration was varied from 1 to 33 mg/ml and average molecular weights of $0.5 \times 10^6$ to $2 \times 10^6$ were obtained; at the highest concentration we also found small amounts of still higher aggregates of molecular weight $4 \times 10^6$.

While the radius of gyration and molecular weight varied greatly in the concentration range under investigation, the cross-section curves showed almost identical slopes. These curves are shown in Fig. 39; the slopes of the Guinier straight lines and therefore also the radii of gyration of the cross-sections agree within the accuracy of measurement as shown in Table VIII. The masses per 1 Å length ($M/1$ Å) calculated from the intensities at zero angle were also identical (Table VIII). In Fig. 40, the $M/1$ Å and $R_q$ values are plotted as a function of glutamate dehydrogenase concentration. It can be seen that both the radii of gyration and masses per 1 Å length of the cross-section are independent of the concentration and thus independent of the size of the associated molecules. This proves that with this enzyme a linear aggregation takes place in the direction of the long axis while the cross-section remains unchanged.

By comparing the shape of the experimental cross-section curves with

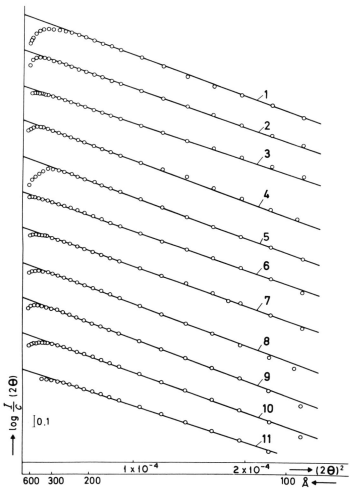

Fig. 39. Cross-section factors of glutamate dehydrogenase in $M/15$ phosphate buffer pH 7.6 in Guinier plot. The concentrations $c$ of the solutions are given in Table VIII. The values of the radii of gyration and of the mass per unit length $(M/1 \text{ Å})$ obtained are also summarized in the table and plotted *versus* the concentration in Fig. 40 (Sund et al., 1969).

theoretical curves for elliptical cross-sections of varying axial ratios it is also possible to obtain information on the shape of the cross-section. Figure 41 shows this comparison for solutions of various concentrations and it is seen that regardless of the concentration there is always a largely isotropic cross-section which is either circular or slightly elliptical. For

TABLE VIII

RADIUS OF GYRATION AND MASS PER UNIT LENGTH OF THE CROSS-SECTION OF GLUTAMATE DEHYDROGENASE. MEASUREMENTS IN PHOSPHATE BUFFER, pH 7.6[a]

| 1<br>No. | 2<br>$c$ (mg/ml) | 3<br>$T$ (°C) | 4<br>$R_q$ (Å) | 5<br>$M/1$ Å |
|---|---|---|---|---|
| 1 | 1.00 | 12.5 | 30.8 | (2200)[b] |
| 2 | 2.00 | 12.5 | 29.9 | — |
| 3 | 2.40 | 21.0 | 29.5 | 2300 |
| 4 | 5.05 | 12.5 | 30.7 | 2320 |
| 5 | 5.08 | 12.5 | 30.8 | 2350 |
| 6 | 10.10 | 12.5 | 30.2 | 2300 |
| 7 | 11.80 | 21.0 | 30.3 | 2300 |
| 8 | 20.50 | 12.5 | 30.7 | 2350 |
| 9 | 20.60 | 12.5 | 30.9 | 2400 |
| 10 | 29.70 | 12.5 | 30.5 | (2560) |
| 11 | 33.00 | 21.0 | 29.3 | 2340 |

[a] Sund et al. (1969).
[b] Values in parentheses are afflicted with larger errors.

a circular cross-section a diameter of 86 Å may be calculated from the value of $R_q$.

The data which can be obtained on the size and shape of the whole molecule are, because of the association–dissociation equilibrium, much less accurate; nevertheless, it could be shown from these data, too, that a linear association must take place. From the radii of gyration $R$ obtained for the whole particles and the known $R_q$ value, the average length of the aggregates could be calculated (see Section II,E). If these lengths are plotted against the average molecular weights (Fig. 42) it can be seen that there is a linear association; that is, the lengths increase in proportion to the molecular weights of the aggregates. Comparing the experimental scattering curves for the different degrees of aggregation with theoretical scattering curves for spherical cylinders of various axial ratios (Fig. 43) again shows that the association takes place in a way similar to that indicated in Fig. 42. The axial ratios correspond, to a first approximation, to the models shown in this figure.

Exact information on the shape of the glutamate dehydrogenase can, of course, only be obtained by preparing monodisperse solutions of the enzyme. This can be done by splitting the enzyme molecule into its smallest enzymatically active units by adding NADH and GTP. For these units, Pilz and Sund (1971) found a radius of gyration of 47 Å, a molecular weight of 300,000, and a ratio of length to diameter of 1.5:1. The

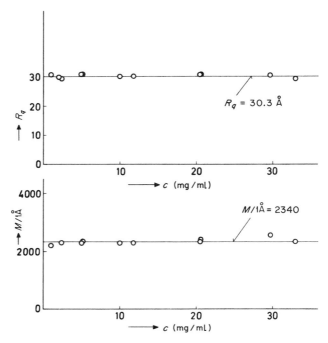

Fig. 40. Radius of gyration $R_q$ of the cross-section and mass per 1 Å length ($M/1$ Å) of glutamate dehydrogenase particles as a function of the protein concentration. Measurements were made in $1/15\,M$ phosphate buffer, pH 7.6 (Sund et al., 1969).

scattering curve of these units is in good agreement with the theoretical curve for the model suggested by Eisenberg (1970) as a result of electron microscope studies. In this model, the smallest enzymically active unit is composed of six ellipsoids, each representing a subunit. These six ellipsoids are arranged in such a way that they form a short rodlike particle, the mean diameter, radius of gyration of the cross-section, and mass per unit length of which correspond to the values given in Table VIII.

### E. Radial Electron Density Distribution by Fourier Transformation

For particles with spherical symmetry it is possible to calculate the radial electron density distribution by Fourier inversion of the scattered amplitudes (see Section II,D) as well as the usual data. Since the determination of this density distribution gives useful information about the conformation of a particle, some details and difficulties of this procedure are described in the following section.

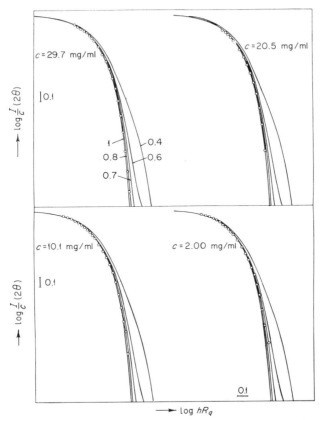

Fig. 41. Comparison of the experimental cross-section curves for glutamate dehydrogenase (○) with theoretical cross-section curves (full lines) for elliptical cylinders with axial ratios ($2a:2b$) between 1:1 and 1:0.4 in log–log plots. The concentrations $c$ of glutamate dehydrogenase are shown (Sund et al., 1969).

### 1. Protein Shells of Spherical Viruses

a. *General.* Anderegg (Anderegg et al., 1961, 1963), Fischbach et al. (1965), and White (1962) have studied a number of spherical viruses in solution and determined their radial density distribution to obtain information about the shapes and positions of their protein and RNA components. As already mentioned (see Figs. 5 and 7), the theoretical curves for particles with spherical symmetry (spheres and hollow spheres) show a number of subsidiary maxima and their intensity curves decrease to zero at their minima. For a Fourier inversion it is also necessary to measure the tail end of the scattering curve, where the intensity has

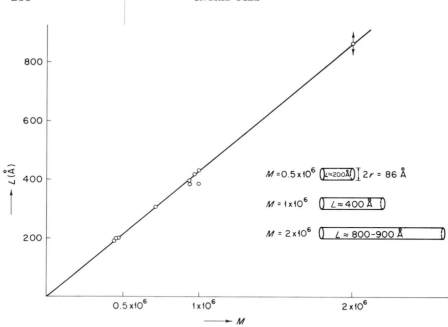

Fig. 42. Mean length $L$ of glutamate dehydrogenase particles in phosphate buffer, pH 7.6, as a function of the average molecular weight $M$ (Sund et al., 1969).

dropped by several orders of magnitude; the accuracy should be sufficient to include as many maxima as possible—at least four to five such maxima being desirable.

In Fig. 44, the slit-corrected scattering curve of bacteriophage R17 is compared with the scattering curve for a uniform sphere over a large angle range. The intensity of the numerous subsidiary maxima is, as is seen from the figure, 1000 to 10,000 times smaller than the intensity of the main maximum. It is therefore necessary, when studying this weak scattering at large angles, to use very concentrated solutions (the authors used concentrations up to 200 mg/ml) and wide slits; only under these conditions can sufficiently high intensities be obtained.

The scattering amplitudes necessary for the Fourier inversion can be obtained from the square root of the intensity curve. To decide whether the positive or negative square root at any given point is to be taken, however, additional information is necessary. Since it is known that spheres and hollow spheres always have a positive main maximum and subsidiary maxima which alternate in sign, the same choice of signs $(+, -, +, -, +, -, \ldots)$ is generally used for the scattering amplitudes.

Another difficulty in these calculations is as follows. The small devia-

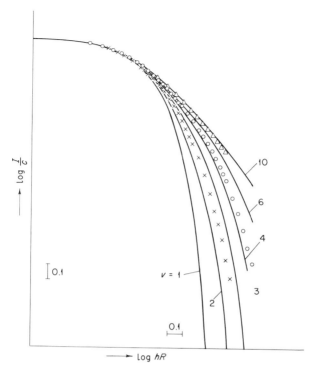

Fig. 43. Theoretical scattering curves of circular cylinders (full lines) in log–log plot; the ratios ($v = L/2r$) of length $L$ to the diameter $2r$ of the cylinder are 1, 2, 3, 4, 6, and 10. These curves are compared with experimental scattering curves for glutamate dehydrogenase particles with the following approximate molecular weights: $0.5 \times 10^6$ (×), $1 \times 10^6$ (○), and $2 \times 10^6$ (△) (Sund et al., 1969).

tions from exactly spherical symmetry, which occur usually with spherical macromolecules, have the effect of preventing the minima in the intensity curves from decreasing to zero, as shown in Fig. 44. In calculating the Fourier transform, only curves whose intensities are zero between the maxima can be used; that is to say, if the amplitude curve is to have successive sections which alternate in sign, then the amplitude curve and the intensity curve must pass through zero between these sections.

There are different procedures for obtaining curves which pass through zero between maxima. One method, used by Anderegg and his co-workers, is illustrated in Fig. 45. The amplitude curve is plotted both positively and negatively and a curve is drawn which fits the data near the maxima and passes through zero between them. This smoothed curve is then used in calculating the Fourier transform.

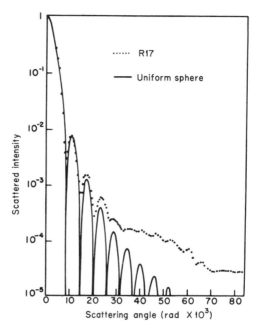

FIG. 44. Slit-corrected scattering curve for the bacteriophage R17 (Fischbach et al., 1965). The dots are representative points from the slit-corrected curve; the solid line is the theoretical scattering curve for a uniform sphere of 133 Å radius (Anderegg, 1967).

Another way is to draw a curve, as smoothly as possible, which connects the minima of the intensity curve. This curve is subtracted from the whole curve and only the residual curve is used for the Fourier transform. Zipper et al. (1971a), who also investigated bacteriophages R17 and fr, calculated radial density distributions and carefully compared the results using both procedures. In addition, they used scattering curves of models to study the influence on the radial density distribution of various procedures and approximations used in calculating results.

The abrupt cutoff in the amplitude curve is a further difficulty. This cutoff arises because it is not possible to measure more than five or eight maxima with sufficient accuracy, since their intensities become too weak. This abrupt cutoff in the amplitude curve produces errors in the transform calculated from it. The amplitude curve is therefore usually multiplied by an artificial temperature factor consisting of a suitable Gaussian curve.

Because of the many approximations required, the electron density distribution cannot be obtained with great accuracy; useful information

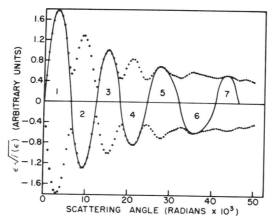

Fig. 45. Amplitude curve ($\epsilon \sqrt{i(\epsilon)}$) for the "top" component of wild cucumber mosaic virus (WCMV). The dots represent the slit-corrected data plotted as both positive and negative numbers. The solid line is the smooth curve that was used to calculate the Fourier transform; $\epsilon$ is the scattering angle (Anderegg, 1967).

about the conformation of macromolecular particles can nevertheless be obtained, as we shall now see.

*b. Squash Mosaic Virus.* Small plant viruses consist of a shell of protein surrounding a core of RNA. It has been known for a long time that plants infected with certain viruses produce not only the complete viruses but also other particles, which consist only of the protein shell and contain no nucleic acid. Mazzone et al. (1962) found, for example, that plants infected with squash mosaic virus (SMV) produce three kinds of macromolecular particles, which contain 0 (top component), $1.6 \times 10^6$ (middle component) and $2.4 \times 10^6$ (bottom component) Daltons of RNA, respectively. The components were called top, middle, and bottom components according to their sedimentation velocities, the bottom component being the complete virus particle, and the top component consisting only of the protein shell.

The scattering curves for the three components are shown in Fig. 46a. It is clear that with decreasing RNA content the height of the subsidiary maxima and thus the hollowness of the particle increases, indicating that the RNA is localized inside the virus particle. We can see this more clearly from the radial electron density distributions in Fig. 46b. The protein forms the outer shell and the top component (which contains no RNA) corresponds to a hollow sphere, with an inner radius of about 95 Å and an outer radius of 140 Å. If the particle contains RNA (middle and

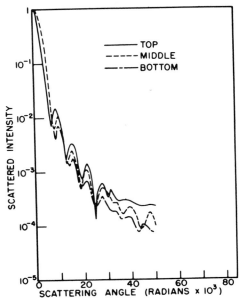

FIG. 46 (a) Slit-corrected scattering curve for "top," "middle," and "bottom" components of squash mosaic virus (SMV) (Anderegg, 1967). (b) The radial density distributions for "top," "middle," and "bottom" components of squash mosaic virus (SMV) normalized so that the ordinate represents "effective electron density." Six maxima with alternating signs were used in calculating the transform, giving a resolution of about 40 Å. An artificial temperature factor was used which consisted of multiplying the amplitude curve by a Gaussian so that the data at the cutoff angle were multiplied by 0.1 (Anderegg, 1967).

bottom component) there must be a region of RNA overlapping the protein between at least 95 Å and 110 Å.

Similar results were obtained by the authors with other viruses (Anderegg *et al.*, 1961, 1963; Anderegg, 1967) and by Zipper *et al.* (1971a).

c. *Bacteriophages fr and R17.* As already mentioned, Bacteriophages consist of a protein shell, which is composed of a number of protein subunits, surrounding a core of nucleic acid. Recently, small-angle scattering studies have made an important contribution to the clarification of a problem concerning simple icosahedral bacteriophages. Proceeding from the bacteriophages fr and R17 viruslike particles which are free of nucleic acid can be obtained in various ways. These particles differ considerably in their sedimentation behavior.

Thus, for particles which were formed by self-assembly from fr protein subunits, a sedimentation coefficient of 60 was found, while empty

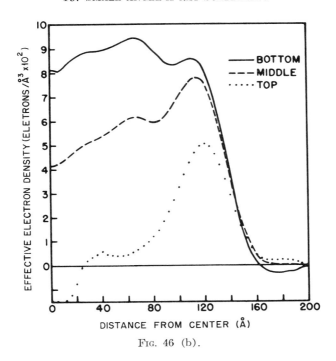

Fig. 46 (b).

protein shells, produced by alkaline treatment of the bacteriophages, had a sedimentation coefficient of 41. Shells could also be obtained by freezing and thawing and the sedimentation coefficients obtained in this way were 40–50.

The high sedimentation coefficient of the reconstituted particles, together with electron microscopic studies, led to the supposition that the reaggregated particles, in contrast to the empty shells produced by freezing and thawing or by alkaline treatment, contained no nucleic acid but had a core consisting of protein. The protein shell of the whole bacteriophage consisted of 180 subunits and it was presumed that the protein core inside the reaggregated particles consisted of 60 protein subunits.

It has recently been possible, using comparative small-angle scattering measurements on reconstituted particles and also on the empty shells produced by alkaline treatment of the bacteriophage fr, to refute this theory. As can be seen from the radial electron density distributions in Fig. 47 the reconstituted particles show only a slightly higher electron density inside than do the empty shells obtained by the alkaline method. It can be estimated roughly that the mass within the interior of the reaggregated particles is only about 5 to 6% of the total mass, corresponding to only about ten protein subunits (Zipper et al., 1971b).

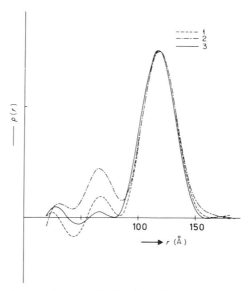

FIG. 47. Radial electron density distribution of empty protein shells obtained by alkaline treatment of phage fr (curve 1) and of viruslike particles aggregated from fr protein subunits (curve 2). Curve 3 represents the theoretical electron density distribution of a protein shell with a mean inner radius of 104.9 Å and a mean outer radius of 131.7 Å. The curves are normalized arbitrarily (Zipper *et al.*, 1971b).

Since, however, this small additional mass is insufficient to explain the high sedimentation coefficient of the reaggregated particles, there must be some other origin for the high S value. The latter may be due to differences in the effect of the outer hydration sheaths, which have a great influence on sedimentation coefficients but have very little effect on small-angle scattering. It is, of course, well known that sedimentation coefficients cannot be interpreted in terms of molecular weight alone (see Chapter 10 in Part B of this series).

Curve 3 in Fig. 47 corresponds to the theoretical electron density distribution for a protein shell with the dimensions determined from recent measurements on bacteriophages fr and R17 (inner radius, 104.9 Å; outer radius, 131.7 Å). Comparison with the two experimentally determined curves shows very good agreement.

### 2. *Ferritin and Apoferritin*

Ferritin, as is well known, consists of a hydrated iron oxide core $(FeOOH)_8 \cdot (FeO \cdot OPO_3H_2)$, surrounded by a protein shell. The investigation of this macromolecule by small-angle scattering is of great

interest, because the whole ferritin molecule, the protein shell by itself—
the so-called apoferritin—and finally the iron oxide core can each be
investigated. The inorganic core of the molecule cannot be obtained in
isolation without its protein shell, but it can nevertheless be investigated
by means of small-angle scattering.

Since the scattering of X-rays can be observed only when the electron
density of the solute particle is different from that of the solvent,
either of the two regions in the macromolecule (which differ in electron
density) can, as it were, be made invisible by making the electron density
of the solvent equal to it. In the case under discussion the electron
density of the aqueous solution can be increased to equal that of the protein shell by adding sucrose. The X-rays then "see" only the hydrated
iron oxide core.

These investigations were carried out first by Kratky and co-workers
(Kratky, 1960b; Bielig et al., 1963). They investigated ferritin first in
aqueous solution and then in 66% sucrose solution (66 gm sucrose in
100 ml) in which the scattering of the protein shell was suppressed.
It was shown that the radius of gyration of the core was substantially
smaller than that of the whole ferritin molecule, which was to be expected in view of the fact that the inorganic component formed the
inner core. The authors calculated a radius of gyration of 37 Å for
ferritin and 29 Å for the hydrated iron oxide core.

Bielig et al. (1964, 1966) also carried out extensive investigations on
apoferritin. The scattering curve showed a series of maxima between
which the intensity decreased nearly to zero, indicating a radial electron
density distribution. A Fourier inversion of the scattering amplitudes
could therefore be carried out as described above and the radial electron

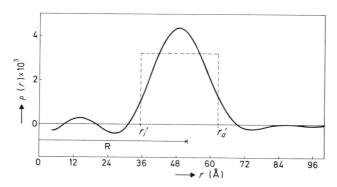

FIG. 48. Electron density distribution $\delta(r)$ for apoferritin obtained by Fourier
transformation. $R$ is the calculated radius of gyration (Bielig et al., 1966).

density distribution could be calculated. The result is illustrated in Fig. 48; the apoferritin corresponds to a hollow sphere with a radius of gyration of 56 Å, an inner radius $r_i$ of 35.5 Å, and an outer radius $r_a$ of 63 Å. These values agree quite closely with those obtained from the X-ray crystallographic studies by Harrison (1963) ($r_i = 37$ Å, $r_a = 61$ Å) and those values obtained later by Fischbach and Anderegg (1965) also using small-angle X-ray scattering ($r_i = 37$ Å, $r_a = 61$ Å). The molecular weight and volume also showed close agreement with the values found by other methods.

Fischbach and Anderegg (1965) also investigated the hydrated iron oxide core of ferritin in the same manner as Kratky and co-workers using 66% sucrose solution, which has an electron density matching that of the protein so that no scattering of the protein takes place. In Fig. 49, the scattering amplitude of apoferritin is plotted as a function of the electron density of the sucrose solution. It can be clearly seen how the scattering of the apoferritin disappears as the concentration or electron density of the sucrose solution increases. When the whole ferritin molecule is investigated in a solution of electron density 0.41 electrons per cubic angstrom (corresponding to 66 gm sucrose in 100 ml), the scattering curve observed will be derived solely from the ferric hydroxide core.

On the basis of electron micrographs, many authors have suggested that the ferric hydroxide core (or micelle), the diameter of which is 27–30 Å, is composed of four to six spherical subunits. Fischbach and Anderegg (1965) compared the experimental scattering curve of the micelle with theoretically calculated curves for such models and also

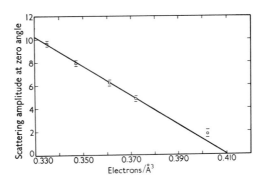

Fig. 49. Amplitude of forward scattering from apoferritin in sucrose solutions of varying sucrose concentration, plotted as a function of the electron density of the sucrose solution and extrapolated so as to determine the electron density at which the scattering amplitude becomes zero (Fischbach and Anderegg, 1965).

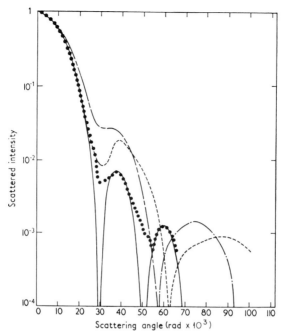

Fig. 50. The scattering from the full ferritin micelle compared to the scattering curves for several theoretical models. The solid dots represent the scattering from full ferritin in a 66% sucrose solution; in such a solution the scattering from the protein is negligible and the scattering is due only to the micelle. The solid line (———) represents the theoretical scattering from a uniform sphere of 36.6 Å radius. The dashed line (– – –) represents the theoretical scattering from six spheres at the vertices of an octahedron. The dash–dot line (– · –) represents the theoretical scattering from four spheres at the vertices of a tetrahedron. The size of each model has been adjusted to have the same radius of gyration as the full ferritin micelle.

for a uniform sphere with a radius of 36.6 Å. Figure 50 shows that the experimental scattering curve of the ferric hydroxide core fits the scattering of a uniform sphere much better than it fits either subunit model.

If the ferric hydroxide micelle, which can best be simulated by a uniform sphere of radius 37 Å, were composed of subunits, then these must, in the author's opinion, be asymmetrical and show an approximately overall spherical shape only when assembled together. The outer diameter of 73 Å for the micelle corresponds to the inner diameter of the apoferritin. As a point of interest it might be mentioned that Fischbach *et al.* (1969) also investigated ferritin crystals and found that the shape of the ferric hydroxide micelle is clearly determined by the form of the inner protein wall; this gives rise to the notion that the ferric oxide core grows only in the interior of the protein.

## F. CHARACTERISTICS OF HOLLOW BODIES

In Section II,H it was explained in some detail how small-angle scattering could be used to determine whether a macromolecule had sponge-like geometry with very small holes or alternatively whether it contained macroscopic hollow spaces. Studies on hemocyanin and fatty acid synthetase are given below as examples; both proteins can be described as hollow bodies.

### 1. *Hemocyanin Helix Pomatia*

Hemocyanin, the blue dye from snails, is stable between pH 4.8 and 6.9. Above and below this pH range the molecule splits into fragments corresponding to about one half, one tenth and one twentieth of the whole particle, as Lontie and Witters (1966) were able to determine with the ultracentrifuge. Small-angle X-ray scattering studies (Pilz *et al.*, 1970c) on hemocyanin in $H_2O$ and in 0.05 $M$ borate buffer at pH 8.2 yielded the following results:

*a. The Whole Molecule in $H_2O$.* The scattering curve illustrated in Fig. 51b exhibits a larger number of subsidiary maxima from which one may deduce that this protein molecule has a symmetrical geometry. Comparison with the theoretical scattering curves of simple bodies shows almost perfect agreement with a circular cylinder, the ratio of whose diameter to height is 1:1.1. Hemocyanin *Helix pomatia* thus has the shape of a highly isotrope circular cylinder.

From the scattering curve and the absolute intensity the following data could also be obtained:

Radius of gyration $R = 164$ Å
Molecular weight $M = 9 \times 10^6$
Volume $V = 1.74 \times 10^7$ Å$^3$
Swelling factor $f_q = 1.51$ (this is 0.44 gm $H_2O$ per 1 gm protein)

From these values it can be quite simply calculated that the hemocyanin molecule cannot have a uniform density. Using the radius of gyration $R$ and the ratio of diameter $2r$ to height $h$, the dimensions and volume of the corresponding solid cylinder can be calculated. Such a cylinder has a volume of $3.66 \times 10^7$ Å$^3$ and this is twice as large as the experimentally determined volume of $1.74 \times 10^7$ Å$^3$. The latter value corresponds, as stated above, to the swollen protein with 0.44 gm $H_2O$ per 1 gm hemocyanin.

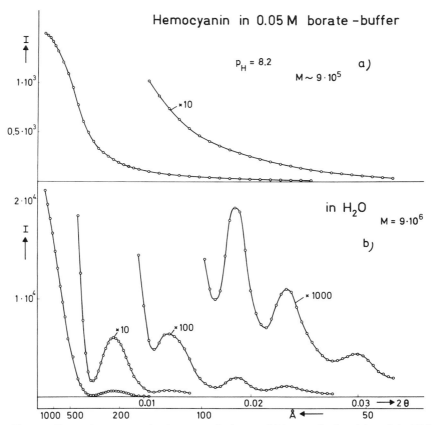

Fig. 51. Scattering curves for hemocyanin in 0.05 $M$ borate buffer (a) and in $H_2O$ (b). The intensity of the tail end of the curves is enlarged 10, 100, and 1000 times. The molecular weight of hemocyanin in water is $9 \times 10^6$ and in borate buffer is approximately $9 \times 10^5$ (Pilz et al., 1970c).

This contradiction can be resolved only if one assumes that there is a hollow space in the interior of the cylinder and that the scattering protein mass is concentrated on the outside. Since the radius of gyration is, as already mentioned, the root-mean-square of the distances of all the scattering electrons from the electronic center of gravity of the particle, a hollow body has a large $R$ value even though the protein volume is relatively small.

A hollow cylinder with a height of 360 Å, an inner radius of 74 Å, and an outer radius of 164 Å would, for instance, provide a model agreeing very well with the small-angle scattering data. Whether or not the shape of the hemocyanin can adequately be described by a hollow

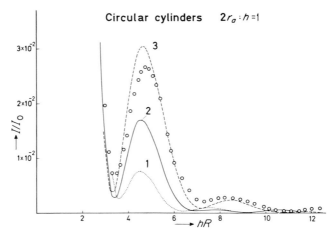

FIG. 52. Comparison of the experimental scattering curve (○) for hemocyanin in water with theoretical curves for different circular cylinders, the ratio of diameter to length of which is 1:1. Curve 1 (dotted line) corresponds to a solid cylinder, while curve 2 (solid line) corresponds to a hollow cylinder with a ratio of inner radius to outer radius $r_i/r_a$ of 0.3 and curve 3 (dashed line) to a hollow cylinder with a ratio of $r_i/r_a$ of 0.5 (Pilz et al., 1970c).

cylinder can be tested by comparing the experimentally found subsidiary maxima with those of the corresponding theoretical scattering curves.

While the scattering curves for hollow and solid cylinders with identical axial ratios show hardly any differences in their main maxima, the hollow structure is clearly distinguishable by the height of its subsidiary maximum (see Section II,C,1c and Fig. 7). In Fig. 52, the middle portion of the experimental curve for hemocyanin is compared with theoretical curves for spherical cylinders. Curve 1 corresponds to a solid cylinder and curves 2 and 3 to hollow cylinders with ratios of inner to outer radius $(r_i/r_a)$ of 0.3 and 0.5. The height of the subsidiary maxima increases greatly with hollowness and the experimental curve clearly indicates a hollow structure for hemocyanin.

While the subsidiary maxima shown in Figs. 51 and 52 are indicative of an overall shape for the hemocyanin of *Helix pomatia* which is best described by a hollow cylinder, there is also a periodicity appearing at very large angles arising from the relatively small subunits of this giant protein. When a protein consists of a large number of more or less identical subunits a long-wave periodicity (minima) of this kind can be observed at large angles (see Section II,C,2,*c*). From the position of these minima the radius of gyration of the subunits can be calculated. Assuming, as a first approximation, that the subunits are spherical, their diameter is found to be 40 Å. The number of subunits can be roughly estimated by comparing the volume of protein in the whole hemocyanin molecule with the volume of a single subunit; this leads to an estimate of 400 subunits in the giant molecule. Various models can be considered in which these subunits are arranged in different ways; models which do not agree with the scattering curve and other data found for the whole molecule can be excluded and in this way a model equivalent in scattering can be arrived at (Pilz et al., 1972).

Electron micrographs of *Helix pomatia* hemocyanin also show a hollow cylinder in which a subunit structure can be clearly recognized (van Bruggen, 1968). From these micrographs van Bruggen has suggested a model with 180 identical spherical subunits. This model can be excluded, since its scattering curve differs clearly from the experimental one (Pilz et al., 1972).

*b. Subunits of Hemocyanin (tenths, halves).* Until recently, it has not been possible to split the hemocyanin molecule into the smallest subunits from which it appears to be constructed, although it is possible to split the molecule into larger subunits. And it is helpful when calculating curves for models to determine the shape and size of all the fragments and subunits into which the molecule can be split. At pH 8.2 we have obtained fractions with a molecular weight of about $9 \times 10^5$, which is only one-tenth that of the whole molecule. The scattering curve of these fractions is shown in Fig. 51 (top curve). All of the subsidiary maxima have disappeared, showing that the subunits present at this pH do not have a uniform isotropic shape.

Recently a homogenous solution of the halves of the hemocyanin helix pomatia was investigated at pH 6.2. All data obtained (molecular weight, radius of gyration, volume) agree exactly with a model of a circular hollow cylinder possessing the same inner and outer diameter as the whole molecule but only half the height. Further, the form of the scattering curve of this model agrees very well in the main maximum and three subsidiary maxima found for the experimental one. These results indicate clearly that the whole molecule, best described by a hollow cylinder (height 360 Å, outer radius 164 Å and inner radius 74 Å), splits into halves by dividing the cylinder vertically to the cylinder axis, obtaining two flat cylinders of only half the height. (Pilz et al., 1973c.)

## 2. *Fatty Acid Synthetase*

Lynen *et al.* (1964) succeeded in isolating the multienzyme complex of the fatty acid synthetase from yeast and in describing its mechanism of action. The synthesis of long fatty acid chains takes place by the cooperative action of at least seven different enzymes. The conformation of this enzyme complex is therefore of special interest.

Electron micrographs obtained by Hofschneider and Hagen (1965) show slightly oval particles with a regular structure. The interpretation of these electron micrograph images in terms of a three-dimensional model has not been possible until recently. Small-angle scattering studies on fatty acid synthetase have been carried out by Pilz et al. (1970a) with this object in view.

The inner portions of the log $I$ vs. $(2\theta)^2$ plot of the scattering curves were linear over a large range of angles, indicating a largely isotrope shape. Disturbances due to interference effects were very small, so that an accurate extrapolation to zero concentration could be made (as already shown in Figs. 19 and 20). The detailed data obtained are summarized in Table IX. The molecular weight of $2.2 \times 10^6$ is in close agreement with the value determined by ultracentrifugation ($2.3 \times 10^6$). Comparison with theoretical scattering curves also showed that the overall shape of fatty acid synthetase is largely isotropic; closest agreement was found with the theoretical curve for an ellipsoid with an axial ratio of 1:1:1.2 (Fig. 53). From this axial ratio and the other data one may calculate overall dimensions of $230 \text{ Å} \times 230 \text{ Å} \times 280 \text{ Å}$ which agree fairly well with the values of $210 \times 250 \text{ Å}$ obtained from the electron micrographs.

Just as with hemocyanin, a simple combination of these data leads one to the conclusion that there are hollow spaces in the fatty acid synthetase complex. From the calculated volume of $3.5 \times 10^6 \text{ Å}^3$ and the axial ratio of 1:1:1.2, one arrives at a radius of gyration (for a solid particle) of $R = 70 \text{ Å}$, which is much smaller than the experimental value of 108 Å.

On the other hand, if the value of 108 Å is used together with the molecular weight, the volume calculated for the solid particle is three times too large ($11 \times 10^6 \text{ Å}^3$ instead of $3.5 \times 10^6 \text{ Å}^3$) and the swelling is ten times too great (2.2 gm $H_2O$ instead of only 0.22 gm $H_2O$ per 1 gm protein).

These contradictions can be eliminated only if one presumes that there are hollow spaces containing solvent in the inner portion of the particle.

TABLE IX
Data for Fatty Acid Synthetase Obtained by Small-Angle Scattering[a]

| | | |
|---|---|---|
| Molecular weight | $M$ | $2.2 \times 10^6$ |
| Volume | $V$ | $3.5 \times 10^6 \text{ Å}^3$ |
| Swelling factor | $f_q$ | 1.3; (corresponding to 0.22 gm $H_2O$ per gram protein) |
| Radius of gyration | $R$ | 108 Å |

[a] Pilz et al. (1970a).

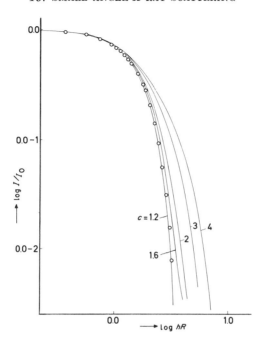

FIG. 53. Comparison of the experimental scattering curve (○) for fatty acid synthetase in 0.1 $M$ phosphate buffer with theoretical curves for ellipsoids with the axial ratios $1:1:c$; the values of $c$ are as indicated (Pilz et al., 1970a).

This supposition is borne out by the finding that the experimental subsidiary maximum occurs at the same position as those for isotropic models (spheres or slightly isotropic ellipsoids of revolution) but that the height is considerably greater than that for isotropic models and this phenomenon is typical of hollow bodies (Fig. 54).

A further indication that fatty acid synthetase must possess hollow spaces is shown by the appearance of thickness factors (Fig. 55). In general, thickness factors are characteristic of flat particles in which two dimensions are clearly larger than the third (see Section II,E,2). The fact that thickness factors (from which an average wall thickness of about 20 Å can be calculated) appear in a particle which is isotropic, can be explained only if we assume that the complex is not compact but contains hollow spaces.

The multienzyme complex, however, cannot be described simply in terms of a hollow ellipsoid of revolution, since hollow spheres or slightly anisotrope hollow ellipsoids of revolution possess more clearly defined subsidiary maxima. The enzyme must have a more complex structure. Some indication of the complexity is seen in the electron micrographs

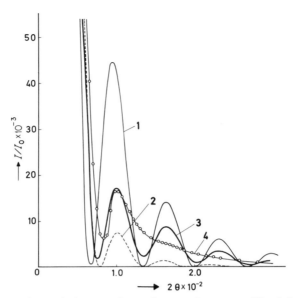

Fig. 54. Comparison of the experimental scattering curve (4) of fatty acid synthetase in 0.1 $M$ phosphate buffer, pH 6.5, with theoretical scattering curves for the following models: (1) hollow sphere with an inner radius of 105 Å and an outer radius of 125 Å; (2) solid sphere; (3) scattering curve for the model described in the text. The models have the same radius of gyration as was determined experimentally for the fatty acid synthetase (Pilz et al., 1970a).

in which one or more walls in the ovals of the ellipse can be observed.

We have therefore tried to arrive at a spatial model for fatty acid synthetase which agrees with small-angle scattering data and electron microscopy. Our model, which is only schematic, consists essentially of a hollow sphere with an average wall thickness of 20 Å. The hollow sphere is divided into two halves by a planar wall also 20 Å thick, as seen in the electron micrographs. The scattering curve for such a model is shown in Fig. 54. The height of the first subsidiary maximum agree well with the experimental curve; in its tail end the theoretical curve oscillates around the experimental curve. This shows that the model is not entirely adequate, though agreement can be improved by modifying the conformation in detail.

We have also demonstrated (Pilz et al., 1970a) that thickness factors are observed not only in the scattering curves of flat lamellar particles (Porod, 1949) but also in bent lamellar regions, such as those occurring in the walls of hollow spheres. Thickness factors are not normally observable with hollow spheres because too many subsidiary maxima appear

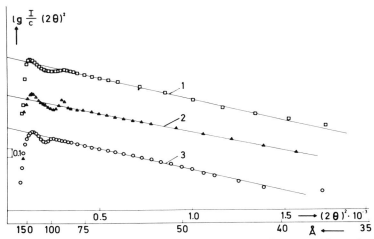

FIG. 55. Thickness factors for the scattering curves of fatty acid synthetase in 0.1 $M$ phosphate buffer, pH 6.5. The concentrations of the enzyme are: 9.84 mg/ml (curve 1), 18.3 mg/ml (curve 2), and 43.0 mg/ml (curve 3) (Pilz et al., 1970a).

as a result of their highly symmetrical shape. If symmetry is sufficiently disturbed, however, thickness factors occur, as can be clearly seen in Fig. 56. Sufficient disturbance of the symmetry was achieved in the following way: scattering curves were calculated for hollow spheres with a constant wall thickness (outer radius $r_a$ less inner radius $r_i$) of 20 Å, while the outer radius was varied from 100 Å up to 200 Å.

The figure shows the Guinier plot of the thickness factor $\log [I \cdot (2\theta)^2]$ vs. $(2\theta)^2$; in this plot the radius of gyration of the thickness $R_t$ can be obtained from the slope of the straight line; by multiplying $R_t$ by $12^{1/2}$ the thickness $t$ itself can be obtained from Eq. (10). Curve 1 shows the scattering curve for an ideal hollow sphere ($r_a = 125$ Å, $r_i = 105$ Å); because of the high symmetry this scattering curve shows definite maxima between which the intensity goes to zero and no thickness factor is seen. In curve 2 the scattering curves for five different hollow spheres were superimposed, with equal weights for each. In curve 3 the scattering curves for 11 and in curve 4 for 20 different hollow spheres were superimposed in the same way; the radii of all the hollow spheres are given in Table X. In curve 4 the symmetry has been sufficiently disturbed to calculate the radius of gyration of the thickness $R_t$ from the slope of the straight line which has been sketched in. The thickness itself is obtained by multiplying by $12^{1/2}$ and leads to the exact value of 20 Å for the wall thickness of the hollow spheres.

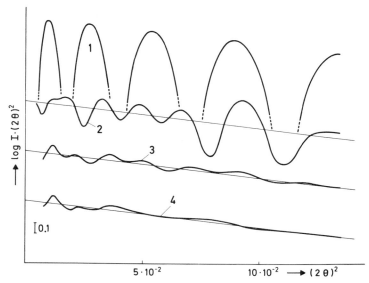

Fig. 56. Thickness factors of theoretical scattering curves for hollow spheres. (Curve 1) Hollow sphere with an outer radius $r_a = 125$ Å and an inner radius $r_i = 105$ Å. In curves 2, 3, and 4, scattering curves for hollow spheres of constant wall thickness $(r_a - r_i = 20$ Å$)$ and different radii are superimposed; the radii are summarized in Table X. In the superposition equal amounts by weight of the different hollow spheres were used. The thin solid lines correspond to theoretical thickness factors of lamellar particles which have a thickness of 20 Å (Pilz et al., 1970a).

## G. Allosteric Effects and Similar Changes in Conformation

As already pointed out, small-angle X-ray scattering is specially suitable for investigating changes in the conformation of macromolecules. Improvements in measurement techniques in recent years have increased their precision, making it possible to observe relatively small changes in molecular structure. As Durchschlag et al. (1969, 1971) have recently demonstrated, allosteric effects can now be investigated by small-angle scattering.

### 1. Allosteric Effect with GPDH

As a rule, the conformational changes occurring in allosteric mechanisms are so small that the accuracy of measurement is seldom sufficient to determine them by absolute methods. By using comparative measurements, however, an appreciably higher degree of accuracy can be obtained if the investigations are carried out under absolutely identical conditions.

TABLE X

OUTER RADIUS ($r_a$) AND INNER RADIUS ($r_i$) FOR HOLLOW SPHERES ($r_a - r_i = 20$ Å): THE SCATTERING CURVES OF THESE SPHERES WERE SUPERIMPOSED AND THE THICKNESS FACTORS WERE PLOTTED IN FIG. 56 IN CURVES 1 TO $4^a$

| Curve 1 | | Curve 2 | | Curve 3 | | Curve 4 | |
|---|---|---|---|---|---|---|---|
| $r_a$ (Å) | $r_i$ (Å) | $r_a$ (Å) | $r_i$ (Å) | $r_a$ (Å) | $r_i$ (Å) | $r_a$ (Å) | $r_i$ (Å) |
| 125 | 105 | 100 | 80 | 100 | 80 | 100 | 80 |
| | | 120 | 100 | 110 | 90 | 105 | 85 |
| | | 150 | 130 | 120 | 100 | 110 | 90 |
| | | 170 | 150 | 130 | 110 | 115 | 95 |
| | | 200 | 180 | 140 | 120 | 120 | 100 |
| | | | | 150 | 130 | 125 | 105 |
| | | | | 160 | 140 | 130 | 110 |
| | | | | 170 | 150 | 135 | 115 |
| | | | | 180 | 160 | 140 | 120 |
| | | | | 190 | 170 | 145 | 125 |
| | | | | 200 | 180 | 150 | 130 |
| | | | | | | 155 | 135 |
| | | | | | | 160 | 140 |
| | | | | | | 165 | 145 |
| | | | | | | 170 | 150 |
| | | | | | | 175 | 155 |
| | | | | | | 180 | 160 |
| | | | | | | 185 | 165 |
| | | | | | | 190 | 170 |
| | | | | | | 200 | 180 |

$^a$ Pilz et al. (1970a).

Durchschlag et al. studied the allosteric behavior (Kirschner et al., 1961) of yeast glyceraldehyde 3-phosphate dehydrogenase (GPDH) at pH 8.5 and 40°C. The coenzyme NAD was added to the apoenzyme and small-angle measurements were made at the following degrees of saturation with NAD: 0.0 (apoenzyme), 0.23, 0.46, 0.72, and 0.99 (holoenzyme). In this way, a volume contraction of the GPDH was observed which depended on the degree of saturation. The volume was determined from the scattering intensity at zero angle and the invariant was determined from Eq. (13), as described in Section II,F. The areas under the curves in Fig. 57 are due to the values of the invariant for samples with the four different degrees of NAD saturation. It is seen that the area and, therefore, the invariant and the volume change with the degree of saturation. In Table XI the volumes for the different degrees of saturation with NAD are summarized and the change in volume from the apoenzyme (degree of saturation zero) to the holoenzyme (degree of

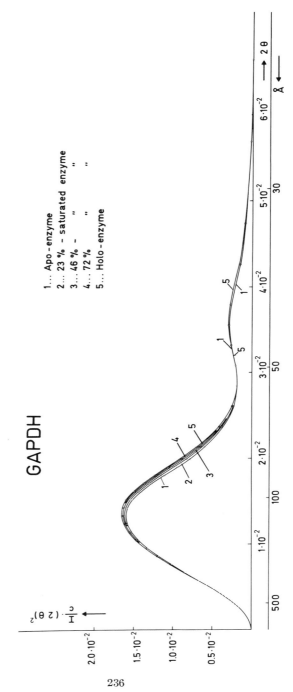

Fig. 57. Invariants of glyceraldehyde 3-phosphate dehydrogenase (GPDH) at pH 8.5 and 40°C as a function of the indicated degrees of saturation with NAD (Durchschlag et al., 1971).

TABLE XI

VOLUME OF GLYCERALDEHYDE 3-PHOSPHATE DEHYDROGENASE IN DEPENDENCE OF THE DEGREE OF SATURATION WITH NAD[a]

| Degree of saturation with NAD | Volume (Å³) | Degree of volume contraction (%) |
|---|---|---|
| 0 (apoenzyme) | $2.64 \times 10^5$ | 0 |
| 0.23 | $2.58 \times 10^5$ | 32 |
| 0.46 | $2.53 \times 10^5$ | 60 |
| 0.72 | $2.49 \times 10^5$ | 82 |
| 0.99 (holoenzyme) | $2.45 \times 10^5$ | 100 |

[a] Durchschlag et al. (1969).

saturation 0.99) is expressed as a percentage. In Fig. 58, the volume contraction is plotted as a function of saturation with NAD and it may be noted that the change in volume or the degree of structural change is not linearly related to the degree of saturation. Since a simple sequential mechanism (Koshland et al., 1966) would require such a linear relationship, this mechanism is excluded by the authors (Durchschlag et al., 1969) and an allosteric mechanism (Monod et al., 1965) is assumed, in which a linear change with the degree of saturation is required.

## 2. Changes in Conformation of Antibodies upon Reaction with Antigen

The interaction of the determinant group on the antigen with the combining site of the antibody represents a highly specific recognition at a

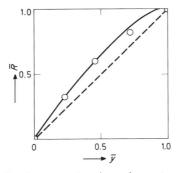

FIG. 58. The degree of volume contraction of yeast apoglyceraldehyde 3-phosphate dehydrogenase as a function of saturation with NAD. $\bar{R}$ = degree of structural change, $\bar{Y}$ = degree of saturation with NAD. The diagonal represents the predicted relationship for the sequential mechanism of cooperative binding.

molecular level. It is known that antibodies are capable of inducing profound conformational changes in the antigens upon binding.

Recently it could be shown by X-ray small angle scattering that, similarly to the induction of changes in antigen conformation by antibodies, there occur also a change in the antibody conformation on the binding of antigen. Specific antibodies (IgG-globulins) were studied in the presence and absence of the antigen; the antigen was a poly-D-alanine (monovalent hapten), a small molecule with a molecular weight of only 350. Since the molecular weight of the antibody was determined to be 150,000 no change in the small angle data would be expected upon the binding of two haptens on the two combining sites (compare Fig. 30) showed clearly, however, that the radius of gyration and the volume of the antibody become smaller as a result of the reaction with the hapten. If 90% of the antibody combining sites were occupied by the hapten there was a decrease in the R-value of 8% and a decrease in the volume of 10% (Pilz et al., 1973b). On the other hand no change in the typical cross-section curves of IgG-globulins could be observed; the course of all curves was similar to that shown in Fig. 29 for other IgG-globulins. This indicates that a contraction of the antibody molecule occurs upon interaction with antigen, without changing the typical T-shape (see Fig. 30 and Fig. 33, model 9) of the molecule. The same effect was observed with another specific antibody and another antigen (Pilz et al., 1973a).

## V. Prospects

Small-angle X-ray scattering is a very young method. Only in the past decade has its development reached a stage at which very dilute protein solutions may be investigated accurately. Progress is rapid and it is likely that during the next few years the method will become a standard one for investigating macromolecular solutes.

Some important points for current and future development may be summarized as follows:

1. The service and adjustment of small-angle cameras require simplification.

2. Scattering data, which are recorded automatically, should be used as input data in computer programs which correct them as far as possible for all collimation and polychromatic effects and then calculate the various data of the dissolved macromolecule directly and compare its scattering curve with curves for various models. For these purposes, it is necessary to correct automatically for the variations in intensity which may arise during the measurement of scattering curves or to eliminate

these from the start by making combined measurements, for instance, using a small-angle X-ray diffractometer with a monitor. Fluctuations in the intensity of the incident beams are eliminated using a second radiation detector which measures a quantity proportional to the intensity of the incident beam.

3. As it is now possible, using computers, to calculate not only scattering curves of triaxial models but models of any shape rapidly and systematically it is feasible to make an accurate study of the influence of changes in shape and subunit structure on the form of the scattering curve. This will make it possible to get more detailed information about the nature of the changes in conformation occurring as a result of denaturation or allosteric effects.

ACKNOWLEDGMENTS

The author is indebted to Prof. O. Kratky for his valuable advice and discussion. She also wishes to thank Miss Müller for drawing the figures.

REFERENCES

Alexander, L. E. (1969). "X-Ray Diffraction Methods in Polymer Science." Wiley (Interscience), New York.
Anderegg, J. W. (1967). In "Small-Angle X-Ray Scattering" (H. Brumberger, ed.), p. 243. Gordon and Breach, New York.
Anderegg, J. W., Geil, P. H., Beeman, W. W., and Kaesberg, P. (1961). *Biophys. J.* **1**, 657.
Anderegg, J. W., Wright, M., and Kaesberg, P. (1963). *Biophys. J.* **3**, 175.
Beeman, W. W. (1967). In "Small-Angle X-Ray Scattering" (H. Brumberger, ed.), p. 197. Gordon & Breach, New York.
Beeman, W. W., Kaesberg, P., Anderegg, J. W., and Webb, M. B. (1957). In "Handbuch der Physik" (S. Flügge, ed.), Vol. 32, p. 321. Springer-Verlag, Berlin and New York.
Bielig, H. J., Kratky, O., Steiner, H., and Wawra, H. (1963). *Monatsh. Chem.* **94**, 989.
Bielig, H. J., Kratky, O., Rohns, G., and Wawra, H. (1964). *Tetrahedron Lett.* **38**, 2701.
Bielig, H. J., Kratky, O., Rohns, G., and Wawra, H. (1966). *Biochim. Biophys. Acta* **112**, 110.
Blake, C. C. F., Koenig, D. F., Mair, G. A., North, A. C. T., Phillips, D. C., and Sarma, V. R. (1965). *Nature (London)* **206**, 757.
Bonse, U., and Hart, M. (1965). *Appl. Phys. Lett.* **6**, 155.
Bonse, U., and Hart, M. (1966). *Z. Phys.* **189**, 151.
Bonse, U., and Hart, M. (1967). In "Small Angle X-Ray Scattering" (H. Brumberger, ed.), p. 121 Gordon & Breach, New York.
Brew, K., Vanaman, T. C., and Hill, R. L. (1967). *J. Biol. Chem.* **242**, 3747.
Burge, R. E., and Draper, J. C. (1967). *Acta Crystallogr.* **22**, 6.
Chonacky, N. J., and Beeman, W. W. (1969). *Acta Crystallogr., Sect. A* **25**, 564.
Cleeman, J. C., and Kratky, O. (1960). *Z. Naturforsch.* **15b**, 525.

Cohen, G., and Eisenberg, H. (1968). *Biopolymers* **6**, 1077.
Conrad, H., Mayer, A., Schwaiger, S., and Schneider, R. (1969a). *Hoppe-Seyler's Z. Physiol. Chem.* **350**, 845.
Conrad, H., Mayer, A., Thomas, H. P., and Vogel, H. (1969b). *J. Mol. Biol.* **41**, 225.
Cullis, A. F., Muirhead, H., Perutz, M. F., Rossmann, M. G., and North, A. C. T. (1962). *Proc. Roy. Soc., Ser. A* **265**, 161.
Damaschun, G. (1964). *Naturwissenschaften* **51**, 378.
Damaschun, G. (1965). *Exp. Tech. Phys.* **13**, 224.
Damaschun, G., and Müller, J. J. (1965). *Z. Naturforsch.* **20a**, 1274.
Damaschun, G., and Pürschel, H. V. (1968). *Acta Biol. Med. Ger.* **21**, 567; *ibid.* **21**, 401.
Damaschun, G., and Pürschel, H. V. (1969). *Monatsh. Chem.* **100**, 510.
Damaschun, G., Kley, G., Müller, J. J., and Pürschel, H. V. (1968a). *Acta Biol. Med. Ger.* **20**, 409.
Damaschun, G., Kley, G., and Müller, J. J. (1968b). *Acta Phys. Austr.* **28**, 233.
Debye, P. (1925). *J. Math. Phys.* **4**, 133.
Debye, P., and Bueche, A. M. (1949). *J. Appl. Phys.* **20**, 518.
Du Mond, J. W. M. (1947). *Phys. Rev.* **72**, 83.
Durchschlag, H., Puchwein, G., Kratky, O., Schuster, I., and Kirschner, K. (1969). *FEBS Lett.* **4**, 75.
Durchschlag, H., Puchwein, G., Kratky, O., Schuster, I., and Kirschner, K. (1971). *Eur. J. Biochem.* **19**, 9.
Echols, G. H., and Anderegg, J. W. (1960). *J. Amer. Chem. Soc.* **82**, 5085.
Edelman, G. M., Cunningham, B. A., Gall, W. E., Gottlieb, P. D., Rutishauser, U., and Waxdal, M. J. (1969). *Proc. Nat. Acad. Sci. U. S.* **63**, 78.
Ehrenberg, W., and Franks, A. (1952). *Nature (London)* **170**, 1076.
Eisenberg, H. (1970). *In* "Pyridine Nucleotide Dependent Dehydrogenases" (H. Sund, ed.), p. 293. Springer-Verlag, Berlin and New York.
Eisenberg, H., and Cohen, G. (1968). *J. Mol. Biol.* **37**, 355.
Federov, B. A. (1968). *Kristallografiya* **13**, 763.
Federov, B. A., Andreeva, N. A., Volkova, L. A., and Voronin, L. A. (1968). *Kristallografiya* **133**, 770.
Feinstein, A., and Rowe, A. J. (1965). *Nature (London)* **205**, 147.
Filmer, D. L., and Kaesberg, P. (1962). *Enzyme Models and Enzyme Struct.* **15**, 210.
Fischbach, F. A., and Anderegg, J. W. (1965). *J. Mol. Biol.* **14**, 458.
Fischbach, F. A., Harrison, P. M., and Anderegg, J. W. (1965). *J. Mol. Biol.* **13**, 638.
Fischbach, F. A., Harrison, P. M., and Hoy, T. G. (1969). *J. Mol. Biol.* **39**, 235.
Franks, A. (1955). *Proc. Phys. Soc., London, Sect. B* **68**, 1054.
Franks, A. (1958). *J. Appl. Phys.* **9**, 349.
Gerold, V. (1957). *Acta Crystallogr.* **10**, 287.
Glatter, O. (1972). *Acta Phys. Austr.* **36**, 307.
Glatter, O. (1973). *J. Appl. Cryst.* (submitted for publication).
Green, D. W., and Aschaffenburg, R. (1959). *J. Mol. Biol.* **1**, 54.
Guinier, A. (1937). *C. R. Acad. Sci.* **204**, 1115.
Guinier, A. (1939). *Ann. Phys. (Paris)* [11] **12**, 161.
Guinier, A. (1943). *J. Chim. Phys.* **40**, 133.
Guinier, A. (1946). *C. R. Acad. Sci.* **223**, 31.
Guinier, A., and Fournet, G. (1947). *J. Phys. Radium* **8**, 345.
Guinier, A., and Fournet, G. (1955). "Small Angle Scattering of X-rays." Wiley, New York.

Haager, O. (1970). Thesis Universitat Graz.
Harrison, P. M. (1963). *J. Mol. Biol.* **6**, 404.
Heine, S., and Roppert, J. (1962). *Acta Phys. Austr.* **15**, 148.
Hendricks, R. W. (1971). *J. Appl. Cryst.* (in press).
Hermans, P. H., Heikens, D., and Weidinger, A. (1959). *J. Polym. Sci.* **35**, 145.
Hill, W. E., Thompson, J. D., and Anderegg, J. W. (1969). *J. Mol. Biol.* **44**, 89.
Hofschneider, P., and Hagen, A. (1965). *Proc. Reg. Conf. (Eur.) Electron Microsc., 3rd, 1964* pp. 69-70.
Hossfeld, F. (1967). Thesis, Technische Hochschule Aachen.
Hossfeld, F. (1968). *Acta Crystallogr. Sect. A* **24**, 643.
Hyman, A., and Vaughan, P. A. (1967). *In* "Small Angle X-Ray Scattering" (H. Brumberger, ed.), p. 477. Gordon & Breach, New York.
Jagodzinsky, H., and Wohlleben, K. (1960). *Z. Elektrochem.* **64**, 212.
Johansson, J. (1933). *Z. Phys.* **82**, 587.
Katz, L. (1958). Ph.D. Thesis, University of Wisconsin, Madison.
Kendrew, J. C., Dickerson, R. E., Strandberg, B. E., Hart, R. G., and Davies, D. R. (1960). *Nature (London)* **185**, 422.
Kent, P., and Brumberger, H. (1964). *Acta Phys. Austr.* **17**, 263.
Kirschner, K., Eigen, M., Bittmann, R., and Voigt, B. (1961). *Proc. Nat. Acad. Sci. U. S.* **56**, 1661.
Kirste, R. G., and Stuhrmann, H. B. (1967). *Z. Phys. Chem. (Frankfurt am Main)* [N. S.] **56**, 334.
Kirste, R. G., Schulz, G. V., and Stuhrmann, H. B. (1960). *Z. Naturforsch.* **24b**, 1385.
Klimanek, P. (1964). *Ann. Phys. (Paris)* [13] **14**, 71.
Koshland, D. E., Jr., Nemethy, G., and Filmer, D. (1966). *Biochemistry* **5**, 365.
Kratky, C., and Kratky, O. (1964). *Z. Instrumentenk.* **72**, 302.
Kratky, O. (1947). *Monatsh. Chem.* **76**, 325.
Kratky, O. (1954). *Z. Elektrochem.* **58**, 49.
Kratky, O. (1955a). *Naturwissenschaften* **42**, 237.
Kratky, O. (1955b). *Kolloid-Z.* **144**, 110.
Kratky, O. (1956). *Z. Elektrochem.* **60**, 245.
Kratky, O. (1958). *Z. Elektrochem.* **62**, 66.
Kratky, O. (1960a). *Angew. Chem.* **72**, 467.
Kratky, O. (1960b). *Makromol. Chem.* **35A**, 12.
Kratky, O. (1963). *Progr. Biophys. Biophys. Chem.* **13**, 105.
Kratky, O. (1964). *Z. Anal. Chem.* **201**, 161.
Kratky, O. (1967). *In* "Small Angle X-ray Scattering" (H. Brumberger, ed.), p. 70. Gordon & Breach, New York.
Kratky, O., and Pilz, I. (1955). *Z. Naturforsch.* **10b**, 389.
Kratky, O., and Porod, G. (1948). *Acta Phys. Austr.* **2**, 133.
Kratky, O., and Porod, G. (1953). *In* "Die Physik der Hochpolymeren" (H. A. Stuart, ed.), Vol. 2, p. 515. Springer-Verlag, Berlin and New York.
Kratky, O., and Skala, Z. (1958). *Z. Elektrochem.* **62**, 73.
Kratky, O., and Wawra, H. (1963). *Monatsh. Chem.* **94**, 981.
Kratky, O., Porod, G., and Kahovec, L. (1951). *Z. Elektrochem.* **55**, 53.
Kratky, O., Pilz, I., Sekora, A. (1955a). *Z. Naturforsch.* **10b**, 510.
Kratky, O., Porod, G., Sekora, A., and Paletta, B. (1955b). *J. Polym. Sci.* **16**, 163.
Kratky, O., Porod, G., and Skala, Z. (1960). *Acta Phys. Austr.* **13**, 76.
Kratky, O., Pilz, I., Schmitz, P. J., and Oberdorfer, R. (1963). *Z. Naturforsch.* **18b**, 180.

Kratky, O., Wawra, H., Pilz, I., Sekora, A., and van Deinse, A. (1964). *Monatsh. Chem.* **95**, 359.
Kratky, O., Pilz, I., and Schmitz, P. J. (1966). *J. Colloid Interface Sci.* **21**, 24.
Kratky, O., Leopold, H., and Stabinger, H. (1969). *Z. Angew. Phys.* **27**, 273.
Kraut, J., Wright, H. T., Kellerman, M., and Freer, S. T. (1967). *Proc. Nat. Acad. Sci. U. S.* **58**, 304.
Krigbaum, W. R., and Godwin, R. W. (1968). *Biochemistry* **7**, 3126.
Krigbaum, W. R., and Kügler, F. R. (1970). *Biochemistry* **9**, 1216.
Krivacic, J., and Rupley, J. A. (1968). *J. Mol. Biol.* **35**, 483.
Lake, J. A. (1967). *Acta Crystallogr.* **23**, 191.
Lely, J. A., and van Ryssel, T. W. (1951). *Philips Tech. Rev.* **13**, 96.
Leonard, B. R. (1951). Ph.D. Thesis, University of Wisconsin, Madison.
Leopold, H. (1965). *Elektronik* **14**, 359.
Lontie, R., and Witters, R. (1966). *In* "Biochemistry of Copper" (J. Peisach, P. Aisen, and W. E. Blumberg, eds.), p. 455. Academic Press, New York.
Luzzati, V. (1958). *Acta Crystallogr.* **11**, 843.
Luzzati, V. (1960). *Acta Crystallogr.* **13**, 939.
Luzzati, V. (1963). *In* "Third International Symposium on X-Ray Optics and X-Ray Microanalysis" (H. H. Pattee, J. E. Cosslett, and A. Engström eds.), p. 133. Academic Press, New York.
Luzzati, V., and Baro, R. (1961). *J. Phys. Radium* [8] **22**, 186a.
Luzzati, V., Witz, J., and Nicolaieff, A. (1961). *J. Mol. Biol.* **3**, 367.
Luzzati, V., Witz, J., and Nicolaieff, A. (1961b). *J. Mol. Biol.* **3**, 379.
Lynen, F., Hopper-Kessel, I., and Eggerer, H. (1964). *Biochem. Z.* **340**, 95.
Malmon, A. G. (1957a). *Acta Crystallogr.* **10**, 639.
Malmon, A. G. (1957b). *Biochim. Biophys. Acta* **26**, 233.
Mazzone, H. M., Incardona, N. L., and Kaesberg, P. (1962). *Biochim. Biophys. Acta* **14**, 1.
Mittelbach, P. (1964). *Acta Phys. Austr.* **19**, 53.
Mittelbach, P., and Porod, G. (1961a). *Acta Phys. Austr.* **14**, 185.
Mittelbach, P., and Porod, G. (1961b). *Acta Phys. Austr.* **14**, 405.
Mittelbach, P., and Porod, G. (1962). *Acta Phys. Austr.* **15**, 122.
Mittelbach, P., and Porod, G. (1965). *Kolloid-Z.* **202**, 40.
Monod, J., Wyman, J., and Changeux, J. P. (1965). *J. Mol. Biol.* **12**, 88.
Muirhead, H., Cox, J. M., Mazzarella, L., and Perutz, M. F. (1967). *J. Mol. Biol.* **28**, 117.
Noelken, M. E., Nelson, C. A., Buckley, C. E., and Tanford, C. (1965). *J. Biol. Chem.* **240**, 218.
Oster, G., and Riley, D. P. (1951). *Acta Crystallogr.* **5**, 1.
Peret, R., and Ruland, W. (1971). *Bien. Conf. Carbon, 10th,* 1971 Abstract SS-91.
Perutz, M. F. (1963). *Angew. Chem.* **75**, 589.
Perutz, M. F. (1970). *Nature (London)* **228**, 726 and 734.
Pilz, I. (1969). *J. Colloid Interface Sci.* **30**, 140.
Pilz, I. (1970). *Allg. Prakt. Chem.* **21**, 21.
Pilz, I., and Kratky, O. (1967). *J. Colloid Interface Sci.* **24**, 211.
Pilz, I., and Sund, H. (1971). *Eur. J. Biochem.* **20**, 561.
Pilz, I., Herbst, M., Kratky, O., Oesterhelt, D., and Lynen, F. (1970a). *Eur. J. Biochem.* **13**, 55.
Pilz, I., Kratky, O., Cramer, F., van der Haar, F., and Schlimme, E. (1970b). *Eur. J. Biochem.* **15**, 401.

Pilz, I., Kratky, O., and Moring-Claesson, I. (1970c). *Z. Naturforsch.* **25b**, 600.
Pilz, I., Puchwein, G., Kratky, O., Herbst, M., Haager, O., Gall, W. E., and Edelmann, G. M. (1970d). *Biochemistry* **9**, 211.
Pilz, I., Glatter, O., and Kratky, O. (1972). *Z. Naturforsch.* **27b**, 518.
Pilz, I., Kratky, O., and Karush, F., (1973a). *Eur. J. Biochem.* (in preparation).
Pilz, I., Kratky, O., Licht, A., and Sela, M. (1973b). *Biochemistry* (in preparation).
Pilz, I., Lontie, R., and Engelborghs, Y. (1973c). (In preparation).
Porod, G. (1948). *Acta Phys. Austr.* **2**, 255.
Porod, G. (1949). *Z. Naturforsch.* **6b**, 401.
Porod, G. (1951a). *Kolloid-Z.* **124**, 83.
Porod, G. (1951b). *Kolloid-Z.* **125**, 51.
Porod, G. (1960). *Makromol. Chem.* **35**, 1.
Porteus, J. O. (1968). *J. Appl. Phys.* **39**, 163.
Ritland, H. N., Kaesberg, P., and Beeman, W. W. (1950a). *J. Appl. Phys.* **21**, 838.
Ross, P. A. (1928). *J. Opt. Soc. Amer.* **16**, 433.
Rupley, J. A. (1968). *J. Mol. Biol.* **35**, 455.
Rupley, J. A., and Gates, V. (1968). *J. Mol. Biol.* **35**, 477.
Sarma, R., Silverton, E. W., Davies, D. R., and Terry, W. D. (1971). *J. Biol. Chem.* **246**, 3753.
Schelten, J., and Hossfeld, F. (1971). *J. Appl. Cryst.* **4**, 210.
Schmidt, P. W. (1955). *Acta Crystallogr.* **8**, 772.
Schmidt, P. W. (1965). *Acta Crystallogr.* **19**, 938.
Schmidt, P. W., and Hight, R. (1959). *J. Appl. Phys.* **30**, 866.
Schmidt, P. W., and Hight, R. (1960). *Acta Crystallogr.* **13**, 480.
Shaffer, L. (1963). Ph.D. Thesis, University of Wisconsin, Madison.
Shehedrin, B. M., and Feigin, L. A. (1966). *Kristallografiya* **11**, 159.
Shull, C. G., and Roess, L. C. (1947). *J. Appl. Phys.* **18**, 295.
Sparks, C. J. (1966). Metals and Ceramics Div. Ann. Progr. Rept., ORNL-3970, p. 57.
Stabinger, H., Leopold, H., and Kratky, O. (1967). *Monatsh. Chem.* **98**, 436.
Stuhrmann, H. B. (1970a). Habilitation paper, University of Mainz.
Stuhrmann, H. B. (1970b). *Acta Crystallogr., Sect. A* **26**, 297.
Stuhrmann, H. B., and Kirste, R. G. (1965). *Z. Phys. Chem. (Frankfurt am Main)* [N.S.] **46**, 247.
Stuhrmann, H. B., and Kirste, R. G. (1967). *Z. Phys. Chem. (Frankfurt am Main)* [N.S.] **56**, 334.
Sund, H., Pilz, I., and Herbst, M. (1969). *Eur. J. Biochem.* **7**, 517.
Syneček, V. (1960). *Acta Crystallogr.* **13**, 378.
Thomas, H. P. (1967). *Z. Phys.* **208**, 338.
Valentine, R. C., and Green, N. M. (1967). *J. Mol. Biol.* **27**, 615.
van Bruggen, E. F. J. (1968). *In* "Physiology and Biochemistry of Haemocyanins" (F. Ghiretti, ed.), p. 37. Academic Press, New York.
Vonk, C. G. (1971). *J. Appl. Cryst.* **4**, 340.
von Nordstrand, R. A., and Hach, K. M. (1953). *123rd Meet., Amer. Chem. Soc. Chicago* p. 112.
White, R. A. (1962). Ph.D. Thesis, University of Wisconsin, Madison.
Witz, J., Timasheff, S. N., and Luzzati, V. (1964). *J. Amer. Chem. Soc.* **86**, 168.
Worthington, C. R. (1956). *J. Sci. Instrum.* **33**, 66.
Zipper, P. (1969). *Acta Phys. Austr.* **30**, 143.
Zipper, P., Kratky, O., Herrmann, R., and Hohn, T. (1971a). *Eur. J. Biochem.* **18**, 1.
Zipper, P., Kratky, O., and Schubert, D. (1971b). *FEBS Lett.* **14**, 219.

# 20 □ Pulsed Nuclear Magnetic Resonance

W. J. O'SULLIVAN,

K. H. MARSDEN, and J. S. LEIGH, Jr.

|  |  |
|---|---|
| Glossary of Symbols | 246 |
| I. Introduction | 247 |
| II. Scope of the Technique | 247 |
|    A. Background | 248 |
|    B. Some Definitions | 250 |
|    C. Information Potentially Available from PRR Studies | 251 |
|    D. Limitations of the PRR Method | 253 |
| III. Theory of Relaxation | 253 |
|    A. Basic Concepts | 253 |
|    B. Phenomenological Description of Relaxation | 257 |
|    C. Principles of Measurement of $T_1$ and $T_2$ | 258 |
|    D. Relaxation Theory Applied to Paramagnetic Systems | 262 |
| IV. Experimental Methods | 267 |
|    A. Apparatus | 267 |
|    B. The Measurement of $T_1$ | 269 |
|    C. The Measurement of $T_2$ | 271 |
| V. Experimental Design | 273 |
|    A. Preliminary Considerations | 273 |
|    B. Experiments with Phosphoenolpyruvate Carboxykinase | 275 |
|    C. Assessment of Data | 276 |
|    D. Temperature Dependence of PRR | 279 |
|    E. Frequency (Field) Dependence of PRR | 281 |
|    F. Formation of a Binary Complex: Mn–Bovine Serum Albumin | 282 |
|    G. Formation of a Ternary Complex: MnADP–Creatine Kinase | 283 |
|    H. Protein Modification | 284 |
|    I. Spin Labels | 285 |
| VI. Individual Enzymes | 286 |
|    A. Creatine Kinase | 286 |
|    B. Pyruvate Kinase | 290 |
|    C. Alcohol Dehydrogenase | 291 |
|    D. Application of Variable Frequency Measurements | 293 |
| VII. Absorption of Water by Diamagnetic Molecules | 294 |
|    A. Scope of the Problem | 295 |

B. Experimental Aspects . . . . . . . . . . 295
C. The Keratin–Water System . . . . . . . . . 296
References . . . . . . . . . . . . . 298

## Glossary of Symbols

| | |
|---|---|
| $T_1$ | longitudinal proton relaxation time |
| $T_2$ | spin–spin proton relaxation time |
| $\epsilon$ | enhancement factor |
| $T_{1,p}$ | contribution of paramagnetic species to observed $T_1$ of water molecules |
| $p$ | mole fraction of water molecules in the first coordination sphere of a paramagnetic species |
| $T_{1,M}$ | relaxation time of water protons in the first coordination sphere of a paramagnetic species |
| $T_{1,0}$ | relaxation time of water protons in the bulk solvent |
| $T_{1,os}$ | outer sphere contribution to the relaxation time |
| $\tau_M$ | residence time of a water molecule in the first coordination sphere of a paramagnetic species |
| $\tau_c$ | correlation time |
| $\tau_r$ | rotational correlation time |
| $\tau_s$ | electron spin relaxation time |
| $\epsilon_a$ | enhancement of a binary metal ion complex (small molecule) |
| $\epsilon_b$ | enhancement of a binary metal–enzyme complex |
| $\epsilon_t$ | enhancement of a ternary enzyme–metal–substrate complex |
| $I$ | angular momentum quantum number |
| $\mu$ | magnetic moment |
| $H_0$ | external magnetic field |
| $M_0$ | magnetization |
| $\nu$ | frequency |
| $\omega$ | angular frequency |
| $\gamma$ | gyromagnetic ratio |
| $t$ | time |
| $\tau_E$ | correlation time for scalar interaction |
| $S$ | electronic spin quantum number |
| $r$ | ion–proton internuclear distance |
| $\beta$ | Bohr magneton |
| $g$ | electronic "g" factor |
| $A$ | hyperfine coupling constant |
| M | metal ion |
| MS | metal–substrate complex |
| EM | metal–enzyme complex |
| ES | enzyme–substrate complex |
| EMS | ternary complex, containing enzyme, metal, and substrate |
| $K_D$ | constant for dissociation of EM |
| $K_1$ | constant for dissociation of MS |
| $K_s$ | constant for dissociation of ES |
| $K_2$ | constant for dissociation of MS from EMS |
| $\beta$ | width parameter |
| $\tau_v$ | correlation time for electron relaxation |

## I. Introduction

Nuclear magnetic resonance (NMR) represents probably the most readily accessible spectroscopic technique capable of observing individual atoms within a molecule. It has the potential advantage over the X-ray method that it can provide not only structural but also dynamic information about the motion of molecules in solution. In its familiar form, the recording of proton spectra, it has been widely applied to the study of small molecules of biological importance (Kowalsky and Cohn, 1964). The large number of protons present, giving rise to overlapping signals, has somewhat limited its application to the study of protein molecules, though these limitations are being overcome by improvements in instrumentation and experimental technique (Bradbury and Crane-Robinson, 1968; Sheard and Bradbury, 1970; McDonald and Phillips, 1970; Metcalfe, 1970).

The broad scope of NMR techniques, with the emphasis on the continuous-wave method, was described by Metcalfe (1970) in Part B of this series. The present chapter is intended to focus attention on a different aspect of NMR, the use of pulsed techniques to measure the nuclear spin relaxation time. Pulsed techniques are used to obtain direct measurements of the longitudinal and spin–spin relaxation times $T_1$ and $T_2$, respectively, of solution protons, usually water protons. While both of these parameters can be obtained from continuous-wave NMR, the ease of measurement and greatly increased accuracy with the pulsed technique, particularly for $T_1$, has made it an important tool in the study of the interaction of paramagnetic species, particularly metal ions, with proteins. Some recent discussions of the applications of pulsed NMR have been presented by Mildvan and Cohn (1970) and Cohn (1970).

## II. Scope of the Technique

The usefulness of the pulsed technique lies in the fast and accurate estimate of one parameter $T_1$, the time constant for a nuclear (usually proton) population that is disturbed in a magnetic field, to reach a new equilibrium state. Relatively speaking, a large population of protons is needed to observe a meaningful signal. The importance of the technique, from a biochemist's point of view, is that the medium to which it is most easily applied is water.

The value of $T_1$ is changed by changing the environment of the water molecules. This may be done in one of two ways; by adsorbing the water molecules at an interface or trapping them in interstices, or by perturbing the environment of the water protons by the introduction of a powerful local field, viz., a paramagnetic entity. A paramagnetic species in solution has a profound effect on $T_1$ of water protons and much of this chapter will be concerned with the use of paramagnetic probes though some attention will be directed to the effects of immobilization of water molecules by binding to diamagnetic macromolecules.

In the case of the paramagnetic species, one is usually dealing with changes at a very specific site on the protein corresponding to the active site of an enzyme. The effects are fairly short range, being principally concerned with the water molecules that are in direct contact with the paramagnetic species, e.g., in the first coordination sphere of a metal ion.

Two classes of paramagnetic species are relevant to the discussion.

(1) One class is the paramagnetic metal ions. In principle, any transition metal ion would have a detectable effect, provided it has, bound in its first coordination sphere, water molecules which exchange rapidly with the bulk solvent. In practice, the metal ion that has proved most useful is manganous ion because of its large effect on $T_1$ when it is complexed (see Section II,B) and because of its ability to act as the activating metal ion in a large number of enzymic reactions, particularly those that utilize adenosine triphosphate (ATP) as a substrate.

(2) The other class is stabilized free radicals. Originally developed by McConnell and his colleagues (Ohnishi and McConnell, 1965; McConnell and McFarland, 1970) to observe changes in electron paramagnetic resonance (EPR) spectra, spin labels have often been found to have a large effect on the value of $T_1$ for water. This has been taken advantage of by attaching a spin label to the reactive thiol groups of creatine kinase (Taylor *et al.*, 1969) and in studies on the binding of a spin-labeled analog of nicotinamide adenine dinucleotide (NAD) to alcohol dehydrogenase (Weiner, 1969; Mildvan and Weiner, 1969a,b). The last few years have seen a considerable increase in the number of spin labels available and this should be an area of increasing importance. Naturally occurring free radicals, such as semiquinones and flavin radicals, are also potentially useful.

A. BACKGROUND

The background of the utility of the measurement of $T_1$ stems from the observation that the binding of some paramagnetic metal ions to macromolecules resulted in a decreased value of $T_1$, that is, an increase in the proton relaxation rate (PRR), rather than the decreased PRR which would be expected if the only change were due to the replacement of water molecules in the first coordination sphere of the metal ion by other ligands. Thus, Eisinger et al. (1962) observed that the binding of $Mn^{2+}$ to DNA produced a substantial increase in the PRR of the solution. They described the effect as an enhancement $\epsilon$ of the PRR of water protons due to the formation of the Mn–DNA complexes. At about the same time, Cohn and Leigh (1962) reported similar effects for the binding of $Mn^{2+}$ to proteins. They demonstrated a basic distinction between two metal-activated enzymes, enolase and creatine kinase. It was found that the addition of enolase to a manganous chloride solution caused a large increase in the PRR, i.e., an increased enhancement. However, the addition of phosphoenolpyruvate (PEP), the substrate of the enzyme, to a solution containing $Mn^{2+}$ and enzyme caused a decrease in the enhancement. That is to say, the enhancement $\epsilon_b$ of the binary Mn–E complex was greater than the enhancement $\epsilon_t$ of the ternary complex containing enzyme, metal, and substrate. The behavior of creatine kinase was quite different. In this case, very little change was seen with metal ion and enzyme alone. However, on the addition of either of the nucleotide substrates, ADP or ATP, a substantial increase in the PRR was observed; i.e., in this case $\epsilon_t$ was greater than $\epsilon_b$. (ADP and ATP both

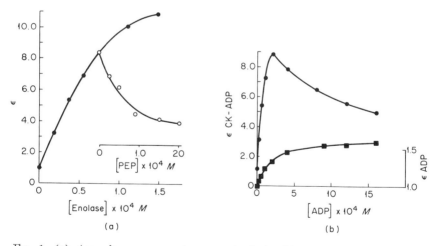

Fig. 1. (a) Ascending curve, enhancement of the binary complex Mn–enolase. Descending curve, enhancement of the ternary Mn–enolase–phosphoenolpyruvate complex with enolase concentration constant at $0.74 \times 10^{-4} M$. Experiments were carried out in $0.05 M$ Tris–HCl, pH 7.5, $0.5 M$ KCl with $1 \times 10^{-4} M$ $MnCl_2$ in a total volume of 0.1 ml. (b) Enhancement as a function of ADP concentration in the absence and presence of rabbit muscle creatine kinase. Experiments were carried out in $0.05 M$ Tris–HCl, pH 7.5, with $1 \times 10^{-4} M$ $MnCl_2$ and $0.78 \times 10^{-4} M$ creatine kinase. [Reproduced with permission from Cohn, *Biochemistry* **2**, 623 (1963).]

produced small effects with $Mn^{2+}$.) These results are illustrated in Fig. 1, taken from Cohn (1963).

The distinction between these two enzymes has formed the basis of classifying metal-activated enzymes into type I ($\epsilon_t > \epsilon_b$), typified by creatine kinase, and type II ($\epsilon_b > \epsilon_t$), typified by enolase and pyruvate kinase (Cohn, 1963). This distinction is elaborated in Section II,C.

## B. Some Definitions

The parameter most often measured is $T_1$, the longitudinal proton relaxation time. However, it has become customary to speak of changes in the proton relaxation rate (PRR), i.e., changes in $1/T_1$. When dealing with solutions containing paramagnetic species, comparisons are made between the paramagnetic contributions to the PRR. A correction for the bulk solution must always be made, thus

$$1/T_{1,p} = 1/T_1 - 1/T_{1,0} \tag{1}$$

where $1/T_1$ is the measured PRR and $1/T_{1,0}$ is the PRR of the bulk solvent.

The enhancement $\epsilon$ observed on the complexing of $Mn^{2+}$ is defined as

$$\epsilon = \frac{(1/T_{1,p})^*}{1/T_{1,p}} = \frac{(1/T_1 - 1/T_{1,0})^*}{1/T_1 - 1/T_{1,0}} \tag{2}$$

where the asterisk indicates the presence of the complexing agent.

Three factors determine the PRR in solutions containing paramagnetic species; (1) the relative number $p$ of water molecules in the first coordination sphere, (2) the time of residence $\tau_M$ of a particular water molecule in the first coordination sphere (the inverse of the chemical exchange rate) and (3) the relaxation time $T_{1,M}$ of a water proton in the first coordination sphere. These terms are related by the expression (O'Reilly and Poole, 1963; Luz and Meiboom, 1964)

$$1/T_{1,p} = p/(T_{1,M} + \tau_M) \tag{3}$$

$T_{1,M}$ is inversely proportional to the correlation time, $\tau_c$, which for manganous ion, is dominated by the rotational correlation time $\tau_r$ that is

$$1/\tau_c = 1/\tau_r + 1/\tau_s + 1/\tau_M \tag{4}$$

as $\tau_s$ is of the order of $10^{-8}$ sec and $\tau_r$ is of the order of $10^{-11}$ sec (Bernheim et al., 1959).

The binding of $Mn^{2+}$ to a macromolecule can lead to changes in all three terms of Eq. (3). $p$ will always decrease, which would tend to decrease the PRR. $\tau_M$ may either decrease or increase but usually stays within one order of magnitude of the value of free manganous ion. However, $\tau_c$ may vary by as much as three orders of magnitude when the manganous ion is bound to a macromolecular site which is tumbling slowly. It is the term $T_{1,M}$ which has changed drastically when a significant increase in enhancement for a Mn–protein complex is observed. For a complex with a high enhancement, the apparent equality of $\tau_c$ and $\tau_r$ may not hold and $\tau_s$ is sometimes the determining correlation time (Reuben and Cohn, 1970).

Further information may come from studies at variable temperature and/or frequency. Thus, $\tau_r$ has a positive temperature coefficient, while $\tau_s$ and $1/\tau_M$ have negative temperature coefficients. The latter two can be distinguished because $1/\tau_M$ is independent of frequency while $\tau_s$ is not; further, $1/\tau_M$ usually has a much higher energy of activation than $\tau_s$.

The theoretical aspects are dealt with in more detail in Section III.

## C. Information Potentially Available from PRR Studies

It is pertinent to indicate the type of information that may, under favorable circumstances, be obtained from PRR studies. The discussion below is principally concerned with enzymes.

1. Evidence for the specific binding of the paramagnetic species to the protein may be obtained.

2. Where manganese is concerned, evidence as to whether the binding takes place directly or indirectly, via the substrate, to an enzyme can also be obtained. In practice, three modes of binding of substrate relative to the metal ion may be observed with manganese-activated enzymes (Mildvan and Cohn, 1970; Cohn, 1963). These are type I, where the metal ion is bound via the substrate, so that there is a substrate bridge, e.g., creatine kinase. The enhancement of the ternary complex containing enzyme, metal, and substrate will be greater than that with enzyme and metal alone. Type II, as with pyruvate kinase, occurs where the metal ion is bound directly and the enhancement is decreased on the addition of substrate. Type III, as with citrate lyase (Ward and Srere, 1965), occurs where $Mn^{2+}$ is bound to the enzyme and is unaffected by the binding of the substrate to the enzyme.

These modes of binding may be diagrammatically represented as in Fig. 2 and some enzymes that fall into these three classes are collected in Table I. A more comprehensive compilation is to be found in the review by Mildvan and Cohn (1970).

3. Binding constants and the number of binding sites for the paramagnetic species and/or its complexes can also be obtained from PRR data. In general, values obtained have compared favorably with those

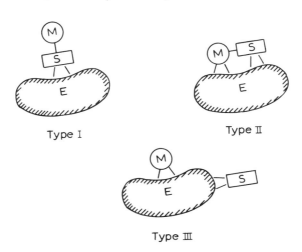

Fig. 2. Diagrammatic representation of ternary enzyme-metal-substrate complexes according to their enhancement behavior.

from more standard techniques, such as equilibrium dialysis. The PRR experiment has the advantage that data can usually be obtained in a relatively short time period, minimizing deterioration of enzyme and substrates.

4. Evidence for the definite existence of particular complexes contain-

TABLE I
Classification of Metal-Activated Enzymes According to Enhancement Behavior

| Type | Type I | Type II | Type III |
|---|---|---|---|
| Description | substrate bridge (E—S—M) | metal bridge $\left( \begin{array}{c} E-M-S \text{ or } E \diagdown_{S}^{M} \end{array} \right)$ | enzyme bridge (S—E—M) |
| Enzymes | $\epsilon_b < \epsilon_t$<br>creatine kinase<br>arginine kinase<br>adenylate kinase<br>3-phosphoglycerate kinase<br>tetrahydrofolate synthetase | $\epsilon_b > \epsilon_t$<br>pyruvate kinase<br>pyruvate carboxylase<br>phosphoenolpyruvate carboxykinase<br>enolase<br>phosphoglucomutase<br>carboxypeptidase | $\epsilon_b = \epsilon_t$<br>citrate lyase<br>dopamine hydroxylase |

ing the enzyme, a paramagnetic species, and various substrates or modifiers in different combinations can often be obtained.

5. Evidence for subtle substrate-induced conformation changes in an enzyme may be forthcoming. This often follows from paragraph 4 above.

6. It provides a very useful technique for studying the binding of substrates to specifically inactivated enzymes. This has been demonstrated for the binding of MnADP$^-$, etc., to derivatives of creatine kinase, specifically inactivated by substitution at the essential thiol group (O'Sullivan and Cohn, 1968; O'Sullivan, 1971).

7. Information about the behavior of the water molecules remaining in the first coordination sphere of the manganese complexes may be obtained from studies at varying temperature and frequency. In some cases, it is possible to obtain a reasonable estimate for the number of water molecules still bound to the manganese (Reuben and Cohn, 1970).

8. Information about changes in the oxidation state of a metal ion, e.g., $Cu^{2+}$ to the diamagnetic $Cu^+$, can also be gathered, though in this respect it is somewhat ancillary to EPR.

D. LIMITATIONS OF THE PRR METHOD

Apart from the necessity for a paramagnetic species, relatively high concentrations of highly purified material are required for PRR. This requirement is partially offset by the fact that small volumes can be used; for routine measurements, 0.1 ml is satisfactory and, under favorable conditions, volumes as small as 10 $\mu$l can be used.

The other principal limitation is that one must be able to observe significant differences in the PRR in the test system. Thus, a complex must be formed with an enhancement value significantly different from that of free manganese and at a concentration sufficient to affect the overall PRR. Some prior knowledge of the order of magnitude of the relevant binding constants is of distinct value in planning an experiment.

Stringent precautions must also be taken to remove any traces of chelating agents, e.g., ethylene diamine tetraacetic acid (EDTA), that might have been used in the enzyme purification, as these would seriously interfere with PRR measurements using paramagnetic metal ions.

### III. Theory of Relaxation

A. BASIC CONCEPTS

Although a complete and rigorous description of NMR phenomena requires the use of quantum mechanics, a classical model will prove most useful in explaining the basic principles and is particularly convenient for describing experimental procedures. A further simplification is obtained by restricting the discussion to nuclei, such as the proton, which has a spin angular momentum quantum number $I$ equal to $\frac{1}{2}$.

According to the quantum theory, nuclei having $I$ equal to $\frac{1}{2}$ and magnetic moment $\mu$, when placed in an external magnetic field $H_0$ applied in the $+z$ direction, distribute themselves between two energy levels corresponding to the $z$-component of spin $I_z$ being quantized parallel or antiparallel to the field $H_0$. The equilibrium distribution is such that there is a small excess of nuclei having $I_z$ parallel to $H_0$: this gives rise to a net macroscopic magnetic moment or magnetization $M_0$ in the direction of $H_0$. The equilibrium distribution may be disturbed by causing transitions between levels through the application of electromagnetic radiation at the appropriate frequency $\nu_0$ such that $h\nu_0$ is equal to $\Delta E$, the energy difference between levels.

This resonance condition may also be written

$$\omega_0 = \gamma H_0 \tag{5}$$

where

$$\omega_0 = 2\pi\nu_0 \tag{6}$$

and $\gamma$ is the gyromagnetic ratio of the resonant nuclear species.

The value of $\gamma$ for the proton is such that, in a typical laboratory magnetic field, the resonant frequency is in the radio frequency (rf) band; e.g., for a field of 10 kG, the resonant frequency is 42.6 MHz. This electromagnetic radiation at the resonant frequency will be subsequently referred to as the $H_1$ or rf field.

In the classical model (see Fig. 3a), an isolated nucleus in a magnetic field $H_0$ experiences a mechanical torque $\mu \times H_0$ which causes it to precess about the field direction with an angular frequency given by Eq. 5. This motion is similar to that of a top which, when set spinning with its axis at an angle to the vertical, precesses about a vertical axis passing through its point of support.

If, instead of to a coordinate system $(x, y, z)$ which is fixed with respect to the laboratory the motion of the nucleus is referred to a coordinate frame $(x', y', z')$ which rotates about the $z$-axis of the laboratory frame in the same sense and at the same frequency as the precessing nucleus, then viewed in this rotating reference frame, the magnetic moment will appear stationary.

The application of magnetic field $H_1$, which is at right angles to $H_0$ and is stationary in the rotating frame, will cause the nuclear moment to precess about $H_1$ with angular frequency $\omega_1$ equal to $\gamma H_1$ in this rotating frame (Fig. 3b).

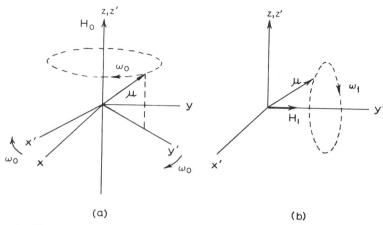

Fig. 3. Classical motion of a nuclear spin having a magnetic moment $\mu$, in a static magnetic field $H_0$. (a) The precession in the laboratory $(x, y, z)$ and rotating $(x', y', z')$ coordinate frames. (b) The nutation, caused by the rf field $H_1$, viewed in the rotating frame.

In the theory of the spinning top, this type of motion is called *nutation*. Although stationary in the rotating frame, in the laboratory reference frame, $H_1$ is a rotating magnetic field at the precessional angular frequency $\omega_0$.

In a macroscopic sample of, say, 1 ml there are about $10^{22}$ nuclei. In the equilibrium state the $x$- and $y$-components are randomly distributed and average out to zero, but in the presence of the external magnetic field $H_0$, the slight preferential alignment of the nuclear moments in the field direction produces a net magnetization $M_0$ parallel to $H_0$. The behavior of $M_0$, which is the resultant of the $10^{22}$ nuclear spins, is similar to that of an individual spin and the application of a resonant rf field $H_1$ causes $M_0$ to nutate about $H_1$ so that it is no longer parallel to $H_0$.

Experimentally, the sample of interest is placed within a coil (Fig. 4a), which has its axis at right angles to $H_0$ and which is energized at the resonant frequency. The oscillating field within the coil may be decomposed into two components, one of which rotates in the $xy$-plane in the same sense as the precessing nuclei and corresponds to the rotating field $H_1$.

In the rotating reference frame $(x', y', z')$, $H_1$ is stationary and in this frame all nuclei and, hence, their resultant $M_0$ nutate about $H_1$ with angular frequency $\omega_1$. If at zero time the resonant rf field $H_1$ is switched on, then at a time $t$, later, the angle through which $M_0$ nutates is $\gamma H_1 t$. Conversely, the time for $M_0$ to nutate through an angle $\theta$ is $\theta/H_1$. Of

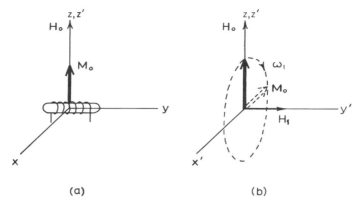

FIG. 4. (a) The magnetization $M_0$ of a macroscopic sample in the static field $H_0$. (b) The nutation of $M_0$ by the rf field $H_1$ as viewed in the rotating frame.

particular importance is the 90° pulse which, as its name implies, is a pulse of resonant rf energy, the duration of which is such that $M_0$ nutates through exactly 90° as shown in Fig. 4b. Similarly a pulse of twice this duration, or a 180° pulse, causes $M_0$ to nutate through 180°,

At the end of a 90° pulse, $M_0$ lies in the $x'y'$-plane of the rotating reference frame but in the laboratory frame it is rotating (in the $xy$-plane) at the resonant frequency. This motion produces changes in the magnetic flux linked with the coil so that a voltage is induced in the coil at the resonant frequency. This signal (referred to as the free precessional signal, or f.p.s.) may be suitably amplified, detected, and displayed on a cathode ray oscilloscope. The free precessional signal is a maximum following a 90° pulse and is zero following a 180° pulse, for then the magnetization $M_0$ is in the negative $z$-direction and there is no component in

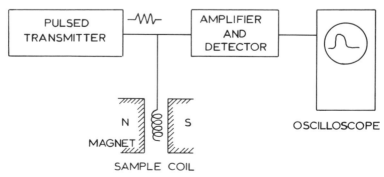

FIG. 5. Simplified block diagram of the essential apparatus required for a pulsed NMR experiment.

the $xy$-plane to induce a signal in the coil. A simplified block diagram showing the essential components for a pulsed NMR experiment is shown in Fig. 5.

## B. Phenomenological Description of Relaxation

The free precessional signal following a 90° pulse does not persist indefinitely but decays away to zero due to two distinct types of relaxation process.

First, following a 90° pulse when the magnetization is in the $xy$-plane, the sample is in a nonequilibrium condition: the equilibrium state corresponds to that of the magnetization $M_0$, being parallel to $H_0$, i.e., in the $+z$-direction. Since any perturbed system tends to revert to its equilibrium state, there occurs a gradual growth of magnetization parallel to $H_0$ and a corresponding decay of magnetization in the $xy$-plane.

This return to equilibrium is governed by a characteristic time $T_1$, the longitudinal or spin–lattice relaxation time, and the process may be represented by an exponential growth equation

$$M_z(t) = M_0[1 - \exp(-t/T_1)] \qquad (7\text{a})$$

where $M_z(t)$ is the magnetization in the $z$-direction at time $t$. The $xy$-component of magnetization $M_{xy}$ (and hence the free precessional signal which is proportional to $M_{xy}$) decays away with a similar time constant $T_2$

$$M_{xy} = M_0 \exp(-t/T_2) \qquad (7\text{b})$$

Relaxation processes arise from the presence of local magnetic fields $H_L$, which are produced by the nuclear magnetic dipoles themselves (and by paramagnetic ions, if present). The actual magnetic field at the site of a nucleus is then $(H_0 + H_L)$ and its precessional frequency is $\gamma(H_0 + H_L)$. Although $H_L$ is small compared to $H_0$, its value varies from one nuclear site to another. This spread in local fields produces a corresponding spread in precessional frequencies so that, following a 90° pulse, some nuclear spins precess at a faster rate and others at a slower rate than the average. Thus, the spins, initially in phase immediately after a 90° pulse, after a while begin to get out of step (i.e., they dephase) so that the signal induced in the coil decreases, eventually becoming zero when the spins have achieved a completely random-phase distribution. The characteristic time $T_2$ which describes the decay of the free precessional signal is referred to as the *transverse* or *spin–spin relaxation time*.

An additional effect is that due to the inhomogeneity or spatial variation of the applied magnetic field over the sample volume. This again causes a spread in the precessional frequencies of the nuclear spins, resulting in a rapid decay of the free precessional signal. Unless magnets of exceptional field homogeneity are used (e.g., as in high-resolution continuous-wave NMR spectroscopy) the signal decay is dominated by the field inhomogeneity so that the natural $T_2$ decay is obscured and cannot be measured directly. However, the effects of field inhomogeneity may be overcome by employing a special technique which produces an echo signal (see Section III,C).

## C. Principles of Measurement of $T_1$ and $T_2$

Consider a sample, initially in an equilibrium state, with the magnetization $M_0$ parallel to $H_0$. At zero time, a 90° pulse of resonant rf is applied, which causes $M_0$ to nutate into the $xy$-plane, thus inducing a signal in the sample coil (see Fig. 6a; note that all the diagrams in Fig. 6 are drawn in the rotating reference frame). Due to the field inhomogeneity and consequent spread in precessional frequencies, the nuclear

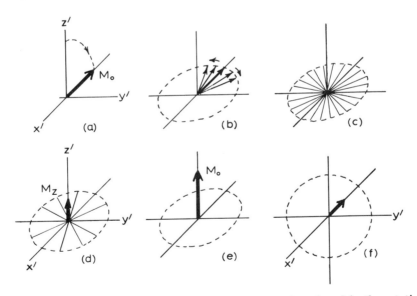

Fig. 6. Evolution of the spin system following a 90° pulse, viewed in the rotating frame. (a) The static magnetization $M_0$ is tipped into the $x'y'$-plane by the 90° pulse; (b) the individual spins begin to dephase; (c) the magnetization is completely dephased; (d) the growth of magnetization $M$ in the direction of $H_0$; (e) the system has returned to its equilibrium state with static magnetization $M_0$; (f) a 90° pulse applied at the time (d) produces a coherent magnetization $M$ in the $x'y'$-plane.

spins dephase or fan out as in Fig. 6b. In a short time the spins become completely dephased (Fig. 6c) so that zero signal is induced in the sample coil, although the magnetization still lies in the $xy$-plane. The nuclear spins continue to relax toward their equilibrium state and the situation some time later is shown in Fig. 6d, where the spins which have relaxed contribute to a magnetization parallel to $H_0$, of value $M_z(t)$ less than $M_0$. After a sufficient length of time, the relaxation process is complete and the magnetization has returned to its equilibrium value $M_0$ (Fig. 6e).

If a second 90° pulse had been applied at a time corresponding to Fig. 6d, all the spins would nutate through 90° and the situation would be that represented in Fig. 6f. As there is now a component of magnetization in the $xy$-plane, a signal will be induced in the sample coil although its magnitude will be less than that following the first 90° pulse. It will be noted that the signal height following a 90° pulse is always proportional to (and therefore a measure of) the magnetization $M_z(t)$ existing in the sample at the time of application of the 90° pulse. This fact is the basis of various techniques for the measurement of $T_1$.

### 1. *Measurement of $T_1$*

In the 90°-$t$-90° method, two 90° pulses are applied with a time interval $t$ between them. When the system has returned to an equilibrium state (a period of $5T_1$ will allow the spin system to achieve more than 99% of its equilibrium magnetization), the sequence is repeated but with a different time interval between the two pulses. The height of the signal following the second 90° pulse is recorded as a function of the time interval $t$ and its growth as $t$ increases is given by Eq. (7a). $T_1$ may then

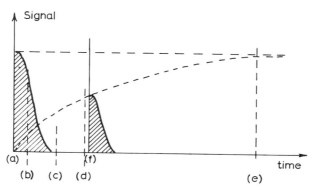

FIG. 7. Representation of the observed signals in the 90°-$t$-90° method of measuring $T_1$. The condition of the spin system at the times denoted by the letters (a), (b), etc., is depicted in the correspondingly lettered diagram of Fig. 6.

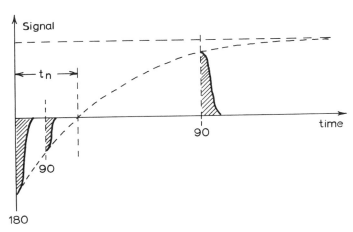

FIG. 8. Representation of the observed signals in the 180°-$t$-90° method of measuring $T_1$. The signal following the 90° pulse is zero at a particular time $t_n$.

be extracted by suitable mathematical or graphical means. The sequence of events in this method is shown in Fig. 7.

A more direct method employs a 180° pulse followed at some variable time interval later by a 90° pulse (the 180°-$t$-90° method). There is no received signal following the 180° pulse since the magnetization is then $-M_0$, i.e., parallel to the $-z$-axis. The magnetization still relaxes back with a characteristic time $T_1$ toward its equilibrium value of $+M_0$. At some particular time $t_n$ after the 180° pulse, the magnetization is zero and a 90° pulse applied at this time will give zero signal. At any other time the magnetization is nonzero (either positive or negative) and a signal is observed following the 90° pulse (see Fig. 8). The essential feature of this method is then to vary the time between pulses until the null condition (zero signal following the 90° pulse) is obtained. The time interval $t_n$ for this condition is related to the spin–lattice relaxation time $T_1$ by the simple expression

$$T_1 = t_n / \ln 2 \tag{8}$$

## 2. *Measurement of $T_2$ by the Spin Echo Method*

The principle of the formation of an echo signal may be understood from a consideration of Fig. 9a, in which the vectors $p$ and $q$ represent two spins in a rotating reference frame. Because they have slightly different precessional frequencies, they appear to be diverging. The application of a 180° pulse $H_1$ along the $y'$-axis causes all spins to nutate in the rotating reference frame to 180° about this axis. The two spins are now in the positions marked $p'$ and $q'$ but, because their relative preces-

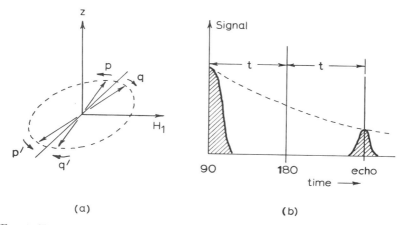

Fig. 9. Echo formation. (a) In the rotating frame, two dephasing spins $p$ and $q$ are tipped by a 180° pulse to new positions $p'$ and $q'$, respectively, where they now rephase; (b) the signals observed in the 90°-$t$-180° method of measuring $T_2$. The application of the 180° pulse at time $t$ causes the dephased spin to regroup so as to form an echo signal at a time $2t$.

sional rates have not changed, they are now converging or rephasing.[1]

The sequence of events in a spin echo experiment is illustrated in Fig. 9b. Following the 90° pulse, the signal decays away rapidly due to the dephasing effect of the magnet inhomogeneity. At time $t$ later, a 180° pulse is applied which causes the spins to begin rephasing and, after a further time interval $t$ (i.e., at a time $2t$ from the 90° pulse), the spins are back in phase and a signal is again induced in the sample coil. Following the formation of the echo the continuing differential precessional rates produce a dephasing and consequent decay of the signal.

The 180° pulse recovers the loss of phase due to the static field inhomogeneity but not that brought about by the varying local fields which produce the spin–spin relaxation $T_2$. The echo height therefore is less than the initial free precession signal and decreases further as the interval $t$ increases. $T_2$ may be found from the decay of the echo signal which, for liquid samples, follows the experimental decay equation

$$y(t) = y_0 \exp(-t/T_2) \tag{9}$$

[1] As an elementary illustration of the formation of spin echoes, a parallel is sometimes drawn between the precessing nuclear moments and runners at a race track. These runners, starting together at time $t = 0$, will progressively spread out on the racetrack because of their different speeds, and if the race goes on for a sufficiently long time they will eventually be uniformly distributed all round the track. If, then, at time $\tau$ they reverse their steps and run in the opposite direction but with the same speeds, they will end up bunched together at time $t = 2\tau$. (Abragam, 1961, p. 62.)

where $y$ represents the echo height and $t$ the time of formation of the echo.

When $T_2$ is long and the magnet homogeneity is poor, the 90°-$t$-180° pulse method of measuring $T_2$ may lead to an erroneous result if the molecule containing the resonant nuclear spin undergoes significant diffusional motion. This motion in an inhomogeneous magnetic field produces corresponding changes in the frequency of the precessing spins, which leads to an incomplete refocusing of the dephased spins by the 180° pulse and a consequent reduction in the height of the echo signal. The decay of the echo signal caused by the diffusional motion is nonexponential and this fact can be used to test whether diffusional effects are present. Experimentally, the effects of self-diffusion may be reduced to negligible proportions by the application of a 90°-180°-180°-180°-. . . or Carr-Purcell sequence of pulses. The 180° pulses are applied at times $t$, $3t$, $5t$, . . . and the echoes which occur at times $2t$, $4t$, $6t$, . . . have an exponential decay characterized by $T_2$ (Carr and Purcell, 1954).

Alternatively, using the 90°-180° method and applying magnetic field gradients of known strength, the effect can be utilized to measure self-diffusion constants in liquids and absorbed systems (Stejskal and Tanner, 1965; Tanner and Stejskal, 1968).

## D. Relaxation Theory Applied to Paramagnetic Systems

### 1. *Introduction*

For nuclei with spin $\frac{1}{2}$, the relaxation processes, both longitudinal and transverse, may be explained in terms of the local magnetic fields which exist at each nuclear site. In pure liquid water, the local fields arise from the magnetic moments of the protons. In solutions containing paramagnetic substances, additional local fields will arise from the unpaired electrons. Since the magnetic moment of an electron is approximately 1000 times greater than that of a proton, even small concentrations of paramagnetic ions have an appreciable effect. For example, the effect of $Mn^{2+}$ ions on the $T_1$ value of 3 sec for pure water is illustrated in Fig. 10.

The local fields fluctuate both in magnitude and direction in a random manner due to the various atomic or molecular motions that take place in the sample. The first effect of the fluctuations is to reduce the time-averaged value of the local field so that its dephasing effect is also reduced; i.e., $T_2$ is increased. Second, the fluctuating field has some frequency component, however small, which coincides with the nuclear resonance frequency and which can therefore induce transitions between the magnetic energy levels. This is the origin of the $T_1$ relaxation mechanism.

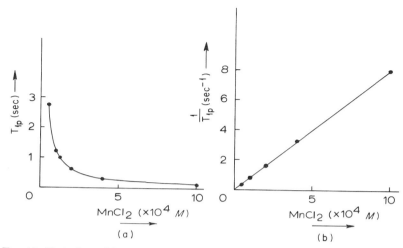

Fig. 10. Variation of longitudinal relaxation time $T_{1,p}$ with manganese concentration, at 22°C.

The fluctuating nature of the local field is described quantitatively in the theory of random processes by a correlation time $\tau_c$. This may be interpreted qualitatively as the average time for the local field experienced by a nucleus to change by a significant amount due to the motion either of the nucleus itself or of its neighboring nuclear or electronic spins.

The theoretical expressions for the longitudinal relaxation time, which will be given later in full, contain terms having a similar form and which may be expressed as

$$\frac{1}{T_1} = A \frac{\tau_c}{1 + \omega_0^2 \tau_c^2} \tag{10}$$

where $A$ is a constant involving various nuclear parameters and internuclear distances which vary according to the relaxation mechanism. When a number of mechanisms contribute to the relaxation process, it is the relaxation rates $1/T_1$, rather than the relaxation times, that are additive.

The value of $T_1$ is very much dependent on the relative magnitudes of the correlation time and the resonant frequency and, as may be seen from inspecting the above expression, three distinct situations can arise.

a. If the correlation time is very short, so that $\omega_0 \tau_c < 1$, the $\omega_0^2 \tau_c^2$ term in the denominator can be ignored and the expression simplifies to $(T_1) \sim 1/\tau_c$.

b. If, on the other hand, $\tau_c$ is very long, so that $\omega_0 \tau_c > 1$, the expression reduces to $(T_1) \sim \omega_0^2 \tau_c$. Since $\omega_0$ is equal to $\gamma H_0$, this means that under these conditions $T_1$ is field dependent.

c. $1/(T_1)$ exhibits a maximum value (and $T_1$ a minimum) when $\omega_0 \tau_c$ is equal to 1. This situation corresponds to the fluctuating local field component at the resonant frequency being a maximum.

Molecular motions are very temperature dependent and in many instances the correlation time exhibits an Arrhenius-type behavior [i.e., $\tau_c = \tau_0 \exp(E/RT)$] where $E$ is an appropriate activation energy. For this reason, the functional behavior of $T_1$ is as illustrated in Fig. 11, where $\ln T_1$ is plotted against the reciprocal of absolute temperature. If the Arrhenius law is obeyed, the limiting cases a and b appear as straight lines of negative and positive slope, respectively. It should be noted that the above discussion is strictly applicable only to the case of a single relaxation mechanism. In general, several distinct mechanisms may be operative so that the observed relaxation rate is the sum of a number of terms similar to Eq. (10), with some additional modification when chemical exchange must be considered. However, in many cases, at a given temperature or in a given temperature range, one mechanism is dominant, the other mechanisms providing negligible contribution to the relaxation rate. In such cases, the behavior of $T_1$ as a function of temperature and of applied field can provide evidence to help decide which, out of a number of possible alternatives, is the dominant relaxation process. This problem will be dealt with in greater detail in Section V,D.

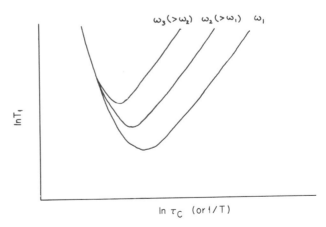

FIG. 11. Behavior of $T_1$ as a function of the correlation time $\tau_c$ or of reciprocal absolute temperature. It is assumed that there is one dominant relaxation mechanism having a contribution of the form $1/T_1 = A\tau_c/(1 + \omega^2 \tau_c)$; $\omega_1$, $\omega_2$, and $\omega_3$ represent three different values of the resonant frequency.

## 2. Effect of Paramagnetic Species

As was mentioned previously, paramagnetic species may have a profound effect on the relaxation time of water protons. The paramagnetic contribution itself $1/T_{1,p}$ [see Eq. (1)], may be considered in two parts; first, a direct effect on the protons in the first hydration sphere surrounding the paramagnetic ion, and second, an effect produced by the exchange between the solvent water molecules and the water molecules in the first hydration sphere [cf. Eq. (3)]. Because the interaction responsible for the relaxation decreases rapidly with distance according to an inverse sixth-power law, the direct or outer-sphere relaxation is negligible and may safely be ignored. By the same token, there is a strong interaction between the paramagnetic ion and the protons of the water molecules of the hydration sphere so that the relaxation rate of these protons is fast, and when a molecule from the hydration sphere undergoes exchange, it thereby contributes to the relaxation of the solution as a whole.

The expression for $1/T_{1,p}$ given in Eq. (3) is repeated for convenience

$$1/T_{1,p} = \frac{p}{T_{1,M} + \tau_M} \tag{3}$$

A term $1/T_{1,os}$ representing the outer-sphere relaxation has been suppressed on the right-hand side since this is usually of negligible consequence.[2]

By inspection, it can be seen that two limiting cases arise. When $\tau_M > T_{1,M}$, the paramagnetic relaxation rate becomes $p/\tau_M$, as would be expected, for under these circumstances the protons of the water molecules exchanging from the hydration sphere have all relaxed, so that the relaxation rate in the bulk solution is determined directly by the exchange rate. When $T_{1,M} > \tau_M$, the paramagnetic relaxation rate becomes $p/T_{1,M}$. This is the case of fast exchange where the average water molecule, during the time of the NMR experiment, makes many exchanges between the hydration sphere and the bulk solution.

In dealing with the magnitude of the relaxation time $T_{1,M}$ of the protons in the first hydration sphere, there are two mechanisms to consider. A direct interaction between the paramagnetic ion and the proton magnetic moment and an indirect or scalar interaction which acts through the agency of the neighboring electrons. A further complication arises from the fact that the correlation time, which describes the fluctuating nature of the local field experienced by a given proton and which, as we have already seen, plays an important role in the relaxation mechanism,

---

[2] The outer-sphere term is usually small but can become dominant if $\tau_M$ is very long, i.e., if water exchange into the first hydration sphere is very slow.

is compounded from three terms. First, there is the fluctuation due to the water molecules exchanging in and out of the hydration sphere. The correlation time appropriate to this motion is the previously defined $\tau_M$. Second, there is the rotation of the water molecule–metal ion vector as the whole complex tumbles. This is described by a rotational correlation time $\tau_r$. Third, the paramagnetic spin flips between parallel and antiparallel alignment with the applied magnetic field and creates corresponding changes in its local field. The characteristic time $\tau_s$, which describes this spin-flipping motion is referred to as the *electronic relaxation time*.

It can be shown that, in combining these motions, it is the rates which are additive. The rate constant which describes the effective correlation time for the direct interaction is therefore given by[3]:

$$1/\tau_c = 1/\tau_M + 1/\tau_r + 1/\tau_s \tag{12}$$

For the scalar interaction, the rotational motion is ineffective and the appropriate correlation time $\tau_E$, is given by

$$1/\tau_E = 1/\tau_M + 1/\tau_s \tag{13}$$

In both cases the dominant contribution is provided by that motion which has the shortest correlation time.

The paramagnetic contribution to the longitudinal relaxation of the protons within the first hydration sphere of a paramagnetic ion is given, for the general case, by the Solomon-Bloembergen equation (Solomon, 1955; Bloembergen, 1957, 1961)

$$\frac{1}{T_{1,M}} = \frac{2}{15} \cdot \frac{S(S+1)\gamma_I^2 g^2 \beta^2}{r^6} \left[ \frac{3\tau_c}{1+\omega_I^2\tau_c^2} + \frac{7\tau_c}{1+\omega_s^2\tau_c^2} \right] + \frac{2}{3} \cdot \frac{S(S+1)A^2}{\hbar^2} \left[ \frac{\tau_E}{1+\omega_s^2\tau_E^2} \right] \tag{14}$$

where $S$ is the electronic quantum number for the paramagnetic ion, $\gamma_I$ is the nuclear gyromagnetic ratio for the proton, $r$ is the ion–proton internuclear distance, $\beta$ is the Bohr magneton, and $g$ the electronic "g" factor; $\omega_I$ and $\omega_s$ are the angular precession frequencies of the proton and the electronic spins, respectively, and $A$ is the hyperfine coupling constant. The two terms in this formidable equation correspond to the direct and scalar interactions previously mentioned. Fortunately, some simplification is possible since, for all cases of interest, $\tau_E$ is sufficiently large to

---

[3] It should be noted that in the $\tau_s$ term of Eq. (12), the $T_1$ of the electron spin is appropriate for the term containing $\omega_I$ in Eq. (14) and $T_2$ of the electron spin is to be used in the term containing $\omega_s$ in Eq. (14) (Abragam, 1961; Connick and Fiat, 1966; Reuben et al., 1970).

make $\omega_s\tau_E \gg 1$. Thus, the scalar term is negligibly small. Furthermore, since $\omega_s \sim 10^3\omega_I$, the second fractional term within the first square bracket is negligible unless $\omega_s\tau_c < 1$, which is not normally the case in enhanced systems. Thus, the expression for $T_{1,M}$ reduces to

$$\frac{1}{T_{1,M}} = \frac{2}{15} \cdot \frac{S(S+1)\gamma_I^2 g^2 \beta^2}{r^6} \frac{3\tau_c}{1+\omega_I^2\tau_c^2} \qquad (15)$$

In considering the paramagnetic contribution to the transverse relaxation of the bulk solution, arguments similar to that advanced in the above discussion on the longitudinal relaxation apply. Eqs. (12) and (13) referring to the correlations times are unmodified while the analogous expression to Eq. (1), defining the paramagnetic contribution to the transverse relaxation in terms of the experimentally measured quantities, is

$$\frac{1}{T_{2,p}} = \frac{1}{T_2} - \frac{1}{T_{2,0}} \qquad (16)$$

In most paramagnetic systems of interest an equation similar to Eq. (3) gives the theoretical expression for $T_{2,p}$ in terms of the transverse relaxation time $T_{2,M}$ of the protons in the water molecules within the first hydration sphere

$$\frac{1}{T_{2,p}} = \frac{p}{\tau_M + T_{2,M}} \qquad (17)$$

Finally, the Solomon-Bloembergen equation for $T_{2,M}$, evaluated under the same simplifying conditions that yield Eq. (15), may be expressed as

$$\frac{1}{T_{2,M}} = \frac{1}{15} \cdot \frac{S(S+1)\gamma_I^2 g^2 \beta^2}{r^6} \left[ 4\tau_c + \frac{3\tau_c}{1+\omega_I^2\tau_c^2} \right]$$
$$+ \frac{1}{3} \cdot \frac{S(S+1)}{\hbar^2} \cdot A^2 \cdot \tau_E \qquad (18)$$

It should be noticed that, in contrast with $T_{1,M}$ given by Eq. (15), the scalar interaction is not in general negligible but has a residual effect which is represented in Eq. (18) by the term involving $\tau_E$.

## IV. Experimental Methods

### A. Apparatus

The following section gives a brief description of the apparatus required to perform a pulsed NMR experiment. A detailed account would not only be out of place here but also, in view of the continuing development of solid-state electronics, soon outdated.

The major single item is a magnet system which should be capable of producing a field of several thousand gauss (the higher the field, the greater the NMR signal, other things being equal). Whether a permanent magnet or an electromagnet is used, it is vital that the field be of an adequate homogeneity. For $T_1$ measurements, it is advisable that the inhomogeneity decay of the free precessional signal be not much less than 1 msec and preferably longer. This corresponds to a field homogeneity over the sample volume of one part in $10^5$ or better. Likewise the magnetic field must be stable in time to the same order of magnitude. For an electromagnet this means that a power supply of high stability is required and because of the heat dissipated, so is adequate cooling of the magnet itself. The advantages of an electromagnet compared with a permanent magnet are its obvious flexibility and its capability of achieving much higher fields than the permanent magnet.

The next important item is a transmitter which must be capable of providing radio frequency pulses of several hundred volts across the sample coil. The simplest version is that of a pulsed power oscillator which is turned on by gating pulses. It suffers the disadvantage that it must be carefully tuned to, and maintained at, the resonant frequency appropriate to the static magnetic field throughout the experiment.

An alternative form of transmitter uses a low-level crystal oscillator at the desired frequency. Its output is passed through a gated preamplifier before final power amplification. Besides eliminating the need for frequency tuning, a further advantage is that the crystal oscillator may be used to inject a frequency reference signal into the receiver producing a "beat" with the NMR signal. Adjustment of the magnetic field to the exact resonance condition is achieved when zero beat is obtained. Additionally and with appropriate circuitry, the frequency reference signal may be used to provide phase-sensitive detection in the receiver.

The gating pulses for the transmitter can be obtained from pulse generators of variable width, which are in turn triggered by timing pulses having variable separation and sequence repetition rate. This pulse programing unit may be conveniently assembled from commercially available waveform and pulse generators or may be fabricated at a more basic level from micrologic elements.

The receiver takes the form of a high-gain tuned radio frequency receiver with a good noise figure and with design attention directed to the recovery of the receiver from the overload conditions caused by the transmitter pulse. This entails a wide bandwidth for the receiver and the use of short time-constant filters in the detector and elsewhere.

## B. The Measurement of $T_1$

The experimental $T_1$ results discussed in Section V,B were obtained using a 30 MHz crystal oscillator-gated amplifier type of transmitter with two separate gated channels. A block diagram of the apparatus is given in Fig. 12. A most useful auxilliary item is the interval timer which gives a digital readout of the time between two pulses. It obviates the necessity for making time measurements from the face of an oscilloscope and, in the case of $T_1$ experiments using the 180°-90° pulse sequence, it provides the sole reading which need be taken.

The static magnetic field was produced by an electromagnet having a 12 inch pole diameter and a 2 inch gap. The inhomogeneity over the sample volume was such that the decay of the free precessional signal was approximately 2 msec.

The sample coil, 3.0 cm in length, was wound on a glass tube of 7 mm inner diameter and was positioned, with its axis vertical, in the center of the magnet gap. The sample tubes, which were 6 mm outer diameter and contained approximately 100 μl of the liquid sample, were lowered into the first tube and located in position by a constriction so that the liquid sample occupied the central portion of the coil. This insured that the radio frequency ($H_1$) field was uniform over the sample volume. The coil system was enclosed in a chamber through which nitrogen gas, cooled

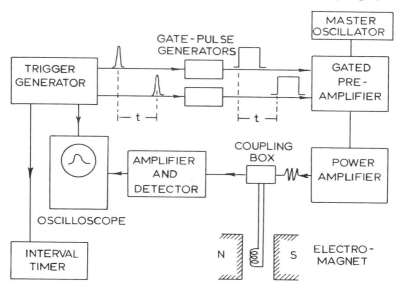

Fig. 12. Block diagram of the apparatus used for the measurement of $T_1$ and $T_2$ as described in the text.

or heated as desired, could be circulated for the measurement of relaxation times at different temperatures.

The experimental procedure was as follows. After allowing an initial warm-up period for the magnet and electronic equipment generally, a sample of a standard paramagnetic solution of known $T_1$ ($\sim$100 msec) was placed in the sample coil for the purpose of making preliminary adjustments. The magnetic field was set to the exact resonance condition; transmitter output and sample-tuned circuits were trimmed and the two gates of the transmitter were individually adjusted to the 180° pulse and 90° pulse conditions respectively.

The measurement of the $T_1$ of an unknown sample then follows in a straightforward manner. Using a 180°-90° pulse sequence, the time interval between the two pulses was varied until a minimum signal following the 90° pulse was obtained, as shown in Fig. 13. Then, in order to obtain this null condition more accurately, the oscilloscope was triggered from the 90° pulse (instead of the 180° pulse as previously), the gain was increased, and the time base was suitably adjusted. This procedure allows a more detailed inspection of the signal following the 90° pulse and is illustrated in Fig. 14. The pulse separation $t_n$ corresponding to the null condition was read directly from the interval timer. $T_1$ is then given by $t_n/\ln 2$ (Eq. 8).

The 180°-90° method provides a convenient and rapid means of measuring $T_1$ and the null condition can generally be located with an error of less than $\pm 2\%$. However, a systematic error will be introduced unless the nuclear magnetization acquires the value $-M_0$ immediately after

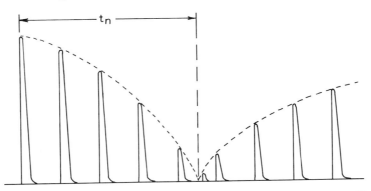

Fig. 13. Representation of the signals observed on the screen of an oscilloscope during the measurement of $T_1$ by the 180°-$t$-90° pulse method. The receiver has linear detection and the oscilloscope is triggered by the 90° pulse of the sequence. The height of the signal produced by the 90° pulse varies with the interval $t$ of the sequence and is zero for an interval $t_n = T_1 \ln 2$.

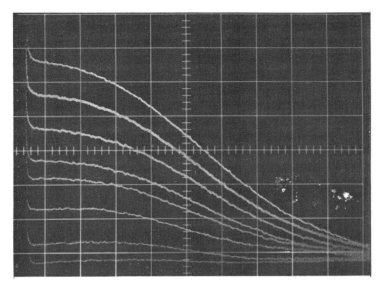

Fig. 14. Measurement of $T_1$ by the 180°-$t$-90° pulse method. The photograph (multiple exposure) shows the signals appearing on the screen of an oscilloscope as the interval $t$ is varied. The oscilloscope is triggered from the 90° pulse; the gain of the Y-amplifier has been increased and the time base expanded to provide an effective enlargement of the signal due to the 90° pulse as the null condition is approached.

the first pulse. For this to be achieved the first requirement is that magnetization have the equilibrium value $+M_0$ prior to the first pulse. This may be attained by insuring that the repetition time for the two pulse sequence is $5T_1$ or larger. The second requirement is that the first pulse tip the magnetization through exactly 180°. This necessitates (a) that the radio frequency ($H_1$) field be uniform over the sample volume, (b) that the static magnetic field obey the resonance condition, and (c) that the width of the first pulse be accurately adjusted to the 180° condition. When measurements are to be made on a large number of samples, it is expedient to make periodic checks of conditions (b) and (c).

C. The Measurement of $T_2$

The apparatus required for the measurement of $T_2$ is identical with that described above for the measurement of $T_1$ though the technique may vary according to the experimental conditions.

When $T_2$ of the sample under test is short compared with the time of dephasing of spins due to the magnetic field inhomogeneity, a single 90°

pulse only is required. $T_2$ is then measured directly from the decay of the free precessional signal.

If $T_2$ is long compared with the inhomogeneity decay, echo techniques must be used. In the 90°-$t$-180° pulse sequence the height of the echo ($y$) is measured on the oscilloscope screen as a function of the time of its formation at $2t$. By varying the time interval $t$, the envelope of the echo decay may be mapped out. This is illustrated in Fig. 15 which is a multiple-exposure photograph of a number of 90°-$t$-180° sequences with progressively increasing value of $t$.

The Carr-Purcell method employs a 90°-$t$-180°-$3t$-180°- . . . sequence of pulses, and echoes are formed, with decaying heights, at times $2t$, $4t$, $6t$, . . . . This is illustrated in the single-exposure photograph of Fig. 16. $T_2$ is evaluated either numerically or graphically from the decay of the echo envelope which is given by $y_{(t)} = y_{(0)} \exp(-t/T_2)$.

The Carr-Purcell sequence must be used whenever diffusional effects are present as these cause additional attenuation of the single echo formed by the 90°-$t$-180° sequence (see Section III,C). Figures 15 and 16 represent measurements taken on the same sample and a diffusional effect may be discerned by noting that the echo decay is slightly faster in the former case. It should be noted that, if the Carr-Purcell sequence

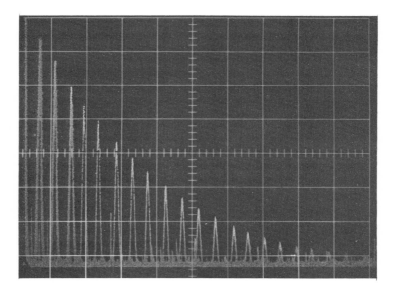

FIG. 15. Determination of $T_2$ by the 90°-$t$-180° pulse method. A photograph (multiple exposures) of the echo signals observed on an oscilloscope screen as the interval $t$ is progressively increased (time scale, 20 msec/cm).

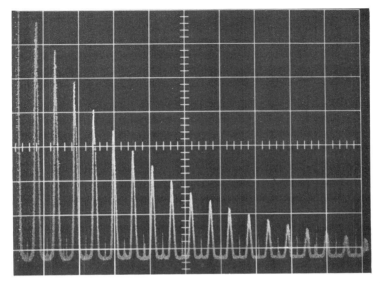

Fig. 16. Determination of $T_2$ by the Carr-Purcell method. The photograph (single exposure) shows the multiple echoes formed by a single 90°-180°-90°- . . . pulse sequence (time scale 20 msec/cm).

is used, it is essential that precise 180° pulses be employed; otherwise imperfect refocusing of the echoes occurs in an accumulative manner and gives rise to an erroneous measurement.

## V. Experimental Design

The essence of the PRR experiment is that one is trying to observe a significant change in $T_1$ caused by the addition of a complexing species, a protein, to a solution containing a paramagnetic moiety and which can be interpreted at least qualitatively. The following remarks are particularly directed to the use of $Mn^{2+}$ as a probe of enzyme/protein structure. They are illustrated by examples from the literature on bovine serum albumin and creatine kinase, respectively.

### A. Preliminary Considerations

In designing a particular experiment such factors as concentration of $Mn^{2+}$, pH, nature of the buffer or other supporting medium, nature of the substrates, purity of material(s), and temperature need to be considered. Each of these will be treated in turn.

### 1. Concentration of $Mn^{2+}$

This is particularly relevant and some previous indication as to what order of magnitude to expect for the binding of $Mn^{2+}$ or Mn–substrate as obtained, for example, from enzyme kinetics, is very useful. Further, shorter values of $T_1$ are relatively easier to determine; a range of $Mn^{2+}$ from $5 \times 10^{-5} M$ to $5 \times 10^{-4} M$ has been the most often employed (see Fig. 10). A useful guide for preliminary experiments is to use a concentration of enzyme of about the same order as the metal ion. To obtain meaningful results, it is necessary not only to have a species formed with an enhancement significantly different from that of the manganous aquocation but there should be sufficient of the species present for it to have a significant effect on the measured value of $T_1$.

### 2. pH

The choice of pH is governed on the alkaline side by the hydrolysis of $Mn^{2+}$ and on the acid side by the nature of the enzyme and/or substrate.

a. The p$K$ for the hydrolysis of the manganous aquocation has been reported as 10.6 (Chaberek et al., 1952). However, $Mn^{2+}$ tends to start precipitating before this pH and although measurements have been carried out above 9.0 (Mildvan and Cohn, 1963) experiments in this region should be approached with caution.

b. In dealing with kinase enzymes, in particular, one has ATP and ADP as substrates, with p$K_a$'s in the region of 6.8–7.0. Thus, it is preferable to do experiments sufficiently above pH 7 that only one ionic species of the nucleotide, and thus of the Mn–nucleotide complex, need be considered.

c. Obviously, the choice of pH will also be guided by the stability properties of the enzyme.

### 3. Nature of the Buffer

The buffer or other supporting medium should be such that there is minimal interaction with $Mn^{2+}$. The metal ion is normally at a concentration of the order of $10^{-4} M$ and as the buffer concentration is expected to be greater than this by at least two orders of magnitude, it can be seen that even weakly complexing buffer species could cause serious perturbation. $N$-Ethyl morpholine and triethanolamine would appear to be quite satisfactory and there does not appear to be serious interference with tris or imidazole. Some of the buffers, e.g., HEPES, described by Good et al. (1966), appear to be satisfactory.

### 4. *Nature of the Substrate*

It is important not only to determine as accurately as possible the dissociation constant of any Mn-substrate complexes formed but also the characteristic enhancement of such complexes, so that their contribution to the observed enhancement can be assessed (see Eq. 21). Both ATP and ADP bind manganous ion strongly and the Mn-complexes, which are the probable substrates of kinase reactions, would be the dominant species in equimolar solutions of metal ion and nucleotide.

### 5. *Purity of the Sample*

Obviously, the protein species under consideration should be as pure as possible and most studies have been done on crystalline preparations. However, meaningful information can be obtained with, say, protein that is only 60–90% pure, provided that the contaminating species does not interfere with the binding of $Mn^{2+}$. Small molecules introduced in purification procedures may also perturb the results and, as mentioned above (Section II,D) stringent precautions are essential to remove traces of chelating agents, such as EDTA.

### 6. *Temperature*

Variation of temperature can influence the PRR in a complex fashion as, apart from the position of the various chemical equilibria in solution, $\tau_M$, $\tau_r$ and $\tau_s$ can be affected. The temperature dependence of the PRR is discussed in detail below (Section V,D).

## B. Experiments with Phosphoenolpyruvate Carboxykinase

As an example of type II, we will describe the calculation of some enhancement values obtained with manganous ion and phosphoenolpyruvate carboxykinase (PEPCK) from sheep kidney mitochondria (R. Barns, D. B. Keech and W. J. O'Sullivan, unpublished results, 1972, cf. Barns *et al.*, 1972). The sheep kidney enzyme appears to be similar to that from pig liver mitochondria (Miller *et al.*, 1968) and exhibits type II behavior.

The calculation of enhancement at two points on a titration of variable enzyme concentration against constant manganese concentration is demonstrated in Table II (a complete titration curve of this type, for bovine serum albumin, is shown in Fig. 19). The effect of the addition of one of the substrates of the reaction, phosphoenolpyruvate (PEP), is also included. The experiments were carried out in 0.04 $M$ $N$-ethylmorpholine–HCl, pH 7.5, at 22°C. The data in Table II demonstrate, successively, the calculation of $T_1$ values from null times [Eq. (8)] conversion to PRR values, correction for the contribution $(1/T_{1,0})$ from the buffer system,

## TABLE II
### Calculation of Enhancements for Mn–PEPCK Complexes[a]

| Components | Null time (sec) | $T_1$ (sec) | $1/T_1$ (sec$^{-1}$) | $1/T_1 - 1/T_0$ (sec$^{-1}$) | $\epsilon$ |
|---|---|---|---|---|---|
| Buffer system | 1.600 | 2.256 | 0.44 | — | — |
| Buffer system + MnCl$_2$ (1.2 × 10$^{-4}$ $M$) | 0.520 | 0.733 | 1.36 | 0.92 | — |
| Buffer system + MnCl$_2$ + PEPCK (0.71 × 10$^{-4}$ $M$) | 0.252 | 0.355 | 2.82 | 2.38 | 2.59 |
| Buffer system + MnCl$_2$ + PEPCK (1.48 × 10$^{-4}$ $M$) | 0.150 | 0.212 | 4.72 | 4.28 | 4.65 |
| Buffer system + MnCl$_2$ + PEPCK (1.48 × 10$^{-4}$ $M$) + PEP (10$^{-4}$ $M$) | 0.205 | 0.289 | 3.46 | 3.02 | 3.28 |

[a] Taken from unpublished results of R. Barns, D. B. Keech, and W. J. O'Sullivan (1972).

and the calculation of the enhancement from the corrected PRR values [see Eq. (2)].

## C. Assessment of Data

It is pertinent here to make some remarks about the analysis of enhancement data. In most of the work in the literature, estimates of the enhancements of various complexes and the respective dissociation constants have been obtained by graphical extrapolation procedures (e.g., Mildvan and Cohn, 1965, 1966; O'Sullivan and Cohn, 1966a). The fact that such procedures appeared to yield values for dissociation constants that were in satisfactory agreement with estimates obtained by other methods (viz., enzyme kinetics, substrate protection kinetics), was to a large extent responsible for their acceptance.

Subsequent analyses (Danchin, 1969; Reed et al., 1970) have thrown considerable doubt on the validity of the graphical extrapolations, even for simple binary complexes. This has led to the development of computer techniques, basically of an iterative nature, for the solution of enhancement data. The theoretical considerations have been discussed in some detail by Deranleau (1969a,b). Most important of Deranleau's conclusions is that approximately 75% of the saturation curve for complex formation is required to define the system accurately. This is a general requirement, relevant to any complex studied by any spectroscopic method. It should, in principle, be fairly easy to meet for the formation of binary complexes but, as discussed below, it may be somewhat more difficult for higher complexes.

## 1. Binary Complexes

For a simple binary complex, involving the formation of manganese–enzyme complex, there are two species in solution and the measured enhancement is given by

$$\epsilon^* = \sum \frac{[M]_i \epsilon_i}{[M]_T} = \frac{[M]}{[M]_T} \cdot \epsilon_0 + \frac{[EM]}{[M]_T} \cdot \epsilon_b \qquad (19)$$

$\epsilon_0$, the enhancement of free manganese M, is 1.0 by definition. $\epsilon_b$ is the enhancement of the EM complex. The dissociation constant for the equilibrium between metal ion and enzyme is given by

$$K_D = \frac{[E][M]}{[EM]} \qquad (20)$$

It can be seen by inspection of Eq. (19) that for a fixed value of $[M]_T$, as $[E]$ is increased, $[EM]$ will tend to $[M]_T$, $[M]$ to zero, and thus $\epsilon^*$ to $\epsilon_b$. Thus, linear extrapolations of $1/\epsilon^*$ against $1/[E]$ have been used to determine $\epsilon_b$ and $K_D$. This is somewhat hazardous as such plots are theoretically nonlinear (Danchin, 1969) and one must rely upon obtaining $\epsilon^*$ values sufficiently close to saturation that the extrapolation does not introduce any significant error. However, as demonstrated by Fig. 1 of Deranleau (1969a), while this may yield a satisfactory value for $\epsilon_b$, the estimation of $K_D$ is much less reliable. Estimation of the latter should be obtained from a range of approximately 20–80% of the theoretical saturation curve and, at greater saturations, the relative error in $K_D$ increases markedly. In Fig. 17, plots of $\epsilon^*$ vs. $[E]$ and of $1/\epsilon^*$ vs. $1/[E]$ are shown, for theoretical data, constructed with $\epsilon_b = 10.0$ for the EM complex and $K_D$ values of $10^{-4}\,M$ and $10^{-5}\,M$ respectively. Provided that the right portion of the curve is selected, the graphical analysis will yield reasonable results for $K_D$ of $10^{-4}\,M$ but is unlikely to do so for $K_D$ of $10^{-5}\,M$. The answer to the problem of analysis appears to lie in the use of nonlinear regression methods, for which computer programs are available (see Reed et al., 1970). A combination of EPR determinations of free manganese and PRR measurements of $\epsilon^*$ allows a direct evaluation of the enhancement for the binary complex (Mildvan and Cohn, 1963, 1965).

Should there be reason to suspect more than one equivalent site, this is most reliably determined by a Scatchard plot. A good example of this (from EPR data) has recently been published (Reuben and Cohn, 1970). For nonequivalent sites, the mathematical treatment may become formidable and the reader is referred to Danchin (1969) and Cohn et al. (1969) where this is treated for the interaction of $Mn^{2+}$ and tRNA.

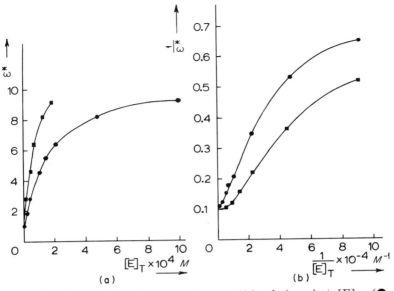

FIG. 17. (a) Plot of theoretical data, for $\epsilon_b = 10.0$, of $\epsilon^*$ against $[E]_T$; (●—●) $K_D = 10^{-4} M$; (■—■) $10^{-5} M$. (b) Double reciprocal plots of the same data.

### 2. Ternary Complexes

The analysis of systems containing ternary complexes becomes considerably more complex. In this case one must consider contributions from three forms of complexed metal ion, Mn–substrate (MS), Mn–enzyme, and the ternary complex (EMS); i.e., the measured enhancement is

$$\epsilon^* = \sum \frac{[M]_i \epsilon_i}{[M]_T} = \frac{[M]\epsilon_0}{[M]_T} + \frac{[EM]\epsilon_b}{[M]_T} + \frac{[MS]\epsilon_a}{[M]_T} + \frac{[EMS]\epsilon_t}{[M]_T} \quad (21)$$

$\epsilon_a$ refers to the enhancement of the MS complex and $\epsilon_t$ to the enhancement of the ternary complex.

Apart from $K_D$, it is necessary to consider the dissociation constants

$$K_1 = \frac{[M][S]}{[MS]} \quad (22)$$

$$K_2 = \frac{[E][MS]}{[EMS]} \quad (23)$$

$$K_s = \frac{[E][S]}{[ES]} \quad (24)$$

and the conservation equations

$$[M]_T = [M] + [EM] + [MS] + [EMS] \tag{25}$$
$$[E]_T = [E] + [ES] + [EM] + [EMS] \tag{26}$$
$$[S]_T = [S] + [MS] + [ES] + [EMS] \tag{27}$$

Consider a type I enzyme with $\epsilon_t$ significantly greater than both $\epsilon_a$ and $\epsilon_b$. Increasing [E] will tend to increase [EMS]; increasing [S] will have the effect of first increasing and then decreasing [EMS] as S starts to compete with MS. Thus, a plot of $\epsilon^*$ vs. [S] should reach a maximum somewhat below the value of $\epsilon_t$ and then decrease (see Fig. 1 and compare Fig. 15). Extrapolation of double reciprocal plots of data on the ascending portion of such curves has been used to estimate values of $\epsilon_t$ and $K_2$. These have yielded satisfactory results in some cases but have been misleading in others (see Reed et al., 1970). The difficulties in interpreting this type of data have now largely been met by computer adaptations of standard successive approximation methods.

Examples of the interpretation of data for binary and ternary complexes are given in later sections (Section V,F, and G).

### 3. Quaternary Complexes

The PRR method has provided unequivocal evidence for the formation of various quaternary complexes, i.e., of the type $EMS_1-S_2$ (Mildvan and Cohn, 1966; O'Sullivan and Cohn, 1968; O'Sullivan et al., 1969). Though attempts have been made to analyze the experimental data, an exact analysis poses formidable problems, with the possible introduction of a variety of mixed complexes. In some situations, partial solutions may be possible. Thus, (1) if $S_2$ saturates all forms of E at concentrations at which it has no significant effect on the other components in the system, then it would be possible to determine $K_2$ for the equilibrium between MS and $ES_2$; and (2) if the equilibria between $S_2$ and the various forms E, ES, and EMS are assumed to be similar, then a reasonable value for its dissociation constant can be obtained by graphical procedures.

The former appears to be potentially applicable to the MnADP–arginine kinase–L-arginine system, though in practice the observed enhancements have been too low to allow accurate analysis. The latter appears to have given reasonable values for the association between creatine and MnADP–creatine kinase (O'Sullivan and Cohn, 1968) and between L-arginine and MnADP–L-arginine kinase (O'Sullivan et al., 1969).

## D. Temperature Dependence of PRR

The effects of temperature variation on the measured PRR arise from a multitude of causes. If we restrict the discussion to the case of Mn

(II), the terms in Eqs. (3), (12), and (19) which may vary significantly with temperature include $[Mn]_i$, $\tau_M$, $\tau_r$, and $\tau_s$. The distribution of chemical species $[Mn]_i$ in a solution may be expected to vary with temperature as the equilibrium constant(s)

$$K_{eq} = K_0 \exp(\Delta H/RT - \Delta S/R), \tag{28}$$

where $\Delta H$ and $\Delta S$ refer to the normal enthalpy and entropy of complex formation.

The terms $\tau_M$ and $\tau_r$, the mean residence time of water in the first hydration sphere and the rotational correlation time, vary similarly with temperature

$$\tau = \tau_0 \exp(\Delta H^*/RT - \Delta S^*/R) \tag{29}$$

where $\Delta H^*$ and $\Delta S^*$ are interpreted as enthalpies and entropies of activation. It is expected that $\Delta H^*$ for $\tau_M$ variation will be in the range 6–9 kcal/mole[4] and that $\Delta H^*$ for $\tau_r$ variation will be 2–5 kcal mole. The variation of $\tau_s$ with temperature is a little more complicated, as will be discussed in Section V,E. However, over a narrow temperature range one may expect $\tau_s$ to show an apparent activation energy of from $-4$ to $+4$ kcal/mole with a dependence on the magnetic field strength.

Sample concentrations may usually be adjusted so that influence on PRR of the temperature dependence of binding constants may be neglected (or corrected).

Relative values of $\tau_M$ and $T_{1,M}$ will then determine the temperature dependence of PRR. When exchange is rapid $\tau_M \ll T_{1,M}$, and the effects of temperature will be contained in the $\tau_c$ terms of Eq. (14). For very long correlation times $(\tau_c \omega_I)^2 \gg 1$, the PRR will vary as $\tau_c^{-1}$. Conversely, for short correlation times, $(\tau_c \omega_I)^2 \ll 1$, the PRR will vary directly as $\tau_c$. For conditions of slow exchange, $\tau_M > T_{1,M}$, the PRR will vary directly as $\tau_M$. If the exchange rate is much slower than $1/T_{1,os}$, the contribution from outer-sphere relaxation, then only the temperature dependence of $T_{1,os}$ will be observed.

Variable temperature data are usually presented as a plot of $\ln 1/T_{1,M}$ vs. $1/T°$. The exponential variation of the various terms will then be exhibited as straight lines with slope(s) equal in magnitude to the corresponding enthalpy change(s). These considerations are illustrated in Fig. 18.

The regions A, B, and C correspond to so-called fast exchange, $\tau_M$ limited, and outer-sphere regions, respectively. The borderline between

---

[4] For ions other than Mn(II) the values of $\Delta H^*$ for proton exchange may be as high as ~15 kcal/mole (Swift and Connick, 1962).

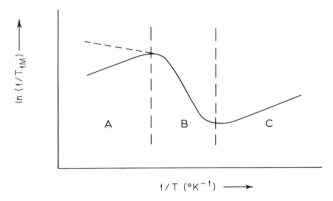

FIG. 18. Variation of ln $1/T_{1,M}$ with respect to reciprocal temperature.

regions A and B is the temperature at which $\tau_M \simeq T_{1,M}$ and the division of regions B and C is at the temperature where $\tau_M \simeq T_{1,os}$.

E. FREQUENCY (FIELD) DEPENDENCE OF PRR

In many ways the effects of frequency variation on PRR are much simpler than are those of temperature variation. First of all, frequency (magnetic field) variation can change only the magnetic parameters in the relaxation equations, $\omega_I$, $\omega_s$, and $\tau_s$. For this reason, frequency independence is a critical test for $\tau_M$-limited PRR.

For any complex with a significant enhancement ($\epsilon > 1$), then $(\tau_c \omega_s)^2 \gg 1$ so that the term in Eq. (14) containing $\tau_s$ is negligible and Eq. (15) applies. Variable frequency data may be usefully presented as a plot of $T_{1,M}$ vs. $\omega_I{}^2$. For cases where $\tau_s$ is not the dominant correlation time, a straight line is obtained with a horizontal intercept at $-\omega_I{}^2 = (\tau_c)^{-2}$. If the relevant correlation time is $\tau_s$, then frequently a nonlinear plot is obtained. The electron relaxation time $\tau_s$ is generally field dependent. The field dependence of $\tau_s$ may be assessed at each field value from the ratio of slope to vertical intercept of the $T_{1,M}$ vs. $\omega_I{}^2$ plot.

Bloembergen and Morgan (1961) have derived an approximate expression for the frequency dependence of the electron relaxation time $\tau_s$ of Mn.

$$1/\tau_s = B[\tau_r/(1 + \omega_s{}^2\tau_r{}^2) + 4\tau_r/(1 + 4\omega_s{}^2\tau_r{}^2)] \qquad (30)$$

where B is a constant, and $\tau_v$ is the correlation time for electron relaxation and represents a characteristic time for symmetry distortions of the Mn complex. Values of $\tau_r$ at 27°C range from $2 \times 10^{-12}$ sec (Bloembergen and Morgan, 1961; Reed et al., 1971; Reuben and Cohn, 1970) for

$Mn(H_2O)_6$ to $6 \times 10^{-12}$ sec for Mn–enzyme complexes (Reuben and Cohn, 1970).

## F. Formation of a Binary Complex: Mn–Bovine Serum Albumin

The first detailed report of the use of PRR measurements to study metal ion–protein interactions was that of Mildvan and Cohn (1963) on the binding of manganous ion to bovine serum albumin. This paper laid the foundation for many of the procedures used in PRR studies. Titrations were carried out with constant manganous ion and varying protein concentrations, and vice versa. In some cases supplementary measurements of free $Mn^{2+}$ were made using an electron paramagnetic resonance spectrometer.

An example of the titration of variable protein with constant $Mn^{2+}$ concentration is illustrated in Fig. 19 [data taken from Table I of Mildvan and Cohn (1963)]. Graphical analysis [see Fig. 1 of Mildvan and Cohn (1963)] of this data yielded an estimate of $\epsilon_b$ of $11.7 \pm 1.1$ with $K_D = 3.4 \times 10^{-5} M$. These values are in good agreement with the results of more recent computer analyses, viz., $\epsilon_b$, 10.2 and $K_D$, $2.0 \times 10^{-5} M$, with a standard deviation of 5.1%.

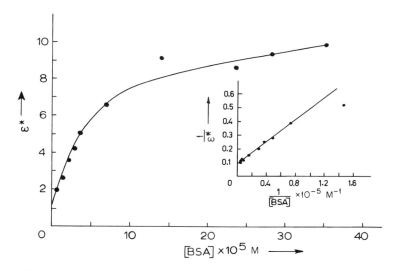

Fig. 19. Variation of enhancement with bovine serum albumin concentration. Data taken from Table 1 of Mildvan and Cohn, *Biochemistry* **2**, 912 (1963). Experiments were carried out in $0.05 M$ Tris–HCl, pH 7.5, and $0.5 M$ tetramethylammonium chloride with $5 \times 10^{-5} M$ $MnCl_2$ at 24°C. Inset: double reciprocal plot of data [reproduced with permission from Mildvan and Cohn, *Biochemistry* **2**, 910 (1963)].

## G. FORMATION OF A TERNARY COMPLEX: MnADP–CREATINE KINASE

The type of titration obtained with creatine kinase, ADP, and manganous ion was illustrated in Fig. 1b. This system was studied in some detail by O'Sullivan and Cohn (1966a), who used graphical procedures to estimate $\epsilon_t$ as 19.4 and $K_2$ as 50 $\mu M$. Subsequent computer analysis revealed that a considerable degree of degeneracy existed in the fit for different $\epsilon_t$ and $K_2$ values. To minimize the degeneracy it was necessary to go to much higher concentrations of enzyme than had been used previously, with correspondingly high concentrations of ADP, in order to cover a sufficient portion of the saturation curve. This introduced two complications. First, the high values for [E] meant that [EM] could not be neglected, particularly as the enhancement for this complex was found to be higher than had been previously recognized.[5] Second, at high [ADP], competition between $ADP^{3-}$ and $MnADP^-$, and thus the value of $K_s$, became very important and it was necessary to extend the titration to take this into account.

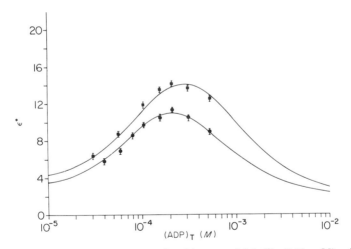

FIG. 20. PRR titration data for creatine kinase and $MnCl_2$ (0.10 m$M$) with ADP at 24°C. Upper curve: $[E]_T = 0.389$ m$M$ active sites; lower curve $[E]_T = 0.205$ m$M$ active sites. Solid curves drawn with $K_D = 0.5$ m$M$, $\epsilon_b = 6.5$, $K_1 = 0.03$ m$M$, $\epsilon_a = 1.7$, $K_2 = 0.064$ m$M$, $\epsilon_t = 20.5$, $K_s = 0.112$ m$M$. Values of $K_2$, $K_s$, and $\epsilon_t$ taken from minimum % S.D. (2.9) in regression analysis. [Reproduced, with permission, from Reed et al., J. Biol. Chem. **245**, 6547 (1970).]

[5] Recent investigations (G. H. Reed and M. Cohn, private communication) have indicated a value of approximately 6 for $\epsilon_b$ of Mn–CrK and of 0.5 m$M$ for $K_D$. O'Sullivan and Cohn (1966a) had obtained 1.6 for $\epsilon_b$. It is not clear whether or not the more recent result indicates weak binding of the Mn to the active site of the enzyme.

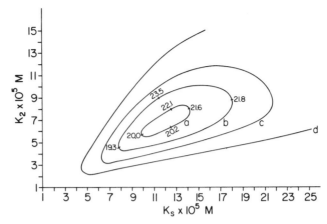

Fig. 21. Constant percentage standard deviation contours as functions of $K_2$ and $K_s$ for data given in Fig. 20. (a) 3.0%; (b) 3.5%; (c) 4.0%; (d) 5.0%. [Reproduced, with permission, from Reed et al., J. Biol. Chem. 245, 6547 (1970).]

The titrations shown in Fig. 20 were designed to include the region of ADP concentrations which theoretical curves had shown to be optimal for distinguishing between possible values of $K_2$, $K_s$, and $\epsilon_t$. The results of calculation from this data are illustrated in Fig. 21 in the form of contour lines of constant error (percentage standard deviation) as functions of $K_2$ and $K_s$. The best fit was obtained at $K_2$ and $K_s$ equal to 64 and 112 $\mu M$, respectively, and $\epsilon_t$ equal to 20.5. In this case, the graphical method had given results in reasonable agreement with those of the computer analysis, though it may be noted that there was very considerable discrepancy in the case of adenylate kinase (Reed et al., 1970).

## H. Protein Modification

Experiments by Mildvan and Cohn (1963) on bovine serum albumin illustrate the use of PRR to follow changes in protein modification. First, the thiol group of bovine serum albumin had been implicated by Holeysovosoka (1961) as the binding site for manganese, from equilibrium dialysis studies. However, Mildvan and Cohn (1963) showed that the reaction of the thiol group with p-mercuribenzoate had no effect on the binding of $Mn^{2+}$. Rather, experiments at variable pH values indicated that dissociable groups, presumably ligands for manganese, with p$K_a$'s consistent with imidazole groups, were involved with the binding of manganese.

Second, the effects on the enhancement parameters for Mn–bovine serum albumin, of various concentrations of the denaturing agents, urea, guanidinium chloride, and decyl sulfate were studied. The effects were complex, particularly with the guanidinium chloride, which appeared to compete directly for some of the manganese binding sites on the protein, apart from disturbing the protein structure. However, to quote from the authors, "it should be pointed out that effects on enhancement of the PRR of the manganese–albumin complex occur at concentrations of urea and guanidinium below those required to produce changes in rotatory dispersion, indicating that the former is a more sensitive indicator of local changes of protein structure."

PRR has been used extensively to study the binding of substrates to modified forms of creatine kinase as is described in Section VI,A.

I. Spin Labels

The use of PRR measurements, in addition to EPR, to study spin-labeled compounds is discussed for creatine kinase in Section VI,A and for alcohol dehydrogenase in Section VI,C. These studies have demonstrated that PRR changes brought about by the paramagnetic probe can provide very useful information concerning the binding of substrates and substrate analogs to enzymes, including the detection of small changes which could be attributed to local conformation changes in the enzyme. With the wide variety of spin label probes now available (McConnell and McFarland, 1970), it appears possible that PRR measurements could be much more widely applied, provided that one is dealing with an aqueous environment.

We would note however that it may be difficult to extend this type of study so that it is possible to estimate the factors [see Eqs. (3), (4), and (15)] that determine the PRR. Some recent experiments by Cohn and McLaughlin on frequency and temperature dependence of the PRR of spin-labeled creatine kinase complexes have not yielded quantitative data on these factors. It appears that a stable hydration sphere does not form around the spin label on the enzyme, so that two forms of water can contribute to the PRR, "free" water in the active site and "bound" water in the hydration sphere of the bound substrate and/or amino acids. Thus, one always has a component of an ill-defined outer-sphere relaxation. The problem could be a general one and would thus limit the usefulness of PRR measurements in the study of spin label compounds.

## VI. Individual Enzymes

In this section, the type of information available from PRR studies is illustrated by reference to four specific cases, chosen because they have been the most extensively studied or because they illustrate certain points. The discussion is extended beyond information obtained only from PRR studies to include information obtained from other magnetic resonance studies where this is complementary or gives a more complete picture.

The proteins which will be treated in further detail are creatine kinase, which is the prototype of type I enzymes and also illustrates some potentialities of the use of spin labels; pyruvate kinase, which is the prototype of type II enzymes; liver alcohol dehydrogenase, which illustrates the information available from a spin-labeled substrate analog; and apotransferrin and transferrin, which illustrate the type of information available from measurements at variable temperature and frequency. A comprehensive review of enzymes studied by nuclear relaxation techniques has recently appeared (Mildvan and Cohn, 1970).

### A. Creatine Kinase

Creatine kinase (ATP: creatine phosphotransferase E.C. 2.7.3.2) catalyzes the reversible transfer of a phosphoryl group from ATP to creatine. The reaction shows an absolute requirement for a divalent metal ion, the primary function of which is to form the active complexed species, $M-ATP^{2-}$ and $M-ADP^{-}$, respectively, of the nucleotide substrates (Kuby and Noltmann, 1962; Morrison and O'Sullivan, 1965). The metal ion requirement is met by $Mg^{2+}$, $Mn^{2+}$, $Ca^{2+}$, and $Co^{2+}$. There does not appear to be any significant variation in the enzymic mechanism with different metal ions (Morrison and Uhr, 1966; O'Sullivan and Cohn, 1966a).

Creatine kinase has been subjected to extensive study by magnetic resonance techniques (see, in particular, Cohn, 1970) and may be discussed as the prototype of type I enzymes with $\epsilon_t > \epsilon_b$ and where there appears to be a substrate bridge between enzyme and metal ion. The early report of Cohn and Leigh (1962), which established the mode of binding of the metal ion, was extended by O'Sullivan and Cohn (1966a) to obtain enhancement parameters for various manganese-containing complexes of the enzyme and the relevant dissociation constants. As already stated, this work has subsequently been subjected to more rigorous analysis using the computer techniques of Reed et al. (1970), leading to a greater degree of precision for the respective parameters, though not to any qualitative changes in the overall picture. We discuss below the types of studies that have been carried out with creatine kinase.

### 1. Determination of Enhancement Parameters

Graphical analysis of the results obtained by O'Sullivan and Cohn (1966a) yielded values of 19.4 and 9.8 for the MnADP–enzyme and MnATP–enzyme complexes, respectively, at 24°C in 0.05 $M$ $N$-ethyl morpholine–HCl, pH 8.0 and 7.5.[6] The relevant dissociation constants for the dissociation of MnADP⁻ and MnATP²⁻ were determined as 50 $\mu M$ and 130 $\mu M$, respectively.

The values for MnADP⁻ have subsequently been subjected to computer analysis, as described in Section V,E (Reed et al., 1970).

The complex formed with MnATP²⁻ posed more formidable problems. Assessment of the data of O'Sullivan and Cohn (1966a) and of unpublished work showed that the experiments were carried out at much too low concentrations of enzyme and ATP to be analyzed meaningfully. Unfortunately, attempts to overcome this were thwarted by the ATPase activity of the enzyme (Sasa and Noda, 1964). With increasing concentrations of enzyme and ATP, measurements of $T_1$ could not be made within the time for negligible breakdown of ATP.[7]

### 2. Studies with Different Substrates

Apart from ADP (or ATP), creatine kinase can utilize a number of nucleotide diphosphates as substrates. This includes not only the "naturally occurring" compounds, such as dADP and IDP (James and Morrison, 1966), but also synthetic analogs, such as 2-Cl-ADP (O'Sullivan et al., 1972). Investigation of the first group by O'Sullivan and Cohn (1966b) showed that there appeared to be a very close correlation between the enhancement of the ternary complexes formed with ADP, 2'-dADP, 3'-dADP, IDP, and GDP and their ability to act as substrates of the reaction. A third parameter, the ability of the metal–nucleotide to potentiate the reaction between iodoacetate and the essential thiol group at the active site of creatine kinase, also followed the same order very closely. Such a three-way correlation would be a powerful argument in favor of a graded degree of substrate-induced conformation change (O'Sullivan and Cohn, 1966b). Unfortunately this interpreta-

---

[6] The lower pH was used for MnATP-enzyme to minimize the ATPase activity of the enzyme.

[7] Cohn et al. (1971) have recently published estimates of $\epsilon_t = 13.5$ and $K_2 = 0.13$ m$M$ for the binding of MnATP²⁻ to creatine kinase.

tion must be open to question, as the analysis of the data was carried out using the approximate graphical procedures, so that extrapolations were made from different portions of the curve of enhancement against substrate concentration and thus indirectly reflected differences in $K_2$. Nevertheless, recent measurements from both the Sydney and the Philadelphia laboratories have confirmed a lower $\epsilon_t$ value with 2′-dADP than with ADP. It has also been found (Cohn, 1970) that, in common with other manganous ion complexes, the correlation time of the MnADP–creatine kinase complex is frequency dependent; i.e., $\tau_s$ rather than $\tau_r$ appears to be the dominant term in Eq. (4). Possibly, different $\epsilon_t$ values could arise from different $\tau_s$ values for the respective MnNDP–enzyme complexes, which would be compatible with different Mn–nucleotides inducing different enzyme conformations. It is pertinent to note that a number of synthetic analogs of ADP, viz., 2Cl-ADP, 2Cl-$N$-6-CH$_3$-ADP, and 2-CH$_3$-S-ADP, give enhancements similar to that of ADP ($\epsilon_t \sim 20$) though these all bind more strongly to the enzyme. The 2Cl-ADP appears to be as good as, or better than ADP as a substrate (O'Sullivan et al., 1972). A final decision will have to await the accurate determination of all the parameters determining the PRR. Similar reservations apply to the interpretation of the data for ATP analogs with adenylate kinase (O'Sullivan and Noda, 1968).

### 3. Formation of a Quaternary Complex

The addition of creatine to a solution containing a significant concentration of MnADP–creatine kinase can produce a significant change in the PRR of the system, presumably due to the formation of the complex MnADP–enzyme–creatine. At low temperatures, $1/T_{1,p}$ is decreased and at high temperatures it is increased, the crossover point occurring at 25°–30°C, depending on the concentration of the various compounds. Such a change, the "creatine effect," is dependent on the integrity of the two essential —SH groups of the enzyme. If they are chemically modified which results in loss of enzymic activity, the creatine effect is lost, though the binding of MnADP$^-$ and MnATP$^{2-}$ appears to be only slightly affected (O'Sullivan and Cohn, 1968; O'Sullivan, 1971) and creatine can still bind to the modified enzyme (Cohn, 1970).

The apparent difference in temperature behaviors of the ternary and quaternary complexes led to the postulate that while the former was $\tau_c$ dominated, the latter was probably $\tau_M$ dominated; i.e., the effect of creatine was to induce a conformation change so as to increase the residence time of the water molecules in the first coordination sphere of the bound manganese at the active site. While a conformation change at the active site does indeed appear to take place in going from the ternary to the quaternary complex (O'Sullivan et al., 1966; Lui and Cunningham, 1966; Cohn, 1970) the data is not sufficient to analyze it in terms of the various PRR parameters.[8]

It is interesting to note that arginine kinase, which is similar in many respects to creatine kinase, demonstrates an analogous arginine effect with the substrate L-arginine. No effect is observed with D-arginine, which is a weak inhibitor of the enzyme (O'Sullivan et al., 1969).

### 4. Spin-Labeled Creatine Kinase

Taylor et al. (1969) described the introduction of a stabilized nitroxide derivative of iodoacetamide at the thiol groups of creatine kinase from both rabbit muscle and chicken heart. Three types of magnetic resonance measurement were applied to the study of the environment of the spin label.

1. The EPR signal from the free radical was relatively stabilized on attachment to the enzyme. Small but significant changes were seen on titrating with ADP or diamagnetic metal complexes. On titrating with MnADP-, the signal "collapsed," indicating that the two paramagnetic species were relatively close. It was calculated that the distance between the unpaired electron of the nitroxide moiety and the bound manganese ion was $\sim 8$ Å in MnADP–enzyme and $\sim 11$ Å in MnATP–enzyme complexes.

2. The spin label contributed to the PRR of water and its attachment to the enzyme increased the effect by a factor of approximately six. This changed further on titration with ADP or diamagnetic metal–ADP and the change in the PRR proved to be a more sensitive parameter than the change in the EPR signal.

3. Further experiments using continuous-wave NMR spectroscopy were carried out with the spin-labeled chicken heart enzyme, which retains activity after complete alkylation of the —SH groups (Hooton, 1968). Again, an approximate calculation for the free radical–proton distance could be made. This distance increased (i.e., the broadening effect decreased) on the addition of $Mg^{2+}$ and still further on the addition of creatine. Fuller details of the experiments with spin-labeled creatine kinase have now appeared (Taylor et al., 1971; Cohn et al., 1971).

[8] Recent experiments on the frequency and temperature dependence of the PRR, together with ancillary EPR measurements, have provided strong evidence that there are no water molecules remaining in the first coordination sphere of the metal ion in the quaternary MnADP-creatine kinase–creatine complex (G. H. Reed, H. Diefenbach and M. Cohn (1972), J. Biol. Chem. **247**, 3066; G. H. Reed and M. Cohn (1972), J. Biol. Chem. **247**, 3073).

## B. Pyruvate Kinase

Pyruvate kinase catalyzes the reaction

$$\text{ATP} + \text{pyruvate} \underset{}{\overset{M^{2+},\ M^+}{\rightleftharpoons}} \text{ADP} + \text{phosphoenolpyruvate}$$

and also two nonphysiological reactions (acting then as a fluorokinase or a hydroxylamine kinase) with fluoride ion and hydroxylamine, respectively, as alternate substrates for pyruvate, provided that bicarbonate ion is also present (Boyer, 1962). The divalent metal may be $Mg^{2+}$ or $Mn^{2+}$ but not $Ca^{2+}$. The most comprehensive NMR studies have been carried out on the rabbit muscle enzyme, a tetramer of M.W. 237,000, with apparently identical subunits (Cottam et al., 1969).

Pyruvate kinase, isolated from rabbit muscle, has served as the prototype of type II enzymes. A large enhancement is seen with manganous ion and enzyme alone and for the binary E–Mn complex, $\epsilon_b$ was estimated by Mildvan and Cohn (1965) to be 33 ± 3 at pH 7.5, 29°C, in the presence of 0.1 $M$ KCl. The enhancement was decreased in the presence of any of the substrates, ADP, ATP, phosphoenolpyruvate, or pyruvate. A comparison of the enhancements of the E–MnADP and E–MnATP complexes, namely $\epsilon_t$ values of 20 and 13, respectively, led the authors to propose that the difference could represent the removal of an extra water molecule from the first coordination sphere of the bound metal ion in the latter complex. In other words, the metal ion was at least bound to the terminal phosphate of the ATP molecule (i.e., to the phosphoryl group undergoing transfer), providing a bridge between the enzyme and the substrate. In this case, it would be expected that the phosphate moiety of phosphoenolpyruvate should also provide a ligand group to the bound manganese ion. Indirect but very convincing evidence for this proposal was obtained from parallel observations on the fluorine resonance of $FPO_3^{2-}$ (Mildvan et al., 1967). Fluorophosphate, the substrate of the fluorokinase reaction catalyzed by pyruvate kinase, binds at the same site as phosphoenolpyruvate and with approximately the same enhancement. The magnetic resonance data, particularly the change in the relaxation time of the fluorine resonance due to the bound manganese, was used to calculate the distance between the metal and the fluorine atoms on the enzyme surface. These calculations were consistent with the binding of $FPO_3^{2-}$ to manganese through an oxyanion.

Compared with the ADP and ATP complexes, much larger decreases in enhancement were calculated for the E–Mn–phosphoenolpyruvate and E–Mn–pyruvate complexes, $\epsilon_t$ values being 2.2 and 11, respectively. These could not be explained in terms of a change in the number of coordinated water molecules alone (though this could possibly be partially responsible for the difference between the two). It was suggested that some conformational change had taken place in each case and that this was reflected in the denominator terms $T_{1,M}$ and $\tau_M$ of Eq. (3). Evidence for conformation changes has also been forthcoming from ultraviolet spectroscopy (Kayne and Suelter, 1965). Further support has come from the difference in the temperature behavior of the PRR for the E–Mn complex as compared with the E–Mn–phosphoenolpyruvate complex (Cohn, 1967). It would appear that the PRR for the latter complex is completely determined by $T_{1,M}$, whereas $\tau_M$ is also of significance for the binary complex.

A recent study by Reuben and Cohn (1970) is of particular interest, as it represents the most detailed analysis to date of the effect of both temperature and frequency on the PRR of manganese–enzyme complexes. Experiments were carried out at four frequencies, 8.13, 24.3, 40.0, and 60.0 MHz, and over the temperature range 5°–37°C. It was found that the PRR did not vary in the same way at all frequencies. At the two lower frequencies $1/T_{1,p}$ for E–Mn decreased monotonically with increasing temperature. However, at the higher frequencies, a maximum with respect to temperature was observed. As a maximum was not observed at the frequency at which the enhancement was greatest, the authors concluded that the PRR was principally determined by $T_{1,M}$, i.e., by $\tau_c$. Experiments at constant temperature with variable frequency demonstrated that the PRR was frequency dependent, so that the term in $\tau_c$ was dominated by $\tau_s$, the electron spin relaxation time. The authors suggested that other macromolecular systems that bind manganous ions might also prove to be $\tau_s$ dominated (cf. Peacocke et al., 1969) rather than $\tau_r$ dominated as had been previously thought (Eisinger et al., 1962; Cohn, 1963) and as is the case for the manganous aquocation (Bloembergen and Morgan, 1961).

The data could also be analyzed to obtain an estimate of the number of water molecules remaining in the first coordination sphere of the bound manganese. Reuben and Cohn (1970) reached the conclusion that a value of 3 best fitted the experimental data. Though it is to be anticipated that such experiments for ternary and higher complexes will be much more difficult to unravel, the returns in terms of obtaining accurate information about the active site of metal–activated enzymes would appear to be great.

C. Alcohol Dehydrogenase

Two possibilities offer themselves for the use of spin labels in the study of enzyme mechanisms. First, the spin label can be attached to a specific portion of the enzyme as was described in Section VI,A,4 for creatine kinase. Second, there is the possibility of incorporating the spin label into a substrate analog. This has been successfully achieved for an analog of NAD (nicotinamide adenine dinucleotide) and thus has led to a new approach to the study of the large group of enzymes, the dehydrogenases, for which NAD serves as a substrate (Weiner, 1969).

The structure of the analog, adenosine diphosphate 4-(2,2,6,6-tetramethylpiperidine-1-oxyl) (abbreviated to ADP-R°) is compared with that of NAD in Fig. 22. The unpaired electron is localized in a region corresponding to the ribotide bond between the pyridine nitrogen and the C-1 atom of the ribose group. As anticipated, ADP-R° in solution increased the PRR of water protons and the PRR was enhanced on the binding of ADP-R° to the enzyme, liver alcohol dehydrogenase. Extensive magnetic resonance studies were carried out with this analog and the liver alcohol dehydrogenase (Weiner, 1969; Mildvan and Weiner, 1969a,b), including electron paramagnetic resonance, PRR studies, and continuous-wave NMR measurements. ADP-R° was shown to be a

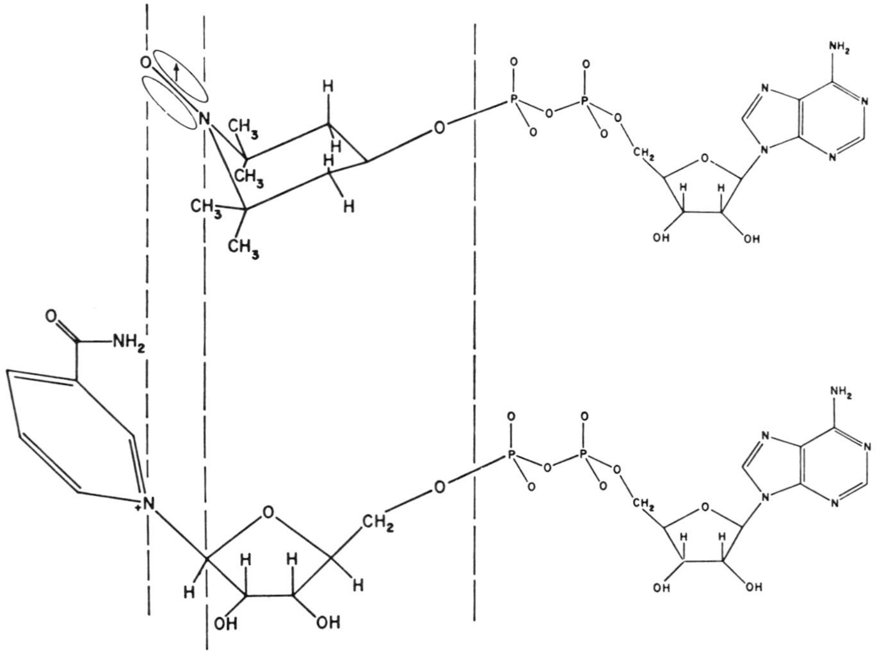

Fig. 22. Comparison of the structures of NAD and ADP-R°. [Reproduced, with permission, from Mildvan and Weiner, *Biochemistry* **8**, 552 (1969).]

competitive inhibitor with respect to the NAD substrate, so that the magnetic resonance experiments could be considered relevant to the active site of the enzyme.

Two types of binding sites for ADP-R° on the enzyme could be determined from both EPR and PRR experiments, namely two tight binding sites and five or six weak binding sites. The enhancement $\epsilon_b$ was found to decrease from $\sim 80$ to $\sim 13$ as the occupancy of the ADP-R° binding sites increased from 0 to 2, indicating site–site interaction. This was attributed to a decrease in $\tau_r$, consistent with opening up of the site. A decrease in $\epsilon_b$ was observed on the addition of ethanol, isobutyramide, or acetaldehyde to the binary E–ADP-R° complex; that is, the enzyme exhibited type II behavior. The decrease was consistent with the formation of E–ADP-R°–S complexes in which the substrates and inhibitor diminished the accessibility of water molecules to the unpaired electron of the bound ADP-R°. Dissociation constants calculated for these respective compounds were again in satisfactory agreement with kinetic estimates.

Experiments were also carried out with the apoenzyme in which the essential zinc atoms had been removed. The apoenzyme retained the two tight binding sites but not the weak binding sites and the enhancements for the former were considerably reduced, though the site–site interaction could still be observed.

## D. Application of Variable Frequency Measurements

Though potentially of great value in the estimation of the parameters that determine the PRR, particularly when carried out in conjunction with measurements at variable temperature (Reuben and Cohn, 1970), few investigations of the frequency dependence of PRR have been carried out. This may be partially due to the relatively sophisticated instrumentation needed, though measurements at two or three frequencies can yield useful information, particularly as to the importance of $\tau_s$ in determining the PRR for paramagnetic systems.

The most extensive investigations of the frequency dependence of PRR have been made by Koenig and co-workers (Koenig and Schillinger, 1969a,b; Gaber et al., 1970; Fabry et al., 1970). Using a pulsed magnetic field technique developed by Redfield and co-workers (Anderson and Redfield, 1959; Redfield et al., 1968), variable frequency studies have been carried out with a high degree of precision over a wide range of frequencies; typically, a range from $\sim 10$ kHz to 50 MHz has been covered. Both diamagnetic and paramagnetic systems have been investigated.

Studies (Koenig and Schillinger, 1969a) with apotransferrin (i.e., transferrin with the metal ions removed) demonstrated the existence of a few water molecules bound rigidly to the protein molecule. These water molecules had enhanced relaxation rates due to proton–proton dipolar interactions, with a correlation time which corresponded to the rotational time of the whole protein molecule. It was thus possible to study the hydrodynamic properties of the protein as a function of pH, temperature, etc. Lifetimes of water molecules bound to protein in the range 0.1–10 μsec were determined.

A number of investigations of systems containing paramagnetic metal ions has been carried out. The systems included transferrin containing $Fe^{3+}$, $Co^{2+}$, and $Cu^{2+}$ and carbonic anhydrase with $Co^{2+}$ substituted for the native $Zn^{2+}$ ion (Koenig and Schillinger, 1969b; Gaber et al., 1970; Fabry et al., 1970).

These studies have all demonstrated additional water bound at the paramagnetic metal ion. In the case of $Co^{2+}$-carbonic anhydrase two types of bound $Co^{2+}$ were indicated. Additions of enzymic inhibitors, such as azide ion or sulfonamides, abolished the enhancement of one class of sites. The authors proposed that this probably represented the active site cobalt.

These studies have clearly demonstrated frequency dispersion of various terms in the relaxation equations. In those cases in which the correlation time itself is frequency independent, an unambiguous determination of $\tau_c$ is possible. However, unlike the studies by Reuben and Cohn (1970) on $Mn^{2+}$-pyruvate kinase described in Section VI,B, the problem of frequency-dependent correlation times ($\tau_s$) was not considered in any detail in the work described above.

## VII. Absorption of Water by Diamagnetic Molecules

Any change in the freedom of motion of a water molecule should lead to some change in values of $T_1$ and $T_2$. In principle, this has the potential of yielding information about the state of water in any biological system (see Tait and Franks, 1971, for a general discussion of water in biological systems). In practice, the difficulty of working with many biological systems and of interpreting the often small changes in $T_1$ and $T_2$ has limited the effective use of magnetic resonance in this area, though evidence for a minimum of two phases of ordered water in skeletal muscle has come from continuous-wave NMR measurements (Hazelwood et al., 1969).

The PRR method has been applied with some success to the problem of elucidating the state of water which has been absorbed into a solid protein matrix, such as keratin in the form of wool and hair. As an illustration of the use of PRR which might serve as an approximate model for the biological situation, some experiments carried out in this laboratory (Physics Department, University of New South Wales) on the absorption of water by various keratins are described.

A. Scope of the Problem

Keratin fibers can absorb up to 34% of their own weight of water at a relative humidity of 100%. It is found that $T_1$ and $T_2$ for the absorbed water protons are unequal and less than the value of 3 sec normally observed for both in water free of paramagnetic species. A number of questions about the nature of the absorbed water are raised. In particular, how does the absorbed water compare with the normal thermodynamic states of pure water. Does it have any "icelike" properties? What is its degree of hydrogen bonding? Does it exist in discrete states or in a continuum of states? If the former, what is the rate of chemical exchange of water between the various states? (Lynch and Marsden, 1971).

B. Experimental Aspects

The wool and horse-hair samples used in these investigations were in the form of cylindrical bundles tightly packed into 8 mm internal diameter glass tubes, with the fibers parallel to the cylinder axis. The bundles were 25 mm long and weighed approximately 1 gm. The rhinoceros horn samples were cut from the bulk horn and machined to a cylindrical shape of similar dimensions. Since rhinoceros horn possesses a discernible grain, which has been shown by X-ray measurements to coincide with the $\alpha$-helix direction of the keratin macromolecules it was possible to prepare samples with the grain either parallel or at right angles to the cylinder axis. Prior to the sample preparation, the bulk wool and hair had been cleaned by washing in ether (to remove natural oils) and in distilled water. In the case of rhinoceros horn, the cleaning took place after the sample was cut. The cleaned and dried specimens were conditioned to a nominal water content by injecting a predetermined amount of water into the sample tubes, which were then sealed and left to equilibrate for at least one week. After the NMR experiments had been completed, the actual water content was determined by drying under vacuum at 105°C to constant weight.

The NMR measurements were carried out at 30 MHz using apparatus similar to that described above, the only essential difference being the larger resonant coil required to accommodate the sample tubes. With this arrangement, however, the rf $H_1$ field was not sufficiently uniform over the sample volume to permit the use of the 180°-$t$-90° pulse sequence for the measurement of $T_1$ or the Carr-Purcell method for $T_2$.

$T_1$ was measured by the 90°-$t$-90° method and the single echo 90°-$t$-180° sequence was used to measure $T_2$, except in those cases where $T_2$ was very short (less than 0.5 msec) when the decay of the free precessional signal could be used.

## C. The Keratin–Water System

Measurements of $T_1$ and $T_2$ of the water absorbed by keratin in the form of wool (Lynch and Marsden, 1969) have been made over a temperature range +90°C to −50°C and for equilibrium water contents ranging up to the saturation value (e.g., 34% equilibrium water content for wool keratin at 100% relative humidity).

The observed NMR signal has two components; a fast-decaying component ($T_2 \sim 25$ μsec) which is attributed to the protons of the keratin molecules and a slower decaying component which is attributed to the absorbed water molecules. At room temperature, $T_2$ of the absorbed water in wool keratin varies from 0.1 msec at 2% equilibrium water content to 8 msec at saturation equilibrium water content, with the rate of change of $T_2$ increasing rapidly as saturation is approached (see Fig. 23a). In general terms, this behavior indicates a lowering of the average binding energy and an increase in the "mobility" of the average water molecule as the equilibrium water content is increased.

Attempts to fit the results to either a two-phase model of the absorbed water (a bound and a free state: Cassie, 1945) or a three-phase model ("bound," "intermediate," and "free" states: Windle, 1956) were unsuccessful. It was possible, however, to obtain a reasonable fit of the room-temperature experimental $T_2$ values to a model of the absorbed water proposed by Feughelman and Haly (1962). In this model the absorbed water molecules are classified into four states according to whether they make 0, 1, 2, or 3 hydrogen bonds with other water molecules or with the protein.

Experimental $T_1$ and $T_2$ values for a given equilibrium water content are illustrated as a function of reciprocal temperature in Fig. 23b. It will be noted that $T_1$ exhibits a rather shallow minimum and that the ratio of $T_1$ to $T_2$ at the $T_1$ minimum is numerically large. Both of these facts are consistent with a theory of nuclear magnetic relaxation based on a continuous distribution of correlation times. By assuming (1) a log-normal probability distribution for $\tau_c$ (i.e., log $\tau_c$ has a normal or

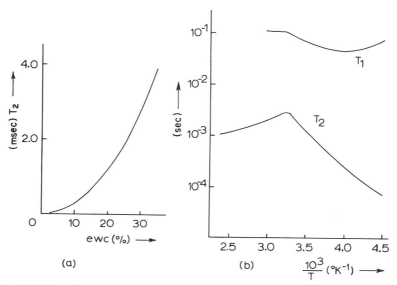

FIG. 23. Relaxation times of the protons of water absorbed by wool keratin. (a) Variation of $T_2$ with equilibrium water content (e.w.c.) at 25°C. (b) Temperature dependence of $T_1$ and $T_2$ in a wool sample having an e.w.c. of 25% approximately.

gaussian distribution) with a width parameter $\beta$ and centered on a mean value $\tau_c^*$ and (2) that $\tau_c^*$ obeys a thermal activation law, it is possible to correlate the theoretical and experimental values of both $T_1$ and $T_2$ below $-40°C$ using only one adjustable parameter (Lynch and Marsden, 1969). According to this analysis the width of the distribution narrows as the equilibrium water content is increased but remains independent of temperature. The activation energy for the mean correlation time also decreases with increasing equilibrium water content.

As the temperature is increased above 36°C (i.e., $1/T$ less than $3.24 \times 10^{-3}$ °K$^{-1}$), $T_2$, instead of increasing begins to decrease. This phenomenon may be explained on the basis of an exchange process taking place between the water protons and protons in a relatively immobile state having a very short $T_2$, e.g., the protons of keratin.

These experimental results refer to wool and hair samples in the form of a cylindrical bundle with the fibers packed parallel to the cylinder axis. With rhinoceros horn, as stated in Section VII,B, it is possible to cut cylindrical samples with the $\alpha$-helix of the keratin macromolecule at right angles to the cylindrical axis so that by rotation of the sample, NMR measurements can be made as a function of the angle between the magnetic field and the direction of the $\alpha$-helix.

In such samples the measured values of $T_2$ show a slight angular dependence which becomes more marked at low temperatures and for low equilibrium water content values. The $T_2$ decay remains exponential with no evidence of multiphase behavior (Lynch and Haly, 1970). These results are consistent with the existence of a state in which the water molecule is bound to the keratin but is free to rotate and for which the lifetime is short so that the conditions of fast exchange apply.

Angular dependence of $T_2$ has also been reported in the collagen–water system (Berendsen, 1962) and DNA–water system (Migchelsen et al., 1968).

### Acknowledgments

We would like to acknowledge support from the National Health and Medical Research Council of Australia (W. J. O'S.) and the Australian Wool Board (K. H. M.).

W. J. O'S. and J. S. L. would also like to acknowledge their debt to Dr. Mildred Cohn for her guidance in the application of magnetic resonance techniques to the study of enzymes. We are grateful to Mr. A. Berzins for his meticulous drawing of many of the figures and to Miss R. Shoebridge for her patient typing of the manuscript.

### References

Abragam, A. (1961). "The Principles of Nuclear Magnetic Resonance" pp. 295 and 311. Oxford Univ. Press, London and New York.
Anderson, A. G., and Redfield, A. G. (1959). *Phys. Rev.* **116,** 583.
Barns, R. J., Keech, D. B., and O'Sullivan, W. J. (1972). *Biochim. Biophys. Acta* **289,** 212.
Berendsen, H. J. C. (1962). *J. Chem. Phys.* **36,** 3297.
Bernheim, R. A., Brown, T. H., Gutowsky, H. S., and Woessner, D. E. (1959). *J. Chem. Phys.* **30,** 950.
Bloembergen, N. (1957). *J. Chem. Phys.* **27,** 572.
Bloembergen, N. (1961). "Nuclear Magnetic Relaxation." Benjamin, New York.
Bloembergen, N., and Morgan, L. O. (1961). *J. Chem. Phys.* **34,** 842.
Boyer, P. D. (1962). In "The Enzymes" (P. D. Boyer, H. Lardy, and K. Myrbäck, eds.), 2nd rev. ed., Vol. 6, p. 95. Academic Press, New York.
Bradbury, E. M., and Crane-Robinson, C. (1968). *Nature (London)* **220,** 1079.
Carr, H. Y., and Purcell, E. M. (1954). *Phys. Rev.* **94,** 630.
Cassie, A. B. D. (1945). *Trans. Faraday Soc.* **41,** 458.
Chaberek, S., Courtney, R. C., and Martell, A. E. (1952). *J. Amer. Chem. Soc.* **74,** 5057.
Cohn, M. (1963). *Biochemistry* **2,** 623.
Cohn, M. (1967). In "Magnetic Resonance in Biological Systems" (A. Ehrenberg, B. G. Malmström, and T. Vänngard, eds.), p. 101. Pergamon, Oxford.
Cohn, M. (1970). *Quart. Rev. Biophys.* **3,** 61.
Cohn, M., and Leigh, J. S., Jr. (1962). *Nature (London)* **193,** 1037.
Cohn, M., Danchin, A., and Grunberg-Manago, M. (1969). *J. Mol. Biol.* **39,** 199.
Cohn, M., Diefenbach, H., and Taylor, J. S. (1971). *J. Biol. Chem.* **246,** 6037.

Connick, R. E., and Fiat, D. (1966). *J. Chem. Phys.* **44**, 4103.
Cottam, G. L., Hollenberg, P. F., and Coon, M. J. (1969). *J. Biol. Chem.* **244**, 1481.
Danchin, A. (1969). *J. Theor. Biol.* **25**, 317.
Deranleau, D. A. (1969a). *J. Amer. Chem. Soc.* **91**, 4044.
Deranleau, D. A. (1969b). *J. Amer. Chem. Soc.* **91**, 4050.
Eisinger, J., Shulman, R. G., and Szymanski, B. M. (1962). *J. Chem. Phys.* **36**, 1721.
Fabry, M. E., Koenig, S. H., and Schillinger, W. E. (1970). *J. Biol. Chem.* **245**, 4256.
Feughelman, M., and Haly, A. R. (1962). *Text. Res. J.* **32**, 996.
Gaber, B., Schillinger, W. E., Koenig, S. H., and Aisen, P. (1970). *J. Biol. Chem.* **245**, 4251.
Good, N. E., Winget, G. D., Winter, W., Connolly, T. N., Izawa, S., and Singh, R. M. M. (1966). *Biochemistry* **5**, 467.
Hazelwood, C. F., Nicholls, B. L., and Chamberlain, N. F. (1969). *Nature (London)* **222**, 747.
Holeysovosoka, H. (1961). *Collect. Czech. Chem. Commun.* **26**, 3074.
Hooton, B. R. (1968). *Biochemistry* **7**, 2063.
James, E, and Morrison, J. F. (1966). *J. Biol. Chem.* **241**, 4758.
Kayne, F. J., and Suelter, C. H. (1965). *J. Amer. Chem. Soc.* **87**, 897.
Koenig, S. H., and Schillinger, W. E. (1969a). *J. Biol. Chem.* **244**, 3283.
Koenig, S. H., and Schillinger, W. E. (1969b). *J. Biol. Chem.* **244**, 6520.
Kowalsky, A., and Cohn, M. (1964). *Annu. Rev. Biochem.* **33**, 481.
Kuby, S. A., and Noltmann, E. A. (1962). *In* "The Enzymes" (P. D. Boyer, H. Lardy, and K. Myrbäck, eds.), 2nd rev. ed., Vol. 6, p. 515. Academic Press, New York.
Lui, N. S. T., and Cunningham, L. (1966). *Biochemistry* **5**, 144.
Luz, Z., and Meiboom, S. (1964). *J. Chem. Phys.* **40**, 2686.
Lynch, L. J., and Haly, A. R. (1970). *Kolloid-Z. Z. Polym.* **239**, 581.
Lynch, L. J., and Marsden, K. H. (1969). *J. Chem. Phys.* **51**, 5681.
Lynch, L. J., and Marsden, K. H. (1971). *Search* **2**, 95.
McConnell, H. M., and McFarland, B. G. (1970). *Quart. Rev. Biophys.* **3**, 91.
McDonald, C. C., and Phillips, W. D. (1970). *In* "Biological Macromolecules" (S. N. Timasheff and G. Fasman, eds.), Vol. 4, p. 1. Dekker, New York.
Metcalfe, J. C. (1970). *In* "Physical Principles and Techniques of Protein Chemistry" (S. J. Leach, ed.), Part B, p. 275. Academic Press, New York.
Migchelsen, C., Berendsen, H. J. C., and Rupprecht, A. (1968). *J. Mol. Biol.* **37**, 235.
Mildvan, A. S., and Cohn, M. (1963). *Biochemistry* **2**, 910.
Mildvan, A. S., and Cohn, M. (1965). *J. Biol. Chem.* **240**, 238.
Mildvan, A. S., and Cohn, M. (1966). *J. Biol. Chem.* **241**, 1178.
Mildvan, A. S., and Cohn, M. (1970). *Advan. Enzymol.* **33**, 1.
Mildvan, A. S., and Weiner, H. (1969a). *Biochemistry* **8**, 552.
Mildvan, A. S., and Weiner, H. (1969b). *J. Biol. Chem.* **244**, 2465.
Mildvan, A. S., Leigh, J. S., and Cohn, M. (1967). *Biochemistry* **6**, 1805.
Miller, R. S., Mildvan, A. S., Chang, H. C., Easterday, R. L., Maruyama, H., and Lane, M. D. (1968). *J. Biol. Chem.* **243**, 6030.
Morrison, J. F., and O'Sullivan, W. J. (1965). *Biochem. J.* **97**, 37.
Morrison, J. F., and Uhr, M. L. (1966). *Biochim. Biophys. Acta* **122**, 57.
Ohnishi, S., and McConnell, H. M. (1965). *J. Amer. Chem. Soc.* **87**, 2293.
O'Reilly, D. E., and Poole, C. P. Jr. (1963). *J. Phys. Chem.* **67**, 1762.

O'Sullivan, W. J. (1971). *Int. J. Protein Res.* **3,** 131.
O'Sullivan, W. J., and Cohn, M. (1966a). *J. Biol. Chem.* **241,** 3104.
O'Sullivan, W. J., and Cohn, M. (1966b). *J. Biol. Chem.* **241,** 3116.
O'Sullivan, W. J., and Cohn, M. (1968). *J. Biol. Chem.* **243,** 2737.
O'Sullivan, W. J., and Noda, L. (1968). *J. Biol. Chem.* **243,** 1424.
O'Sullivan, W. J., Reed, G. H., Marsden, K. H., Gough, G., and Lee, C. S. (1972). *J. Biol. Chem.* **247,** 7839.
O'Sullivan, W. J., Virden, R., and Blethen, S. B. (1969). *Eur. J. Biochem.* **8,** 562.
O'Sullivan, W. J., Diefenbach, H., and Cohn, M. (1966). *Biochemistry* **5,** 2666.
Peacocke, A. R., Richards, R. E., and Sheard, B. (1969). *Mol. Phys.* **16,** 177.
Redfield, A. G., Fite, W., and Bleich, H. E. (1968). *Rev. Sci. Instrum.* **39,** 710.
Reed, G, H., Cohn, M., and O'Sullivan, W. J. (1970). *J. Biol. Chem.* **245,** 6547.
Reed, G. H., Leigh, J. S., Jr., and Pearson, J. E. (1971). *J. Chem. Phys.* **55,** 3311.
Reuben, J., and Cohn, M. (1970). *J. Biol. Chem.* **245,** 6539.
Reuben, J., Reed, G. H., and Cohn, M. (1970). *J. Chem. Phys.* **52,** 1617.
Sasa, T., and Noda, L. (1964). *Biochim. Biophys. Acta* **81,** 270.
Sheard, B., and Bradbury, E. M. (1970). *Progr. Biophys. Mol. Biol.* **20,** 187.
Solomon, I. (1955). *Phys. Rev.* **99,** 559.
Stejskal, E. O., and Tanner, J. E. (1965). *J. Chem. Phys.* **42,** 288.
Swift, T. J., and Connick, R. E. (1962). *J. Chem. Phys.* **37,** 307.
Tait, M. J., and Franks, F. (1971). *Nature (London)* **230,** 91.
Tanner, J. E., and Stejskal, E. O. (1968). *J. Chem. Phys.* **49,** 1768.
Taylor, J. S., Leigh, J. S., and Cohn, M. (1969). *Proc. Nat. Acad. Sci. U. S.* **64,** 219.
Taylor, J. S., McLaughlin, A., and Cohn, M. (1971). *J. Biol. Chem.* **246,** 6029.
Ward, R. L., and Srere, P. A. (1965). *Biochim. Biophys. Acta* **99,** 270.
Weiner, H. (1969). *Biochemistry* **8,** 526.
Windle, J. J. (1956). *J. Polym. Sci.* **21,** 103.

# 21 ☐ The Use of Least Squares in Data Analysis

R. D. B. FRASER and E. SUZUKI

|   |   |
|---|---|
| Glossary of Symbols | 301 |
| I. Introduction | 302 |
| II. Outline of Theory | 304 |
|     A. General Considerations | 304 |
|     B. The Gauss-Newton Method | 305 |
|     C. Iterative Refinement | 307 |
|     D. Constrained Optimization | 307 |
| III. The Iteration Process | 309 |
|     A. Measures of Goodness of Fit | 310 |
|     B. Damped Least Squares | 311 |
|     C. Scaled Parameter Adjustments | 312 |
|     D. Temporary Constraints | 314 |
| IV. Statistical Aspects | 316 |
|     A. Choice of Weighting Function | 316 |
|     B. Errors in Parameter Estimates | 317 |
|     C. Significance Tests | 317 |
| V. Model Functions | 321 |
|     A. Analytical Representation | 321 |
|     B. Digital Standard Shape Functions | 329 |
|     C. Baseline Functions | 331 |
| VI. Computational Procedure | 331 |
|     A. Flow Chart | 331 |
|     B. Computer Program | 334 |
|     C. Input | 334 |
|     D. Output | 348 |
| VII. Applications | 352 |
| References | 353 |

## Glossary of Symbols

$Y_i(X_i)$     digital data element
$n$     number of data elements
$F_i$     value predicted by model function for $X = X_i$
$\delta_i$     deviation between $F_i$ and $Y_i$

| | |
|---|---|
| $W_i$ | weight assigned to $\delta_i^2$ |
| $S$ | sum of weighted squares of deviations |
| $\sigma$ | weighted standard deviation |
| $R$ | ratio of $\sigma$ values for two optimizations |
| $R_{p,q,\alpha}$ | quantity used to test significance |
| $P_j$ | parameter of model function |
| $m$ | number of parameters |
| $\Delta P_j$ | parameter adjustment |
| $c_{kj}$ | element of $m \times m$ matrix |
| $d_k$ | element of $m \times 1$ matrix |
| $A_t, B_t$ | components of model function |
| $A_0$ | maximum of model band |
| $X_0$ | $X$-value of model band maximum |
| $\Delta X_{1/2}$ | width of model band at half height |
| $n_b$ | number of model bands |
| $G_k$ | constraining function |
| $D$ | damping factor |
| $h$ | scaling factor |

## I. Introduction

In many of the techniques of protein chemistry described in this and the preceding volumes of this series, data are collected with a view to estimating the values of a series of parameters which characterize the properties of a protein. Frequently, the parameters are those of a theoretical model and the problem arises of finding the set of parameter values which gives the best fit to the observed data. To take a simple example, suppose that incompletely resolved bands are observed in a chromatographic study of the hydrolyzate of a protein (Fig. 1) and we wish to estimate the concentrations of the five components known to be present in the mixture. If the shape of the individual bands could be assumed to be Gaussian, the model function for the system would be a set of five such bands, each characterized by the three parameters peak height, peak position, and bandwidth at half peak height, together with a baseline.

A frequently used criterion of best fit is that of "least squares," in which the parameter values are selected so as to minimize the sum of the squared deviations between the observed data and values calculated from the model. In general, the amount of calculation required to optimize the parameter values is prodigious but with the advent of high-speed digital computers and the development of methods for securing rapid convergence the computation time is usually on the order of minutes.

Many examples of the successful use of computers for the least-squares analysis of data have been reported (Section VII) but so far the method

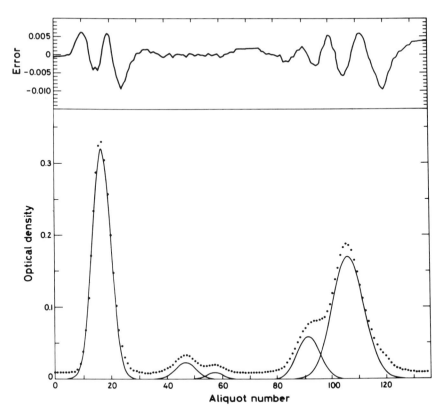

Fig. 1. Typical example of a situation encountered in protein chemistry where the method of least squares can be used in data analysis. The peaks in the output from an amino acid analyzer (dotted) are incompletely resolved. The individual band shapes (full lines) are assumed to be Gaussian and a simulated output is constructed in such a way as to minimize the sum of the squares of the deviations between the observed and simulated data points. Residual deviations are shown in the upper curve. A better fit may be obtained by using more elaborate simulating functions (Section V).

has not been used widely in protein chemistry even though the advantages to be gained by using this powerful technique are self-evident. We believe that this may be due in part to the rather formidable nature of the standard mathematical texts on the subject and to the fairly complex nature of the computer program that is required to translate the theory into a useful analytical tool. The aim in the present treatment is to provide an introduction to the relevant theory and to describe in detail a computer program (Section VI) which can readily be tailored to deal with any specific problem in protein chemistry to which the reader may wish to apply it. An illustrative example is given in Section VI.

The theory of the procedure of parameter optimization by least squares is usually formulated in the elegant language of matrix algebra. In our experience this is a barrier for many protein chemists to obtaining the understanding so essential for the successful adaptation of the method to specific problems. Hence, we have chosen to write out equations in full in order that they may be comprehensible to the widest audience possible.

A discussion of the relative merits of different schemes for parameter optimization would be out of place in the present context and reference may be made to an excellent monograph by Kowalik and Osborne (1968) and to papers by Duke and Gibb (1967), Pitha and Jones (1966), and Box (1966) for discussions of this topic. The method presented here is applicable to a wide variety of problems and the computer program listed has been written in a form such that it can readily be adapted to the individual needs of the reader. In order to assist such adaptation, each step in the procedure has been explicitly described.

Once the parameter values have been optimized according to the least-squares criterion it is often pertinent to inquire whether they are consistent with values predicted by a theoretical treatment or with results obtained by other methods. For example, a particular hypothesis might require that the areas beneath two of the component bands in Fig. 1 bear some fixed ratio. Methods of using the results obtained in least-squares optimizations to test such hypotheses are discussed in Section IV.

The procedures described in the present chapter refer to the general case where the number of data exceeds the number of parameters to be optimized and the function to be fitted is nonlinear with respect to the parameters. In these circumstances iterative methods must be used. When the function is linear the problem is greatly simplified and only involves solving a set of simultaneous linear equations (Kowalik and Osborne, 1968).

## II. Outline of Theory

### A. GENERAL CONSIDERATIONS

The most frequently encountered form of data is that in which measurements of a quantity $Y$, such as absorbance or count, are recorded at intervals of a second quantity $X$, such as wavelength or aliquot number.

For the purposes of illustration, the data will therefore be assumed to be in the form of a list of $n$ pairs of values $(X_i, Y_i)$, $i = 1, \ldots, n$.

The model function $F$ which is to be fitted to these data will be formulated in such a way that for each value of $X_i$, a corresponding $Y$ value conveniently denoted by $F_i$ can be calculated. The difference, or deviation, between $F_i$ and $Y_i$ will be denoted by $\delta_i$ and the least-squares criterion seeks to choose values $P_1, \ldots, P_m$ for the parameters of the model function which minimize the weighted sum of squares

$$S = \sum_{i=1}^{n} W_i(F_i - Y_i)^2 = \sum_{i=1}^{n} W_i \delta_i^2 \tag{1}$$

The weighting factor $W_i$ is included to allow for the possibility that some observations may be more reliable than others. In many instances $W_i$ is taken as unity for all $i$ so that all data points are accorded equal weight.

In many cases, for example with data in which there is overlap of bands, the model function $F$ has the form

$$F_i = \sum_{t=1}^{nb} A_{ti} + B_i \tag{2}$$

where each $A$ term contains parameters representing a component band and $B$ contains additional parameters representing a baseline. No special form for the model function is assumed, however, in the following analysis.

The condition that $S$ in Eq. (1) be a minimum requires that the partial derivatives of $S$ with respect to each of the parameters are zero and so $m$ simultaneous equations for the $m$ parameter values can be obtained from Eq. (1). In general, these equations are nonlinear with respect to the parameters and methods involving successive approximations must be used to evaluate the optimum parameter values.

## B. The Gauss-Newton Method

The most generally useful way of finding the set of parameter values which minimize $S$ in Eq. (1) is termed the Gauss or Gauss-Newton method (Kowalik and Osborne, 1968; Hamilton, 1964) and depends upon the fact that it is often possible to estimate approximate values for the parameters.

Let $P_1, \ldots, P_m$ be the parameter values which minimize $S$, and let $P_1^{(0)}, \ldots, P_m^{(0)}$ be the approximate values. Using Taylor's expansion and neglecting second- and higher-order terms

$$F_i = F_i^{(0)} + \sum_{j=1}^{m} \left(\frac{\partial F_i}{\partial P_j}\right)^{(0)} \Delta P_j \qquad (i = 1, \ldots, n) \qquad (3)$$

where

$$F_i = F_i(P_1, \ldots, P_m)$$
$$F_i^{(0)} = F_i(P_1^{(0)}, \ldots, P_m^{(0)})$$
$$\left(\frac{\partial F_i}{\partial P_j}\right)^{(0)} = \frac{\partial F_i}{\partial P_j}(P_1^{(0)}, \ldots, P_m^{(0)})$$
$$\Delta P_j = P_j - P_j^{(0)}$$

The superscript (0) is used to distinguish quantities calculated from the initial guesses of the parameter values and the minimized deviation $\delta_i$ will from Eq. (3) be given by

$$\delta_i = \delta_i^{(0)} + \sum_{j=1}^{m} \left(\frac{\partial F_i}{\partial P_j}\right)^{(0)} \Delta P_j \qquad (i = 1, \ldots, n) \qquad (4)$$

The condition that $S$ be a minimum requires that

$$\frac{\partial S}{\partial P_k} = 2 \sum_{i=1}^{n} W_i \delta_i \frac{\partial \delta_i}{\partial P_k} = 0 \qquad (k = 1, \ldots, m) \qquad (5)$$

Using the value for $\delta_i$ obtained in Eq. (4) and approximating $\partial \delta_i/\partial P_k = \partial F_i/\partial P_k$ by $(\partial F_i/\partial P_k)^{(0)}$ we obtain a set of $m$ simultaneous linear equations

$$\sum_{j=1}^{m} \Delta P_j \sum_{i=1}^{n} W_i \left(\frac{\partial F_i}{\partial P_j}\right)^{(0)} \left(\frac{\partial F_i}{\partial P_k}\right)^{(0)} = -\sum_{i=1}^{n} W_i \delta_i^{(0)} \left(\frac{\partial F_i}{\partial P_k}\right)^{(0)}$$
$$(k = 1, \ldots, m) \qquad (6)$$

or

$$\begin{aligned}
c_{11}\Delta P_1 + c_{12}\Delta P_2 + \cdots + c_{1m}\Delta P_m &= d_1 \\
c_{21}\Delta P_1 + c_{22}\Delta P_2 + \cdots + c_{2m}\Delta P_m &= d_2 \\
&\vdots \\
c_{m1}\Delta P_1 + c_{m2}\Delta P_2 + \cdots + c_{mm}\Delta P_m &= d_m
\end{aligned} \qquad (7)$$

where

$$c_{jk} = c_{kj} = \sum_{i=1}^{n} W_i \left(\frac{\partial F_i}{\partial P_j}\right)^{(0)} \left(\frac{\partial F_i}{\partial P_k}\right)^{(0)} \qquad (8)$$

and

$$d_k = -\sum_{i=1}^{n} W_i \, \delta_k{}^{(0)} \left(\frac{\partial F_i}{\partial P_k}\right)^{(0)} \tag{9}$$

Computer routines for solving such a set of simultaneous linear equations are readily available. They require the arrays $[c_{kj}]$ and $[d_k]$ as input and return the required array of parameter adjustments $[\Delta P_j]$.

## C. Iterative Refinement

In general, the model function $F$ is not linearly dependent on the parameters and the value of $F_i$ calculated from Eq. (3) will only be an approximation to the optimum value. As a result, the computed parameter adjustments will not be correct, but in favorable cases the set of parameter values $P_j{}^{(1)} = P_j{}^{(0)} + \Delta P_j{}^{(0)}$ will lead to a lower value of $S$ than the set $P_j{}^{(0)}$ ($j = 1, \ldots, m$), where $\Delta P_j{}^{(0)}$ indicates a value calculated from Eq. (6). In these circumstances, the procedure outlined in Section II,B can be used iteratively using the new parameter values as the starting set for a further cycle of refinement. This process is continued until the computed parameter adjustments become negligible. The iteration is then said to have converged and the set of parameter values corresponds to a minimum of $S$.

In some instances, for example, when the initial parameter values are badly chosen, the value of $S^{(1)}$ calculated from the adjusted parameter values will be greater than $S^{(0)}$ calculated from the initial set and the iteration diverges. Methods of overcoming this situation are discussed in Section III.

## D. Constrained Optimization

It sometimes happens that the values of the parameters in the model function $F$ must be optimized to give the best fit to the observed data subject to certain conditions, or constraints, on the values which the parameters may assume. For example, if the frequency of an absorption band maximum is known accurately, it should not be treated as an independent parameter. One method of introducing a constraint of this kind is to recast $F$ in such a way that the frequency is eliminated as an independent variable. A simpler and more versatile method of introducing constraints on the parameters is to express them as a series of relationships

$$G_l(P_1, \ldots, P_m) = 0 \qquad (l = 1, \ldots, r) \tag{10}$$

For example if $P_7$ in our model function were the frequency of an absorption band maximum known to occur at 716 cm$^{-1}$ we would set

$$G_1(P_7) = P_7 - 716 \tag{11}$$

Suppose further that $P_6$ were the peak absorbance of the absorption band at 716 cm$^{-1}$ and we knew that it was always one-half the peak absorbance, say $P_9$, of a second band in the model. Then we would set

$$G_2(P_6, P_9) = 2P_6 - P_9 \tag{12}$$

Quite apart from providing a better model, the introduction of constraints reduces the number of independent parameters and thus increases the likelihood of obtaining a convergent iteration.

The constraints can be introduced into the optimization procedure (Meiron, 1965; Kowalik and Osborne, 1968) by minimizing

$$S = \sum_{i=1}^{n} W_i \delta_i^2 + \sum_{l=1}^{r} \lambda_l G_l \tag{13}$$

where the $\lambda_l$'s, which are termed Lagrange's undetermined multipliers, are to be treated as independent variables.

The condition that $S$ be a minimum requires that $\partial S/P_k = 0$ for all $k$, giving

$$2 \sum_{i=1}^{n} W_i \delta_i \frac{\partial \delta_i}{\partial P_k} + \sum_{l=1}^{r} \lambda_l \frac{\partial G_l}{\partial P_k} = 0 \qquad (k = 1, \ldots, m) \tag{14}$$

and $\partial S/\partial \lambda_l = 0$ for all $l$ giving

$$G_l = 0 \qquad (l = 1, \ldots, r) \tag{15}$$

as required by Eq. (10). Taken together, Eqs. (14) and (15) constitute a set of $m + r$ simultaneous equations for the $m$ parameter values and $r$ values of $\lambda$. A Taylor expansion for $\delta_i$ is given in Eq. (4) and a similar approximation for $G$ is

$$G_l = G_l^{(0)} + \sum_{j=1}^{m} \left(\frac{\partial G_l}{\partial P_j}\right)^{(0)} \cdot \Delta P_j \tag{16}$$

where the (0) again refers to an evaluation using the initial parameter estimates.

By substituting the value for $\delta_i$ obtained from Eq. (4) in Eq. (14) we obtain

$$\sum_{j=1}^{m} \Delta P_j \sum_{i=1}^{n} W_i \left(\frac{\partial F_i}{\partial P_j}\right)^{(0)} \left(\frac{\partial F_i}{\partial P_k}\right)^{(0)} + \sum_{l=1}^{r} \tfrac{1}{2}\lambda_l \left(\frac{\partial G_l}{\partial P_k}\right)^{(0)}$$
$$= - \sum_{i=1}^{n} W_i \delta_i^{(0)} \left(\frac{\partial F_i}{\partial P_k}\right)^{(0)} \qquad (k = 1, \ldots, m) \tag{17}$$

and by substituting the value for $G_l$ obtained from Eq. (16) into Eq. (15), we obtain

$$\sum_{j=1}^{m} \Delta P_j \left(\frac{\partial G_l}{\partial P_j}\right)^{(0)} = -G_l^{(0)} \qquad (l = 1, \ldots, r) \qquad (18)$$

Eqs. (17) and (18) constitute a set of $(m + r)$ equations for the $m$ parameter adjustments and the $r$ semimultipliers; the two arrays of coefficients needed to solve these equations are given by

$$\left.\begin{aligned} c_{jk} = c_{kj} &= \sum_{i=1}^{n} W_i \left(\frac{\partial F_i}{\partial P_j}\right)^{(0)} \left(\frac{\partial F_i}{\partial P_k}\right)^{(0)} \qquad 1 \leqslant j \leqslant m \\ &= \left(\frac{\partial G_{j-m}}{\partial P_k}\right)^{(0)} \qquad m < j \leqslant m + r \\ d_k &= -\sum_{i=1}^{n} W_i \delta_i^{(0)} \left(\frac{\partial F_i}{\partial P_k}\right)^{(0)} \end{aligned}\right\} \quad 1 \leqslant k \leqslant m \quad (19a)$$

$$\left.\begin{aligned} c_{jk} = c_{kj} &= \left(\frac{\partial G_{k-m}}{\partial P_j}\right)^{(0)} \qquad 1 \leqslant j \leqslant m \\ &= 0 \qquad m < j \leqslant m + r \\ d_k &= -(G_{k-m})^{(0)} \end{aligned}\right\} \quad m < k \leqslant m + r \quad (19b)$$

In the computer program given in Section VI, the coefficient $c_{jk}$ is stored in an array element called CMAT and the coefficient $d_k$ in an array element called DMAT.

### III. The Iteration Process

The Gauss-Newton method for minimizing the function $S$ in Eq. (1) possesses the property of very rapid convergence provided certain conditions are satisfied. The commonest causes of divergence are:

a. Poor initial estimates of the trial parameters.
b. Highly nonlinear dependence of $F$ upon one or more parameters.
c. A high degree of correlation between parameters, that is, the effect on $F$ of changing one parameter is very similar to that of changing another parameter.

A combination of a and b is particularly likely to lead to divergence as the truncated Taylor expansion in Eqs. (3) and (4) will give very inaccurate values for $F_i$ and $\delta_i$, respectively. Detailed discussions of the convergence of nonlinear least-squares optimizations have been given by

Hamilton (1964), Kowalik and Osborne (1968), and Marquardt (1963). In the present section measures of goodness of fit are discussed and three methods of improving the convergence properties are described.

## A. Measures of Goodness of Fit

The progress of an iteration can be followed by calculating the value $S$ of the sum of squares after each cycle and comparing successive values. The physical significance of a particular numerical value of $S$ is not easily grasped and a more convenient measure is the weighted standard deviation

$$\sigma = \left[ \frac{S}{\sum_{i=1}^{n} W_i} \right]^{1/2} \tag{20}$$

which with unit weights reduces to

$$\sigma = \left[ \frac{S}{n} \right]^{1/2} \tag{21}$$

and is sometimes termed the root-mean-square deviation. If the weighting functions are dimensionless, the weighted standard deviation has the same units as the physically observed quantity $Y$.

For some purposes, a dimensionless measure is desirable and a coefficient of variation defined by

$$v = \frac{\sigma}{\bar{Y}} \tag{22}$$

where $\bar{Y}$ is the mean value of $Y_i$. This coefficient is sometimes used in statistical studies but is unsuitable, for example, in electron spin resonance studies where $\bar{Y} \sim 0$. A more convenient measure is a generalized weighted standard deviation (Hamilton, 1964, 1965) defined by

$$\sigma_g = \left[ \frac{\sum_{i=1}^{n} W_i \delta_i^2}{\sum_{i=1}^{n} W_i Y_i^2} \right]^{1/2} \tag{23}$$

If the superscript $(r)$ is applied to quantities associated with the $r^{\text{th}}$ cycle of iteration we may characterize a convergent cycle of iteration by the result

$$\sigma^{(r)} < \sigma^{(r-1)} \tag{24}$$

and a divergent cycle by the opposite result.

A criterion of the attainment of a stationary or quasistationary value of $S$ is that the fractional change in $\sigma$ between successive cycles should be less than some small number $\epsilon$. In the program listed in Section VI the related condition

$$\left| \frac{\sigma^{(r-1)} - \sigma^{(r)}}{\sigma^{(r-1)}} \right| \leqslant 10^{-6} \tag{25}$$

is used to decide when the iteration should be terminated. One objection to this procedure is that, if the minimum of $S$ is ill defined due to high correlations between the parameters, small changes may still be taking place in the parameter values even though the inequality in Eq. (25) is satisfied. In an attempt to overcome this occasional difficulty, some authors have devised tests which require that the fractional change in each parameter be less than some prescribed minimum. The selection of suitable minima is difficult, however, and a test of the fractional change in $|\sigma|$ has proved to be the more convenient.

## B. Damped Least Squares

When $\sigma^{(r)} \geqslant \sigma^{(r-1)}$ the iteration procedure is potentially divergent and this situation is usually associated with large values of the calculated parameter adjustments. Levenberg (1944) suggested that if the magnitudes of the parameter adjustments were also minimized the errors introduced by truncating the Taylor series for $F_i$ in Eq. (3) would be reduced. The function to be minimized takes the form shown in Eq. (26) and the damping of the parameter adjustments generally insures convergence.

$$S = \sum_{i=1}^{n} W_i \delta_i^2 + \sum_{l=1}^{r} \lambda_l G_l + D^2 \sum_{j=1}^{m} \mu_j^2 (\Delta P_j)^2 \tag{26}$$

Marquardt (1963) and Meiron (1965) have discussed criteria for choosing the weights $\mu_j$ and the optimum choice given by Meiron is

$$\mu_j = \left[ \sum_{i=1}^{n} W_i \left( \frac{\partial F_i}{\partial P_j} \right)^2 \right]^{1/2} \tag{27}$$

which from Eq. (8) reduces to

$$\mu_j^2 = c_{jj} \tag{28}$$

With this choice of $\mu_j$ it may be shown that the only modification required to Eq. (19) is to multiply the diagonal terms $c_{jj}$ by $(1 + D^2)$. The formulation given by Marquardt (1963) requires additional modifications to these equations but is mathematically equivalent to Meiron's method (Pitha and Jones, 1966).

The coefficient $D$ in Eq. (26) regulates the amount of damping and is referred to as the *damping factor*. If very large values of $D$ are used, it can be shown (Marquardt, 1963) that the optimization process approaches the "steepest-descent" method, which is slow and inefficient.

Two methods of choosing $D$ have been suggested. Marquardt (1963) advocates using the smallest value of $D$ which will insure $\sigma^{(r+1)} < \sigma^{(r)}$, whereas other authors have recommended that $D$ be chosen so as to maximize $\sigma^{(r)} - \sigma^{(r+1)}$ (Levenberg, 1944; Papoušek and Plíva, 1965; Pitha and Jones, 1966; Fraser and Suzuki, 1970a). This latter method is used in the program given in Section VI and is based on the following algorithm.

If $D^{(r)}$ is the damping factor used in the $r^{\text{th}}$ cycle of refinement, a series of trial refinements for the $(r + 1)^{\text{th}}$ cycle are carried out using $k^{\nu-1} \cdot D^{(r)}$, $k^{\nu} D^{(r)}$, and $k^{\nu+1} \cdot D^{(r)}$ as damping factors with $\nu = 0$ and $k > 1$. The corresponding values of $\sigma^{(r+1)}$ are compared and $\nu$ is increased or decreased by 1 iteratively until a minimum is detected by the condition that

$$\sigma^{(r+1)}(k^{\nu} \cdot D^{(r)}) < \sigma^{(r+1)}(k^{\nu-1} \cdot D^{(r)}) \tag{29a}$$

and

$$\sigma^{(r+1)}(k^{\nu} \cdot D^{(r)}) < \sigma^{(r+1)}(k^{\nu+1} \cdot D^{(r)}) \tag{29b}$$

The optimum value of $D^{(r+1)}$ is then taken as $k^{\nu} \cdot D^{(r)}$. A suitable value for $k$ has been found to be 2 and $D^{(0)}$ may be taken as unity.

In some instances, $\sigma^{(r+1)}$ becomes insensitive to the value of $D$ and a test must be made for this condition during the optimization of the damping factor. Under certain conditions, the optimum value of the damping factor becomes very large or very small and it is advantageous to restrict the range of search (Section VI,B).

An example of the use of damped least squares to secure convergence in an otherwise divergent system is given in Table I for the problem illustrated in Fig. 2. A feature of the method, which is apparent from Table I, is a rapid rate of convergence in the initial stages followed by a rather slow convergence as the solution is approached.

## C. Scaled Parameter Adjustments

Another method by which the Gauss-Newton method may be modified to improve convergence is to scale the calculated parameter adjustments by a constant factor $h$ in such a way that $\sigma^{(r)} - \sigma^{(r+1)}$ is maximized (Hamilton, 1964).

TABLE I

EXAMPLES OF THE USE OF MODIFIED PROCEDURES TO SECURE
CONVERGENCE IN THE PROBLEM IN FIG. 2

| Cycle | Standard deviation $\sigma$ ($\times 100$) | | | |
|---|---|---|---|---|
| | Unmodified iteration | Scaled adjustments | Damped adjustments | Damped and scaled |
| 0 | 6.0720 | 6.0720 | 6.0720 | 6.0720 |
| 1 | 4.2231 | 3.6306 | 1.0264 | 0.9515 |
| 2 | 32.2796 | 3.0181 | 0.1155 | 0.0627 |
| 3 | Diverging | 2.3935 | 0.0292 | 0.0286 |
| 4 | | 0.3592 | 0.0254 | 0.0253 |
| 5 | | 0.0271 | 0.0223 | 0.0208 |
| 6 | | 0.0029 | 0.0195 | 0.0139 |
| 7 | | 0.0028 | 0.0168 | 0.0071 |
| 8 | | | 0.0144 | 0.0040 |
| 9 | | | 0.0123 | 0.0029 |
| 10 | | | 0.0104 | 0.0028 |
| 11 | | | 0.0082 | |
| 12 | | | 0.0043 | |
| 13 | | | 0.0031 | |
| 14 | | | 0.0028 | |

An algorithm similar to that employed with damped least squares may be employed with trials being carried out for the $(r + 1)$th cycle using $h = k^{\nu-1}$, $k^{\nu}$, and $k^{\nu+1}$ with $\nu$ initially zero and searching for a value of $\nu$ such that

$$\sigma^{(r+1)}(k^\nu) < \sigma^{(r+1)}(k^{\nu-1}) \tag{30a}$$

and

$$\sigma^{(r+1)}(k^\nu) < \sigma^{(r+1)}(k^{\nu+1}) \tag{30b}$$

The optimum value of $h$ is taken as $k^\nu$. A convenient value for $k$ has been found to be 1.2 and, as with the optimization of the damping factor, it is advantageous to limit the range of search (Section VI,B).

An example of the use of scaled parameter adjustments to secure convergence is given in Table I. In this particular example, the method of scaled parameter adjustments requires fewer cycles of refinement than the method of damped least squares but this is not always the case.

Kowalik and Osborne (1968) recommend that the parameter adjustments calculated by damped least squares should also be scaled to maximize $\sigma^{(r)} - \sigma^{(r+1)}$ and the effect of optimizing both $D$ and $h$ is also given in Table I. The rate of convergence of the damped least-squares method is improved by this modification. Once the parameters have been re-

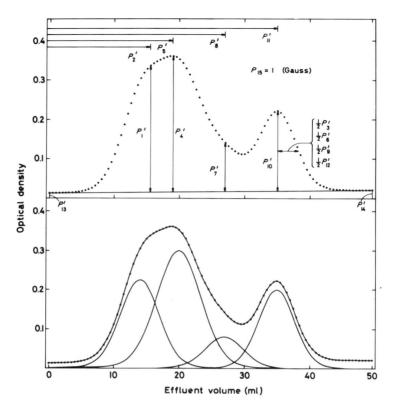

Fig. 2. Example of poorly resolved set of bands (upper curve). The initial estimates of $P_1$ through $P_{14}$ are indicated and $P_{15}$ corresponding to $f$ in Eq. (40) was assigned a value of 1.0. The combination of poor resolution and a poor set of initial estimates led to divergence when Eq. (6) was used (Table I). However, convergent solutions were obtained (lower curve) by the methods described in Section III. The relative efficiencies of the various methods are compared in Table I.

fined to the point where the unmodified least-squares procedure no longer diverges, it will normally give a rate of convergence which is superior to any of the modified procedures.

### D. Temporary Constraints

A third method of improving convergence which may be used in conjunction with damped least squares and scaled parameter adjustments is that of introducing temporary constraints on certain parameters (Fraser and Suzuki, 1970a).

Boulton (1968) describes an automated procedure in which the correla-

TABLE II

CORRELATION COEFFICIENTS BETWEEN PARAMETERS IN THE EXAMPLE OF DIVERGENT ITERATION GIVEN IN TABLE I

| k \ j | Band 1 | | | Band 2 | | | Band 3 | | | Band 4 | | | Baseline | | |
|---|---|---|---|---|---|---|---|---|---|---|---|---|---|---|---|
| | 1 | 2 | 3 | 4 | 5 | 6 | 7 | 8 | 9 | 10 | 11 | 12 | 13 | 14 | 15 |
| 1 | 1.00 | 1.00 | 0.98 | −1.00 | −0.83 | 0.99 | 0.94 | −0.90 | 0.78 | −0.57 | 0.40 | −0.23 | 0.19 | 0.37 | 0.49 |
| 2 | | 1.00 | 0.99 | −0.99 | −0.86 | 0.97 | 0.91 | −0.87 | 0.74 | −0.55 | 0.37 | −0.21 | 0.25 | 0.42 | 0.55 |
| 3 | | | 1.00 | −0.98 | −0.89 | 0.94 | 0.87 | −0.82 | 0.69 | −0.52 | 0.33 | −0.17 | 0.29 | 0.47 | 0.60 |
| 4 | | | | 1.00 | 0.79 | −0.99 | −0.95 | 0.92 | −0.81 | 0.61 | −0.45 | 0.28 | −0.20 | −0.37 | −0.49 |
| 5 | | | | | 1.00 | −0.73 | −0.58 | 0.50 | −0.31 | 0.19 | 0.01 | −0.12 | −0.41 | −0.52 | −0.67 |
| 6 | | | | | | 1.00 | 0.98 | −0.96 | 0.85 | −0.63 | 0.48 | −0.30 | 0.12 | 0.29 | 0.39 |
| 7 | | | | | | | 1.00 | −0.99 | 0.92 | −0.70 | 0.58 | −0.40 | 0.06 | 0.23 | 0.31 |
| 8 | | | | | | | | 1.00 | −0.95 | 0.73 | −0.62 | 0.43 | 0.00 | −0.17 | −0.24 |
| 9 | | | | | | | | | 1.00 | −0.82 | 0.80 | −0.63 | −0.07 | 0.11 | 0.14 |
| 10 | | | | | | | | | | 1.00 | −0.74 | 0.53 | −0.09 | −0.28 | −0.21 |
| 11 | | | | | | | | | | | 1.00 | −0.80 | −0.14 | −0.03 | −0.04 |
| 12 | | | | | | | | | | | | 1.00 | 0.18 | −0.04 | 0.13 |
| 13 | | | | | | | | | | | | | 1.00 | 0.74 | 0.88 |
| 14 | | | | | | | | | | | | | | 1.00 | 0.85 |
| 15 | | | | | | | | | | | | | | | 1.00 |

tion between parameters is tested when the iteration fails to converge and highly correlated parameters are temporarily fixed. This facility is not provided in the program listed in Section VI because it has not been found to be as effective as the methods described in Sections III,B and III,C. However, the correlation coefficients are printed out if the iteration diverges so that a suitably designed iteration procedure can be devised.

In the example shown in Fig. 2, the iteration diverges after one cycle and the correlation matrix is given in Table II. It will be seen that the correlation between certain of the parameters of the badly overlapping pair of bands are very high.

The correlation coefficients may be calculated from the inverse of the matrix of coefficients $[c_{kj}]$ in Eq. (7). If $(c^{-1})_{kj}$ is used to denote an element of the inverse array, the correlation between $P_j$ and $P_k$ is

$$\rho_{kj} = \frac{(c^{-1})_{kj}}{\{(c^{-1})_{jj}(c^{-1})_{kk}\}^{1/2}} \quad (31)$$

The calculation of the inverse of the matrix $[c_{kj}]$ may be performed by the same subroutine MATINV that is used to solve Eqs. (17) and (18).

The use of temporary constraints can of course be combined with other methods of securing convergence and provision for fixing up to $m - 1$ of the $m$ parameters is provided in the program listed in Section VI.

## IV. Statistical Aspects

A. Choice of Weighting Function

Factors affecting the choice of an appropriate weight $W_i^{1/2}$ for the deviation $\delta_i$ have been discussed by Hamilton (1964) and Blackburn (1970). A commonly used scheme is to assume that the errors of different observations are uncorrelated and to use a value of $W_i$ which is inversely proportional to the estimated variance of the $i^{th}$ observation. In some applications, for example, spectroscopic and chromatographic, the use of unit weight functions gives satisfactory results, while in others, for example, when counting techniques are used, careful choice of weighting function is essential. In the program given in Section VI unit weights are assumed.

If the errors in the data are correlated this may be taken into account by an appropriate modification of the theory outlined in Section II (Hamilton, 1964).

## B. Errors in Parameter Estimates

When the sum of squares $S$ has been minimized, an unbiased estimate of the standard deviation is given (Marquardt et al., 1961) by

$$\hat{\sigma} = \left[\frac{S_{\min}}{n-m}\right]^{1/2} \qquad (32)$$

where $S_{\min}$ is the minimum value of $S$ and $m$ is the number of independent parameters in the fitting function $F$. The quantity $n - m$ is the number of degrees of freedom for the optimization.

Unbiased estimates of the standard deviations of the parameter values may be calculated from the relation

$$\hat{\sigma}_j = \hat{\sigma} \cdot (c^{-1})_{jj} \qquad (33)$$

Useful discussions of the topic have been given by Hamilton (1964), Luenberger and Dennis (1966), and Ederer (1969).

## C. Significance Tests

Hypothesis testing on the basis of least-squares optimization has been discussed in some detail by Hamilton (1964). When the fitting function $F$ is known to be capable of representing error-free data, exactly precise tests of hypotheses concerning the parameters can be made. Various examples from X-ray crystallography have been given by Hamilton (1965) and the way in which his treatment may be used in problems relevant to protein chemistry will be illustrated by some specific examples.

The method consists in formulating a hypothesis of dimension $p$ concerning the parameters. The standard deviation $\sigma_{\min}$, obtained when the $n$ observations are fitted using $m$ independent parameters (a refinement with $n - m = q$ degrees of freedom), is then compared with the value $\sigma$, obtained when the optimization is constrained on the assumption that the hypothesis is true. The level of significance ($\alpha$) at which the hypothesis can be accepted may be determined by comparing the ratio of the two standard deviations with a tabulated quantity $R_{p,q,\alpha}$. Extracts from the tables given by Hamilton (1965) for $\alpha = 0.5$, 0.05, and 0.005 are given in Table III. The following questions are frequently of interest in practice.

1. Is the optimized set of parameters consistent with a second set derived from another investigation or predicted by a theoretical treatment?

Let $\sigma_{\min}$ be the standard deviation calculated using the optimized parameter values and let $\sigma$ be the value calculated using the second set. We then test the hypothesis that the second set of parameters is

## TABLE III
### Values of $R_{p,q,\alpha}$ for Significance Tests[a]

$\alpha = 0.005$

| q \ p | 1 | 2 | 3 | 4 | 5 | 6 | 7 | 8 | 9 | 10 | 12 | 15 | 20 |
|---|---|---|---|---|---|---|---|---|---|---|---|---|---|
| 1 | * | * | * | * | * | * | * | * | * | * | * | * | * |
| 2 | 10.012 | 14.142 | 17.313 | 19.987 | 22.344 | 24.474 | 26.434 | 28.257 | 29.971 | 31.591 | 34.605 | 38.688 | 44.671 |
| 3 | 4.418 | 5.848 | 6.962 | 7.912 | 8.755 | 9.522 | 10.231 | 10.894 | 11.517 | 12.109 | 13.212 | 14.711 | 16.917 |
| 4 | 2.972 | 3.761 | 4.381 | 4.915 | 5.392 | 5.828 | 6.232 | 6.611 | 6.969 | 7.309 | 7.944 | 8.811 | 10.031 |
| 5 | 2.357 | 2.885 | 3.304 | 3.667 | 3.992 | 4.291 | 4.569 | 4.831 | 5.078 | 5.314 | 5.755 | 6.359 | 7.253 |
| 6 | 2.026 | 2.418 | 2.731 | 3.003 | 3.249 | 3.475 | 3.686 | 3.884 | 4.073 | 4.252 | 4.590 | 5.053 | 5.741 |
| 7 | 1.822 | 2.132 | 2.380 | 2.597 | 2.793 | 2.974 | 3.144 | 3.304 | 3.456 | 3.602 | 3.875 | 4.251 | 4.812 |
| 8 | 1.684 | 1.939 | 2.144 | 2.324 | 2.488 | 2.639 | 2.781 | 2.915 | 3.042 | 3.164 | 3.394 | 3.712 | 4.186 |
| 9 | 1.585 | 1.802 | 1.976 | 2.130 | 2.269 | 2.399 | 2.521 | 2.636 | 2.746 | 2.851 | 3.050 | 3.325 | 3.736 |
| 10 | 1.511 | 1.695 | 1.850 | 1.984 | 2.106 | 2.220 | 2.326 | 2.427 | 2.524 | 2.617 | 2.792 | 3.034 | 3.398 |
| 11 | 1.453 | 1.619 | 1.753 | 1.871 | 1.980 | 2.080 | 2.175 | 2.265 | 2.352 | 2.434 | 2.591 | 2.808 | 3.135 |
| 12 | 1.407 | 1.555 | 1.675 | 1.781 | 1.879 | 1.969 | 2.055 | 2.136 | 2.214 | 2.289 | 2.430 | 2.627 | 2.924 |
| 13 | 1.369 | 1.503 | 1.612 | 1.708 | 1.796 | 1.879 | 1.957 | 2.031 | 2.102 | 2.170 | 2.299 | 2.479 | 2.751 |
| 14 | 1.338 | 1.460 | 1.559 | 1.647 | 1.728 | 1.804 | 1.875 | 1.943 | 2.008 | 2.071 | 2.190 | 2.356 | 2.607 |
| 15 | 1.311 | 1.424 | 1.515 | 1.596 | 1.671 | 1.740 | 1.806 | 1.869 | 1.929 | 1.987 | 2.098 | 2.252 | 2.485 |
| 16 | 1.289 | 1.393 | 1.477 | 1.552 | 1.621 | 1.686 | 1.747 | 1.806 | 1.862 | 1.916 | 2.019 | 2.162 | 2.381 |
| 17 | 1.269 | 1.366 | 1.444 | 1.514 | 1.579 | 1.639 | 1.696 | 1.751 | 1.803 | 1.854 | 1.950 | 2.085 | 2.290 |
| 18 | 1.252 | 1.342 | 1.416 | 1.481 | 1.542 | 1.598 | 1.652 | 1.703 | 1.752 | 1.800 | 1.890 | 2.017 | 2.211 |
| 19 | 1.237 | 1.322 | 1.391 | 1.452 | 1.509 | 1.562 | 1.613 | 1.661 | 1.707 | 1.752 | 1.838 | 1.957 | 2.140 |
| 20 | 1.224 | 1.303 | 1.368 | 1.426 | 1.480 | 1.530 | 1.578 | 1.624 | 1.667 | 1.710 | 1.791 | 1.904 | 2.078 |
| 21 | 1.212 | 1.287 | 1.349 | 1.403 | 1.454 | 1.502 | 1.547 | 1.590 | 1.632 | 1.672 | 1.749 | 1.857 | 2.022 |
| 22 | 1.201 | 1.272 | 1.331 | 1.383 | 1.431 | 1.476 | 1.519 | 1.560 | 1.600 | 1.638 | 1.711 | 1.814 | 1.972 |
| 23 | 1.191 | 1.259 | 1.315 | 1.364 | 1.410 | 1.453 | 1.494 | 1.533 | 1.571 | 1.607 | 1.677 | 1.775 | 1.926 |
| 24 | 1.182 | 1.247 | 1.300 | 1.347 | 1.391 | 1.432 | 1.471 | 1.508 | 1.545 | 1.579 | 1.646 | 1.740 | 1.885 |
| 25 | 1.174 | 1.236 | 1.287 | 1.332 | 1.374 | 1.413 | 1.450 | 1.486 | 1.521 | 1.554 | 1.618 | 1.708 | 1.847 |
| 26 | 1.167 | 1.226 | 1.274 | 1.318 | 1.358 | 1.395 | 1.431 | 1.465 | 1.499 | 1.531 | 1.592 | 1.679 | 1.812 |
| 27 | 1.160 | 1.217 | 1.263 | 1.305 | 1.343 | 1.379 | 1.414 | 1.447 | 1.478 | 1.509 | 1.568 | 1.652 | 1.780 |
| 28 | 1.154 | 1.208 | 1.253 | 1.293 | 1.330 | 1.364 | 1.397 | 1.429 | 1.460 | 1.489 | 1.546 | 1.627 | 1.750 |
| 29 | 1.148 | 1.200 | 1.243 | 1.282 | 1.317 | 1.351 | 1.382 | 1.413 | 1.443 | 1.471 | 1.526 | 1.604 | 1.723 |
| 30 | 1.143 | 1.193 | 1.234 | 1.271 | 1.306 | 1.338 | 1.369 | 1.398 | 1.427 | 1.454 | 1.507 | 1.582 | 1.698 |
| 40 | 1.105 | 1.142 | 1.172 | 1.199 | 1.224 | 1.248 | 1.270 | 1.292 | 1.313 | 1.334 | 1.373 | 1.429 | 1.516 |
| 60 | 1.068 | 1.092 | 1.112 | 1.130 | 1.146 | 1.162 | 1.176 | 1.191 | 1.205 | 1.218 | 1.244 | 1.282 | 1.340 |
| 120 | 1.034 | 1.045 | 1.055 | 1.063 | 1.071 | 1.079 | 1.086 | 1.093 | 1.100 | 1.107 | 1.120 | 1.139 | 1.168 |
| ∞ | 1.000 | 1.000 | 1.000 | 1.000 | 1.000 | 1.000 | 1.000 | 1.000 | 1.000 | 1.000 | 1.000 | 1.000 | 1.000 |

[a] From Hamilton (1965).
* $R$ exceeds a value of 100.

correct. The dimension $p$ of the hypothesis is $m$ and the number of degrees of freedom for the refinement is $q = n - m$. If we wish to test at a level of significance equal to $\alpha$ we compare $R_{m,n-m,\alpha}$ with

$$R = \frac{\sigma}{\sigma_{\min}} \qquad (34)$$

If $R \leqslant R_{m,n-m,\alpha}$ we can accept the hypothesis at a $100\alpha\%$ level of significance; if $R > R_{m,n-m,\alpha}$ we reject the hypothesis at that level.

As an example, consider the analysis of an output from an automatic amino acid analyzer using a digital standard shape function (Section VI). The number of data $n = 135$, the number of parameters $m = 17$, and $\sigma_{\min} = 1.77 \times 10^{-3}$. Suppose investigator Z uses an analog curve resolver to fit the same data and produces a different set of parameter values which he claims are correct. In order to test his claim, we calculate a standard deviation using his set of parameter values and obtain $\sigma = 1.98 \times 10^{-3}$. From Eq. (34) this leads to a value of $R = 1.12$ which is to be compared with $R_{17,118,\alpha}$. From Table III we obtain interpolated values of 1.066, 1.114, and 1.151 corresponding to $\alpha = 0.5, 0.05$, and

## 21. LEAST SQUARES IN DATA ANALYSIS

### TABLE III (Continued)

$\alpha = 0.05$

| q \ p | 1 | 2 | 3 | 4 | 5 | 6 | 7 | 8 | 9 | 10 | 12 | 15 | 20 |
|---|---|---|---|---|---|---|---|---|---|---|---|---|---|
| 1 | 12.746 | 20.000 | 25.458 | 29.989 | 33.938 | 37.483 | 40.723 | 43.727 | 46.539 | 49.191 | 54.110 | 60.747 | 70.436 |
| 2 | 3.203 | 4.472 | 5.454 | 6.284 | 7.017 | 7.680 | 8.291 | 8.859 | 9.393 | 9.898 | 10.839 | 12.113 | 13.981 |
| 3 | 2.092 | 2.714 | 3.206 | 2.627 | 4.003 | 4.345 | 4.662 | 4.959 | 5.238 | 5.503 | 5.998 | 6.672 | 7.664 |
| 4 | 1.711 | 2.115 | 2.438 | 2.718 | 2.970 | 3.201 | 3.415 | 3.617 | 3.808 | 3.989 | 4.328 | 4.792 | 5.478 |
| 5 | 1.524 | 1.821 | 2.061 | 2.270 | 2.460 | 2.634 | 2.798 | 2.951 | 3.097 | 3.236 | 3.497 | 3.854 | 4.385 |
| 6 | 1.413 | 1.648 | 1.838 | 2.006 | 2.158 | 2.299 | 2.431 | 2.555 | 2.674 | 2.787 | 3.000 | 3.293 | 3.730 |
| 7 | 1.341 | 1.534 | 1.692 | 1.832 | 1.959 | 2.077 | 2.188 | 2.293 | 2.393 | 2.489 | 2.670 | 2.919 | 3.293 |
| 8 | 1.290 | 1.454 | 1.589 | 1.708 | 1.818 | 1.920 | 2.016 | 2.107 | 2.194 | 2.277 | 2.434 | 2.652 | 2.979 |
| 9 | 1.252 | 1.395 | 1.512 | 1.617 | 1.713 | 1.803 | 1.887 | 1.967 | 2.044 | 2.118 | 2.258 | 2.452 | 2.743 |
| 10 | 1.223 | 1.349 | 1.453 | 1.546 | 1.632 | 1.712 | 1.787 | 1.859 | 1.928 | 1.995 | 2.120 | 2.295 | 2.559 |
| 11 | 1.200 | 1.313 | 1.407 | 1.490 | 1.567 | 1.640 | 1.708 | 1.773 | 1.836 | 1.896 | 2.010 | 2.170 | 2.411 |
| 12 | 1.181 | 1.284 | 1.368 | 1.444 | 1.515 | 1.581 | 1.643 | 1.703 | 1.760 | 1.815 | 1.920 | 2.067 | 2.289 |
| 13 | 1.166 | 1.259 | 1.337 | 1.406 | 1.471 | 1.532 | 1.589 | 1.644 | 1.697 | 1.748 | 1.845 | 1.981 | 2.187 |
| 14 | 1.153 | 1.239 | 1.310 | 1.374 | 1.434 | 1.490 | 1.543 | 1.594 | 1.643 | 1.691 | 1.781 | 1.908 | 2.100 |
| 15 | 1.141 | 1.221 | 1.287 | 1.347 | 1.403 | 1.455 | 1.504 | 1.552 | 1.598 | 1.642 | 1.726 | 1.845 | 2.026 |
| 16 | 1.132 | 1.206 | 1.268 | 1.324 | 1.375 | 1.424 | 1.471 | 1.515 | 1.558 | 1.600 | 1.679 | 1.790 | 1.961 |
| 17 | 1.123 | 1.193 | 1.251 | 1.303 | 1.351 | 1.397 | 1.441 | 1.483 | 1.523 | 1.562 | 1.637 | 1.742 | 1.904 |
| 18 | 1.116 | 1.181 | 1.236 | 1.285 | 1.331 | 1.374 | 1.415 | 1.455 | 1.493 | 1.530 | 1.600 | 1.700 | 1.853 |
| 19 | 1.109 | 1.171 | 1.222 | 1.269 | 1.312 | 1.353 | 1.392 | 1.429 | 1.465 | 1.501 | 1.568 | 1.662 | 1.808 |
| 20 | 1.103 | 1.162 | 1.210 | 1.254 | 1.295 | 1.334 | 1.371 | 1.407 | 1.441 | 1.474 | 1.538 | 1.629 | 1.768 |
| 21 | 1.098 | 1.153 | 1.200 | 1.241 | 1.280 | 1.317 | 1.352 | 1.386 | 1.419 | 1.451 | 1.512 | 1.598 | 1.731 |
| 22 | 1.093 | 1.146 | 1.190 | 1.230 | 1.267 | 1.302 | 1.336 | 1.368 | 1.399 | 1.430 | 1.488 | 1.570 | 1.698 |
| 23 | 1.089 | 1.139 | 1.181 | 1.219 | 1.255 | 1.288 | 1.320 | 1.351 | 1.381 | 1.410 | 1.466 | 1.545 | 1.667 |
| 24 | 1.085 | 1.133 | 1.173 | 1.209 | 1.243 | 1.276 | 1.306 | 1.336 | 1.365 | 1.393 | 1.446 | 1.522 | 1.640 |
| 25 | 1.082 | 1.127 | 1.166 | 1.201 | 1.233 | 1.264 | 1.294 | 1.322 | 1.350 | 1.376 | 1.428 | 1.501 | 1.614 |
| 26 | 1.078 | 1.122 | 1.159 | 1.192 | 1.224 | 1.253 | 1.282 | 1.309 | 1.336 | 1.362 | 1.411 | 1.482 | 1.591 |
| 27 | 1.075 | 1.117 | 1.153 | 1.185 | 1.215 | 1.244 | 1.271 | 1.297 | 1.323 | 1.348 | 1.396 | 1.464 | 1.569 |
| 28 | 1.072 | 1.113 | 1.147 | 1.178 | 1.207 | 1.235 | 1.261 | 1.286 | 1.311 | 1.335 | 1.381 | 1.447 | 1.549 |
| 29 | 1.070 | 1.109 | 1.142 | 1.172 | 1.200 | 1.226 | 1.252 | 1.276 | 1.300 | 1.323 | 1.368 | 1.431 | 1.530 |
| 30 | 1.067 | 1.105 | 1.137 | 1.166 | 1.193 | 1.218 | 1.243 | 1.267 | 1.290 | 1.312 | 1.355 | 1.417 | 1.513 |
| 40 | 1.050 | 1.078 | 1.101 | 1.123 | 1.143 | 1.162 | 1.180 | 1.198 | 1.216 | 1.233 | 1.265 | 1.312 | 1.385 |
| 60 | 1.033 | 1.051 | 1.067 | 1.081 | 1.094 | 1.107 | 1.119 | 1.131 | 1.143 | 1.154 | 1.176 | 1.208 | 1.258 |
| 120 | 1.016 | 1.025 | 1.033 | 1.040 | 1.047 | 1.053 | 1.059 | 1.065 | 1.071 | 1.077 | 1.088 | 1.104 | 1.130 |
| ∞ | 1.000 | 1.000 | 1.000 | 1.000 | 1.000 | 1.000 | 1.000 | 1.000 | 1.000 | 1.000 | 1.000 | 1.000 | 1.000 |

0.005, respectively. Thus, we can reject Z's claim at the 5% level but we cannot reject it at the 0.5% level. A more familiar way of expressing this result would be to say that the odds are more than 20 to 1 against Z's results being correct.

2. There are grounds for believing that one of the parameters has a certain value. Are the data consistent with this value?

Let $\sigma_{\min}$ be the standard deviation obtained when the parameter is allowed to vary and let $\sigma$ be the value obtained when it is fixed at the theoretical value. We then test the hypothesis that the parameter has the theoretical value. The dimension of the hypothesis is $p = 1$ and, as before, $q = n - m$. We accept the hypothesis at a $100\alpha\%$ level of significance if

$$R = \sigma/\sigma_{\min} \leqslant R_{1, n-m, \alpha} \tag{35}$$

3. There are grounds for supposing that fixed linear relationships exist between the parameters. Is the data consistent with such relationships?

Let $\sigma$ be the standard deviation obtained when the optimization is carried out subject to $p$ fixed linear relationships. We formulate the hypothesis that these linear relationships exist at a $100\alpha\%$ level of significance and test whether

## TABLE III (Continued)

$\alpha = 0.5$

| q \ p | 1 | 2 | 3 | 4 | 5 | 6 | 7 | 8 | 9 | 10 | 12 | 15 | 20 |
|---|---|---|---|---|---|---|---|---|---|---|---|---|---|
| 1 | 1.414 | 2.000 | 2.475 | 2.879 | 3.236 | 3.557 | 3.853 | 4.127 | 4.385 | 4.628 | 5.080 | 5.692 | 6.586 |
| 2 | 1.155 | 1.414 | 1.644 | 1.848 | 2.032 | 2.202 | 2.359 | 2.507 | 2.647 | 2.779 | 3.028 | 3.366 | 3.864 |
| 3 | 1.093 | 1.260 | 1.414 | 1.555 | 1.684 | 1.805 | 1.918 | 2.025 | 2.127 | 2.224 | 2.406 | 2.656 | 3.028 |
| 4 | 1.066 | 1.189 | 1.306 | 1.414 | 1.515 | 1.610 | 1.700 | 1.785 | 1.867 | 1.945 | 2.092 | 2.296 | 2.600 |
| 5 | 1.051 | 1.149 | 1.243 | 1.331 | 1.414 | 1.493 | 1.568 | 1.639 | 1.708 | 1.774 | 1.899 | 2.072 | 2.333 |
| 6 | 1.042 | 1.122 | 1.201 | 1.276 | 1.347 | 1.414 | 1.479 | 1.540 | 1.600 | 1.657 | 1.766 | 1.918 | 2.148 |
| 7 | 1.035 | 1.104 | 1.172 | 1.237 | 1.298 | 1.358 | 1.414 | 1.469 | 1.521 | 1.572 | 1.669 | 1.805 | 2.012 |
| 8 | 1.031 | 1.091 | 1.150 | 1.207 | 1.262 | 1.315 | 1.365 | 1.414 | 1.461 | 1.507 | 1.595 | 1.718 | 1.906 |
| 9 | 1.027 | 1.080 | 1.133 | 1.184 | 1.234 | 1.281 | 1.327 | 1.371 | 1.414 | 1.456 | 1.536 | 1.649 | 1.821 |
| 10 | 1.024 | 1.072 | 1.120 | 1.166 | 1.211 | 1.254 | 1.296 | 1.336 | 1.376 | 1.414 | 1.488 | 1.592 | 1.752 |
| 11 | 1.022 | 1.065 | 1.109 | 1.151 | 1.192 | 1.232 | 1.270 | 1.308 | 1.344 | 1.380 | 1.448 | 1.545 | 1.694 |
| 12 | 1.020 | 1.059 | 1.099 | 1.138 | 1.176 | 1.213 | 1.249 | 1.284 | 1.317 | 1.350 | 1.414 | 1.505 | 1.645 |
| 13 | 1.018 | 1.055 | 1.092 | 1.128 | 1.163 | 1.197 | 1.231 | 1.263 | 1.295 | 1.326 | 1.385 | 1.470 | 1.602 |
| 14 | 1.017 | 1.051 | 1.085 | 1.119 | 1.152 | 1.184 | 1.215 | 1.245 | 1.275 | 1.304 | 1.360 | 1.440 | 1.565 |
| 15 | 1.016 | 1.047 | 1.079 | 1.111 | 1.142 | 1.172 | 1.201 | 1.230 | 1.258 | 1.285 | 1.338 | 1.414 | 1.532 |
| 16 | 1.015 | 1.044 | 1.074 | 1.104 | 1.133 | 1.161 | 1.189 | 1.216 | 1.243 | 1.268 | 1.319 | 1.391 | 1.504 |
| 17 | 1.014 | 1.042 | 1.070 | 1.098 | 1.125 | 1.152 | 1.178 | 1.204 | 1.229 | 1.254 | 1.302 | 1.370 | 1.478 |
| 18 | 1.013 | 1.039 | 1.066 | 1.093 | 1.119 | 1.144 | 1.169 | 1.193 | 1.217 | 1.241 | 1.286 | 1.352 | 1.454 |
| 19 | 1.012 | 1.037 | 1.063 | 1.088 | 1.112 | 1.137 | 1.160 | 1.183 | 1.206 | 1.229 | 1.272 | 1.335 | 1.433 |
| 20 | 1.012 | 1.035 | 1.059 | 1.083 | 1.107 | 1.130 | 1.152 | 1.175 | 1.196 | 1.218 | 1.260 | 1.320 | 1.414 |
| 21 | 1.011 | 1.034 | 1.057 | 1.079 | 1.102 | 1.124 | 1.145 | 1.167 | 1.188 | 1.208 | 1.248 | 1.306 | 1.397 |
| 22 | 1.011 | 1.032 | 1.054 | 1.076 | 1.097 | 1.118 | 1.139 | 1.159 | 1.179 | 1.199 | 1.238 | 1.293 | 1.381 |
| 23 | 1.010 | 1.031 | 1.052 | 1.073 | 1.093 | 1.113 | 1.133 | 1.153 | 1.172 | 1.191 | 1.228 | 1.281 | 1.366 |
| 24 | 1.010 | 1.029 | 1.049 | 1.070 | 1.089 | 1.109 | 1.128 | 1.147 | 1.165 | 1.183 | 1.219 | 1.271 | 1.352 |
| 25 | 1.009 | 1.028 | 1.048 | 1.067 | 1.086 | 1.104 | 1.123 | 1.141 | 1.159 | 1.176 | 1.211 | 1.261 | 1.340 |
| 26 | 1.009 | 1.027 | 1.046 | 1.064 | 1.082 | 1.101 | 1.118 | 1.136 | 1.153 | 1.170 | 1.203 | 1.251 | 1.328 |
| 27 | 1.009 | 1.026 | 1.044 | 1.062 | 1.079 | 1.097 | 1.114 | 1.131 | 1.148 | 1.164 | 1.196 | 1.243 | 1.317 |
| 28 | 1.008 | 1.025 | 1.042 | 1.060 | 1.077 | 1.093 | 1.110 | 1.126 | 1.142 | 1.158 | 1.189 | 1.235 | 1.307 |
| 29 | 1.008 | 1.024 | 1.041 | 1.058 | 1.074 | 1.090 | 1.106 | 1.122 | 1.138 | 1.153 | 1.183 | 1.227 | 1.297 |
| 30 | 1.008 | 1.023 | 1.040 | 1.056 | 1.072 | 1.087 | 1.103 | 1.118 | 1.133 | 1.148 | 1.178 | 1.220 | 1.288 |
| 40 | 1.006 | 1.017 | 1.030 | 1.042 | 1.054 | 1.066 | 1.078 | 1.089 | 1.101 | 1.112 | 1.135 | 1.168 | 1.221 |
| 60 | 1.004 | 1.012 | 1.020 | 1.028 | 1.036 | 1.044 | 1.052 | 1.060 | 1.068 | 1.076 | 1.091 | 1.114 | 1.151 |
| 120 | 1.002 | 1.006 | 1.010 | 1.014 | 1.018 | 1.022 | 1.026 | 1.030 | 1.034 | 1.038 | 1.046 | 1.058 | 1.078 |
| ∞ | 1.000 | 1.000 | 1.000 | 1.000 | 1.000 | 1.000 | 1.000 | 1.000 | 1.000 | 1.000 | 1.000 | 1.000 | 1.000 |

$$R = \sigma/\sigma_{\min} \leqslant R_{p,n-m,\alpha} \tag{36}$$

If this condition holds we accept the hypothesis at that level.

4. When the optimized parameter values are used, the calculated contribution of a component of the system to the model function $F$ is very small. How sure can we be that it really exists?

Let $\sigma$ be the standard deviation obtained when the component is omitted from the optimization. We then test the hypothesis that the component is not present at significance level $\alpha$. The dimension is equal to the number of parameters which are needed to describe the component, say $p$, and we test

$$R = \sigma/\sigma_{\min} \quad \text{and} \quad R_{p,n-m,\alpha}$$

and accept the hypothesis if $R \leqslant R_{p,n-m,\alpha}$.

5. We do not know the precise form to take for the fitting function $F$. We try function $F1$ with $m_1$ independent parameters and obtain a value $\sigma_1$, and function $F2$ with $m_2$ independent parameters and obtain a value $\sigma_2$. Which is the better model function?

In this instance, we cannot assume that *either* of the model functions would adequately represent error-free data and so cannot make quantitative tests of significance. If $m_2 > m_1$ we expect $\sigma_2$ to be less than $\sigma_1$ because the number of degrees of freedom has been reduced. The quantity $\hat{\sigma}$ defined in Eq. (32) may be used to compare the models and if

$$\frac{\hat{\sigma}_2}{\hat{\sigma}_1} = \frac{\sigma_2}{\sigma_1}\left[\frac{n-m_1}{n-m_2}\right]^{1/2} > 1 \qquad (37)$$

we may conclude that the function $F1$ is a better model. How much better we cannot say in any precise terms.

## V. Model Functions

In many instances, the form of the function $F$ which is to be fitted to the experimental data can be predicted from theoretical considerations. In other cases, the theory may be imperfectly understood or instrumental factors may modify the idealized form and a certain degree of empiricism must be introduced. One method of approximating $F$ is to use combinations of simple analytical functions; another is to use the observed data obtained from a very simple system to analyze that of a more complex system.

### A. Analytical Representation

In spectroscopic and chromatographic methods the data frequently consists of a series of overlapping bands and the problem is to estimate the individual components. A number of functions have been employed for this purpose and in selecting models for $F$ it should be borne in mind that the number of independent parameters should be chosen carefully in relation to the number of data (Margulies, 1968).

#### 1. Gaussian Bands

The Gaussian band shape, illustrated in Fig. 3, is given by

$$A(X, A_0, X_0, \Delta X_{1/2}) = A_0 \exp\left\{-\ln 2 \cdot \left[\frac{2(X-X_0)}{\Delta X_{1/2}}\right]^2\right\} \qquad (38)$$

where $A_0$ is the peak height, $X_0$ is the $X$ coordinate of the peak, and $\Delta X_{1/2}$ is the bandwidth at half height. If these quantities differ from band to band then there are three parameters per component, but if all bands may be assumed to have the same halfwidth this reduces to two. Expressions for the area beneath a Gaussian band and for the partial derivatives required in Eq. (19) are given in Table IV.

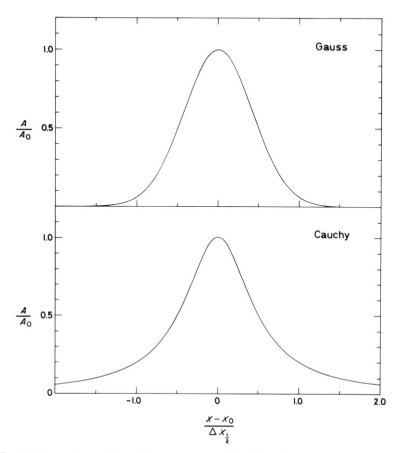

FIG. 3. Comparison of band shapes generated by Gaussian function (upper curve) and Cauchy function (lower curve). The Cauchy band has much longer "tails" which contain a considerable proportion of the total area beneath the curve. For a given peak height and half bandwidth the area beneath a Cauchy band is about 50% greater.

## 2. *Cauchy Bands*

Another symmetrical band shape, which differs from the Gaussian in having long "tails," is the Cauchy or Lorentz function (Fig. 3)

$$A(X, A_0, X_0, \Delta X_{1/2}) = \frac{A_0}{1 + [2(X - X_0)/\Delta X_{1/2}]^2} \quad (39)$$

Expressions for the area and the derivatives are given in Table V, and it will be seen that the tails contain a significant fraction of the total area.

### TABLE IV
#### The Gaussian Function

$$A(X, A_0, X_0, \Delta X_{1/2}) = A_0 \exp\left\{-\ln 2 \cdot \left[\frac{2(X-X_0)}{\Delta X_{1/2}}\right]^2\right\}$$

*Partial Derivatives*

$$\frac{\partial A}{\partial A_0} = \exp\left\{-\ln 2 \cdot \left[\frac{2(X-X_0)}{\Delta X_{1/2}}\right]^2\right\}$$

$$\frac{\partial A}{\partial X_0} = \frac{8 \ln 2 \cdot A_0 (X-X_0)}{\Delta X_{1/2}^2} \frac{\partial A}{\partial A_0}$$

$$\frac{\partial A}{\partial \Delta X_{1/2}} = \frac{(X-X_0)}{\Delta X_{1/2}} \frac{\partial A}{\partial X_0}$$

*Area*

$$\frac{1}{2}\left(\frac{\pi}{\ln 2}\right)^{1/2} A_0 \Delta X_{1/2}$$

### 3. *Sum and Product Bands*

A great variety of symmetrical band shapes may be generated by combining Gaussian and Cauchy functions as sums or products (Fraser and Suzuki, 1966, 1969; Pitha and Jones, 1966, 1967). One of the most useful, which only involves one additional parameter, is the sum of a Gaussian and a Cauchy band with equal half-widths in the proportions $f$ to $(1-f)$ giving

$$A = f A_0 \exp\left\{-\ln 2 \cdot \left[\frac{2(X-X_0)}{\Delta X_{1/2}}\right]^2\right\} + \frac{(1-f)A_0}{1+[2(X-X_0)/\Delta X_{1/2}]^2} \quad (40)$$

Often $f$ may be assumed to have the same value for all bands and so

### TABLE V
#### The Cauchy Function

$$A(X, A_0, X_0, \Delta X_{1/2}) = \frac{A_0}{1+[2(X-X_0)/\Delta X_{1/2}]^2}$$

*Partial Derivatives*

$$\frac{\partial A}{\partial A_0} = \frac{1}{1+[2(X-X_0)/\Delta X_{1/2}]^2}$$

$$\frac{\partial A}{\partial X_0} = \frac{8 A_0 (X-X_0)}{\Delta X_{1/2}^2}\left(\frac{\partial A}{\partial A_0}\right)^2$$

$$\frac{\partial A}{\partial \Delta X_{1/2}} = \frac{(X-X_0)}{\Delta X_{1/2}} \frac{\partial A}{\partial X_0}$$

*Area*

$$\frac{1}{2}\pi A_0 \Delta X_{1/2}$$

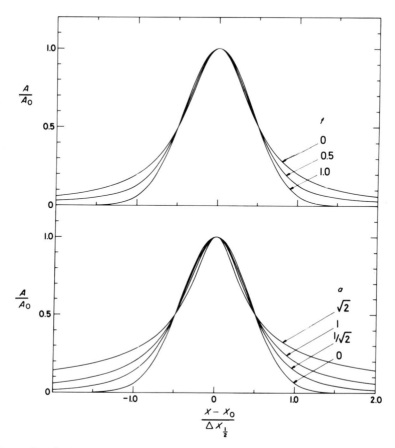

Fig. 4. Band shapes generated by using linear combinations of Gauss and Cauchy bands (upper set) and by varying the value of the parameter $a$ in the four-parameter symmetrical band shape given in Eq. (43) (lower set). [Reprinted from *Analyt. Chem.* **41**, 1969, pp. 37-39. Copyright (1969) by the American Chemical Society. Reprinted by permission of the copyright owner.]

only one additional parameter needs to be introduced into the optimization procedure. In this case, the partial derivatives with respect to $A_0$, $X_0$, and $\Delta X_{1/2}$ are simply linear combinations of those given in Tables IV and V in the proportions $f:(1-f)$. The partial derivative $\partial F_i/\partial f$ is obtained by summing a term of the type

$$A_0 \left[ \exp\left\{ -\ln 2 \cdot \left[ \frac{2(X-X_0)}{\Delta X_{1/2}} \right]^2 \right\} - \left\{ 1 + \left[ \frac{2(X-X_0)}{\Delta X_{1/2}} \right]^2 \right\}^{-1} \right] \quad (41)$$

for each band. The area beneath the band is given by

$$\tfrac{1}{2} A_0 \Delta X_{1/2} \left[ f \left( \frac{\pi}{\ln 2} \right)^{1/2} + \pi(1-f) \right] \tag{42}$$

Computer subroutines for fitting this type of band to experimental data are included in the program given in Section VI. The sum function in Eq. (40) is extremely versatile as fixing $f = 1$ gives a Gaussian shape, fixing $f = 0$ gives a Cauchy shape, and using $f$ as an independent parameter optimizes the shape of the tail (Fig. 4).

### 4. A Further Four-Parameter Symmetrical Band Shape

An alternative to the four-parameter sum function is provided by the expression (Fraser and Suzuki, 1969)

$$A = \frac{A_0}{\{1 + [2^{a^2} - 1][2(X - X_0)/\Delta X_{1/2}]^2\}^{1/a^2}} \qquad (a \neq 0) \tag{43}$$

which is illustrated in Fig. 4 for several values of $a$ between 0 and $\sqrt{2}$. As $a \to 0$ the band shape reduces to a Gaussian and for $a = 1$ it reduces to the Cauchy shape. Values of $a$ greater than 1 produce shapes with tails even more pronounced than the Cauchy band shape but when $a \geq \sqrt{2}$ the area beneath the band becomes infinite. Expressions for the area and the partial derivatives are given in Table VI.

### 5. Asymmetric Band Shapes

Empirical fitting of asymmetric band shapes with analytical functions is not usually successful and the digital standard shape method described

**TABLE VI**
**A Four-Parameter Function for Representing Symmetrical Band Shapes**

$$A(X, A_0, X_0, \Delta X_{1/2}, a) = \frac{A_0}{\{1 + (2^{a^2} - 1)[2(X - X_0)/\Delta X_{1/2}]^2\}^{1/a^2}} \qquad 0 < a < \sqrt{2}$$

*Partial Derivatives*

$$\frac{\partial A}{\partial A_0} = \frac{1}{\{1 + (2^{a^2} - 1)[2(X - X_0)/\Delta X_{1/2}]^2\}^{1/a^2}}$$

$$\frac{\partial A}{\partial X_0} = \frac{8 A_0 (2^{a^2} - 1)(X - X_0)}{(a \Delta X_{1/2})^2} \left( \frac{\partial A}{\partial A_0} \right)^{1+a^2}$$

$$\frac{\partial A}{\partial \Delta X_{1/2}} = \frac{X - X_0}{\Delta X_{1/2}} \frac{\partial A}{\partial X_0}$$

$$\frac{\partial A}{\partial a} = \left( \frac{-2 A_0}{a} \right) \left\{ \ln \left( \frac{\partial A}{\partial A_0} \right) + 4 \ln 2 \left[ \frac{X - X_0}{\Delta X_{1/2}} \right]^2 \left( 2 \frac{\partial A}{\partial A_0} \right)^{a^2} \right\} \frac{\partial A}{\partial A_0}$$

*Area*

$$\tfrac{1}{2} A_0 \Delta X_{1/2} \pi^{1/2} \Gamma \left( \frac{1}{a^2} - \frac{1}{2} \right) \left[ (2^{a^2} - 1)^{1/2} \Gamma \left( \frac{1}{a^2} \right) \right]^{-1}$$

where $\Gamma(x)$ is the gamma function of argument $x$

at the end of this section is to be preferred. Small amounts of asymmetry may sometimes be dealt with by using the log-normal band shape

$$A = A_0 \exp\left[-\ln 2 \cdot \left\{\frac{\ln\left[1 + 2b(X - X_0)/\Delta X_{1/2}\right]}{b}\right\}^2\right]$$

$$(2b(X - X_0)/\Delta X_{1/2} > -1) \quad (44a)$$

$$A = 0 \qquad (2b(X - X_0)/\Delta X_{1/2} \leqslant -1) \quad (44b)$$

where $b$ is a parameter which determines the degree of asymmetry (Fig. 5). As $b \to 0$, the band shape reduces to a symmetrical Gaussian shape while negative values of $b$ produce negatively skewed band shapes. The parameter $\Delta X_{1/2}$ in Eq. (44a) is related to the halfwidth of the skewed band by

$$\text{Actual bandwidth} = (\Delta X_{1/2} \cdot \sinh b)/b \quad (45)$$

and the area beneath the band is given by

$$\frac{1}{2}\left(\frac{\pi}{\ln 2}\right)^{1/2} A_0 \Delta X_{1/2} \exp\left(\frac{b^2}{4 \ln 2}\right) \quad (46)$$

and expressions for the partial derivatives have been given by Fraser and Suzuki (1970a).

A simple method of introducing *small* amounts of asymmetry into a symmetrical function is to multiply it by a factor

$$\{1 + \tanh\left[2b(X - X_0)/\Delta X_{1/2}\right]\} \quad (47)$$

where $b$ is an adjustable parameter. The effect of multiplying the expression for a Gaussian shape in Eq. (38) by this factor is illustrated in Fig. 5. The parameters $A_0$, $X_0$, and $\Delta X_{1/2}$ no longer have their original significance but an advantage of the procedure is that the formula for the area beneath the band in terms of these parameters remains unchanged.

More complex functions have been used to represent asymmetric chromatographic bands (Levy and Martin, 1968; Grushka et al., 1969; Gladney et al., 1969; Anderson et al., 1970b; Buys and de Clerk, 1972) and in polarographic studies (Gutknecht and Perone, 1970) but in the authors' experience it is usually simpler and more effective to use the digital standard shape function method rather than to attempt finding analytical functions to represent complicated band shapes.

## 6. Derivative Functions

In some techniques, the data is collected in the form of derivative spectra and the functions above can be used in derivative form to fit the ob-

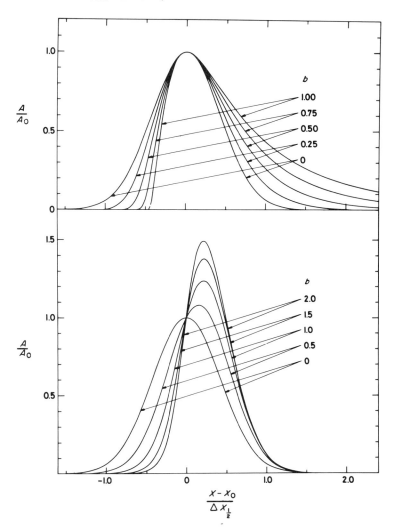

Fig. 5. Asymmetric band shapes generated by using nonzero values of the parameter $b$ in Eq. (44) (upper set) and in the product of the expression in Eq. (47) with a Gaussian function (lower set). [Reprinted from *Analyt. Chem.* **41**, 1969, pp. 37–39. Copyright (1969) by the American Chemical Society. Reprinted by permission of the copyright owner.]

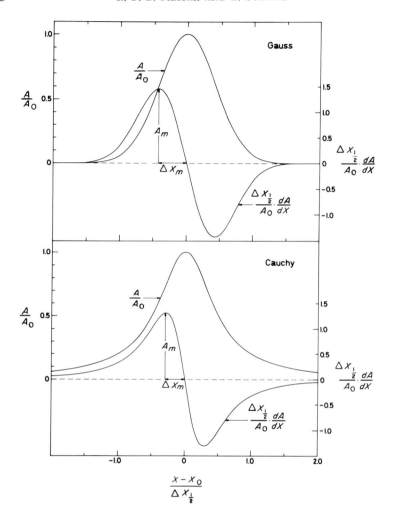

Fig. 6. Gaussian band and its first derivative (upper curve) and Cauchy band and its first derivative (lower curve). The quantities $A_m$ and $\Delta X_m$ may be related to the values $A_0$ and $\Delta X_{1/2}$ (see text) in the parent band. [Reprinted from *Analyt. Chem.* **41,** 1969, pp. 37–39. Copyright (1969) by the American Chemical Society. Reprinted by permission of the copyright owner.]

served data. It is convenient to use the same parameters and these are related to the observable quantities illustrated in Fig. 6, as follows

$$\begin{array}{ll} \text{Gauss} & \text{Cauchy} \\ A_0 = A_m \Delta X_m \exp(1/2) & A_0 = 8A_m \Delta X_m / 3 \\ A(X_0) = 0 & A(X_0) = 0 \\ \Delta X_{1/2} = 2(2 \ln 2)^{1/2} \Delta X_m & \Delta X_{1/2} = 2\sqrt{3} \Delta X_m \end{array} \quad (48)$$

The partial derivatives for these two types of derivative band are given in Table VII.

## B. Digital Standard Shape Functions

In cases in which the form of $F$ cannot be predicted from theoretical considerations and attempts to construct an empirical model using analytical functions are not successful, the method whereby a digital standard shape function is fitted can be used (Keller et al., 1966; Anderson et al., 1970b; Trombka and Schmadebeck, 1970).

In the case of a complex of overlapping bands, a pure component is

### TABLE VII
### Derivative Functions

*First Derivative of the Gaussian Function*

$$A'(X, A_0, X_0, \Delta X_{1/2}) = - \frac{8 \ln 2 A_0 (X - X_0) \exp\{-\ln 2[2(X - X_0)/\Delta X_{1/2}]^2\}}{\Delta X_{1/2}^2}$$

*Partial Derivatives*

$$\frac{\partial A'}{\partial A_0} = - \frac{8 \ln 2 (X - X_0) \exp\{-\ln 2[2(X - X_0)/\Delta X_{1/2}]^2\}}{\Delta X_{1/2}^2}$$

$$\frac{\partial A'}{\partial X_0} = - \frac{\partial A'}{\partial A_0} \frac{A_0\{1 - 2 \ln 2[2(X - X_0)/\Delta X_{1/2}]^2\}}{X - X_0}$$

$$\frac{\partial A'}{\partial \Delta X_{1/2}} = -2 \frac{\partial A'}{\partial A_0} \frac{A_0\{1 - \ln 2[2(X - X_0)/\Delta X_{1/2}]^2\}}{\Delta X_{1/2}}$$

*First Derivative of the Cauchy Function*

$$A'(X, A_0, X_0, \Delta X_{1/2}) = - \frac{8A_0(X - X_0)}{\Delta X_{1/2}^2 \{1 + [2(X - X_0)/\Delta X_{1/2}]^2\}^2}$$

*Partial Derivatives*

$$\frac{\partial A'}{\partial A_0} = - \frac{8(X - X_0)}{\Delta X_{1/2}^2 \{1 + [2(X - X_0)/\Delta X_{1/2}]^2\}^2}$$

$$\frac{\partial Y'}{\partial X_0} = - \frac{\partial A'}{\partial A_0} \frac{A_0\{1 - 4/[1 + \frac{1}{4}\Delta X_{1/2}^2/(X - X_0)^2]\}}{X - X_0}$$

$$\frac{\partial A'}{\partial \Delta X_{1/2}} = -2 \frac{\partial A'}{\partial A_0} \frac{A_0\{1 - 2/[1 + \frac{1}{4}\Delta X_{1/2}^2/(X - X_0)^2]\}}{\Delta X_{1/2}}$$

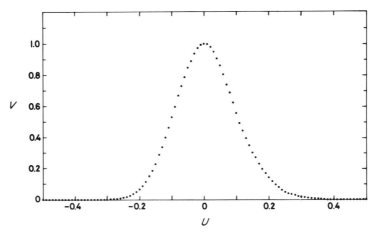

Fig. 7. An example of a standard shape function used in resolving overlapping components in the output from an automatic amino acid analyzer (Fig. 9).

run and the band shape is stored in the computer as a series of readings of $Y$ measured at equispaced $X$ values in the range $X_{min}$ to $X_{max}$ (Fig. 7).

It is convenient to use the variables $U$ and $V$ to describe the standard shape; they are related to the measured variables by

$$U = \frac{X - X_{min}}{X_{max} - X_{min}} - \frac{1}{2} \tag{49}$$

$$V = Y/(Y_{max} - Y_{min}) \tag{50}$$

where $Y_{max}$ and $Y_{min}$ are the maximum and minimum values of $Y$ in the standard shape. Bands in the complex spectrum may then be specified by reference to the standard shape $V(U)$ by

$$A = A_0 V\left(\frac{X - X_0}{\Delta X}\right) \tag{51}$$

where $A_0$ has the same significance as previously, $X_0$ represents the $X$-axis displacement of the point $U = 0$ in the standard shape, and $\Delta X$ is the scale expansion of the $U$-axis, i.e., the $X$ range in the observed component corresponding to $U = -0.5$ to $+0.5$ in the standard shape.

It is essential to remove both baseline and baseline drift from the observations used to obtain the standard shape and frequently $V(U) = 0$, for $U \leqslant -0.5$ and $U \geqslant 0.5$. Values within the range $0.5 \geqslant U \geqslant -0.5$ may be obtained by using a suitable interpolation formula. The partial derivatives of the digital shape function parameters must be calculated numerically from the relationship

$$\frac{\partial F_i}{\partial P_j} = \frac{F_i(P_1, \ldots, P_j(1 + \epsilon), \ldots, P_m) - F_i(P_1, \ldots, P_j, \ldots, P_m)}{\epsilon P_j}$$
$$(P_j \neq 0) \quad (52)$$

where $\epsilon$ is a small quantity, say $\sim 10^{-4}$. If values of $P_j$ very close to zero are likely to be encountered, the term $P_j(1 + \epsilon)$ in the numerator should be replaced by $(P_j + \epsilon)$ and the denominator becomes $\epsilon$.

In many applications it is the area beneath the band which is the quantity of interest. When a digital standard shape function is used the area beneath a component band is simply $A_0 \cdot \Delta X$ times the area beneath the standard shape function.

## C. Baseline Functions

The function $B(X)$ in Eq. (2) represents the baseline contribution and this also can be optimized during the curve fitting procedure.

A linear baseline may be specified by two parameters $B_1$ and $B_n$ equal to baseline ordinates at $X = X_1$ and $X_n$, respectively. For intermediate values

$$B = B_1 + \frac{(X - X_1)(B_n - B_1)}{X_n - X_1} \quad (53)$$

and the partial derivatives required in Eq. (19) are

$$\frac{\partial B}{\partial B_1} = 1 - (X - X_1)/(X_n - X_1) \quad (54)$$

$$\frac{\partial B}{\partial B_n} = (X - X_1)/(X_n - X_1) \quad (55)$$

Curved baselines may be approximated by polynomials (Stone, 1962; Marshall et al., 1965) and special conditions, such as a flat baseline with $B_1 = B_n$, can be readily incorporated (Fraser and Suzuki, 1970a).

## VI. Computational Procedure

In this section a procedure for parameter optimization is described which is applicable to a wide variety of problems.

## A. Flow Chart

A flow chart for the procedure is given in Fig. 8. The operation of the program is regulated by means of control cards bearing the words JOB, DATA, RUN, or END.

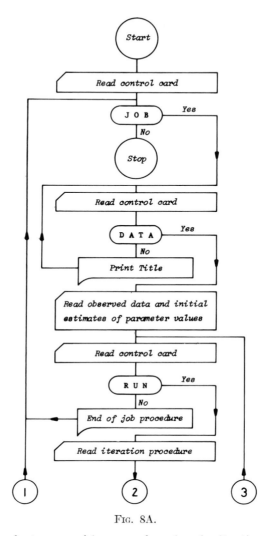

Fig. 8A.

Fig. 8. Flow chart summarizing procedure for the iterative nonlinear least-squares optimization procedure.

Each complete problem is signaled by a JOB card followed by a series of cards bearing title material to be reproduced on the output medium. The end of the title is signaled by a DATA card and subsequent cards supply information about the data to be fitted, the digital standard shape function (if applicable), the number of parameters, and the initial estimates of the parameter values.

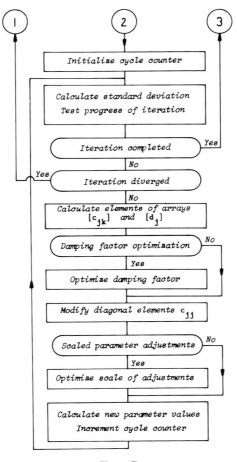

Fig. 8B.

A RUN card signals that a refinement is to be carried out according to the procedure specified on subsequent cards and the iteration then proceeds until the specified number of cycles has been completed or the iteration converges or diverges. If the latter condition obtains, the job is aborted, the parameter correlation coefficients are printed out, and the program then proceeds to the next job. If the operation is normal and the iteration does not diverge, a further card is read. If it is a RUN card, further refinement is carried out according to the procedure specified on subsequent cards; if it is a JOB or an END card, the optimized parameter values and standard deviations are printed out together with the observed and fitted values and an unbiased estimate of the standard deviation. If a JOB card was read the entire procedure is repeated; if an END card was read the program terminates.

B. Computer Program

The procedure outlined in the flow chart may be illustrated by a program designed to deal with the resolution of multicomponent systems. A listing of this program, which is written in Fortran IV, is given in Table VIII. The subroutines FCALC and DFCALC refer to a set of Gauss-Cauchy sum bands (Fig. 4) and a linear baseline. The only modification required for other applications is to replace these subroutines by appropriate ones for calculating $F_i$ ($i = 1, \ldots, n$) and $\partial F_i / \partial P_j$ ($i = 1, \ldots, n, j = 1, \ldots, m$), respectively.

If a digital standard shape function is to be used, the initial C's in instructions 31 through 35 are removed and instructions 390 through 431 are deleted and replaced by the alternative block 390 through 452 listed at the end of Table VIII.

The detailed operation of the program is indicated by the insertion of comments in the listing. Various additions to the subroutine JOBEND may be made for particular applications; for example, band areas may be printed and the observed data and fitted function can be plotted out as illustrated in Fig. 10.

Logical unit 5 is the standard input medium (usually a card reader) and logical unit 6 is the standard output medium (usually a line printer).

C. Input

A typical input relating to the problem illustrated in Fig. 9 is listed in Table IX. The commencement of a job is signaled by a control card bearing the word JOB in columns 1–3 followed by any number of cards bearing title material to be reproduced on the output medium (repeating format 20A4/). A control card bearing the word DATA signals the end of the title and the following card indicates the number of data $n$ (format I4). Then follows the observed data in the order $X_1, Y_1; X_2, Y_2; X_3, Y_3; \ldots ; X_n, Y_n$ (repeating format 8F10.0/). Advantage may be taken of the fact that the inclusion of a decimal point in the data overrides the decimal portion of the specified format.

In the digital standard shape function version, the next thirteen cards contain 101 values of $V$ at equispaced intervals of $U$ in the range $-0.5 \leqslant U \leqslant 0.5$. The format for the list of $V$ values is similar to that for the

## TABLE VIII
### Listing of a Computer Program for Analyzing Overlapping Components[a]

●●●●●●●●●●●●●●●●●●●●●●●●●●●●●●●●●●●●●●●●●●●●●●●●●●●●●●●●●●●●●●●●●●●●●●●●●

THIS LISTING IS SUITABLE FOR USE WITH THE CDC 3600. FOR USE WITH THE IBM 7040 THE C IN COLUMN 1 OF CARD 10 SHOULD BE REMOVED AND C INSERTED IN COLUMN 1 OF CARDS 1 AND 11

●●●●●●●●●●●●●●●●●●●●●●●●●●●●●●●●●●●●●●●●●●●●●●●●●●●●●●●●●●●●●●●●●●●●●●●●●

```
            PROGRAM LSQ                                                     1
  C         OPTIMIZES M PARAMETERS TO GIVE BEST FIT OF MODEL FUCTIONS        2
  C         F(1)-F(N) TO OBSERVED DATA Y(1)-Y(N). UP TO L=M-1 OF THE         3
  C         PARAMETERS MAY BE CONSTRAINED TO RETAIN PRESELECTED VALUES.      4
  C         IN THIS VERSION EACH Y(I) IS ASSOCIATED WITH A SECOND VARIABLE   5
  C         X(I) AND F IS A FUNCTION OF X AND THE M PARAMETERS.              6
            DIMENSION X(200),Y(200),F(200),P(20),G(19),DF(200,20),IFP(19),   7
           1DG(19,20),NAME(20),CMAT(39,39),DMAT(39),TCMAT(39,39),TDMAT(39)   8
            COMMON V(101)                                                    9
  C         DATA NAME1/4HJOB /,NAME2/4HDATA/,NAME3/4HRUN /,NAME4/4HEND /    10
            DATA (NAME1=4HJOB ),(NAME2=4HDATA),(NAME3=4HRUN ),(NAME4=4HEND) 11
            READ(5,1) NAME                                                  12
          1 FORMAT(20A4)                                                    13
          2 IF(NAME(1).NE.NAME1) STOP                                       14
          3 READ(5,1) NAME                                                  15
            IF(NAME(1).EQ.NAME2) GO TO 5                                    16
            WRITE(6,4) NAME                                                 17
          4 FORMAT(1H ,20A4)                                                18
            GO TO 3                                                         19
  C         READ NUMBER OF OBSERVATIONS                                     20
          5 READ(5,6) N                                                     21
          6 FORMAT(13I4)                                                    22
  C         READ OBSERVED DATA                                              23
            READ(5,7) (X(I),Y(I),I=1,N)                                     24
          7 FORMAT(8F10.0)                                                  25
            WRITE(6,8)                                                      26
          8 FORMAT(/14H OBSERVED DATA/3(4X,1HI,4X,4HX(I),8X,4HY(I),4X))     27
            WRITE(6,9) (I,X(I),Y(I),I=1,N)                                  28
          9 FORMAT(3(I5,2E12.4))                                            29
  C         READ STANDARD SHAPE FUNCTION                                    30
  C         READ(5,7) V                                                     31
  C         WRITE(6,10)                                                     32
  C      10 FORMAT(/24H STANDARD SHAPE FUNCTION)                            33
  C         WRITE(6,11) (I,V(I),I=1,101)                                    34
  C      11 FORMAT(6(I5,F8.4))                                              35
            WRITE(6,12)                                                     36
         12 FORMAT(/25H INITIAL PARAMETER VALUES)                           37
  C         READ NUMBER OF PARAMETERS                                       38
            READ(5,6) M                                                     39
  C         READ TRIAL VALUES OF PARAMETERS                                 40
            READ(5,7) (P(J),J=1,M)                                          41
            WRITE(6,14)                                                     42
         13 WRITE(6,14)                                                     42
         14 FORMAT(5(4X,1HJ,4X,4HP(J),4X))                                  43
            WRITE(6,15) (J,P(J),J=1,M)                                      44
         15 FORMAT(5(I5,E12.4))                                             45
         16 READ(5,1) NAME                                                  46
  C         SET FLAG IF JOB COMPLETED                                       47
            IF(NAME(1).EQ.NAME3) GO TO 17                                   48
            IT=4                                                            49
            GO TO 24                                                        50
  C         READ INSTRUCTIONS FOR ITERATION                                 51
         17 READ(5,18) NCYCLE,MODE,D                                        52
  C         MODE 0=AUTOD+AUTOS, 1=AUTOD, 2=AUTOS, 3=NONAUTO                 53
         18 FORMAT(2I4,F10.0)                                               54
  C         READ NUMBER OF FIXED PARAMETERS                                 55
            READ(5,6) L                                                     56
            LM=L+M                                                          57
  C         READ INDICES OF FIXED PARAMETERS                                58
            IF(L.NE.0) READ(5,6) (IFP(I),I=1,L)                             59
            WRITE(6,19) NCYCLE,MODE                                         60
```

[a] In this application the components are assumed to be capable of representation by Gauss-Cauchy sum functions with the same value of $f$ for all components. By altering the program as indicated a digital standard shape function may be used in place of the sum function.

## TABLE VIII (Continued)

```
      19 FORMAT(/4H RUN,I3,13H CYCLES   MODE,I2)               61
         IF(L.EQ.0) WRITE(6,20) L                              62
         IF(L.NE.0) WRITE(6,20) L,(IFP(I),I=1,L)               63
      20 FORMAT(I3,17H FIXED PARAMETERS,1913)                  64
         ICYCLE=0                                              65
C        CALCULATE F FOR X=X(1) TO X(N)                        66
      21 CALL FCALC(N,X,M,P,F)                                 67
C        CALCULATE RMS DEVIATION AND TEST PROGRESS OF ITERATION 68
         SD=DRMS(N,F,Y)                                        69
         CALL TEST(ICYCLE,NCYCLE,SD,IT,NAME,NAME1,NAME4,D,M,L,IFP, 70
        1TCMAT,TDMAT)                                          71
         GO TO(24,22,13),IT                                    72
C        ITERATION DIVERGED ABORT JOB                          73
      22 WRITE(6,23)                                           74
      23 FORMAT(1H1)                                           75
         GO TO 2                                               76
C        CALCULATE G FOR THE L FIXED PARAMETERS                77
      24 CALL GCALC(L,G)                                       78
C        CALCULATE PARTIAL DERIVATIVES OF F WRT THE M PARAMETERS 79
C        FOR X=X(1) TO X(N)                                    80
         CALL DFCALC(N,X,F,M,P,DF)                             81
C        CALCULATE THE PARTIAL DERIVATIVES OF G WRT THE M PARAMETERS 82
C        FOR EACH OF THE L FIXED PARAMETERS                    83
         CALL DGCALC(M,L,IFP,DG)                               84
C        LOAD DMAT ELEMENTS 1 TO M                             85
         DO 25 J=1,M                                           86
         DMAT(J)=0.                                            87
         DO 25 I=1,N                                           88
      25 DMAT(J)=DMAT(J)-DF(I,J)*(F(I)-Y(I))                   89
         J=M                                                   90
C        LOAD DMAT ELEMENTS M+1 TO M+L                         91
         DO 26 K=1,L                                           92
         J=J+1                                                 93
      26 DMAT(J)=-G(K)                                         94
C        INITIALIZE ALL CMAT ELEMENTS TO ZERO                  95
         DO 27 J=1,LM                                          96
         DO 27 K=1,LM                                          97
      27 CMAT(J,K)=0.                                          98
C        LOAD CMAT ELEMENTS IN RANGE J=1 TO M, K=1 TO M        99
         DO 29 J=1,M                                          100
         DO 29 K=J,M                                          101
         DO 28 I=1,N                                          102
      28 CMAT(J,K)=CMAT(J,K)+DF(I,J)*DF(I,K)                  103
      29 CMAT(K,J)=CMAT(J,K)                                  104
C        LOAD CMAT ELEMENTS IN RANGE J=M+1 TO M+L, K=1 TO M   105
C        AND J=1 TO M, K=M+1 TO M+L                           106
         DO 30 I=1,L                                          107
         J=M+I                                                108
         DO 30 K=1,M                                          109
         CMAT(J,K)=DG(I,K)                                    110
      30 CMAT(K,J)=DG(I,K)                                    111
C        STORE CMAT AND DMAT ARRAYS FOR CALCULATION OF INVERSE IF REQUIRED 112
         DO 31 J=1,LM                                         113
         TDMAT(J)=DMAT(J)                                     114
         DO 31 K=1,LM                                         115
      31 TCMAT(J,K)=CMAT(J,K)                                 116
C        TEST IF FLAG SET FOR JOB COMPLETED                   117
         IF(IT.NE.4) GO TO 32                                 118
         CALL JOBEND(N,X,Y,F,M,L,SD,TCMAT,TDMAT)              119
         WRITE(6,23)                                          120
         GO TO 2                                              121
C        OPTIMIZE DAMPING FACTOR IF MODE = 0 OR 1             122
      32 IF(MODE.LT.2) CALL AUTOD(N,X,Y,F,M,P,LM,D,CMAT,DMAT) 123
C        MULTIPLY DIAGONAL ELEMENTS OF CMAT BY (1+DAMPING FACTOR**2) 124
         DO 33 J=1,LM                                         125
      33 CMAT(J,J)=(1.+D*D)*CMAT(J,J)                         126
C        SOLVE FOR PARAMETER ADJUSTMENTS                      127
         CALL MATINV(CMAT,DMAT,LM,1)                          128
C        OPTIMIZE SCALE OF PARAMETER ADJUSTMENTS IF MODE =0 OR 2 129
         IF(MODE.EQ.0.OR.MODE.EQ.2) CALL AUTOS(N,X,Y,F,M,P,DMAT) 130
C        ADJUST PARAMETER VALUES                              131
         DO 34 J=1,M                                          132
      34 P(J)=P(J)+DMAT(J)                                    133
         ICYCLE=ICYCLE+1                                      134
         GO TO 21                                             135
         END                                                  136
```

## TABLE VIII (Continued)

```
      SUBROUTINE TEST(ICYCLE,NCYCLE,SD,IT,NAME,NAME1,NAME4,D,M,L,IFP,      137
     1TCMAT,TDMAT)                                                          138
C     MONITORS ITERATION                                                    139
      DIMENSION NAME(20),TCMAT(39,39),TDMAT(39),IFP(19)                     140
      WRITE(6,1) ICYCLE,SD,D                                                141
    1 FORMAT(6H CYCLE,I3,6H    SD=,E13.6,5H    D=,E13.6)                   142
      IF(ICYCLE.NE.0) GO TO 4                                               143
C     SET FLAG FOR ITERATION INCOMPLETE                                     144
    2 IT=1                                                                  145
C     STORE STANDARD DEVIATION                                              146
    3 SDO=SD                                                                147
      RETURN                                                                148
C     TEST IF ITERATION CONVERGED                                           149
    4 IF((SDO-SD)/SDO.GT.1.E-6) GO TO 16                                   150
C     TEST IF ITERATION DIVERGED                                            151
      IF((SD-SDO)/SDO.GT.1.E-6) GO TO 6                                    152
      WRITE(6,5)                                                            153
    5 FORMAT(20H ITERATION COMPLETED//25H REFINED PARAMETER VALUES)        154
      GO TO 18                                                              155
    6 WRITE(6,7)                                                            156
    7 FORMAT(19H ITERATION DIVERGED//25H CORRELATION COEFFICIENTS)         157
C     CALCULATE CORRELATION COEFFICIENTS                                    158
      CALL MATINV(TCMAT,TDMAT,L+M,0)                                        159
      DO 9 J=2,M                                                            160
      KK=J-1                                                                161
      DO 9 K=1,KK                                                           162
      TCMAT(J,K)=0.0                                                        163
      DO 8 I=1,L                                                            164
      IF(J.EQ.IFP(I).OR.K.EQ.IFP(I)) GO TO 9                               165
    8 CONTINUE                                                              166
      TCMAT(J,K)=TCMAT(K,J)/SQRT(ABS(TCMAT(J,J)*TCMAT(K,K)))               167
    9 CONTINUE                                                              168
      WRITE(6,10) (J,J=1,M)                                                 169
   10 FORMAT(5X,20I6)                                                       170
      DO 11 J=1,M                                                           171
      TCMAT(J,J)=1.0                                                        172
   11 WRITE(6,12) J,(TCMAT(J,K),K=1,J)                                      173
   12 FORMAT(I3,2X,20F6.2)                                                  174
      DO 14 I=1,100                                                         175
      READ(5,13) NAME                                                       176
      IF(NAME(1).EQ.NAME1.OR.NAME(1).EQ.NAME4) GO TO 15                    177
   13 FORMAT(20A4)                                                          178
   14 CONTINUE                                                              179
C     SET FLAG FOR JOB ABORTED                                              180
   15 IT=2                                                                  181
      RETURN                                                                182
   16 IF(ICYCLE.NE.NCYCLE) GO TO 2                                         183
      WRITE(6,17)                                                           184
   17 FORMAT(17H CYCLES COMPLETED//25H REFINED PARAMETER VALUES)           185
C     SET FLAG FOR CYCLES COMPLETED                                         186
   18 IT=3                                                                  187
      GO TO 3                                                               188
      END                                                                   189

      SUBROUTINE AUTOD(N,X,Y,TF,M,P,LM,D,CMAT,DMAT)                        190
C     OPTIMIZES DAMPING FACTOR (D)                                          191
      DIMENSION X(200),Y(200),TF(200),P(20),CMAT(39,39),DMAT(39),          192
     1TCMAT(39,39),TDMAT(39),E(3),TP(20)                                   193
      DMIN=1.0E-6                                                           194
      DMAX=1.0E2                                                            195
      DINC=2.0                                                              196
      IF(D.LT.DMIN)  D=1.0                                                  197
      IF(D.GT.DMAX)  D=DMAX                                                 198
      K=1                                                                   199
      L=3                                                                   200
C     CALCULATE STANDARD DEVIATION FOR 0.5*D,D,2.0*D                        201
    1 DO 5 I=K,L                                                            202
      TD=D*DINC**(I-2)                                                      203
      DO 3 J=1,LM                                                           204
      TDMAT(J)=DMAT(J)                                                      205
      DO 2 KK=1,LM                                                          206
    2 TCMAT(J,KK)=CMAT(J,KK)                                                207
    3 TCMAT(J,J)=TCMAT(J,J)*(1.+TD*TD)                                      208
      CALL MATINV(TCMAT,TDMAT,LM,1)                                         209
      DO 4 J=1,M                                                            210
```

## TABLE VIII (Continued)

```
      4 TP(J)=P(J)+TDMAT(J)                                      211
        CALL FCALC(N,X,M,TP,TF)                                  212
      5 E(I)=DRMS(N,TF,Y)                                        213
        K=1                                                      214
C       TEST IF D OUTSIDE RANGE                                  215
        IF(D.LT.DMIN.OR.D.GT.DMAX)  GO TO 6                      216
C       TEST IF STANDARD DEVIATION INSENSITIVE TO D              217
        IF(ABS((E(3)-E(2))/E(2)).LE.1.E-6)   RETURN              218
        IF(ABS((E(1)-E(2))/E(2)).LE.1.E-6)   RETURN              219
C       TEST IF MINIMUM DETECTED                                 220
        IF(E(2).LT.E(1).AND.E(2).LT.E(3))  RETURN                221
C       TEST IF BEST D GT CURRENT D                              222
        IF(E(3).LT.E(1))  K=3                                    223
        L=4-K                                                    224
        E(L)=E(2)                                                225
        E(2)=E(K)                                                226
C       CHOOSE BETTER D                                          227
        D=D*DINC**(2-L)                                          228
        L=K                                                      229
        GO TO 1                                                  230
      6 WRITE(6,7) D                                             231
      7 FORMAT(3H D=,E10.3,27H  OUTSIDE RANGE DMIN - DMAX)       232
        IF(D.LT.DMIN)  D=DMIN                                    233
        RETURN                                                   234
        END                                                      235

        SUBROUTINE AUTOS (N,X,Y,F,M,P,DMAT)                      236
C       OPTIMIZE SCALE FACTOR (H) OF PARAMETER ADJUSTMENTS       237
        DIMENSION X(200),Y(200),F(200),P(20),DMAT(39),E(3)       238
       1,PTEMP(20)                                               239
        K=1                                                      240
        L=3                                                      241
        H=1.0                                                    242
        HMIN=0.01                                                243
        HMAX=100.0                                               244
        HINC=1.2                                                 245
C       CALCULATE STANDARD DEVIATION FOR RANGE OF H VALUES       246
      1 DO 3 I=K,L                                               247
        DO 2 J=1,M                                               248
      2 PTEMP(J)=DMAT(J)*H*HINC**(I-2)+P(J)                      249
        CALL FCALC(N,X,M,PTEMP,F)                                250
      3 E(I)=DRMS(N,F,Y)                                         251
        K=1                                                      252
C       TEST IF H OUTSIDE SELECTED RANGE                         253
        IF(H.LT.HMIN.OR.H.GT.HMAX) GO TO 4                       254
C       TEST IF STANDARD DEVIATION INSENSITIVE TO H              255
        IF(ABS((E(1)-E(2))/E(2)).LE.1.E-6) RETURN                256
        IF(ABS((E(3)-E(2))/E(2)).LE.1.E-6) RETURN                257
C       TEST IF MINIMUM DETECTED                                 258
        IF(E(2).LT.E(1).AND.E(2).LT.E(3)) GO TO 6                259
C       TEST IF BEST H GT CURRENT H                              260
        IF(E(3).LT.E(1))  K=3                                    261
        L=4-K                                                    262
        E(L)=E(2)                                                263
        E(2)=E(K)                                                264
C       CHOOSE BETTER H                                          265
        H=H*HINC**(2-L)                                          266
        L=K                                                      267
        GO TO 1                                                  268
      4 WRITE(6,5) H                                             269
      5 FORMAT(3H H=,F10.6,26H EXEEDS RANGE HMIN - HMAX )        270
C       MULTIPLY PARAMETER ADJUSTMENTS BY BEST H                 271
      6 DO 7 J=1,M                                               272
      7 DMAT(J)=DMAT(J)*H                                        273
        RETURN                                                   274
        END                                                      275

        SUBROUTINE GCALC(L,G)                                    276
        DIMENSION G(19)                                          277
        DO 1 I=1,L                                               278
      1 G(I)=0.                                                  279
        RETURN                                                   280
        END                                                      281

        SUBROUTINE DGCALC(M,L,IFP,DG)                            282
```

## 21. LEAST SQUARES IN DATA ANALYSIS

### TABLE VIII (Continued)

```
      DIMENSION IFP(19),DG(19,20)                              283
      DO 1 K=1,L                                                284
      DO 1 J=1,M                                                285
      DG(K,J)=0.                                                286
      IF(J.EQ.IFP(K)) DG(K,J)=1.                                287
    1 CONTINUE                                                  288
      RETURN                                                    289
      END                                                       290

      SUBROUTINE MATINV(A,B,N,L)                                291
C     IF L=0 RETURNS INVERSE OF A IN A, IF L=1 SOLUTION OF AX=B IN B  292
      DIMENSION A(39,39),B(39),IP(39),IN(39,2)                  293
      D=1.                                                      294
      DO 1 I=1,N                                                295
    1 IP(I)=0                                                   296
      DO 12 I=1,N                                               297
      AMAX=0.                                                   298
      DO 3 J=1,N                                                299
      IF(IP(J).GT.0) GO TO 3                                    300
      IF(IP(J).LT.0) GO TO 4                                    301
      DO 2 K=1,N                                                302
      IF(IP(K).EQ.1) GO TO 2                                    303
      IF(IP(K).GT.1) GO TO 4                                    304
      IF(ABS(A(J,K)).LE.AMAX) GO TO 2                           305
      IR=J                                                      306
      IC=K                                                      307
      AMAX=ABS(A(J,K))                                          308
    2 CONTINUE                                                  309
    3 CONTINUE                                                  310
      IP(IC)=IP(IC)+1                                           311
      IF(AMAX.GT.1.E-30) GO TO 6                                312
    4 WRITE(6,5)                                                313
    5 FORMAT(/16H SINGULAR MATRIX)                              314
      STOP                                                      315
    6 IF(IR.EQ.IC) GO TO 8                                      316
      D=-D                                                      317
      DO 7 K=1,N                                                318
      AMAX=A(IR,K)                                              319
      A(IR,K)=A(IC,K)                                           320
    7 A(IC,K)=AMAX                                              321
      IF(L.EQ.0) GO TO 8                                        322
      AMAX=B(IR)                                                323
      B(IR)=B(IC)                                               324
      B(IC)=AMAX                                                325
    8 IN(I,1)=IR                                                326
      IN(I,2)=IC                                                327
      AMAX=A(IC,IC)                                             328
      D=D*AMAX                                                  329
      A(IC,IC)=1.                                               330
      DO 9 K=1,N                                                331
    9 A(IC,K)=A(IC,K)/AMAX                                      332
      IF(L.EQ.0) GO TO 10                                       333
      B(IC)=B(IC)/AMAX                                          334
   10 DO 12 J=1,N                                               335
      IF(J.EQ.IC) GO TO 12                                      336
      AMAX=A(J,IC)                                              337
      A(J,IC)=0.                                                338
      DO 11 K=1,N                                               339
   11 A(J,K)=A(J,K)-A(IC,K)*AMAX                                340
      IF(L.EQ.0) GO TO 12                                       341
      B(J)=B(J)-B(IC)*AMAX                                      342
   12 CONTINUE                                                  343
      IF(L.EQ.1) RETURN                                         344
      DO 14 I=1,N                                               345
      J=N+1-I                                                   346
      IF(IN(J,1).EQ.IN(J,2)) GO TO 14                           347
      IR=IN(J,1)                                                348
      IC=IN(J,2)                                                349
      DO 13 K=1,N                                               350
      AMAX=A(K,IR)                                              351
      A(K,IR)=A(K,IC)                                           352
   13 A(K,IC)=AMAX                                              353
   14 CONTINUE                                                  354
      RETURN                                                    355
      END                                                       356

      FUNCTION DRMS(N,F,Y)                                      357
```

## TABLE VIII (Continued)

```
          DIMENSION F(200),Y(200)                                    358
          DRMS=0.                                                    359
          DO 1 I=1,N                                                 360
        1 DRMS=DRMS+(F(I)-Y(I))**2                                   361
          DRMS=SQRT(DRMS/FLOAT(N))                                   362
          RETURN                                                     363
          END                                                        364

          SUBROUTINE JOBEND(N,X,Y,F,M,L,SD,TCMAT,TDMAT)              365
C         END OF JOB PROCEDURE                                       366
          DIMENSION X(200),Y(200),F(200),TCMAT(39,39),TDMAT(39),E(200) 367
C         CALCULATE STANDARD DEVIATIONS OF PARAMETER ESTIMATES       368
          LM=L+M                                                     369
          CALL MATINV(TCMAT,TDMAT,LM,0)                              370
          DO 1 J=1,M                                                 371
        1 TDMAT(J)=SD*SQRT(ABS(TCMAT(J,J))*FLOAT(N)/FLOAT(N+L-M))    372
          WRITE(6,2)                                                 373
        2 FORMAT(/43H STANDARD DEVIATIONS OF PARAMETER ESTIMATES)    374
          WRITE(6,3) (J,TDMAT(J),J=1,M)                              375
        3 FORMAT(5(I5,E12.4))                                        376
C         CALCULATE DEVIATIONS                                       377
          DO 4 I=1,N                                                 378
        4 E(I)=F(I)-Y(I)                                             379
          WRITE(6,5)                                                 380
        5 FORMAT(/14H FITTED VALUES/2(4X,1HI,4X,4HY(I),8X,4HF(I),9X,4HE(I), 381
         13X))                                                       382
          WRITE(6,6) (I,Y(I),F(I),E(I),I=1,N)                        383
        6 FORMAT(2(I5,2E12.4,E12.3))                                 384
          SD=SD*SQRT(FLOAT(N)/FLOAT(N+L-M))                          385
          WRITE(6,7) SD                                              386
        7 FORMAT(/42H UNBIASED ESTIMATE OF STANDARD DEVIATION = ,E12.4) 387
          RETURN                                                     388
          END                                                        389

          SUBROUTINE FCALC(N,X,M,P,F)                                390
C         SHAPE=F*GAUSS + (1-F)*CAUCHY                               391
          DIMENSION X(200),F(200),P(20)                              392
          NB=(M-3)/3                                                 393
          Q1=P(M-2)                                                  394
          Q2=(P(M-1)-Q1)/(X(N)-X(1))                                 395
          DO 1 I=1,N                                                 396
        1 F(I)=Q1+Q2*(X(I)-X(1))                                     397
          Q1=ALOG(2.)                                                398
          DO 2 J=1,NB                                                399
          Q2=4./P(3*J)**2                                            400
          Q3=P(3*J-2)                                                401
          Q4=P(3*J-1)                                                402
          DO 2 I=1,N                                                 403
          Q5=Q2*(X(I)-Q4)**2                                         404
        2 F(I)=F(I)+Q3*(P(M)*EXP(-Q1*Q5)+(1.-P(M))/(1.+Q5))          405
          RETURN                                                     406
          END                                                        407

          SUBROUTINE DFCALC(N,X,F,M,P,DF)                            408
C         FOR SHAPE= F*GAUSS + (1-F)*CAUCHY                          409
          DIMENSION X(200),F(200),P(20),DF(200,20)                   410
          NB=(M-3)/3                                                 411
          Q1=ALOG(2.)                                                412
          DO 1 I=1,N                                                 413
          DF(I,M)=0.                                                 414
          DF(I,M-1)=(X(I)-X(1))/(X(N)-X(1))                          415
          DF(I,M-2)=1.-DF(I,M-1)                                     416
          DO 1 K=1,NB                                                417
          J1=3*K-2                                                   418
          J2=3*K-1                                                   419
          J3=3*K                                                     420
          Q2=(X(I)-P(J2))/P(J3)                                      421
          Q3=EXP(-4.*Q1*Q2*Q2)                                       422
          Q4=1./(1.+4.*Q2*Q2)                                        423
          Q5=8.*Q1*Q2*Q3*P(J1)/P(J3)                                 424
          Q6=8.*Q2*Q4*Q4*P(J1)/P(J3)                                 425
          DF(I,J1)=P(M)*Q3+(1.-P(M))*Q4                              426
          DF(I,J2)=P(M)*Q5+(1.-P(M))*Q6                              427
          DF(I,J3)=Q2*DF(I,J2)                                       428
        1 DF(I,M)=DF(I,M)+P(J1)*(Q3-Q4)                              429
          RETURN                                                     430
          END                                                        431
```

## 21. LEAST SQUARES IN DATA ANALYSIS 341

### TABLE VIII (Continued)

```
FOR OPERATION WITH A STANDARD SHAPE FUNCTION
(1) REMOVE C FROM COLUMN 1 IN CARDS 31 - 35 INCLUSIVE
(2) REMOVE ALL CARDS AFTER 389 AND REPLACE WITH FOLLOWING

      FUNCTION STD(X,H,T,W)                                           390
C     COMPUTES A(X) FROM DIGITAL STANDARD SHAPE FUNCTION IN V(1)-V(101) 391
C     H=HEIGHT, T=X SHIFT OF V(51), W=X SCALE EXPANSION                392
      COMMON V(101)                                                    393
      U=100.*(X-T)/W+50.                                               394
C     TEST IF STORED RANGE OF V EXEEDED                                395
      IF(U.GT.0.) GO TO 1                                              396
      STD=H*V(1)                                                       397
      RETURN                                                           398
    1 J=INT(U)+1                                                       399
C     TEST IF STORED RANGE OF V EXEEDED                                400
      IF(J.LT.101) GO TO 2                                             401
      STD=H*V(101)                                                     402
      RETURN                                                           403
C     INTERPOLATE A(X) FROM STORED VALUES OF V                         404
    2 D=U-FLOAT(J-1)                                                   405
      STD=V(J)+D*(V(J+1)-V(J))                                         406
      IF(J.EQ.1) GO TO 3                                               407
      IF(J.EQ.100) GO TO 4                                             408
      STD=H*(STD+.25*D*(D-1.)*(V(J+2)-V(J+1)-V(J)+V(J-1)))              409
      RETURN                                                           410
    3 STD=H*(STD+.25*D*(D-1.)*(V(J+2)-V(J+1)))                         411
      RETURN                                                           412
    4 STD=H*(STD+.25*D*(D-1.)*(V(J-1)-V(J)))                           413
      RETURN                                                           414
      END                                                              415

      SUBROUTINE FCALC(N,X,M,P,F)                                      416
      DIMENSION X(200),F(200),P(20)                                    417
      NB=(M-2)/3                                                       418
      R=P(M-1)                                                         419
      S=(P(M)-R)/(X(N)-X(1))                                           420
      DO 2 I=1,N                                                       421
      T=R+S*(X(I)-X(1))                                                422
      DO 1 K=1,NB                                                      423
      J1=3*K-2                                                         424
      J2=3*K-1                                                         425
      J3=3*K                                                           426
    1 T=T+STD(X(I),P(J1),P(J2),P(J3))                                  427
    2 F(I)=T                                                           428
      RETURN                                                           429
      END                                                              430

      SUBROUTINE DFCALC(N,X,F,M,P,DF)                                  431
      DIMENSION X(200),F(200),P(20),DF(200,20)                         432
      NB=(M-2)/3                                                       433
      S=1.E-4                                                          434
      T=1.+0.5*S                                                       435
      U=1.-0.5*S                                                       436
      DO 2 I=1,N                                                       437
      R=(X(I)-X(1))/(X(N)-X(1))                                        438
      DO 1 K=1,NB                                                      439
      J1=3*K-2                                                         440
      J2=3*K-1                                                         441
      J3=3*K                                                           442
      DF(I,J1)=(STD(X(I),P(J1)*T,P(J2),P(J3))-STD(X(I),P(J1)*U,P(J2),  443
     1P(J3)))/(S*P(J1))                                                444
      DF(I,J2)=(STD(X(I),P(J1),P(J2)*T,P(J3))-STD(X(I),P(J1),P(J2)*U,  445
     1P(J3)))/(S*P(J2))                                                446
    1 DF(I,J3)=(STD(X(I),P(J1),P(J2),P(J3)*T)-STD(X(I),P(J1),P(J2),P(J3) 447
     1*U))/(S*P(J3))                                                   448
      DF(I,M-1)=1.-R                                                   449
    2 DF(I,M)=R                                                        450
      RETURN                                                           451
      END                                                              452
```

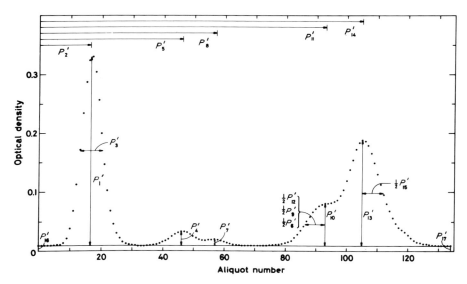

Fig. 9. Output from Spinco automatic amino acid analyzer with partially resolved components. The method of estimating the initial values of parameter 1 through 17 is shown for a model consisting of five bands plus a linear baseline.

observed data (Table X). In this example, the digital standard shape function is that illustrated in Fig. 7.

The number of parameters ($m$) is defined on the next card (format I4). This is equal to three times the number of bands plus three with the sum function version and one less with the alternative version. In the present example, these numbers are $(3 \times 5) + 3 = 18$ (Table IX) and $(3 \times 5) + 2 = 17$ (Table X). The following cards contain the initial estimates of the parameter values in similar format to the observed data. The parameters are listed in the order $A_0$, $X_0$, $\Delta X_{1/2}$ for each band in turn followed by the baseline estimates $B_1$ and $B_n$ (Section V,C) and, in the case of the sum function version, the parameter $f$ in Eq. (40).

The method used to obtain the initial estimates of the parameter values is illustrated in Fig. 9. The value of $P_{18}$ was arbitrarily taken as 1.0; that is, a Gaussian shape was assumed. The initial estimates for operation with the digital standard shape function version are the same for band heights and positions but the halfwidths estimated from the observed data must be divided by the halfwidth of the standard band, which in the case depicted in Fig. 7 is 0.21.

In choosing the initial estimates of parameter values, care must be taken to avoid values for which $\partial F_i/\partial P_j$ is zero for all $i$. For example,

## TABLE IX

Listing of the Input Deck for the Problem Illustrated in Fig. 9 Used
In Conjunction with the Program Listed in Table VIII

```
JOB
ANALYSIS OF AN OUTPUT FROM SPINCO 120B AUTOMATIC AMINO ACID ANALYSER, USING
(1) GAUSS SHAPE   (2) GAUSS + CAUCHY SHAPE
DATA
 135
      0.000    0.009     1.000    0.009     2.000    0.009     3.000    0.009
      4.000    0.009     5.000    0.009     6.000    0.010     7.000    0.010
      8.000    0.013     9.000    0.021    10.000    0.038    11.000    0.068
     12.000    0.112    13.000    0.171    14.000    0.234    15.000    0.287
     16.000    0.324    17.000    0.330    18.000    0.304    19.000    0.257
     20.000    0.200    21.000    0.147    22.000    0.102    23.000    0.068
     24.000    0.046    25.000    0.030    26.000    0.022    27.000    0.015
     28.000    0.012    29.000    0.011    30.000    0.009    31.000    0.009
     32.000    0.008    33.000    0.008    34.000    0.008    35.000    0.009
     36.000    0.009    37.000    0.010    38.000    0.011    39.000    0.013
     40.000    0.015    41.000    0.017    42.000    0.020    43.000    0.025
     44.000    0.028    45.000    0.031    46.000    0.033    47.000    0.033
     48.000    0.032    49.000    0.028    50.000    0.025    51.000    0.022
     52.000    0.019    53.000    0.018    54.000    0.018    55.000    0.019
     56.000    0.019    57.000    0.020    58.000    0.019    59.000    0.017
     60.000    0.016    61.000    0.014    62.000    0.012    63.000    0.011
     64.000    0.010    65.000    0.010    66.000    0.009    67.000    0.009
     68.000    0.009    69.000    0.009    70.000    0.009    71.000    0.009
     72.000    0.009    73.000    0.009    74.000    0.009    75.000    0.010
     76.000    0.010    77.000    0.011    78.000    0.011    79.000    0.012
     80.000    0.013    81.000    0.015    82.000    0.017    83.000    0.023
     84.000    0.028    85.000    0.035    86.000    0.042    87.000    0.051
     88.000    0.058    89.000    0.066    90.000    0.071    91.000    0.076
     92.000    0.079    93.000    0.081    94.000    0.081    95.000    0.082
     96.000    0.084    97.000    0.086    98.000    0.094    99.000    0.104
    100.000    0.119   101.000    0.138   102.000    0.154   103.000    0.172
    104.000    0.183   105.000    0.188   106.000    0.186   107.000    0.179
    108.000    0.165   109.000    0.149   110.000    0.130   111.000    0.112
    112.000    0.095   113.000    0.080   114.000    0.068   115.000    0.056
    116.000    0.047   117.000    0.040   118.000    0.036   119.000    0.032
    120.000    0.027   121.000    0.022   122.000    0.018   123.000    0.015
    124.000    0.014   125.000    0.012   126.000    0.012   127.000    0.011
    128.000    0.010   129.000    0.010   130.000    0.009   131.000    0.009
    132.000    0.009   133.000    0.009   134.000    0.009
 18
 .320        16.5      8.0       .024     46.0      13.5        .011      57.0
 13.5         .041    93.0      13.5       .179    105.0       13.9       .009
 .009         1.0
RUN
  40      0  0.0
   1
  18
RUN
  40      0  0.0
   0
END
```

$\partial A/\partial X_0$ for the Gaussian function (Table IV) is zero for $A_0 = 0$ and if this value were used as an initial estimate for the height of a weak band the quantity $S$ in Eq. (1) would not be a function of $X_0$ and the method would break down. It would be permissible to fix $A_0 = 0$ but if this parameter were subsequently released the method would again break down. Provided values of zero are avoided for initial estimates of the heights or half-widths of bands, this problem is rarely encountered in practice.

A control card bearing the word RUN signals that a refinement is to be carried out and the next card specifies the maximum number of cycles,

## TABLE X
### Listing of the Input Deck for the Problem Illustrated in Fig. 9 Used in Conjunction with the Alternative Form of the Program Listed in Table VIII

```
JOB
ANALYSIS OF AN OUTPUT FROM SPINCO 120B AUTOMATIC AMINO ACID ANALYSER, USING
DIGITIZED STANDARD SHAPE FUNCTION
DATA
  135
     0.000      0.009      1.000      0.009      2.000      0.009      3.000      0.009
     4.000      0.009      5.000      0.009      6.000      0.010      7.000      0.010
     8.000      0.013      9.000      0.021     10.000      0.038     11.000      0.068
    12.000      0.112     13.000      0.171     14.000      0.234     15.000      0.287
    16.000      0.324     17.000      0.330     18.000      0.304     19.000      0.257
    20.000      0.200     21.000      0.147     22.000      0.102     23.000      0.068
    24.000      0.046     25.000      0.030     26.000      0.022     27.000      0.015
    28.000      0.012     29.000      0.011     30.000      0.009     31.000      0.009
    32.000      0.008     33.000      0.008     34.000      0.008     35.000      0.009
    36.000      0.009     37.000      0.010     38.000      0.011     39.000      0.013
    40.000      0.015     41.000      0.017     42.000      0.020     43.000      0.025
    44.000      0.028     45.000      0.031     46.000      0.033     47.000      0.033
    48.000      0.032     49.000      0.028     50.000      0.025     51.000      0.022
    52.000      0.019     53.000      0.018     54.000      0.018     55.000      0.019
    56.000      0.019     57.000      0.020     58.000      0.019     59.000      0.017
    60.000      0.016     61.000      0.014     62.000      0.012     63.000      0.011
    64.000      0.010     65.000      0.010     66.000      0.009     67.000      0.009
    68.000      0.009     69.000      0.009     70.000      0.009     71.000      0.009
    72.000      0.009     73.000      0.009     74.000      0.009     75.000      0.010
    76.000      0.010     77.000      0.011     78.000      0.011     79.000      0.012
    80.000      0.013     81.000      0.015     82.000      0.017     83.000      0.023
    84.000      0.028     85.000      0.035     86.000      0.042     87.000      0.051
    88.000      0.058     89.000      0.066     90.000      0.071     91.000      0.076
    92.000      0.079     93.000      0.081     94.000      0.081     95.000      0.082
    96.000      0.084     97.000      0.086     98.000      0.094     99.000      0.104
   100.000      0.119    101.000      0.138    102.000      0.154    103.000      0.172
   104.000      0.183    105.000      0.188    106.000      0.186    107.000      0.179
   108.000      0.165    109.000      0.149    110.000      0.130    111.000      0.112
   112.000      0.095    113.000      0.080    114.000      0.068    115.000      0.056
   116.000      0.047    117.000      0.040    118.000      0.036    119.000      0.032
   120.000      0.027    121.000      0.022    122.000      0.018    123.000      0.015
   124.000      0.014    125.000      0.012    126.000      0.012    127.000      0.011
   128.000      0.010    129.000      0.010    130.000      0.009    131.000      0.009
   132.000      0.009    133.000      0.009    134.000      0.009
   .000       .000       .000       .000       .000       .000       .000       .000
   .000       .000       .000       .000       .000       .000       .000       .000
   .001       .001       .002       .002       .003       .004       .005       .006
   .010       .014       .018       .026       .039       .053       .068       .091
   .116       .152       .187       .231       .290       .341       .402       .466
   .533       .599       .673       .734       .791       .850       .893       .940
   .977       .997      1.00        .998       .982       .952       .909       .864
   .807       .743       .690       .627       .559       .495       .447       .398
   .352       .306       .273       .235       .201       .172       .144       .120
   .101       .082       .063       .052       .043       .037       .031       .025
   .021       .016       .015       .013       .010       .009       .006       .005
   .003       .001       .001       .0000      .0000      .000       .000       .000
   .000       .000       .000       .000       .000
  17
 .32       16.5        32.0       0.024      46.0       53.0        0.011     57.0
 53.0       0.041      93.0       53.0        0.179    105.0        53.0       0.009
  0.009
RUN
   40   0  0.0
    0
END
```

## TABLE XI
### Listing of the Output from the Program Given in Table VIII When Used with the Input Listed in Table X

ANALYSIS OF AN OUTPUT FROM SPINCO 120B AUTOMATIC AMINO ACID ANALYSER, USING DIGITIZED STANDARD SHAPE FUNCTION

OBSERVED DATA

| I | X(I) | Y(I) | I | X(I) | Y(I) | I | X(I) | Y(I) |
|---|------|------|---|------|------|---|------|------|
| 1 | 0.0000+000 | 9.0000-003 | 2 | 1.0000+000 | 9.0000-003 | 3 | 2.0000+000 | 9.0000-003 |
| 4 | 3.0000+000 | 9.0000-003 | 5 | 4.0000+000 | 9.0000-003 | 6 | 5.0000+000 | 9.0000-003 |
| 7 | 6.0000+000 | 1.0000-002 | 8 | 7.0000+000 | 1.0000-002 | 9 | 8.0000+000 | 1.3000-002 |
| 10 | 9.0000+000 | 2.1000-002 | 11 | 1.0000+001 | 3.8000-002 | 12 | 1.1000+001 | 6.8000-002 |
| 13 | 1.2000+001 | 1.1200-001 | 14 | 1.3000+001 | 1.7100-001 | 15 | 1.4000+001 | 2.3400-001 |
| 16 | 1.5000+001 | 2.8700-001 | 17 | 1.6000+001 | 3.2400-001 | 18 | 1.7000+001 | 3.3000-001 |
| 19 | 1.8000+001 | 3.0400-001 | 20 | 1.9000+001 | 2.5700-001 | 21 | 2.0000+001 | 2.0000-001 |
| 22 | 2.1000+001 | 1.4700-001 | 23 | 2.2000+001 | 1.0200-001 | 24 | 2.3000+001 | 6.8000-002 |
| 25 | 2.4000+001 | 4.6000-002 | 26 | 2.5000+001 | 3.0000-002 | 27 | 2.6000+001 | 2.2000-002 |
| 28 | 2.7000+001 | 1.5000-002 | 29 | 2.8000+001 | 1.2000-002 | 30 | 2.9000+001 | 1.1000-002 |
| 31 | 3.0000+001 | 9.0000-003 | 32 | 3.1000+001 | 9.0000-003 | 33 | 3.2000+001 | 8.0000-003 |
| 34 | 3.3000+001 | 8.0000-003 | 35 | 3.4000+001 | 8.0000-003 | 36 | 3.5000+001 | 9.0000-003 |
| 37 | 3.6000+001 | 9.0000-003 | 38 | 3.7000+001 | 1.0000-002 | 39 | 3.8000+001 | 1.1000-002 |
| 40 | 3.9000+001 | 1.3000-002 | 41 | 4.0000+001 | 1.5000-002 | 42 | 4.1000+001 | 1.7000-002 |
| 43 | 4.2000+001 | 2.0000-002 | 44 | 4.3000+001 | 2.5000-002 | 45 | 4.4000+001 | 2.8000-002 |
| 46 | 4.5000+001 | 3.1000-002 | 47 | 4.6000+001 | 3.3000-002 | 48 | 4.7000+001 | 3.3000-002 |
| 49 | 4.8000+001 | 3.2000-002 | 50 | 4.9000+001 | 2.8000-002 | 51 | 5.0000+001 | 2.5000-002 |
| 52 | 5.1000+001 | 2.2000-002 | 53 | 5.2000+001 | 1.9000-002 | 54 | 5.3000+001 | 1.8000-002 |
| 55 | 5.4000+001 | 1.8000-002 | 56 | 5.5000+001 | 1.9000-002 | 57 | 5.6000+001 | 1.9000-002 |
| 58 | 5.7000+001 | 2.0000-002 | 59 | 5.8000+001 | 1.9000-002 | 60 | 5.9000+001 | 1.7000-002 |
| 61 | 6.0000+001 | 1.6000-002 | 62 | 6.1000+001 | 1.4000-002 | 63 | 6.2000+001 | 1.2000-002 |
| 64 | 6.3000+001 | 1.1000-002 | 65 | 6.4000+001 | 1.0000-002 | 66 | 6.5000+001 | 1.0000-002 |
| 67 | 6.6000+001 | 9.0000-003 | 68 | 6.7000+001 | 9.0000-003 | 69 | 6.8000+001 | 9.0000-003 |
| 70 | 6.9000+001 | 9.0000-003 | 71 | 7.0000+001 | 9.0000-003 | 72 | 7.1000+001 | 9.0000-003 |
| 73 | 7.2000+001 | 9.0000-003 | 74 | 7.3000+001 | 9.0000-003 | 75 | 7.4000+001 | 9.0000-003 |
| 76 | 7.5000+001 | 1.0000-002 | 77 | 7.6000+001 | 1.0000-002 | 78 | 7.7000+001 | 1.1000-002 |
| 79 | 7.8000+001 | 1.1000-002 | 80 | 7.9000+001 | 1.2000-002 | 81 | 8.0000+001 | 1.3000-002 |
| 82 | 8.1000+001 | 1.5000-002 | 83 | 8.2000+001 | 1.7000-002 | 84 | 8.3000+001 | 2.3000-002 |
| 85 | 8.4000+001 | 2.8000-002 | 86 | 8.5000+001 | 3.5000-002 | 87 | 8.6000+001 | 4.2000-002 |
| 88 | 8.7000+001 | 5.1000-002 | 89 | 8.8000+001 | 5.8000-002 | 90 | 8.9000+001 | 6.6000-002 |
| 91 | 9.0000+001 | 7.1000-002 | 92 | 9.1000+001 | 7.6000-002 | 93 | 9.2000+001 | 7.9000-002 |
| 94 | 9.3000+001 | 8.1000-002 | 95 | 9.4000+001 | 8.1000-002 | 96 | 9.5000+001 | 8.2000-002 |
| 97 | 9.6000+001 | 8.4000-002 | 98 | 9.7000+001 | 8.6000-002 | 99 | 9.8000+001 | 9.4000-002 |
| 100 | 9.9000+001 | 1.0400-001 | 101 | 1.0000+002 | 1.1900-001 | 102 | 1.0100+002 | 1.3800-001 |
| 103 | 1.0200+002 | 1.5400-001 | 104 | 1.0300+002 | 1.7200-001 | 105 | 1.0400+002 | 1.8300-001 |
| 106 | 1.0500+002 | 1.8800-001 | 107 | 1.0600+002 | 1.8600-001 | 108 | 1.0700+002 | 1.7900-001 |
| 109 | 1.0800+002 | 1.6500-001 | 110 | 1.0900+002 | 1.4900-001 | 111 | 1.1000+002 | 1.3000-001 |
| 112 | 1.1100+002 | 1.1200-001 | 113 | 1.1200+002 | 9.5000-002 | 114 | 1.1300+002 | 8.0000-002 |
| 115 | 1.1400+002 | 6.8000-002 | 116 | 1.1500+002 | 5.6000-002 | 117 | 1.1600+002 | 4.7000-002 |
| 118 | 1.1700+002 | 4.0000-002 | 119 | 1.1800+002 | 3.6000-002 | 120 | 1.1900+002 | 3.2000-002 |
| 121 | 1.2000+002 | 2.7000-002 | 122 | 1.2100+002 | 2.2000-002 | 123 | 1.2200+002 | 1.8000-002 |
| 124 | 1.2300+002 | 1.5000-002 | 125 | 1.2400+002 | 1.4000-002 | 126 | 1.2500+002 | 1.2000-002 |
| 127 | 1.2600+002 | 1.2000-002 | 128 | 1.2700+002 | 1.1000-002 | 129 | 1.2800+002 | 1.0000-002 |
| 130 | 1.2900+002 | 1.0000-002 | 131 | 1.3000+002 | 9.0000-003 | 132 | 1.3100+002 | 9.0000-003 |
| 133 | 1.3200+002 | 9.0000-003 | 134 | 1.3300+002 | 9.0000-003 | 135 | 1.3400+002 | 9.0000-003 |

STANDARD SHAPE FUNCTION

| 1 | 0.0000 | 2 | 0.0000 | 3 | 0.0000 | 4 | 0.0000 | 5 | 0.0000 | 6 | 0.0000 |
|---|--------|---|--------|---|--------|---|--------|---|--------|---|--------|
| 7 | 0.0000 | 8 | 0.0000 | 9 | 0.0000 | 10 | 0.0000 | 11 | 0.0000 | 12 | 0.0000 |
| 13 | 0.0000 | 14 | 0.0000 | 15 | 0.0000 | 16 | -0.0000 | 17 | 0.0010 | 18 | 0.0010 |
| 19 | 0.0020 | 20 | 0.0020 | 21 | 0.0030 | 22 | 0.0040 | 23 | 0.0050 | 24 | 0.0060 |
| 25 | 0.0100 | 26 | 0.0140 | 27 | 0.0180 | 28 | 0.0260 | 29 | 0.0390 | 30 | 0.0530 |
| 31 | 0.0680 | 32 | 0.0910 | 33 | 0.1160 | 34 | 0.1520 | 35 | 0.1870 | 36 | 0.2310 |
| 37 | 0.2900 | 38 | 0.3410 | 39 | 0.4020 | 40 | 0.4660 | 41 | 0.5330 | 42 | 0.5990 |
| 43 | 0.6730 | 44 | 0.7340 | 45 | 0.7910 | 46 | 0.8500 | 47 | 0.8930 | 48 | 0.9400 |
| 49 | 0.9770 | 50 | 0.9970 | 51 | 1.0000 | 52 | 0.9980 | 53 | 0.9820 | 54 | 0.9520 |
| 55 | 0.9090 | 56 | 0.8640 | 57 | 0.8070 | 58 | 0.7430 | 59 | 0.6900 | 60 | 0.6270 |
| 61 | 0.5590 | 62 | 0.4950 | 63 | 0.4470 | 64 | 0.3980 | 65 | 0.3520 | 66 | 0.3060 |
| 67 | 0.2730 | 68 | 0.2350 | 69 | 0.2010 | 70 | 0.1720 | 71 | 0.1440 | 72 | 0.1200 |
| 73 | 0.1010 | 74 | 0.0820 | 75 | 0.0630 | 76 | 0.0520 | 77 | 0.0430 | 78 | 0.0370 |
| 79 | 0.0310 | 80 | 0.0250 | 81 | 0.0210 | 82 | 0.0160 | 83 | 0.0150 | 84 | 0.0130 |
| 85 | 0.0100 | 86 | 0.0090 | 87 | 0.0040 | 88 | 0.0050 | 89 | 0.0030 | 90 | 0.0010 |
| 91 | 0.0010 | 92 | 0.0000 | 93 | 0.0000 | 94 | 0.0000 | 95 | 0.0000 | 96 | 0.0000 |
| 97 | 0.0000 | 98 | 0.0000 | 99 | 0.0000 | 100 | 0.0000 | 101 | 0.0000 | | |

## TABLE XI (*Continued*)

```
INITIAL PARAMETER VALUES
   J      P(J)         J       P(J)         J      P(J)         J       P(J)         J       P(J)
   1   3.2000-001      2   1.6500+001       3   3.2000+001      4   2.4000-002       5   4.6000+001
   6   5.3000+001      7   1.1000-002       8   5.7000+001      9   5.3000+001      10   4.1000-002
  11   9.3000+001     12   5.3000+001      13   1.7900-001     14   1.0500+002      15   5.3000+001
  16   9.0000-003     17   9.0000-003

RUN 40 CYCLES  MODE 0
 0 FIXED PARAMETERS
 CYCLE  0    SD= 1.052793-002    D=-0.000000+000
 CYCLE  1    SD= 2.810305-003    D= 2.500000-001
 CYCLE  2    SD= 1.792608-003    D= 1.250000-001
 CYCLE  3    SD= 1.776847-003    D= 1.250000-001
 CYCLE  4    SD= 1.776707-003    D= 2.500000-001
 CYCLE  5    SD= 1.776704-003    D= 2.500000-001
 CYCLE  6    SD= 1.776704-003    D= 2.500000-001
 ITERATION COMPLETED

REFINED PARAMETER VALUES
   J      P(J)         J       P(J)         J      P(J)         J       P(J)         J       P(J)
   1   3.2302-001      2   1.6657+001       3   3.4918+001      4   2.3734-002       5   4.6454+001
   6   4.2325+001      7   9.0699-003       8   5.7384+001      9   2.8150+001      10   6.4789-002
  11   9.1617+001     12   5.2967+001      13   1.7168-001     14   1.0562+002      15   5.7815+001
  16   8.0715-003     17   1.1422-002

STANDARD DEVIATIONS OF PARAMETER ESTIMATES
   1   1.0268-003      2   1.1536-002       3   1.4545-001      4   9.6101-004       5   1.9369-001
   6   2.4897+000      7   1.1235-003       8   4.1646-001      9   4.3777+000      10   8.7258-004
  11   1.3728-001     12   1.3564+000      13   9.0768-004     14   5.3553-002      15   5.1116-001
  16   4.8218-004     17   4.9202-004

FITTED VALUES
   I      Y(I)         F(I)         E(I)        I      Y(I)         F(I)         E(I)
   1   9.0000-003   8.0715-003   -9.285-004     2   9.0000-003   8.0965-003   -9.035-004
   3   9.0000-003   8.1215-003   -8.785-004     4   9.0000-003   8.1465-003   -8.535-004
   5   9.0000-003   8.1715-003   -8.285-004     6   9.0000-003   8.5196-003   -4.804-004
   7   1.0000-002   9.0025-003   -9.975-004     8   1.0000-002   9.1811-003   -8.190-005
   9   1.3000-002   1.3009-002    9.180-006    10   2.1000-002   2.1208-002    2.076-004
  11   3.8000-002   3.7188-002   -8.117-004    12   6.8000-002   6.6379-002   -1.621-003
  13   1.1200-001   1.1294-001    9.363-004    14   1.7100-001   1.7029-001   -7.107-004
  15   2.3400-001   2.3384-001   -1.574-004    16   2.8700-001   2.8674-001   -2.639-004
  17   3.2400-001   3.2512-001    1.116-003    18   3.3000-001   3.3091-001    9.113-004
  19   3.0400-001   3.0444-001    4.445-004    20   2.5700-001   2.5448-001   -2.517-003
  21   2.0000-001   1.9852-001   -1.476-003    22   1.4700-001   1.4602-001   -9.783-004
  23   1.0200-001   1.0412-001    2.117-003    24   6.8000-002   7.1959-002    3.959-003
  25   4.6000-002   4.7244-002    1.244-003    26   3.0000-002   2.9643-002   -3.575-004
  27   2.2000-002   2.1100-002   -8.997-004    28   1.5000-002   1.6001-002    1.001-003
  29   1.2000-002   1.3344-002    1.344-003    30   1.1000-002   1.1366-002    3.664-004
  31   9.0000-003   9.6267-003    6.267-004    32   9.0000-003   8.8410-003   -1.590-004
  33   8.0000-003   8.8918-003    8.918-004    34   8.0000-003   8.9441-003    9.441-004
  35   8.0000-003   9.0065-003    1.007-003    36   9.0000-003   9.0865-003    8.655-005
  37   9.0000-003   9.3275-003    3.275-004    38   1.0000-002   9.8104-003   -1.896-004
  39   1.1000-002   1.0648-002   -3.518-004    40   1.3000-002   1.2118-002   -8.823-004
  41   1.5000-002   1.4268-002   -7.322-004    42   1.7000-002   1.7347-002    3.471-004
  43   2.0000-002   2.0937-002    9.367-004    44   2.5000-002   2.4842-002   -1.585-004
  45   2.8000-002   2.8242-002    2.417-004    46   3.1000-002   3.1030-002    3.002-005
  47   3.3000-002   3.2864-002   -1.361-004    48   3.3000-002   3.2858-002   -1.423-004
  49   3.2000-002   3.1230-002   -7.701-004    50   2.8000-002   2.8456-002    4.558-004
  51   2.5000-002   2.5235-002    2.351-004    52   2.2000-002   2.1737-002   -2.631-004
  53   1.9000-002   1.9500-002    4.998-004    54   1.8000-002   1.8135-002    1.350-004
  55   1.8000-002   1.7956-002   -4.399-005    56   1.9000-002   1.8539-002   -4.614-004
  57   1.9000-002   1.9363-002    3.626-004    58   2.0000-002   1.9742-002   -2.576-004
  59   1.9000-002   1.9225-002    2.251-004    60   1.7000-002   1.7540-002    5.402-004
  61   1.6000-002   1.5435-002   -5.651-004    62   1.4000-002   1.3504-002   -4.962-004
  63   1.2000-002   1.2086-002    8.562-005    64   1.1000-002   1.0988-002   -1.151-005
  65   1.0000-002   1.0323-002    3.231-004    66   1.0000-002   1.0029-002    2.945-005
  67   9.0000-003   9.8831-003    8.831-004    68   9.0000-003   9.8361-003    8.361-004
  69   9.0000-003   9.8047-003    8.047-004    70   9.0000-003   9.7964-003    7.964-004
  71   9.0000-003   9.8218-003    8.218-004    72   9.0000-003   9.8468-003    8.468-004
  73   9.0000-003   9.8718-003    8.718-004    74   9.0000-003   9.8948-003    8.948-004
  75   9.0000-003   9.9866-003    9.866-004    76   1.0000-002   1.0076-002    7.640-005
  77   1.0000-002   1.0200-002    1.996-004    78   1.1000-002   1.0335-002   -6.648-004
  79   1.1000-002   1.0745-002   -2.546-004    80   1.2000-002   1.1285-002   -7.153-004
  81   1.3000-002   1.2658-002   -3.420-004    82   1.5000-002   1.4453-002   -5.467-004
  83   1.7000-002   1.7356-002    3.564-004    84   2.3000-002   2.1628-002   -1.372-003
```

TABLE XI (*Continued*)

| | | | | | | |
|---|---|---|---|---|---|---|
| 85 | 2.8000-002 | 2.7480-002 | -5.203-004 | 86 | 3.5000-002 | 3.4234-002 | -7.665-004 |
| 87 | 4.2000-002 | 4.2294-002 | 2.943-004 | 88 | 5.1000-002 | 5.0740-002 | -2.600-004 |
| 89 | 5.8000-002 | 5.8888-002 | 8.884-004 | 90 | 6.6000-002 | 6.6275-002 | 2.753-004 |
| 91 | 7.1000-002 | 7.2095-002 | 1.095-003 | 92 | 7.6000-002 | 7.7015-002 | 1.015-003 |
| 93 | 7.9000-002 | 7.8740-002 | -2.605-004 | 94 | 8.1000-002 | 8.0047-002 | -9.534-004 |
| 95 | 8.1000-002 | 7.9321-002 | -1.679-003 | 96 | 8.2000-002 | 7.9340-002 | -2.660-003 |
| 97 | 8.4000-002 | 8.2322-002 | -1.678-003 | 98 | 8.6000-002 | 8.6603-002 | 6.033-004 |
| 99 | 9.4000-002 | 9.6328-002 | 2.328-003 | 100 | 1.0400-001 | 1.0859-001 | 4.591-003 |
| 101 | 1.1900-001 | 1.2329-001 | 4.290-003 | 102 | 1.3800-001 | 1.3993-001 | 1.928-003 |
| 103 | 1.5400-001 | 1.5397-001 | -2.900-005 | 104 | 1.7200-001 | 1.6736-001 | -4.643-003 |
| 105 | 1.8300-001 | 1.7835-001 | -4.647-003 | 106 | 1.8800-001 | 1.8492-001 | -3.077-003 |
| 107 | 1.8600-001 | 1.8469-001 | -1.306-003 | 108 | 1.7900-001 | 1.7920-001 | 1.975-004 |
| 109 | 1.6500-001 | 1.6700-001 | 2.002-003 | 110 | 1.4900-001 | 1.5176-001 | 2.761-003 |
| 111 | 1.3000-001 | 1.3368-001 | 3.676-003 | 112 | 1.1200-001 | 1.1521-001 | 3.207-003 |
| 113 | 9.5000-002 | 9.5611-002 | 6.113-004 | 114 | 8.0000-002 | 8.1178-002 | 1.178-003 |
| 115 | 6.8000-002 | 6.7260-002 | -7.400-004 | 116 | 5.6000-002 | 5.6321-002 | 3.207-004 |
| 117 | 4.7000-002 | 4.5698-002 | -1.302-003 | 118 | 4.0000-002 | 3.7165-002 | -2.835-003 |
| 119 | 3.6000-002 | 3.0204-002 | -5.796-003 | 120 | 3.2000-002 | 2.4595-002 | -7.405-003 |
| 121 | 2.7000-002 | 2.0183-002 | -6.817-003 | 122 | 2.2000-002 | 1.7822-002 | -4.178-003 |
| 123 | 1.8000-002 | 1.6077-002 | -1.923-003 | 124 | 1.5000-002 | 1.4686-002 | -3.139-004 |
| 125 | 1.4000-002 | 1.3761-002 | -2.388-004 | 126 | 1.2000-002 | 1.3147-002 | 1.147-003 |
| 127 | 1.2000-002 | 1.2635-002 | 6.350-004 | 128 | 1.1000-002 | 1.2107-002 | 1.107-003 |
| 129 | 1.0000-002 | 1.1524-002 | 1.524-003 | 130 | 1.0000-002 | 1.1392-002 | 1.392-003 |
| 131 | 9.0000-003 | 1.1322-002 | 2.322-003 | 132 | 9.0000-003 | 1.1347-002 | 2.347-003 |
| 133 | 9.0000-003 | 1.1372-002 | 2.372-003 | 134 | 9.0000-003 | 1.1397-002 | 2.397-003 |
| 135 | 9.0000-003 | 1.1422-002 | 2.422-003 | | | | |

UNBIASED ESTIMATE OF STANDARD DEVIATION = 1.9004-003

the mode of operation (0, 1, 2, or 3), and the initial value for the damping factor $D$ (format 2I4,F10.0). Mode 1 gives automatic optimization of the damping factor (Section III,B); mode 2 gives automatic scaling of parameter adjustments (Section III,C); and mode 0 gives both. In general, the unmodified least-squares procedure (mode 3 plus $D = 0$) gives the most rapid convergence but when the iteration diverges modes 0, 1, or 2 may be used. If a damping factor other than zero is specified with mode 2 or 3, it remains fixed at this initial value for all cycles.

The following card defines the number of parameters which are to be held at their initial values (format I4) and if this is nonzero a further card is read bearing the indices of the fixed parameters (multiple format I4). In the example in Table IX, an initial iteration was carried out with the function parameter $f(P_{18})$ fixed at a value of 1.0 (i.e., Gaussian band shape) yielding a value of $\sigma = 3.223 \times 10^{-3}$. The result of this analysis is shown in Fig. 1.

The following card in the input deck listed in Table IX contains the word RUN, indicating that further refinement is required, while that in Table X contains the word END, indicating that the end of the job has been reached. If, instead, the word JOB had been encountered in either case, the program would have returned to the starting point to deal with a completely new job.

Following the second RUN card in Table IX, the description of the required method of iteration follows the same pattern as before and the termination of the job is signaled by an END card.

## D. Output

The output obtained from the digital standard shape version of the program with the input listed in Table X is given in Table XI. The various sections are self-explanatory and it will be seen that after six cycles, using combined optimization of damping factor and scale of parameter adjustments, the iteration converges, giving a value of $\sigma = 1.78 \times 10^{-3}$ (Fig. 10). This may be compared with a value of $\sigma = 3.22 \times$

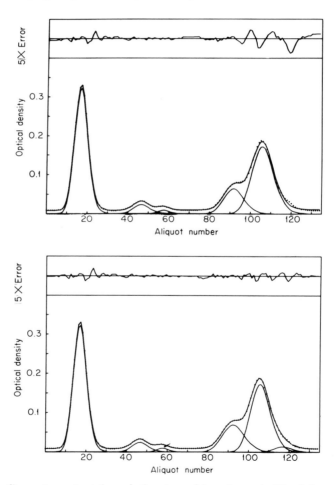

Fig. 10. Computer output for solution to problem shown in Fig. 9 for a five-band model (upper) and a six-band model (lower), using the digital standard shape function version of the program LSQ. The routine for plotting the graphical output was added to the subroutine JOBEND.

TABLE XII
SELECTED EXAMPLES OF THE APPLICATION OF ITERATIVE NONLINEAR
LEAST-SQUARES IN DATA ANALYSIS

| Application | Remarks |
| --- | --- |
| Activation analysis | |
| DeVoe (1969) | Proceedings of international conference |
| Kerrigan (1970) | Review |
| Tunnicliff et al. (1970) | Decomposition of spectral data with up to six different decay times |
| Chromatography, gas | |
| Anderson et al. (1970a) | Use of digital standard shape function for resolving mixture of benzene derivatives |
| Anderson et al. (1970b) | Use of asymmetric analytical functions for hydrocarbon mixtures |
| Gladney et al. (1969) | Skew function applied to mixture of $n$-alcohols and to petroleum ethers |
| Littlewood et al. (1969) | Gauss shape used for aliphatic hydrocarbon mixtures, also a description of a peak-finding method using moments |
| Roberts et al. (1970) | Skew Gauss shape used for mixtures of benzene, 2,2,4-trimethyl pentane, and $n$-heptane |
| Westerberg (1969) | Comparison with graphical methods |
| Chromatography, ion exchange | |
| Fraser and Suzuki (1966) | Gauss shape used for amino acid analysis |
| Fraser and Suzuki (this chapter) | Digital standard shape function used for amino acid analysis |
| Chromatography, paper | |
| Boulton (1968) | Review on paper chromatography of fluorescent substances |
| Countercurrent distribution | |
| Kirdani and Priore (1968) | Mixture of estriol derivatives |
| Kirdani and Priore (1970a) | Identification of estrogen conjugates and determination of partition coefficients |
| Kirdani and Priore (1970b) | Isotope effect on CCD of estriols |
| Priore and Kirdani (1968) | Description of least-squares method for countercurrent distribution with application to estriol–glucosiduronates mixture |
| Electron spin resonance | |
| Ibers and Swalen (1962) | Description of line shapes and application to polycrystalline substances |
| Johnston and Hecht (1965) | Determination of g-values of tetrakis (pyridine)silver(II) peroxydisulfate |
| Leaver et al. (1969) | Indoles, phenols, disulfides, thiols |

TABLE XII (*Continued*)

| Application | Remarks |
|---|---|
| Leaver (1971) | Xanthene dyes |
| Marquardt et al. (1961) | Description of algorithm and application to $N$-ethyl propionamide |
| **Electronic spectra** | |
| Bell and Biggers (1965) | Separation of overlapping peaks in uranyl perchlorate spectrum in 3200–5200 Å region using Gaussian band shape |
| Ederer (1969) | Analysis of Beutler-Fano Resonance profile |
| Guschlbauer et al. (1965) | Use of component spectra for resolving nucleoside spectrum |
| Lee et al. (1965) | Multicomponent spectrophotometric analysis of nucleoside mixture |
| Roos (1964) | Use of Gauss band shape for Cu(II) dimethyl glyoxime–water spectrum |
| **Enzyme kinetics** | |
| Garfinkel et al. (1970) | Review on computer application to biochemical kinetics |
| Morrison (1969) | Kinetic studies on allosteric enzyme reaction mechanism |
| Schwartz and Bodansky (1968) | Review |
| **Gamma-ray and pulse height spectra** | |
| Heath (1962) | Use of Gaussian function |
| Trombka (1970) | Review |
| Trombka and Schmadebeck (1970) | Review |
| **Infrared spectra** | |
| Burns and Law (1970) | Use of Gauss band shape in metal silicate spectrum |
| Fraser and Suzuki (1970b) | Resolution of $\beta$-keratin dichroism spectrum, using Gauss-Cauchy sum function |
| Fraser and Suzuki (1970c) | Review (Part B, Chapter 13, of this series) |
| Fraser et al. (1971) | Feather keratin polarized spectrum |
| Papoušek and Plíva (1965) | $n$-Decane spectrum |
| Pitha and Jones (1966) | Comparison of optimization methods |
| Pitha and Jones (1967) | Study of band shapes |
| Stone (1962) | Use of Cauchy band shape for benzene spectrum in near-infrared region, also use of second-order Taylor approximation |
| Suzuki (1967) | *Bombyx mori* silk, polarized spectrum |
| **Mass spectra** | |
| Brauman (1970) | Review |
| Luenberger and Dennis (1966) | Feasibility study on the resolution of overlapping peaks with high noise level |

TABLE XII (*Continued*)

| Application | Remarks |
|---|---|
| **Mössbauer spectra** | |
| Duke and Gibb (1967) | Optimization procedures in relation to Mössbauer spectra |
| Marshall *et al.* (1965) | Use of Cauchy band shapes and polynomial baseline |
| **Nuclear magnetic resonance** | |
| Keller *et al.* (1966) | Use of digital standard shape function for 3-chlorothietane spectrum |
| **Optical rotatory dispersion and circular dichroism** | |
| Blout *et al.* (1967) | Review of application to homopolypeptides and proteins |
| Carver *et al.* (1966a) | $\alpha$-Helical polypeptides and polyproline |
| Carver *et al.* (1966b) | Cotton effect in polypeptides |
| Madison and Schellman (1970) | Proline derivatives |
| Magar (1968) | Least-squares estimation of $\alpha$, $\beta$, and random contents of lysozyme and myoglobin |
| Moscowitz (1960a) | Carbonyl chromophore |
| Moscowitz (1960b) | Ketones |
| Tinoco and Cantor (1970) | Review |
| **Polarography** | |
| Gutknecht and Perone (1970) | Use of fitted skew Gauss shape to multicomponent systems |
| **Sedimentation equilibrium** | |
| Haschemeyer and Bowers (1970) | Analysis of concentration spectra and concentration difference spectra of self-associating systems |
| Reinhardt and Squire (1965) | Analysis of sedimentation equilibrium constant of ICS hormone |
| Teller (1965) | Dimerization of $\alpha$-chymotrypsin |
| **Titration** | |
| Sato and Momoki (1970) | End point detection of chelatometric titration |
| **X-Ray diffraction** | |
| Ahmed *et al.* (1970) | Collection of papers on least-squares refinement of atomic parameters |
| Arnott (1968) | Structure of poly-L-alanine and polyproline II |
| Arnott and Wonacott (1966) | Structure of poly-L-alanine |
| Fraser *et al.* (1969) | Structure of $\beta$-keratin |
| Hindeleh and Johnson (1972) | Structure of cellulose |
| Kasper and Lonsdale (1959) | Description of methods and application to X-ray crystallography |
| Pepinsky *et al.* (1961) | Computing methods in X-ray crystallography |

TABLE XII (*Continued*)

| Application | Remarks |
|---|---|
| Rollett (1965) | Applications in X-ray crystallography |
| Warwicker (1970) | Structure of nylon |
| General accounts | |
| Anderssen and Osborne (1969) | Proceedings of symposium on least squares in data analysis |
| Cooper and Steinberg (1970) | General description |
| Hamilton (1964) | Deals with least squares, estimation, and hypothesis testing |
| Kowalik and Osborne (1968) | Theory of unconstrained optimization |

$10^{-3}$ for a Gaussian shape (Fig. 1) and $\sigma = 2.80 \times 10^{-3}$ for a Gauss-Cauchy sum function.

Inspection of the residual error curve for the standard shape optimization (Fig. 10) suggests that a small band has been overlooked near $X = 119$ and a further refinement including a band near this position gives $\sigma = 0.93 \times 10^{-3}$ (Fig. 10). From the discussion given in Section IV,C,4 we may test whether a band is in fact present by calculating $R = 1.78/0.93 = 1.92$ and comparing it with $R_{3,135-20,\alpha}$ extracted from Table III. The value corresponding to $\sigma = 0.005$ is 1.055 and so we may accept the hypothesis that the additional band is genuine at the 0.5% level. The value of $R$ is actually very much greater than $R_{3,115,0.05}$ and the result is far more significant than indicated. It should be emphasized that in making a test of significance we are assuming first that the weights of the observations are equal and second that the model, i.e., six bands each having the specified standard shape, together with a linear baseline, is an adequate description of the system.

### VII. Applications

Although the advantages of using least-squares procedures for parameter optimization and model testing are self-evident, it is only very recently that there has been any significant use of the method in protein chemistry. The aim of the present contribution has been to outline the relevant theory and to give a detailed account of how it can be applied to a typical problem in chromatographic analysis. The program given may readily be adapted to other problems and to assist in this respect selected examples of applications to a wide variety of techniques are listed in Table XII. Most of the examples given are from fields of study other than protein chemistry but the procedures used are in most instances of general interest and applicability.

#### Acknowledgments

We are indebted to Mr. R. J. Rowlands for his valuable comments on the first draft of the contribution and to the late Dr. W. C. Hamilton for useful discussion of significance testing. The subroutine MATINV (Table VIII) was adapted from a subroutine of similar name provided by the CSIRO Users Group. The output from the Spinco automatic amino acid analyzer was kindly supplied by Dr. C. M. Roxburgh. We are grateful to the International Union of Crystallography for permission to reproduce the material in Table III from Hamilton (1965).

#### References

Ahmed, F. R., Hall, S. R., and Huber, C. P. (1970). "Crystallographic Computing." Munksgaard, Copenhagen.
Anderson, A. H., Gibb, T. C., and Littlewood, A. B. (1970a). *Anal. Chem.* **42**, 434.
Anderson, A. H., Gibb, T. C., and Littlewood, A. B. (1970b). *J. Chromatogr. Sci.* **8**, 640.
Anderssen, R. S., and Osborne, M. R., eds. (1969). "Least Squares Methods in Data Analysis." Australian National University, Canberra, Australia.
Arnott, S. (1968). *In* "Symposium on Fibrous Proteins, Australia, 1967" (W. G. Crewther, ed.), p. 26. Butterworth, Sydney, Australia.
Arnott, S., and Wonacott, A. J. (1966). *J. Mol. Biol.* **21**, 371.
Bell, J. T., and Biggers, R. E. (1965). *J. Mol. Spectrosc.* **18**, 247.
Blackburn, J. A. (1970). *In* "Spectral Analysis" (J. A. Blackburn, ed.), pp. 35–66. Dekker, New York.
Blout, E. R., Carver, J. P., and Shechter, E. (1967). *In* "Optical Rotatory Dispersion and Circular Dichroism in Organic Chemistry" (G. Snatzke, ed.), pp. 224–313. Heyden, London.
Boulton, A. A. (1968). *Methods Biochem. Anal.* **16**, 327–393.
Box, M. J. (1966). *Computer J.* **9**, 67.
Brauman, J. I. (1970). *In* "Spectral Analysis" (J. A. Blackburn, ed.), pp. 235–257. Dekker, New York.
Burns, R. G., and Law, A. D. (1970). *Nature (London)* **226**, 73.
Buys, T. S., and de Clerk, K. (1972). *Anal. Chem.* **44**, 1273.
Carver, J. P., Shechter, E., and Blout, E. R. (1966a). *J. Amer. Chem. Soc.* **88**, 2550.
Carver, J. P., Shechter, E., and Blout, E. R. (1966b). *J. Amer. Chem. Soc.* **88**, 2562.
Cooper, L., and Steinberg, D. (1970). "Introduction to Methods of Optimization." Saunders, Philadelphia.
DeVoe, J. R. (1969). *Nat. Bur. Stand. (U. S.), Spec. Publ.* **312**, Vol. II.
Duke, B. J., and Gibb, T. C. (1967). *J. Chem. Soc., A* p. 1478.
Ederer, D. (1969). *Appl. Opt.* **8**, 2315.
Fraser, R. D. B., and Suzuki, E. (1966). *Anal. Chem.* **38**, 1770.
Fraser, R. D. B., and Suzuki, E. (1969). *Anal. Chem.* **41**, 37, 935.
Fraser, R. D. B., and Suzuki, E. (1970a). *In* "Spectral Analysis" (J. A. Blackburn, ed.), pp. 171–211. Dekker, New York.
Fraser, R. D. B., and Suzuki, E. (1970b). *Spectrochin. Acta,* **26A**, 423.
Fraser, R. D. B., and Suzuki, E. (1970c). *In* "Physical Principles and Techniques of

Protein Chemistry" (S. J. Leach, ed.), Part B, pp. 213–273. Academic Press, New York.

Fraser, R. D. B., MacRae, T. P., Parry, D. A. D., and Suzuki, E. (1969). *Polymer* **10**, 810.

Fraser, R. D. B., MacRae, T. P., Parry, D. A. D., and Suzuki, E. (1971). *Polymer* **12**, 35.

Garfinkel, D., Garfinkel, L., Pring, M., Green, S. B., and Chance, B. (1970). *Annu. Rev. Biochem.* **39**, 728.

Gladney, H. M., Dowden, B. F., and Swalen, J. D. (1969). *Anal. Chem.* **41**, 883.

Grushka, E., Myers, M. N., Schettler, P. D., and Giddings, J. C. (1969). *Anal. Chem.* **41**, 889.

Guschlbauer, W., Richards, E. G., Beurling, K., Adams, A., and Fresco, J. R. (1965). *Biochemistry* **4**, 964.

Gutknecht, W. F., and Perone, S. P. (1970). *Anal. Chem.* **42**, 906.

Hamilton, W. C. (1964). "Statistics in Physical Science." Ronald Press, New York.

Hamilton, W. C. (1965). *Acta Crystallogr.* **18**, 502.

Haschemeyer, R. H., and Bowers, W. F. (1970). *Biochemistry* **9**, 435.

Heath, R. L. (1962). *Nucleonics* **20**, 67.

Hindeleh, A. M., and Johnson, D. J. (1972). *Polymer* **13**, 423.

Ibers, J. A., and Swalen, J. D. (1962). *Phys. Rev.* **127**, 1914.

Johnston, T. S., and Hecht, H. G. (1965). *J. Mol. Spectrosc.* **17**, 98.

Kasper, S., and Lonsdale, K., eds. (1959). "International Tables for Crystallography," Vol. II, pp. 92 and 326. Kynoch Press, Birmingham.

Keller, W. D., Lusebrink, T. R., and Sederholm, C. H. (1966). *J. Chem. Phys.* **44**, 782.

Kerrigan, F. J. (1970). *In* "Spectral Analysis" (J. A. Blackburn, ed.), pp. 213–233. Dekker, New York.

Kirdani, R. Y., and Priore, R. L. (1968). *Anal. Biochem.* **24**, 377.

Kirdani, R. Y., and Priore, R. L. (1970a). *Anal. Biochem.* **35**, 1.

Kirdani, R. Y., and Priore, R. L. (1970b). *Anal. Biochem.* **35**, 23.

Kowalik, J., and Osborne, M. R. (1968). "Methods for Unconstrained Optimization Problems." Amer. Elsevier, New York.

Leaver, I. H. (1971). *Aust. J. Chem.* **24**, 753.

Leaver, I. H., Ramsay, G. C., and Suzuki, E. (1969). *Aust. J. Chem.* **22**, 1891.

Lee, S., McMullen, D., Brown, G. L., and Stokes, A. R. (1965). *Biochem. J.* **94**, 314.

Levenberg, K. (1944). *Quart. Appl. Math.* **2**, 164.

Levy, E. J., and Martin, A. J. (1968). Paper presented at *Pittsburgh Conf. Anal. Chem. Appl. Spectrosc., 1968*.

Littlewood, A. B., Gibb, T. C., and Anderson, A. H. (1969). *In* "Gas Chromatography 1968, Proceedings of Seventh International Symposium" (C. L. A. Harbourn, ed.), p. 297. Institute of Petroleum, London.

Luenberger, D. G., and Dennis, U. E. (1966). *Anal. Chem.* **38**, 715.

Madison, V., and Schellman, J. (1970). *Biopolymers* **9**, 511.

Magar, M. E. (1968). *Biochemistry* **7**, 617.

Margulies, S. (1968). *Rev. Sci. Instrum.* **39**, 478.

Marquardt, D. W. (1963). *J. Soc. Ind. Appl. Math.* **11**, 431.

Marquardt, D. W., Bennett, R. G., and Burrell, E. J. (1961). *J. Mol. Spectrosc.* **7**, 269.

Marshall, S. W., Nelson, J. A., and Wilenzick, R. M. (1965). *Comm. Ass. Comput. Mach.* **8**, 313.

Meiron, J. (1965). *J. Opt. Soc. Amer.* **55**, 1105.
Morrison, J. F. (1969). *In* "Least Squares Methods in Data Analysis" (R. S. Anderssen and M. R. Osborne, eds.), pp. 63–69. Australian National University, Canberra, Australia.
Moscowitz, A. (1969a). *In* "Optical Rotatory Dispersion" (C. Djerassi, ed.), pp. 150–177. McGraw-Hill, New York.
Moscowitz, A. (1960b). *Rev. Mod. Phys.* **32**, 440.
Papoušek, D., and Plíva, J. (1965). *Collect. Czech. Chem. Commun.* **30**, 3007.
Pepinsky, R., Robertson, J. M., and Speakman, J. C. (1961). "Computing Methods and the Phase Problem in X-Ray Crystal Analysis." Pergamon, Oxford.
Pitha, J., and Jones, R. N. (1966). *Can. J. Chem.* **44**, 3031.
Pitha, J., and Jones, R. N. (1967). *Can. J. Chem.* **45**, 2347.
Priore, R. L., and Kirdani, R. Y. (1968). *Anal. Biochem.* **24**, 360.
Reinhardt, W. P., and Squire, P. G. (1965). *Biochim. Biophys. Acta* **94**, 566.
Roberts, S. M., Wilkinson, D. H., and Walker, L. R. (1970). *Anal. Chem.* **42**, 886.
Rollett, J. S. (1965). "Computing Methods in Crystallography." Pergamon, Oxford.
Roos, B. (1964). *Acta Chem. Scand.* **18**, 2186.
Sato, H., and Momoki, K. (1970). *Anal. Chem.* **42**, 1477.
Schwartz, M. K., and Bodansky, O. (1968). *Methods Biochem. Anal.* **16**, 183–218.
Stone, H. (1962). *J. Opt. Soc. Amer.* **52**, 998.
Suzuki, E. (1967). *Spectrochim. Acta* **23A**, 2302.
Teller, D. C. (1965). Ph.D. Thesis, University of California.
Tinoco, I., Jr., and Cantor, C. R. (1970). *Methods Biochem. Anal.* **18**, 81.
Trombka, J. I. (1970). *In* "Spectral Analysis" (J. A. Blackburn, ed.), pp. 259-282. Dekker, New York.
Trombka, J. I., and Schmadebeck, R. L. (1970). *In* "Spectral Analysis" (J. A. Blackburn, ed.), pp. 121-170. Dekker, New York.
Tunnicliff, D. D., Bowers, R. C., and Wyld, G. E. A. (1970). *Anal. Chem.* **42**, 1048.
Warwicker, J. O. (1970). *J. Soc. Dyers Colour.* **86**, 303.
Westerberg, A. W. (1969). *Anal. Chem.* **41**, 1770.

# 22 □ Optical Rotatory Dispersion and the Main Chain Conformation of Proteins

KAZUTOMO IMAHORI and N. A. NICOLA

|   |   |
|---|---|
| Glossary of Symbols . . . . . . . . . . . | 358 |
| I. Introduction . . . . . . . . . . . . | 360 |
| II. The Basic Relations for Optically Active Molecules . . . . | 362 |
|    A. Phenomenological Description . . . . . . . . | 364 |
|    B. Quantum Mechanical Description . . . . . . . | 366 |
|    C. Description of ORD and CD Bands and Their Interrelationships . . . . . . . | 369 |
| III. Visible Rotatory Dispersion . . . . . . . . . | 371 |
|    A. $[m']$ or $[\alpha]_D$ Method . . . . . . . . . . | 373 |
|    B. $\lambda_c$ Method . . . . . . . . . . . . | 373 |
|    C. The Two-Term Drude Equation . . . . . . . | 375 |
|    D. Moffitt Equation for Proteins Containing $\alpha$-Helix and Unordered Structures . . . . . . . | 380 |
|    E. Relations Between the Use of the Moffitt Equation and Other Methods . . . . . . . . . . | 388 |
|    F. The Moffitt Equation Applied to Proteins Containing $\beta$-Structure . . . . . . . . . . | 390 |
| IV. Peptide Cotton Effects . . . . . . . . . . | 394 |
|    A. Proteins Containing Helix and Unordered Form Only . . | 397 |
|    B. Proteins Containing $\beta$-Structure . . . . . . . | 399 |
|    C. Discussion of the Assumptions and Limitations of the Various Methods for Estimating Protein Secondary Structure from the ORD . . . . . . | 405 |
| V. Cotton Effects Due to Side-Chain Chromophores . . . . | 412 |
|    A. Effect of Side-Chain Optical Activity on Estimates of Helical Content . . . . . . . . | 414 |
|    B. Near-UV Cotton Effects . . . . . . . . . | 417 |
| VI. Other Secondary Structures . . . . . . . . . | 421 |
|    A. Collagen and Polyproline . . . . . . . . | 421 |
|    B. The Screw Sense of the $\alpha$-Helix . . . . . . . | 426 |
| VII. Conformational Transitions . . . . . . . . | 427 |
| VIII. Experimental Considerations . . . . . . . . | 430 |
|    A. Instrumentation . . . . . . . . . . | 430 |

B. Sensitivity . . . . . . . . . . . . . 434
C. Instrumental Factors . . . . . . . . . . 434
D. Choice of Solution and Cells . . . . . . . . 435
E. Scanning Speed . . . . . . . . . . . 435
F. Turbidity and Orientation of the Sample . . . . . 435
IX. The Relative Advantages of ORD and CD . . . . . . 436
Appendix . . . . . . . . . . . . . . 439
References . . . . . . . . . . . . . 440

## Glossary of Symbols[1]

$\varphi'$     optical rotation in radians per centimeter
$\alpha$     angle of rotation of plane polarized light in degrees
$[\alpha]$     specific rotation
$[\Phi]$     molar rotation

[1] The expressions for optical rotation used in this chapter are as follows:
$\varphi'$ is the angle of rotation of plane-polarized light in radians per unit optical path length (centimeters) by the sample and is given by:

$$\varphi' = \frac{\pi}{\lambda}(n_L - n_R) \qquad \text{radians per centimeter}$$

where $\lambda$ is the wavelength of the light used, $n_L$ and $n_R$ are the refractive indices of the left and right circularly polarized components, respectively, at this wavelength.

The observed rotation in degrees and for variable path length is

$$\alpha = \frac{180}{\pi} l \varphi' \qquad \text{degrees}$$

where $l$ is in centimeters.

The specific rotation is defined as

$$[\alpha] = 100\alpha/cd \qquad \text{degrees-cm}^2 \text{ per decagram}$$

where $c$ is the concentration in grams per 100 ml and $d$ is the path length in decimeters. This may then be converted to a molar rotation

$$[\Phi] = [\alpha]M/100 \qquad \text{degrees-cm}^2 \text{ per decimole}$$

where $M$ is the molecular weight. However, most data for proteins are reported as mean residue rotations defined in

$$[m] = [\alpha]M_0/100 \qquad \text{degrees-cm}^2 \text{ per decimole}$$

where $M_0$ is the mean residue molecular weight of the protein. For a protein or polypeptide, $[m]$ is thus an average rotation per mole of peptide bonds so that this value is comparable between proteins of different molecular weights. In an attempt to eliminate the contribution of solvent to the optical rotation, mean residue rotations are sometimes reported as reduced mean residue rotations defined as

$$[m'] = 3/(n^2 + 2)[m] \qquad \text{degrees-cm}^2 \text{ per decimole}$$

where $n$ is the bulk solvent refractive index at the specified wavelength. $3/(n^2 + 2)$ is the Lorentz internal field correction and reduces the rotations to those expected

| | |
|---|---|
| $[m]$ | mean residue rotation |
| $[\Phi']$, $[m']$ | reduced molar rotation or reduced mean residue rotation, respectively |
| $M$ | molecular weight |
| $M_0$ | mean residue weight (for a protein this is the molecular weight divided by the number of amino acid residues in the protein) |
| $n_\lambda$, $n_L$, $n_R$ | isotropic refractive index and refractive indices for the left and right components of polarized light, respectively (wavelength dependent) |
| $l$, $d$ | optical path length in centimeters and decimeters, respectively |
| $c$ | concentration in grams per 100 ml |
| $K_\lambda$, $K_L$, $K_R$ | absorption indices (subscripts defined as for refractive indices) |
| $I_0$, $I$ | intensity of the incident and transmitted light beams, respectively |
| $\epsilon_\lambda$, $\epsilon_L$, $\epsilon_R$ | molar absorptions (subscripts defined as for refractive indices) |
| $R_{0a}$ | component rotational strength for the electronic transition from state 0 to state $a$. Units are Debye Bohr Magnetons (DBM). 1 DBM = $0.9273 \times 10^{-38}$ cgs units |
| $R_{n\pi^*}$, $R_{\pi\pi^*}$ | rotational strengths of the $n$–$\pi^*$ and $\pi$–$\pi^*$ electronic transitions of the peptide chromophore, respectively |
| $N_0$ | Avogadro's number, $6.0230 \times 10^{26}$ (kg-mole)$^{-1}$ |
| $h$ | Planck's constant, $6.6252 \times 10^{-34}$ joule-sec |
| $c$ | velocity of light in vacuum, $2.99793 \times 10^8$ m/sec |
| $\mathbf{\mu}_{0a}$ | electric transition moment integral from state 0 to state $a$ |
| $\mathbf{m}_{0a}$ | magnetic transition moment integral from state 0 to state $a$ |
| $m$ | mass of the electron, $9.108 \times 10^{-31}$ kg |
| $e$ | electronic charge of the electron, $1.6021 \times 10^{-19}$ coulomb |
| $\mathbf{r}_i$, $\mathbf{p}_i$ | position and momentum vectors for electron $i$ from a given coordinate system |
| $\psi_0$, $\psi_a$ | electronic wave functions for a molecule in states 0 and $a$, respectively |
| $d\tau$ | a volume element |
| $V_{ij}$ | interaction potential between the transition dipole moments of the $i^{th}$ and $j^{th}$ transitions |
| $\Delta_i$ | halfwidth between the two extrema of the $i^{th}$ ORD Cotton effect |
| $\Delta_i^0$ | halfwidth of a Gaussian CD band (i.e., wavelength interval over which the ellipticity falls to $1/e$ of its maximum value) |
| $[\theta]$ | mean residue ellipticity |
| $[\theta_i^0]$ | maximum ellipticity of an isolated or resolved CD band |
| $\lambda_i^0$ | wavelength at which maximum ellipticity occurs |
| $a_i$ | constant in the multiterm Drude equation simply related to the rotational strength |
| $\lambda_i$ | characteristic wavelength of the $i^{th}$ electronic transition |
| $\lambda_c$ | constant in the one-term Drude equation related to helix content |
| $a_1$, $a_2$ | constants in the two-term Drude equation |
| $\lambda_0$ | constant in the Moffitt-Yang equation, usually 212 nm |
| $X^H$, $X^\beta$, $X^R$ | fractions of helical, $\beta$-, or unordered structures, respectively |

*in vacuo*. However, the applicability of this correction for proteins, where a variety of microenvironments all different from the external solvent may occur, has recently been questioned. It is now becoming more common to make no assumptions about solvent effects and report rotations as $[m]$.

| | |
|---|---|
| $(\varphi, \psi)$ | dihedral angles N–C$\alpha$ and C$\alpha$–C', respectively |
| $a_0, b_0$ | Moffitt Parameters for 100% $\alpha$-helix |
| $a'_0, b'_0$ | Moffitt Parameters obtained experimentally for sample |
| $a_0^R, b_0^R$ | Moffitt Parameters for 100% random coil |
| $a_\beta, b_\beta$ | Moffitt Parameters for 100% $\beta$-structure |
| $a_0^H = a_0 - a_0^R$ | change in $a'_0$ in going from 100% random coil to 100% $\alpha$-helix |
| $a_0^\beta = a_\beta - a_0^R$ | change in $a'_0$ in going from 100% random coil to 100% $\beta$-structure |
| $b_0^H, b_0^\beta$ | similar definitions to above |

## I. Introduction

Optical rotatory dispersion (ORD) and its companion technique, circular dichroism (CD), have found increasing use in chemistry and biology ever since Biot first discovered the optical rotation of quartz in 1815. The particular power of these techniques lies in their ability to provide specific stereochemical information which is often difficult or impossible to obtain from most other methods.

In the study of small organic molecules ORD has been used in the determination of absolute configurations and the relative positions of functional groups; in analytical problems, such as the separation of racemic mixtures; in the detection of weak absorption bands (magnetically allowed, electrically forbidden) or bands which are poorly resolved; in the stereochemistry of complex formation; and, finally, in solvation problems and the study of conformational equilibria with temperature.

In the last 20 years or so ORD has also developed into a powerful tool for the characterization of the conformations of biopolymers. The main applications have been toward estimating the types and contents of secondary structure of proteins or following conformational changes of proteins under a variety of conditions. However, nucleic acids are also being thoroughly studied by ORD and CD with a view to determining base stacking characteristics. Furthermore, recent developments in theory make it more likely that specific environments in proteins may be probed either by binding symmetric molecules and observing their induced optical activity or by looking more carefully at near-ultraviolet ORD and CD in the region of protein chromophore absorption. Finally, magnetic ORD (MORD) and magnetic CD (MCD) open up the possibility of inducing optical activity in any molecule (by the Faraday effect), and these techniques have already been used to estimate the tryptophan content of proteins.

Despite these varied uses for ORD one may still question the relevance of such techniques at a time when the structure of several proteins is known virtually to atomic resolution from X-ray crystallographic studies.

In the past ORD has played an important part in the actual development of concepts relating to protein structure but this role now seems to be superseded by the use of X-ray crystallography. Although this is to some extent true, ORD and CD do have certain advantages over the X-ray method even in the development of ideas about protein structure in general. First, they are solution techniques which require an absolute minimum of perturbation to the system. This means that experimental conditions can be chosen so that physiological relevance is ensured. Furthermore, in the solution state, desired perturbations such as change in concentration (of protein or any other component, such as substrate, salt, effector, or denaturant) can be effected more readily than in the crystal state. Second, measurements can be made rapidly, making it possible to study the dynamics of conformational changes in protein structures (e.g., unfolding–refolding reactions, allosteric binding). Clearly, the length of time required to obtain structural information from X-ray crystallography prohibits its use in this way.

The measurement of ORD and CD in protein chemistry is becoming extremely general and these properties are often among the first reported properties of new proteins and particularly in studies which involve conformational change, such as denaturation, binding, aggregation, and chemical modification. Consequently, there is an immense amount of literature on ORD which cannot properly be reviewed here. Fortunately, however, there are several excellent reviews on specialized applications of ORD to which the reader will be referred. This chapter will be primarily concerned with the use of ORD in the determination of protein secondary structure and will discuss the various methods employed, the basic theory behind them, and their limitations.

For reviews on the optical activity of small organic molecules the reader is referred to three books on the subject (Djerassi, 1960; Crabbé, 1965; Snatzke, 1967). Snatzke's book also contains a review by Blout et al. on a comparison of the various methods for calculating secondary structure in proteins and polypeptides. An excellent early introduction to the ORD of proteins is that by Schellman and Schellman (1964). For many useful practical points the review by Fasman (1963) is recommended. More recent reviews are available by Yang (1969) and Tinoco and Cantor (1970), the latter including discussion of theoretical advances and the ORD of nucleic acids. The reader interested in optical activity of nucleic acids is also referred to the review by Bush (1971). For a comprehensive survey of theoretical advances in the field of protein optical activity the reader should consult Chapter 23 in this volume. The review by Urry (1969a) covers some of the general theory for optically active molecules, discusses the optical activity of protein-related model compounds (see also Urry, 1968), and also the optical activity of suspensions and turbid solutions of optically active molecules. A review of the optical activity of symmetric molecules (cofactors, metals, heme, etc.) when they are bound to proteins is to be found in Ulmer and Vallee (1965) and Vallee and Wacker (1970). Finally, Jirgensons' (1969) book on the optical activity of proteins catalogs the uses of the techniques in this field up to 1969, although there is little discussion of theory.

## II. The Basic Relations for Optically Active Molecules

An optically active molecule rotates the plane of polarization of polarized light. The variation of the magnitude of this rotation with wavelength is termed optical rotatory dispersion (ORD). In spectral regions where the molecule absorbs light, the closely related phenomenon of circular dichroism (CD) also occurs, this being a distortion of the plane-polarized beam so that the resultant electric vector of the light wave no longer oscillates in one plane but traces out an ellipse. In such regions of the spectrum the optical rotation may be thought of as a rotation of the major axis of the ellipse. The circular dichroism is a measure of the ellipticity of the light wave.

Pasteur first pointed out that for a molecule to display optical activity it must be nonsuperimposable on its mirror image (Pasteur, 1848). Voigt (1905) restated this basic postulate in terms of the stereochemical requirements needed for such a condition to hold. These are that the molecule must not possess a center of inversion, nor a plane of symmetry, nor an alternating rotation–reflection axis of symmetry. In view of recent results, it is now clear that these criteria should be applied not only to the isolated chromophore under study but to the whole system of the molecule plus its environment (including the solvation sphere).

Thus, optical activity is displayed by inherently disymmetric chromophores (such as hexahelicene), symmetric chromophores in asymmetric molecules (such as the carbonyl group in steroids), and symmetric molecules which are asymmetrically perturbed by their environment (such as heme in hemoglobin).

In the accessible spectral range the only inherently disymmetric chromophore which commonly occurs in proteins is the disulfide group. This group can exist in either of two skewed conformations of opposite chirality because of the restricted rotation about the —S—S— bond. Inherently disymmetric chromophores of this type generally display more intense optical activity than do symmetric chromophores in asymmetric fields.

However, the major source of protein optical activity arises from the second situation mentioned above. In the first place, each amino acid, whether incorporated into a protein or as the free amino acid, contains an asymmetric center in the tetrahedral α-carbon atom. Although this center does not provide any chromophores in the accessible spectral range its inherent asymmetry perturbs the carboxyl group in free amino acids or the amide group in proteins to give optical activity. This also applies to other chromophores linked to the α-carbon atom such as the benzyl, indolyl, phenolic, and histidyl chromophores of the aromatic amino acids. This source of optical activity in proteins essentially depends on amino acid composition only. Second, the protein backbone chain can take up regular conformations which are nonsuperimposable on their mirror images (helices, β-structures, polyproline structures, etc.). This form of asymmetry also perturbs the symmetric protein chromophores, especially the amide group. In these regular structures the perturbations of the amide group are regular and repeating so that it is possible to estimate the number of amide groups in such an array from the amide optical activity. Finally, proteins also take up specific three-dimensional structures where any chromophore may find itself in an asymmetric field. It is much more difficult to predict this type of structure from the optical activity of the chromophores.

Some proteins also contain symmetric molecules which are not covalently linked to any asymmetric centers. In these cases optical activity of the symmetric molecule may arise if the molecule is bound to the protein in such a way that the specific arrangement of dipoles and charges around it form an asymmetric field. This is the source of the optical activity of heme in the hemoproteins, and of metals and cofactors which are bound to enzymes.

It is thus important to realize that protein optical activity arises from several component effects. For the peptide chromophore the major sources are point asymmetry of the individual amino acids and form asymmetry of the backbone chain. Tertiary structure effects are less important because the amide chromophore may be to some extent "shielded" from the rest of the protein when it occurs in ordered structures. However, the optical activity of the aromatic chromophores depends much more strongly on the specific three-dimensional structure within which it occurs. This can serve to increase, decrease, or leave unchanged the inherent optical activity of the amino acid (see Section V).

Both ORD and CD are closely related to the absorption spectrum of the molecule under study. This is because it is the same electronic transitions which are involved in all three phenomena. Some of the electronic transitions interact with their local environment and became optically active, each of these then having a component ORD and CD spectrum. However, whereas electronic transitions always give rise to positive absorption bands the ORD and CD bands can be positive or negative in sign and need not bear any relation to the size of the absorption band which gives rise to them. The experimental ORD or CD curve is a sum of components from all the optically active transitions.

## A. Phenomenological Description

Plane-polarized light can be considered as the vector sum of two circularly polarized components. If one electric vector component rotates clockwise and the other counterclockwise at exactly the same velocity then the vector sum of these will be a vector which oscillates in magnitude but remains in one plane. The frequency of the light will correspond to the frequency of rotation of these electric (or magnetic) vectors.

All the phenomena of optical activity can then be rationalized if we assume that anisotropic but not isotropic matter interacts preferentially with one of the circular components. Fresnel in 1825 first showed that a difference in the indices of refraction of the left and right circularly polarized components would result in rotation of the plane of polarization because in the medium the two vectors would rotate at different velocities and hence be out of phase when they emerged. On this basis it is easy to show that the rotation per unit length (centimeters), in radians, is

$$\varphi' = \pi/\lambda(n_L - n_R) \qquad (1)$$

where $\lambda$ is the wavelength used (*in vacuo*) and $n_L$ and $n_R$ are the respective indices of refraction for the left and right circularly polarized components.

Similarly, one can explain circular dichroism as a difference in absorption of the left and right circularly polarized components by an optically active molecule. This results in the component vectors being of unequal size so that their vector sum traces out an ellipse which is tilted relative to the original plane of polarization due to the accompanying optical rotation (see Chapter 23, this volume, Section II,A). The ellipticity (ratio of the minor to the major axis) per unit length is given as

$$\theta' = \pi/\lambda(K_L - K_R) \qquad (2)$$

where $K_L$ and $K_R$ are the absorption indices of the left and right circularly polarized beams.

$K$ is defined in

$$I = I_0 e^{-(4\pi K/\lambda)l} \qquad (3)$$

where $I_0$ is the intensity of the incident light beam and $I$ is the intensity of the transmitted beam.

Now, since both ORD and CD are seen to be the difference between two quantities, they are generally very small compared to isotropic refraction and absorption[2] and, moreover, they can be positive or negative. (Clearly, if one optical isomer interacts preferentially with the left component, the other will interact preferentially with the right.) Also, since the CD is a difference in absorption, it will be shaped like an absorption band and be confined to the same discrete wavelength interval as the responsible absorption band. The ORD, on the other hand, will show the anomalous dispersion that the refractive index shows near absorption bands and, like the refractive index, will occur over a much larger wavelength interval extending well outside the range of the absorption band. The shapes of ORD and CD spectra and their relation to the dispersion of the refractive index and absorption spectrum are

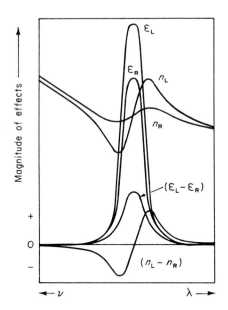

FIG. 1. A positive Cotton effect for an isolated optically active absorption band. $\epsilon_L$ and $\epsilon_R$ are the absorption curves, and $n_L$ and $n_R$ the dispersion of the refractive index for the left and right circularly polarized components of plane-polarized light, respectively. The CD band is related to a difference in absorption and the ORD Cotton effect is related to a difference in refractive indices for these two components. (From Woldbye, 1967.)

[2] Defined as $(n_L + n_R)/2$ and $(\epsilon_L \pm \epsilon_R)/2$.

shown in Fig. 1 for an isolated transition. The term *Cotton effect* is applied to the characteristic ORD and CD curves in the region of the responsible absorption band.

## B. QUANTUM MECHANICAL DESCRIPTION[3]

Although the above description tells us what optically active molecules do to light to produce optical rotation, it does not tell us how or why they do this. We need to know the structural requirements necessary for a molecule to display optical activity if we are to use ORD to elucidate protein structures.

Rosenfeld (1928) first attempted a quantum mechanical description of optical activity. His treatment gave the following equation

$$[\Phi]_\lambda = 96\pi N_0/hc \sum_i R_i \lambda_i/(\lambda^2 - \lambda_i^2) \tag{4}$$

where $[\Phi]_\lambda$ is the reduced molar ellipticity $[=3/(n^2 + 2) \cdot (\alpha M/cd)]$, $N_0$ is Avogadro's number, $\lambda$ is the experimental wavelength, and $\lambda_i$ is the characteristic wavelength of the $i^{th}$ transition. $R_i$ is termed the rotational strength of the $i^{th}$ transition and is given by

$$R_i = Im(\mathbf{\mu}_i \cdot \mathbf{m}_i) \tag{5}$$

i.e., it is equal to the imaginary part of the dot-product of the electric transition moment ($\mathbf{\mu}_i$) and the magnetic transition moment ($\mathbf{m}_i$) of the $i^{th}$ transition.[4] It is thus similar to the dipole strength in an electronic transition. However, unlike the dipole strength, the rotational strength

---

[3] Some of the theoretical treatment in this section, particularly in Sections II,B and II,C, overlaps with that in the following chapter on circular dichroism. This permits an appreciation of basic ORD theory without immediate reference to the CD chapter.

[4] These are quantum mechanical quantities defined by

$$\mathbf{\mu}_{0a} = e \int \Psi_0^* \sum_i \mathbf{r}_i \cdot \Psi_a d\tau$$

$$\mathbf{m}_{a0} = (e/2mc) \int \Psi_a^* \sum_i (\mathbf{r}_i \times \mathbf{p}_i) \cdot \Psi_0 d\tau$$

where $\Psi_0$ and $\Psi_a$ are the electronic wave functions of the molecule in the ground and excited states, respectively; $\Psi_0^*$ and $\Psi_a^*$ are their complex conjugates; $\mathbf{r}_i$ and $\mathbf{p}_i$ are the position and momentum vectors, respectively, of electron $i$ from a coordinate system; $m$ and $e$ are the electron mass and charge; and $d\tau$ is a volume element. The rotational strength for the transition from state 0 to state $a$ is then given by

$$R_{0a} = Im(\mathbf{\mu}_{0a} \cdot \mathbf{m}_{a0})$$

may be positive or negative giving rise to positive or negative Cotton effects, respectively.

Another important property of the rotational strength is that the sum of the rotational strengths over all the transitions of a molecule must equal zero (Kuhn, 1930; Condon, 1937). This is often difficult to test experimentally because of the possible contributions from transitions which lie in the inaccessible far ultraviolet. It should also be borne in mind that, as Urry has pointed out (Urry, 1969a), changes in ORD and CD patterns can occur when, say, heme is removed from myoglobin without necessarily implying any change in conformation, since removal of heme rotational strength must be accompanied by a compensating change in the protein rotational strength elsewhere in the spectrum by virtue of the above sum rule.

From the definition of the rotational strength, it is clear that only those electronic transitions which have associated with them a magnetic moment which is not perpendicular to the electric moment can display optical activity.[5] The basic requirement for optical activity is thus a movement of charge simultaneously along and around a given axis upon electronic excitation so that the electron displacement becomes associated with an electric moment in the direction of linear displacement and an induced magnetic moment in this same direction due to the simultaneous circular motion of the electron about this direction. There are two recognized ways in which this type of electron motion may be achieved. The first is the coupled-oscillator mechanism of Kuhn (1930) and Kirkwood (1937). In this case two strong electronic transitions which have transition dipoles which are noncoplanar and separated in space can couple through their dipole moments so that they move in phase. This type of interaction can cause the displacement of charge along and around an axis required for optical activity. It appears to be the source of optical rotation observed for strong absorption bands in symmetrical molecules such as the $\pi-\pi^*$ transitions of the peptide bonds. The rotational strength due to such coupling is given by

$$R_{ij} = 2\pi/hc(V_{ij}\lambda_i\lambda_j\mathbf{r}_{ij} \cdot \mathbf{\mu}_i \times \mathbf{\mu}_j)/(\lambda_i^2 - \lambda_j^2) \qquad (6)$$

where $R_{ij}$ is the rotational strength of the $i^{\text{th}}$ transition due to coupling with the $j^{\text{th}}$ transition and $V_{ij}$ is the interaction potential between the transition dipole moments. $\mathbf{\mu}_i$ and $\mathbf{\mu}_j$ are the respective transition dipole

---

[5] The physical basis for this is that the scattered wavelet produced when light interacts with matter recombines with the incident beam. If the scattered wavelets have component electric and magnetic vectors in the same direction, then recombination with the incident beam must result in rotation of the plane of polarization of the incident beam (see Kauzmann, 1957, for a full discussion).

moment vectors and $\mathbf{r}_{ij}$ is the vector distance between the transition moment origins.[6] The rotational strength of the $j^{\text{th}}$ transition due to coupling with the $i^{\text{th}}$ is

$$R_{ji} = -R_{ij}$$

in accordance with the sum rule for rotational strengths for any molecule. This is the basis of the reciprocal relations experimentally observed by Miles and Urry (1967) in the case of adenosine—5'-mononicotinate and other model systems (Urry, 1969a).

However, this mechanism cannot explain the inherent optical activity of molecules with single chromophores nor the strong optical activity sometimes shown by very weak magnetically allowed but electrically forbidden transitions (in an inherently symmetric chromophore, such as the $n$–$\pi^*$ transition of the carbonyl group). The second mechanism by which optical activity may occur is sometimes called the one-electron theory originated by Condon et al. (1937) which is successful in explaining the above observations. In this theory the static effect of asymmetrically placed dipoles and charges on the appropriate orbitals of the molecule causes the transition to acquire a small electric moment on excitation.

A magnetic transition normally involves circular motion of charge but the asymmetric environment causes a small linear displacement of charge as well. This mechanism is the major source of the optical activity of the $n$–$\pi^*$ transition of the peptide chromophore. For static charges the induced optical activity is proportional to $r^{-3}$ or $r^{-4}$ (Schellman and Oriel, 1962; Caldwell and Eyring, 1963) so that this mechanism is most important only for short-range interactions where the charges are not in a symmetry plane of the peptide bond. This is in contrast to the coupled-oscillator or exciton mechanism where the whole exciton system must be considered. Furthermore, Schellman has indicated that a quadrant rule applies to the carbonyl chromophore and that for a large Cotton effect to occur the molecule must be rigid, be in an environment of low dielectric constant, and have vicinal charges properly distributed close around it. This situation does not often occur with small amides but certainly does in the case of the peptide $n$–$\pi^*$ transition in the $\alpha$-helix (Litman and Schellman, 1965).

Urry (1968b) has carried out model compound studies of the peptide bond with L-pyrrolid-2-one derivatives and other compounds to determine the effect of vicinal groups on the magnitude and sign of the peptide $n$–$\pi^*$ rotational strengths. He has introduced the concept of partial molar

---

[6] Note that because of the denominator $(\lambda_i^2 - \lambda_j^2)$ this rotational strength will be greatest when coupling occurs between two transitions of comparable energies.

rotatory powers of vicinal groups. The partial molar rotatory power depends on a geometric factor (distance and orientation of the vicinal group from the $n$–$\pi^*$ transition) and a factor which reflects the physical properties of the vicinal group (see Urry, 1968b, for a review).

Tinoco (1962) has formulated a coherent general theory of the optical activity of polymers in which the above two mechanisms are unified within one theoretical framework. His theory also revealed a third mechanism for generating optical activity in symmetric chromophores. This is through the coupling of strong electric transition moments with magnetic transition moments. An example of this is the coupling of the $\pi$–$\pi^*$ transitions with the $n$–$\pi^*$ transition in the peptide group which is thought to be the major mechanism for the optical activity of the $\beta$-structure in proteins and polypeptides (Madison and Schellman, 1972). Like the one-electron mechanism, magnetic–electric coupling is a short-range interaction depending on $r^{-4}$ so that it is most sensitive to local geometry. Tinoco's first-order perturbation theory of the optical activity of polymers is thus of great importance and along with the later matrix formulation of Bayley et al. (1969) has formed the basis of many of the recent calculations on the optical activity of proteins and polypeptides [see Sears and Beychok, this volume, for a detailed discussion of these theories (their Sections II,E and II,F) and their application to protein optical activity (their Sections II,G,1–II,G,7)].

## C. Description of ORD and CD Bands and Their Interrelationships

The theoretically derived Rosenfeld equation cannot be successful in predicting the optical rotation near the responsible absorption band because when $\lambda = \lambda_i$ it predicts an optical rotation which tends towards infinity. Furthermore, it is difficult to predict the optical rotation at any particular wavelength because of the summation over all transitions. However, for an isolated transition this equation adequately describes the optical rotation in regions distant from the absorption band.

To eliminate the discontinuity at $\lambda = \lambda_i$ Moscowitz (1962) introduced an empirical term into the Rosenfeld equation which is analogous to the damping factor in absorption bands. By analogy with ordinary dispersion of the refractive index the Rosenfeld equation then becomes

$$[\Phi]_\lambda = 96\pi N_0/hc \sum_i R_i(\lambda^2 - \lambda_i^2)\lambda_i^2/[(\lambda^2 - \lambda_i^2)^2 + 2\Delta_i^2\lambda_i^2] \tag{7}$$

where $\Delta_i$ is the halfwidth between the two ORD extrema. However, a good fit of this equation to an ORD curve is not possible over the whole wavelength range.

Urry (1969b) has found a good fit for the data for $d$-camphor 10-sulfonate over the whole range using a series of the following form

$$[m_i]_\lambda = 96\pi N_0/hc \sum_x^{\text{odd}} \frac{R_i 2^x}{(2x)!} \frac{\lambda_i^{2x}(\lambda^2 - \lambda_i^2)^x}{(\lambda^2 - \lambda_i^2)^{2x} + (2\Delta_i)^{2x}\lambda_i^{2x}} \tag{8}$$

where $x = 1$ and 3. The rotational strength thus obtained also agrees well with that obtained from CD. Another way of generating an ORD curve is by application of the Kronig-Kramers transforms. Just as these transforms interrelate absorption and dispersion they also interrelate ORD and CD. The ORD can be calculated from the CD from the equation

$$[m_i]_\lambda = 2/\pi \oint_0^\infty [\theta_i]_{\lambda'} \lambda'/(\lambda^2 - \lambda'^2) d\lambda' \tag{9}$$

and the CD can be calculated from the ORD from the equation

$$[\theta_i]_\lambda = -2/\pi\lambda \oint_0^\infty [m_i]_{\lambda'} \lambda'^2/(\lambda^2 - \lambda'^2) d\lambda' \tag{10}$$

The integration from 0 to $\infty$ presents difficulties unless we integrate over each component ORD and CD curve separately. The total contribution of an isolated ORD or CD curve can usually be measured or approximated by an appropriate relation describing the band shape. Another problem of the integration is the discontinuity at $\lambda = \lambda'$. Thus we have indicated that the Cauchy principal value of the integral is to be taken. If one assumes a Gaussian shape for the CD band[7] (by analogy with absorption bands) we obtain for the $i^{\text{th}}$ CD band

$$[\theta_i]_\lambda = [\theta_i^0] e^{-(\lambda - \lambda_i^0)^2/(\Delta_i^0)^2} \tag{11}$$

where $\theta_i^0$ is the extremum value of $\theta_i$ at $\lambda_i^0$ and $\Delta_i^0$ is the half bandwidth (wavelength interval over which $\theta_i$ falls to $1/e$ of $\theta_i^0$).

If one substitutes this into the Kronig-Kramers transform, one obtains

$$[m_i]_\lambda = 2[\theta_i^0]/\sqrt{\pi} \left\{ e^{-c^2} \int_0^c e^{x^2} dx - \Delta_i^0/2(\lambda + \lambda_i^0) \right\} \tag{12}$$

where $c = (\lambda - \lambda_i^0)/\Delta_i^0$. Numerical values of $e^{-c^2} \int_0^c e^{x^2} dx$ have been listed for various values of $c$ in the literature (see, e.g., Yang, 1969).[8]

[7] Studies on the isolated $n-\pi^*$ transition of cyclopentanone show that the CD band is skewed to shorter wavelengths, this skew increasing if the data are plotted against wave number. Thus it may be preferable to resolve CD spectra against wavelength (Urry, 1968b).

[8] The ORD and CD can also be calculated without making any assumptions about band shape and band number by employing direct integration of the total ORD or CD curve. Thièry (1969) and Emeis et al. (1967) have developed numerical

The above equation thus provides a description of the ORD associated with a Gaussian CD band and can be used to resolve ORD spectra if the CD data are available. Alternatively, it may be possible to fit the ORD spectrum with a minimum number of these components empirically using a least-squares analysis of goodness of fit. This method has been successful in resolving the far-UV ORD of the α-helix, the results agreeing with the data obtained from CD (Blout et al., 1967).

When $\lambda - \lambda_i^0 \gg \Delta_i^0$, i.e., far from the absorption band, Eq. (12) reduces to

$$[m_i]_\lambda = 2[\theta_i^0]/\pi^{-\frac{1}{2}} \cdot \Delta_i^0 \lambda_i^0/(\lambda^2 - \lambda_i^{02}) \tag{13}$$

which is identical to the Rosenfeld equation (also valid in this range) if

$$R_i = hc[\theta_i^0]\Delta_i^0/48N_0\pi^{-3/2}\lambda_i^0$$
$$= 1.09 \times 10^{-42} 2[\theta_i^0]\Delta_i^0/\pi^{1/2}\lambda_i^0 \text{ erg-cm}^3 \tag{14}$$

Thus one can obtain the rotational strength from the band parameters of a Gaussian CD band. Even if the Gaussian assumption is not made the rotational strength may be obtained directly from the CD by integration using

$$R_i = hc/48\pi^2 N_0 \int_0^\infty [\theta_i]_{\lambda'}/\lambda' d\lambda' \text{ erg-cm}^3 \tag{15}$$

The rotational strength may also be obtained from the isolated or resolved ORD band either by fitting the data at $\lambda \gg \lambda_i$ to the Rosenfeld equation or by using certain approximate relations between the ORD band parameters and rotational strengths (Urry, 1969a; Schellman and Schellman, 1964). However, the CD approach is usually more satisfactory for this purpose. As has already been mentioned, experimental determination of rotational strengths is the most critical information that can be extracted from optical activity measurements since it is this quantity which carries the structural information and which can be compared with theoretical predictions.

## III. Visible Rotatory Dispersion

Since we have seen that the Rosenfeld equation adequately describes the ORD in regions far removed from absorption bands it should be possible to apply this equation to the analysis of the rotatory dispersion of

---

methods for these integrations. Although these methods are generally successful (see, e.g., Cassim and Yang, 1970) they are not strictly valid because the integration is not from 0 to ∞ but only over the accessible spectral range.

proteins in the visible spectral range. Even before the derivation of the Rosenfeld equation, it had been known that Drude-type equations could describe the rotation outside of absorption bands. In the general case a multiterm Drude equation of the form

$$[m']_\lambda = \sum_i a_i \lambda_i^2 / (\lambda^2 - \lambda_i^2) \tag{16}$$

may be used. Here $a_i$ is a constant and $\lambda_i$ is the wavelength at the maximum of the $i^{\text{th}}$ absorption band. Comparison of this equation with the quantum mechanically derived Rosenfeld equation indicates that $a_i$ is closely related to the rotational strength of the $i^{\text{th}}$ transition and thus embodies most of the structural information of interest.

Since both $a_i$ and $\lambda_i$ reflect the nature of the chromophore $[m']_\lambda$ and especially $[m']_D$ have been used as physical constants to identify the compound (see also Section IX). In the case of polypeptides and proteins both $a_i$ and $\lambda_i$ of the peptide chromophore vary with the conformation of the backbone chain since, as we have already pointed out, the peptide bond itself is symmetric (Section II).

If the following assumptions are made:

1. the protein structure can be divided into homogeneous units of ordered secondary structure each with component sets of $a_i$ and $\lambda_i$ for their peptide bond rotations;

2. the contributions from each type of secondary structure are additive; and

3. the visible rotatory dispersion of the protein consists solely of contributions from peptide bond rotations;

then Eq. (16) may be written as

$$[m']_\lambda = \sum_i \sum_j x_j a_{ij} \lambda_{ij}^2 / (\lambda^2 - \lambda_{ij}^2) \tag{17}$$

Here $x_j$ is the fractional content of the $j^{\text{th}}$ conformation and $\lambda_{ij}$ and $a_{ij}$ correspond to the wavelength and rotatory strength of the $i^{\text{th}}$ absorption band of the $j^{\text{th}}$ conformation.

It is thus possible, in principle at least, to estimate $a_i$ and $\lambda_i$ for each conformation and by comparing these values with those for model systems of known structure to obtain the contents $(x_j)$ of each conformation in the protein. In practice, however, this is not possible even with the aid of a computer because of the difficulty of obtaining a unique solution and also because the summation over all bands is usually impracticable, some of the bands lying outside the range of all instruments.

For these reasons Eqs. (16) and (17) cannot be generally used and various modifications of these equations have been tried in order to make them usable in practice. The various simplified forms all eliminate the summation over all bands and approximate the true ORD to a summation over only a few averaged transitions. They are all thus empirical equations with adjustable parameters. The three above assumptions as well as the approximations used to obtain workable equations will be discussed in a later section. The various modified equations will now be discussed.

## A. $[m']$ OR $[\alpha]_D$ METHOD

The simplest method is simply to measure the optical rotation at one wavelength (historically, this was usually the sodium D line). Since the rotation at any wavelength is comprised of contributions from all of the optically active transitions in the protein, it is characteristic for the total conformation of the protein. Thus any change in conformation [i.e., the $x_j$ value in Eq. (17)] can be followed by this method although the actual type of conformational change involved cannot usually be determined. In particular, sharp transitions in conformations, such as helix to coil (Doty et al., 1957) and coil to $\beta$-structure (Fasman and Blout, 1960), can be easily followed by changes in molecular rotation at a single wavelength. Sometimes, the demonstration of the presence of a transition is all that is required but quantitative interpretation of such changes in $[m']_\lambda$ requires a correlation with the results of other physicochemical methods since $[m']_\lambda$ values vary, sometimes being dependent on the nature of the solvent and independent of any conformational changes.

## B. $\lambda_c$ METHOD

If we assume that for all $\lambda_{ij}$'s, $|\lambda_{ij}^2 - \lambda_c^2| \ll |\lambda^2 - \lambda_c^2|$, where $\lambda_c$ is a constant to be determined, then Eq. (17) reduces to the well-known one-term Drude equation[9]

$$[m']_\lambda = k/(\lambda^2 - \lambda_c^2) \qquad (18)$$

where $k = \Sigma_i \Sigma_j x_j a_{ij} \lambda_{ij}^2$. Making this assumption means that we consider that all the transitions of each conformation are spaced so closely that they can be approximately represented by a single wavelength $\lambda_c$. The particular value of $\lambda_c$ which is found to best fit the data should then represent some sort of weighted sum average of the transition wavelengths of the $\alpha$-helix, $\beta$-structure, random coil, etc.

Many proteins appear to obey this simple Drude equation and since

---

[9] Sometimes written as $[m']_\lambda = k\lambda_c^2/(\lambda^2 - \lambda_c^2)$ in which case $k = \Sigma_i \Sigma_j x_j a_{ij}$, i.e., an average rotational strength for the whole protein.

$\lambda_c$ decreases on denaturation (Linderstrøm-Lang and Schellman, 1954) it is expected that $\lambda_c$ will reflect the conformation of the protein (Yang and Doty, 1957).

In fact, as illustrated in Fig. 2, where a series of hypothetical mixtures of helices and coils are calculated in accordance with Eq. (18), the rotatory dispersion of helical contents up to 40 or 50% is indistinguishable from that predicted by the one-term Drude equation (Yang and Doty, 1957). It will be seen from Fig. 2 that $\lambda_c$ values decrease in accordance with the decrease in helix content. Thus, $\lambda_c$ provides a means of estimating the helical content of proteins.

Jirgensons (1961), however, found that for some proteins $\lambda_c$ does not decrease but sometimes increases slightly upon denaturation. This seems to be related to a high content of $\beta$-structure in the native protein (see Section III,F).

$\lambda_c$ will provide an estimate of helix content of the protein semiquantitatively only under the restricted conditions where helix content does not

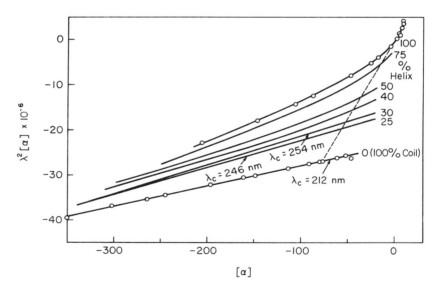

Fig. 2. Yang's plot of the optical rotatory dispersion data for poly-L-glutamic acid in the helical (pH 4.7) and randomly coiled (pH 6.6) conformations in dioxan:0.2 $M$ sodium chloride (1:2). Dispersion curves have also been calculated for various proportions of helical and coiled structures by combining the above two curves in the appropriate ratios. The dashed line represents the values at the sodium D line (589.3 nm). [Reprinted from Yang and Doty (1957), *J. Am. Chem. Soc.* **79**, 761. Copyright (1957) by the American Chemical Society. Reprinted by permission of the copyright owner.]

exceed 50%, the content of $\beta$-structure is not appreciable, and the analysis is carried out on ORD very far away from any of the absorption bands. Although this limits the applicability of the $\lambda_c$ method it has often been used because the analysis of ORD curves by this method is much simpler than by the other methods.

There are two methods for plotting data according to Eq. (18). If Eq. (18) is written in the form

$$[m']_\lambda \lambda^2 = [m']_\lambda \lambda_c^2 + k \qquad (19)$$

then it is evident that a plot of $[m']_\lambda \lambda^2$ against $[m']_\lambda$ should give a straight line, the slope being equal to $\lambda_c^2$. This kind of plot is called *Yang's plot* (Yang and Doty, 1957) and is demonstrated in Fig. 2.

Another modification of Eq. (18) may be written as

$$1/[m']_\lambda = \lambda^2/k - \lambda_c^2/k \qquad (20)$$

Thus a plot of $-1/[m']_\lambda$ against $\lambda^2$ will give a straight line and $\lambda_c^2$ can be calculated as (intercept on ordinate) ÷ (slope) or as the intercept on the abscissa. This is called *Lowry's plot* (1935). An example of such a plot is given in Fig. 3. The $\lambda_c$ values obtained for some proteins are given in Table I of the Appendix.

## C. The Two-Term Drude Equation

Since the one-term Drude equation can be applied with some success for many proteins it is reasonable to assume that a two-term Drude equation will be even more useful.

When one looks at the ORD curve of the pure helical form (Fig. 4) it is evident that the curve is quite anomalous, that is, the curve passes

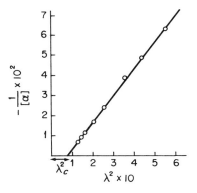

Fig. 3. Lowry's plot of the optical rotatory dispersion data for bovine serum albumin. $\lambda_c^2$ is the intercept on the abscissa.

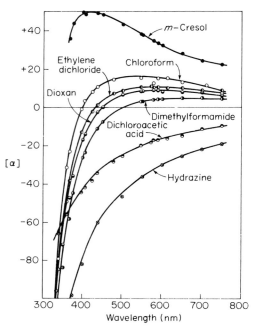

FIG. 4. The rotatory dispersion of poly-γ-benzyl-L-glutamate in several solvents. All are helix promoting solvents except dichloroacetic acid and hydrazine which favor the random coil. [Reprinted from Yang and Doty (1957), *J. Am. Chem. Soc.* **79**, 761. Copyright (1957) by the American Chemical Society. Reprinted by permission of the copyright owner.]

through a maximum, crosses over the zero-rotation axis, and changes sign at lower wavelengths. This kind of anomalous dispersion curve can be approximated by a two-term Drude equation:

$$[m'] = a_1\lambda_1^2/(\lambda^2 - \lambda_1^2) + a_2\lambda_2^2/(\lambda^2 - \lambda_2^2) \tag{21}$$

If $a_1$ is positive, $a_2$ is negative and, $|a_1| > |a_2|$ and $\lambda_1 < \lambda_2$ we anticipate such anomalous dispersion curves.

Imahori (1963) first reported that the ORD of poly-L-glutamic acid in its helical form fits Eq. (21) with $\lambda_1 = 190$ nm and $\lambda_2 = 220$ nm. Later, Yamaoka (1964) introduced the same type of equation with $\lambda_1 = 193$ nm and $\lambda_2 = 226$ nm. Schechter and Blout (1964a,b) proposed that $\lambda_1 = 193$ nm and $\lambda_2 = 225$ nm.

The four parameters $a_1$, $a_2$, $\lambda_1$, and $\lambda_2$ can be evaluated by plotting the data in accordance with the following equation which is a modified form of Eq. (21).

$$[m'](\lambda^2/\lambda_1^2 - 1) = a_1 + a_2(\lambda^2/\lambda_1^2 - 1)/(\lambda^2/\lambda_2^2 - 1) \tag{22}$$

Thus, by plotting the left side of Eq. (22) against $(\lambda^2/\lambda_1^2 - 1)/(\lambda^2/\lambda_2^2 - 1)$ for trial values of $\lambda_1$ and $\lambda_2$ until a straight line was obtained, Schechter and Blout (1964a,b) obtained $a_1 = +2900$ and $a_2 = -1930$ for helical polypeptides.

The ORD of the random-coil conformation can also be described by Eq. (21) and assuming 193 and 225 nm for $\lambda_1$ and $\lambda_2$, respectively, values of $a_1 = -750$ and $a_2 = +60$ may be obtained.

Using $a_1^H$ and $a_2^H$ for the helical contributions to $a_1$ and $a_2$, respectively, and $a_1^R$ and $a_2^R$ for the random-coil contributions then the ORD of a helix–coil mixture can be given by

$$[m']_\lambda = \frac{[x^H a_1^H + (1 - x^H)a_1^R]\lambda_1^2}{\lambda^2 - \lambda_1^2} + \frac{[x^H a_2^H + (1 - x^H)a_2^R]\lambda_2^2}{\lambda^2 - \lambda_2^2}$$

$$= \frac{[x^H(a_1^H - a_1^R) + a_1^R]\lambda_1^2}{\lambda^2 - \lambda_1^2} + \frac{[x^H(a_2^H - a_2^R) + a_2^R]\lambda_2^2}{\lambda^2 - \lambda_2^2}$$

$$= \frac{a_1 \lambda_1^2}{\lambda^2 - \lambda_1^2} + \frac{a_2 \lambda_2^2}{\lambda^2 - \lambda_2^2} \tag{23}$$

where, $x^H$ is the fraction of helix conformation.

Equation (23) leads to the following relations

$$a_1 = x^H(a_1^H - a_1^R) + a_1^R \tag{23a}$$
$$a_2 = x^H(a_2^H - a_2^R) + a_2^R \tag{23b}$$

or

$$x^H = \frac{a_1 - a_1^R}{a_1^H - a_1^R} \tag{23c}$$

$$x^H = \frac{a_2 - a_2^R}{a_2^H - a_2^R} \tag{23d}$$

By inserting $a_1^H$, $a_2^H$, $a_1^R$, and $a_2^R$ values as obtained above into Eqs. (23c) and (23d) we obtain

$$x^H = \frac{a_1 + 750}{3650} \tag{23e}$$

$$x^H = \frac{-(a_2 + 60)}{1990} \tag{23f}$$

from which $x^H$ may be estimated.

Thus, if one obtains $a_1$ and $a_2$ values by analyzing the ORD curve according to Eq. (22), then the helical content can be estimated from Eq. (23e) or Eq. (23f).

It should be noted, however, that Eqs. (23e) and (23f) can be applied only to polypeptides and proteins in aqueous solutions since both $a_1$ and $a_2$ are sensitive to the refractive index of the solvent (Schechter and

Blout, 1964b). Further, no contributions from $\beta$-structure have been considered in deriving Eqs. (21) and (23) and consequently Eqs. (23e) and (23f) can only be applied to polypeptides or protein molecules which contain only helical and coiled conformations.

From the position of the resolved absorption and CD bands of helical and randomly coiled polypeptides (see Section IV) it is apparent that $a_1$ and $a_2$ at $\lambda_1 = 193$ nm and $\lambda_2 = 225$ nm are not actual rotatory strengths but empirical parameters which are a function of the rotatory strength as well as of the locations of all optically active absorption bands.

Thus, if there is a shift in the location of any absorption band of a conformation because of solvent effects, this will result in a significant change in $a_1$ or $a_2$ or both. This sometimes causes a discrepancy between helical contents obtained from $a_1$ and from $a_2$.

It is equally probable that this internal disagreement between helical contents calculated from $a_1$ and $a_2$ is due to the existence of other structures in the molecule, such as $\beta$-structure. In fact, lysozyme, which contains a large amount of $\beta$-structure, gives 610 and $-670$ for $a_1$ and $a_2$, respectively (Schechter and Blout, 1964b). This would suggest 37 and 31% helical content by Eqs. (23e) and (23f).

Thus, there are at least two cases where some discrepancy between helical contents obtained by $a_1$ and $a_2$ would be expected. Schechter and Blout (1964b) examined how $a_1$ and $a_2$ varied with the dielectric constant of the solvent. They came to the conclusion that solvents can be grouped into two categories. In high dielectric constant solvents (aqueous) the helix content can be estimated by Eq. (23e) or (23f). However, in low dielectric constant solvents the helical content $x^H$ may be obtained by the following equations

$$x^H = \frac{a_1 + 600}{36.2} \tag{23g}$$

$$x^H = -\frac{a_2}{19.0} \tag{23h}$$

Furthermore, they reported that the variation of $a_1$ due to solvent effects can be compensated for by an opposite variation of $a_2$. By combining Eqs. (23e) and (23f) or Eqs. (23g) and (23h) into one to yield $(a_1 - a_2)$ they proposed an equation which is nearly independent of the dielectric constant of the solvent

$$x^H = \frac{a_1 - a_2 + 650}{55.8} \tag{23i}$$

Table I shows a comparison of the helical contents of several proteins estimated from Eqs. (23e), (23f), (23g), (23h), and (23i). It should be

TABLE I[a]

HELICAL CONTENTS OF SOME PROTEINS OBTAINED FROM DIFFERENT FORMS
OF THE TWO TERM DRUDE EQUATION

| | | $100 \times x^H$ from Equations: | | | | |
|---|---|---|---|---|---|---|
| Proteins | Solvent | 23e | 23f | 23g | 23h | 23i |
| Tropomyosin | water | *87* | *85* | *83* | 92 | 86 |
| Tropomyosin | chloroethanol | 99 | 91 | *96* | *99* | 97 |
| Bovine serum albumin | water | *55* | *55* | 52 | 61 | 55 |
| Bovine serum albumin | chloroethanol | 55 | 52 | *61* | *58* | 60 |
| Ribonuclease | water | *27* | *26* | 23 | 30 | *26* |
| Ribonuclease | chloroethanol | 67 | 59 | *64* | *65* | *64* |

[a] Some values in this table are calculated by the author, based on the data of Schechter and Blout (1964a,b). The values in italics are calculated by Schechter and Blout.

noted that when the protein is in aqueous solution $x^H$ obtained from Eq. (23i) agrees best with $x^H$ obtained from Eqs. (23e) and (23f) but agrees better with $x^H$ obtained from Eqs. (23g) and (23h) when the protein is in solvents of low dielectric constant.

The table also shows that $a_1$ is more sensitive than $a_2$ to the nature of the solvent. Furthermore, the $x^H$ values obtained from Eq. (23f) in organic solvents are generally underestimated and those from Eq. (23h) in aqueous solution are overestimated.

Thus, the best way to estimate the helix content of proteins from the two-term Drude equation seems to be as follows. The ORD data are plotted as $[m'](\lambda^2 - \lambda_1^2)/\lambda_1^2$ versus $(\lambda^2 - \lambda_1^2)\lambda_2^2/\lambda_1^2(\lambda^2 - \lambda_2^2)$ (Fig. 5).

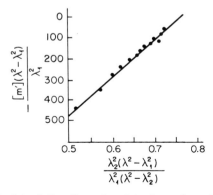

FIG. 5. A modified plot of the dispersion data for poly-L-glutamic acid in accordance with the two-term Drude equation. The slope of this curve is $-a_2$ and the intercept on the ordinate at $\lambda_2^2(\lambda^2 - \lambda_1^2)/\lambda_1^2(\lambda^2 - \lambda_2^2) = 0$ is $-a_1$. (From Imahori, 1963.)

The $a_1$ and $a_2$ values thus obtained are put into Eq. (23i) and $x^H$ is estimated.

### D. Moffitt Equation for Proteins Containing $\alpha$-Helix and Unordered Structures

#### 1. *General Principles*

A simple ORD curve which is characteristic of a random-coil conformation can be easily approximated by a one-term Drude equation. However, as seen in Fig. 4, the ORD curve of purely helical forms cannot be represented by a one-term Drude equation. The shapes of the curves suggest that they may be approximated by the following type of equation

$$[m']_\lambda = a_0\lambda_0^2/(\lambda^2 - \lambda_0^2) + b_0\lambda_0^4/(\lambda^2 - \lambda_0^2)^2 \qquad (24)$$

If we assume $a_0$ is positive and $b_0$ is negative then at wavelengths much greater than $\lambda_0$ the first term on the right-hand side of Eq. (24) will be dominant. The resulting ORD curve will correspond to the tail of the positive Cotton effect which is observed at longer wavelengths (see Fig. 4). However, as $\lambda$ approaches $\lambda_0$ the second term becomes appreciable and overcomes the contribution of the first term so that the rotation decreases steeply.

It should be noted that Eq. (24) was developed theoretically by Moffitt (1956) who applied Kirkwood's (1937) polarizability theory of optical activity coupled with the theory of molecular excitons (Davydov, 1962) to the $\pi-\pi^*$ transitions of the $\alpha$-helical peptide bonds. He predicted that the monomer transitions would give rise to a splitting of the polymer transition into two components, one polarized parallel to the helix axis and the other perpendicular to it.

Although Moffitt's treatment was adequate for the absorption spectrum of the $\alpha$-helix, it was later shown (Moffitt *et al.*, 1957) that important contributions to the optical activity of the $\alpha$-helix had been missed because the whole exciton system had not been considered (he did not treat degenerate transitions). In addition to this, the significant rotational strength of the $n-\pi^*$ transition of the peptide bond could not be considered in terms of the coupled-oscillator mechanism of Kirkwood (cf. Section II).

Thus, the theoretical basis for the Moffitt equation is no longer considered adequate and it is usually used as an empirical equation with $\lambda_0$, $a_0$, and $b_0$ adjustable parameters. In fact, it is easy to show that the Moffitt equation is a simple derivative of the general multiterm Drude equation. If Eq. (16) is expanded in a Taylor's series about $(\lambda^2 - \lambda_0^2)^{-1}$ where $\lambda_0$ is an arbitrary constant we obtain

$$[m']_\lambda = \sum_i [a_i\lambda_i^2/(\lambda^2 - \lambda_0^2) + a_i\lambda_i^2(\lambda_i^2 - \lambda_0^2)/(\lambda^2 - \lambda_0^2)^2 + \cdots] \quad (24\text{a})$$

This formulation has the advantage that the denominators for all transitions $i$ are the same. Thus the summation need only be considered in the numerators to give a constant. Furthermore, this series converges rapidly if $\lambda_i \ll \lambda$, i.e., if $[m']_\lambda$ is restricted to wavelengths far from any absorption bands.

If we consider only the first term of the expansion we obtain the one term Drude equation used in the $\lambda_c$ method with $\lambda_0 = \lambda_c$ and $k = \Sigma_i a_i \lambda_i^2$. Since we consider only the first term of the expansion, the restriction that the rotation be very far from any absorption bands must be obeyed if the one-term Drude equation is to be obeyed.

If we consider the first two terms of the expansion we obtain

$$[m']_\lambda = a_0\lambda_0^2/(\lambda^2 - \lambda_0^2) + b_0\lambda_0^4/(\lambda^2 - \lambda_0^2)^2$$

which is Eq. (24) where

$$a_0\lambda_0^2 = \sum_i a_i\lambda_i^2 \quad (24\text{b})$$

and

$$b_0\lambda_0^4 = \sum_i a_i\lambda_i^2(\lambda_i^2 - \lambda_0^2) \quad (24\text{c})$$

Furthermore, the theoretical derivation of the Moffitt equation yielded a two-term Drude equation which was then converted to the Moffitt equation. This again emphasizes that all these equations are special cases of the general multiterm Drude equation. The interrelationship of the parameters in these different equations will be discussed in Section III,E.

Experimental tests of Eq. (24) were quite successful since it fitted well with the data then available on poly-$\gamma$-benzyl-L-glutamate and poly-L-glutamic acid.

Graphic solution of Eq. (24) is achieved by rearranging Eq. (24) in the form

$$[m']_\lambda(\lambda^2 - \lambda_0^2)/\lambda_0^2 = a_0 + b_0\lambda_0^2/(\lambda^2 - \lambda_0^2) \quad (24\text{d})$$

Thus, by plotting $[m']_\lambda(\lambda^2 - \lambda_0^2)/\lambda_0^2$ against $\lambda_0^2/(\lambda^2 - \lambda_0^2)$ for trial values of $\lambda_0$ until a straight line was obtained, Moffitt and Yang (1956) obtained $\lambda_0 = 212$ nm and this value is now widely accepted within the uncertainty of $\pm 2$ nm. This kind of plot is generally called a *Moffitt–Yang plot* (Fig. 6). For poly-$\gamma$-benzyl-L-glutamate, poly-L-glutamic acid, and other polypeptides in several helix-forming solvents $b_0$ gave an average

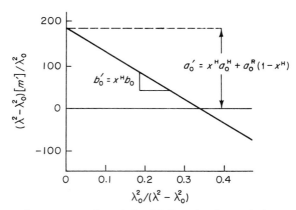

Fig. 6. Schematic representation of the Moffitt plot for optical rotatory dispersion data. The meaning of the symbols is discussed in the text. The slope shown ($b'_0$) is that for 100% α-helix. (From Imahori, 1963.)

value of $-630$. While this value was almost independent of the solvent the $a_0$ value varied widely.

Since the Moffitt equation is derivable from the multiterm Drude equation, which should apply for any optically active molecule in regions distant from absorption bands, we can carry out a similar expansion of this equation for the random coil, arriving at the same equation as that for the helix but with different values of $a_0$ and $b_0$. When the fraction of helical content is $x^H$ the rotatory dispersion of a helix–coil mixture should be given by the following equation assuming the additivity of rotatory contributions from helix and coil structures

$$[m']_\lambda = \frac{[(1 - x^H)a_0^R + x^H a_0]\lambda_0^2}{\lambda^2 - \lambda_0^2} + \frac{[(1 - x^H)b_0^R + x^H b_0]\lambda_0^4}{(\lambda^2 - \lambda_0^2)^2}$$
$$= \frac{(a_0^R + x^H a_0^H)\lambda_0^2}{\lambda^2 - \lambda_0^2} + \frac{(b_0^R + x^H b_0^H)\lambda_0^4}{(\lambda^2 - \lambda_0^2)^2}$$
$$= \frac{a'_0 \lambda_0^2}{\lambda^2 - \lambda_0^2} + \frac{b'_0 \lambda_0^4}{(\lambda^2 - \lambda_0^2)^2} \quad (25)$$

where

$$a_0^H = a_0 - a_0^R \quad (25\text{a})$$
$$b_0^H = b_0 - b_0^R \quad (25\text{b})$$

Again by rearranging Eq. (25) in the form of Eq. (24d) one obtains $a'_0$ and $b'_0$ from the intercept and the slope of the Moffitt–Yang plot, respectively. Thus it is possible to obtain the fractional helical content from the following equations

$$x^H = \frac{a'_0 - a_0^R}{a_0^H} = \frac{a'_0 - a_0^R}{a_0 - a_0^R} \quad (26a)$$

$$x^H = \frac{b'_0 - b_0^R}{b_0^H} = \frac{b'_0 - b_0^R}{b_0 - b_0^R} \quad (26b)$$

The symbols $a_0$, $a_0^H$, $b_0$, and $b_0^H$ are sometimes confusing and have different meanings for different authors. To avoid such confusion, it is important to define these parameters as used in this text. $a_0$ and $b_0$ are the intercept and slope, respectively, obtained from a Moffitt-Yang plot of pure helical conformations. $a_0^H$ and $b_0^H$ are defined by Eqs. (25a) and (25b) and represent the contribution of helical conformations over and above that of random coil to $a'_0$ and $b'_0$, respectively. $a'_0$ and $b'_0$ are the slope and intercept obtained experimentally for a sample in a Moffitt-

TABLE II

Parameters of the One-Term Drude and Moffitt Equations for Poly-L-Glutamic Acid (pH 7.3) above 270 mµ[a]

| Solvent (moles/liter) | $\lambda_c$ (nm) | $k \times 10^8$ (deg-cm$^4$/dg) | $b'_0$ (deg-cm$^2$/dmole) | $a'_0$ (deg-cm$^2$/dmole) |
|---|---|---|---|---|
| KF | | | | |
| 0 | 208 | −37.5 | 50 | −820 |
| 0.2 | 204 | −34.2 | 50 | −750 |
| 0.5 | 205 | −32.9 | 50 | −730 |
| 1.0 | 205 | −32.1 | 40 | −710 |
| 2.0 | 208 | −29.1 | 20 | −650 |
| 3.0 | 211 | −25.9 | 10 | −580 |
| 4.0 | 219 | −22.6 | −20 | −510 |
| 6.0 | 223 | −20.3 | −50 | −460 |
| LiBr | | | | |
| 2.0 | 206 | −28.7 | 40 | −620 |
| 4.0 | 206 | −24.7 | 30 | −550 |
| 6.0 | 211 | −20.4 | 0 | −450 |
| 7.5 | 217 | −15.6 | −20 | −350 |
| 8.0 | 221 | −15.4 | −40 | −340 |
| Dioxan (%, v/v) | | | | |
| 10 | 203 | −34.7 | 30 | −770 |
| 20 | 211 | −32.0 | −30 | −700 |
| 30[b] | 221 | −25.4 | −60 | −560 |
| 40[b] | | Nonlinear | −340 | −230 |
| 50[b] | | Nonlinear | −670 | 140 |
| 10(+0.2 M KF) | 203 | −31.6 | 50 | −700 |

[a] Data taken from Iizuka and Yang (1965).

[b] At and above 30% dioxan (pH 7.3), poly-L-glutamic acid undergoes a coil-to-helix transition.

Yang plot. Since the $\lambda_i$ value of Eq. (24c) is close to $\lambda_0$, the $b_0{}^R$ value is generally close to zero. As shown in Table II poly-L-glutamic acid at neutral pH under various solvent conditions consistently gives $b_0{}^R \approx 0$ (Iizuka and Yang, 1965). Thus, Eq. (26b) can be reduced to

$$x^H = \frac{-b'_0}{630} \tag{26c}$$

which has been used as a standard method to estimate the helical content of many proteins (Urnes and Doty, 1961).

Although $b_0{}^R$ obtained for randomly coiled polypeptides is nearly equal to zero it is not certain whether this value will still apply to the nonhelical, unordered, but nevertheless rigid conformation involved in native protein molecules. Some workers prefer to determine $b_0{}^R$ for the protein in its denatured state. However, although this corrects to some extent for the different intrinsic rotations of different amino acids, the structure of a denatured protein is still expected to be that of a random coil rather than the more rigid structure of the unordered regions of native proteins.

Even for the pure $\alpha$-helical structure there is not complete agreement as to the value of $b_0$. The generally accepted value of $-630$ was obtained from polypeptides in organic solvents. More recently Cassim and Yang (1970) in careful studies of poly-L-glutamic acid and poly-L-lysine in aqueous solution found $b_0$ values around $-670$ to $-680$. Approximately the same values are found by analyzing ORD curves which have been calculated from Kronig-Kramers transforms of the CD data for these polypeptides. This puts the $b_0$ value close to $-700$ for an infinite, perfect $\alpha$-helix in water.

The actual value of $b'_0$ depends on the length of the helix, the number of helices (end-effects), the nature of the side chains, the precise geometry of the helix, and solvent effects. Since it is the peptide Cotton effects which give rise to the visible rotatory dispersion of proteins, these considerations will be discussed in Section IV, on peptide Cotton effects. We will only mention here that, *a priori*, one would expect that polypeptides of uniform helix length, side chains, and geometry would not necessarily be good models for proteins where short and irregular helices can often occur. In fact, Woody and Tinoco (1967) have predicted from their calculations that $b'_0$ decreases for short helices compared to the infinite helix. Epand and Scheraga (1968) noted that for a helical sequence of 15 residues $b_0$ decreases by 10% compared to the infinite helix. Thus, the use of $b_0$ obtained from data on infinite model polypeptide helices is expected to underestimate the helical content in proteins if short helices occur. In an attempt to overcome this problem

Chen and Yang (1971) and Chen et al. (1972) have used proteins whose structure is known from X-ray crystallography to calibrate $b'_0$ and $a'_0$ for the main types of protein secondary structure.

The proteins they used were myoglobin, lysozyme, ribonuclease, papain, and lactic dehydrogenase. Experimental $b'_0$'s and $a'_0$'s were obtained for the proteins and the fractions of helix, $\beta$-, and unordered structure obtained from the published X-ray results. The data were then inserted into a set of simultaneous equations of the form

$$X = f_H X_H + f_\beta X_\beta + (1 - f_H - f_\beta) X_R$$

where $X$ is $b'_0$ or $a'_0$ for the protein; $f_H$ and $f_\beta$ are the fractional contents of helix and $\beta$-structure in the protein; and $X_H$, $X_\beta$, and $X_R$ are the values of $a'_0$ or $b'_0$ to be determined for the pure helical, $\beta$-, or unordered structures. Using $\lambda_0 = 212$ nm these authors found

| | | | | | |
|---|---|---|---|---|---|
| $b'_0$ (helix) | = | $-580 \pm 20$ | $a'_0$ (helix) | = | $330 \pm 190$ |
| $b'_0$ (beta) | = | $60 \pm 30$ | $a'_0$ (beta) | = | $-810 \pm 400$ |
| $b'_0$ (unordered) | = | $-10 \pm 20$ | $a'_0$ (unordered) | = | $-420 \pm 180$ |

Criticism of this approach will be discussed in Section IV,B but for $b_0$ at least we see that the results agree fairly well with the values estimated from studies with synthetic polypeptides and also indicate that the $a'_0$ value varies much more than the $b'_0$ value. The variation of $b_0$ in going from polypeptides to proteins is, in fact, in quantitative agreement with that expected in going from an infinite helix to a short helix (Chen et al., 1972). The use of $b'_0$ for estimation of $\beta$-structure will be discussed in Section III,F.

Despite the limitations mentioned above in using the $b_0$ method, the helical contents estimated from Eq. (26c) for myoglobin, hemoglobin, lysozyme, ribonuclease, chymotrypsin, and tropomyosin agree well with those obtained by X-ray diffraction studies. Thus, the $b_0$ method is still a very popular and reliable technique for characterizing protein conformations in solution, especially in estimating their $\alpha$-helical content.

## 2. The Effects of Solvent

Although Eq. (26c) is usually applicable for proteins in aqueous solution one should be aware that $b_0^H$ and $b'_0$ are somewhat dependent on the nature of the solvent.

By the extensive and careful examination of poly-$\gamma$-benzyl-L-glutamate in various organic solvents Cassim and Taylor (1965) found the following relation to hold between $b_0$ and the refractive indices of the solvents, $n$ (measured at the sodium D line)

$$b_0 = 730.3n - 1701 \tag{27}$$

The variation of $b_0$ with refractive index is probably due to the small blue-shifts of $n-\pi^*$ and red-shifts of $\pi-\pi^*$ transitions in solvents of increasing refractive index. This should change the values used for $\lambda_i$ and $\lambda_0$.

Experimental confirmation of such shifts has not been obtained because of the strong absorption of organic solvents in the far UV. Since no helical polypeptides other than poly-$\gamma$-benzyl-L-glutamate have been the subject of such detailed examination it is not known whether Eq. (27) is of general applicability.

However, one should be careful in applying Eq. (26b) or Eq. (26c) to proteins in organic solvents. One might apply Eq. (27) for proteins in those organic solvents which have far larger refractive indices than water but in aqueous solution or any solvent of $n = 1.3$ to $1.4$ one should use Eq. (26c) and not Eq. (27). Equation (27) gives too large a magnitude for $b_0$ ($-730$) for helical polypeptides in aqueous solution and this value would lead to underestimates of the helical content of many proteins causing a serious deviation from the content obtained from X-ray analysis.

Equation (26a) provides an alternative means of estimating helical content which can be termed the $a_0$ method. However, the uncertainty in using Eq. (26a) lies in the determination of $a_0^H$. As is evident from Eq. (25a) $a_0^H$ can be obtained by subtracting the rotation of polypeptides in the denatured state from that in their helical state and assuming that the residue rotations remain unchanged during denaturation. A similar procedure is required for the native and denatured proteins in order to obtain $x^H$ from Eq. (26a). Unlike the nonhelical $b_0^R$ in Eq. (26b), which is very small, $a_0^R$ is very large and varies with the nature of the solvent, as shown in Table II. Thus, any uncertainty in the determination of $a_0^R$ would result in serious errors in the estimation of helical content. Thus, the $a_0$ method is not used to the same extent as the $b_0$ method in quantitative estimates of helical content.

The value of $a_0^H$ has been obtained from the ORD of mesohelices on the assumption that residue rotations of L-residues would cancel that of D-residues (Doty and Lundberg, 1957). A mean value of about 650 has been suggested for $a_0^H$ from this treatment. By averaging the $a_0^H$ values of many kinds of polypeptides obtained from Eq. (25a) a value of 680 has been proposed for $a_0^H$ (Doty et al., 1958; Urnes and Doty, 1961). The helical content of proteins can then be obtained by the following equation.

$$x^H = (a'_0 - a_0^R)/680 \tag{26d}$$

It should be noted that $a_0^R$ in Eq. (26d) should be obtained from the ORD of the protein in its denatured state since $a_0^R$ varies with the protein as well as the solvent.

Although the $a_0$ method is somewhat less reliable than the $b_0$ method in estimating helical contents of proteins it is still a useful parameter especially when the helical content is low. Table III shows the helical contents of several proteins as estimated from both the $a_0$ and $b_0$ methods. It is evident that the estimates from $a'_0$ are in reasonable agreement with those from $b'_0$ except for $\beta$-lactoglobulin which is thought to have a high content of $\beta$-structure. This is discussed further in Section III,F and the use that $a_0$ may be put to in estimating $\beta$-structure from visible rotatory dispersion is also discussed in Section III,F.

Changes in the $a'_0$ value may also be used to follow conformational changes in the protein if the solvent has not been markedly changed in the process. In fact the drastic changes observed in $[\alpha]_D$ during protein denaturation are primarily a reflection of changes in $a'_0$ rather than $b'_0$.

### 3. The Moffitt Equation Applied at Wavelengths Below 310 nm

Equations (26c) and (26d) are applicable for the ORD in the visible and near-UV regions only. However, myoglobin, for example, presents a problem in this region because of the Cotton effects induced in the heme group and we are forced to analyze the ORD below 300 nm.

Urnes et al. (1961) have studied the ORD of copolymers of L-tyrosine with L-glutamic acid in the wavelength region between 240 nm and 310 nm and found that $a_0^H = 610$ and $b_0^H = -535$ with $\lambda_0 = 216$ nm. The ORD

TABLE III
HELICAL CONTENTS OF SOME PROTEINS OBTAINED FROM THEIR MOFFITT-YANG PARAMETERS

| Proteins | Helical content obtained from: | |
| --- | --- | --- |
| | $b'_0$ | $a'_0$ |
| Tropomyosin | 0.90 | 0.90 |
| Hemoglobin | 0.70–0.80 | |
| Myoglobin | 0.70–0.80 | |
| Serum albumin | 0.46 | 0.56 |
| Insulin | 0.40 | 0.55 |
| Ovalbumin | 0.31 | 0.48 |
| Lysozyme | 0.29 | 0.37 |
| Histone | 0.20 | 0.28 |
| Ribonuclease | 0.16 | 0.16 |
| Chymotrypsinogen | 0.10 | 0.10 |
| Lactoglobulin | 0–0.10 | 0.70 |

of myoglobin results in values of $a'_0 - a_0^R = 420$ and $b'_0 = -390$. In this way, the helical contents obtained from $a_0$ and $b_0$ were 70% and 73%, respectively, which were in fairly good agreement with the value obtained by X-ray analysis. Later, Urnes and Doty (1961) examined several water-soluble polypeptides and increased $\lambda_0$ to 220 nm to linearize the Moffitt-Yang plot in the region 240–310 nm. $a_0$ and $b_0$ in this region varied from $-130$ to $-170$ and from $-340$ to $-400$ with the averages $a_0 = -150$ and $b_0 = -360$. The corresponding average values for the coiled form were $a_0^R = -710$ and $b_0^R = 70$.

Thus the helical contents may be obtained from the following equations

$$x^H = \frac{a'_0 - a_0^R}{560} \tag{28a}$$

$$x^H = \frac{70 - b'_0}{430} \tag{28b}$$

It should be noted that these equations can only be applied for $a'_0$ and $b'_0$ values which have been determined from a Moffitt-Yang plot using $\lambda_0 = 220$ nm for ORD data obtained in the limited wavelength region 240 to 310 nm. The experimental $b'_0$ value for myoglobin was $-260$, which according to Eq. (28b) gives 77% helical content, in good agreement with X-ray results.

E. RELATIONS BETWEEN THE USE OF THE MOFFITT EQUATION AND OTHER METHODS

### 1. $[\alpha]_D$ and $\lambda_c$

Finally, it may be helpful to compare the Moffitt equation with other methods. It has already been mentioned that the $[m]_\lambda$ method or the $[\alpha]_D$ method correspond, in principle, to the $a_0$ method. The ambiguity caused by solvent effects or intrinsic residue rotations handicaps both the $a_0$ and $[\alpha]_D$ methods. The parallelism between $a_0$ and $[\alpha]_D$ holds only for helix and coil mixtures. As will be explained in the following section, when $\beta$-structure exists in addition to helix and coil structures the $a_0$ method seems to be superior to the $[\alpha]_D$ method. It should also be noted that the parallelism between the $a_0$ and $[m]_\lambda$ methods holds only at longer wavelengths where the first term of the Moffitt equation is much larger than the second term.

Downie (1960) found the following relation between $\lambda_c$ and Moffitt's parameters

$$\lambda_c^2 = \frac{\lambda_0[(\lambda^2 - \lambda_0^2) + (b'_0/a'_0)(\lambda^2 + \lambda_0^2)]}{[(\lambda^2 - \lambda_0^2) + 2(b'_0/a'_0)\lambda_0^2]} \tag{29}$$

or

$$\lambda_c \approx \lambda_0[1 + (b'_0/a'_0)]^{1/2} \qquad (29a)$$

It is evident from Eq. (25) that when $x^H$ gets close to zero $b'_0$ is almost equal to zero while $a'_0$ reduces to $a_0{}^R$. Thus, it becomes clear that $\lambda_c$ equals $\lambda_0$ when the protein is completely denatured.

Furthermore, as the helical content increases $b'_0$ decreases, $a'_0$ increases, and hence $\lambda_c$ increases in accord with the data in Fig. 2 (Yang and Doty, 1957).

In others words, $\lambda_c$ is a function of both $a'_0$ and $b'_0$ so that it is expected that it would reflect the conformation of the protein. It is possible to estimate helical contents from the $\lambda_c$ value using Fig. 2 or Eq. (29a). However, the problem is that $\lambda_c$ is a function of $a'_0$, which depends on the nature of the solvent and is influenced by the presence of other conformations, such as $\beta$-structure. In the Moffitt equation one is able to cancel these effects to some extent by combining the use of $a'_0$ and $b'_0$ values.

## 2. The Moffitt Equation and the Two-Term Drude Equation

So far we have suggested two equations for analyzing the ORD curves of proteins—the Moffitt equation and the two-term Drude equation. There has been much discussion in the literature as to which is the superior method, and we do not intend to state a preference here. However, one should be aware that both equations are abbreviated and approximate forms of Eq. (17). $\lambda_1$, $\lambda_2$, or $\lambda_0$ of the two equations are selected so as to allow rapid convergence of the Taylor expansion of Eq. (17). Thus, neither equation contains any more information than the empirical equation.

In fact, a set of $a'_0$ and $b'_0$ values can be related to a set of $a_1$ and $a_2$ values by the following equations (Imahori, 1963; Moffitt and Yang, 1956)

$$a'_0 = \sum_{i=1,2} (a_i + a_i\delta_i + a_i\delta_i{}^2) \qquad (30a)$$

$$b'_0 = \sum_{i=1,2} (2a_i\delta_i + 5a_i\delta_i{}^2) \qquad (30b)$$

where $\delta_i = (\lambda_i - \lambda_0)/\lambda_0$. If the $a_i$'s are a linear function of the helical content, $a'_0$ and $b'_0$ must also vary with the helical content. Conversely, either $a_1$ or $a_2$ can be expressed in terms of $a'_0$ and $b'_0$.

It is true that the $a'_0$ term of the Moffitt equation is very sensitive to solvent while $(a_1 - a_2)$ of the two-term Drude equation is not. This can be easily understood from Eqs. (30a) and (30b). Among the terms on the right-hand side of Eq. (30a), $(a_1 + a_2)$ is dominant in determining the value of $a'_0$. If, as mentioned previously, $a_1$ and $a_2$ are both dependent on the solvent the sensitivity would be exaggerated by adding these two terms. However, in Eq. (30b) the magnitude of $b'_0$ is determined primarily by $(a_1\delta_1 + a_2\delta_2)$. Since $\delta_1$ and $\delta_2$ are almost equal in magnitude but opposite in sign $(a_1\delta_1 + a_2\delta_2) \approx (a_1 - a_2)|\delta_1|$ may be obtained. It is evident now that the $b_0$ term is almost equivalent to $(a_1 - a_2)$. Thus, the $(a_1 - a_2)$ method recommended by Schechter and Blout is similar to a determination of the $b'_0$ value.

## F. The Moffitt Equation Applied to Proteins Containing $\beta$-Structure

Although the existence of $\beta$-structure was recognized a long time ago, the beauty of the $\alpha$-helical conformation proposed by Pauling and Corey drew attention away from the $\beta$-structure for some time. It had already been noted by Yang and Doty (1957) that the ORD curve for $\beta$-aggregates of the oligomers of $\gamma$-benzyl-L-glutamate did not obey the one-term Drude equation. Later, Imahori (1960) first discovered that it could be described by the Moffitt equation. This is to be expected since the ORD of the $\beta$-structure should obey Eq. (16) and Moffitt's equation simply corresponds to the first two terms of its Taylor expansion. However, it may be surprising to see that the $\lambda_0$ value which had been selected from the Taylor series expansion of the ORD of polypeptides in their $\alpha$-helical conformation is also valid for the ORD of the $\beta$-structure.

Thus, the Moffitt equation for pure $\beta$-structure may be given by

$$[m']_\lambda = a_\beta \lambda_0^2/(\lambda^2 - \lambda_0^2) + b_\beta \lambda_0^4/(\lambda^2 - \lambda_0^2)^2 \qquad (31)$$

where $a_\beta$ and $b_\beta$ are the values of $a'_0$ and $b'_0$ for 100% $\beta$-structure.

With a similar treatment to that for Eq. (25) the ORD curve for a mixture of helix, coil, and $\beta$-structure may be expressed as follows

$$[m']_\lambda = \frac{a_0^R \lambda_0^2}{\lambda^2 - \lambda_0^2} + \frac{[x^\beta a_0^\beta + x^H a_0^H]}{\lambda^2 - \lambda_0^2} \cdot \lambda_0^2 + \frac{[x^\beta b_0^\beta + x^H b_0^H]}{(\lambda^2 - \lambda_0^2)^2} \cdot \lambda_0^4 \qquad (32)$$

where

$$a_0^\beta = a_\beta - a_0^R \qquad (32\text{a})$$
$$b_0^\beta = b_\beta - b_0^R \qquad (32\text{b})$$

and $x^\beta$ denotes the fractional $\beta$-structure content. Since

$$a'_0 = a_0^R + x^\beta a_0^\beta + x^H a_0^H \qquad (32\text{c})$$

and

$$b'_0 = x^\beta b_0^\beta + x^H b_0^H \qquad (32\text{d})$$

both $x^\beta$ and $x^H$ can be obtained from $a'_0$ and $b'_0$.

However, the problem is to estimate $a_\beta$ and $b_\beta$ which are necessary for solution of Eqs. (32a) and (32b). The $a_\beta$ and $b_\beta$ values of several polypeptides and proteins are listed in Table IV. Imahori (1960) first reported that on denaturation bovine serum albumin could show transitions to $\beta$-structure with positive $a_0^\beta$ and $b_0^\beta$. However, the reference value for 100% $\beta$-form could not be obtained since the amount of $\beta$-structure could not be determined. Wada et al. (1961) investigated the oligomer of $\gamma$-benzyl-L-glutamate at various concentrations and they estimated $a_0^\beta = 840$ and $b_0^\beta = 420$ by extrapolating their results to

TABLE IV

The Moffitt Parameters for Several Peptides and Proteins Containing $\beta$-Structures

| Substance | $a'_0$ | $b'_0$ |
|---|---|---|
| Oligomers of $\gamma$-benzyl-L-glutamate | | |
| 6.3% in $CHCl_3$ | 85 | 55 |
| 1.9% in $CH_2ClCH_2Cl$ | 210 | 30 |
| 0.8% in $CH_2ClCH_2Cl$ | 110 | −10 |
| 4.2% in dioxan | 60 | 60 |
| 0.8% in dioxan | −60 | 0 |
| Poly-O-acetyl-L-serine | | |
| in 1:3 $CHCl_2COOH$–$CHCl_3$ | 685 | 0 |
| in $CHCl_2COOH$[a] | + | −80 |
| Poly-O-benzyl-L-serine | | |
| in 90:10 chloroform-dichloroacetic acid | | |
| Initial | 680 | 100 |
| Final (after 10 days) | 600 | 190 |
| in $CHCl_2COOH$[a] | 200 | −50 |
| Poly-S-carbobenzoxymethyl-L-cysteine in | | |
| $CHCl_2COOHCHCl_3$ (0.5:95.5) | 385 | 0 |
| Poly-L-lysine in water | | |
| 0.01%, pH 11, 50°C, 10 min | −340 | −150 |
| 0.01%, 1 $M$ NaCl, pH 11, 50°C, 15 min | −60 | −240 |
| neutral pH[a] | −980 | 0 |
| Bovine serum albumin | | |
| exposed to pH 12, then pH 6 | −220 | 230 |
| in 60% $n$-propanol, 100°C, 10 min | −300 | 0 |
| in 8 $M$ urea[a] | −620 | 0 |
| Ovalbumin in 8 $M$ urea | | 270 |
| Silk fibroin in aqueous solution | | |
| 0.01% in 50% dioxan, 9 days | −260 | 90 |
| 0.01% in 93% $CH_3OH$ | −40 | 30 |
| 0.01% in 50% $CH_3OH$ | Nonlinear | |
| in 8 $M$ urea[a] | −260 | 10 |

[a] The polymer is in the disordered form.

100% β-structure. On the other hand, Fasman and Blout (1960) found that the β-structure in poly-$O$-acetyl-L-serine in a dichloroacetic acid–chloroform mixture had a positive $a'_0$ but $b'_0$ was zero. Imahori extended this work and proposed that the conformation of this polymer was intrachain cross-β since both parameters were independent of polymer concentration (Imahori and Yahara, 1964). If we assume that poly-$O$-acetyl-L-serine in 1:7 dichloroacetic acid–chloroform mixtures has 100% β-structure, $a_0^\beta$ and $b_0^\beta$ are found to be $+490$ and $0$, respectively. A similar examination of poly-$O$-benzyl-L-serine resulted in $a_0^\beta = 300$ and $b_0^\beta = 100$ (Bradbury et al., 1962b). On the other hand, Sarker and Doty (1966) and Davidson et al. (1966) have independently arrived at values of $a_0^\beta = 920$ and $b_0^\beta = -150$ to $-240$ for poly-L-lysine.

Thus, the data for the values of $a_0^\beta$ and $b_0^\beta$ vary with different investigators, $b_0^\beta$ varying even in sign. One reason for these results is that it is very difficult to attain 100% β-structure. Although more than 90% of helical poly-L-lysine was converted into β-structure it is possible that a small amount of helical form may have remained, leading to the negative $b_0^\beta$ value.

Another reason for the confusion may arise from the wavelength regions used for the ORD measurements. As shown in Fig. 7, the $b'_0$ value of β-aggregates of the oligomers of γ-benzyl-L-glutamate shows a remarkable dependence on the wavelength region used (Yahara, 1966). The longer the wavelength region used the more positive the value for $b_0^\beta$, the shorter wavelength region giving zero for $b_0^\beta$. Thus, the positive

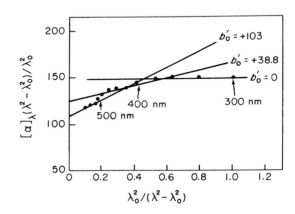

FIG. 7. Moffitt plot of the dispersion data for low-molecular-weight poly-γ-benzyl-L-glutamate at a concentration of 5% in chloroform. In this solvent the polymer exists as a β-aggregate. Note that the data cannot be fitted with one straight line over the whole wavelength range. (From Yahara, 1966.)

value for $b_0{}^\beta$ obtained for oligomers of γ-benzyl-L-glutamate will tend to zero if the wavelength region used is limited to that below 400 nm.

A similar phenomenon is observed for proteins. Iizuka and Yang (1966) repeated Imahori's experiments on denatured bovine serum albumin but at wavelengths far below 400 nm and obtained zero values for $b_0{}^\beta$.

The value of the other parameter necessary for obtaining the content of β-structure in the Moffitt analysis, $a_0{}^\beta$, also varies widely as indicated in Table IV.

It is important to realize at this stage that at least part of the reason for these discrepancies is the fact that there are at least three main types of β-structure which are expected to have widely different ORD parameters (see Section IV,C,3). These are the parallel, antiparallel, and intrachain cross-β-structures. Although the antiparallel structure is the most common in polypeptides and proteins the other structures have also been found in some proteins.

If a single value for $a_0{}^\beta$ for β-structures were to be assigned, then, in principle, the content of helix, β-structures, and unordered structure could be estimated from Eq. (32). However, even without making such an assumption it is still possible to detect the presence of β-structures by applying both the $a_0$ and $b_0$ methods and seeing if they agree in their estimates of helix content. For example, native β-lactoglobulin has an $a'_0$ and $b'_0$ of $-160$ and $-68$ and the denatured protein has $-663$ and $-51$, respectively (Tanford et al., 1960). Thus the $b'_0$ is essentially zero and indicates the absence of much α-helical structure. In contrast, the $a_0$ method gives a helical content of about 70%. Such a discrepancy may be attributed to the presence of structures other than α-helices and coils and probably indicates the presence of β-structure. It was on this basis that Hamaguchi and Imahori (1964) estimated the content of β-structure in lysozyme before its verification by X-ray analysis (Blake et al., 1965).

If a value for $a_0{}^\beta$ were to be established it would be possible to estimate the content of β-structure from a modified form of Eq. (32)

$$x^\beta = \frac{a'_0 - a_0{}^R - a_0{}^H(b'_0/b_0{}^H)}{a_0{}^\beta} = \frac{a'_0 - a_0{}^R + 1.1 b'_0}{a_0{}^\beta} \qquad (33)$$

In deducing Eq. (33) it was assumed that $b_0{}^R = b_0{}^\beta = 0$, $b_0{}^H = -630$, and $a_0{}^H = +680$. Taking lysozyme as an example, an $a'_0$ and $b'_0$ of $-270$ and $-140$, respectively, gave 15% of β-structure assuming 840 for $a_0{}^\beta$. The helical content was about 25%.

The fair agreement of these data with those of X-ray analysis makes it tempting to apply Eq. (33) for other cases. However, one should be aware that the estimates of $a_0^\beta$ are tentative and Eqs. (32) and (33) can be used only qualitatively or semiquantitatively at best.

## IV. Peptide Cotton Effects

Although the rotatory dispersion of proteins in the visible can be represented by Eq. (16) and its derivatives we have already seen (Fig. 1) that in the vicinity of any $\lambda_i$ anomalous dispersion occurs and that this has been termed a Cotton effect. These Cotton effects are observed at the near-UV absorption bands and yield information on the content and conformation of the chromophoric residues in the protein (Section V).

Cotton effects are also observed in the far-UV (185–250 nm) and for most proteins these are mainly due to amide electronic transitions of the peptide bond. Since the peptide bond itself is a symmetric chromophore, its optical activity arises through interactions of its transitions with other transitions and dipoles provided by the asymmetrically folded backbone chain of the protein in a particular secondary structure. The optical activity of the $\pi-\pi^*$ transition of the peptide bond arises mainly from strong coupling of the transition moments between the monomer units. However, since the $n-\pi^*$ transition of the peptide bond is electrically forbidden, it must derive its optical activity through magnetic–electric coupling of monomer transition moments or through the perturbing effect of the polymer electric field. These mechanisms have been discussed in Section II.

Early calculations on the optical activity of helical structures have been discussed in Section III,D,1.

More recently, extensive calculations on the optical activity of the $\alpha$-helix using molecular exciton theory have yielded a high degree of success in reproducing the main features of ORD and CD curves of $\alpha$-helical polypeptides in the accessible range (Schellman and Oriel, 1962; Woody and Tinoco, 1967; Woody, 1968; Madison and Schellman, 1972; see, also, Chapter 23, this volume, Section II,G). In general these results predict a negative Cotton effect at about 207 nm which corresponds to the parallel polarized component of the $\pi-\pi^*$ exciton system and a strong positive band at about 191 nm which is a composite of many bands, including the perpendicular–polarized component of the exciton system. The negative rotational strength of the $n-\pi^*$ transition is also predicted at about 222 nm. A strong negative band is predicted to lie just outside the accessible range.

Holzwarth and Doty (1965) showed that the absorption spectrum and CD spectrum of $\alpha$-helical poly-$\gamma$-methyl-L-glutamate in trifluoroethanol could be simultaneously resolved with three bands centered at 190 nm, 206 nm, and 222 nm. They assigned these bands to the $\pi-\pi^*$ transition polarized perpendicular to the helix axis, the $\pi-\pi^*$ transition polarized parallel to the helix axis, and the $n-\pi^*$ transition, respectively. Furthermore, Blout et al. (1967) showed that the far-UV ORD of the $\alpha$-helix could only be fitted with a minimum of three Cotton effects. Using a least-squares fitting program and a modification of Eq. (12) to describe the Cotton effect shape, they showed that the data for poly-$\gamma$-morpholinoethyl-$\alpha$-L-glutamide in methanol–water (9:1) gave a positive Cotton effect centered at 192 nm and two negative Cotton effects centered at 209 nm and 224 nm. Gaussian bands centered at the same wavelength also fit the CD data for this compound. An interesting point is that the expression used in this study also contained a Drude-type term to allow for background contributions from bands centered outside the spectral region studied. Blout et al. found that no contribution from other bands need be considered since the fit with just three Cotton effects was within the experimental error. Finally, Cassim and Yang (1970), in a very careful study specifically designed to test the molecular exciton model for the $\alpha$-helix, have applied the Kronig-Kramers transforms to the experimental CD curves for $\alpha$-helical polypeptides and have compared the ORD curves thus calculated with the experimental ORD curves. They find essentially perfect agreement between the two curves.

All these results suggest that there is essentially no contribution to the accessible far-UV ORD of the $\alpha$-helix from Cotton effects centered in the inaccessible far UV. Although this does not disprove the predictions of the $\alpha$-helix optical activity based on the molecular exciton model (Moffitt et al., 1957; Woody and Tinoco, 1967; Woody, 1968; see, also, Chapter 23, this volume, Section II,G,1) it does indicate that contributions from the inaccessible spectral region to the far-UV Cotton effects must cancel each other out.[10] Thus, in considering the optical

---

[10] The recent vacuum-ultraviolet CD spectra for aqueous solutions of helical poly-L-glutamic acid measured by Johnson and Tinoco (1972) show a positive shoulder and not a large negative band at 180 nm. Mandel and Holzwarth (1972) have measured the unoriented and oriented circular dichroism as well as the absorption and linear dichroism of poly-$\gamma$-methyl-L-glutamate in hexafluoro*iso*propanol between 183 nm and 240 nm. Although, in their resolution of these curves they do find evidence for the predicted band (which has the shape of a derivative of an absorption band) it is only about ¼ the size of that predicted theoretically. Furthermore, comparison of their fitted spectrum with the experimental spectrum of Johnson and Tinoco in the 160 to 183 nm region reveals that there is a positive band near 175 nm which is not included in their analysis. This again suggests that there is cancellation of the vacuum-UV Cotton effects so that they do not significantly affect the ORD seen in the accessible spectral range.

activity of the α-helix, we need only discuss effects on the three observed Cotton effects.

Blout et al. (1967), using the same empirical procedure discussed above, have also resolved the far-UV ORD of the random-coil conformation of poly-L-glutamic acid in water at pH 7. They find that a two Cotton effect solution adequately represents the experimental ORD although a three Cotton effect solution improves the fit to within experimental error. In the two Cotton effect solution a negative band is centered at 198 nm and a positive band is centered at 218 nm. In the three Cotton effect solution a very weak negative Cotton effect is included centered at 235 nm. Again no significant background rotation is required for a good fit. However, theoretical results (Zubkov et al., 1971; see, also, Chapter 23, this volume, Section II,G,5) are much more difficult to correlate with the experimental results. Generally the shapes of the curves agree with experiment but they are shifted by 10–15 nm to lower wavelengths.

The calculations on $\beta$-structure (antiparallel) were more successful (Woody, 1969; Chapter 23, this volume, Section II,G,3 and references cited therein). The $n-\pi^*$ transition was associated with a small negative Cotton effect centered at 225 nm, the size agreeing approximately with experiment. The $\pi-\pi^*$ transition is split into three main components. One ($y$) is polarized in the direction of the chain and gives rise to a positive Cotton effect centered at 198–200 nm of large rotational strength. The second ($x$) is polarized in the direction of the interchain hydrogen bonds and gives rise to a smaller positive Cotton effect centered at 194–196 nm. The third ($z$) is polarized perpendicular to the plane of the sheet and gives rise to a large negative Cotton effect centered at 182–185 nm. There is some disagreement about the sign of the Cotton effect of the $y$ component (Pysh, 1966; Woody, 1969) and the inaccessible $z$ component (Pysh, 1966; Zubkov and Vol'kenshtein, 1970). However, all the results indicate that the antiparallel $\beta$-structure is characterized by negative rotational strength at about 220 nm and positive rotational strength at about 200 nm from the addition of nearly overlapping $x$ and $y$ components, whether they are positive or negative (Madison and Schellman, 1972). The prediction of positive rather than negative rotational strength centered at about 180 nm is in better agreement with experiment (Timasheff et al., 1967). Madison and Schellman (1972) also conclude that the major contribution to $\beta$-structure optical activity comes from the electric–magnetic coupling mechanism because there is a large cancellation of $\pi-\pi^*$ exciton rotatory strength and because the H-bonding occurs in a plane of symmetry of the peptide group thus reducing optical activity from the one-electron mechanism. This means that "perfect" $\beta$-

structure optical activity is not very dependent on chain length or width but is extremely sensitive to distortions which destroy the symmetry.

## A. Proteins Containing Helix and Unordered Form Only

The far-UV ORD of polypeptides in their $\alpha$-helical and random-coil conformations are shown in Fig. 8 (Blout et al., 1962). The CD of the $\alpha$-helix shows two negative peaks at 222 nm and 206 nm and one positive peak at 190 nm (Holzwarth and Doty, 1965; and Chapter 23, this volume, Fig. 5). Since these bands overlap, the corresponding ORD curve displays

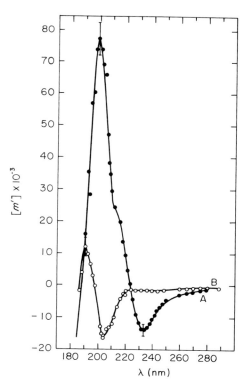

Fig. 8. Curve A (●—●): the ultraviolet optical rotatory dispersion of the helical form of poly-L-glutamic acid in water (pH 4.3). Curve B (○—○): the ultraviolet optical rotatory dispersion of the random coil form of poly-L-glutamic acid in water (pH 7.1). $[m']$ is the reduced mean residue rotation calculated using the refractive index of water at 240 nm. [Reprinted from Blout et al. (1962), J. Am. Chem. Soc. 84, 3193. Copyright (1962) by the American Chemical Society. Reprinted by permission of the copyright owner.]

a trough and a peak at 233 nm and 198 nm, respectively, with a crossover point near 224 nm, and a weak shoulder near 210 nm.

The random-coil ORD shows a trough and a peak at 205 and 190 nm, respectively; an inflection point at 197; and a very small Cotton effect near 230 nm.

Thus, the 233 nm trough of the $\alpha$-helix ($n$–$\pi^*$ transition) is well separated from the major contributions of the random coil to the ORD and has been used to estimate helical contents in proteins. The problem is to find a suitable reference value for the magnitude of this trough for the purely $\alpha$-helical conformation. Poly-L-glutamic acid has been used as a standard but $[m']_{233}$ values vary from $-13,000$ to $-16,000$ (Yang and McCabe, 1965). Values of $[m']_{233}$ (helix) obtained from different polypeptides may be summarized as follows: Poly-L-lysine in 1 $M$ NaBr, $-13,500$; poly-$\gamma$-benzyl-L-glutamate in dioxan, $-14,400$; poly-$\delta$-hydroxy-$O$-acetyl $\alpha$-amino valerate in trifluoroacetone trihydrate, $-15,400$; poly-$\gamma$-ethyl-L-glutamate in trifluoroethanol, $-18,500$. The discrepancies in this value may be partially due to solvent effects as well as to the factors discussed in Section IV,C. It should also be remembered that poly-L-glutamic acid aggregates below pH 4 with a consequent increase in the magnitude of $[m']_{233}$. The coiled form of poly-L-glutamate gives $[m']_{233} = -1800$.

The contributions due to solvent effects and side-chain Cotton effects may be partially cancelled by subtracting $[m']_{233}$ of the denatured protein from that of the native one.

Tentatively we may assume $[m']_{233}$ (helix) $= -15,000$ and $[m']_{233}$ (coil) $= -2000$. We can then obtain a rough estimate of the helical content from the following equation:

$$x^H = \frac{-\{[m']_{233} \text{ (native)} - [m']_{233} \text{ (denatured)}\}}{13,000} \tag{34}$$

A formula similar to Eq. (34) can be deduced for the data at the 198 nm peak. However, this peak has not been used as often to estimate helical content because of the difficulties cited below. Cotton effects of the coiled form with a trough at 205 nm make a large negative contribution at this wavelength. Although $[m']_{198}$ for the coiled form has been estimated to be around $-5000$ from poly-L-glutamic acid data, the Cotton effects of the coiled form show a steep gradient at 198 nm. Accordingly, spectral shifts due to solvent perturbation, etc., will result in a large change in the value of $[m']_{198}$ (coil). This makes the estimation of the reference value for $[m']_{198}$ (coil) very uncertain. Second, as will be described later, some side-chain chromophores, especially the aromatic side chains, have large Cotton effects around 198 nm and their contributions to $[m']_{198}$ vary greatly with the conformation of the backbone chain. For example, the Cotton effect due to tyrosyl residues changes its sign depending on whether it is incorporated in a helical region or a coiled region (Shiraki and Imahori, 1967). Thus, we have no way of cancelling out the side-chain contribution unless we know whether each side-chain residue is located in a helical or a coiled segment. Furthermore, Woody and Tinoco (1967) have predicted on theoretical grounds that the peak at 198 nm should change by a factor of two in going from a helical segment five residues long to the infinite helix. On the other hand, the trough at 233 nm is relatively insensitive to chain length (see Section IV,C).

Another difficulty exists in the limitations of the instrument used. Strong absorption of both the solute and solvent at this wavelength make the measurements unreliable. A small signal-to-noise ratio and artifacts due to stray light also add to the unreliability of the data.

If one does use the $[m']_{198}$ method for the estimation of helical content we may assume $[m']_{199}$ (helix) = 75,000 and $[m']_{198}$ (coil) = −5000. By neglecting the variation of side-chain contributions the fraction of helical content can be obtained from

$$x^H = \frac{[m']_{198} \text{ (native)} - [m']_{198} \text{ (denatured)}}{80,000} \tag{35}$$

### B. Proteins Containing $\beta$-Structure

The Cotton effects of the $\beta$-structure are more difficult to measure. Those polypeptides known to contain $\beta$-structure are usually insoluble in water and until recently it was not possible to measure Cotton effects for such polypeptides. The $\beta$-structure of axially oriented poly-L-isoleucine in a solid film was reported to have a peak around 207 nm (Blout and Schechter, 1963). More recently, however, Cotton effects for the $\beta$-structure in aqueous solution have been observed for poly-L-lysine (Sarker and Doty, 1966; Davidson et al., 1966) and silk fibroin (Iizuka and Yang, 1966). Figure 8 shows an example: A trough at 229 to 230 nm and a peak at 204 to 205 nm were observed. Another trough should exist around 190 nm. The two curves in Fig. 8 for the ORD profiles of $\alpha$-helix and $\beta$-structure are notably similar, the main difference being that no shoulder appears near 210 nm for the $\beta$-structure. As will be described below this similarity makes the estimation of $x^H$ and $x^\beta$ very difficult, if the protein contains the helical, $\beta$, and coiled conformations.

In principle it is possible to estimate $x^H$ and $x^\beta$ from the following equation

$$[m'] = x^H[m']_H + x^\beta[m']_\beta + (1 - x^H - x^\beta)[m']_R \tag{36}$$

The content of each conformation can be determined by employing two simultaneous equations, say at 230 and 205 nm. However, the reference value for $\beta$-structure is uncertain. Sarker and Doty (1966) measured the ORD of poly-L-lysine after heating at 50°C for 10 min at pH 11 and found a trough and a peak at 230 and 205 nm, respectively, with $[m']_{230} = -6000$ and $[m']_{205} = 22,000$. The same polymer was studied simultaneously by Davidson et al. (1966), who found that $[m']_{230} = -6000$ and $[m']_{205} = 20,000$. Iizuka and Yang (1966) studied the $\beta$-structure of silk fibroin and reported that both $[m']_{230}$ and $[m']_{205}$ varied depending on the solvent composition: $[m']_{230}$ was $-6000$, $-5000$, and $-3000$ and $[m']_{205}$ was 20,000, 24,000 and 27,000 in 1:1 dioxane–water, 1:1 methanol–water, and 93:7 methanol–water, respectively. Sarker and Doty (1966) reported that the $\beta$-structure of poly-L-lysine can be produced at neutral pH with the addition of sodium dodecyl sulfate, but the magnitude of the 230 nm trough was much shallower than that in water at pH 11.

The discrepancy cannot be attributed to the different samples since both poly-L-lysine and silk fibroin showed variations in $[m']_{230}$ and $[m']_{205}$ with the different solvents. Furthermore, both the peak and trough for the $\beta$-structure were located close to, but with smaller magnitudes than, those for the helical conformation. Thus, rotations due to the $\beta$-structure can easily be overshadowed by those of the helices if they coexist in a protein molecule.

The content of $\beta$-structure in $\beta$-lactoglobulin has been estimated by the Moffitt equation, the two-term Drude equation, and $[m']_{230}$ (Timasheff et al., 1966). These authors observed unusual optical rotatory behavior in $\beta$-lactoglobulin and tried to explain it by the presence of some specific conformation different from $\alpha$-helix or random coil. From the infrared absorption spectrum they attributed this conformation to $\beta$-structure although they did not try to estimate its content from the infrared spectrum. Rather, they tried to estimate it from Eqs. (32c) and (32d). The problem was to assign values to $a_0^\beta$ and $b_0^\beta$. They used $a_0^\beta = +400$ and $b_0^\beta = 0$, which are somewhat different from the average literature values, $a_0^\beta = +600$ and $b_0^\beta = 100$. They obtained $x^H = 0.13$ and $x^\beta = 0.40$. These results have been partly supported by the fact that $x^H$ obtained from Eqs. (23c) and (23f) gave 0.28 and 0.14, respectively, although these results only suggested the existence of $\beta$-structure rather than its content. They tested for the presence of $\beta$-structure by subtracting the rotatory contributions of 13% $\alpha$-helix and 47% random coil from the experimental curves. The resultant curve with its positive maximum at 207 nm ($[m'] = 6500$) was similar to that obtained for polyisoleucine in the $\beta$-structure (Blout and Schechter, 1963). If we assume $[m']_{207} = 22,000$ for 100% $\beta$-structure, the above results suggest the existence of

30% β-structure in β-lactoglobulin. Although this value is a little smaller than that obtained by the Moffitt equation, we can accept their results as indicating the limits for estimation of β-structure by the two different methods.

More recently a self-consistent set of far-UV ORD curves has been obtained for poly-L-lysine in the pure α-helical, antiparallel β-, and random-coil conformations (Greenfield et al., 1967). These curves are similar to those in Fig. 8 but do show some differences principally between the random-coil conformations (Fig. 9). These authors have also computed a series of ORD spectra for varying proportions of α-helix, β-structure, and random coil from linear combinations of the three "pure" spectra.

Their ORD spectrum for 100% α-helix shows a trough at 233 nm

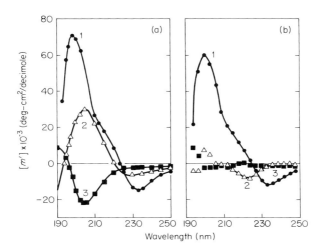

FIG. 9. Two models for the far-ultraviolet rotatory dispersion curves of the three main types of protein secondary structure. (a) Polypeptide model using poly-L-lysine. (1) The α-helix (poly-L-lysine in water at pH 11) (2) the antiparallel β-structure (poly-L-lysine in water pH 11, heated at 50°C for 12–25 min, then cooled to 22°C) (3) the random coil (poly-L-lysine in water at pH 5). Data replotted from Greenfield et al. (1967). (b) Protein model based on the ORD of five proteins whose structure is known from X-ray crystallography. Curves (1), (2), and (3) are the resolved ORD curves for pure α-helix, β-structure, and unordered structures, respectively, in the proteins. The data is that of Chen et al. (1972) and has been replotted after multiplying each value by Lorentz's factor using the refractive index of water at 240 nm. Below 210 nm a smooth line cannot be drawn through the data points for the β-structure and unordered structure. [Reprinted from Greenfield et al. (1967). Biochemistry **6**, 1630 and Chen et al. (1972). Biochemistry **11**, 4120. Copyright (1967 and 1972) by the American Chemical Society. Reprinted by permission of the copyright owner.]

$([m']_{233} = -14{,}600)$, a shoulder at 210 nm, and a peak at 198 nm $([m']_{198} = 63{,}000)$. The random coil shows a trough at 205 nm $([m']_{205} = -21{,}700)$ and a peak at 190 nm $([m']_{190} = 18{,}500)$. The $\beta$-structure shows a trough at 230 nm $([m']_{230} = -6200)$ and a peak at 205 nm $([m']_{205} = 27{,}000)$.

Using these curves the equation corresponding to Eq. (34) becomes

$$x^{\mathrm{H}} = \frac{-\{[m']_{233}\ (\text{native}) - [m']_{233}\ (\text{denatured})\}}{12{,}200} \tag{37}$$

for proteins containing only helix and coil, using $[m']_{233}$ (denatured) = $-2400$. The curves computed for varying amounts of helix and coil also showed that the trough at 233 nm decreases in size but does not shift in going from 100% helix to 100% coil. Sarker and Doty (1966) have also shown from studies with a copolymer of poly-L-lysine with tyrosine that at a level of 4% tyrosine the Cotton effects were the same as for pure poly-L-lysine. However, this may not be of general significance (see Section V).

Estimation of $\beta$-structure content from these curves also presents difficulties. Fasman and Potter (1967) have shown that different polymers, all presumably in the $\beta$-form, show different ORD curves. This may be partly due to the presence of different types of $\beta$-structure (interchain parallel $\beta$, antiparallel $\beta$, or intrachain cross-$\beta$) as well as the larger solvent dependence of $\beta$-structures compared to the $\alpha$-helix. Furthermore, because of the similarity of the ORD for the $\alpha$-helical and $\beta$-structures, $\beta$-structure can only be estimated when the protein does not contain a large amount of $\alpha$-helix. Jirgensons (1969) suggests that a protein containing 80% helix and 20% $\beta$-structure cannot be distinguished from one containing 80% helix and 20% random coil on the basis of its far-UV ORD.

Greenfield et al., applying their curve fitting method, find the best fit to the ORD data for lysozyme with 20% $\alpha$-helix, 32% $\beta$-structure and 48% random coil, although this fit is still well outside experimental error. X-ray results give 35% helix, 10% $\beta$-structure, and 55% random coil. For myoglobin these authors find 54% $\alpha$-helix, 36% $\beta$, and 10% random coil compared with 77% $\alpha$-helix and 23% random coil from X-ray studies. In both cases curves having the composition given by the X-ray structure gave a less satisfactory fit to the experimental data. The conclusions were thus that their procedure overestimates $\beta$-content and underestimates both helical and unordered content when applied to proteins. The main problem was that adding random-coil ORD curves to the $\alpha$-helical curve tended to shift the peak at 199 nm to the blue so that a compensating amount of $\beta$-curve had to be added to shift it back to the red again.

Perhaps the most satisfactory way of using the poly-L-lysine data is to arrive at an estimate of $\alpha$-helical content by the $[m']_{233}$ method and then refine this value, as well as to estimate the content of other structures by comparison of the experimental ORD with the computed curves of Greenfield et al., over the whole available spectral range.

There is still, however, a major objection to this method and that is that the uniform secondary structures found in polypeptide models do not necessarily correspond to the type of structures found in proteins. The structures in polypeptides are expected to have uniform geometry over long chain lengths because of the uniformity in the nature of the side chains. In proteins only short and somewhat distorted structures are expected and in particular the unordered structure in a protein will not have the large degree of freedom of movement that polypeptide random coils or even denatured proteins have.

Chen et al. (1972) (see Section III,D,1 for details) have also calculated the far-UV ORD curves expected for 100% $\alpha$-helical, 100% $\beta$-sheet, and 100% unordered structures from an analysis of the ORD curves of proteins.

Although these resolved curves do bear some resemblance to those obtained from polypeptide models there are significant differences in all the curves, especially in those for $\beta$-structure and "random coil" at the lower wavelengths (see Fig. 9).

The resolved curve for the $\alpha$-helix is very similar in shape to that obtained for $\alpha$-helical polypeptides but is smaller. These authors find $[m]_{233}$ (helix) $= -15{,}300 \pm 550$. When this value is reduced with Lorentz's factor we find that it is about 20% lower than the average value for polypeptides. This result has been predicted on the basis of theoretical calculations on short, distorted helices (see Section IV,C,1) which might be expected to occur in proteins. In fact, Chen et al. showed that the difference between their $[m]_{233}$ (helix), which is for an average of eleven residues in a helical array, and the $[m]_{233}$ for an infinite number of residues in a helical array (the experimental data for poly-L-glutamic acid and poly-L-lysine) was the same as that predicted theoretically by Woody and Tinoco (1967). For any number of residues in a helical array they suggest the following relationship

$$[m]_{233} = [m]_{233}^{\infty} \text{ (helix) } (1 - k/n)$$

where $n$ is the number of residues in the array and $k$ is a constant. Such a relationship can also be found to hold experimentally for oligopeptides of L-lysine (Yarron et al., 1971).

Using the result of Chen et al. the equation corresponding to Eq. (34) becomes

$$x^H = \frac{-\{[m']_{233} \text{ (native)} - [m']_{233} \text{ (denatured)}\}}{9758} \tag{38}$$

where $[m']_{233}$ (denatured) $= -1924$.

Using this equation for the five proteins employed to derive it as well as on another five proteins whose structure is known from X-ray crystallography (insulin, nuclease, cytochrome C, α-chymotrypsin, and chymotrypsinogen A) gave very satisfactory results in terms of helical contents, in most cases better than those obtained from Eq. (37).

For the β-structure Chen et al. show a resolved curve with the ORD minimum located below the 230 nm found for poly-L-lysine at 223 nm. They found $[m']_{233}$ (β-structure) $= -9900$, whereas for poly-L-lysine in the β-conformation (antiparallel) $[m']_{230} = -6000$. At lower wavelengths the Chen et al. curve is much smaller than that for poly-L-lysine. The curve obtained for the unordered structures of proteins was found to be greatly different from that obtained for the randomly coiled polypeptides. Although at 233 nm the difference was not great, ($[m']_{233} = -1925$ for the proteins and $[m']_{233} = -2400$ for the polypeptides) the protein curve is much smaller than that for polypeptides at lower wavelengths.

Using the resolved curves shown in Fig. 9b Chen et al. attempted to find the best fit to the experimental ORD spectra of the ten proteins mentioned before. In most cases the fit obtained was quite good and the proportions of each type of secondary structure thus calculated were usually in reasonable agreement with the values obtained from the X-ray crystallographic data.

The authors also carried out a similar analysis with the CD data of these proteins and found that, in general, the CD analysis provided answers closer to the actual proportions of secondary structures than the ORD did. This was thought to be a result of the curve fitting procedure and the steepness of the ORD curve compared to the CD curve for the α-helix in the 205 to 230 nm range.

Although the results obtained with this approach were encouraging one should be aware of the assumptions behind this method. In the first place, perhaps artificially, it classifies all protein secondary structures into three types: α-helix, β-structure, and unordered structure. No distinction was made between residues occurring in α or $3_{10}$ helical segments or residues occurring in distorted or short helical segments. (Lactate dehydrogenase, for example, contained each of these types of residues.) Furthermore, no distinction was made between parallel, antiparallel, or distorted β-structure. (Again, these different types occur in lactate dehydrogenase and lysozyme.) The authors also realize that the unordered spectra calculated represent a statistical average and may be different for each protein. In Section IV,C we will discuss the differences in the ORD spectra for these different structures. Thus the procedure has been one of averaging out the differences in the structures of the individual proteins and its validity must rest in how representative the averaged structures are of those found in most proteins. Chen *et al.* give a partial answer to this by calculating the resolved $[m]_{233}$ and $b_0$ for the three secondary structures from different sets of three proteins (different combinations of the five proteins used in the final analysis). They find quite large variations among the different sets [for example, $[m]_{233}$ (helix) varied from $-9990$ to $-15,300$ and $b_0$ (helix) varied from $-548$ to $-786$].

Finally, this procedure also averages out nonpeptide Cotton effects (see Section V) the assumption being that these are small compared to the peptide Cotton effects.

Certainly as more and more X-ray results become available the analyses just mentioned should be continually repeated to test the generality of the results. At present it seems that each protein possesses both a unique structure and a unique ORD spectrum. The choice of models for comparison is still an open question.

C. Discussion of the Assumptions and Limitations of the Various Methods for Estimating Protein Secondary Structure from the ORD

The assumptions made when the types and amounts of secondary structure in a protein are estimated from a rotatory parameter have been listed by Carver *et al.* (1966). A given rotatory parameter ($b_0$, $a_0$, $[m']_{233}$, etc.) obtained for the protein is compared with that for model systems assuming that the contribution from each type of secondary structure is proportional to its fractional content in the protein. This procedure implies that the following criteria hold:

1. the peptide bonds of the backbone chain are the only source of optical activity contributing to the rotatory parameter,

2. for residues located within a given ordered segment of the molecule the rotatory parameter is independent of the number of residues in that segment,

3. the rotatory parameter is determined only by the average peptide bond conformation of the array and is insensitive to differences in side chains, to small distortions in geometry away from the ideal or average conformation of the array, and to changes in the local environment which do not reflect conformational changes from one type of structure to another,

4. the experimental error in the determination of the rotatory param-

eter is much less than the maximum possible contribution of each type of structure to the parameter,

5. the structure for which the rotatory parameter is obtained in the model system corresponds to the structure inferred to exist in the protein. This assumption becomes most important when the optical activity of random coils in polypeptides is used as the model for the optical activity of unordered regions in a protein (see Section IV,C,2).

It is clear that assumptions (1) to (5) are serious oversimplifications when rotatory parameters obtained for the protein are compared with those obtained for synthetic polypeptides in different solvents. It is important to realize, however, that the assumptions are not made valid by obtaining rotatory parameters calibrated against the X-ray structures of proteins. The latter process merely serves to average out the parameters over the variations of a particular type of structure in the group of proteins used.

Assumption (1) does not hold for the $b_0$ or the other visible ORD methods when extrinsic Cotton effects occur (e.g., heme in the hemoproteins) or when Cotton effects of the aromatic side chains (and cystine) are large enough to interfere with the visible or far-UV ORD. These will be discussed in more detail in the next section.

The remaining assumptions will be discussed in terms of the optical activity of specific secondary structures.

### 1. The $\alpha$-Helix

Theoretical calculations by Woody and Tinoco (1967) on the rotational strengths of the $\alpha$-helix have predicted that $[m']_{198}$ would be very sensitive to the number of residues in the helical segment, decreasing by nearly 50% in going from the infinite helix to helices only ten residues in length. $[m']_{233}$, which primarily reflects the rotational strength of the $n-\pi^*$ transition of the peptide bond, is much less sensitive to chain length, as observed also by Vournakis et al. (1968), the corresponding change being only 20%. The most insensitive parameters as regards chain length, however, were $b_0$ and $a_1$ (193 nm), the changes being only 15% for these parameters. Vournakis et al. pointed out that the major effect of short helices on reducing $R_{n\pi*}$ or the optical activity per peptide bond is the "end-effect" since short segments within an infinite helix gave rotational strengths much closer to that of the infinite helix. Madison and Schellman (1972) also concluded that $[m']_{233}$ or $[\theta]_{222}$ were the least sensitive to helix length and suggested that the choice of $\lambda_0$ at 212 nm (suppressing contributions from the hypersensitive 207 nm band) was the basis for the success of the $b_0$ method.

The reason for the insensitivity of the 222 nm Cotton effect to helix length as compared with the other bands is that the former arises from the one-electron mechanism which depends on short-range interactions only, while the latter depends on the entire exciton array because exciton interactions have a much less steep dependence on distance. Litman and Schellman (1965), however, point out that, for this very reason, residues which are not in helical arrays but nevertheless have α-helical ($\varphi$, $\psi$) angles may still make (smaller) contributions to the $n-\pi^*$ Cotton effect at 220 nm. This is a potential source of error in estimates of helical content in proteins.

Vournakis et al. also calculated $R_{n\pi*}$ of the infinite α-helix as a function of the exact conformation of the backbone chain [as expressed by a uniform set of ($\varphi$, $\psi$) angles for each residue] and the nature of the side chain. They showed that almost the whole variation of $R_{n\pi*}$ with conformation was due to contributions from the perturbations of the N and H atoms of the peptide bond. There was, in fact, a good correlation between the magnitude of $R_{n\pi*}$ and the energy of the peptide hydrogen bond, both of these being determined by the degree of nonlinearity of the peptide hydrogen bond. The main effect of side chains on $R_{n\pi*}$ was through their influence in determining which conformation the backbone would take up but also through their contribution to the perturbation field of the $n-\pi^*$ transition. Litman and Schellman (1965) have also demonstrated the dependence of $R_{n\pi*}$ on the conformation of cyclic dipeptides (see, also, Bayley et al., 1969). Very small distortions of the conformation (a few degrees in $\varphi$ or $\psi$; can have very large effects on $R_{n\pi*}$.

On this basis Schellman and Lowe (1968) have suggested an interpretation of the CD and ORD data for ribonuclease. In ribonuclease the $n-\pi^*$ Cotton effect is centered at 217 nm instead of 222 nm for the α-helix. The CD and ORD spectra cannot be fitted with any combination of the Greenfield and Fasman curves. However, if the spectrum for the α-helix is shifted 5 nm to the blue then the combination of the curves in accordance with the X-ray results gives a good fit to the experimental curve. They suggested this blue-shift was due to the presence of short helices in ribonuclease causing exposure of the carbonyl group at the carboxyl end of the helix to the surroundings. Also, end distortion effects causing nonlinearity of the N—H· ·O—C hydrogen bond means that the $n$ orbital on the oxygen atom can partially participate in the hydrogen bond, resulting in a blue-shift of the $n-\pi^*$ transition. Nagy and Strzelecka-Gołaszewska (1972) have offered a similar explanation for the smaller blue-shift they observe for the $n-\pi^*$ transition in G-actin.

Goodman and Rosen (1964) have also given experimental evidence that rotatory properties depend on chain length from a study of carbobenzoxy glutamate oligomers and Epand and Scheraga (1968) have observed that the $n$–$\pi^*$ Cotton effect for an L-valine oligomer sandwiched between two D,L-lysine oligomers is shifted about 2 nm to the blue. In 98% methanol only about nine residues of the valine block were found to be in the right-handed $\alpha$-helical form, this short and distorted helix being judged the cause of the shift.

Furthermore, calculations on the helical segments in myoglobin and lysozyme from the X-ray coordinates indicate that the ORD for the short and irregular helices is only in qualitative agreement with the ORD for idealized helices (Madison and Schellman, 1972). All these results suggest that there may be no general model for the $\alpha$-helix in proteins since the optical activity of helical segments in proteins can vary considerably with helix length, conformation, nature of side chain, etc. It would appear that careful measurement of the wavelength of the $n$–$\pi^*$ transition should accompany measurements of the rotation at the extremum in estimates of helix content. Even then only general statements of the helix content rather than specific estimates can be made. $[m']_{233}$ and $b_0$ or $a_1$ measurements, give better average helical contents than do measurements at other wavelengths. These methods are also probably better than curve fitting the whole far-UV ORD or CD since the 207 and 198 nm Cotton effects are hypersensitive to helix geometry while the 233 nm Cotton effect is not. However, the disparity between the various Cotton effect sizes and the estimates from visible rotatory dispersion may yield valuable information on the types of helices present in the protein. The ORD of "perfect" helices should be relatively independent of solvent because the tightly packed helix itself provides the uniform local environment for the $n$–$\pi^*$ transition. This is not the case for distorted helices which will be more environment dependent.

## 2. *The Random Coil*

The fraction of protein structure which is not in a recognizable segment of ordered structure does not necessarily correspond to statistical random coils or even the random-coil structures of synthetic polypeptides. Although it is not ordered in a regular array it is still held fairly rigid by the other units of ordered structure in the protein and has a much lower entropy than a truly random coil.

Recent calculations on the optical activity of aperiodic or "random-coil" structures have been less successful than calculations on the ordered regions of polypeptides. Aebersold and Pysh (1970) combined optical theory with configurational statistics to determine population averaged rotational strengths of randomly generated oligopeptides of alanine. They considered only exciton interactions and could not obtain agreement with experiment unless the favored regions in the conformational energy map were changed from those usually accepted as such. On the other hand, Zubkov et al. (1971) have used statistical procedures to obtain convergence of their CD calculations for random oligomers and have considered both exciton and nonexciton components of the optical activity. Again, good agreement with experiment was not possible (the calculated curves were blue-shifted by 10–15 nm relative to experiment) but when the region of conformational space was restricted to a forbidden region the resulting CD curve bore no resemblance at all to experimental curves.

Ronish and Krimm (1972) have calculated the CD of random oligopeptides by summing dipeptide CD spectra. The dipeptide CD spectra were calculated from a Boltzmann distribution over the conformational energy map, the CD calculations for each $(\varphi,\psi)$ angle (10° intervals) including consideration of both $n-\pi^*$ and $\pi-\pi^*$ transitions. Although the sum of the dipeptide spectra showed consistent differences with the spectra calculated directly for the equivalent oligopeptides (the summed dipeptide spectra were smaller in magnitude and slightly red-shifted) the basic features were the same. Their calculated CD spectrum for the random coil disagrees with the previous calculations in not exhibiting a long-wavelength positive band but only a negative band at $\sim$213 nm and a positive band at $\sim$195 nm. This spectrum is more in agreement with their experimental assignment of the random-coil CD spectrum to that of poly-L-proline in concentrated aqueous $CaCl_2$ solution (Tiffany and Krimm, 1968) and other systems. These authors believe that charged poly-L-glutamic acid or charged poly-L-lysine do not exist as true random coils but as extended helices (Krimm and Mark, 1968).

The lack of agreement of the above calculations with the experimental optical activity of disordered structures in polypeptides and denatured proteins emphasizes the difficulty of predicting the properties of nonperiodic structures which have freedom of motion. The difficulty may lie in the statistical weighting and averaging of conformations (through energetic considerations), in the choosing of wave functions and spectral parameters or in the incompleteness of the actual calculations in considering all possible mechanisms.

The unordered regions of native proteins, however, do not have the freedom of motion required to establish the weighted statistical average of the optical activity. Because of the more or less fixed conformation of residues in these regions it is interesting to compare the predicted optical activity with the experimental. Madison and Schellman (1972) have done this for the unordered regions in chymotrypsin, lysozyme, and ribonuclease S using the X-ray coordinates. They found that although the predicted curves were nearly all similar they did not even qualitatively resemble those found for disordered polypeptides or denatured proteins. Furthermore, by subtracting out the β-structure and α-helix components from the experimental CD spectra for the whole proteins they were able to show that their calculated curves did not agree with these either. (The predicted curves contained a small negative band at about 220 nm and a large positive and negative band at 200 and 180 nm, respectively, while experimentally there was a small positive band at 220 nm and a large negative one at 200 nm). They suggest this may be due to the influence of environmental effects since these regions of the protein, if fairly rigid, can interact strongly with their local specific surroundings. This would be a possible explanation for the larger optical activity displayed by these regions than is expected theoretically. However, this argument does not apply to the α-helix and β-structure because the ordered packing and internal hydrogen bonds supply their own consistent environment. In fact, because of the lack of regular repeating units in the unordered segments of a protein it is possible to obtain any type of ORD spectrum or to observe a cancellation of rotational strengths. Nagy and Strzelecka-Gołaszewska (1972), on the basis of curve fitting to Greenfield curves, have suggested that 60% of the residues in G-actin contribute nothing to the optical activity of the protein (as opposed to a 60% random-coil contribution).

It is clear, therefore, that the "random-coil" part of a protein structure is the most difficult to evaluate. Usually the recognizable α-helical contributions and β-structure contributions to the optical activity are estimated and the random coil estimated by subtraction. However, where the three contributions overlap considerably, random-coil contributions may seriously affect the estimations of ordered structures.

### 3. *The β-Structure*

Theoretical calculations on the optical activity of the β-conformation have been carried out by Pysh (1966, 1970a), Rosenheck and Sommer (1967), Zubkov and Vol'kenshtein (1967a,b), Urry (1968a), Woody (1969), and Madison and Schellman (1972). For the much more common antiparallel β-structure most of the calculations agree fairly well with experimental spectra.

Woody also considered two different proposed conformations of the β-structure, namely, that of Fraser and MacRae (1962) for β-keratin and that of Arnott et al. (1967) for poly-L-alanine in the β-conformation. The first point that Woody makes is that the average wavelength of the $\pi-\pi^*$ exciton system is shifted from 195 nm to about 189 nm in going from the α-helical to the β-conformation. He attributes this to the different environment of the $\pi-\pi^*$ transition with respect to the dipole field of the polymer.

A very interesting feature of his calculations is that the $x$-polarized component of the $\pi$–$\pi^*$ transition is very sensitive to the width of the $\beta$-sheet. Since this component is responsible for most of the absorption intensity of the $\beta$-structure the wavelength maximum of the absorption spectrum is found to shift from 189 nm to 196 nm on increasing the number of strands in the antiparallel $\beta$-structure from one to six. This prediction was borne out by the experimental absorption spectra of Quadrifoglio and Urry (1968) for low-molecular-weight poly-L-serine ($\lambda_{max}$ = 187 nm) and of Rosenheck and Doty (1961) for high-molecular-weight poly-L-lysine ($\lambda_{max}$ = 194 nm).

The rotational strengths also depend on the number of residues in each strand of the $\beta$-sheet, up to eight residues in length, the dependence being different for the $x$, $y$, and $z$ components of the $\pi$–$\pi^*$ system.

The $n$–$\pi^*$ transition of the antiparallel $\beta$-structure had a negative rotational strength and occurred at about 220 nm. The rotational strength of this transition increased in magnitude as the number of residues in each strand increased but decreased in magnitude when the number of strands in the structure increased.

On the basis of these calculations Woody suggests that all the polypeptide models known to be in the $\beta$-structure have the antiparallel conformation. The proteins lysozyme and ribonuclease contain antiparallel $\beta$-structure; subtilisin BPN' contains some parallel $\beta$-structure; and carboxypeptidase A contains a segment of $\beta$-structure which runs in both parallel and antiparallel senses.

Calculations on the parallel $\beta$-structure (Woody, 1969) also suggested a means of distinguishing between parallel and antiparallel structures. In the antiparallel form the strongest absorption band has a position depending on sheet width but it is always associated with a small positive CD band. In the parallel form the major absorption band is always blue-shifted relative to its large negative CD band. Furthermore, the difference in $\lambda_{max}$ of the absorption and CD bands is much more sensitive to sheet width for the parallel (maximum $\Delta\lambda$ = 13 nm) than the antiparallel structure (maximum $\Delta\lambda$ = 5 nm).

Although there is qualitative agreement between the predicted optical properties of $\beta$-structures and experiment, Madison and Schellman (1972) have pointed out that the calculations have all been carried out on ideal $\beta$-structures of varying width and length. The $\beta$-structures in the proteins mentioned above are very irregular and distorted and vary greatly in width and length. When this disorder is considered there are strong qualitative and quantitative changes in the shape of the ORD and CD spectrum. In fact Madison and Schellman (1972) suggest that the two-fold screw axes in an antiparallel $\beta$-array cause a considerable cancellation of rotatory strength. Thus distortions which destroy this symmetry have large effects. Experimental curves do in fact show a wide range of dependence on the solvent used and the side chain of the polypeptide (Iizuka and Yang, 1966; Stevens et al., 1968; Timasheff, 1970).

The above calculations have helped clarify many points which would be very difficult to obtain experimentally. All the calculations show that the assumptions made in calculating secondary structure from ORD are not strictly valid. When the array is small, the ORD for each type of secondary structure depends on the size of the array and on the exact ($\varphi$, $\psi$) angles of the residues. These calculations have not shown that it is impractical to use ORD for characterizing protein structure but have paved the way to a more critical examination of ORD and CD spectra. Criteria such as the wavelength of the Cotton effects, their relative sizes, and even their number which, in the past, have been ignored may now reveal the presence of short and distorted helices and distorted $\beta$-sheets, and may distinguish between parallel and antiparallel $\beta$-sheets and so forth. Certainly the calculations indicate that estimates of helix, $\beta$-, and "random" structures are unlikely to be accurate to more than 10–20% but they also reveal which parameters are least sensitive to specific effects and suggest means for obtaining a much more critical qualitative picture of protein structure.

## V. Cotton Effects Due to Side-Chain Chromophores

In Section IV we discussed the ORD of the peptide chromophore and how this relates to the conformation of the backbone chain. However, proteins also contain aromatic side chains which can display optical activity. This optical activity can in some cases hinder the estimation of secondary structure in proteins but it can also provide specific information about the conformation of particular side chains in the protein.

Side-chain optical activity has occasionally been picked up as irregularities in the ORD curve in the near UV (250–300 nm) where aromatic residues absorb light. Schellman and Schellman (1956) observed such small irregularities in the ORD curve of bovine serum albumin near 290 nm and similar Cotton effects were observed in human serum albumin (Jirgensons, 1962). Simmons and Blout (1960) observed a similar peak for tobacco mosaic virus protein near 290 nm which disappeared upon denaturation. Although Urness and Doty (1961) had suggested that such effects could be due to artifacts caused by strong absorption by aromatic residues in this region, Myers and Edsall (1965) showed, by

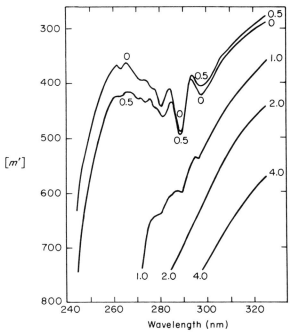

Fig. 10. Rotatory dispersion curves for carbonic anhydrase B in the native state at pH 7.0 (curve marked 0) and in guanidine hydrochloride solutions at molar concentrations indicated by the numbers adjoining the curves. All solutions were in 0.1 $M$ phosphate buffer and the guanidine hydrochloride solutions had stood at 0° for 48 hr before the measurements were made. (From Myers and Edsall, 1965.)

varying the concentration and light path, that at least in carbonic anhydrase the Cotton effects were real (Fig. 10). They attributed the positive Cotton effect with peak and trough at 293 and 289 nm, respectively, to tryptophanyl residues and the one with peak and trough at 285 and 280 nm to tyrosyl residues. Several small Cotton effects between 260 and 270 nm probably arise from phenylalanyl residues.

Tryptophan, tyrosine, phenylalanine, and cystine may display optical activity in the near UV as well as in the far UV along with methionine, histidine and, of course, the peptide bond. Tyrosine shows absorption maxima at 275, 222, and 193 nm, the bands at 222 and especially 193 nm being very intense. Tryptophan shows bands at 280, 220, and 195 nm, the last two bands also being intense, although not as strong as the tyrosyl 193 nm band. Phenylalanine shows bands at 257, 206, and 188 nm, the 188 nm band being even more intense than the tyrosine 193 nm band. Methionine and histidine show relatively weaker bands at 205 and 211 nm, respectively. Cystine shows bands at 250 and 210 nm, both relatively weak.[11]

It is clear that many of these bands will overlap in proteins so that CD will be the preferable technique for resolving the component Cotton effects. This has been the method used by Horwitz et al. (1970) in low-temperature resolution studies of model compounds and proteins containing aromatic chromophores. Thus, specific aspects of side-chain optical activity will be left to the chapter on circular dichroism, and we will only make some general comments.

The protein chromophores are asymmetric by virtue of their tetrahedral α-carbon group and will display optical activity even as the isolated amino acids. Cystine is a special case in that a second asymmetric center is present, since disulfides can exist as right-handed or left-handed screws due to restricted rotation about the —S—S— bond. These intrinsic rotations of the aromatic amino acids can be modified in the protein through interaction of the side-chain transition moments with the specific environment in the protein or the peptide transition moments in the α-helical backbone. We would thus expect that those residues that are fixed in geometry would experience the largest induced Cotton effects, although we cannot predict, without knowing the specific three-dimensional structure, how this will modify the intrinsic rotation of the residues (since complex arrangements of charges in the asymmetric field can completely alter the intrinsic optical activity of the residue or even cancel it).

A. Effect of Side-Chain Optical Activity on Estimates of Helical Content

The presence of large Cotton effects due to aromatic transitions may present serious problems in the estimation of secondary structure from the far-UV ORD and also the visible ORD. This is especially true for ORD because of the diffuseness of ORD Cotton effects. However, a number of model polypeptide studies have helped clarify this problem (e.g., see Fasman et al., 1964, 1965; Shiraki and Imahori, 1967).

Poly-L-tyrosine, in the helical form, shows multiple Cotton effects with two peaks at 286 and 254 nm and a trough at 238 nm (Beychok and Fasman, 1964). In a thorough theoretical study on poly-L-tyrosine Chen and Woody (1971) were able to reproduce the main features

[11] In addition to the above residues, ionized carboxyls (aspartic and glutamic acids) show a weak band at 200 nm and asparagine and glutamine show typical amide spectra with bands at 225 nm and 190 nm. Finally, lysine and arginine show weak bands at 213 nm.

of the far-UV CD of this polymer only if it existed in the right-handed
α-helical conformation (see also Chapter 23, this volume, Section II,G,2).
Thus, the 233 nm trough characteristic of the right-handed α-helix has
shifted to 238 nm. Nonhelical poly-L-tyrosine at the same pH (11.2)
displayed a positive Cotton effect at 225 nm. Beychok and Fasman were
in fact able to resolve the CD for helical poly-L-tyrosine into four component Gaussian bands at 270 and 248 nm for tyrosine and at 224 and
190 nm for the α-helical Cotton effects. Kronig-Kramers transformations
of these components into ORD curves yielded a resultant ORD curve
which agreed with experiments. This then indicated that the two
anomalous properties of poly-L-tyrosine in the α-helical form—a smaller
than usual trough shifted to 238 nm and a positive $b'_0$—could be explained simply by adding positive Cotton effects that correspond to the
tyrosyl chromophore.

Damle (1970) has also presented ORD and CD data for poly-L-tyrosine in trimethyl phosphate showing, from polarized infrared spectra, that
the conformation of the polymer in this solvent is α-helical. In the
study mentioned above, some of the tyrosyl residues may have been
ionized and influenced the optical activity. Damle found that in his
system there was evidence for side-chain–side-chain interactions and obtained similar curves to Beychok and Fasman.

Goodman and Tonioli (1968) have carried out a CD study of poly-$O$-carbobenzoxy-L-tyrosine in trimethyl phosphate and also found large
Cotton effects at 220 and 199 nm attributable to tyrosine transitions.
These shifted the α-helix CD Cotton effect from 222 to 232 nm.

According to the work of Fasman et al. (1964) on copolymers of
L-tyrosine and L-glutamic acid, aromatic groups in proteins would not
be expected to display significant Cotton effects if these groups comprise
less than 10% of the total amino acid residues. The ORD of poly-L-tryptophan films revealed two positive peaks at about 280 nm and a
large negative trough at 233 nm (Fasman et al., 1965). Again, Moffitt-Yang plots of the visible ORD of this polymer give a positive $b'_0$. X-Ray
studies of poly-L-tryptophan films (Peggion et al., 1968) indicated that
this polymer also exists in the right-hand α-helical conformation. Furthermore, copolymers of γ-ethyl-L-glutamate and L-tryptophan in ethylene
glycol monomethyl ether reveal a gradual change in the CD, as the
tryptophan content is lowered, toward that typical for the α-helix. Below 30% tryptophan content the CD curve was not distorted but the
ellipticity was small. The results also indicated that interaction among
chromophores could occur in the helical arrays and this could enhance
the optical activity of the side chains (see, also, Cosani et al., 1968).

Poly-L-phenylanine is insoluble in water but a block copolymer of

D,L-glutamic acid–L-phenylalanine–D,L-glutamic acid is soluble and showed Cotton effects in the region 246–258 nm (Sage and Fasman, 1966). This polymer also showed dextrorotation at the higher wavelengths and a small negative trough at 237 nm. It should be remembered in this context that although the near-UV absorption bands of phenylalanine are very small the far-UV bands are very intense and the CD of model compounds shows positive Cotton effects (see Chapter 23, this volume, Fig. 22).

Poly-L-histidine at low pH, where it shows properties of the random coil, displays a positive Cotton effect at about 220 nm. However, in the helical form at pH 5.78, this becomes a negative Cotton effect. Thus the $n$–$\pi^*$ Cotton effect of the $\alpha$-helix may overshadow the side-chain Cotton effects (see, e.g., Beychok, 1967). This probably explains why the levorotation increases in the coil-to-helix transition of this polymer (Norland et al., 1963).

Coleman and Blout (1967), in an extensive study of the optical activity of the disulfide bond, showed that in the cyclic disulfide-containing peptides, arginine vasotocin, and 8-L-ornithine vasopressin, the ORD in the far UV was dominated by a strong negative Cotton effect at 200 nm which was assigned to cystine. Similar Cotton effects were found in model compounds containing cystine. However, with copolymers of glutamic acid and cystine containing up to 10% half-cystine (inter- and intrachain), no distortion of the usual $\alpha$-helical far-UV ORD spectra or evidence for cystine Cotton effects at 200 nm were found, although there were consistently negative Cotton effects in the near UV.

In general, most derivatives of tyrosine, phenylalanine, histidine, and probably tryptophan, whether monomers or polymers, show positive Cotton effects at about 200 nm. The size of these Cotton effects, however, may vary in different situations. It would seem that these contributions would, in general, cause an underestimation of helical contents and would create the most serious errors when there was little $\alpha$-helix in the protein. For example, by analyzing the ORD curve of helical poly-L-tyrosine, one can determine $a_i$ and $\lambda_i$ of Eq. (16). By inserting these data into Eq. (17) and rearranging them into the Moffitt equation, we can see how $b_0$ varies with the content of tyrosyl residues. If the tyrosine content were less than 10% the effect on $b_0$ would be of the same order of magnitude as the experimental error in evaluating it. Also, the far-UV ORD spectra for helical polypeptides containing 10% of tyrosine was calculated (Shiraki, 1968), and this curve was the same within experimental error as that for helical polypeptides lacking aromatic side chains.

Nevertheless, in some proteins, greatly enhanced optical activity of aromatic chromophores and low levels of $\alpha$-helical or other ordered structures can give rise to very different ORD curves. For example, the ORD of avidin shows positive Cotton effects centered at 274 and 228 nm, and these disappear when the protein is dissociated into subunits in 6 $M$ guanidine hydrochloride. Streptavidin shows similar Cotton effects, both proteins exhibiting a peak instead of a trough at 233 nm. Furthermore, carbonic anhydrase, which has a chromophoric content of less than 10%, displays marked side-chain Cotton effects which must have a large induced component. Again, denaturation causes the near-UV Cotton effects to disappear and in the far-UV ORD the minimum shifts from 223 nm to 229 nm (Green and Melamed, 1966).

## B. Near-UV Cotton Effects

For many proteins, the Cotton effects due to aromatic groups vanish upon denaturation of the protein. In these proteins, it would seem that the Cotton effects (which are much larger than one would expect from studies on the isolated amino acid model compounds) due to aromatic groups must have a large component induced by the secondary and tertiary structures of the protein. In fact, model polymers, such as poly-L-tyrosine, show a large decrease in the amplitude of aromatic Cotton effects as a result of the helix-to-coil transition (Myers and Edsall, 1965; Fasman et al., 1964). One would thus expect that induced Cotton effects would arise as a result of a tertiary structure in which rotation of the side-chain chromophores is strongly hindered (Hashizume et al. 1967).

It is clear that side-chain Cotton effects do not only depend on secondary structure as for the model polymers but also on the specific three-dimensional arrangement of groups around the chromophore. For example, the Cotton effects in ribonuclease are much larger than one would expect from studies on poly-L-tyrosine and Hashizume et al. (1967) showed that upon denaturation the decrease in amplitude of the tyrosyl Cotton effects does not parallel the decrease in $[m']_{233}$.

Another example of this is given by lysozyme. This protein reveals Cotton effects near 280 nm which probably originate mainly from tryptophan (Glazer and Simmons, 1965; Nalper et al., 1971; Cowburn et al., 1972) and to a lesser extent from ionizable tyrosine residues (Ikeda et al., 1967; Nalper et al., 1971). These Cotton effects disappear on exposure of the enzyme to sodium dodecyl sulfate although there is virtually no change in the 233 nm trough. Furthermore, enhancement of the near-UV Cotton effects in aqueous ethylene glycol or in the presence of the competitive inhibitor N-acetyl-D-glucosamine indicates that some of the residues giving rise to the Cotton effects are at least partially exposed to solvent. The displacement of water or direct interactions which may occur by the treatments just mentioned may cause small conformational changes which enhance the optical activity of the side chains. It becomes clear then that the induced optical activity of aromatic residues disappears on denaturation not only because the secondary structure is destroyed but also because no rigid tertiary conformation exists to interact with the chromophore. This means that it is very difficult to predict the magnitudes and even signs of aromatic Cotton effects in particular proteins even if the secondary structure of the protein is known. The near-UV peaks for tyrosine and tryptophan, as opposed to the far-UV peaks, vary enormously in size and sign between different proteins and work by Strickland's group on model compounds indicates that even different vibrational bands of the same chromophore may differ in sign.

Assignments of near-UV Cotton effects are very difficult because of the overlapping of transitions and the large shifts in wavelength maximum which can sometimes occur. Also the different contributions from different vibronic components of a transition further complicates the assignments. Any assignments to be made for the near-UV are nearly always better from a CD than from an ORD spectrum.

Even from a CD spectrum assignments should be backed up by other studies, such as the pH dependence of the Cotton effect in the absence of conformational changes. The most common use of this is the red-shift that accompanies the ionization of tyrosine. For example, as the pH of a lysozyme solution is raised a new band appears at 298 nm in the CD spectrum and increases in size in proportion to the degree of ionization of the tyrosyl residues. In addition a positive band near 250 nm increases in size. Both these bands are assigned to ionized tyrosyl residues (Ikeda et al., 1967; Halper et al., 1971). Furthermore the 298 nm band appears to be conformation dependent while that at 250 nm persists even at pH 12.6 where the protein is extensively denatured (Ikeda et al., 1967). In this way it is possible to separate out the contributions of tyrosine and tryptophan to the native CD spectrum in the near-UV.

Another method for assigning Cotton effects is to examine the effect of chemical modification of various aromatic chromophores using group-specific reagents. McCubbin et al. (1972) have used tetranitromethane, cyanuric fluoride, $N$-acetylimidazole, 2-hydroxy-5-nitrobenzyl bromide, alkaline hydrogen peroxide–dioxan treatment, and $N$-bromosuccinimide to separate out overlapping Cotton effects of tryptophan and tyrosine in concanavalin A.

Simpson and Vallee (1966) modified the three accessible tyrosyl groups in ribonuclease and found little effect on the near-UV Cotton effects. Dioxan (45%), which should perturb accessible tyrosines, also had little effect, so these authors concluded that major contributions to the Cotton effects come from buried residues.

A problem with these types of studies, however, is that chemical modification of vicinal groups or small conformational changes may have large effects on the induced Cotton effects without indicating which residues are responsible.

Horwitz et al. (1970); Horwitz and Strickland, (1971) have overcome this problem for ribonuclease by resolving the absorption and CD spectra, both at liquid nitrogen temperatures. They have assigned four main vibronic bands to each type of tyrosine in ribonuclease and analyzed the absorption spectrum into three spectral types. The first type corresponds to one tyrosine group with its 0–0 transition at 288.5 nm, the second type to two tyrosines with their 0–0 transitions at 286 nm, and the third to three tyrosines with 0–0 transitions at 283.5 nm (the 0–0 transition is the longest wavelength band). From the X-ray data on ribonuclease and the effects of solvents on model compounds the first spectral type was assigned to a fully buried tyrosine, the second also to buried tyrosines and the third to tyrosines with their hydroxyl groups exposed to solvent.

Resolution of the CD spectra using the same bands indicated that the tyrosine with its 0–0 transition at 288.5 nm contributed 15–20% of the CD spectrum. The two tyrosines at 286 nm contributed nothing and the three exposed tyrosines contributed 35–45% of the spectrum. There was also a large contribution to the CD (40–50%) from the disulfide groups, centered at about 270 nm. All of the contributions were negative in the near-UV. From the X-ray evidence all of the tyrosines in ribonuclease are constrained in unique sites and for the tyrosines at 286 nm their unique conformation apparently causes their rotational strengths to cancel out. In ribonuclease S these authors found that the only change observed relative to ribonuclease A was that the tyrosine at 288.5 nm moved to 286 nm and ceased to contribute to the CD spectrum.

More recently, Strickland (1972) has calculated the near-UV optical activity of the tyrosyl chromophores in ribonucleases S and A. He concludes that the major source of the optical activity is dynamic coupling of the $^1L_b$ near-ultraviolet transitions with $\pi$–$\pi^*$ transitions of other moieties including other tyrosines. The major contribution is, in fact, coupling between two of the exposed tyrosines which also results in addition of a small exciton component to the CD. The rotation of the buried tyrosines is small in ribonuclease S but the altered orientation of tyrosine 25 in ribonuclease A may account for the increased rotation in the CD spectrum since it is then close enough to interact with other aromatic transitions. The inherent disymmetry of the disulfide bonds was also considered and is consistent with the negative contribution experimentally observed in the near-UV. It will be recognized that this theoretical study has been very successful in reproducing the major features of the experimental optical activity of ribonuclease in the near-UV.

Although ribonuclease is a special case in that it contains no tryptophan, this study is probably the most complete analysis of near-UV Cotton effects to date and reveals the special powers of the technique in revealing subtle conformational changes which involve interactions with chromophoric residues. The sign and position of Cotton effects of disulfides are also capable of revealing the dihedral angle of the —S—S— bond (see, e.g., Chapter 23, this volume, their Section IV,D).

In native proteins aromatic residues may show enhanced optical activity which depends on the particular conformation of and around the residues. There has also been some evidence that optical activity of different residues may tend to cancel each other out but on denaturation the side-chain optical activity becomes that of an equivalent mixture of aromatic amino acids. For example, in the case of peroxidase C, Strickland et al. (1970) could find no phenylalanyl fine structure in the near-UV CD spectrum of the native protein even though it contains 23 phenylalanines. However, when the protein was denatured with guanidine hydrochloride the fine structure bands appeared in the CD spectrum and were consistent with the expected contribution from 23 unoriented phenylalanines. Thus, in this case it would appear that the particular orientations of the phenylalanines in the native structure cause a net cancellation of rotational strengths.

We will mention briefly some other nonpeptide Cotton effects which can occur in proteins. The reviews by Ulmer and Vallee (1965) and Vallee and Wacker (1970) discuss the extrinsic Cotton effects of coenzymes in enzymes and metals in metalloproteins in some detail. Extrinsic Cotton effects have also been observed in all hemoproteins studied. Heme itself is symmetric and optically inactive but it becomes optically active on binding to polypeptides (Stryer, 1961) or the apoprotein. Studies with dyes binding to polypeptides (Stryer and Blout, 1961) had suggested that these extrinsic Cotton effects were due to their interaction with the helix, since opposite-sense Cotton effects were obtained when the binding occurred with helices of opposite sense. Furthermore, the Cotton effects disappeared on denaturation. However, the positive Soret Cotton effect for heme in hemoglobin and myoglobin had the wrong sign for interaction with right-handed helices. Furthermore, it is now apparent that some hemoproteins have negative Soret

Cotton effects, although the peptide α-helices are all right handed; see for example, *Chironomus* hemoglobin (Gersonde et al., 1972), lamprey hemoglobin (Lampe et al., 1972), and leghemoglobins (Minasian, Nicola, and Leach, unpublished). In one case at least (that of lamprey hemoglobin) there is even a reversal of the sign of the heme Cotton effect with different ligands (Lampe et al., 1971).

Hsu and Woody (1971) have carried out a theoretical study on the Soret Cotton effects of myoglobin and hemoglobin to explain the cause, sign, and magnitude of the induced optical activity. They conclude that the main source of the induced Cotton effect is the coupling of the heme transitions with nearby protein aromatic transition moments. This has interesting consequences because it means that the reciprocal relations mentioned in Section II,B will hold between any changes in the heme optical activity and aromatic optical activity. Many authors have reported changes in near-UV Cotton effects when different ligands bind to hemoglobin or myoglobin, although there is usually little change in the far-UV ORD. Since different ligands alter the spectral properties of heme (and hence its transition moment) then these must also alter the near-UV Cotton effects in the opposite sense, without necessarily implying any conformational change at all.

This mechanism of induction of optical activity of symmetric molecules need not be the only one and, indeed, cannot be the mechanism for the Cotton effects of dyes bound to polypeptide helices which lack aromatic residues. However, the existence of extrinsic Cotton effects may open up a new approach to probing local conformations of protein structure. Symmetric molecules which are also specific modifying agents, e.g., dinitrophenyl fluoride and 1-dimethyl aminonapthaline-5-sulfonyl (dansyl) chloride, can also be attached to certain sites in the protein and then display optical activity. Such studies may become more meaningful, as will side-chain Cotton effects, when more precise information is available on the relation between sign and magnitude of Cotton effects and the chromophore environment. At present these studies are still useful for detecting changes in environment which would prove very difficult to detect by other techniques.

## VI. Other Secondary Structures

### A. Collagen and Polyproline

The fibrous protein collagen has been shown to possess a highly ordered and rigid structure by the use of electron microscopy, flow birefringence, and other hydrodynamic properties (Boedtker and Doty, 1956).

Cohen (1963) found that over the wavelength range of 589 to 436 nm the dispersion of a collagen solution obeyed the one-term Drude equation

$$[\alpha]_\lambda = \frac{A}{\lambda^2 - \lambda_c^2} \tag{39}$$

with $\lambda_c = 205 \pm 15$ nm; $[\alpha]_D$ at 11°C was about $-350°$. This large negative rotation in the visible region was found to be typical for a variety of native collagens. The structure of collagen has been proposed to be a triple-stranded left-handed helix from the analysis of X-ray diffraction patterns. When a collagen solution is transformed into its disordered form (gelatin) by heating, the specific rotation rises to about $-110°$ while $\lambda_c$ remains essentially unchanged (Cohen, 1963).

Thus the characteristic ORD properties of collagen can be summarized as follows:

1. a large negative $[\alpha]_D$ in the native state,
2. an ORD which obeys a one-term Drude equation,
3. a large decrease in levorotation upon denaturation,

All three properties are quite different from the corresponding properties of the $\alpha$-helix and the $\beta$-structure, and provide a method of identification. The fact that the data fit a one-term Drude equation with $\lambda_c = 205$ nm suggests that the far-UV ORD may be approximated by a single Cotton effect.

Two regular conformations are possible for poly-L-proline. One is termed poly-L-proline II and is a left-handed helix with three residues per turn—the displacement along the chain between successive repeating units involving a rotation about the main chain axis of 120° and a translation along this axis of 3.12 Å (Cowan and McGavin, 1955). The $\alpha$-carbons are taken to be in the trans position relative to the N—$C_\alpha$ bond (Fig. 11). The other form, termed poly-L-proline I, is a right-handed helix containing between three and four residues per turn and a translation of 1.85 Å per residue (Traub and Shmueli, 1963). This form has a cis peptide bond configuration with rotation about the $C_\alpha$—$C'$ bond such that the oxygen atom is nearly trans to the $C_\alpha$ hydrogen. Although poly-L-proline I is not a strict mirror image of poly-L-proline II, the different structures with their opposite helix sense and peptide configuration suggest that the ORD curve of the former should be similar to the mirror image of the latter.

This is verified by the ORD curves presented in Fig. 12 (Fasman and Blout, 1963). The poly-L-proline II ORD curve obeys Eq. (39) with $\lambda_c = 210$ nm in water and 206 nm in other solvents (Gratzer et al., 1963; Blout and Fasman, 1957). It is also characterized by a large

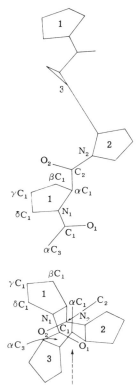

FIG. 11. The conformation of a single chain of poly-L-proline II seen in projection, below, along the fiber axis and above, perpendicular to this axis and along the direction indicated by the arrow. (From Cowan and McGavin, 1955.)

negative specific rotation with $[\alpha]_{546}$ ranging from $-500$ to $-700$ (Kurtz et al., 1958; Blout and Fasman, 1957). Harrington and Sela (1958) have studied changes in the optical rotatory properties during heat denaturation of poly-L-proline II. From the characteristic rotation of form II ($[\alpha]_D = -500$ at room temperature) the rotation decreased abruptly in magnitude to $[\alpha]_D = -100$ close to 60°.

All of these rotatory properties show marked similarities to those of collagen. Furthermore, in the far UV the Cotton effects for poly-L-proline II (Fig. 13a) (Blout et al., 1963) are very similar to those for collagen (Fig. 13b) although the Cotton effects for collagen occur at shorter wavelengths.

We will therefore discuss the optical activity of poly-L-proline II as a model for collagen because it has been extensively studied and analyzed.

The far-UV ORD spectrum resembles a single negative Cotton effect

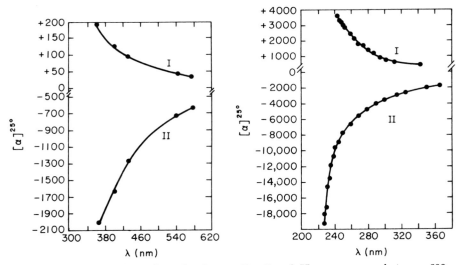

FIG. 12. Rotatory dispersion of poly-L-proline I and II, upper curve between 600 and 360 nm, lower curve between 360 and 230 nm. Poly-L-proline I is 0.5% in propionic acid in a 1 mm cell and poly-L-proline II is 0.14% in water in a 1 mm cell. (From Fasman and Blout, 1963.)

with a trough at 216 nm and a peak at 195 nm centered at about 207 nm. However, Blout et al. (1967) have shown that a double Cotton effect plus nonzero background contributions are required to fit the data. They find a large negative Cotton effect centered at 207 nm and a smaller positive Cotton effect centered at 221 nm. This is in agreement with the two bands seen in the CD.

Theoretical calculations on poly-L-proline II predict a splitting of the $\pi-\pi^*$ transition so that a parallel-polarized component would produce a positive Cotton effect at 206 nm and a perpendicular-polarized component would produce a large negative Cotton effect at 191 nm. This does not agree at all with experiment. However, Madison and Schellman (1970) showed that reasonable agreement with experiment could be obtained if small variations in the conformation were allowed. The conformations which agreed with the experimental ORD data were similar to the structure of computed minimum conformation energy and differed from the conformation determined by X-ray diffraction studies on fibers. It has been pointed out, however, that the X-ray results did not unequivocally define a unique structure (see Chapter 23, this volume, Section II,G,4).

Pysh (1967) has also shown that a small positive $n-\pi^*$ rotational strength can be produced by poly-L-proline II where every third residue is a glycine. From the position of the observed 221 nm Cotton effect this

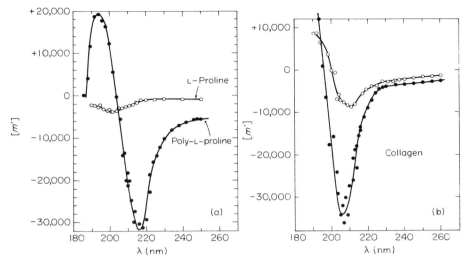

FIG. 13. (a) The far-ultraviolet rotatory dispersion of poly-L-proline II (●) and L-proline (○) in water. (b) The far-ultraviolet rotatory dispersion of native calf skin collagen (●) in 0.01 M acetic acid and the same preparation heated at 50°C for 30 min, cooled to 25°C, and measured immediately (○). The latter preparation is at least partly denatured. [Reprinted from Blout et al. (1963), J. Am. Chem. Soc. **85**, 644. Copyright (1963) by the American Chemical Society. Reprinted by permission of the copyright owenr.]

would appear to represent a positive $n-\pi^*$ rotational strength in contrast to the results for the $\alpha$-helix.

The similarity of the Cotton effects for poly-L-proline II and collagen suggests that the latter's ORD may have a similar explanation. At any rate, the unique rotatory properties of collagen furnish an easy method for following changes in its structure during any physical and chemical treatments.

From a theoretical point of view, it is interesting to see if the singular ORD of collagen results from the three stranded structure or merely from the conformation of the individual strands. Poly-L-proline II is a good model for collagen, as we have mentioned, because two thirds of the total amino acids in collagen are proline or hydroxyproline. However, since poly-L-proline II does not contain amide hydrogen atoms, there is no possibility that hydrogen bonds like those that stabilize the three-stranded helix can be formed. Thus poly-L-proline II should exist as a single-stranded helix. This already suggests that the major contributions to the optical activity of collagen arise from the strands themselves and not from the triple helix structure which is right handed. Furthermore, the calculations by Madison and Schellman (1970) on (Gly-Pro-Gly)$_8$ in different conformations indicated that the major features of the optical activity were the same whether individual chains or the associated chains in the triple helix were considered. There was also very little effect of chain length.

For poly-L-proline I the data in Fig. 12 indicate a positive Cotton effect. Blout and Fasman (1958) reported that this dispersion did not obey a one-term Drude equation or the Moffitt equation but Harrington and Sela (1958) reported that a one-term Drude equation was adequate. In propanol, poly-L-proline I showed a negative Cotton effect (Blout et al., 1963) but in thin oriented films it showed a positive Cotton effect (Blout and Schechter, 1963). It is now accepted that this polymer shows an ORD spectrum in the far UV that resembles a positive Cotton effect with a peak and trough at 221 and 204 nm, respectively, and a cross-over point at about 211 nm (Timasheff et al., 1967).

The calculations of Pysh (1967, 1970b) and Madison and Schellman (1972) on this structure are in agreement with experiment. The transition is again split through exciton interactions to give a low-wavelength negative Cotton effect at 190 nm and a positive Cotton effect at 209 nm. The parallel-polarized component is predicted to give rise to most of the absorption intensity and this is also in agreement with experiment, the absorption maximum occurring at 210 nm. Linear dichroism measurements indicate that this band is polarized parallel to the helix axis. The $n-\pi^*$ transition is predicted to give a weak negative rotational strength.

## B. The Screw Sense of the $\alpha$-Helix

In the previous sections the estimation of helical content by the various methods (including $[\alpha]_D$, $\lambda_c$, two-term Drude equation, Moffitt equation, and $[m']_{233}$ methods) was based upon the assumption that the protein contains only one type of helix, namely the right-handed $\alpha$-helix. If a protein contains both right- and left-handed helices it is not possible to determine the absolute amount of helix from ORD alone. If the left-handed helix is a mirror image of the right-handed helix the optical rotations should be equal in magnitude but opposite in sign. Thus the apparent fraction of helix ($X_{app}{}^H$) which is the real value obtained by the various methods described before is simply given by

$$X_{app}{}^H = X_R{}^H - X_L{}^H \qquad (40)$$

where $X_R{}^H$ and $X_L{}^H$ represent the fractions of right-hand and left-hand helices, respectively. Strictly speaking Eq. (40) is not valid since Tinoco and Woody (1960) have pointed out that the interactions of L-residues in a right-handed and left-handed helix will be different and may not cancel.

However, it is certain that the helical content will be underestimated by all the ORD methods if the sample contains left-handed helices as well as the right.

Since infrared and far-ultraviolet absorption techniques do not discriminate between left- and right-handed helices they may be used to obtain the sum of helical contents of both senses if the presence of left-handed helices is suspected. ORD studies should then reveal the amounts of each type of helix sense.

The experimental data to date, however, show that right-handed helices are overwhelmingly preferred to left-handed ones when polymers of L-amino acids take up a helical conformation. For example, Blout *et al.* (1957) first studied the D- and L-copolymers of γ-benzylglutamate. The rotations increased with the introduction of small fractions of D-isomers into L-polypeptides as long as they measured these samples in chloroform, a helix-promoting solvent. On the other hand, in dichloracetic acid, a coil-promoting solvent, the incorporation of D-isomers reduced proportionally the magnitude of the configurational rotations. This is consistent with the idea that the helices of L-isomers retain their right handedness even with the inclusion of a small amount of D-isomers.

Work on the stepwise synthesis of oligopeptides can also be understood in terms of the formation of right-handed helices for the pentamer or higher polymers of L-amino acid residues (Goodman *et al.*, 1962). As has been cited above, helical L-polypeptides always display a trough at 233 nm and a peak at 198 nm having approximately equal relative magnitudes among the several polypeptides so far studied which strongly confirms that right-handed helices are favoured by L-amino acid polymers. Lastly, all the *proteins* so far studied by X-ray analysis reveal the presence of only right-handed helical segments.

One of the few examples of left-handed helical conformations is that described by Bradbury *et al.* (1962a), who showed that poly-β-benzyl-aspartate can take up the left-handed ω-helix conformation. Although this conformation could be stabilized in proteins by intermolecular packing there has been little evidence for the existence of such structures to date. All these data suggest that the possibility of left-handed helices occurring in proteins and causing errors in helix estimations by ORD methods is remote.

### VII. Conformational Transitions

Because ORD Cotton effects show measurable rotation over a very large wavelength range the rotation at any one wavelength represents contributions from all the secondary structures in the protein. This means that the rotation at a single wavelength (usually in the visible) is often more sensitive to *any* conformational change involving secondary structure than the circular dichroism at a single wavelength. Thus, in studies which follow the transition itself in detail and under different conditions the rotation at one wavelength may be conveniently used rather than the CD. For studies of the nature of the conformational transition it is preferable to study the far-UV Cotton effects in more detail.

The most thoroughly characterized conformational transition, both theoretically and experimentally, is the helix-to-coil transition which occurs with polypeptides. The first study of this type on synthetic polypeptides was that of Yang and Doty (1957) who showed that poly-$\gamma$-benzyl-L-glutamate undergoes a sharp helix-to-coil transition as the volume percent of dichloroacetic acid in ethylene dichloride increases to 80% (at room temperature). This transition could be followed both by changes in $[\alpha]_D$ and in the intrinsic viscosity.

For water-soluble polypeptides, such as poly-L-glutamic acid, the helix–coil transition may be induced simply by changing the pH (Doty *et al.*, 1957). This transition was centered near pH 6 (again both by $[\alpha]$ and by intrinsic viscosity) and corresponded to the midpoint of the carboxyl group ionization. Incorporation of amino acid residues with hydrophobic side chains into polyglutamic acid shifted the transition pH to the alkaline side. With respect to this observation it is interesting that addition of organic solvents, such as dioxan, to aqueous solutions of polyglutamic acid also shifts the pH of the transition to the alkaline side reaching a value near 8 at 50% v/v dioxan (Iizuka and Yang, 1965). These effects seem to be primarily related to solvent effects on the p$K$ of the carboxylate groups. When two opposing ionizations are involved as with the helix–coil transition of copolymers of glutamic acid and lysine the effect of pH is more complicated (Doty *et al.*, 1958).

The transition from coil to $\beta$-structure has been observed for poly-$O$-acetyl-L-serine (Fasman and Blout, 1960), poly-$O$-benzyl-L-serine (Bradbury *et al.*, 1962b) and silk fibroin (Iizuka and Yang, 1966). These changes are brought about by changing the solvent composition. It has also been reported that helical poly-L-lysine at pH above 11 can be converted to $\beta$-structure by heating at 50°C for a short time (Sarkar and Doty, 1966; Davidson *et al.*, 1966; Townend *et al.*, 1966). The coiled form of poly-L-lysine at pH 7 can also be converted to the $\beta$-structure if a small amount of sodium dodecyl sulfate is added (Sarkar and Doty, 1966). The presence of $\beta$-structure in these cases was established by infrared spectral studies.

Although ORD is very sensitive to changes in secondary structure, it is less sensitive to changes in higher-order structures. Thus, the use of other methods which do reflect changes in tertiary or quaternary structure should accompany ORD measurements wherever possible. Among such methods we may include ultraviolet difference spectroscopy near 280 nm, infrared spectroscopy, group-specific chemical modifications, sedimentation in the ultracentrifuge, light scattering, and perhaps intrinsic viscosity. An example of such a study is that of Hermans and Scheraga (1961) in which the thermal transition of ribonuclease A was studied at various pH's using both difference spectroscopy at 287 nm and optical rotation measurements at 436 nm. Although the thermal transition temperature varied markedly with pH, both techniques showed the same transition temperature at a given pH, and the thermal melting profiles obtained by the two methods agreed quite well.

A similar but more extensive study has been carried out by Bello (1969). He added various amounts of polyols to aqueous solutions of ribonuclease A and from ultraviolet difference spectroscopy, circular dichroism, and optical rotation studies was able to observe thermal transitions at all concentrations of polyol up to 97%. Ethylene glycol lowered the melting temperature at all its concentrations but glycerol raised the transition temperature when its concentration was lower than 80%. At higher concentrations of glycerol the transition temperature began to decrease. At all concentrations of polyols the melting temperatures obtained by the three methods were in good agreement. However, there were differences in the shapes of the curves. The melting profiles obtained from UV difference spectra were sigmoid at 0, 75, and 97% ethylene glycol concentrations. However, the melting profiles obtained from the ORD were sigmoidal below 75% glycerol, showed no transition at all at 75% glycerol, and were bell-shaped above 75% glycerol with [$\alpha$] becoming first more positive and then more negative. It should be noted that light scattering measurements of the sample gave a similar bell-shaped curve when plotted against temperature.

In these cases the transition temperature was taken from the descending limb of the bell-shaped curve. At 75% glycerol concentrations the ascending and descending limbs appeared to coincide at the same temperature, thus cancelling each other out and exhibiting no transition at all. Circular dichroism measurements at 277 nm which give similar information to that obtained from UV difference spectroscopy also gave sigmoidal melting curves at all concentrations of glycerol.

Although further work may be required to resolve the above discrepancies it is tempting to assign the first rise in both [$\alpha$] and scattering power (ascending limb of the bell-shaped curve) to a combined conformational change of the main chain and aggregation which has essentially no net effect on the conformational status of tyrosyl residues and the descending limb to a thermal denaturation which exposes all the tyrosines.

In fact, the melting curve obtained for ribonuclease from $[\theta]_{222}$ measurements also showed a biphasic transition while that obtained from $[\theta]_{280}$ was sigmoidal (Hashizume et al., 1967). It is clear therefore that it is possible to discuss conformational changes in much more detail when various physical methods are combined.

Biologically, one of the most important phenomena associated with conformational transitions is the allosteric effect. If the cooperativity of ligand binding is associated with a conformational transition of the protein molecule, then a plot of the conformational change against the ligand concentration should be sigmoidal rather than hyperbolic. Unfortunately, such a curve obtained for hemoglobin, which is a typical allosteric protein, was hyperbolic. However, the conformational change involved here has been shown by Perutz (1970) to involve a change in tertiary and quaternary structure with little change in secondary structure.

A sigmoidal curve has been obtained in the case of glyceraldehyde 3-phosphate dehydrogenase obtained from baker's yeast (YGPD) (Jaenicke, 1968). It is well known that this enzyme shows a sigmoidal curve for the rate and binding plotted against concentration of the coenzyme nicotinamide adenine dinucleotide (NAD). The binding of NAD to the apoenzyme resulted in a change of conformation of the protein which could be observed in changes of $a'_0$, $b'_0$, $[m']_{233}$, sedimentation constant, and viscosity. The change in the S-value reflects a change in shape, both holoenzyme and apoenzyme showing essentially the same molecular weight. When $[\alpha]_{360}$, $b'_0$, or the S-value were plotted against the concentration of NAD, sigmoidal curves were obtained only when the experiments were carried out at 40°C and pH 8.5. These three sigmoidal curves could be fitted to each other and to the kinetic curve by adjusting the ordinate scale. Thus, it could be concluded that the cooperative behavior observed in the kinetics of NAD binding was mediated by conformational changes in the protein.

## VIII. Experimental Considerations

A. INSTRUMENTATION

There are several kinds of automatic recording spectropolarimeters of which the most commonly used is probably the Cary 60. This instrument has been discussed briefly by Yang (1969) and in more detail by Cary et al. (1964). We will refer here to the Jasco ORD/UV-5, a block

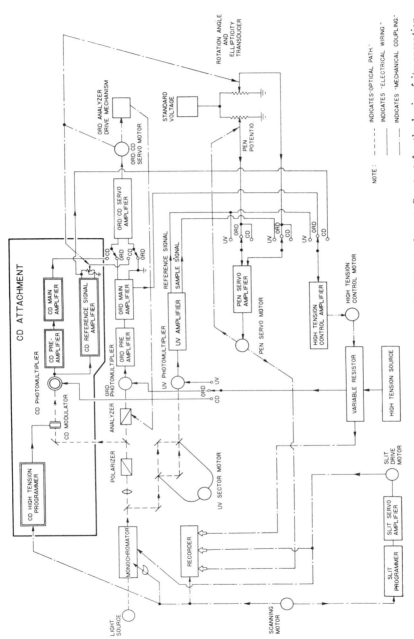

Fig. 14. Block diagram for the JASCO ORD/UV-5 instrument with CD attachment. See text for the basis of its operation.

diagram of which is shown in Fig. 14. Light from a xenon lamp passes through a monochrometer, which can be driven by a synchronous motor at different speeds, is converted into a parallel beam by a lens and then passed through a polarizer. The polarizer is vibrated symmetrically to the left and right about a mean reference position. The light passing through the analyzer thus takes the form shown in Fig. 15. As long as the direction of the analyzer is perpendicular to the reference direction of the polarizer, the light (which is transformed into an electric current by a photomultiplier), follows a sine wave as shown in Fig. 15 (curve $a'$).

If this alternating current is submitted to a synchronous rectifier, the direct current component becomes zero and no servo-output signal appears. However, if an optically active substance is placed in between the polarizer and analyzer the photomultiplier output will vary with the rotation angle $\theta$ as shown in curve $b'$ of Fig. 15. The rectifying output of this signal then produces an output in the servoamplifier and this eventually rotates the servomotor. The servomotor operates so as to rotate the analyzer, yielding a photomultiplier output shown by curve $a'$ of Fig. 15. The rotation angle of the analyzer, which is equal to the angle of optical rotation, is recorded on chart paper with an appropriate mechanical and electrical amplification. The slit opening can be adjusted

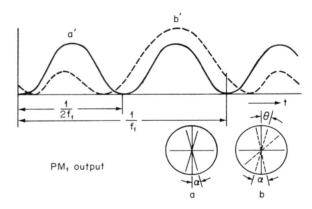

FIG. 15. Output of the photomultiplier in the JASCO ORD/UV-5. The curve marked $a'$ represents the photomultiplier output when there is no sample between the polarizer and the analyzer. In this case the direction of the analyzer [horizontal arrows in (a)] is perpendicular to the reference direction of the polarizer [vertical arrows in diagram (a)]. The curve marked $b'$ represents the photomultiplier output when an optically active sample has been inserted. The reference direction of the plane-polarized light has rotated through $\theta°$ with respect to its original direction so that the analyzer is no longer crossed with the polarizer.

Fig. 16. Photographs of the JASCO ORD/UV-5 spectropolarimeter. (A) whole view of the instrument. (B) The servomotor seen from the left-hand side. The center of the photograph shows the mechanical amplifier of the ORD signal. (C) The gear assembly for wavelength scale expansion and scanning speed.

either by a program or manually and the photomultiplier output signal is maintained at a constant level throughout the whole wavelength range by means of the high-voltage amplifier. This voltage is also recorded on the chart. The recorded voltage can be used to judge whether the instrument is performing properly and whether the ORD curve is reliable. If the voltage is too high, artifacts may be obtained, as will be described in the following section. Pictures of the instrument are shown in Fig. 16.

## B. Sensitivity

The automatic recording polarimeters which are commercially available should have an accuracy of $\pm 0.001°$. However, this accuracy can be achieved only when enough light reaches the analyzer. This means that a solution which has low light absorption but large rotation can be measured with a high accuracy. However, a sample which has large absorption but small rotation is very difficult to measure accurately and the resulting ORD curve will be noisy. Thus, the trough at 233 nm for helical polypeptides can be measured with a greater accuracy than the peak at 198 nm.

Besides absorption of the solution itself, oxygen absorbs light especially below 200 nm. Thus it is advisable to purge the sample chamber with dry nitrogen to obtain better results in this wavelength region. In this way one can make measurements down to 185 nm as long as the sample does not absorb too strongly. Purging with nitrogen has another advantage. Ozone is easily formed within the instrument by the strong xenon light source and will oxidize the mirror, resulting in a large reduction in light intensity. Purging with dry nitrogen is thus important also in the maintenance of the instrument.

## C. Instrumental Factors

Many of the considerations outlined in the chapter on ultraviolet absorption in Part A of this series are valid also for ORD measurements. However, special attention should be given to the rotatory artifacts which may appear in regions of high absorbance. Many unusual Cotton effects have been attributed to stray light polarized within the cell housing and which dominate the measured ORD spectra because of the opacity of the solution.

In fact, if one places $p$-cresol in one cell in series with polyglutamic acid in another, Cotton effects appear at 276 nm as the absorbance of the $p$-cresol solution is raised above 2 (Urnes and Doty, 1961).

As a paramount criterion of the authenticity of Cotton effects it is suggested that they are submitted to a Beer's law type test. If the value of the specific rotation or molecular rotation of the Cotton effects does not depend on the concentration or length of the sample solution cell one can safely say that the observed Cotton effects are genuine. This kind of caution is specifically needed for cases where the specific rotations are small but the molar extinction coefficients are large.

### D. Choice of Solution and Cells

For the above reasons the concentration of the solution and the length of the cell should be chosen so that the optical density does not exceed 2. It should also be remembered that the light intensity obeys Beer's law

$$I = I_0 \times 10^{-\epsilon Cl}$$

where $\epsilon$ is the molar extinction coefficient, $C$ is the concentration in moles per liter and $l$ is the length of the cell. $I$ and $I_0$ denote the light intensity after and before the light enters the cell. However, the observed rotation is linearly proportional to the concentration and cell length. Thus, reducing the cell length greatly increases the signal-to-noise ratio in the ORD curve. Although both concentration and cell length are variable, changing the cell length is recommended, since this will decrease the absorption of the solvent as well.

As standard choices the cell length may be selected as follows: for the visible region, 50 mm; for the UV region above 250 nm, 10 mm; between 250 and 215 nm, 1 mm cell; for below 215 nm, 0.1 mm cell.

It is important to choose cells which have no distortions in the windows, since distortion will cause a large optical rotation. If this condition is fulfilled, both rectangular and cylindrical cells may be used. Distortions in the cell windows can be detected by making measurements with the cell filled only with water.

### E. Scanning Speed

It should be noted that ORD measurements are carried out under conditions where polarizer and analyzer almost cross. This means that only a very small fraction of the initial light can pass through the analyzer and necessarily makes the response of the recording pen to the signal very poor. Consequently, it is important to scan the spectrum sufficiently slowly to allow the pen to attain its equilibrium position. Scanning speeds depend on the wavelength as well as the optical density of the sample solution. However, for measurements below 300 nm and especially below 250 nm, very slow scanning is recommended.

### F. Turbidity and Orientation of the Sample

When the solution is turbid some scattered light reaches the photomultiplier. It should be noted that depolarization and optical rotation accompany light scattering. Because of this depolarization, the intensity of the signal is sometimes exceeded by that of the scattered light. It is therefore most important to remove even small amounts of precipitate either by ultracentrifugation or by filtration before making a measurement. However, when the molecular weight of the sample becomes too high (say about a million), the scattering effect due to the sample itself is unavoidable, and the measurement of ORD becomes very difficult, especially in the shorter wavelength regions. In such cases one is forced to analyze ORD curves at longer wavelength regions or to utilize CD. CD has a great advantage in this regard. This can be demonstrated by inserting a Ludox solution in the light beam: there is great difficulty in making ORD measurements but a much lesser effect on the CD results. The effect of light scattering on optical activity has been discussed by Urry and Krivacic (1970).

If the concentration of the sample is too high, gels may form which can cause significant errors in the ORD. It should be remembered that the various equations we have given for the ORD are only valid for solutions where the solute is randomly oriented. When poly-$\gamma$-benzyl-L-glutamate becomes oriented as a gel several thousand degrees may be obtained for $[\alpha]_D$.

## IX. The Relative Advantages of ORD and CD

Throughout this chapter, we have emphasized that ORD and CD are very closely related to each other. Since either of these curves can be obtained from the other through the general Kronig-Kramer transforms, it is clear that both measurements must contain essentially the same amount and type of information. This information is a set of component rotational strengths and characteristic wavelengths which may be related to the structure of the molecule under study. It is therefore quite reasonable to ask which technique is best to use and whether in fact both techniques are needed.

Historically, optical rotation at a single wavelength and later ORD are the much older techniques, their popularity almost certainly being due to the availability of a commercial spectropolarimeter. As Djerassi (1967) has pointed out, this historical "accident" was probably very lucky since the subtleties of ORD might otherwise never have been uncovered. In recent years the development of commercial circular dichrometers has seen a large swing away from ORD to CD.

The relative advantages of the two techniques stem from two basic facts. The first is the characteristic band shapes for ORD and CD Cotton effects—namely CD Cotton effects are discrete and can be approximated by Gaussian curves over the same wavelength range as the responsible absorption band, whereas ORD Cotton effects have a complex shape and extend over very large wavelength ranges. The second is that none of the present commercial instruments are capable of recording ORD or CD below about 185 nm.

This means that CD cannot be used at all for those chromophores whose absorption bands are centered below 185 nm, whereas the ORD spectrum will exhibit the tail of the corresponding Cotton effect in the accessible spectral range. By fitting these data with a Drude-type equation the rotational strength and even the wavelength of the transition can be estimated.

The optical rotation at any one wavelength represents contributions from all optically active transitions, whereas the CD at any one wavelength contains contributions from only a few of the transitions. Thus $[\alpha]$ can be used much more efficiently to characterize different compounds than $[\theta]$, especially if some of the chromophores contained in the molecules being compared are the same. The finite contributions to the ORD from all secondary structures in a protein also mean that the optical rotation at a single wavelength is much more sensitive to *any* conformational change that may occur than is the CD at one wavelength. With ORD it is also possible to follow conformational transitions in the visible wavelength range with equipment which is inexpensive and easy to operate. This is the reason that most studies involving extensive measurements of denaturation or other conformational transitions employ $[\alpha]$ to follow the transition. Furthermore, the use of ORD (especially in the visible wavelength range) can circumvent a common problem encountered in using CD to characterize conformation and conformational changes. The problem is that some molecules have very intense absorption bands which give rise to only weak optical activity. In these cases, since CD can only be measured at the same wavelength as the absorption bands, the concentration of the molecule cannot be increased to improve the signal without a concomittant increase in the noise level or even the production of absorption artifacts in the CD pattern (see Sections VIII,B and C). However, since the ORD can be measured outside the absorption bands, and this measurement can be used to follow conformational changes, then the concentration can be increased greatly to obtain a good signal without increasing the noise level. Finally, the visible ORD methods (Section III) produce parameters (e.g., $b'_0$) which represent average contributions from *all* the far-ultraviolet Cotton effects. Compared with the far-ultraviolet ORD and CD methods, which measure individual Cotton effects, the visible ORD parameters are not so sensitive to such factors as solvent effects (which can cause red-shifts of $\pi-\pi^*$ transitions and blue-shifts of $n-\pi^*$ transitions with increasing refractive index) and the exact geometries and lengths of the secondary structures [which can differentially affect $R_{n\pi*}$ and $R_{\pi\pi*}$ (see Section IV,C)]. Thus, visible ORD may give better "average" values for the secondary structure.

Circular dichroism is invariably preferred over ORD whenever one wishes to separate out component transitions. Their approximately Gaussian band shape, discreteness, and simple relationship to absorption bands makes it much simpler to resolve CD curves into their components than to resolve ORD curves. Once resolution has been achieved, it is easy to obtain the wavelength and rotational strength of each transition. Thus, theoretical calculations of optical activity are always compared with the CD spectrum. It is also easier to characterize conformational changes by CD since the Cotton effects for different structures are more easily separable than for ORD. It is, in fact, the easier resolution of CD curves into component transitions and the greater sensitivity of this technique in detecting small bands that are adjacent to large ones which are the special advantages of this technique and which have caused its greatly increased popularity. However, Djerassi (1967) has pointed out that sometimes the ORD can be more successful than the CD in detecting overlapping Cotton effects (e.g., in the case of cholestan-3α-ol nitrite).

We have seen that the two techniques have their own special advantages depending on the particular situation and on the question being asked by the experimenter. However, it is not always simply a question of choosing between the two techniques. The two techniques used together with the Kronig-Kramer transforms can become powerful investigatory tools. First, simultaneous resolution of absorption, CD and ORD spectra may help in the unambiguous resolution of all three curves into component curves. Second, the absence of primary standards for calibrating circular dichrometers has seen the development of methods which obtain an ORD curve using a primary standard (quartz or sucrose) followed by the transformation into a CD curve, using the Kronig-Kramers transforms (see, e.g., Cassim and Yang, 1970). Finally, it is possible to detect transitions centered outside the range of both instruments using the combined techniques. This requires the measurement of both the ORD and CD spectra of the same sample with both instruments calibrated as described above. The CD is then transformed into an ORD curve and this calculated curve is subtracted from the experimental ORD. The difference between these two curves then represents the contributions to the ORD from transitions centered in the inaccessible spectral region. It was in this way that Cassim and Yang (1970) tested the molecular exciton model for the α-helix (Section IV). The information obtained from such a procedure is greater than that which may be obtained from either technique alone.

## APPENDIX: TABLE I
### Optical Rotatory Parameters For Several Proteins in Their Native And Denatured States

| Proteins | Native | | | | Denatured | | |
|---|---|---|---|---|---|---|---|
| | $[\alpha]_D$ | $\lambda_c{}^a$ | $b'_0$ | Solvent | $[\alpha]_D$ | $\lambda_c$ | Solvent |
| Paramyosin | −11 | NL | −600 | 0.6$M$ KCl | −63 | | 9.5$M$ urea |
| Tropomyosin | −12 | NL | −650 | pH 7, 0.6$M$ KCl | −109 | 210 | pH 7.5, 8$M$ urea |
| L-Meromyosin-I | −13 | NL | −660 | 0.6$M$ KCl | −118 | 212 | 9.5$M$ urea |
| L-Meromyosin | −20 | NL | −490 | 0.6$M$ KCl | −107 | 215 | 9.5$M$ urea |
| Ovalbumin | −28 | 266 | −450 | pH 7.1 | −98 | 226 | 8.6$M$ urea |
| β-Lactoglobulin | −28 | 245 | | 0.1$M$ NaCl, pH 5.5 | −117 | 225 | 8.5$M$ urea |
| Myosin | −29 | NL | −370 | 0.6$M$ KCl | −108 | 218 | 9.5$M$ urea |
| Insulin | −29 | 266 | −264 | pH 1.8 | −88 | 226 | 9$M$ urea |
| H-Meromyosin | −35 | NL | −300 | 0.6$M$ KCl | −103 | 215 | 9.5$M$ urea |
| Lysozyme | −50 | 257 | | | | | |
| Bovine serum albumin | −59 | 265 | −370 | pH 5.5 | −109 | 221 | 8.5$M$ urea pH 5.5 |
| α-Chymotrypsin | −66 | 241 | | | −112 | 220 | 8$M$ urea pH 3 |
| Ribonuclease IA | −74 | 233 | | pH 5.4, 0.1$M$ KCl | −109 | 220 | 8$M$ urea pH 5.6 |
| Chymotrypsinogen | −78 | 239 | | pH 6.7 | −117 | 224 | 8$M$ urea pH 6.5 |

[a] NL: Nonlinear in Yang's plot.

## Acknowledgments

One of the authors (N.A.N.) wishes to thank the Commonwealth Scientific and Industrial Research Organization (Australia) for financial support during the preparation of this manuscript.

## References

Aebersold, D., and Pysh, E. S. (1970). *J. Chem. Phys.* **53**, 2156.
Arnott, S., Dover, S. D., and Elliot, A. (1967). *J. Mol. Biol.* **30**, 201.
Bayley, P. M., Nielsen, E. B., and Schellman, J. A. (1969). *J. Phys. Chem.* **73**, 228.
Bello, J. (1969). *Biochemistry* **8**, 4535.
Beychok, S. (1965). *Proc. Nat. Acad. Sci. U. S.* **53**, 999.
Beychok, S. (1967). *In* "Poly-$\alpha$-Amino Acids" (G. D. Fasman, ed.), Chapter 7, p. 293. Dekker, New York.
Beychok, S., and Fasman, G. D. (1964). *Biochemistry* **3**, 1675.
Blake, C. C. F., Koenig, D. F., Mair, G. A., North, A. C. T., Phillips, D. C., and Sarma, V. R. (1965). *Nature (London)* **206**, 757.
Blout, E. R., and Fasman, G. D. (1957) *Recent Advan. Gelatin Glue. Res., Proc. Conf., 1957.* Vol. 1, p. 122.
Blout, E. R., and Schechter, E. (1963). *Biopolymers* **1**, 568.
Blout, E. R., Doty, P., and Yang, J. T. (1957). *J. Amer. Chem. Soc.* **79**, 749.
Blout, E. R., Schmier, I., and Simmons, N. S. (1962). *J. Amer. Chem. Soc.* **84**, 3193.
Blout, E. R., Carver, J. P., and Gross, J. (1963). *J. Amer. Chem. Soc.* **85**, 644.
Blout, E. R., Carver, J. P., and Schechter, E. (1967). *In* "Optical Rotatory Dispersion and Circular Dichroism in Organic Chemistry" (G. Snatzke, ed.), p. 224. Heyden, London.
Boedtker, H., and Doty, P. (1956). *J. Amer. Chem. Soc.* **78**, 4267.
Bovey, F. A., and Hood, F. P. (1967). *Biopolymers* **5**, 325.
Bradbury, E. M., Brown, L., Downie, A. R., Elliot, A., Fraser, R. D. B., and Hanby, W. E. (1962a). *J. Mol. Biol.* **5**, 230.
Bradbury, E. M., Elliot, A., and Hanby, W. E. (1962b). *J. Mol. Biol.* **5**, 487.
Bush, C. A. (1971). *In* "Physical Techniques in Biological Research," (G. Oster, ed.), 2nd ed., Vol. 1, Part A, p. 347. Academic Press, New York.
Caldwell, D. J., and Eyring, H. (1963). *Rev. Mod. Phys.* **35**, 577.
Carver, J. P., Schechter, E., and Blout, E. R. (1966). *J. Amer. Chem. Soc.* **88**, 2562.
Cary, H., Hawes, R. C., Hooper, P. B., Duffield, J. J., and George, K. P. (1964). *Appl. Opt.* **3**, 329.
Cassim, J. Y., and Taylor, E. W. (1965). *Biophys. J.* **5**, 553.
Cassim, J. Y., and Yang, J. T. (1970). *Biopolymers* **9**, 1475.
Chen, A. K., and Woody, R. W. (1971). *J. Amer. Chem. Soc.* **93**, 29.
Chen, Y.-H., Yang, J. T., and Martinez, H. M. (1972). *Biochemistry* **11**, 4120.
Cohen, C. (1963). *J. Biophys. Biochem. Cytol.* **1**, 3.
Coleman, D. L., and Blout, E. R. (1967). *In* "Conformation of Biopolymers" (G. N. Ramachandran, ed.), Vol. 1, p. 123. Academic Press, New York.
Condon, E. U. (1937). *Rev. Mod. Phys.* **9**, 432.
Condon, E. U., Altar, W., and Eyring, H. (1937). *J. Chem. Phys.* **5**, 753.
Cosani, A., Peggion, E., Verdini, A. S., and Terbojevich, M. (1968). *Biopolymers* **6**, 963.

Cowan, P. M., and McGavin, S. (1955). *Nature (London)* **176**, 501.
Cowburn, D. A., Brew, K., and Gratzer, W. B. (1972). *Biochemistry* **11**, 1228.
Crabbé, P. (1965). "Optical Rotatory Dispersion and Circular Dichroism in Organic Chemistry." Holden-Day, San Francisco, California.
Damle, V. N. (1970). *Biopolymers* **9**, 937.
Davidson, B., Tooney, N., and Fasman, G. D. (1966). *Biochem. Biophys. Res. Commun.* **23**, 156.
Davydov, A. S. (1962). "Theory of Molecular Excitons." McGraw-Hill, New York.
Djerassi, C. (1960). "Optical Rotatory Dispersion: Applications to Organic Chemistry." McGraw-Hill, New York.
Djerassi, C. (1967). *In* "Optical Rotatory Dispersion and Circular Dichroism in Organic Chemistry" (G. Snatzke, ed.), p. 16. Heyden, London.
Doty, P., and Lundberg, R. D. (1957). *Proc. Nat. Acad. Sci. U. S.* **43**, 213.
Doty, P., Wada, A., Yang, J. T., and Blout, E. R. (1957). *J. Polym. Sci.* **23**, 851.
Doty, P., Imahori, K., and Klemperer, E. (1958). *Proc. Nat. Acad. Sci. U. S.* **44**, 424.
Downie, A. R. (1960). Cited by Todd (1960).
Emeis, C. A., Osterhofh, L. J., and de Vries, G. (1967) *Proc. Roy. Soc., Ser A* **297**, 54.
Epand, R. F., and Scheraga, H. A. (1968). *Biopolymers* **6**, 1551.
Fasman, G. D. (1963). *In* "Methods in Enzymology" (S. P. Colowick and N. O. Kaplan, eds.), Vol. 6, p. 928. Academic Press, New York.
Fasman, G. D., and Blout, E. R. (1960). *J. Amer. Chem. Soc.* **82**, 2262.
Fasman, G. D., and Blout, E. R. (1963). *Biopolymers* **1**, 3.
Fasman, G. D., and Potter, J. (1967). *Biochem. Biophys. Res. Commun.* **27**, 209.
Fasman, G. D., Bodenheimer, E., and Lindblow, C. (1964). *Biochemistry* **3**, 1665.
Fasman, G. D., Landsberg, M., and Buchwald, M. (1965). *Can. J. Chem.* **43**, 1588.
Fraser, R. D. B., and MacRae, T. P. (1962). *J. Mol. Biol.* **5**, 457.
Fresnel, A. (1825). *Ann. Chim. Phys.* [2] **28**, 147.
Gersonde, K., Sick, H., Wollmer, A., and Buse, G. (1972). *Eur. J. Biochem.* **25**, 181.
Glazer, A. N., and Simmons, N. S. (1965). *J. Amer. Chem. Soc.* **87**, 2287.
Goodman, M., and Rosen, I. G. (1964). *Biopolymers* **2**, 537.
Goodman, M., and Tonioli, C. (1968). *Biopolymers* **6**, 1673.
Goodman, M., Schmitt, E. E., and Yphantis, D. A. (1962). *J. Amer. Chem. Soc.* **84**, 1283.
Gratzer, W. B., Rhodes, W., and Fasman, G. D. (1963). *Biopolymers* **1**, 319.
Green, N. M., and Melamed, M. D. (1966). *Biochem. J.* **100**, 614.
Greenfield, N., Davidson, B., and Fasman, G. D. (1967). *Biochemistry* **6**, 1630.
Halper, J. P., Latovitzki, N., Bernstein, M., and Beychok, S. (1971). *Proc. Nat. Acad. Sci. U. S.* **68**, 517.
Hamaguchi, K., and Imahori, K. (1964). *J. Biochem. (Tokyo)* **55**, 388.
Harrington, W. F., and Sela, M. (1958). *Biochim. Biophys. Acta* **27**, 24.
Hashizume, H., Shiraki, M., and Imahori, K. (1967). *J. Biochem. (Tokyo)* **62**, 543.
Hermans, J., Jr., and Scheraga, H. A. (1961). *J. Amer. Chem. Soc.* **83**, 3293.
Holzwarth, G. M., and Doty, P. (1965). *J. Amer. Chem. Soc.* **87**, 218.
Holzwarth, G. M., Gratzer, W. B., and Doty, P. (1962). *J. Amer. Chem. Soc.* **84**, 3194.
Horwitz, J., and Strickland, E. H. (1971). *J. Biol. Chem.* **246**, 3749.
Horwitz, J., Strickland, E. H., and Billups, C. (1970). *J. Amer. Chem. Soc.* **92**, 2119.
Hsu, M.-C., and Woody, R. W. (1971). *J. Amer. Chem. Soc.* **93**, 3515.

Iizuka, E., and Yang, J. T. (1965). *Biochemistry* **4,** 1249.
Iizuka, E., and Yang, J. T. (1966). *Proc. Nat. Acad. Sci. U. S.* **55,** 1175.
Ikeda, K., Hamaguchi, K., Imanishi, M., and Amano, T. (1967). *J. Biochem. (Tokyo)* **62,** 315.
Imahori, K. (1960). *Biochim. Biophys. Acta* **37,** 336.
Imahori, K. (1963). *Kobunshi* **12,** Suppl. 1, 34.
Imahori, K., and Yahara, I. (1964). *Biopolym. Symp.* **1,** 421.
Jaenicke, R. (1968). *Abstr. Int. Symp. Macromol. Chem. 1968* B 1-1.
Jirgensons, B (1961) *Makromol. Chem.* **44/46,** 123.
Jirgensons, B. (1962). *Arch. Biochem. Biophys.* **96,** 314.
Jirgensons, B. (1969). "Optical Rotatory Dispersion of Proteins and Other Macromolecules." Springer-Verlag, Berlin and New York.
Johnson, C., and Tinoco, I. (1972). *J. Amer. Chem. Soc.* **94,** 4389.
Kauzmann, W. (1957). "Quantum Chemistry." Academic Press, New York.
Kirkwood, J. G. (1937). *J. Chem. Phys.* **5,** 479.
Krimm, S., and Mark, J. E. (1968). *Proc. Nat. Acad. Sci. U. S.* **60,** 1122.
Kuhn, W. (1930). *Trans. Faraday Soc.* **26,** 293.
Kurtz, J., Berger, A., and Katchalski, E. (1957). *Recent Advan. Gelatin Glue Res., Proc. Conf., 1957* p. 731.
Lampe, J., Rein, N., and Schelen, W. (1972). *FEBS Lett.* **23,** 282.
Linderstrøm-Lang, K., and Schellman, J. A. (1954). *Biochim. Biophys. Acta* **15,** 156.
Litman, B. J., and Schellman, J. A. (1965). *J. Phys. Chem.* **69,** 978.
Lowry, T. M. (1935). "Optical Rotatory Power." Longmans, Green, New York.
McCubbin, W. D., Oikawa, K., and Kay, C. M. (1972). *FEBS Lett.* **23,** 100.
Madison, V., and Schellman, J. A. (1970). *Biopolymers* **9,** 569.
Madison, V., and Schellman, J. (1972). *Biopolymers* **11,** 1041.
Madison, V., Bayley, P., and Schellman, J. A. (1970). *Abstr. Proc. Int. Congr. Biochem., 8th, 1969,* p. 67.
Mandel, R., and Holzwarth, G. (1972). *J. Chem. Phys.* **57,** 3469.
Miles, D. W., and Urry, D. W. (1967). *J. Phys. Chem.* **71,** 4448.
Moffitt, W. (1956). *J. Chem. Phys.* **25,** 467.
Moffitt, W., and Yang, J. T. (1956). *Proc. Nat. Acad. Sci. U. S.* **42,** 596.
Moffitt, W., Fitts, D., and Kirkwood, J. G. (1957). *Proc. Nat. Acad. Sci. U. S.* **43,** 723.
Moscowitz, A. (1962). *Advan. Chem. Phys.* **4,** 67.
Myers, D. V., and Edsall, J. T. (1965). *Proc. Nat. Acad. Sci. U. S.* **53,** 169.
Nagy, B., ad Strzelecka-Gołaszewska, H. (1972). *Arch. Biochem. Biophys.* **150,** 428.
Norland, K., Fasman, G. D., Katchalski, E., and Blout, E. R. (1963). *Biopolymers* **1,** 277.
Pasteur, L. (1848). *Ann. Chim. Phys.* [2] **24,** 443.
Peggion, E., Cosani, A., Verdini, A. S., Del Pra, A., and Mammi, M. (1968). *Biopolymers* **6,** 1477.
Perutz, M. F. (1970). *Nature (London)* **228,** 726.
Pysh, E. S. (1966). *Proc. Nat. Acad. Sci. U. S.* **56,** 825.
Pysh, E. S. (1967). *J. Mol. Biol.* **23,** 587.
Pysh, E. S. (1970a). *Science* **167,** 290.
Pysh, E. S. (1970b). *J. Chem. Phys.* **52,** 4723.
Quadrifoglio, F., and Urry, D. W. (1968). *J. Amer. Chem. Soc.* **90,** 275.
Ronish, E. W., and Krimm, S. (1972). *Biopolymers* **11,** 1919.

Rosenfeld, V. L. (1928). *Z. Phys.* **52**, 161.
Rosenheck, K., and Doty, P. (1961). *Proc. Nat. Acad. Sci. U. S.* **47**, 1755.
Rosenheck, K., and Sommer, B. (1967). *J. Chem. Phys.* **46**, 532.
Sage, H. J., and Fasman, G. D. (1966). *Biochemistry* **5**, 286.
Sarkar, P. K., and Doty, P. (1966). *Proc. Nat. Acad. Sci. U. S.* **55**, 981.
Saxena, V. P., and Wetlaufer, D. B. (1971). *Proc. Nat. Acad. Sci. U. S.* **68**, 969.
Schechter, B., Schechter, I., Ramachandran, J., Conway-Jacobs, A., and Sela, M. (1971). *Eur. J. Biochem.* **20**, 301.
Schechter, E., and Blout, E. R. (1964a). *Proc. Nat. Acad. Sci. U. S.* **51**, 695.
Schechter, E., and Blout, E. R. (1964b). *Proc. Nat. Acad. Sci. U. S.* **51**, 794.
Schellman, J. A., and Lowe, M. J. (1968). *J. Amer. Chem. Soc.* **90**, 1070.
Schellman, J. A., and Oriel, P. J. (1962). *J. Chem. Phys.* **37**, 2114.
Schellman, J. A., and Schellman, C. (1964). *In* "The Proteins" (H. Neurath, ed.), 2nd ed., Vol. 2, p. 1. Academic Press, New York.
Schellman, J. A., and Schellman, C. G. (1956). *Arch. Biochem. Biophys.* **65**, 58.
Shiraki, M. (1968). Ph.D. Thesis, University of Tokyo.
Shiraki, M., and Imahori, K. (1967). *Sci. Pap. Coll. Gen. Educ., Univ. Tokyo* **16**, 215.
Simmons, N. S., and Blout, E. R. (1960). *Biophys. J.* **1**, 55.
Simpson, R. T., and Vallee, B. L. (1966). *Biochemistry* **5**, 2531.
Snatzke, G., ed. (1967). "Optical Rotatory Dispersion and Circular Dichroism in Organic Chemistry." Heyden, London.
Stevens, L., Townend, R., Timasheff, S. N., Fasman, G. D., and Potter, J. (1968). *Biochemistry* **7**, 3717.
Strickland, E. H. (1972). *Biochemistry* **11**, 3465.
Strickland, E. H., Kay, E., and Shannon, M. (1970). *J. Biol. Chem.* **245**, 1233.
Stryer, L. (1961). *Biochim. Biophys. Acta* **54**, 395.
Stryer, L., and Blout, E. R. (1961). *J. Amer. Chem. Soc.* **83**, 1411.
Tanford, C., De, P. K., and Taggart, V. G. (1960). *J. Amer. Chem. Soc.* **83**, 6028.
Thiery, J. M. (1969). Ph.D. Thesis, University of California, Berkeley.
Tiffany, M. L., and Krimm, S. (1968). *Biopolymers* **6**, 1767.
Timasheff, S. N. (1970). *Accounts Chem. Res.* **3**, 62.
Timasheff, S. N., Townend, R., and Mescaanti, L. (1966). *J. Biol. Chem.* **241**, 1863.
Timasheff, S. N., Susi, H., Townend, R., Stevens, L., Gorbunoff, M. J., and Kumosinski, T. F. (1967). *In* "Conformation of Biopolymers" (G. N. Ramachandran ed.), Vol. 1, p. 173. Academic Press, New York (see, also, Bovey and Hood, 1967).
Tinoco, I., Jr. (1962). *Advan. Chem. Phys.* **4**, 113.
Tinoco, I., Jr., and Cantor, C. R. (1970) *Methods Biochem. Anal.* **18**, 81.
Tinoco, I., Jr., and Woody, R. W. (1960). *J. Chem. Phys.* **32**, 461.
Todd, A. (1960). *In* "A Laboratory Manual of Analytical Methods of Protein Chemistry" (P. Alexander and R. J. Block, eds.), Vol. 2, Chapter 8, p. 246. Pergamon, Oxford.
Townend, R., Kumosinski, T. F., Timasheff, S. N., Fasma, G. D., and Davidson, B. (1966). *Biochem. Biophys. Res. Commun.* **23**, 163.
Traub, W., and Shmueli, U. (1963). *Nature (London)* **198**, 1165.
Ulmer, D. D., and Vallee, B. L. (1965). *Advan. Enzymol.* **27**, 37.
Urnes, P., and Doty, P. (1961). *Advan. Protein Chem.* **16**, 401.
Urnes, P., Imahori, K., and Doty, P. (1961). *Proc. Nat. Acad. Sci. U. S.* **47**, 1635.

Urry, D. W. (1968a). *Proc. Nat. Acad. Sci. U. S.* **60**, 394.
Urry, D. W. (1968b). *Annu. Rev. Phys. Chem.* **19**, 477.
Urry, D. W., ed. (1969a). "Spectroscopic Approaches to Biomolecular Conformation." A.M.A.
Urry, D. W. (1969b). *Proc. Nat. Acad. Sci. U. S.* **63**, 261.
Urry, D. W., and Krivacic, J. (1970). *Proc. Nat. Acad. Sci. U. S.* **65**, 845.
Vallee, B. L., and Wacker, W. E. C. (1970). *In* "The Proteins" (H. Neurath, ed.), 2nd ed., Vol. 5, p. 104. Academic Press, New York.
Voigt, W. (1905). *Ann. Phys. (Leipzig)* [4] **18**, 645.
Vournakis, J. N., Yan, J. F., and Scheraga, H. A. (1968). *Biopolymers* **6**, 1531.
Wada, A., Tsuboi, M., and Konishi, E. (1961). *J. Phys. Chem.* **65**, 1119.
Woldbye, F. (1967). *In* "Optical Rotatory Dispersion and Circular Dichroism in Organic Chemistry" (G. Snatzke, ed.), Chapter 5, Fig. 1, p. 85. Heyden, London.
Woody, R. W. (1968). *J. Chem. Phys.* **49**, 4797.
Woody, R. W. (1969). *Biopolymers* **8**, 669.
Woody, R. W., and Tinoco, I., Jr. (1967). *J. Chem. Phys.* **46**, 4927.
Yahara, I. (1966). Ph.D. Thesis, University of Tokyo.
Yamaoka, K. K. (1964). *Biopolymers* **2**, 219.
Yang, J. T. (1967). *In* "Poly-$\alpha$-Amino Acids" (G. D. Fasman, ed.), p. 239. Dekker, New York.
Yang, J. T. (1969). *In* "A Laboratory Manual of Analytical Methods of Protein Chemistry" (P. Alexander and H. P. Lundgren, eds.), Vol. 5, p. 25. Pergamon, Oxford.
Yang, J. T., and Doty, P. (1957). *J. Amer. Chem. Soc.* **79**, 761.
Yang, J. T., and McCabe, W. J. (1965). *Biopolymers* **3**, 209.
Yaron, A., Katchalski, E., Berger, A., Fasman, G. D., and Sober, H. A. (1971). *Biopolymers* **10**, 1107.
Zubkov, V. A., and Vol'kenshtein, M. V. (1967a). *Biopolymers* **5**, 465.
Zubkov, V. A., and Vol'kenshtein, M. V. (1967b). *Dokl. Akad. Nauk SSSR* **175**, 942.
Zubkov, V. A., and Vol'kenshtein, M. V. (1970). *Mol. Biol.* **4**, 2829.
Zubkov, V. A., Birshtein, T. M., Milevskaya, I. S., and Vol'kenshtein, M. V. (1971). *Biopolymers* **10**, 2051.

# 23 □ Circular Dichroism

## DUANE W. SEARS and SHERMAN BEYCHOK

| | |
|---|---|
| Glossary | 446 |
| I. Introduction | 447 |
| II. Theory of Optical Activity and Its Applications | 449 |
|    A. Phenomenological Description of Optical Activity | 449 |
|    B. ORD and CD | 454 |
|    C. Rotational Strength | 457 |
|    D. CD Spectra | 460 |
|    E. Quantum Mechanical Theory of Optical Activity | 471 |
|    F. Symmetry Rules and the Origins of Optical Activity | 488 |
|    G. Calculations | 506 |
| III. Secondary Structures of Proteins | 533 |
|    A. CD Spectra of Polypeptides in $\alpha$-Helical, $\beta$-Pleated Sheet and Aperiodic Conformations | 533 |
|    B. Estimates of Protein Secondary Structure | 536 |
| IV. Side-Chain Optical Activity in Model Compounds and Proteins | 541 |
|    A. Tyrosyl Residues | 541 |
|    B. Tryptophanyl Residues | 547 |
|    C. Phenylalanyl Residues | 550 |
|    D. Cystinyl Residues (Disulfide) | 552 |
| V. Selected Proteins | 554 |
|    A. Hemoglobin and Myoglobin | 554 |
|    B. Ribonuclease | 558 |
|    C. Lysozymes | 562 |
|    D. Insulin | 566 |
|    E. Immunoglobulins | 569 |
|    F. Elastase | 574 |
|    G. Lactate Dehydrogenase | 576 |
|    H. Staphylococcal Nuclease | 578 |
|    I. Neurophysins | 579 |
|    J. Concanavalin A | 581 |
| VI. Concluding Remarks | 583 |
|    Appendix: Electromagnetic Units | 583 |
|    References | 585 |

## Glossary

| | |
|---|---|
| $\lambda, \lambda', \lambda_0, \lambda_a, \lambda_a^0$ | wavelength of light |
| $A(\lambda)_L, A(\lambda)_R$ | optical density for left and right circularly polarized light, respectively |
| $\epsilon(\lambda)_L, \epsilon(\lambda)_R$ | molar extinction coefficient for left and right circularly polarized light, respectively |
| $n(\lambda)_L, n(\lambda)_R$ | index of refraction for left and right circularly polarized light, respectively |
| $C$ | concentration in moles per liter |
| $l$ | optical path length in centimeters |
| $\mathbf{E}_L^0, \mathbf{E}(\lambda)_L, \mathbf{E}_R^0, \mathbf{E}(\lambda)_R$ | electric field vector for left and right circularly polarized light, respectively |
| $\phi(\lambda)$ | optical rotation in radians per centimeter |
| $\delta(\lambda)$ | optical rotation in radians |
| $\alpha(\lambda)$ | optical rotation in degrees |
| $\theta'(\lambda)$ | ellipticity in radians |
| $\psi(\lambda)$ | ellipticity in degrees |
| $\theta(\lambda)$ | ellipticity in radians per centimeter |
| $[\alpha(\lambda)]$ | specific rotation |
| $[\phi(\lambda)], [m(\lambda)], m_\lambda$ | molecular or molar rotation |
| $C_0$ | concentration in grams per cubic centimeter |
| $d$ | optical path length in decimeters |
| $[\theta(\lambda)], \theta_\lambda$ | molecular or molar ellipticity |
| $\epsilon(\lambda)_{0a}, \epsilon_{0a}$ | partial molar extinction coefficient |
| $[\phi(\lambda)_{0a}]$ | partial molecular rotation |
| $[\theta(\lambda)_{0a}], [\theta_{0a}]$ | partial molecular ellipticity |
| $|0\rangle, |a\rangle$ | molecular ground and excited state vectors, respectively |
| $[m'], [\theta']$ | reduced molecular rotation and ellipticity, respectively |
| $\boldsymbol{\mu}, \boldsymbol{\mu}_i$ | electric dipole moment operator |
| $\boldsymbol{\mu}_{0a}, \boldsymbol{\mu}_{i0a}, \boldsymbol{\mu}_{0AK}$ | electric transition dipole moment |
| $\mathbf{m}, \mathbf{m}_i, \mathbf{M}$ | magnetic dipole moment operator |
| $\mathbf{m}_{0a}, \mathbf{m}_{i0a}, \mathbf{M}_{AK0}$ | magnetic transition dipole moment |
| $\phi_0, \phi_a, \psi_{1a}, \psi_{2a}, \phi_{i0}, \phi_{ia}, \psi_{ia}, \psi_{AK}, \psi_\pm$ | molecular wavefunctions |
| $\mathbf{r}_s, \mathbf{r}_{is}$ | position operator of the $s^{th}$ electron |
| $e$ | charge of the proton |
| $m$ | mass of the electron |
| $c$ | vacuum speed of light |
| $h$ | Planck's constant |
| $i$ | imaginary square root of minus one |
| $\mathbf{p}_s$ | momentum of the $s^{th}$ electron |
| $\mathbf{p}, \mathbf{p}_i$ | total momentum operator |
| $\nabla_s$ | gradient operator of the $s^{th}$ electron |
| $\nabla$ | total gradient operator |
| $R_a, R_{0a}, R_\pm, R_{0AK}, R'_{0AK}$ | rotational or rotatory strength |
| $\nu_a, \nu_{AK}, \nu_\pm$ | transition frequency |
| $N_0$ | Avogadro's number |
| $D_{0a}$ | dipole strength |
| $n(\lambda)$ | average index of refraction |

$\Delta_a$ absorption band halfwidth
$\Delta_a^0$ circular dichroism band halfwidth
$\mathbf{R}_i, \mathbf{R}, \mathbf{R}_{12}$ distance vector
$\mathbf{\mu}_{iaa}, \mathbf{\mu}_{i00}$ permanent dipole moment
$V$ potential energy operator
$V_{i0a;k0b}, V_{12}$ potential energy
$\delta_{ij}$ Kronecker delta function
$\underline{\underline{H}}$ Hamiltonian matrix
$\underline{\underline{I}}$ unit matrix
$E, E_a, E_b, E_0$ energy
$\mathbf{\mu}, \mathbf{\mu}'$ electric transition dipole moment row vector
$\mathbf{m}, \mathbf{m}', \mathbf{M}, \mathbf{M}'$ magnetic transition dipole moment row vector
$R$ distance
$\underline{\underline{Q}}_1$ quadrupole tensor
$Q_{xy}$ transition quadrupole moment
$\mathbf{R}_{21}$ distance row vector
$x, X, y, Y, z, Z$ cartesian coordinates
$\mu_{1z}, \mu_{1x}, \mu_{1y}$ components of electric transition dipole moment
$m_{1z}, m_{1x}, m_{1y}$ components of magnetic transition dipole moment
$\underline{\underline{C}}$ unitary transformation matrix
$C_{iaK}, C_{11}$ exciton coefficient or element of the unitary transformation matrix

## I. Introduction

Our objective in this chapter is to describe several recent applications of circular dichroism (CD) measurements to studies of protein conformation and to review a few theoretical advances. When one considers that the first CD measurements on protein were reported as recently as 1962 (Holzwarth et al., 1962), it is somewhat surprising that reviewers must apologize for being incomplete and selective so few years later, but we find ourselves in that predicament. One reason for this is that CD measurements have become fairly routine, in protein investigations, so that often the CD spectrum is one figure among many in a paper dealing with diverse aspects of a particular protein. A more significant reason, however, is that optical activity, and increasingly CD, in particular, remains a unique tool for analyzing secondary structures of dissolved proteins. If one wants to compare two conformations in solution with respect to secondary structure or to estimate secondary structure elements of a newly isolated protein or polypeptide, the most direct and frequently used method is the measurement of the CD spectrum in the wavelength interval 240–190 nm.

No solution technique, now or in the foreseeable future, has the overall power, definiteness, and completeness of X-ray diffraction analysis of crystalline proteins. However, even that extraordinary method of structure determination is not always applicable, or possible, or practical. It is thus especially important that the accuracy and limitations of a method such as CD be continuously evaluated and tested by comparison with X-ray diffraction results whenever these are available.

Among specialists and nonspecialists, in varying degrees, there is uncertainty as to the confidence which may be placed in current interpretations of far-ultraviolet CD spectra of proteins. Understandably, this has led to some ambivalence and controversy, ranging from utter disillusionment with the usefulness of far-UV CD measurements all the way to noncritical acceptance of any value cited for content of $\alpha$-helix or $\beta$-pleated sheet. To the extent that it is possible, we shall endeavor to cite experiments and calculations which suggest that neither of the extreme positions is valid. In addition, theoretical work bearing on this will be presented and evaluated.

CD measurements in the spectral interval 320 to 250 nm fall into a somewhat different category. Here, the technique is but one of a class of spectral measurements which, taken together, provide valuable information about interactions of certain side-chain residues. The expectations and claims with respect to these measurements have not been as great as with the far-UV CD. Moreover, a number of theoretical approaches to calculating the CD of known structures appear possible but, at the time of this writing, there has yet been little exploitation of recent theoretical work. Accordingly, we will survey these theoretical aspects and discuss possible future applications or tests of theory.

Induced optical activity in small bound substances is still another area in which the experimental results have far outpaced the theoretical or semiempirical understanding of the phenomenon. Here again, there is a good deal of casting about to see whether more information about tertiary structure can be gained from CD measurements than has heretofore been possible. While awaiting a more secure theoretical basis for interpretation of the CD bands, many investigators find these measurements of significant diagnostic value in monitoring binding, conformational change, altered reactivity, and related interactions. The same may be said of side-chain optical activity in the absence of bound substances. From an experimental point of view, such measurements, at present, represent a significant power of optical activity and, accordingly, we shall review selected cases in some depth.

In consideration of space limitations, as well as the limited competence of the authors, very complex situations, such as membrane CD, are not extensively reviewed. Conjugated proteins, including metalloproteins, are examined only briefly for the same reasons. Moreover, details of instrumentation, evaluation of signal-to-noise ratios, relative performance of different instruments, and optical or electronic aspects of the instruments, as well as data handling, are only touched on as they relate to particular protein investigations. We apologize for these exclusions; fortunately, comprehensive reviews devoted solely or partly to optical activity of biological macromolecules have dealt with the foregoing in a most excellent manner. Our task in the limited scope of this article is thus made vastly simpler by our ability to refer the reader throughout the text to several outstanding recent reviews.

## II. Theory of Optical Activity and Its Applications

Fundamentally, optical activity is a manifestation of the differential interaction of matter with left and right circularly polarized light (LCPL and RCPL henceforth). The interesting, particularly useful feature of this phenomenon is that it relates, in a sensitive way, to molecular geometry itself; that is, the very extent to which a molecule or a molecular structure interacts differently with the two circular polarizations depends directly on both its inherent conformation and its precise orientation relative to other molecules, structures, or charges in its immediate environment. In this section, we focus on the theory which underlies this important stereochemical property of optical activity. Our primary aims here are twofold: to examine in detail some of the more important (from the standpoint of protein chemistry) functional relationships which link the observables in an optical activity experiment to the parameters defining molecular structure; and to review, in the end, the theoretical calculations in which these relationships have been utilized as a means of probing the conformations and the optical activity of certain proteins and related molecules.

### A. Phenomenological Description of Optical Activity

Macroscopically, an optically active medium[1] is one in which monochromatic LCPL and RCPL travel with unequal velocities. An optically

---

[1] It will be assumed throughout that an optically active medium only comprises randomly oriented molecules and is free of external fields. Systems in which the molecules assume preferred orientations and systems to which external, static electric, and magnetic fields are applied can also differentially interact with LCPL and RCPL in a way which is independent of optical activity itself. Such systems are not considered here. The reader is referred to Tinoco (1962), Disch and Sverdlik (1969), Tinoco and Cantor (1970), and Kahn and Beychok (1968) for discussion of the optical activity of linearly birefringent and linearly dichroic substances (such as crystals); and to Tinoco and Bush (1964) and Briat and Djerassi (1968) for discussion of the optical activity of substances placed in static electric and magnetic fields.

active medium is also one which differentially absorbs light of opposite circular polarizations. The former phenomenon is reflected in the respective indices of refraction, $n(\lambda)_L$ and $n(\lambda)_R$, for LCPL and RCPL since the ratio of the speed of light in a vacuum to that in the medium is different in the two cases. The latter phenomenon is reflected in the differences in the respective optical densities (absorbances) $A(\lambda)_L$ and $A(\lambda)_R$ and molar extinction coefficients $\epsilon(\lambda)_L$ and $\epsilon(\lambda)_R$ of the two polarizations. These parameters are defined by the well-known Beer-Lambert law:

$$A(\lambda)_{L,R} = \epsilon(\lambda)_{L,R} Cl = \log_{10}(|\mathbf{E}^0_{L,R}|^2/|\mathbf{E}(\lambda)_{L,R}|^2) \quad (1)$$

Here, $C$ is the concentration of absorbers in moles per liter; $l$ is the optical path length in centimeters; the symbol L,R denotes either LCPL or RCPL; $\mathbf{E}_R^0$ and $\mathbf{E}(\lambda)_R$, for example, are the respective magnitudes of the electric field vectors for RCPL entering and emerging from the medium; and the square of the ratio of these magnitudes, i.e., $|\mathbf{E}_R^0|^2/|\mathbf{E}(\lambda)_R|^2$, equals the ratio of the initial to the final intensity of light. [The symbol $\lambda$ is included as a reminder that the parameter is a function of the experimental wavelength of light.] Optical activity is macroscopically and phenomenologically defined and quantified in terms of these differences: $n(\lambda)_L - n(\lambda)_R$ is the circular birefringence; $A(\lambda)_L - A(\lambda)_R$ is the circular dichroism (CD) (cf. Condon, 1937; Kauzmann et al., 1940; Moscowitz, 1962).

The circular birefringence itself is normally several orders of magnitude smaller than either $n(\lambda)_L$ or $n(\lambda)_R$ alone; and the same is true, but to a somewhat lesser extent, of the circular dichroism as compared with either $A(\lambda)_L$ or $A(\lambda)_R$. Because the conventional instrumentation for measuring the index of refraction and the absorption is usually neither sensitive enough to detect accurately such small differences in these parameters nor conveniently set up to do selective measurements with LCPL and RCPL, it is seldom feasible or practical to determine the circular birefringence and the circular dichroism by a direct measurement of the appropriate quantities. Instead, the optical activity of a sample is almost always determined with special equipment[2] designed to measure certain experimentally defined angles, these being proportional either to the circular birefringence or to the circular dichroism. These angles are defined in the following way.

Consider the light emerging from an optically active sample which is simultaneously irradiated with monochromatic LCPL and RCPL, coherently superposed in the form of plane-polarized light (see Fig. 1). If the wavelength of the light is away from the absorption regions of the

---

[2] The present-day instrumentation used for measuring optical activity is thoroughly discussed by Tinoco and Cantor (1970).

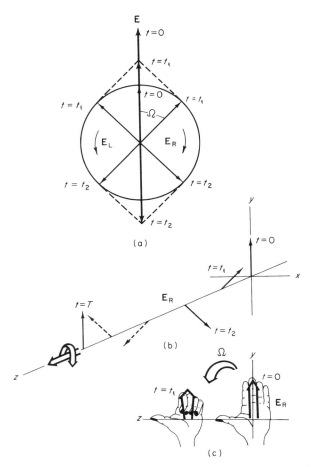

Fig. 1. (a) The oscillating electric field vector **E** of plane-polarized light viewed at various times ($t$) as it moves toward the observer in a plane normal to the plane of the paper. **E** can be regarded as the vector resultant of two coherently superposed electric field vector components, one being $\mathbf{E}_L$ for left circularly polarized light, and the other being $\mathbf{E}_R$, for right circularly polarized light. $\mathbf{E}_L$ appears to rotate counterclockwise as it moves toward the observer and $\mathbf{E}_R$ appears to rotate clockwise. (b) $\mathbf{E}_R$ is illustrated at various times, between $t = 0$ and $t = T$ as it moves down the $z$-axis; $T$ is the period which equals $1/\nu$ where $\nu$ is the frequency of light. The vector arrows represented by dashed lines lie behind the $yz$-plane. (c) Actually, *left* and *right* are misnomers since $\mathbf{E}_L$ describes a right-hand screw as it moves down the $z$-axis and $\mathbf{E}_R$ a left-hand screw. This is illustrated for $\mathbf{E}_R$ with the schematic left hand: with the fingers initially extended in the direction of $\mathbf{E}_R$ (in this case at $t = 0$) and with the thumb pointing along the $z$-axis—in either direction—to the next vector in the series (in this case at $t = t_1$), the fingers are curled through the angle $\Omega$ between the two vectors; a left-hand screw is described if the angle is the smallest of the two possible angles, in this case $\Omega$ as opposed to $2\pi - \Omega$.

sample, the light will emerge still in plane-polarized form but the plane will be rotated relative to the original plane; this is explained by the fact that the velocity difference between LCPL and RCPL in the medium results in a change in the relative phases of these two components. The angle between the planes, the optical rotation, is[3]

$$\phi(\lambda) = [n(\lambda)_L - n(\lambda)_R]\pi/\lambda \text{ (radians per centimeter)} \quad (2)$$

or

$$\delta(\lambda) = \phi(\lambda)l \text{ (radians)} \quad (3)$$

or

$$\alpha(\lambda) = 180\delta(\lambda)/\pi \text{ (degrees)} \quad (4)$$

From Eq. (2) we see that the optical rotation is proportional to the circular birefringence; Eq. (2) is also one-half the phase difference between the LCPL and RCPL components resulting from their passage through the medium.

If the light is in an absorption region of the sample, not only will the relative phases of LCPL and RCPL change but the magnitudes of the electric field vectors of the two polarizations will also change in an unequal fashion, giving rise to elliptically polarized light; the emerging electric field vector, rather than tracing a line as before, traces (with time) an ellipse when viewed head-on (see Fig. 2). The phase differential

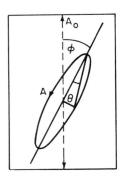

FIG. 2. The definitions of the rotation $\phi$ and the ellipticity $\theta$. The light is approaching the observer. The dashed line labeled $A_0$ represents the plane-polarized light incident on the sample. The transmitted light, the line labeled A, is elliptically polarized. The rotation is positive here, being so defined for clockwise rotation, whereas the ellipticity is negative, this being the convention for left circularly polarized light (indicated by the arrow on line A) (Tinoco, 1965).

[3] The symbols used here for these angles and the ones below are by no means universally accepted by investigators in this field. We have attempted to conform to the most common usage where possible.

between the polarized components causes rotation of the major axis of the ellipse through an angle relative to the plane of the entering beam as defined in Eqs. (2)–(4). The circular dichroism of the sample is found to be approximately proportional to the ellipticity of the ellipse. The ellipticity—defined as the arctangent of the ratio of the minor axis ($|\mathbf{E}(\lambda)_R| - |\mathbf{E}(\lambda)_L|$) to the major axis ($|\mathbf{E}(\lambda)_R| + |\mathbf{E}(\lambda)_L|$)—can be written in the following way, using Eq. (1) above

$$\theta'(\lambda) = \tan^{-1}\left\{\frac{\exp[-2.303A(\lambda)_R/2] - \exp[-2.303A(\lambda)_L/2]}{\exp[-2.303A(\lambda)_R/2] + \exp[-2.303A(\lambda)_L/2]}\right\} \quad (5)$$

Because the CD is usually much smaller than the absorbance, the exponentials in this equation and the arctangent can be expanded in Taylor series to give the following approximate linear relationship between $\theta'(\lambda)$ and the CD

$$\theta'(\lambda) \approx 2.303[A(\lambda)_L - A(\lambda)_R]/4 \text{ (radians)} \quad (6)$$

In other units, the ellipticity is

$$\psi(\lambda) = 180\theta'(\lambda)/\pi \text{ (degrees)} \quad (7)$$

or

$$\theta(\lambda) = \theta'(\lambda)/l \text{ (radians per centimeter)} \quad (8)$$

In most investigations involving proteins, the optical rotation is reported in terms of either the specific rotation

$$[\alpha(\lambda)] = \alpha(\lambda)/dC_0 \text{ (degree-centimeter squared per dekagram)} \quad (9)$$

or, now more commonly, the molecular (molar) rotation

$$[\phi(\lambda)] \equiv [m(\lambda)] = 100\alpha(\lambda)/lC \text{ (degree-centimeter squared per decimole)} \quad (10)$$

$C_0$ is the concentration in grams per cubic centimeter and $d$ is the optical path length in decimeters. Similarly, the ellipticity is usually reported in terms of the molecular (molar) ellipticity:

$$[\theta(\lambda)] = 100\psi(\lambda)/lC$$
$$= 3298[\epsilon(\lambda)_L - \epsilon(\lambda)_R] \text{ (degree-centimeter squared per decimole)} \quad (11)$$

The symbol $C$ in Eqs. (10) and (11) denotes one of several possibilities, depending upon the type of information to be conveyed. It may refer to the concentration, in moles per liter, of either the protein itself or a particular amino acid or prosthetic group (e.g., heme) of the protein; or it may denote the concentration of residues, found by dividing the concentration of the protein in grams per liter by the mean gram molecular weight of its constituent amino acids. When $C$ is the concentration of a particular residue, Eqs. (10) and (11) are expressions of the rotation or ellipticity of the residue itself. When the concentration of residues is used, these equations give, in essence, an average value for the rotation or ellipticity per residue.

Occasionally, in an attempt to take into account the effects of a solvent on an optically active solute, the molecular rotation and the molecular ellipticity are multiplied by a function of the average index of refraction $n(\lambda) = [n(\lambda)_L + n(\lambda)_R]/2$. (The rationale for doing this is discussed in Section II,D,2.) This function is usually the factor $3/[n(\lambda)^2 + 2]$. When treated as such, the molecular rotation and molecular ellipticity are said to be "reduced."

## B. ORD AND CD

The optical rotation and ellipticity are normally measured over a range of wavelengths and the plots of these parameters versus the wavelength (or frequency) are, respectively, the optical rotatory dispersion (ORD) spectrum and the CD spectrum of a sample. ORD and CD spectra, like absorption spectra, are constituted by the sum of spectral bands which attend the individual transitions of the molecule. Stated mathematically: for a homogeneous sample in the presence of monochromatic light

$$\epsilon(\lambda) = \sum_{\{0a\}} [\epsilon(\lambda)_{0a}] \qquad (12)$$

$$[\phi(\lambda)] = \sum_{\{0a\}} [\phi(\lambda)_{0a}] \qquad (13)$$

and

$$[\theta(\lambda)] = \sum_{\{0a\}} [\theta(\lambda)_{0a}] \qquad (14)$$

$\epsilon(\lambda)_{0a}$ is the partial molar extinction coefficient, $[\phi(\lambda)_{0a}]$ is the partial molecular rotation, and $[\theta(\lambda)_{0a}]$ is the partial molecular ellipticity. These functions characterize the bands in the respective spectra which arise from a single transition, $0a$, from the ground state (or from one of several temperature-weighted ground states) $|0\rangle$ to an excited state $|a\rangle$ of the molecule.[4] The former summation includes all transitions of the molecule,

---

[4] Throughout this article the states $|0\rangle$, $|a\rangle$, etc. denote either purely electronic or purely vibronic—electronic plus vibrational—states of the molecule. Although the quantum mechanical theory for the optical activity of pure rotational (Chiu, 1970) and pure vibrational (Hameka, 1964a; Deutsche, 1969; Deutsche and Moscowitz, 1968, 1970) transitions has been developed in detail, the optical activity of such transitions in proteins has not been observed directly, with one exception, and therefore does not concern us here; the one exception to this is found in a recent work by Chirgadze et al. (1971) in which near-infrared ORD measurements on polypeptide α-helical and β-structures were made.

whereas the latter two summations implicitly include only those transitions which are optically active.

Ideal representations of an isolated absorption band and the associated ORD curve of an optically active transition are illustrated in Fig. 3a. The ORD curve, with its minimum, maximum, and inflection point so arranged, is typical of dispersion phenomena. In Fig. 3b, ideal representations of the absorption and CD bands of an optically active transition are depicted. The marked similarities between the two, in shape and position, stem from the fact that both are absorption phenomena. It will be noted in this figure that the inflection point of the ORD curve, the point of maximum (or minimum) CD, and the point of maximum absorption of the same optically active transition all ideally fall at the same wavelength, $\lambda_0$.

The ORD curve illustrated in Fig. 3 characterizes a so-called negative Cotton effect (in honor of its discoverer, the nineteenth century chemist, A. Cotton), whereas the CD band characterizes a positive Cotton effect. The CD band which would accompany the ORD curve depicted in Fig.

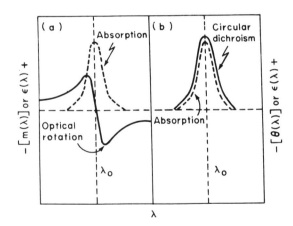

FIG. 3. (a) Ideally represented on the same scale are the absorption and the associated ORD curves of a single optically active transition. With the long-wavelength portion of the ORD curve being negative, the curve is representative of a negative Cotton effect; a positive Cotton effect would appear as the mirror image of this ORD curve as reflected through the horizontal dashed line. (b) The absorption and ellipticity bands of an optically active transition are illustrated on the same scale. Both bands are ideally centered at the same wavelength, $\lambda_0$. The CD band here is positive and hence represents a positive Cotton effect; note that the CD curve accompanying the ORD curve shown in (a) would appear as the mirror image of this CD band reflected through the horizontal dashed line. $\lambda$ increases in value toward the right and decreases toward the left in each half of each figure (Beychok, S., 1966. *Science*, **154**, 1288–1299. Copyright 1966 by the American Association for the Advancement of Science).

3a—that is, the CD band corresponding to a negative Cotton effect—would appear as the mirror image (through the horizontal axis) of the CD curve in Fig. 3b; likewise, the ORD curve associated with a positive Cotton effect would resemble the mirror image of the ORD curve illustrated in Fig. 3a.

A key mathematical relationship exists between the corresponding ORD and CD spectra of a sample. It can be shown (Moffitt and Moscowitz, 1959) that the optical rotation and ellipticity are mathematically interconvertible through the Kronig-Kramers transforms; that is, for each optically active transition

$$[\phi(\lambda)_{0a}] = + (2/\pi) \fint_0^\infty \{[\theta(\lambda')_{0a}]\lambda'/(\lambda^2 - \lambda'^2)\} \, d\lambda' \tag{15}$$

and

$$[\theta(\lambda)_{0a}] = - (2/\pi\lambda) \fint_0^\infty \{[\phi(\lambda')_{0a}]\lambda'^2/(\lambda^2 - \lambda'^2)\} \, d\lambda' \tag{16}$$

where the horizontal bar on the integral sign signifies the principal part of the integral. The significance of Eqs. (15) and (16) is of fundamental importance: in effect, they prove that the information content of the CD and ORD spectra of a substance is one and the same, since one spectrum can, in theory, be completely transformed into the other.

While in principle one need only consider either the ORD or CD spectrum without loss of generality, in practice there are important experimental differences between the two types of measurements. For reasons which will become apparent below, it is imperative to resolve an optical activity spectrum into its constituent bands for a precise assessment of the stereochemical information embodied therein, and to this end the CD spectrum is usually the better suited of the two. As illustrated in Fig. 3, the CD band of an optically active transition is confined to a much narrower region—the absorption region of the spectrum—in comparison to the corresponding partial molecular rotation curve which extends relatively far from $\lambda_0$. Except in regions where optically active transitions are very closely spaced, the band overlap between adjacent bands in an ORD spectrum will tend to be greater, on the average, than in the corresponding CD spectrum. Intuitively then, the CD spectrum is generally expected to be the easier of the two to resolve (see Section II,D,1). The long-ranging tails on the ORD curve, although hindering the resolution, do afford a distinctive advantage to this type measurement: ORD spectra usually provide more information about transitions which lie beyond any experimentally inaccessible absorption regions since there is greater likelihood, in comparison with the CD bands, that the tails of the ORD curves of these transitions will extend from the inaccessible to the accessible regions.

It is now fairly plain that ORD and CD measurements each have certain advantages and disadvantages. These should be weighed according to the goals in an investigation. In addition, one should also weigh the possibility of using both types of measurements in a complementary fashion. The ORD and CD, together with the Kronig-Kramers transforms, constitute a powerful biochemical tool, so to speak, with capabilities and advantages that neither spectrum has alone. In particular, the complementary usage of the two has in the past provided a means of doing the following: (1) simultaneously resolving CD and ORD spectra (Carver et al., 1966); (2) assessing the mutual reliability of ORD and CD data (Holzwarth et al., 1962; Holzwarth and Doty, 1965); (3) testing the theory underlying the optical activity of the $\alpha$-helix (Cassim and Yang, 1970); and (4) calibrating the circular dichrometer (Emeis et al., 1967; Cassim and Yang, 1969; DeTar, 1969; Krueger and Pschigoda, 1971). For further discussion and evaluation of ORD measurements, particularly as they relate to proteins, the reader is referred to Imahori and Nicola (Chapter 22 in this volume).

## C. Rotational Strength

At a molecular level, optical activity manifests itself in electromagnetically induced charge displacements which are simultaneously linear and circular in nature. In the language of quantum mechanics, then, an optically active transition will have an allowed electric transition dipole moment (linear charge displacement)

$$\mathbf{\mu}_{0a} = \langle 0|\mathbf{\mu}|a\rangle = \int \phi_0 \mathbf{\mu} \phi_a \, d\tau \tag{17}$$

as well as an allowed magnetic transition dipole moment (circular charge displacement)

$$\mathbf{m}_{0a} = \langle 0|\mathbf{m}|a\rangle = \int \phi_0 \mathbf{m} \phi_a \, d\tau \tag{18}$$

(The units of the molecular spectroscopic parameters are found in the Appendix.) The wavefunctions $\phi_0$ and $\phi_a$ in these equations respectively represent the ground $|0\rangle$ and excited $|a\rangle$ states of the molecule and both are assumed to be real (the arbitrary phase factor being set equal to zero). The integrals are taken over all space, an elemental volume of which is represented by $d\tau$. The electric dipole moment operator $\mathbf{\mu}$ is defined by the following summation

$$\mathbf{\mu} = -e \sum_s \mathbf{r}_s \tag{19}$$

where $-e$ is the charge on one electron ($e$ itself is positive); $r_s$ is the position vector operator which locates the electron labeled $s$ in a specified coordinate system; and the sum includes all electrons of the molecule. Similarly, the magnetic dipole moment operator $\mathbf{m}$ is constituted by the sum of angular momentum operators $\mathbf{r}_s \times \mathbf{p}_s$ ($\times$ denotes the vector cross-product) of the individual charges

$$\mathbf{m} = (-e/2mc) \sum_s \mathbf{r}_s \times \mathbf{p}_s = (ihe/4\pi mc) \sum_s \mathbf{r}_s \times \boldsymbol{\nabla}_s \qquad (20)$$

where $m$ is the mass of the electron, $c$ is the vacuum speed of light, $h$ is Planck's constant, and $i = \sqrt{-1}$. The last expression in Eq. (20) follows from the definition of the momentum operator of the electron $\mathbf{p}_s$ in terms of the gradient operator $\boldsymbol{\nabla}_s$

$$\mathbf{p} = \Sigma \mathbf{p}_s = \Sigma(-ih/2\pi)\boldsymbol{\nabla}_s = (-ih/2\pi)\boldsymbol{\nabla} \qquad (21)$$

$\mathbf{p}$ and $\boldsymbol{\nabla}$ are, respectively, the total momentum and total gradient operators for the molecule.

For randomly oriented molecules,[5] the optical activity is quantum mechanically quantified in terms of the rotational (rotatory) strength which is defined as the imaginary part[6] of the vector dot-product of the electric and magnetic transition dipole moments.

$$R_{0a} = \text{Im}[\boldsymbol{\mu}_{0a} \cdot \mathbf{m}_{a0}] = -i\boldsymbol{\mu}_{0a} \cdot \mathbf{m}_{a0} \qquad (22)$$

where $\mathbf{m}_{a0} = -\mathbf{m}_{0a}$ ($\boldsymbol{\mu}_{0a} = \boldsymbol{\mu}_{a0}$) for real wavefunctions and Im denotes the imaginary part of the quantity in the brackets. (Because the vector dot-product here is purely imaginary when the wavefunctions are real—$\boldsymbol{\mu}_{0a}$ is real and $\mathbf{m}_{a0}$ is purely imaginary in this case—the notation Im can be replaced by $-i$.) Equation (22) is a direct consequence of the fact that optically active molecules show different transition probabilities for left and right circularly polarized light (Condon, 1937).

The rotational strength is a parameter of central importance in the theory of optical activity. On the one hand, it links together the molecular and spectral phenomena. On the other hand, it embodies the stereochemical information reflected in an optically active transition insofar as the electric and magnetic transition moments can be related to the conformation and environment of the molecule. The first of these features is to be treated presently, while the latter is reserved for discussion in Sections II,E and II,F.

The rotational strength, as it is written in Eq. (22), has the following properties. First, it obeys a sum rule (Condon, 1937)

---

[5] See footnote 1.

[6] Any complex number $N$ can be written as follows: $N = N_R + iN_I$ where $N_R$ is the real part of $N(N_R = \text{Re}[N])$ and $N_I$ is the imaginary part of $N(N_I = \text{Im}[N] = -iN$ if $N_R = 0$). Note that both $N_R$ and $N_I$ are defined as real numbers themselves and that $R_{0a}$ is therefore a real number.

$$\sum_{\{0a\}} R_{0a} = 0 \qquad (23)$$

which, in effect, says that the net rotational strength of all transitions of a molecule is zero. Second, this parameter is origin dependent, its value being contingent upon the frame of reference in which the electrons are located by their respective position vectors ($r_s$). Clearly, the latter property is an undesirable one but it can be circumvented (Moscowitz, 1965) if the electric transition dipole moment is evaluated in the dipole-momentum, or gradient, formalism (Bohm, 1951)

$$\mathbf{\mu}_{0a} = -ihe\mathbf{p}_{0a}/m(E_a - E_0) = -he\mathbf{\nabla}_{0a}/2\pi m\nu_a \qquad (24)$$

where $\mathbf{p}_{0a} = \langle 0|\mathbf{p}|a\rangle$ and $\mathbf{\nabla}_{0a} = \langle 0|\mathbf{\nabla}|a\rangle$ [see Eqs. (17) and (21)]. $E_0$ is the energy of the ground state, arbitrarily set equal to zero here: $E_a = h\nu_a$ is the energy of the excited state; and $\nu_a$ is the frequency of the transition between these two states. Unfortunately, however, two new problems accompany the use of Eq. (24): it is only strictly valid for exact wavefunctions, which are seldom if ever realized, and it leads to the violation (Harris, 1969) of the sum rule (Eq. 23) when substituted into Eq. (22). Gould and Hoffmann (1970) have considered these problems and have concluded that the relative advantages and disadvantages weigh in favor of using Eq. (24) rather than Eq. (17), as it stands, to calculate rotational strengths. (The same conclusion has also been reached informally in several of the recent calculations discussed in Section II,G.)

A pressing question at this point is how is the rotational strength related to the two optical activity spectra. The optical rotatory dispersion is linked to the rotational strength in the following way (Rosenfeld, 1928; Condon, 1937; Moscowitz, 1962)

$$[\phi(\lambda)] = \sum_{\{0a\}} [\phi(\lambda)_{0a}] = (96\pi N_0/hc) \sum_{\{0a\}} \frac{R_{0a}(\lambda^2 - \lambda_a^2)\lambda_a^2}{(\lambda^2 - \lambda_a^2)^2 + 2\Delta_a^2 \lambda_a^2} \qquad (25)$$

Here, $N_0$ is Avogadro's number, $\lambda$ is the experimental wavelength, $\lambda_a$ ($= c/\nu_a$) is the wavelength of the 0a transition, and $\Delta_a$ is the halfwidth of the absorption band of the transition [see Eq. (29)]. Note that the justification for the presence of $\Delta_a$ in Eq. (25) is purely an empirical one[7]; it is included here to eliminate the discontinuities which would

---

[7] The factor $2\Delta_a^2$ does not arise in a natural, rigorous way in the derivation of Eq. (25). Rather, it is included there on intuitive grounds alone: in deriving this expression, Moscowitz (1957, 1962) argued, by analogy with the theory of natural linewidths in absorption spectroscopy, that a damping factor should appear in the denominator and that this factor should approximately equal the square of the full width of the band at half intensity $(2\Delta_a/\sqrt{2})^2 = 2\Delta_a^2$. For further discussion of the implications of this factor the reader is referred to Condon (1937), Moscowitz (1962), Hameka (1965), and Tinoco (1965).

exist, in its absence, when $\lambda = \lambda_a$. When measurements are made far outside the absorption regions of a sample (i.e., when $\lambda \gg \lambda_a$ or $\lambda_a \gg \lambda$) then the $\Delta_a$'s are usually set equal to zero; the resulting equation then is the quantum mechanical analogue of the classical Drude equation (see Chapter 22 in this volume).

The circular dichroism is related to the rotational strength by an expression similar to Eq. (25), as first shown by Condon (1937). However, there exists another more useful functional relationship which is analogous to the relationship existing between the dipole strength and the absorption band. Just as the dipole strength $D_{0a}$ gauges the area under the absorption band of the $0a$ transition (Mulliken, 1939)

$$D_{0a} = \mathbf{\mu}_{0a} \cdot \mathbf{\mu}_{a0} = |\mathbf{\mu}_{0a}|^2 = (6909hc/8\pi^3 N_0) \int_0^\infty \epsilon(\lambda') \, d\lambda'/\lambda' \qquad (26)$$

the rotational strength gauges the area under the CD band (Moffitt and Moscowitz, 1959)

$$R_{0a} = (6909hc/32\pi^3 N_0) \int_0^\infty [\epsilon(\lambda')_{0a,L} - \epsilon(\lambda')_{0a,R}] \, d\lambda'/\lambda' \qquad (27)$$

Or with Eq. (11)

$$R_{0a} = (hc/48\pi^2 N_0) \int_0^\infty [\theta(\lambda')_{0a}] \, d\lambda'/\lambda' \qquad (28)$$

The constants in these equations have the following values: $6909hc/8\pi^3 N_0 = 9.180 \times 10^{-39}$, $6909hc/32\pi^3 N_0 = 2.295 \times 10^{-39}$, and $hc/48\pi^2 N_0 = 0.696 \times 10^{-42}$. If we continue the analogy between the absorption and CD bands we see that since the dipole strength is a measure of the total probability that a transition will occur, then the rotational strength is a measure of the total difference in probability that a transition occurs in LCPL as opposed to RCPL.

D. CD Spectra

1. *Resolution*

It follows from the foregoing discussion that for optimal assessment of the stereochemical information embodied in a complex CD spectrum, one must first resolve the bands constituting the spectral curve. Ideally, this means finding the shape, sign, and position of each contributing band; from the shape, as a function of wavelength, and the sign of the band, the rotational strength can be evaluated [Eqs. (27) and (28)], and from the position of the band in the spectrum one would hope to identify the source among the transitions of the molecule. In practice, resolving the CD spectrum of a large molecule, such as a protein, is usually a difficult and often unsettled problem due to the inherent complexities of such a spectrum, as discussed in this and the following two sections. Our aims in this section are to focus on some of the limitations which these complexities impose on the resolution of a CD spectrum and to consider briefly a few commonly followed procedures for resolving CD spectra. [A more extensive review of these and various other curve fitting procedures is found in Tinoco and Cantor (1970).]

Except for the additional complications which arise from the overlapping of positive and negative CD bands, resolving the CD spectrum and resolving the absorption spectrum of a molecule are basically similar problems; in fact, they are complementary problems in the sense that the resolved form of one spectrum may greatly facilitate the resolution of the other. Recalling that circular dichroism is itself an absorptive phenomenon, one expects the corresponding CD and absorption bands of a transition to lie in essentially the same wavelength range (see Fig. 3) and, in certain instances (see below), one may also expect them to be identically shaped. Clearly then, the resolved bands of one spectrum are potentially revealing of the positions and even the shapes of the bands in the other spectrum, and it is for this reason that the CD spectrum of a molecule is almost always resolved in conjunction with the absorption spectrum. (For example, see Fig. 5 below.)

The most common procedure for resolving absorption and CD spectra is to fit the observed spectral curve with a set of Gaussian functions (Moscowitz, 1960a; Carver et al., 1966; Tinoco and Cantor, 1970). The obvious assumption here is that the real absorption and CD bands are Gaussian, or approximately Gaussian, in shape, being characterized by functions of the following types:

$$\epsilon(\lambda)_{0a} = \epsilon_{0a} \exp[-(\lambda - \lambda_a)/\Delta_a]^2 \quad (29)$$

$$[\theta(\lambda)_{0a}] = [\theta_{0a}] \exp[-(\lambda - \lambda_a^0)/\Delta_a^0]^2 \quad (30)$$

The band halfwidth, $\Delta_a$ or $\Delta_a^0$, is the distance, in wavelength units, from the center of the band at $\lambda_a$ or $\lambda_a^0$ to either point on the band where $\epsilon(\lambda)_{0a}$ or $[\theta(\lambda)_{0a}]$ falls to $e^{-1}$ of its peak value $\epsilon_{0a}$ or $[\theta_{0a}]$. In general, the halfwidths and centers of a CD band and its corresponding absorption band are not necessarily the same. However, $\Delta_a = \Delta_a^0$ and $\lambda_a = \lambda_a^0$ when a transition is allowed (i.e., $\epsilon > 1000$), for in this situation it can be shown (Moscowitz, 1965) that the bands are identically shaped, being different in value at any wavelength by only a constant multiplicative factor. Mathematically,

$$[\theta(\lambda)_{0a}]/\epsilon(\lambda)_{0a} = (3298)\ 4R_{0a}/D_{0a} \quad (31)$$

where the factor $4R_{0a}/D_{0a}$ is the anisotropy (Condon, 1937) constant of the transition.

Apart from providing a relatively simple mathematical means for handling the bands of a spectrum, the Gaussian representations of absorption and CD bands are particularly useful in another respect; they lead to very simple expressions for the dipole strength and the rotational strength of a transition. By substituting Eq. (29) into Eq. (26) and Eq. (30) into Eq. (28), and by making the reasonable assumption that $\lambda_a \gg \Delta_a$ and $\lambda_a^0 \gg \Delta_a^0$, it is found that (Moscowitz, 1960a)

$$D_{0a} \approx 9.180 \times 10^{-39} \sqrt{\pi} \; \epsilon_{0a}\Delta_a/\lambda_a \tag{32}$$

and

$$R_{0a} \approx 0.696 \times 10^{-42} \sqrt{\pi} \; [\theta_{0a}]\Delta_a^0/\lambda_a^0 \tag{33}$$

Although it is frequently the case that the bands associated with electronic transitions closely approximate the Gaussian shape, distinctly non-Gaussian bands are also a likely possibility, especially if a transition is weakly allowed or forbidden in nature. In the latter case, the band may have a skewed appearance in comparison to the Gaussian or it may have a complex, nonuniform shape, possibly even showing a variation in sign over the range of the band; such occurrences are traced to either the effects of temperature and solvent on the conformations and energy states of the molecule (see Section II,D,2), or to the vibronic fine structure of the band itself (see Section II,D,3). If a band is known to be non-Gaussian, it may be possible to represent its shape with another function or series of functions (Moscowitz, 1960b; Tinoco, 1965). If not, the remaining alternative is to represent the entire band, or those portions of the band which are homogeneous in sign, with an "effective" band function which preserves the area under the band but not its shape (Moscowitz, 1960a,b); if an effective Gaussian is used, Eqs. (29) and (30) still obtain but the parameters which define the dimensions of the Gaussian (e.g., $\lambda_a^0$, $\Delta_a^0$) are obviously now no longer necessarily those of the real band itself.

Of the many problems encountered in resolving a spectral curve, the most serious is that of establishing a unique resolution. Several factors come into play here. For instance, one can resolve almost any curve in more than one way depending upon the latitude one is given in varying the number, positions, shapes, and sizes of bands used to fit the curve. Obviously, then, the resolution of a spectrum may be seriously hindered if any doubt exists as to the number or approximate positions of the contributing transitions. Whenever this doubt does exist, as is often the case, the logical criterion for resolution is to achieve a maximal fit of the spectral curve with the minimum number of bands, in a way which is consistent with any prior knowledge concerning the transitions

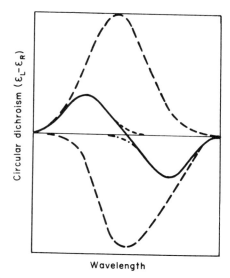

Fig. 4. Two arbitrary resolutions of a circular dichroism spectrum (solid line). The short-dashed lines essentially follow the spectrum, whereas the long-dashed lines give two large bands which nearly cancel (Tinoco, 1965).

of the molecule. However, the nagging possibility always remains with such a procedure that bands will be created which do not correspond to real transitions and, conversely, that some transitions will not be assigned bands. Two other problems which seriously impede the unique resolution of a spectrum involve the overlapping of adjacent bands: spectral data itself is rarely accurate enough to permit the unique resolution of more than three relatively contiguous bands (Carver et al., 1966); and it is easily shown that closely spaced positive and negative CD bands can be resolved in an infinite number of ways, as is partially illustrated in Fig. 4.[8]

As a means of at least partially alleviating the uniqueness problem

---

[8] A further complication accompanies the situation depicted in Fig. 4: if the distance separating the centers of the two overlapping bands is much less than the halfwidth of either band, it can be shown that the rotational strengths of both bands cannot be determined from the spectrum. What can be determined is in general a complex matter but is something essentially intermediate between two extremes. On the one hand, if the bands have identical dimensions, neither the rotational strength nor the separation of the centers of the bands (the splitting) is alone observable, but the product of these two parameters—the rotational couple—is observable (Bayley et al., 1969). On the other hand, if the sizes of the bands are greatly disproportionate, the observable is the difference between the rotational strengths of the two bands.

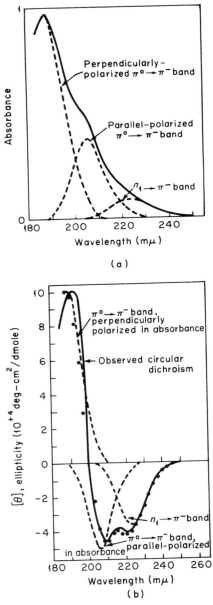

Fig. 5. (a) Absorption spectrum of α-helical poly-γ-methyl-L-glutamate in trifluoroethanol, showing proposed resolution into parallel- and perpendicular-polarized $\pi^0$–$\pi^-$ bands and an $n_1$–$\pi^-$ band. (b) The circular dichroism of α-helical poly-γ-methyl-L-glutamate in trifluoroethanol, showing the proposed resolution of the ob-

described here, a CD spectrum is frequently analyzed in conjunction with the absorption spectrum, or possibly the ORD spectrum, of the same molecule. The results of such an analysis are illustrated in Fig. 5. As shown, the CD and absorption spectra of the α-helix can be resolved with just three bands. Essentially, Holzwarth and Doty (1965) accomplished this by assuming that the shapes of corresponding CD and absorption bands are related by a constant but unknown multiplicative factor [see Eq. (31)]. They determined these constants and thus resolved the spectra by simultaneously adjusting the constants until a maximal fit of both spectra was achieved. On the basis of other theoretical and experimental work (as cited by them), they assigned the bands to the following transitions, as labeled in the figure: the absorption band centered at 189 m$\mu$ ($\lambda_\perp$) and the CD band centered at 190 m$\mu$ to the component of the peptide $\pi^0$–$\pi^-$ transition polarized perpendicular to the helix axis (see Section II,E); the two bands centered at 206 m$\mu$ ($\lambda_\parallel$) to the parallel-polarized component of the $\pi^0$–$\pi^-$ transition; and the two bands at 222 m$\mu$ to the peptide $n_1$–$\pi^-$ transition. It is interesting to note that a similar resolution of the α-helix CD spectrum was also proposed by Carver et al. (1966), who carried out a simultaneous least-squares fit of both the ORD and CD curves of the helix using the Kronig-Kramers transforms, Eqs. (15) and (16).

In spite of the fact that the CD spectrum of the α-helix appears to be satisfactorily resolved in both procedures with only three bands, this alone is not sufficient to rule out the possibility of other bands in this region.[9] In fact, from theoretical considerations it is predicted (Tinoco, 1964) that the perpendicularly polarized component of the $\pi^0$–$\pi^-$ transition will give rise not only to a band such as the one depicted in Fig. 5

---

[9] It has recently been speculated, on the basis of the gas-phase spectra of amides (Basch et al., 1967; Barnes and Rhodes, 1968; and Nagakura and Robin, as cited in the latter), that a third, Rydberg-like, transition also exists in this region of the spectrum. However, more recent work (Basch et al., 1968; Pysh, 1970a, and the references cited therein) indicates that this so-called "mystery band" probably disappears in condensed-phase spectra and is therefore not contributing to the solution spectra of proteins.

---

served data into three CD bands associated with the observed bands in the absorption spectrum. The solid line is the measured circular dichroism. The filled circles indicate the sum of the three CD bands shown. The shape and intensity of each of the component CD bands were obtained from one of the component absorption bands shown in (a) by application of the relationship $[\theta(\lambda)_{0a}] = c_a \, \epsilon(\lambda)_{0a}$ where $c_a$ is an adjustable constant (Holzwarth and Doty, 1965).

but it will also give rise to numerous other closely spaced bands which together form a non-Gaussian spectral envelope.[10] Except for the slight red-shift of the 190 mμ CD band relative to the 189 mμ absorption band, which is consistent with this prediction, there is no other indication of the presence of this envelope in Fig. 5; however, the CD curve could just as well have been resolved including such a curve (Woody, 1968).

## 2. Solvent and Temperature Effects

Considering that circular dichroism is an extremely sensitive function of conformation and environment, it is not surprising to find that the solvent and temperature conditions in an experiment can nontrivially affect the CD spectrum of a molecule. There are literally a variety of ways in which these factors can influence the outcome of a spectral measurement, including the direct induction of optical activity by stereospecific interactions between the solvent and the solute. Although the gross aspects of solvent and temperature effects have been given extensive theoretical treatment (Kirkwood, 1937; Kauzmann et al., 1940; Weigang, 1964), these effects are usually of such overall complexity on an experimental level that the theory is unfortunately of little value in aiding one to account for them in a quantitative fashion. Therefore, one must be satisfied, for the most part at least, with qualitative descriptions of these effects, as the reader will find in several excellent discussions of this matter (cf. Kauzmann et al., 1940; Kauzmann and Eyring, 1941; Donovan, 1969; Urry, 1968a, 1970).

If any attempt at all is made to account for solvent effects in an optical investigation, it is usually done in the following way. The simple assumption is made that the solvent molecules neighboring a solute molecule alter the latter's environment in a way such that the field (of the incoming light) experienced by the solute molecule is increased by the Lorentz factor $[n(\lambda)^2 + 2]/3$, where $n(\lambda)$ is the average index of refraction of the medium (cf. Condon, 1937; Kirkwood, 1937; Moffitt and Moscowitz, 1959). When this assumption is made, the definitions—in terms of molecular parameters—of the ellipticity and rotation [e.g., Eq. (25)] will explicitly include the Lorentz factor whereas the definition of the extinction coefficient will include the factor $[n(\lambda)^2 + 2]^2/9n(\lambda)$. As a means of standardizing the rotation and the ellipticity, then, it is

[10] This envelope is functionally characterized by the sum of a Gaussian function and the derivative of a Gaussian function (Tinoco, 1964). In shape it resembles the derivative of a Gaussian, appearing as a short-wavelength negative band centered at approximately $\lambda_- = \lambda_\perp - \Delta_\perp/\sqrt{2}$, contiguous with a long-wavelength positive band (of comparable size) centered at approximately $\lambda_+ = \lambda_\perp + \Delta_\perp/\sqrt{2}$. $\Delta_\perp$ is found from the approximate width $2\Delta_\perp/\sqrt{2}$ of the perpendicularly polarized absorp- at ½ intensity.

sometimes the practice, as noted earlier, to report experimental values for these parameters which are multiplied by $3/[n(\lambda)^2 + 2]$, the rationale being that this will divide out, so to speak, the contribution of the solvent. For the same reason, one will often see the right-hand sides of the following equations multiplied by these factors: Eqs. (15) and (16) by $[n(\lambda)^2 + 2]/[n(\lambda')^2 + 2]$; Eq. (26) by $9n(\lambda')/[n(\lambda')^2 + 2]^2$; and Eqs. (27) and (28) by $3/[n(\lambda')^2 + 2]$.

As far as protein investigations are concerned, these factors are of questionable practical value for they probably give only a crude, if not wholly inaccurate, accounting of the effects of solvation reflected in the CD and absorption spectra of proteins. Except in fortuitous circumstances, one expects that the chromophores of a protein molecule will "see" different effective microenvironments (e.g., the core as opposed to the surface of the molecule) which do not in general mimic the local environment within the solvent itself. This point is clearly illustrated in Fig. 6, which shows a proposed (Horwitz et al., 1970) resolution of the near-UV

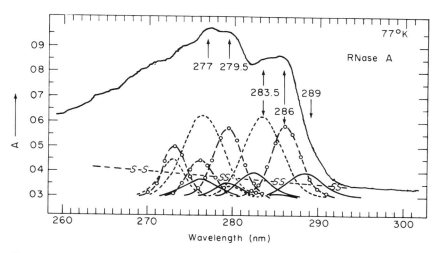

FIG. 6. Instrument trace of the absorption spectrum of 2.8 mM RNase A in water–glycerol (1:1, v/v) with 25 mM sodium phosphate (pH 7) at 77°K. Individual tyrosine and disulfide bands are presented under the RNase A spectrum: —S—S—, disulfide, three types of tyrosine residues are identical in terms of the position of the 0–0 bands; (———) 288.5 nm; (O–O–O) 286 nm; and (– – –) 283.5 nm. The absorption spectrum of each type of tyrosine residue is represented by the 0–0, 0 + 420, 0 + 800, 0 + 1250, and 0 + (2 × 800) cm$^{-1}$ bands, except that the 0 + (2 × 800) cm$^{-1}$ band was omitted from the spectrum having its 0–0 band at 283.5 nm. The remaining short-wavelength transitions were omitted in all cases. The area under the RNase A spectrum at 77°K was the same as that at 298°K, within experimental error of ±10%. The path length was 0.20 mm (Horwitz et al., 1970).

absorption spectrum of RNase A. Here, the environmentally distinct types of tyrosines are identified by the three bands, at 288.5 nm, 286 nm, and 283.5 nm. The observation is even more interesting if one compares the absorption spectrum of RNase A with the absorption spectrum of RNase S, as subsequently recorded by Horwitz and Strickland (1971) under the same solvent and temperature conditions (see caption of Fig. 6). In the resolved spectrum of RNase S, Horwitz and Strickland found that the tyrosine band which was formerly located at 288.5 nm in the spectrum of RNase A appeared shifted approximately 2.5 nm to shorter wavelengths; by direct comparison of the X-ray crystal structures of RNase A (Kartha et al., 1967) and RNase S (Wyckoff et al., 1967, 1970) it can be argued that this shift is due to a slight alteration in the local environment of only one tyrosine, namely residue 25.

Temperature effects in an optical activity experiment can be basically understood by recalling that the temperature determines the equilibrium distribution of molecules in different energy states. Hence, one is forced to conclude (Moscowitz et al., 1963a,b) that an experimental CD spectrum most likely does not represent one molecular species in a particular conformational state, vibrational state, state of solvation, etc. Rather it represents a temperature-weighted average of the CD spectra of several distinct molecular species, each of which is in a different energy state. This immediately poses a general question as to the extent the CD spectrum of a protein represents a population of conformationally different species rather than one, unique conformational state. Several low-temperature CD and absorption measurements have provided new insights on this question.

With the recent improvements in instrumentation (Myer and MacDonald, 1967; Horwitz et al., 1968) and low temperature techniques (Horwitz et al., 1969; Strickland et al., 1969), it has been possible to measure the near-UV spectra of the following proteins and protein-related model compounds at extremely low temperatures (down to 77°K): phenylalanine derivatives (Horwitz et al., 1969; Strickland et al., 1970), tyrosine derivatives (Horwitz et al., 1970; Strickland et al., 1970, 1972a), tryptophan derivatives (Strickland et al., 1969, 1970, 1972b), RNase A (Horwitz et al., 1970), RNase S (Horwitz and Strickland, 1971), chymotrypsinogen A (Strickland et al., 1969), horse radish peroxidase A1 (Strickland et al., 1971), carboxypeptidase A (Fretto and Strickland, 1971a,b), and bovine visual pigment$_{500}$ (Horwitz and Heller, 1971). At such low temperatures, the lower energy states become heavily populated and the low-temperature CD spectra of these compounds, then, are presumably representative of fewer conformational states. Therefore, it was expected, and was usually found to be the case in these experiments, that the low-temperature CD spectra would be sharper and more intense than the counterpart high-temperature spectra. By comparison, the absorption spectra, being relatively insensitive to conformation, were found to intensify and sharpen on cooling to a much lesser extent. Of the many other observations made in these interesting works, perhaps the most significant as far as proteins in general are concerned is that their CD spectra do not drastically change upon cooling; this was in sharp contrast to the observations made concerning the CD spectra of the smaller model compounds. Thus, the conclusion may be drawn that proteins, or at least the proteins studied, exist in relatively few different conformational states at room temperature and that the actual cooling process does not alter their conformations in any gross way.

### 3. *Vibronic Fine Structure*

One particularly interesting finding in the low-temperature studies discussed in the preceding section was that the low temperatures often tended to enhance the vibronic fine structure of the CD and absorption spectra studied. This in turn facilitated the resolution of the spectra as is illustrated, for example, in Fig. 6 where vibronic progressions[11] are assigned for the tyrosine bands of RNase A. The vibronic fine structure of a CD spectrum adds another dimension to the information which the CD conveys and this further requires special theoretical treatment.

The general quantum theory dealing with the vibronic fine structure of electronic CD bands has been developed by Weigang (1965a,b,c, 1966), Harnung et al. (1971), and Lin (1971), all of whom extended an earlier treatment given this subject by Moffitt and Moscowitz (1959). At the heart of this theory is the first-order perturbation expansion (within the limits of the Born-Oppenheimer adiabatic approximation) of the electric and magnetic transition dipole moments. The expanded transition moments lead to an expression for the rotational strength which includes the usual zero-order term for the total electronic rotational strength, Eq. (22), plus first- and second-order terms; the latter are loosely re-

---

[11] The notation 0–0 denotes an electronic transition from the ground vibrational state 0 of the ground electronic state to the ground vibrational state 0 of the excited electronic state. Similarly, $0 + 800$ cm$^{-1}$ represents an electronic transition from the same ground state configuration as previously to the 800 cm$^{-1}$ excited vibrational state of the excited electronic state. In tyrosine for example, one vibronic progression is found by adding multiples of 800 cm$^{-1}$ to the wave number which characterizes the 0–0 band (e.g., in tyrosine the wave number corresponding to the 289 nm band in Fig. 6). Note that the wavelength at which a progression begins depends upon the solvent or the local microenvironment of the chromophore, whereas the spacing between levels in the same progression is constant, depending on the nature of the chromophore itself.

garded as "borrowed" moments from other spectral transitions which perturb the vibrational states. The relative importance of these terms depends on the nature of the transition. When the transition is electrically and magnetically allowed (case I), the zero-order term dominates the expression for rotational strength and it is predicted that the corresponding absorption and CD bands of a transition will be centered at the same wavelength, their shapes being related by the anisotropy factor, Eq. (31). When only one of the two transition moments is allowed (case II), the first-order vibrational terms of the forbidden transition moment may strongly influence the shape of a CD band but make no *net* contribution to the total integrated electronic rotational strength. However, Weigang has pointed out that when only one transition is allowed, second-order vibrational terms (case III) may significantly add with mixed signs to the total rotational strength; the total observed rotational strength in this case then is constituted by electronic and vibrational contributions. Vibrational contributions to the total rotational strength are also very important whenever both transition moments are forbidden (also case III), because the higher-order terms are the only contributions in this case. In case III, and usually to a lesser extent in case II, there may not be, as there is in case I, an obvious relationship between the CD and the absorption bands of an electronic transition. For example, in case II the peaks of corresponding CD and absorption bands may be shifted relative to one another, or, in case III, it is likely that several distinct vibronic bands, each with different signs and intensities, exist rather than just one electronic CD band. In these situations, it is much more difficult to relate molecular chirality to the sign of the rotatory strength: the sign will not only depend on the chirality but it will also depend on the symmetry of the perturbing vibrations, the sign of the borrowed rotational strength, and the degree of electronic–vibrational coupling (Weigang, 1965a).

It follows from the foregoing statements that in order to unambiguously interpret the information conveyed by a CD spectrum it may be desirable or even necessary to consider the vibronic fine structure of the spectrum. The extent to which this is possible—that is, the degree of vibronic fine structure resolution possible—naturally varies greatly from experiment to experiment depending upon several factors. However, it should be borne in mind that, compared to electronic bands, vibronic bands are much more suitable as fundamental informational units upon which to base the definitions of dipole strength, rotational strength, anisotropy, band position, band shape, etc. This stems from the fact that there is true correspondence only between vibronic CD and absorption bands, as pointed out by Strickland *et al.* (1969):

Each vibronic transition in an optically active molecule possesses both absorption intensity and CD intensity. When spectra are displayed in terms of the underlying vibronic transitions, each CD transition must coincide with an absorption transition located at the same wavelength. The positions of vibronic transitions are revealed by the vibrational fine structure superimposed on each electronic transition. Thus each fine structure absorption band gives the position of a possible CD band, and conversely each fine structure CD band indicates the location of an absorption band. The ratio of CD to absorption for each vibronic transition may vary from practically zero to a large positive or negative value. The correspondence between CD and absorption exists *only* for each vibronic transition. Therefore, if the vibrational fine structure is not resolved, or is ignored, an apparent mismatch may occur between CD and absorption bands.

## E. Quantum Mechanical Theory of Optical Activity

The earliest theories of optical activity were founded on classical notions concerning the interaction between electromagnetic radiation and matter. The names most notably associated with the development of the classical theory of optical activity are M. Born, W. Kuhn, and C. W. Oseen. Although the classical theory will not directly concern us here, it does warrant mentioning for it played an influential role in the development of the first quantum mechanical theories of optical activity. That is to say, the classical theory having provided a satisfactory qualitative account of many of the substantive aspects of optical activity phenomena served to guide the course of development of the early quantum theory and provided, as well, a firm conceptual basis for interpreting the subsequent quantum mechanical results. The interested reader will find references to the original classical work and further discussion of the relationship between the classical and the quantum theory in Condon (1937), Condon *et al.* (1937), Kirkwood (1937), and Kauzmann *et al.* (1940).

Rosenfeld (1928) was the first to describe the phenomenon of optical rotation in quantum mechanical terms. By considering the properties of a semiclassical system in which an electron is both constrained to quantized molecular states and under the influence of a classical electromagnetic radiation field, Rosenfeld showed that the optical rotation is directly proportional to the parameter now referred to as the *rotational strength*. Taken in its broadest sense, then, Rosenfeld's theory established the quantum mechanical criterion for optical activity: a transition, in order to be optically active, must be electrically as well as magnetically allowed with nonperpendicular transition moments.

In three papers appearing almost simultaneously, Condon and his colleagues (Condon, 1937; Condon *et al.*, 1937) and Kirkwood (1937) independently reformulated Rosenfeld's theory and then further developed it in terms of first-order perturbation theory. The fundamental result of the Condon-Altar-Eyring (CAE) theory, or the so-called one-electron theory, can be loosely summarized as follows: here, optical activity is shown to arise in situations where an electron moves in an asymmetric potential field. The main point of distinction between Rosenfeld's theory and the CAE theory is that the former (taken at face value) characterizes inherently optically active transitions which are both electrically and magnetically allowed by nature, whereas the latter describes the way in which an otherwise inactive transition becomes optically active by the presence of a static perturbing field. In both theories, the motion of a single electron is implicated in the generation of optical activity, however, and this general mode of optical activity is referred to here as the *one-electron mechanism*.

A fundamentally different result was obtained by Kirkwood in his so-called polarizability theory where he quantum mechanically rederived the classical Born-Oseen-Kuhn coupled-oscillator model of optical activity. Kirkwood described the way in which optical rotation arises from the electromagnetically induced motions of two or more mutually interacting (coupled) electrons residing in different chromophores of the same molecule. Like the one-electron theory, Kirkwood's theory also concerns a perturbation effect but, in contrast, the perturbation here is viewed as being dynamic rather than static. Of the various terms in his equations, Kirkwood emphasized those which involved coupled, electrically allowed transitions; this is the *coupled-oscillator mechanism* of optical activity. He neglected those terms which indicated that the coupling between an electrically allowed transition and a magnetically allowed transition could also result in the production of optical activity; this is the *electric–magnetic coupling mechanism* of optical activity as subsequently formulated by Schellman (1968). Except for brief mention in two earlier papers (Kauzmann et al., 1940; Moffitt et al., 1957), this mechanism was ignored in optical activity considerations until only recently. It is now recognized that in certain situations electric–magnetic coupling is a potentially significant source of optical activity, as has been demonstrated in several of the protein-related calculations which are discussed in Section II,G.

In addition to forming much of the foundation of modern, first-order perturbation optical activity theory (as evidenced below when the expressions which characterize the three mechanisms are considered within a more general theoretical framework) the Kirkwood, Condon, and CAE papers contain a tremendous wealth of ideas concerning optical activity. Together they form a remarkably comprehensive treatment of this subject. Of their many important ideas and contributions in this field (some of which are cited elsewhere in this volume), a few are particularly germane to the topic at hand.

In the Condon papers, one finds the first quantum mechanical treatment of circular dichroism: Condon et al. (1937) demonstrated that the absorption probabilities for LCPL and RCPL are different for an optically active molecule; and Condon (1937) derived the first quantum mechanical expression [similar to Eq. (25)] for the ellipticity, thereby establishing the link between CD and the rotational strength. In the Kirkwood (1937) paper, the ground was laid for the quantum mechanical treatment of the optical properties of large, multielectronic polymers. A brief digression concerning this aspect of Kirkwood's theory will prove useful in later discussion.

Central to Kirkwood's theory is the assumption that each electron of a large molecule is constrained to a well-defined spatial region in that molecule; such a region, for our purposes, is classified either as a chromophoric subunit, as a nonchromophoric subunit, or as a nonchromophoric bond between subunits. Implicit in this "separability condition" (Deutsche et al., 1969) which Kirkwood imposed on the electrons is the neglect of the following phenomena: charge-transfer complexes; the overlap between subunit wavefunctions; and intersubunit, but not intrasubunit, electron exchange. When these phenomena were neglected, it was possible for Kirkwood to construct exceedingly simple wavefunctions for quantum mechanically complex molecules by taking products of the wavefunctions of the subunits. When molecular wavefunctions are so constructed, the interpretation of the optical properties of the molecule as a whole can be made (as seen presently) in terms of the optical properties of the subunits themselves; the advantage here is that the optical parameters (e.g., dipole moments and transition wavelengths) of the subunits are usually easier to measure experimentally or evaluate theoretically than are the same properties of the entire molecule. Although the separability condition has played and continues to play a key role in our understanding of protein optical activity, one should also be aware of its limitations. First, a loss of quantitative accuracy ultimately accompanies the use of this condition (Deutsche et al., 1969); however, such a loss may be acceptable if completely descriptive molecular wavefunctions are too complex or if one's goals can be satisfied with qualitative results. Second, the problem of assigning electrons to specific regions of a molecule is a nontrivial one. As Deutsche et al. (1969) point out, the separability condition is largely a matter of the "nebulous and imperfect body of insight called chemical intuition" and ". . . the extent to which this condition of separability into chromophoric and nonchromophoric moieties is satisfied depends not only on the nature of the chromophore itself, but on the nature and disposition of the extrachromophoric part of the molecule as well."

Further improvements in the general quantum theory of optical activity awaited the ideas set forth by Moffitt (1956a). Intrigued by the discoveries that proteins (Pauling et al., 1951) and nucleic acids (Watson and Crick, 1953) could exist in elegant helical conformations, Moffitt endeavored to characterize the optical properties of polymers constituted by regularly ordered arrays of identical subunits. Moffitt focused his attention on Kirkwood's (1937) theory. Realizing that the wavefunctions used by Kirkwood were limited to nondegenerate systems only, Moffitt turned to solid-state exciton theory (Frenkel, 1936; Davydov, 1948, 1951, 1962) in order to construct excited-state wavefunctions suitable for molecules containing identical subunits and therefore exhibiting degenerate transitions. In essence, Moffitt envisaged the polymer molecule as a one-dimensional solid and its excited states as polymer excitons; he then formulated the optical properties of the polymer accordingly. One important consequence of Moffitt's work was the generalization of Kirkwood's coupled-oscillator mechanism of optical activity so as to include the coupling between degenerate as well as nondegenerate transitions.

Moffitt's unique concept of polymer excitons is key to our understanding of the optical activity of the transitions of proteins; it is best illustrated in a simple example. Consider an ideal system composed of two identical, mutually interacting chromophores (1 and 2) each of which is characterized by a ground-state wavefunction ($\phi_{10}$ and $\phi_{20}$) and only one excited-state wavefunction ($\phi_{1a}$ and $\phi_{2a}$). The absorption of a single photon by this dimer system initially places it in either of two states as symbolized by the following wavefunctions

$$\psi_{1a} = \phi_{1a}\phi_{20} \tag{34}$$

or

$$\psi_{2a} = \phi_{10}\phi_{2a} \tag{35}$$

However, as they stand, neither Eq. (34) nor Eq. (35) alone completely describes the system for all time since there is a finite probability—quantum mechanically arising from the interaction between the chromophores—that the excitation will spontaneously "migrate" from one chromophore to the other. Complete wavefunctions ($\psi_\pm$) for the dimer states then are found by taking linear combinations[12] of $\psi_{1a}$ and $\psi_{2a}$:

---

[12] Because the chromophores are identical and therefore are physically indistinguishable, the absorption of a photon by this system is just one of a general class of quantum mechanical interference phenomena, the wavefunctions for which must be written as normalized, linear combinations of the wavefunctions describing the possible outcomes of an experiment (cf. Feynman et al., 1965).

$$\psi_\pm = (\psi_{1a} \pm \psi_{2a})/\sqrt{2} \tag{36}$$

$\psi_+$ and $\psi_-$ represent "delocalized" dimer excitons, the excitation being distributed between two chromophores; $1/\sqrt{2}$ is the normalization factor needed to conserve probability. If these arguments are now generalized for the polymer with $N$ identical chromophores, the following polymer exciton wavefunctions obtain:

$$\psi_{AK} = \sum_{i=1}^{N} C_{iaK} \psi_{ia} \qquad (K = 1 \text{ to } N) \tag{37}$$

$\psi_{AK}$ is the wavefunction characterizing the $K^{\text{th}}$ exciton band; $\psi_{ia}$, like the wavefunctions defined in Eqs. (34) and (35), represents the polymer when only one chromophore ($i$) is in its excited state ($a$); and the $N$ sets of $N$ coefficients, $C_{iaK}$, are the normalizing factors which are found either by symmetry considerations (Moffitt, 1956a; Loxsom, 1969a) or by solving the secular determinant (Tinoco, 1962). The unique optical properties (cf. Tinoco, 1962; Craig and Wamsley, 1968) which such wavefunctions describe are considered in further detail below (see Section II,F,2).

Moffitt (1956a,b) specifically utilized his theory to characterize the absorption and ORD properties of the $\alpha$-helix. He was led to predict that the $\pi^0$–$\pi^-$ transitions of a right-handed polypeptide helix would together give rise to two distinct polymer absorption bands—one polarized parallel to the helix axis and another, at slightly longer wavelengths, polarized perpendicular to the helix axis. This prediction has subsequently been given strong experimental support: the splitting of the $\pi^0$–$\pi^-$ transition in the $\alpha$-helix absorption spectrum was first reported by Gratzer et al. (1961), Rosenheck and Doty (1961), and Tinoco et al. (1962); and the predicted polarizations of the two bands (see Fig. 5) were confirmed by Gratzer et al. (1961) and Holzwarth and Doty (1965). Moffitt's conclusions concerning the ORD of the $\alpha$-helix were in error, however. In a joint critique of Moffitt's theory by Moffitt, Fitts, and Kirkwood (MFK) (1957), it was shown that Moffitt's use of periodic boundary conditions for the infinite helix, while not inappropriate for calculating the absorption parameters of the helix, resulted in an incomplete expression for the rotational strength. The important missing factor was subsequently shown (Tinoco, 1964) to give rise to the non-Gaussian band described in footnote 9 (see also the discussion at the end of this section).

As a preliminary step to correcting the mistake in Moffitt's theory, MFK rederived Kirkwood's (1937) original theory so that it explicitly included degenerate transitions. The resulting theory was particularly important in that it paved the way for the more general theory of optical activity which was subsequently proposed by Tinoco (1962). Following the basic format of the MFK theory, but utilizing more complete polymer wavefunctions, Tinoco formulated a first-order perturbation theory of optical activity which is of pivotal importance in the history of optical activity theory. In essence, it unified into one theoretical framework the fundamental notions embodied in the earlier theories proposed by Rosenfeld (1928), Kirkwood (1937), Condon et al. (1937), and Moffitt (1956a). Moreover, its appearance marked the beginning of an era in which optical activity calculations have met with increasing success; in particular, the equations in Tinoco's paper or modified forms or extensions of these equations are at the heart of many of the calculations (Section II,G) involving protein-related structures.[13] The salient features of this theory are now briefly outlined.

In defining the electric and magnetic dipole moment operators for a polymer molecule, Tinoco imposed Kirkwood's (1937) separability condition (see above) on the electrons assigning each one to a specific subunit, or "group," in the polymer. Hence, the total electronic dipole moment operator of the polymer can be reduced to the sum of dipole moment operators of the individual subunits of the polymer

$$\mathbf{\mu} = \sum_i \mathbf{\mu}_i \tag{38}$$

where the summation includes all subunits (labeled $i$) of the polymer. Likewise, the total magnetic moment operator of the polymer is constituted from the operators for the subunits themselves:

$$\mathbf{M} = \sum_i [\mathbf{m}_i + (-e/2mc)\, \mathbf{R}_i \times \mathbf{p}_i] \tag{39}$$

Here, $\mathbf{m}_i$ and $\mathbf{p}_i$ are the magnetic dipole moment operator and the momentum operator, respectively, of the $i^{\text{th}}$ subunit. The division of terms in this equation follows from the fact that each electron in the polymer is located by a position vector, $\mathbf{R}_i + \mathbf{r}_{is}$; $\mathbf{R}_i$ locates the $i^{\text{th}}$ subunit relative to an arbitrary origin and $\mathbf{r}_{is}$ locates the $s^{\text{th}}$ electron of subunit $i$ relative to $\mathbf{R}_i$. (Note that the selection of the point in the subunit to which $\mathbf{R}_i$ extends is an important consideration, for, if properly chosen, the expressions for the rotational strength of the polymer are greatly simplified as we shall see presently.)

Now that the necessary operators have been defined, only the ground- and excited-state wavefunctions of the polymer are needed to formulate rotational strengths. Tinoco took two approaches in this matter, construct-

---

[13] This has been true for nucleic acid compounds also (see Bush and Brahms, 1967; Johnson and Tinoco, 1969; Gratzer and Cowburn, 1969; Tinoco and Cantor, 1970; Bush, 1970).

ing first-order perturbation wavefunctions from two different basis sets. One basis set of wavefunctions was formed from the eigenfunctions of the vacuum Hamiltonians of the isolated subunits; this is referred to herein as the *isolated-subunit formalism*. The other basis set was created from the self-consistent field wavefunctions of the subunits. In this case, each subunit wavefunction is an eigenfunction of the Hamiltonian operator formed from two operators: the vacuum Hamiltonian of the subunit plus a potential energy operator representing the time-averaged polymer static field felt by that subunit. This is the *static-field formalism*.

When the polymer wavefunctions of the isolated-subunit formalism (see Tinoco's Eqs. IIIB-14 and IIIB-15) are used to evaluate the electric and magnetic transition dipole moments of the polymer (see Tinoco's Eqs. IIIB-18 and IIIB-20) one finds, in accordance with Eq. (22), the following general expression for the rotational strength[14]

$$R_{0AK} = \text{Im}[\boldsymbol{\mu}_{0AK} \cdot \mathbf{M}_{AK0}] \tag{40}$$

$$= \sum_{i=1}^{N} \sum_{j=1}^{N} C_{iaK} C_{jaK} \left\{ \text{Im}[\boldsymbol{\mu}_{i0a} \cdot \mathbf{m}_{ja0}] \right. \tag{41a}$$

$$+ \mathbf{R}_j \cdot (\boldsymbol{\mu}_{j0a} \times \boldsymbol{\mu}_{i0a}) \pi / \lambda_a \tag{41b}$$

$$+ \sum_{k \neq i}^{N} V_{i0a;k0a} (\text{Im}[\boldsymbol{\mu}_{j0a} \cdot \mathbf{m}_{ka0}] - \text{Im}[\boldsymbol{\mu}_{k0a} \cdot \mathbf{m}_{ja0}]) \lambda_a / 2hc \tag{41c}$$

$$+ \sum_{k \neq i}^{N} V_{i0a;k0a} (\mathbf{R}_k + \mathbf{R}_j) \cdot (\boldsymbol{\mu}_{k0a} \times \boldsymbol{\mu}_{j0a}) \pi / 2hc \tag{41d}$$

$$+ \sum_{k \neq i} \sum_{b \neq a} V_{i0a;k0b} (\text{Im}[\boldsymbol{\mu}_{j0a} \cdot \mathbf{m}_{kb0}] / \lambda_a$$
$$+ \text{Im}[\boldsymbol{\mu}_{k0b} \cdot \mathbf{m}_{ja0}] / \lambda_b) 2\lambda_b^2 \lambda_a^2 / hc(\lambda_b^2 - \lambda_a^2) \tag{41e}$$

[14] This equation is the most general expression for the rotational strength allowed for by the wavefunctions and transition moments developed in Tinoco's (1962) paper. It is also slightly more general than any of the expressions found therein since Tinoco chose to emphasize several special cases of the most general situation. Two aspects of Eq. (41) are not in accord with Tinoco's results, however. The most important discrepancy resides in the exciton term of Eq. (41b) which corresponds to Tinoco's Eq. (IIIB-80a); a positive sign rather than a negative sign as found by Tinoco should precede this term. The other difference centers around Eqs. (41h), (41j), and (41l), which do not appear in Tinoco's expression of a related nature [i.e., Eq. (IIIB-22)]. The reason for this disparity remains unclear; however, one can envisage situations, rare as they may be, in which these terms could give nonnegligible contributions to the rotational strength, just as the other remaining static-field terms, Eqs. (41g), (41i), and (41k), have been shown to contribute nonnegligibly to the rotational strengths of the peptide $n_1-\pi^-$ and $\pi^0-\pi^-$ transitions (Schellman and Oriel, 1962; Woody and Tinoco, 1967).

$$+ \sum_{k \neq i} \sum_{b \neq a} V_{i0a;k0b}(\mathbf{R}_k - \mathbf{R}_j) \cdot (\mathbf{\mu}_{k0b} \times \mathbf{\mu}_{j0a}) 2\pi \lambda_b \lambda_a / hc(\lambda_b{}^2 - \lambda_a{}^2) \quad (41\mathrm{f})$$

$$+ \sum_{k \neq i} \sum_{b \neq a} V_{iab;k00}(\mathrm{Im}[\mathbf{\mu}_{j0a} \cdot \mathbf{m}_{ib0}] + \mathrm{Im}[\mathbf{\mu}_{i0b} \cdot \mathbf{m}_{ja0}]) \lambda_b \lambda_a / hc(\lambda_b - \lambda_a) \quad (41\mathrm{g})$$

$$+ \sum_{k \neq i} \sum_{b \neq a} V_{iab;k00}(\mathbf{R}_i/\lambda_b - \mathbf{R}_j/\lambda_a) \cdot (\mathbf{\mu}_{i0b} \times \mathbf{\mu}_{j0a}) \pi \lambda_b \lambda_a / hc(\lambda_b - \lambda_a) \quad (41\mathrm{h})$$

$$- \sum_{k \neq i} \sum_{b \neq a} V_{i0b;k00}(\mathrm{Im}[\mathbf{\mu}_{j0a} \cdot \mathbf{m}_{iab}] + \mathrm{Im}[\mathbf{\mu}_{iab} \cdot \mathbf{m}_{ja0}]) \lambda_b / hc \quad (41\mathrm{i})$$

$$- \sum_{k \neq i} \sum_{b \neq a} V_{i0b;k00}(\mathbf{R}_j - \mathbf{R}_i + \mathbf{R}_i \lambda_a/\lambda_b) \cdot (\mathbf{\mu}_{j0a} \times \mathbf{\mu}_{iab}) \pi \lambda_b / hc\lambda_a \quad (41\mathrm{j})$$

$$- \sum_{k \neq i} V_{i0a;k00} \mathrm{Im}[(\mathbf{\mu}_{iaa} - \mathbf{\mu}_{i00}) \cdot \mathbf{m}_{ja0}] \lambda_a / hc \quad (41\mathrm{k})$$

$$- \sum_{k \neq i} V_{i0a;k00} \mathbf{R}_j \cdot [\mathbf{\mu}_{j0a} \times (\mathbf{\mu}_{iaa} - \mathbf{\mu}_{i00})] \pi / hc \Big\} \quad (41\mathrm{l})$$

$R_{0AK}$ is the rotational strength of the $K^{\mathrm{th}}$ exciton band ($K = 1$ to $N$) of the polymer transition $0A$ which corresponds to a discrete transition $0a$ in one of $N$ identical chromophores. The outer summations labeled with $i$, $j$, and/or $k$ specify subunits: those with a superscript $N$ include only identical subunits which exhibit the $0a$ transition whereas those summation signs without the superscript $N$ include all subunits of the polymer. The inner summations labeled $b \neq a$ include all but the ground state and the excited states $a$ of the subunit specified by the outer summation. The electric and magnetic transition dipole moments, including those for transitions between excited states (e.g., $\mathbf{\mu}_{iab}$ and $\mathbf{m}_{iab}$), specifically refer to the isolated subunits. $\mathbf{\mu}_{i00}$ and $\mathbf{\mu}_{iaa}$ are, respectively, the permanent dipole moments of the ground state and excited state of the isolated subunit. The potential energy terms $V_{i0a;k0b}$, $V_{iab;k00}$, etc. are defined in the following way:

$$\begin{aligned} V_{i0a;k0b} &= \int \phi_{i0}\phi_{ia} V \phi_{k0}\phi_{kb} \, d\tau \\ V_{iab;k00} &= \int \phi_{ia}\phi_{ib} V \phi_{k0}\phi_{k0} \, d\tau \end{aligned} \quad (42)$$

etc.

where $\phi_{i0}$, etc., are the state wavefunctions of the isolated subunits themselves and $V$ is the potential energy operator which characterizes all intersubunit interactions. [Complete and approximate expressions for $V$ are found in Tinoco (1962), Schellman (1966, 1968), and Hohn and Weigang (1968); certain approximate forms of these potential energy terms are further considered in Section II,F.] Finally, it should be noted in passing that in the course of deriving Eq. (41) it was assumed that the exciton coefficients (i.e., $C_{iaK}$ and $C_{jaK}$) and the basis set wavefunctions are all real, and that the permanent magnetic moments of the subunits are negligibly small.

Although the foregoing expression for the rotational strength is explicitly written for degenerate transitions, it is completely general for either degenerate or nondegenerate transitions. If the transition is degenerate, $C_{iaK}$ and $C_{jaK}$ are found as discussed earlier in connection with the exciton wavefunctions, Eq. (37), and the transition frequency $\nu_{aK}$ of the exciton band is as follows[15]

$$\nu_{AK} = \nu_a + \sum_{i=1}^{N} \sum_{j \neq i}^{N} (C_{iaK} C_{jaK} V_{i0a;j0a})/h \tag{43a}$$

$$+ \left( \sum_{i=1}^{N} C_{iaK}{}^2 \sum_{j \neq i} V_{iaa;j00} \right) \Big/ h - \sum_{j \neq i} V_{i00;j00}/h \tag{43b}$$

where $\nu_a$ is the transition frequency of the isolated subunit itself. If the polymer transition is nondegenerate, then $N = 1$, $C_{iaK} C_{jaK} = \delta_{ij}$ (the Kronecker delta function, which is zero whenever $i \neq j$ and 1 whenever $i = j$), and Eq. (41) can therefore be rewritten as follows

$$R_{0A} = \text{Im}[\mathbf{\mu}_{i0a} \cdot \mathbf{m}_{ia0}] \tag{44a}$$

$$+ \sum_{k \neq i} \sum_{b \neq a} V_{i0a;k0b} (\text{Im}[\mathbf{\mu}_{i0a} \cdot \mathbf{m}_{kb0}]/\lambda_a$$

$$+ \text{Im}[\mathbf{\mu}_{k0b} \cdot \mathbf{m}_{ia0}]/\lambda_b) 2\lambda_a{}^2 \lambda_b{}^2 / hc(\lambda_b{}^2 - \lambda_a{}^2) \tag{44b}$$

$$+ \sum_{k \neq i} \sum_{b \neq a} V_{i0a;k0b} (\mathbf{R}_k - \mathbf{R}_i) \cdot (\mathbf{\mu}_{k0b} \times \mathbf{\mu}_{i0a}) 2\pi \lambda_b \lambda_a / hc(\lambda_b{}^2 - \lambda_a{}^2) \tag{44c}$$

$$+ \sum_{k \neq i} \sum_{b \neq a} V_{iab;k00} (\text{Im}[\mathbf{\mu}_{i0a} \cdot \mathbf{m}_{ib0}] + \text{Im}[\mathbf{\mu}_{i0b} \cdot \mathbf{m}_{ia0}]) \lambda_b \lambda_a / hc(\lambda_b - \lambda_a) \tag{44d}$$

$$- \sum_{k \neq i} \sum_{b \neq a} V_{iab;k00} \mathbf{R}_i \cdot (\mathbf{\mu}_{i0b} \times \mathbf{\mu}_{i0a}) \pi / hc \tag{44e}$$

$$- \sum_{k \neq i} \sum_{b \neq a} V_{i0b;k00} (\text{Im}[\mathbf{\mu}_{i0a} \cdot \mathbf{m}_{iab}] + \text{Im}[\mathbf{\mu}_{iab} \cdot \mathbf{m}_{ia0}]) \lambda_b / hc \tag{44f}$$

$$- \sum_{k \neq i} \sum_{b \neq a} V_{i0b;k00} \mathbf{R}_i \cdot (\mathbf{\mu}_{i0a} \times \mathbf{\mu}_{iab}) \pi / hc \tag{44g}$$

[15] Actually, Eqs. (43) and (44) are valid only if there is strong coupling between subunits. If there is weak coupling, the vibrations of the polymer will alter the transition frequencies, as Tinoco (1962) has shown. Because it is not always easy to decide whether the coupling is strong or weak, it is usually advisable to use experimental transition frequencies whenever possible rather than to attempt to calculate them.

$$-\sum_{k\neq i} V_{i0a;k00}\text{Im}[(\mathbf{\mu}_{iaa} - \mathbf{\mu}_{i00}) \cdot \mathbf{m}_{ia0}]\lambda_a/hc \tag{44h}$$

$$-\sum_{k\neq i} V_{i0a;k00}\mathbf{R}_i \cdot [\mathbf{\mu}_{i0a} \times (\mathbf{\mu}_{iaa} - \mathbf{\mu}_{i00})]\pi/hc \tag{44i}$$

The polymer transition frequency in this case is given by the following expression

$$\nu_A = \nu_a + \sum_{j\neq i} (V_{iaa;j00} - V_{i00;j00})/h \tag{45}$$

Turning to the static-field formalism, one finds that when the static-field polymer wavefunctions are used the most general first-order expression for the rotational strength is identical in form to Eqs. (41a)–(41f):

$$R'_{0AK} = \sum_{i=1}^{N}\sum_{j=1}^{N} C_{iaK}C_{jaK} \left\{\text{Im}[\mathbf{\mu}_{i0a} \cdot \mathbf{m}_{ja0}] \right. \tag{46a}$$

$$+ \mathbf{R}_j \cdot (\mathbf{\mu}_{j0a} \times \mathbf{\mu}_{i0a})\pi/\lambda_a \tag{46b}$$

$$+ \sum_{k\neq i}^{N} V_{i0a;k0a}(\text{Im}[\mathbf{\mu}_{j0a} \cdot \mathbf{m}_{ka0}] - \text{Im}[\mathbf{\mu}_{k0a} \cdot \mathbf{m}_{ja0}])\lambda_a/2hc \tag{46c}$$

$$+ \sum_{k\neq i}^{N} V_{i0a;k0a}(\mathbf{R}_k + \mathbf{R}_j) \cdot (\mathbf{\mu}_{k0a} \times \mathbf{\mu}_{j0a})\pi/2hc \tag{46d}$$

$$+ \sum_{k\neq i}\sum_{b\neq a} V_{i0a;k0b}(\text{Im}[\mathbf{\mu}_{j0a} \cdot \mathbf{m}_{kb0}]/\lambda_a$$

$$+ \text{Im}[\mathbf{\mu}_{k0b} \cdot \mathbf{m}_{ja0}]/\lambda_b)2\lambda_b^2\lambda_a^2/hc(\lambda_b^2 - \lambda_a^2) \tag{46e}$$

$$+ \sum_{k\neq i}\sum_{b\neq a} V_{i0a;k0b}(\mathbf{R}_k - \mathbf{R}_j) \cdot (\mathbf{\mu}_{k0b} \times \mathbf{\mu}_{j0a})2\pi\lambda_b\lambda_a/hc(\lambda_b^2 - \lambda_a^2)\left.\right\} \tag{46f}$$

Terms equivalent to Eqs. (41g)–(41l) are missing in Eq. (46) because the potential energy terms (i.e., $V_{iab;k00}$, $V_{i0b;k00}$, and $V_{i0a;k00}$) of the former characterize polymer static-field effects which, it will be recalled, are already taken into account in the basis set wavefunctions in this formalism. Again, the expression for the rotational strength here is completely valid for both degenerate and nondegenerate transitions, but if the transition is degenerate, $\nu_{AK}$ is given by Eq. (43a) alone and if the transition is nondegenerate, $\nu_A = \nu_a$. In spite of the facts that Eq. (46) is in the same form as Eqs. (41a)–(41f) and that Eq. (43a) is the same in both formalisms, the meanings of the individual parameters which constitute these equations are clearly different in the two cases. The transition and permanent dipole moments, the transition frequencies, the wavefunctions which define the potential energy terms, and the exciton coefficients all refer either to isolated subunits—Eq. (41)—or to subunits as they exist in the polymer static field—Eq. (46).

Thus far the expressions for the rotational strength have been completely general for transitions which are allowed both electrically and magnetically. It is frequently the case, however, that a transition is forbidden in one of these two components. As a consequence, the formula for $R_{0AK}$ can be greatly simplified, as noted earlier, provided that the points to which the $\mathbf{R}_i$, $\mathbf{R}_j$, and $\mathbf{R}_k$ vectors extend are judiciously selected; if possible, each point should be one relative to which either the local electric or magnetic transition dipole moment of a subunit (e.g., $\pmb{\mu}_{i0a}$, $\mathbf{m}_{ia0}$) is zero.[16] Therefore, if Eq. (41) corresponds to a local transition which is magnetically allowed ($\mathbf{m}_{ia0} \equiv \mathbf{m}_{ja0} \equiv \mathbf{m}_{ka0} \neq 0$) but electrically forbidden ($\pmb{\mu}_{i0a} \equiv \pmb{\mu}_{j0a} \equiv \pmb{\mu}_{k0a} = 0$), it simplifies to the following expression

$$R_{0A} = \sum_{K=1}^{N} R_{0AK} \tag{47}$$

$$= \sum_{i=1}^{N} \Big\{ + \sum_{k \neq i} \sum_{b \neq a} V_{i0a;k0b} \text{Im}[\pmb{\mu}_{k0b} \cdot \mathbf{m}_{ia0}] 2\lambda_b \lambda_a^2 / hc(\lambda_b^2 - \lambda_a^2) \tag{48a}$$

$$+ \sum_{k \neq i} \sum_{b \neq a} V_{iab;k00} \text{Im}[\pmb{\mu}_{i0b} \cdot \mathbf{m}_{ia0}] \lambda_b \lambda_a / hc(\lambda_b - \lambda_a) \tag{48b}$$

$$- \sum_{k \neq i} \sum_{b \neq a} V_{i0b;k00} \text{Im}[\pmb{\mu}_{iab} \cdot \mathbf{m}_{ia0}] \lambda_b / hc \tag{48c}$$

$$- \sum_{k \neq i} V_{i0a;k00} \text{Im}[(\pmb{\mu}_{iaa} - \pmb{\mu}_{i00}) \cdot \mathbf{m}_{ia0}] \lambda_a / hc \Big\} \tag{48d}$$

It will be noted here that the exciton bands have been added together using the relationship $\Sigma_{K=1}^{N} C_{iaK} C_{jaK} = \delta_{ij}$; this has been done because the interactions between electrically forbidden transitions are generally weak and are therefore expected to lead to negligibly small exciton splitting and correspondingly small exciton contributions to the rotational strength. Tinoco specifically recommended using this expression (corresponding to his Eq. III-24) for calculating the rotational strengths of electrically forbidden transitions rather than using the corresponding static-field

---

[16] A certain amount of caution must be exercised in using the static-field formalism here. A transition which is either electrically or magnetically forbidden in the isolated subunit may not be necessarily so when the subunit is part of the macromolecule; that is to say, the static field of the polymer may break down the selection rules which govern the transitions of the subunit *in vacuo*. Furthermore, the static field of a molecule may not be uniform throughout. Thus different subunits which are anatomically identical may experience different local fields (as discussed in Section II,D,2) and may therefore have slightly different transition moments and wavelengths as defined in the static-field formalism.

expression [identical to Eq. (48a) alone]. The rationale for this stems from the fact that the directions and magnitudes of magnetic transition dipole moments are impossible to measure by ordinary means and therefore must be quantum mechanically calculated, a process generally easier with isolated-subunit wavefunctions.

If the polymer transition corresponds to a local transition which is magnetically forbidden ($\mathbf{m}_{ia0} \equiv \mathbf{m}_{ja0} \equiv \mathbf{m}_{ka0} = 0$; $\mathbf{\mu}_{j0a} \equiv \mathbf{\mu}_{j0a} \equiv \mathbf{\mu}_{k0a} \neq 0$), Eq. (41) reduces to

$$R_{0AK} = \sum_{i=1}^{N} \sum_{j=1}^{N} C_{iaK} C_{jaK} \left\{ (\mathbf{R}_j - \mathbf{R}_i) \cdot (\mathbf{\mu}_{j0a} \times \mathbf{\mu}_{i0a}) \pi/2\lambda_a \right. \tag{49a}$$

$$+ \sum_{k \neq i}^{N} V_{i0a;k0a}(\mathbf{R}_k + \mathbf{R}_j) \cdot (\mathbf{\mu}_{k0a} \times \mathbf{\mu}_{j0a}) \pi/2hc \tag{49b}$$

$$+ \sum_{k \neq i} \sum_{b \neq a} V_{i0a;k0b} \text{Im}[\mathbf{\mu}_{j0a} \cdot \mathbf{m}_{kb0}] 2\lambda_b^2 \lambda_a / hc(\lambda_b^2 - \lambda_a^2) \tag{49c}$$

$$+ \sum_{k \neq i} \sum_{b \neq a} V_{i0a;k0b}(\mathbf{R}_k - \mathbf{R}_j) \cdot (\mathbf{\mu}_{k0b} \times \mathbf{\mu}_{j0a}) 2\pi \lambda_b \lambda_a / hc(\lambda_b^2 - \lambda_a^2) \tag{49d}$$

$$+ \sum_{k \neq i} \sum_{b \neq a} V_{iab;k00} \text{Im}[\mathbf{\mu}_{j0a} \cdot \mathbf{m}_{ib0}] \lambda_b \lambda_a / hc(\lambda_b - \lambda_a) \tag{49e}$$

$$+ \sum_{k \neq i} \sum_{b \neq a} V_{iab;k00}(\mathbf{R}_i/\lambda_b - \mathbf{R}_j/\lambda_a) \cdot (\mathbf{\mu}_{i0b} \times \mathbf{\mu}_{j0a}) \pi \lambda_b \lambda_a / hc(\lambda_b - \lambda_a) \tag{49f}$$

$$- \sum_{k \neq i} \sum_{b \neq a} V_{i0b;k00} \text{Im}[\mathbf{\mu}_{j0a} \cdot \mathbf{m}_{iab}] \lambda_b / hc \tag{49g}$$

$$- \sum_{k \neq i} \sum_{b \neq a} V_{i0b;k00}(\mathbf{R}_j - \mathbf{R}_i + \mathbf{R}_i \lambda_a / \lambda_b) \cdot (\mathbf{\mu}_{j0a} \times \mathbf{\mu}_{iab}) \pi \lambda_b / hc \lambda_a \tag{49h}$$

$$\left. - \sum_{k \neq i} V_{i0a;k00} \mathbf{R}_j \cdot [\mathbf{\mu}_{j0a} \times (\mathbf{\mu}_{iaa} - \mathbf{\mu}_{i00})] \pi/hc \right\} \tag{49i}$$

where Eq. (49a) is another way of writing Eq. (41b) provided that the coefficients are real. Again, in the corresponding static-field expression

$$R'_{0AK} = \sum_{i=1}^{N} \sum_{j=1}^{N} C_{iaK} C_{jaK} \left\{ (\mathbf{R}_j - \mathbf{R}_i) \cdot (\mathbf{\mu}_{j0a} \times \mathbf{\mu}_{i0a}) \pi/2\lambda_a \right. \tag{50a}$$

$$+ \sum_{k \neq i}^{N} V_{i0a;k0a}(\mathbf{R}_k + \mathbf{R}_j) \cdot (\mathbf{\mu}_{k0a} \times \mathbf{\mu}_{j0a}) \pi/2hc \tag{50b}$$

$$+ \sum_{k \neq i} \sum_{b \neq a} V_{i0a;k0b} \text{Im}[\mathbf{\mu}_{j0a} \cdot \mathbf{m}_{kb0}] 2\lambda_b^2 \lambda_a / hc(\lambda_b^2 - \lambda_a^2) \tag{50c}$$

$$\left. + \sum_{k \neq i} \sum_{b \neq a} V_{i0a;k0b}(\mathbf{R}_k - \mathbf{R}_j) \cdot (\mathbf{\mu}_{k0b} \times \mathbf{\mu}_{j0a}) 2\pi \lambda_b \lambda_a / hc(\lambda_b^2 - \lambda_a^2) \right\} \tag{50d}$$

only those terms with nonstatic potential energy terms are retained. Tinoco specifically recommended using the static-field formalism to calculate rotational strengths for electrically allowed transitions since it should be possible, in principle at least, to measure directly the magnitudes and directions of the electric transition moments of the polymer subunits. In practice, however, the formalism one chooses or the extent to which the formalisms are mixed—as is frequently the case—is largely a matter of necessity, depending upon the available experimental and theoretical information concerning for example, the transition moments, transition frequencies, wavefunctions, etc., of the subunits.

Returning to the most general expression here (Eq. 41), one readily finds that Tinoco's equations totally embody the fundamental principles set forth in the earlier theories of optical activity developed by Rosenfeld (1928), Condon et al. (1937), Kirkwood (1937), and Moffitt (1956a). That part of Eq. (41) which characterizes the contribution to the rotational strength arising from intrinsically optically active transitions (Rosenfeld) is Eq. (41a) when $i = j$. The optical activity which stems from the perturbation of a chromophore by the time-averaged static field of the polymer, the CAE one-electron mechanism, is described by Eqs. (41g)–(41l) when $i = j$. Kirkwood's coupled-oscillator mechanism for nondegenerate, electric transitions is represented by Eq. (41f), whereas Moffitt's extension of this mechanism for degenerate transitions is represented by the exciton term, Eq. (41b). The conventional electric–magnetic coupling mechanism (Schellman, 1968) is represented by Eq. (41e). The remaining terms in Eq. (41) can be viewed, if one so desires, as extensions or combinations of the above mechanisms. For instance, several terms other than Eq. (41e)—Eqs. (41a), $i \neq j$; (41c), $k \neq j$; (41g), $i \neq j$; and (41i), $i \neq j$—involve the coupling of electric and magnetic transitions of different chromophores; and some of these, i.e., Eqs. (41g) and (41i), also contain one-electron-type potential energy terms. As another example, Eqs. (41h) and (41j) contain one-electron-type potential energy terms but their vector products characterize coupled-oscillator effects whenever $i \neq j$.

Although the above equations generally provide a good starting point for an optical activity calculation, as evidenced in many of the calculations reviewed in Section II,G, the recent trend, in protein calculations at least, is to replace these first-order perturbation expressions with expressions based on the matrix formalism of quantum mechanics. In short, this approach, as described in detail by Bayley et al. (1969) and Pysh (1970a), involves solving the secular determinant for the system under consideration. In this way, the eigenvalues and eigenvectors of the system are found and these in turn can be used to calculate rotational strengths. A simple example based on one given in Bayley *et al.* (1969) best illustrates the results of such a procedure.

Consider a system composed of two interacting peptides (labeled *1* and *2*). Restricting our attention to only the $n_1$–$\pi^-$ and $\pi^0$–$\pi^-$ transitions, we find that the Hamiltonian matrix for this system is

$$\underline{\underline{H}} = \begin{pmatrix} \begin{pmatrix} E_a & V_{1ab;200} \\ V_{1ba;200} & E_b \end{pmatrix} & \begin{pmatrix} 0 & V_{10a;20b} \\ V_{10b;20a} & V_{10b;20b} \end{pmatrix} \\ \begin{pmatrix} 0 & V_{20a;10b} \\ V_{20b;10a} & V_{20b;10b} \end{pmatrix} & \begin{pmatrix} E_a & V_{2ba;100} \\ V_{2ab;100} & E_b \end{pmatrix} \end{pmatrix} \qquad (51)$$

Here, the energy terms are as those in the foregoing expressions; $E_a = h\nu_a$; and $E_b = h\nu_b$. It has been assumed that the potential energy of interaction between the magnetic $n_1$–$\pi^-$ transitions $V_{10a;20a}$ is negligibly small, being replaced by zero in the matrix. One now diagonalizes Eq. (51) with the unitary transformation matrix $\underline{\underline{C}}$ solving the secular determinant $|\underline{\underline{H}} - E\underline{\underline{I}}| = 0$ in the process. ($E$ denotes the eigenvalues of the determinant and $\underline{\underline{I}}$ is the unitary matrix.) In this way one can determine the elements (coefficients) of $\underline{\underline{C}}$ and then use these to evaluate the electric and magnetic transition dipole moments for the system.

The electric transition dipole moments of the system are found from the following equation

$$\underline{\mu}' = \underline{\mu}\,\underline{\underline{C}} \qquad (52)$$

where the elements of $\underline{\mu}$ (a row vector) are the transition moments of the isolated peptides and the elements of $\underline{\mu}'$ are transition moments for the system as a whole. For the dipeptide model system then, an element of $\underline{\mu}'$ would appear thus:

$$\underline{\mu}'_{0K} = \underline{\mu}_{10a} C_{1aK} + \underline{\mu}_{10b} C_{1bK} + \underline{\mu}_{20a} C_{2aK} + \underline{\mu}_{20b} C_{2bK} \qquad (53)$$

which further simplifies to

$$\underline{\mu}'_{0K} = \underline{\mu}_{10b} C_{1bK} + \underline{\mu}_{20b} C_{2bK} \qquad (54)$$

since $\underline{\mu}_{10a} \equiv \underline{\mu}_{20a} = 0$ for the $n_1$–$\pi^-$ transition. The element $\underline{\mu}'_{0K}$ is the transition dipole moment of the $0K$ transition from the ground state (0) of the system to the $K^{th}$ energy level; the energy $E_K$ for this transition is the $K^{th}$ eigenvalue; and since there are only four wavefunctions in the basis set, $K = 1$ to 4. The similarity in form between the coefficients $C_{1bK}$, etc. here and the exciton coefficients defined in earlier equations is intentional for they are both basically the same in meaning: that is, they transform the frame of reference from one basis set (e.g., the isolated-subunit wavefunctions) to the basis set which is constituted by the eigenfunctions of the Hamiltonian for the entire system.

The corresponding magnetic transition dipole moment operator matrix $\underline{\mathbf{M}}'$ of the system is found in a similar fashion. An element of $\underline{\mathbf{M}}'$ is written as follows:

$$\mathbf{M}'_{K0} = \mathbf{m}'_{K0} + (\mathbf{R} \times \mathbf{\mu}'_{K0}) i\pi/\lambda_K \tag{55}$$

where $\lambda_K = hc/E_K$; $\mathbf{\mu}'_{K0}$ is found as before; and $\mathbf{m}'_{K0}$ is found with the relationship $\underline{\mathbf{m}}' = \underline{\mathbf{m}}\underline{\underline{\mathbf{C}}}$ (it being assumed here that the coefficients are all real). For simplicity of notation, there is no subscript designation on the position vector $\mathbf{R}$ in Eq. (55); it is assumed that the appropriate subscript is automatically selected as the cross-product of each component of $\mathbf{\mu}'_{K0}$ is formed. For the dipeptide, an element of $\mathbf{M}'_{K0}$ is written as follows (keeping in mind that $\mathbf{m}_{10b} = \mathbf{m}_{20b} = 0$)

$$\mathbf{M}'_{K0} = \mathbf{m}_{1a0} C_{1aK} + (\mathbf{R}_1 \times \mathbf{\mu}_{10b}) i\pi C_{1bK}/\lambda_K + \mathbf{m}_{2a0} C_{2aK}$$
$$+ (\mathbf{R}_2 \times \mathbf{\mu}_{20b}) i\pi C_{2bK}/\lambda_K \tag{56}$$

Finally, the matrix for the rotational strength is

$$R = \text{Im}[\underline{\mathbf{\mu}}' : \tilde{\underline{\mathbf{M}}}'] \tag{57}$$

Here $\tilde{\underline{\mathbf{M}}}'$ is the transpose of $\underline{\mathbf{M}}'$ and the symbol ( : ) means that the scalar products are formed element by element. With Eqs. (54), (56), and (57), then, one finds the following expression for the rotational strength of the $K^{\text{th}}$ transition of the dipeptide system

$$R_{0K} = \text{Im}[\mathbf{\mu}'_{0K} \cdot \mathbf{M}'_{K0}] \tag{58}$$
$$= C_{1bK} C_{1aK} \text{Im}[\mathbf{\mu}_{10b} \cdot \mathbf{m}_{1a0}] + C_{2bK} C_{2aK} \text{Im}[\mathbf{\mu}_{20b} \cdot \mathbf{m}_{2a0}] \tag{59a}$$
$$+ C_{1bK} C_{2aK} \text{Im}[\mathbf{\mu}_{10b} \cdot \mathbf{m}_{2a0}] + C_{2bK} C_{1aK} \text{Im}[\mathbf{\mu}_{20b} \cdot \mathbf{m}_{1a0}] \tag{59b}$$
$$+ C_{1bK} C_{2bK} (\mathbf{R}_2 - \mathbf{R}_1) \cdot (\mathbf{\mu}_{20b} \times \mathbf{\mu}_{10b}) \pi/\lambda_K \tag{59c}$$

As we see from the vector products in Eq. (59), the matrix formalism also renders the three basic mechanisms of optical activity discussed above in connection with perturbation theory: Eq. (59a) characterizes the one-electron mechanism; Eq. (59b) describes the electric–magnetic coupling mechanism; and Eq. (59c) characterizes the coupled-oscillator mechanism. However, it should be noted that the potential energy terms which effect these three mechanisms are now no longer independent of one another, as they were in perturbation theory, but are linked together in the Hamiltonian matrix, Eq. (51). Thus, the mechanisms are also linked to one another through the coefficients and are therefore not simply additive as they were in first-order perturbation theory.

There are basically two advantages to using the matrix formalism as it has been described here. On the one hand, it represents a higher order of approximation than the first-order perturbation approach and it should, in principle at least, lead to more accurate results in optical activity calculations. On the other hand, the matrix formalism circumvents two inherent difficulties associated with perturbation theory: perturbation theory per se is restricted to situations in which the perturbation energies are relatively small, whereas the matrix formalism is not; moreover, with the matrix method one avoids the discontinuities arising from the denominators of Eqs. (41e)–(41h) as $\lambda_b$ approaches $\lambda_a$. In spite of these limitations, though, the perturbation expressions have one distinct advantage over the corresponding matrix expressions: they are usually much simpler to evaluate. Except for the exciton coefficients appearing in Eqs. (41), etc., one determines the rotational strength in the perturbation formalism simply by adding terms, regardless of the number of subunits, transitions, or mechanisms under consideration. In contrast, the Hamiltonian matrix in the matrix formalism becomes larger and increasingly more difficult and expensive (in computer time) to diagonalize as one incorporates more transitions, subunits, etc., into the calculation.

From a purely theoretical point of view, most of the recent interest in optical activity has centered around deriving the theory from principles more general than those upon which the first-order perturbation theory of optical activity is based. One example of this, the matrix formalism or the so-called excited-state configuration interaction method, has already been discussed. Essentially, the remaining nonperturbation approaches to the theory of optical activity can be grouped into three categories[17]: (1) Ando and Nakano (1966), Ando (1968), Nakano and Kimura (1969), Rhodes (1970a,b), Deutsche (1970), and Blum and Frisch (1970, 1971) have derived expressions for ORD following the guidelines of linear response theory, or generalized susceptibility theory. (2) The ORD of polymeric molecules has been formulated in terms of time-dependent Hartree theory by McLachlan and Ball (1964) and Harris (1965). (3) Finally, in several papers (Stephen, 1958; Power and Shail, 1959; Hameka, 1964b, 1965; Gō, 1965, 1966, 1967; Chiu, 1969; Hutchinson, 1968; Lawetz and Hutchinson, 1969; Atkins and Woolley, 1970; Loxsom, 1969a,b, 1970; Loxsom et al., 1971; Philpott, 1972) the theory of optical activity has been formulated within the framework of scattering theory.

While these nonperturbation approaches can be shown to contain implicitly low-order perturbation theory—this is, to lead the perturbation formalism when appropriate limiting approximations are made—

---

[17] With a few exceptions (Lawetz and Hutchinson, 1969; Blum and Frisch, 1970; Loxsom et al., 1971), these works have focused specifically on ORD. The recent surge of interest in CD makes it highly desirable that the findings and conclusions reached in these works be extended to include CD explicitly. It appears, in a few instances at least, that but relatively modest theoretical manipulations are required to achieve this.

they offer as their main advantage over the perturbation method (and the matrix method as well) greater accuracy since they retain the effects of multiply excited configurations. In addition to this, they offer a more rigorous, if not more esthetic, conceptual framework for understanding optical activity. For example, in contrast to the semiclassical theory of optical activity that was conceived by Rosenfeld (1928), Condon (1937), and Kirkwood (1937), scattering theory provides a complete quantum mechanical description of optical activity because: (1) *both* the material and electromagnetic field properties are quantized and (2) the real observables of an optical activity experiment are rigorously defined in terms of the Stokes parameters (cf. McMaster, 1954; Jauch and Rohrlich, 1955; Gō, 1967).

The question is now raised as to what if anything new have these recent improvements in the theory provided concerning polymer molecules, especially proteins. So far, only the optical activity of helical polymers, such as the $\alpha$-helix, has been given consideration (McLachlan and Ball, 1964; Gō, 1965; Loxsom, 1969a,b, 1970; Rhodes, 1970a; Loxsom *et al.*, 1971; Philpott, 1972) but some interesting findings have emerged (see also Section II,G,1). In contrast to the conclusions reached by MFK (1957) concerning Moffitt's (1956a) theory (see p. 475), it has been demonstrated by Ando (1968), Loxsom (1969a,b, 1970), and Deutsche (1970) that periodic boundary conditions (p.b.c.) are in fact valid in formulating the optical activity parameters for the degenerate transitions of a long multimonomeric (and homogeneous) helical polymer provided that the correct level of approximation is retained in the pertinent expressions. Specifically, it has been shown by these authors that Moffitt lost an important term in his equations because he was led to incorrect helix dipole selection rules by utilizing p.b.c. in conjunction with Rosenfeld's (1928) expression for the ORD (essentially Eq. 25). Rosenfeld's expression is valid only for molecules whose overall dimensions are small in comparison with the wavelength of the light. MFK corrected the error by abandoning p.b.c. and taking end-effects into consideration. But, the above authors maintain that p.b.c. are perfectly acceptable for very long helices—and in fact greatly simplify the problem—provided that the ORD is described by a more general expression (cf. Stephen, 1958) than Rosenfeld's. (If Rosenfeld's assumption concerning the relative size of the molecule and the wavelength of light is not made, the simple relationship between the ORD and the electric and magnetic transition moments, as embodied in the rotational strength, no longer holds.) The significance of these findings is summarized by Rhodes (1970a): "For most systems the MFK terms [and hence Tinoco's equations] are sufficiently accurate. The important result is not a quantitative modification of the theory, but is the fact that the applicability of periodic symmetry has been restored to its proper place."

As the reader is likely to perceive, the discussion herein has been limited to but a scant part of an enormous body of literature concerning the theory of optical activity generated over the long history of interest in this field. The emphasis here has been placed on that which is considered to be most fundamental to our understanding of optical activity. In addition, a few recent theoretical advances have been discussed primarily because, in the words of Pysh (1970a), they "should form the foundation for future calculations." Although a few other related theoretical works are discussed in the following section, a great number of the papers which have added valuable insight, improvements, modifications, or simply clarity to the theory have been ignored here for the sake of conciseness, and for this the authors apologize.

## F. Symmetry Rules and the Origins of Optical Activity

In the ensuing discussion, patterned after Schellman's (1968) inspirational treatise on this subject, the major representative terms of the three mechanisms of optical activity are evaluated with the aid of a highly simplified model system. This system, being similar to the two-state exciton and dipeptide model systems developed earlier (in Section II,E), will consist of two interacting groups and will exhibit only two transitions; the set of physical parameters (e.g., wavelength and transition moment) associated with each transition will vary according to the individual mechanism under consideration and, in fact, will determine which mechanism is operative since only one can prevail for a given set. The purposes of using this dimer model are twofold: to present a basic conceptual picture of the origins of optical activity; and to develop for each mechanism simple symmetry rules, or geometrical constructs as Schellman defines them, which correlate rotational strengths with the possible three-dimensional configurations of the two groups relative to one another. Ultimately, of course, it is hoped that these rules can be utilized as a means of predicting either the conformations of a real molecular structure from observed rotational strengths or, in the opposite sense, the order of magnitude and the sign of the rotational strength from the known geometry of a molecular structure. Although these symmetry rules can be useful in certain well-defined situations, as we shall see, their limited applicability, as dictated by the simplicity of the model itself, must be kept in mind: "To apply the results (rules) to real molecules, one must assume either that other groups and transitions are unimportant or that their effects are additive. If this is not true, symmetry rules are of little value and the origins of Cotton effects can be comprehended only on the basis of quantitative calculations. Despite their limitations, symmetry rules have great heuristic value and have provided the solution to many practical structural problems" (Schellman, 1968).

### 1. *The One-Electron Mechanism*

As defined in the previous section, the one-electron mechanism prevails in either of two situations: (1) when a chromophore is inherently optically active and (2) when a chromophore is optically active by virtue of the perturbing effects of a static asymmetric environment. In order for the first case to obtain, the transition itself and the chromophore from which it arises must satisfy several reciprocally related conditions. Obviously, the transition must be inherently allowed both electrically and magnetically since Rosenfeld's expression, Eq. (22), as it applies to a single chromophore, must be nonzero in this situation. This condition, in turn, requires that the chromophore be asymmetric, having no planes or centers of symmetry. This last requirement can be understood by the following series of arguments[18]: For one or more components of the electric and magnetic transition dipole moment integrals [Eqs. (17) and (18)] to be nonzero, one or more vector components (e.g., the $x$ component) of the electric and magnetic dipole moment operators must belong to the same representation (of the symmetry group of the chromophore) to which the excited-state wavefunction belongs. Therefore, Eq. (22) will be nonzero in value only when the excited-state wavefunction, at least one component of the electric dipole moment operator, and the same component of the magnetic dipole moment operator all belong to the same representation. However, this occurs if and only if the chromophore is asymmetric, for it can be shown that a plane or center of symmetry will cause the corresponding components of these operators to segregate according to different representations of the same symmetry

---

[18] Implicit in these arguments are the following: the ground-state wavefunction belongs to the totally symmetric representation of the symmetry group of the chromophore; the electric dipole moment operator and, therefore, the electric transition moment both transform (i.e., in a coordinate transformation) like polar vectors; the magnetic dipole moment operator and, therefore, the magnetic transition moment both transform like axial vectors. Note that because the rotational strength is formed from the dot-product of an axial vector and a polar vector, it is a pseudoscalar, which is in every way identical to a scalar except that it changes sign in a coordinate inversion; this property is expected of the rotational strength since it is a scalar quantity but is oppositely signed for molecular structures which are the inverse of or the mirror image of one another.

group (Schellman, 1966). Since most protein chromophores can be regarded as having planar symmetry, few are expected to exhibit inherent optical activity, to a first approximation at least, and we will focus our attention on the second situation above. It should be noted though that a particularly important exception is the disulfide, the optical activity of which is understood to arise, at least in part, from the inherently asymmetric screw sense of its dihedral angle as discussed in Section IV,D.

More often than not optical activity is found to arise in chromophores which are inherently symmetric. The basic explanation for this is that the chromophore is asymmetrically perturbed by its environment which breaks down its effective symmetry and thereby renders some or all of its transitions optically active. (Note that an inherently optically active transition may acquire additional optical activity in this way.) Such a perturbing environment could be produced by the remainder of the molecule containing the chromophore, another chromophore, an ion, a solvent molecule, etc.

The kind of perturbation which gives rise to the one-electron mechanism is the time-averaged static field in which the chromophore is located. [The "dynamic" perturbations which produce the coupled-oscillator and electric-magnetic mechanisms are discussed in Sections II,F,2 and II,F,3.] The field essentially generates "lending and borrowing" of transition and permanent dipole moments between different states of the same chromophore, as indicated by Eqs. (41g)–(41l) when $i = j$. That is to say, nonzero values for the rotational strength are achieved because the perturbation—as manifested in the static-field energy terms $V_{iab;koo}$, $V_{iob;koo}$, and $V_{ioa;koo}$—"mixes"—through the vector dot- and cross-products—either different transitions of the same chromophore [Eqs. (41g)–(41j)] or a transition with its ground- and excited-state permanent dipole moments [Eqs. (41k) and (41l)]. Of the various one-electron terms, only Eq. (41g) will be considered further for the following reasons: it tends to be the dominant one-electron term when a transition is electrically forbidden [see Eq. (48b)] because of its relatively small denominator; and it provides an adequate qualitative description of the optical activity of a transition of central interest here, the $n_1$–$\pi^-$ transition originating in the carbonyl moiety of ketones and peptides.

Returning to the model outlined above, one finds that the one-electron mechanism obtains under the following set of conditions: both transitions arise from the same group (arbitrarily labeled *1*). One transition (transition $a$ of wavelenth $\lambda_a$) is magnetically allowed ($\mathbf{m}_{1ao} \neq 0$); the other transition (transition $b$ of wavelength $\lambda_b$) is electrically allowed ($\mathbf{\mu}_{1ob} \neq 0$). Finally, the second group acts as a perturber, mixing together transitions $a$ and $b$. Under these conditions, then, Eq. (48b) reduces to

$$R_a = + \frac{V_{1ab;200}\text{Im}[\mathbf{\mu}_{10b} \cdot \mathbf{m}_{1a0}]\lambda_b\lambda_a}{hc(\lambda_b - \lambda_a)} \tag{60}$$

$R_a$ is the rotational strength of the band at $\lambda_a$ and $V_{1ab;200}$ is the potential energy of interaction between the groups. Likewise, the rotational strength $R_b$ at $\lambda_b$ is found to be[19]

$$R_b = - \frac{V_{1ab;200}\text{Im}[\mathbf{\mu}_{10b} \cdot \mathbf{m}_{1a0}]\lambda_b\lambda_a}{hc(\lambda_b - \lambda_a)} \tag{61}$$

which is equal in magnitude to $R_a$ but opposite in sign. Adopting the model to the ketone and peptide chromophores, $R_a$ signifies the rotational strength of the $n_1$–$\pi^-$ transition located at approximately 290 m$\mu$ in ketones, or at approximately 225 m$\mu$ in peptides, and the $b$ transition signifies an as yet unspecified electrically allowed transition.

As Eqs. (60) and (61) reveal, the geometrical dependence of the one-electron mechanism—that is, the dependence on the relative orientation of the groups—lies completely within the nature of the potential energy of interaction $V_{1ab;200}$; clearly, the vector dot-product in these equations has no geometrical dependence since the transition moment vectors are (to a first approximation at least) predetermined by the electronic structure of the chromophore itself. In order that $V_{1ab;200}$ be nonvanishing, it has been shown by Schellman (1966) that the potential energy operator $V$ must belong in part to the pseudoscalar representation of the symmetry group to which the chromophore belongs; Schellman also demonstrated that the higher the symmetry of the chromophore, the higher the multipole (i.e., monopole, dipole) term in $V$ needed to induce optical activity. Both of these findings suggest that any symmetry rule which relates the optical activity of a chromophore to its environment will depend in a nontrivial way on the natural symmetry properties of the chromophore itself. This has in fact also been elegantly demonstrated by Schellman, who concluded from an analysis of the appropriate pseudoscalar potential functions for various symmetry groups that qualitative symmetry rules, or so-called regional rules, can be devised in the following manner (Schellman, 1968): "... use the natural planes of symmetry of the chromophore to divide all space into regions bounded by these planes. The contribution to the rotational strength of the perturbing group changes sign as it passes from one region to another. [$R$ is zero when

---

[13] Note that Eq. (48) (or any of the other perturbation expressions, for that matter) is defined for a transition arbitrarily labeled $a$ which is mixed or coupled with other transitions labeled $b$; therefore, in order to find the rotational strength of a transition labeled $b$, one must interchange labels. The signs of the terms in the numerator of Eq. (60) are insensitive to a label change, whereas the denominator is not, thus accounting for the sign difference here between $R_a$ and $R_b$.

the pertuber lies in a plane dividing two regions.] Once the sign of one region is established (by experimental or detailed calculation), all the others follow." Note that this procedure only gives the minimum number of planes.

Applying this procedure to the ketone chromophore, one is led, upon first consideration, to a quadrant rule: the two orthogonal symmetry planes (in Fig. 7b the $xz$- and the $yz$-planes) of the carbonyl which inter-

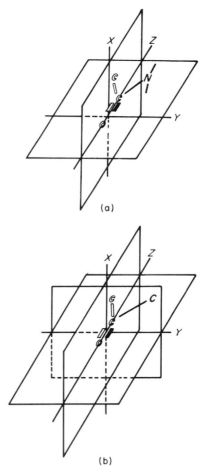

FIG. 7. (a) Quadrant rule for the peptide group. The vertical surface is not planar because of the horizontal distortion of the nonbonding electrons of the C=O group. (b) Octant rule for ketones. The surface separating the C from the O atom is not planar because of the unsymmetrical distribution of the $\pi$ electrons on the C and O atoms. No attempt is made to depict the deviations from planarity, since these are very sensitive to the wave functions assumed (Schellman, 1968).

sect along the CO bond (the $z$-axis) divide the space around the carbonyl into quadrants; the sign of the rotational strength then corresponds either to the sign or the negative of the sign of the product $xy$, where $x$ and $y$ locate the perturbing group in the chosen coordinate system. However, this analysis seemingly fails since a substantial body of experimental evidence (Moffitt et al., 1961; Djerassi, 1960; Bouman and Moscowitz, 1968) decidedly supports octant behavior ($xyz$ sign dependence) rather than quadrant behavior for the ketone $n_1$–$\pi^-$ transition. To resolve this difficulty, Schellman (1968) argues that additional symmetry planes must be used in constructing the regional rule if the wavefunctions or orbitals of the transition states have higher local symmetry than the chromophore itself. In the case of the ketone, the excited-state $\pi^-$ orbital will have a nodal plane which lies between the carbon and oxygen atoms. Inclusion of this plane, which is orthogonal to the other two, yields the desired octant symmetry rule (Fig. 7b). A similar set of considerations (see Fig. 7a) suggests a quadrant rule for the peptide: the physical plane of the peptide and the orthogonal nodal $xz$-plane of the nonbonding, $n_1$, atomic orbital of oxygen divide all space into quadrants.

The construction of regional rules in the manner outlined here raises numerous questions. For example: Under what circumstances do the atomic orbitals of a chromophore change its "effective" symmetry? Is a regional rule dependent upon the nature of the perturbation? In what way does a symmetry rule depend upon the transitions which are thought to mix with the transition in question? The answers to these and other questions can be resolved only through more detailed considerations. For example, Bouman and Moscowitz (1968), extending Moscowitz' (1962) earlier calculations, arrived at the following interesting conclusions concerning the optical activity of the $n_1$–$\pi^-$ transition of ketones in the presence of nonpolar perturbers: (1) A quadrant rule obtains when the $n_1$–$\pi^-$ transition is mixed (by incomplete coulombic screening) with the $\pi^0$–$\pi^-$ transition, assuming that the nonbonding oxygen orbital is localized near the oxygen atom. (2) However, an octant rule obtains—but the signs are wrong—if mixing is with the $\pi^0$–$\pi^-$ transition and the nonbonding oxygen orbital is assumed to be delocalized over the ketone. (3) An octant rule obtains—with correct signs (i.e., negative rotational strength in the $+x$, $+y$, $+z$ octant of the coordinate system of Fig. 7b)—if mixing is with the $n_1$–$d^*$ transition and the perturber is either a C, H, Cl, Br, or I atom. (4) Finally, the contribution to the rotational strength of the latter of these three possibilities appears to outweigh the contributions of the former two by a factor of ten in most instances. Thus, an octant symmetry rule is favored.

In a completely separate set of theoretical considerations, Hohn and Weigang (1968) have also concluded that the ketone should indeed exhibit octant behavior, but for entirely different reasons than those given above. Their calculations indicate that for nonpolar perturbing groups, in particular, the electric–magnetic coupling mechanism may dominate over the one-electron mechanism in this system but still give rise to simple octant behavior with correct signs [see Eq. (77) below]. Also in this work, answers are given to some of the questions raised earlier. Evidence is presented which suggests that the sign and nature of a regional rule depends upon various other factors, such as the type of perturbation (e.g., ion, polar group), its distance from the chromophore, the nature of the perturbation basis set (e.g., whether mixing is with $\pi^0$–$\pi^-$ transitions, $n_1$–$d^*$ transitions, or both), and the assumptions made in the actual calculations (e.g., neglecting three-center integrals).

It is clear now that any regional rule must ultimately be founded upon experimentation accompanied with careful theoretical considerations. Once established, the rule must be applied cautiously; there are, for example, certain known exceptions (Deutsche et al., 1969) to the octant rule for ketones as originally stated by Moffitt et al. (1961). Although it is possible to construct more complicated rules to include the exceptions, this in itself tends to defeat the real purpose of a symmetry rule: simplicity. The more complicated the considerations become, the more quantitative calculations are needed.

In conclusion, the regional rules, as they have been developed here through symmetry considerations alone, do provide a simple, effective starting point for understanding the optical activity of ketones and peptides. An octant rule for ketones seems to be established at least in certain well-defined situations. As far as the peptide quadrant rule is concerned the existing experimental and theoretical evidence (Litman and Schellman, 1965; Schellman and Oriel, 1962; Schellman, 1966; Schellman and Nielsen, 1967a; Urry, 1968b, 1970; Mayers and Urry, 1971) does not unequivocally support quadrant, octant, or perhaps some other more complicated regional behavior for this chromophore. However, it has been brought to our attention in a recent private communication from Schellman that if one is going to use a symmetry rule, "it should be based on wavefunctions" and that with "all the wavefunctions . . . ever tried for the peptide bond, the quadrant rule comes out every time. Octant rule suggestions have always come from an analogy with the ketone case, not from any real wavefunctions." He further pointed out that "the octant versus quadrant rule has scarcely ever been tested anyhow, since the predictions are the same for the position of most perturbants. The violations of the quadrant rule have arisen from an over simplified interpretation of the structure of the molecule or from the wrong assignment of a configuration." It should be stressed in conclusion here that Schellman tempered these remarks by cautioning that "such symmetry rules are highly overrated" since "those . . . who are involved in real calculations of the optical properties of peptides . . . do as complete a calculation as [one] can. This involves two other mechanisms plus the best wave functions [one] can think of for the peptide group and the best charge distribution in the environment."

## 2. *The Coupled-Oscillator Mechanism*

The coupled-oscillator mechanism is the source of optical activity in our model system if both groups have a single electrically allowed (magnetically forbidden) transition. If the transitions are nondegenerate, we find from either Eq. (49d) or Eq. (50d) the following

$$R_{1a} = + \frac{V_{12}\mathbf{R}_{21} \cdot (\mathbf{\mu}_2 \times \mathbf{\mu}_1) 2\pi \lambda_b \lambda_a}{hc(\lambda_b^2 - \lambda_a^2)} \quad (62)$$

where $V_{12} = V_{10a;20b}$; $\mathbf{R}_{21} \cdot (\mathbf{\mu}_2 \times \mathbf{\mu}_1) = (\mathbf{R}_2 - \mathbf{R}_1) \cdot (\mathbf{\mu}_{20b} \times \mathbf{\mu}_{10a})$; and $R_{1a}$ is the rotational strength of the transition $(\lambda_a)$, arising from the group arbitrarily labeled *1*, coupled with the transition $(\lambda_b)$ of the other group. Similarly, the rotational strength of the *b* transition is (see footnote 19)

$$R_{2b} = - \frac{V_{12}\mathbf{R}_{21} \cdot (\mathbf{\mu}_2 \times \mathbf{\mu}_1) 2\pi \lambda_b \lambda_a}{hc(\lambda_b^2 - \lambda_a^2)} \quad (63)$$

which is equal to $-R_{1a}$.

If the transitions are degenerate (i.e., if the groups are identical) the exciton coefficients in Eq. (49) or (50) must be determined. For our dimer here the coefficients are as follows [Tinoco, 1963; see also Eq. (36)]: $C_{11} = +1/\sqrt{2}$ $(C_{iaK}; i = 1; K = 1)$; $C_{21} = +1/\sqrt{2}$; $C_{12} = -1/\sqrt{2}$; $C_{22} = +1/\sqrt{2}$. Thus, from Eq. (49a) or Eq. (50a), the exciton rotational strength is [20]:

$$R_{\pm} = \pm \frac{\pi}{2\lambda_a} \mathbf{R}_{21} \cdot (\mathbf{\mu}_2 \times \mathbf{\mu}_1) \quad (64)$$

where $\lambda_a$ is the transition wavelength of either chromophore in the system and $\mathbf{R}_{21} \cdot (\mathbf{\mu}_2 \times \mathbf{\mu}_1) = (\mathbf{R}_2 - \mathbf{R}_1) \cdot (\mathbf{\mu}_{20a} \times \mathbf{\mu}_{10a})$. $R_{\pm}$ denotes the rotational strength of either the + or − exciton band which is located in the spectrum at frequency $\nu_+$ or $\nu_-$, respectively, as determined by Eq. (43a)

---

[20] Note that $R_+$ and $R_-$ here have signs opposite those assigned them in Tinoco's (1963) paper, this mix-up following from a sign mistake which appeared in an earlier equation (Tinoco, 1962), as discussed in footnote 14. E. H. Strickland in a recent private communication to us has pointed out, however, that in spite of the sign error in Tinoco's (1963) equation corresponding to our Eq. (64), the signs are correct in his equation in which the vector dot- and cross-products have been expanded in terms of spherical polar coordinates.

$$\nu_\pm = \nu_a \pm V_{12}/h \tag{65}$$

where $V_{12} = V_{10a;20a}$.

From Eqs. (62)–(65), it is concluded that the geometrical dependence of the coupled-oscillator mechanism is contingent upon the values of both the potential energy of interaction $V_{12}$ and the optical factor $\mathbf{R}_{21} \cdot (\mathbf{\mu}_2 \times \mathbf{\mu}_1)$. The way in which the optical activity is related to $V_{12}$ depends upon the method by which it is evaluated. Usually, the potential energy is calculated in accordance with either of two approximations: in the monopole approximation (Tinoco, 1962), transition monopole moments are distributed at various points throughout the groups and the resulting potential energy (which resembles the coulombic potential between static point charges) is constituted by the pairwise sum of "electrostatic" interactions between transition monopoles. In the dipole–dipole approximation, the potential energy operator $V$ is expanded (Schellman, 1966, 1968; Hohn and Weigang, 1968; Craig and Wamsley, 1968) to give, as the first nonzero term

$$V_{12} = \frac{\mathbf{\mu}_1 \cdot \mathbf{\mu}_2}{R^3} - \frac{3(\mathbf{\mu}_1 \cdot \mathbf{R}_{21})(\mathbf{\mu}_2 \cdot \mathbf{R}_{21})}{R^5} \tag{66}$$

where $R = |\mathbf{R}_{21}|$. (This expression is identical in form to the potential energy of interaction between two static point dipoles.)

Certain advantages and disadvantages distinguish these approximations. The monopole approximation leads to an expression for the potential energy which is very sensitive to the relative orientations of the groups but the primary disadvantage here is that the monopole positions and charges cannot be experimentally measured and must therefore be calculated using wavefunctions which are themselves usually of a highly approximate nature. The obvious advantage of the dipole–dipole approximation is that, with Eq. (66), either measured or calculated transition dipole moments can be used to evaluate the potential energy. However, as its main disadvantages, the accuracy of the value of $V_{12}$ diminishes as the distance R separating the chromophores becomes roughly comparable to the sizes of the chromophores themselves; and $V_{12}$ as defined by Eq. (66) is less sensitive to the geometrical relationship between groups (as compared with the potential energy in the monopole approximation) because it depends only on the relative orientation of the axes along which the transition dipole moments lie but not on the orientation of the chromophores relative to these axes.

The geometric dependence of the optical factor, like the potential energy of interaction in Eq. (66), again involves only the relative orientation of the two transition moment axes and is therefore not unique in the sense that any configuration of chromophores rotated about their transition moment axes gives the same value for this factor. Therefore, it must be kept in mind that the coupled-oscillator mechanism, as described by the foregoing equations, does not relate the sign and magnitude of the rotational strength to the possible three-dimensional structures of the dimer in a unique, one-to-one-fashion; instead, Eqs. (62)–(69) can only serve to place a limitation on the number of possible conformations.

For the purpose of illustrating the manner in which symmetry rules are developed for this mechanism, we will assume that the model system actually consists of two phenolic chromophores. A symmetry rule for nondegenerate transitions is derived from a consideration of the coupling between the $^1L_b$ transition (at 272 m$\mu$) of one chromophore and the $^1L_a$ transition (at 213 m$\mu$) of the other chromophore (see Fig. 8); similarly, a symmetry rule for the degenerate case is derived from a consideration of the coupling between the $^1L_a$ transitions (Fig. 9). [The notation, the frequencies, and the orientations of the axes along which the transition moments lie in phenol are taken from the work of Hooker and Schellman (1970).] As a further simplification, the transition moment vectors (axes) are constrained to be perpendicular to the vector distance separating the chromophores (i.e., $\mathbf{\mu}_1 \perp \mathbf{R}_{21} \perp \mathbf{\mu}_2$). This constraint results in the elimination of the second term in Eq. (66), leaving

$$V_{12} = \frac{\mathbf{\mu}_1 \cdot \mathbf{\mu}_2}{R^3} \tag{67}$$

for the following analysis where it will be assumed that the dipole–dipole approximation is in effect.

In order to establish the sign of the rotational strength for a given geometry of the dimer, one must be able to determine the sign of the vector dot-product $(\mathbf{\mu}_1 \cdot \mathbf{\mu}_2)$ in Eq. (67) as well as the sign of the triple scalar product $\mathbf{R}_{21} \cdot (\mathbf{\mu}_2 \times \mathbf{\mu}_1)$ of the optical factor. At first glance, this process appears to be complicated by the fact that the transition moment vectors, represented by double-headed arrows in Figs. 8 and 9, may point in either of two directions ("up" or "down"). Since there are two experimentally indistinguishable phases associated with any transition, there are four possible arrangements for two such vectors—both "up," both "down," one "up" and the other "down"—and, hence, four ways to take the vector cross- $(\mathbf{\mu}_2 \times \mathbf{\mu}_1)$ and dot-products. However, it is easily verified for both the nondegenerate and degenerate cases that the sign of the rotational strength is independent of the choice of phases for the transition moments, even though the signs of the vector dot- and cross-products themselves are not. Thus, we will make arbitrary choices as to the phases of the transitions.

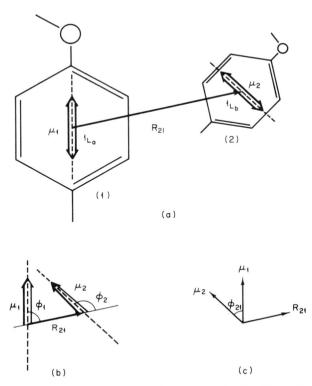

Fig. 8. (a) The $^1L_a$ and $^1L_b$ transition dipole moments (double-headed arrows) of two phenol chromophores. Phenol 2 lies into the paper behind phenol 1. The transition axes (dashed lines) are separated from center to center of the phenol rings by the vector distance $\mathbf{R}_{21} = \mathbf{R}_2 - \mathbf{R}_1$ where $\mathbf{R}_2$ and $\mathbf{R}_1$ refer to an arbitrary origin. (b) One of four possible phase configurations for the transition moment vectors $\boldsymbol{\mu}_1$ and $\boldsymbol{\mu}_2$. Both $\boldsymbol{\mu}_1$ and $\boldsymbol{\mu}_2$ point "up" in this configuration and $\boldsymbol{\mu}_2$ here is tilted counterclockwise through an angle $\theta_{21}$ with respect to $\boldsymbol{\mu}_1$. Assuming that the angles $\phi_1$ and $\phi_2$ are 90°, as is the case for the symmetry rules developed in the text, the transition axes are related to one another in a left-hand screw sense, as are the transition vectors for this choice of phases (see Fig. 1). (c) The vectors are appropriately arranged, tail-to-tail, for evaluating $\mathbf{R}_{21} \cdot (\boldsymbol{\mu}_2 \times \boldsymbol{\mu}_1)$ and $\boldsymbol{\mu}_1 \cdot \boldsymbol{\mu}_2$. Both are clearly positive in value here since the angle (not shown) between $\mathbf{R}_{21}$ and $\boldsymbol{\mu}_2 \times \boldsymbol{\mu}_1$ is zero degrees and the angle ($\theta_{21}$) between $\boldsymbol{\mu}_1$ and $\boldsymbol{\mu}_2$ is less than 90°.

As indicated by Eq. (62), when the $^1L_a$ transition (transition $a$ of group $1$) is coupled to the $^1L_b$ transition (transition $b$ of group $2$), the sign of the rotational strength of the former transition is the negative of the product of signs of $\boldsymbol{\mu}_2 \cdot \boldsymbol{\mu}_1$, $\mathbf{R}_{21} \cdot (\boldsymbol{\mu}_2 \times \boldsymbol{\mu}_1)$, and the wavelength factor $(\lambda_b^2 - \lambda_a^2)$. [Note that the magnitude of the rotational strength is actually twice the value of Eq. (66) since both phenols have $^1L_a$ and $^1L_b$

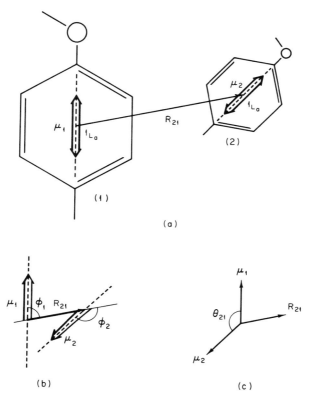

Fig. 9. (a) The same situation as that depicted in Fig. 8 except for the transition of phenol 2 which is now the $^1L_a$ transition. (b) Assuming that $\phi_1 = \phi_2 = 90°$, the transition moment axes are related to one another in a right-hand screw sense whereas the transition moment vectors $\mathbf{\mu}_1$ and $\mathbf{\mu}_2$, for this choice of phases, are related in a left-hand screw sense. (c) Arranging the vectors tail-to-tail, the optical factor $\mathbf{R}_{21} \cdot (\mathbf{\mu}_2 \times \mathbf{\mu}_1)$ is easily seen to be positive (the angle between $\mathbf{R}_{21}$ and $\mathbf{\mu}_2 \times \mathbf{\mu}_1$ being zero degrees) and the dot-product $\mathbf{\mu}_1 \cdot \mathbf{\mu}_2$ is seen to be negative (the angle $\theta_{21}$ being greater than $90°$).

transitions.] The wavelength factor is positive here since $\lambda_a = 213$ m$\mu < \lambda_b = 272$ m$\mu$; and the potential energy and the optical factor are both positive for the situation depicted in Fig. 8. Therefore, for the geometry of the dimer shown in Fig. 8, $R_{1a}$ is positive and $R_{2b}$ is negative. Generalizing from this example, the following can be stated: When the transition moment axes, and not necessarily the transition moment vectors themselves, are related by a left-hand screw (as they are in Fig. 8), the rotational strength of the long-wavelength transition will be negative and the rotational strength of the concomitant short-wavelength transition will be positive; had the axes been related by a right-hand screw, the exact opposite would have been true. This then forms a symmetry rule for nondegenerate coupled oscillators when their transition axes are perpendicular to the vector distance separating them.

The symmetry rule is found to be identical for degenerate coupled oscillators. Referring to Fig. 9, we see that for the choice of phases shown, $\lambda_+ (= c/\nu_+)$ corresponds to the long-wavelength exciton band and its rotational strength, $R_+$, Eq. (64), is positive since the optical factor is positive. Conversely the rotational strength, $R_-$, of the short-wavelength band ($\lambda_-$) is negative. (The CD spectrum for this situation is depicted in Fig. 4 if the wavelength increases from right to left.) Noting the handedness of the axes in Fig. 9, the symmetry rule for the degenerate case, simply stated, says that when the transition axes are related by a right-hand screw, the rotational strength of the long-wavelength exciton band is positive and the rotational strength of the short-wavelength band is negative; the exact opposite obtains when the screw sense of the transition axes is left-handed. (As before, this symmetry rule is independent of the labeling of groups and transitions and the choice of phases for the transition moments; had $\mu_2$ been opposite in phase, for example, $V_{12}$ and $\mathbf{R}_{21} \cdot (\mu_2 \times \mu_1)$ would have both been opposite in sign, and $R_+$, being negative, would have corresponded to the short-wavelength band.)

In summary, for two degenerate or nondegenerate coupled oscillators perpendicular to the vector distance separating them, right-handedness is associated with long-wavelength-positive and short-wavelength-negative rotational strength; conversely, left-handedness is associated with long-wavelength-negative and short-wavelength-positive rotational strength; and, finally, the rotational strength is zero if the two transition moment axes are parallel or perpendicular to one another. If the axes are not perpendicular to $\mathbf{R}_{21}$, then more general expressions for the potential energy and optical factor must be considered: for the angles as they are depicted in Figs. 8 and 9, the general expressions for the potential energy and optical factors are, respectively

$$V_{12} = R^{-3}|\mu_1||\mu_2|(\cos\theta_{21} - 3\cos\phi_1\cos\phi_2) \tag{68}$$

and

$$\mathbf{R}_{21} \cdot (\mu_2 \times \mu_1) = \pm R|\mu_1||\mu_2|\sin\phi_1\sin\phi_2 \\ \times [1 - \csc^2\phi_1 \csc^2\phi_2 (\cos\theta_{21} - \cos\phi_1\cos\phi_2)^2]^{1/2} \tag{69}$$

where $0° < (\theta_{21}, \phi_1, \phi_2) < 180°$. The positive sign in the latter equation is taken if the angle (not shown in the figures) between vectors $\mathbf{R}_{21}$ and $\mu_2 \times \mu_1$ is less than 90° and the negative sign is taken if the angle is greater than 90°. [Other expressions in terms of Cartesian coordinates and spherical polar coordinates are found in Schellman (1968) and Tinoco (1963) (see footnote 20), respectively.]

### 3. *The Electric–Magnetic Coupling Mechanism*

The model system will now consist of two amide chromophores and a symmetry rule for the electric–magnetic coupling mechanism will be developed in terms of the coupling existing between the $n_1$–$\pi^-$ transition on one amide (arbitrarily, transition $a$ of group *1*) and the $\pi^0$–$\pi^-$ transition of the other amide (transition $b$ of group *2*). From Eq. (48a), the appropriate expression for the rotational strength $R_{1a}$ of the $n_1$–$\pi^-$ transition here is[21]

$$R_{1a} = +\frac{V_{12}\mathrm{Im}[\mathbf{\mu}_2 \cdot \mathbf{m}_1]2\lambda_b\lambda_a^2}{hc(\lambda_b^2 - \lambda_a^2)} \qquad (70)$$

where $V_{12} = V_{10a;20b}$ and $\mathbf{\mu}_2 \cdot \mathbf{m}_1 = \mathbf{\mu}_{20b} \cdot \mathbf{m}_{1a0}$. With the aid of Eq. (49c) or Eq. (50c), the rotational strength $R_{2b}$ of the coupled $\pi^0$–$\pi^-$ transition is found (see footnote 19) to be equal in magnitude but opposite in sign to $R_{1a}$. In the ensuing discussion, it is assumed that a coordinate system is located in the dimer such that its origin is centered at the oxygen nucleus of amide *1* and its $z$-axis runs along the CO bond (Fig. 10). Furthermore, it is assumed that the $n_1$–$\pi^-$ transition of amide *1* is polarized along the $z$-axis and that the origin (in space) of $\mathbf{m}_{1a0}$ corresponds to the origin of the coordinate system; implicit in the latter assumption is the problematic (Schellman and Nielsen, 1967a; Bouman and Moscowitz, 1968) assumption that the oxygen nonbonding atomic orbital is localized near the oxygen atom.

Developing a symmetry rule for this mechanism is somewhat difficult because of the nature of the potential energy term $V_{12}$. The question is now raised as to how an electrically allowed transition physically interacts with an electrically forbidden one. The answer is revealed in the symmetry group character table for the amide chromophore, in which it is found that the $n_1$–$\pi^-$ transition has an allowed quadrupole moment. Thus, the potential energy term for this mechanism is constituted by the interaction between the quadrupole charge distribution associated with the $n_1$–$\pi^-$ transition of one amide and the electric dipole charge distribution of the $\pi^0$–$\pi^-$ transition of the other amide.

The transition quadrupole charge distribution is usually evaluated using either of two approximations which are analogous to the monopole and dipole–dipole approximations discussed earlier. In the one approximation, four point transition monopole charges alternating in sign are

---

[21]Note that the corresponding expression in Schellman's (1968) discussion of this mechanism is incorrect in its energy factors.

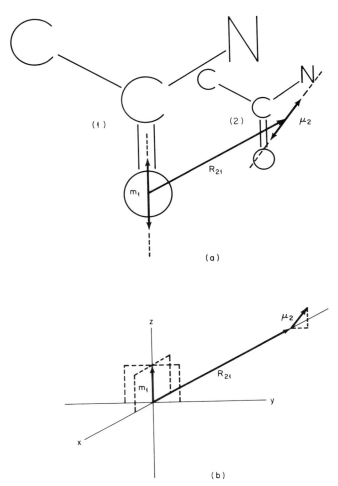

Fig. 10. (a) Two amide chromophores separated by the distance vector $\mathbf{R}_{21} = \mathbf{R}_2 - \mathbf{R}_1$ where $\mathbf{R}_2$ and $\mathbf{R}_1$ refer to an arbitrary origin. $\mathbf{R}_{21}$ points behind the plane of the paper and is assumed to be normal to $\mathbf{\mu}_2$ and $\mathbf{m}_1$ as well as to the plane of amide 1. The following set of assumptions are made: The $n_1$–$\pi^-$ magnetic transition dipole moment $\mathbf{m}_1$ is polarized along the carbonyl CO bond and its origin (in space) is at the oxygen nucleus. The $\pi^0$–$\pi^-$ electric transition dipole moment, $\mathbf{\mu}_2$, is polarized at an angle of 9° in relationship to the line (not shown) connecting the centers of the N and O atoms (Peterson and Simpson, 1957) and its origin is at the point closest to the carbonyl carbon on this imaginary NO line (Woody and Tinoco, 1967). The transition axes (dashed lines) are related to one another in a right-hand screw sense. (b) A coordinate system is superimposed on amide 1 such that its origin coincides with the oxygen nucleus, its z-axis runs parallel to the CO bond, and $\mathbf{R}_{21}$ points in the negative x-direction. The dashed rectangular planes, the xz- and yz-planes, illustrate the two orientations of the plane of amide 1 discussed in the text.

symmetrically distributed around the carbonyl oxygen, above and below the amide plane, on both sides of the carbonyl; each charge is placed in the $xy$-plane since the only allowed component of the $z$-polarized $n_1$–$\pi^-$ transition is the $xy$-quadrupole, and each is located at a distance of 0.4 to 0.6 Å away from the oxygen nucleus (Woody and Tinoco, 1967; Schellman and Nielsen, 1967a; Woody, 1968). In the other approximation, the $n_1$–$\pi^-$ quadrupole charge distribution is a point transition quadrupole moment placed at the oxygen nucleus (Schellman and Oriel, 1962; Hohn and Weigang, 1968; Vournakis et al., 1968) and $V_{12}$ is found by expanding the potential energy operator $V$ in a Taylor series (Schellman, 1968; Hohn and Weigang, 1968). The expansion produces the following dipole–quadrupole interaction term

$$V_{12} = +\frac{\mathbf{R}_{21} \cdot \underline{\underline{Q}}_1 \cdot \underline{\mu}_2}{R^5} - \frac{5(\mathbf{R}_{21} \cdot \underline{\underline{Q}}_1 \cdot \mathbf{R}_{21})(\mathbf{R}_{21} \cdot \underline{\mu}_2)}{2R^7} \quad (71)$$

where $\mathbf{R}_{21}$ and $\underline{\mu}_2$ are row matrices; and $\underline{\underline{Q}}_1 \equiv \underline{\underline{Q}}_{1a0}$ is the quadruple tensor or diadic (of transition $a$ of amide 1) which is represented by the following matrix (Schellman, 1968):

$$\underline{\underline{Q}}_1 = \begin{pmatrix} 2Q_{xx} - Q_{yy} - Q_{zz} & 3Q_{xy} & 3Q_{xz} \\ 3Q_{xy} & 2Q_{yy} - Q_{xx} - Q_{zz} & 3Q_{yz} \\ 3Q_{xz} & 3Q_{yz} & 2Q_{zz} - Q_{xx} - Q_{yy} \end{pmatrix} \quad (72)$$

where each element of the matrix is an integral of the form

$$Q_{xy} = -e\int \phi_{1a} xy \phi_{10} \, d\tau \quad (73)$$

Because the $xy$-quadrupole is the only allowed component for the $z$-polarized $n_1$–$\pi^-$ transition, Eq. (72) simplifies to the following

$$\underline{\underline{Q}}_1 = 3Q_{xy}\begin{pmatrix} 0 & 1 & 0 \\ 1 & 0 & 0 \\ 0 & 0 & 0 \end{pmatrix} \quad (74)$$

Since either approximation will give the same symmetry rule developed below but the latter is somewhat easier to work with, it will be adopted for our purposes here. Using Eq. (74) and setting $\mathbf{R}_{21} = X\mathbf{i} + Y\mathbf{j} + Z\mathbf{k}$, matrix multiplication leads to the following general expression for $R_{1a}$ in terms of Cartesian coordinates

$$R_{1a} = -\frac{6\lambda_b \lambda_a^2 \mu_{2z} i m_{1z} Q_{xy}}{R^7 hc(\lambda_b^2 - \lambda_a^2)}[\mu_{2x}Y(R^2 - 5X^2) + \mu_{2y}X(R^2 - 5Y^2) - 5\mu_{2z}XYZ] \quad (75)$$

(Note that $\underline{\mu}_2 \cdot \mathbf{m}_1 = \mu_{2z} m_{1z}$ since $m_{1x} \equiv m_{1y} = 0$; and that $\text{Im}[m_{1z}] = -im_{1z}$, since the wavefunctions are assumed to be real.)

It is apparent from this equation that a generally complicated relationship exists between the geometry of the dimer and the sign and magnitude of the electric–magnetic coupling rotational strength. In order to further simplify this relationship, the system is now restricted to a very limited portion of the available conformation space; that is, only those geometries are considered in which $\mathbf{R}_{21}$ is colinear with either the $x$- or $y$-axis, amide *1* is coplanar with either the $yz$- or $xz$-plane and $\mathbf{\mu}_2 \perp \mathbf{R}_{21} \perp \mathbf{m}_1$.

For the configuration illustrated in Fig. 10b, in which $\mathbf{R}_{21}$ is parallel to the $x$-axis ($Y = Z = 0$; and $\mu_{2x} = 0$ since $\mathbf{\mu}_2 \perp \mathbf{R}_{21}$), Eq. (75) simplifies to the following

$$R_{1a} = -\frac{6\lambda_b\lambda_a^2\mu_{2y}\mu_{2z}X im_{1z}Q_{xy}}{R^5 hc(\lambda_b^2 - \lambda_a^2)} \tag{76}$$

The sign of $R_{1a}$ here is equal to the product of signs of $X$, $(\lambda_b^2 - \lambda_a^2)$, $\mu_{2y}\mu_{2z}$, and $im_{1z}Q_{xy}$. With $\mathbf{R}_{21}$ colinear with the negative $x$-axis, $X$ is negative. The wavelength factor is negative since the $\pi^0\text{--}\pi^-$ transition is at a shorter wavelength ($\lambda_b \approx 190$ m$\mu$) than the $n_1\text{--}\pi^-$ transition ($\lambda_a \approx 222$ m$\mu$). The sign of the product $\mu_{2y}\mu_{2z}$ is positive since both, as they are depicted in the coordinate system in Fig. 10, are greater than zero; the sign here, it should be noted, is independent of the phase of $\mathbf{\mu}_2$ because the integrals for $\mu_{2y}$ and $\mu_{2z}$, Eq. (17), both involve the same wavefunctions and both will have identical phase signs which cancel—multiply to give a net positive sign—in the product $\mu_{2y}\mu_{2z}$. The sign of $im_{1z}Q_{xy}$ is less obvious; it will depend upon whether amide *1* lies in the $yz$- or the $xz$-plane of the coordinate system. As illustrated in Fig. 10a, the plane of amide *1* is normal to $\mathbf{R}_{21}$ and therefore conforms with the $yz$-plane. Thus, with the nodal plane of the nonbonding oxygen orbital in the $xz$-plane, it can be shown (for amides and ketones as well) that $im_{1z}Q_{xy}$ is positive (Schellman and Oriel, 1962; Vournakis et al., 1968). Again, this result is independent of the phases associated with $m_{1z}$ and $Q_{xy}$ because both integrals, Eqs. (18) and (73), involve the same wavefunctions and therefore have identical phase signs cancelling in the product $im_{1z}Q_{xy}$. Summarizing, $R_{1a}$ is negative for the conformation illustrated in Fig. 10a.

The sign of $R_{1a}$ will change if the geometry is altered by any of the following operations:

(1) The screw sense of the two transition axes is changed from right-handed (as it is in Fig. 10) to left-handed; $\mu_{2y}\mu_{2z}$ changes sign.

(2) The plane of amide *1* is rotated 90° about the $z$-axis, placing it in the $xz$-plane; the integral for $m_{1z}$, but not $Q_{xy}$, changes sign when the nodal plane of the oxygen orbital is designated in the $yz$-plane, as compared with the $xz$-plane, and $im_{1z}Q_{xy}$ becomes negative.

(3) Amide *2* is rotated 90° about the $z$-axis while maintaining the same screw sense of the transition axes; in essence, this operation is equivalent to the preceding one. Clearly any combination of two of these

operations will leave the sign of $R_{1a}$ unchanged. Finally, $R_{1a}$ is zero if $\mathbf{\mu}_2 \perp \mathbf{m}_1$ or $\mathbf{\mu}_2 \parallel \mathbf{m}_1$.

The conclusions just reached can be summarized in terms of a simple symmetry rule for the electric–magnetic coupling mechanism for situations in which $\mathbf{\mu}_2 \perp \mathbf{R}_{21} \perp \mathbf{m}_1$: if $\mathbf{R}_{21}$ is normal to the plane of the chromophore exhibiting the magnetically allowed transition, the rotational strength of the long-wavelength transition is negative when the screw sense of the two transition axes is right-handed and positive when the screw sense is left-handed; conversely, if $\mathbf{R}_{21}$ lies in the plane of this chromophore, the rotational strength of the long-wavelength transition is positive when the screw sense is right-handed and negative when the screw sense is left-handed; the sign of the short-wavelength rotational strength is always opposite to that of the long-wavelength rotational strength. Finally, the rotational strength is zero whenever the transition axes are parallel or perpendicular to one another [or for that matter when $\mathbf{R}_{21} \parallel z$-axis; see Eq. (75)]. The symmetry rule here applies in general not only to a system composed of two amides but also to systems consisting of two ketones, or one amide and one ketone, or one amide (or ketone) and another chromophore with a long- or short-wavelength electrically allowed transition. However, it must be reiterated that this rule applies only to a very limited number of the possible conformations of these systems and is therefore very restricted in its usefulness. In fact, the rule's only obvious value is that it does serve to illustrate that the sign of the electric–magnetic coupling rotational strength has both a regional dependence, in analogy to the one-electron mechanism, and a dependence on the handedness of the two transition axes, in analogy to the coupled-oscillator mechanism.

Naturally, other symmetry rules can be developed for this mechanism. One such rule, developed by Hohn and Weigang (1968), is particularly interesting. They found the following expression [their Eq. (13) in our notation] for the rotational strength of the ketone $n_1$–$\pi^-$ transition coupled with the electrically allowed transitions of a vicinal, nonpolar, isotropic perturber

$$R_{1a} = -15R^{-7}XYZ\bar{\alpha}(\nu_a)im_{1z}Q_{xy} \tag{77}$$

where $\bar{\alpha}(\nu_a)$ is the mean polarizability of the isotropic perturber. The interesting feature of this equation is that it constitutes an octant $(XYZ)$ rule since $\bar{\alpha}(\nu_a)$ is a constant at frequency $\nu_a$ and $im_{1z}Q_{xy}$ is a constant of the transition for a specified coordinate system. Somewhat surprisingly, then, the octant behavior of the rotational strength of the ketone $n_1$–$\pi^-$ transition is supported not only by the one-electron mechanism (Section II,F,1) but by the electric–magnetic coupling mechanism as well. Incidentally, Eq. (77) predicts signs for each octant which are identical to the signs in the octant rule as stated by Moffitt *et al.* (1961).

## G. Calculations

In the past 10 years, commencing with the appearance of Tinoco's (1962) theory, there has been an increasing number of successful efforts —where success is measured more in qualitative than in quantitative terms—to reproduce theoretically and, hence, to interpret the optical activity spectra or spectral parameters of various protein-related structures. An excellent review of these calculations was recently compiled by Deutsche *et al.* (1969) but since that time many papers of a related nature have appeared. It is therefore our intention here to bring the reader up to date on the progress in this growing area of interest.

To a large degree, the success of these calculations can be attributed to Tinoco's (1962) theory of optical activity which has itself either provided the theory or paved the way for subsequent improvements in the theory upon which most of these calculations are based. Since the salient features of the theory have already been discussed in the foregoing sections, we will mainly focus our attention here on the results rather than the theoretical details of these calculations. However, it should be emphasized that the three mechanisms of optical activity as described above do indeed "supply all the ingredients for sophisticated calculations, which consist in the compounding of all three mechanisms for as many groups and excited states as are necessary or feasible" (Schellman, 1968). What is meant by "compounding" of the mechanisms is essentially the following: If the perturbation formalism is involved, the calculation is executed by simply summing the appropriate terms for each mechanism either as they appear in Eqs. (41), etc., or as they appear in a somewhat altered form in the Caldwell-Eyring-Urry (Caldwell and Eyring, 1963; Urry, 1968a) semiempirical extension of Tinoco's theory. If on the other hand the calculation is carried out according to the matrix formalism, compounding the terms characterizing each mechanism involves diagonalization of the Hamiltonian matrix.

### 1. *The α-Helix*

In the first attempt to calculate the rotational strength ($R_{n\pi}$) of the $n_1$–$\pi^-$ transition of the α-helix, Schellman and Oriel (1962) utilized the one-electron terms in Tinoco's (1962) newly proposed theory to estimate the effects of the static field of the polypeptide backbone on the individual peptide residues. It was predicted[22] that $R_{n\pi} = -0.034$ Debye-

Bohr-magnetons (DBM) which is to be compared with the range of experimental values, $-0.202$ to $-0.272$, subsequently obtained from the resolved CD spectra of α-helices lacking chromophoric side chains (Holzwarth and Doty, 1965; Breslow et al., 1965; Carver et al., 1966; Townend et al., 1966; Quadrifoglio and Urry, 1968a; Cassim and Yang, 1970). Clearly the calculated and experimental values agree in sign and order of magnitude, although the former is too small by roughly a factor of seven.

At the same time, Tinoco and his colleagues (Woody, 1962; Tinoco et al., 1963; Bradley et al., 1963; Tinoco, 1964) began a series of investigations concerning the optical activity of helical polymers which culminated. (Woody and Tinoco, 1967) in a detailed calculation of the CD and ORD of the α-helix (and the $3_{10}$-helix as well). The calculation was essentially complete in the following ways: all of the appropriate terms in Tinoco's theory were taken into consideration; and the mixing and coupling of the $n_1$–$\pi^-$ and $\pi^0$–$\pi^-$ transitions with each other as well as with other transitions in the far UV (Tinoco et al., 1962; Barnes and Simpson, 1963) were taken into account.

For the infinite right-handed helix of poly-L-alanine, Woody and Tinoco calculated $R_{n\pi} = -0.028$ DBM; this essentially agreed with the value found by Schellman and Oriel (1962) and was also smaller (roughly an order of magnitude) than the experimental value of $-0.257$ DBM obtained from the 221 mμ CD band of α-helical poly-L-alanine itself (Quadrifoglio and Urry, 1968a).

For the $\pi^0$–$\pi^-$ transition of an α-helix of $N$ residues, Woody and Tinoco showed that the $N$ exciton CD bands arising from the $N$-fold splitting of this transition tend to cluster in three separate spectral regions as $N$ approaches infinity. (1) One region consists mainly of negative CD bands of parallel-polarized (with respect to the helix axis) absorption intensities. The total calculated rotational strength $(R_\parallel)$ in this region was $-1.17$ DBM. By taking the average of the individual wavelengths of the exciton bands weighted by their individual rotational strengths, $R_\parallel$ was found to be centered at approximately 198 mμ. (2) At shorter wavelengths, the perpendicularly polarized exciton transitions give rise to a group of predominantly positive rotational strengths which average to a

---

[22] Because of a mathematical error (J. A. Schellman, as cited in footnote 30 of Woody and Tinoco, 1967) the published value of the rotational strength was too large by a factor of five.

Note that, unless otherwise stated, all values for rotational strengths reported herein are (1) in units of Debye-Bohr-magnetons (see Appendix), (2) on a per residue basis, and (3) uncorrected for solvents (see Section II,D,2).

net value of +6.40 DBM at 188 m$\mu$. In essence, the net rotational strength here can be regarded as the sum of two components: one ($R_\perp$ = +1.17 DBM = $-R_\parallel$) is the "normal" perpendicular component which one expects from the parallel–perpendicular splitting of the $\pi^0$–$\pi^-$ transition of the helix (Moffitt, 1956a; Moffitt et al., 1957); the other component ($R_+$ = +5.23 DBM) is the long-wavelength branch of the non-Gaussian CD envelope, as described by Tinoco (1964), which stems from the term found by Moffitt et al. (1957) to be missing in Moffitt's (1956a) original treatment of this problem. (See discussion in Section II,E and footnote 10.) (3) At still shorter wavelengths, there is a cluster of perpendicularly polarized exciton bands which give a net rotational strength of $-5.23$ DBM at an average wavelength of 186 m$\mu$. The rotational strength here, then, is essentially the negative, short-wavelength branch ($R_-$) of the non-Gaussian CD envelope.

Woody and Tinoco also calculated the total "intrinsic" or nonexciton rotational strength ($R_{ne}$) of the $\pi^0$–$\pi^-$ transition which arises from the mixing and coupling of this transition with other peptide transitions. It was found that this contribution is relatively small in comparison to the exciton rotational strengths; that is, $R_{ne} = -0.11$ DBM. Assuming that the total nonexciton contribution is localized in the region of the spectrum where the monomer $\pi^0$–$\pi^-$ transition lies (see footnote 24), Woody and Tinoco's net calculated rotational strength for this region, as determined by the sum $R_\perp + R_{ne} + R_+$, is +6.29 DBM.

Matching these calculated results with experiment is obviously complicated by the complex nature of the exciton rotational strength. However, if it is assumed that the three somewhat arbitrary groups of rotational strengths here "comprise" three distinct spectral bands,[23] comparisons can be made. The following ranges of experimental rotational strengths are found from the resolved CD bands of right-handed, $\alpha$-helical poly-L-peptides with nonchromophoric side chains (Holzwarth and Doty, 1965; Carver et al., 1966; Townend et al., 1966; Quadrifoglio and Urry, 1968a; Cassim and Yang, 1970): $-0.16$ to $-0.312$ DBM (204 m$\mu$–208 m$\mu$) and +0.432 to +0.872 (189 m$\mu$–192 m$\mu$); the range of the peak wavelengths of the bands is noted in parentheses. (Since CD measurements in and below the 185–190 m$\mu$ region are unreliable, no direct comparison between theory ($R_-$) and experiment can be made for this region.) If one now compares these values with the corresponding computed values in the preceding paragraphs, one finds that the results of the Woody and Tinoco calculation are in qualitative agreement with experiment but

[23] It can be shown that, using least-squares fitting, the system of $\pi^0$–$\pi^-$ exciton rotational strengths can be adequately represented by three Cotton effects (Madison, 1969).

quantitatively overestimate the measured values by roughly an order of magnitude.

Woody and Tinoco also considered the effects of chain length on the optical activity of the $\alpha$-helix. Increasing $N$ from 5 to 40, it was found that the values of $R_{n\pi}$, $R_{\parallel}$, $R_{\perp}$, $R_{+}$, $R_{-}$, and $R_{ne}$, as well as the average positions of these bands, were all sensitive to the value chosen for $N$. In particular, $R_{n\pi}$, $R_{\parallel}$, and $R_{\perp}$, which are the least sensitive, vary 25% over this range of residues. In contrast, $R_{+}$ and $R_{-}$ vary 75%, whereas $R_{ne}$ varies 60%. Finally, $R_{n\pi}$ and $R_{\parallel}$ are found to converge quite rapidly to those values for the infinite helix as $N$ increases from 5 to 40, whereas the remaining rotational strength components are found to converge somewhat less rapidly over this range. In summary, it is concluded that the shapes of the CD and ORD spectra of short helices should differ significantly in the $\pi^0$–$\pi^-$ region from spectra of long helices which more closely resemble the spectra of the infinite $\alpha$-helix itself.

In a subsequent calculation, Woody (1968) significantly improved the agreement between theory and experiment. Relaxing the constraints of the separability condition on the electrons (see Section II,E), Woody tested the importance of charge-transfer complexes between spatially adjacent peptide residues of the $\alpha$-helix; these effects were found to be negligibly small and it was concluded, therefore, that they do not account for the six- to eightfold discrepancy between the experimental values for $R_{n\pi}$ and the earlier calculated values. Using the matrix formalism and making "a more consistent choice of permanent and transition monopole positions" as compared with those used in the earlier Woody-Tinoco calculation, Woody determined the following rotational strengths and band positions for $\alpha$-helical poly-L-alanine: $R_{n\pi} = -0.2217$ DBM (222 m$\mu$), $R_{\parallel} = -0.7251$ DBM (202 m$\mu$), and $R_{\perp} + R_{ne} = +0.3327$ DBM (188 m$\mu$), where $R_{ne} = -0.78$ DBM. The value for $R_{n\pi}$ is in excellent agreement with the experimental range cited earlier, and the value for $R_{\parallel}$ is improved over that calculated by Woody and Tinoco but is still two to four times greater in magnitude than allowed for by the experimental range above. At this point, Woody argued that the apparent discrepancy between experiment and computation concerning the $\pi^0$–$\pi^-$ region may come from the fact that the resolution of the experimental CD spectra is done without incorporating the non-Gaussian envelope predicted by theory. Therefore, Woody calculated the appropriate parameters (Tinoco, 1964) describing the non-Gaussian envelope and used these along with three Gaussian bands, assuming somewhat arbitrary band halfwidths and peak wavelengths, to calculate the CD curve shown in Fig. 11. Although the theoretical curve is remarkably similar to the experimental curve for $\alpha$-helical poly-L-alanine (Quadrifoglio and Urry,

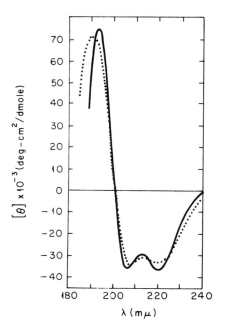

Fig. 11. Comparison of the experimental (Quadrifoglio and Urry, 1968a) and theoretical CD curves for "infinite," right-handed α-helical poly-L-alanine. The dashed line is the experimental curve for poly-L-alanine in trifluoroethanol–trifluoroacetic acid, 98.5:1.5, v/v (Woody, 1968).

1968a), also shown in this figure, Woody cautions that the exact shape of the calculated curve is extremely sensitive to very small variations in the chosen peak wavelengths and band halfwidths and that quantitative agreement between the two curves is therefore probably largely fortuitous. Nonetheless, Woody's results appear to provide at least an excellent qualitative description and, hence, interpretation of the characteristic features of the CD of the α-helix.

Following the basic outline of the Schellman and Oriel (1962) calculation but using improved (Woody, 1968) oxygen atomic orbitals to calculate the quadrupole moment of the $n_1$–$\pi^-$ transition, Vournakis et al. (1968) systematically computed rotational strengths for uniform, homogeneous left- and right-handed α-helices of varying side chains and dimensions. For infinite right-handed α-helices (41-mers), the calculated values for $R_{n\pi}$, in good qualitative agreement with experiment, ranged between −0.081 and −0.243 DBM, depending upon the side chain and the exact dihedral angles defining the backbone conformation. By varying the nature of the side chain, the number of residues in the helix, and the angles defining the backbone conformation, Vournakis *et al.* showed that $R_{n\pi}$ is markedly sensitive to all three of these factors. In addition, it was concluded that the most important effect of the backbone conformation on $R_{n\pi}$ is attributable to the configuration or, more specifically, the nonlinearity, of the hydrogen bonds of the backbone. The finding that $R_{n\pi}$ depends on the nature of the side chain—which in turn is one factor energetically determining the exact conformation of the backbone—is consistent with the notion (Holzwarth and Doty, 1965; Carver *et al.*, 1966) that differences in side chains are at least in part responsible for the observed variations in the 220–224 extrema of the CD spectra of different $\alpha$-helical polypeptides.

Utilizing parameters empirically calibrated from studies concerning various cyclic model amide compounds, Urry (1968a,b) computed— according to a semiempirical theory (Caldwell and Eyring, 1963; Urry, 1968a) based on Tinoco's (1962) perturbation theory—$n_1$–$\pi^-$ rotational strengths for various lengths of right-handed poly-L-alanine $\alpha$-helices. $R_{n\pi}$ was found to be sensitive to chain length, and for an infinite helix (35 residues), $R_{n\pi}$ was calculated as $-0.215$ DBM, in excellent agreement with experiment. Although Urry's computed values for $R_{n\pi}$ are generally within the range of values found in the Schellman and Oriel (1962), Woody and Tinoco (1967), and Vournakis *et al.* (1968) calculations, there is a significant point of difference between Urry's calculation and the latter: whereas the latter workers conclude that the major factor determining the value of $R_{n\pi}$ is the dipole field of the amide groups comprising the helix, Urry concludes that the major factor is the perturbing field of the $C_\alpha$-H groups; in fact, Urry finds that the amide groups make only a small contribution to the $n_1$–$\pi^-$ optical activity which is of a positive rather than a negative sign.

Applying the matrix formalism, Madison (1969) examined extensively the variations in the optical activity of $\alpha$-helical chains as a function of length. It was found that the calculated mean residue ellipticities at 194 m$\mu$ and 204 m$\mu$ are extremely sensitive to chain length, whereas the ellipticity at 222 m$\mu$ is less so. For helices shorter than eight residues, Madison's calculation leads to the following predictions: a positive ellipticity for the longest wavelength band of the $\pi^0$–$\pi^-$ exciton system, in contrast to the large, negative ellipticity observed in the 206–208 m$\mu$ region of the CD spectra of large helices; a positive ellipticity for the band in the 194 m$\mu$ region, but two to ten times lower in magnitude than that observed for long helices; and a negative ellipticity for the band in the 222 m$\mu$ region, this being roughly a fourth to a half the magnitude of the ellipticity observed for long helices. For helices longer than eight residues, Madison found that the calculated ellipticities in the 194 m$\mu$

and 222 mμ regions converge quite rapidly to the values for long helices. In contrast, the calculated ellipticity at 207 mμ—the magnitude of which is very sensitive to the chosen halfwidth and position—shows little sign of converging for values of $N$ up to 18, remaining much smaller (in magnitude) than experimental values for large helices. These results, along with others discussed above (Woody and Tinoco, 1967; Urry, 1968b; Vournakis et al., 1968), clearly raise serious questions as to the limitations and applicability of the procedures which have been developed to estimate the helix content of proteins which usually contain only short helical pieces; in these procedures the CD of any helix piece is assumed to resemble the CD of a large α-helix (see Section III).

Zubkov and Vol'kenshtein (1970) have reexamined the relative importance of the exciton and nonexciton features of the $\pi^0$–$\pi^-$ optical activity of the α-helix. Proceeding in much the same manner and using essentially the same spectral and physical parameters as used in the Woody-Tinoco calculations, Zubkov and Vol'kenshtein computed the following exciton rotational strengths for the helix: $R_\| = -1.08$, $R_\perp + R_+ = +3.45$, and $R_- = -2.37$ DBM at average wavelengths 199 mμ, 191 mμ, and 186 mμ, respectively. These are roughly comparable to those calculated by Woody and Tinoco. The feature distinguishing the Zubkov and Vol'kenshtein calculation from the Woody and Tinoco calculation is the way in which the nonexciton contribution to the rotational strength of the $\pi^0$–$\pi^-$ transition is treated; in the former, the nonexciton contribution is spread over the $\pi^0$–$\pi^-$ exciton system, whereas in the latter it is localized to the region of the spectrum[24] corresponding to the $\pi^0$–$\pi^-$ transition wavelength of the peptide itself. Nevertheless, the conclusions reached are similar, Zubkov and Vol'kenshtein find that the rotational strength of the major component of the nonexciton contribution is $-0.184$ DBM, which is close to the value given $R_{ne}$ by Woody and Tinoco, but

[24] This is explained by the manner in which rotational strengths were evaluated in the two calculations. In evaluating the rotational strength of the $\pi^0$–$\pi^-$ transition due to mixing and coupling with the $n_1$–$\pi^-$ transition or other peptide transitions in the far UV, Woody and Tinoco (1967) used expressions for the rotational strength in which the exciton contribution had been summed. (In the case of the $n_1$–$\pi^-$ transition, their expression is equivalent to our Eqs. (49c), (49e), and (49g) if $K$ is summed from 1 to $N$; in the case of lower-lying transitions, their expression is equivalent to our Eq. (50) if summed from $K = 1$ to $N$.) In contrast, Zubkov and Vol'kenshtein (1970) used expressions in which the exciton coefficients are retained. The difference between the two calculations is summarized as follows: in the former, all of the nonexciton rotational strength is centered at the wavelength used for the $\pi^0$–$\pi^-$ transition of the peptide itself, whereas in the latter, the nonexciton rotational strength is distributed throughout the exciton system since the rotational strength of each individual exciton band now consists of a part solely attributed to nonexciton effects.

that this component is centered near 197 m$\mu$, which is 7 m$\mu$ to the red of the wavelength at which $R_{ne}$ is centered in the Woody-Tinoco calculation. Most importantly, the results of both calculations support the notion that in relation to any nonexciton contributions the exciton splitting and band system of the $\pi^0$–$\pi^-$ transition are the dominant factors in the optical activity of this transition in the $\alpha$-helix.

Recently, Pysh (1970a) has calculated two-state ($n_1$–$\pi^-$ and $\pi^0$–$\pi^-$) and three-state ($n_1$–$\pi^-$, $\pi^0$–$\pi^-$, and $n'$–$\pi^-$) CD spectra for the $\alpha$-helix in accordance with the matrix formalism. The results here are essentially the same as those reported earlier by Woody (1968) and Madison (1969). In both the two-state and three-state calculations, the qualitative features (i.e., the three extrema) of the $\alpha$-helix CD spectrum are reproduced. Quantitatively, though, some discrepancy is found to exist as to the magnitudes of the calculated extrema in comparison with experiment. However, as discussed above in connection with Woody's (1968) calculation and Fig. 11, the extent of quantitative agreement or disagreement between theory and experiment here may largely be a matter of both the choice of band positions and halfwidths for the calculated CD curve and the inclusion of the distinctly non-Gaussian envelope of the perpendicularly polarized component of the $\pi^0$–$\pi^-$ transition.

Although our present understanding of both the spectral properties of polypeptides and the theory of optical activity together seem to provide an excellent qualitative account of the optical activity of the $\alpha$-helix—as judged by the overall success of the foregoing calculations—these have not escaped serious challenge. On the one hand, the question has been raised (as discussed in footnote 9) as to whether or not another amide transition (the mystery band) lies between the $n_1$–$\pi^-$ and $\pi^0$–$\pi^-$ transition. On the other hand, the adequacy of the theory has been seriously questioned in two recent papers. Very carefully calculating (via the Kronig-Kramers transforms) ORD curves from the experimental CD curves of various $\alpha$-helical homopolypeptides, Cassim and Yang (1970) conclude that the negative, non-Gaussian $\pi^0$–$\pi^-$ exciton CD band, which is predicted (Tinoco, 1964; Woody and Tinoco, 1967; see footnote 10) to lie just below the experimentally accessible absorption region at about 185 m$\mu$, is not present because the CD spectra exactly transform into their corresponding ORD spectra without the inclusion of such a band. Two possible explanations are offered: either a positive band cancels this negative band or the theoretical prediction is itself incorrect. Reviewing the available information concerning the transitions of peptides, Cassim and Yang argue that of the two the latter possibility is more likely. It must be stressed, however, that since our knowledge concerning the experimentally inaccessible transitions of peptides is by no means complete, deciding on the likelihood of a transition existing in this region is a highly speculative matter.

In the other challenge to the theory alluded to above, Loxsom et al. (1971) compare the results of first-order perturbation theory with results obtained from a more modern formulation of the rotational strength of the helix which is based on scattering theory (Loxsom, 1970; also see Section II,E). Using the same parameters used in the Schellman and Oriel (1962) and Woody and Tinoco (1967) calculations, Loxsom et al. conclude that perturbation theory, while adequate for calculating absorption parameters of the helix, is inadequate for calculating the CD of the helix *if* the coupling between monomer transitions (i.e., the $\pi^0-\pi^-$ transitions with lower-lying transitions) is sufficiently strong to account for the $\pi^0-\pi^-$ hypochromism observed in the absorption spectrum of the helix. Specifically, it was shown that both the perturbation and nonperturbation approaches (using the same input data) are in basic agreement with each other when only the weak coupling of the $n_1-\pi^-$ and $\pi^0-\pi^-$ transitions is considered; but that the calculated CD in the nonperturbation approach is in much better agreement with experiment as the strength of the coupling between the $\pi^0-\pi^-$ and $\pi^+-\pi^-$ (at 148 m$\mu$) transitions is increased to the point of giving a 30% hypochromism effect. (The coupling is artificially varied by varying the transition charges of the latter transition.) Although the differences between nonperturbation and the perturbation results are substantial in this case, it must be borne in mind first that this criticism levied against perturbation theory is based on an artificial hypochromic effect and second that the real source(s) and nature of the observed hypochromism are not as yet fully understood.

## 2. *Poly*-L-*tyrosine*

Several homopoly-$\alpha$-amino acids with strongly absorbing aromatic side chains exhibit interesting CD and ORD spectra. Of these spectra only the optical activity of poly-L-tyrosine has been examined in a systematic theoretical fashion. A controversy surrounds this particular polypeptide (as fully discussed by Deutsche et al., 1969) for it has been difficult to assess by various techniques (including CD and ORD) whether under certain experimental conditions the polymer exists as a right-handed $\alpha$-helix or some other regularly ordered structure. As far as the optical activity spectra are concerned, the difficulty here stems from the fact that the four transitions of the phenolic group above 185 m$\mu$ plus the peptide $n_1-\pi^-$ and $\pi^0-\pi^-$ transitions together render unique CD and ORD spectra, which bear essentially no resemblance to the spectra for secondary polypeptide structures.

In calculating the ORD of poly-L-tyrosine, Pao et al. (1965) assumed an α-helical configuration for poly-L-tyrosine and considered the exciton splitting of the peptide $n_1$–$\pi^-$ and $\pi^0$–$\pi^-$ transitions and the phenolic $^1L_a$ (225 m$\mu$) and $^1L_b$ (277 m$\mu$) transitions. Certain features of their calculated ORD resemble the experimental ORD curves of Fasman et al. (1964), and on this basis Pao et al. attribute the ORD spectrum and the corresponding CD spectrum (Beychok and Fasman, 1964) to an α-helical conformation.

However, as Chen and Woody (1971) point out in a subsequent investigation of this matter, the agreement between the experimental and calculated ORD of Pao et al. may be largely fortuitous since not all of the potentially important sources of optical activity were considered; the nearly degenerate, strongly allowed $^1B_a$ and $^1B_b$ phenolic transitions (near 193 m$\mu$) and the coupling between nondegenerate transitions were neglected. In an effort to settle the question of whether or not the CD spectrum of poly-L-tyrosine represents a helical structure, then, Chen and Woody carried out a detailed calculation which incorporated all mechanisms of optical activity, all six poly-L-tyrosine transitions lying above 185 m$\mu$, and many of the modifications instituted in Woody's (1968) prior calculation concerning the α-helix. Four conformations of poly-L-tyrosine were specifically considered here—two left-handed and two right-handed helices, the dihedral angles of which were based on the energy considerations of Ooi et al. (1967) and Yan et al. (1968). The primary differences between helices of the same handedness were the values selected for the respective angles $\chi_1$ and $\chi_2$ about the $C_\beta$–$C_\alpha$ bond and the C–O bond of the phenolic OH group. CD curves were calculated for each conformation assuming a band halfwidth of 10 m$\mu$. Each curve was distinct and it was found that the essential qualitative features of each curve were independent of: the chain length ($N$ = 10 to 20 residues); the assumed energy difference between $\pi^0$–$\pi^-$ and B transitions; and small variations (±10°) about the side-chain $\chi_1$ and $\chi_2$ angles.

The main difficulty faced by Chen and Woody was finding an appropriate experimental spectrum with which to compare their calculated results because the only published spectra for poly-L-tyrosine in aqueous solvent (Beychok and Fasman, 1964; Quadrifoglio et al., 1970) were recorded at high pH's (pH = 11.2 and 10.8, respectively) at which the phenol is partially ionized and its transitions are red-shifted 10 to 19 m$\mu$ (Fasman et al., 1964; Legrand and Viennet, 1965). Chen and Woody made comparisons with the CD spectrum of poly-L-tyrosine in methanol, published by Shiraki and Imahori (1966). Of the calculated curves for the four helical structures, only one closely resembled the experimental curve; the structure was one of the right-handed helices (labeled RA), its CD being qualitatively similar to the experimental curve except in the region of the spectrum above 260 mμ, as illustrated in Fig. 12.

In testing the source and seriousness of this discrepancy above 260 mμ, Chen and Woody made the following observations: if the $^1L_b$ transition is omitted in the calculation, the qualitative features below 260 mμ of the RA curve (and the curves for the other conformations as well) are not significantly altered; and, in contrast to the amplitudes of the 220 and 227 mμ bands of the calculated curve, the amplitudes of the 200 and 280 mμ bands vary greatly with modest variations in $\chi_1$ and $\chi_2$. It was therefore argued that the main qualitative differences between the calculated and the experimental curves could be attributed to the effects of solvent in determining the values of $\chi_1$ and $\chi_2$ rather than to some very great structural difference in the RA conformation as it compares with poly-L-tyrosine in methanol. As further support of this argument, unpublished experimental work was cited here (C. Geary and R. W. Woody, see their footnote 34) in which it was found that the CD spectrum of the block copolymer (D,L-glu)$_{50}$(L-tyr)$_{50}$(D,L-glu)$_{50}$ in aqueous solvent (at neutral pH) is qualitatively similar to the Shiraki and Imahori-spectrum, except that a small positive rather than a negative band is

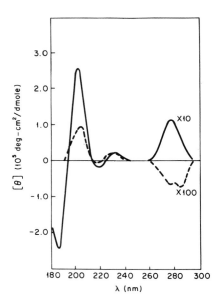

FIG. 12. Comparison of experimental curve (dashed line) for poly-L-tyrosine in methanol (Shiraki and Imahori, 1966) with the calculated curve (solid line) for a decamer of poly-L-tyrosine in the RA conformation (Chen and Woody, 1971).

found in 280 m$\mu$ region.[25] Hence, Chen and Woody conclude that the experimental spectrum in Fig. 11 represents poly-L-tyrosine in a right-handed helix similar to the RA conformation.

### 3. $\beta$-Pleated Sheet

The optical activity of the $\beta$-structure of polypeptides was first theoretically examined by Pysh (1966), who essentially followed the format of the Woody and Tinoco (1967) calculation concerning the $\alpha$-helix. For the $n_1$–$\pi^-$ transition of a large array of peptide chains in antiparallel alignment, Pysh calculated a rotational strength of $-0.035$ DBM, which is in good qualitative agreement with the range of experimental values, $-0.06$ to $-0.156$ DBM (Iizuka and Yang, 1966; Townend et al., 1966; Timasheff et al., 1967; Quadrifoglio and Urry, 1968b), determined from the long-wavelength negative band characteristic of this structure (see Fig. 13). Pysh also predicted that the exciton splitting of the $\pi^0$–$\pi^-$ transition would lead to essentially three bands identifiable by the polarization of their associated absorption intensities: one band, at

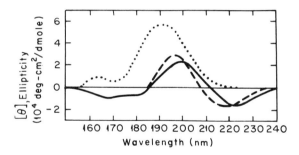

Fig. 13. Three-state solution (including the $n_1$–$\pi^-$, $\pi^0$–$\pi^-$, and $n'$–$\pi^-$ peptide transitions) for the CD spectrum (solid line) of an antiparallel $\beta$-pleated sheet, two chains wide and ten peptide units long. Also shown are the calculated oscillator strength (dotted line), in arbitrary units, and the experimental CD spectrum (dashed line) of preheated, high molecular weight poly-L-lysine as determined by Townend et al. (1966). (Pysh, 1970a.)

---

[25] Recently, several other CD spectra for poly-L-tyrosine in various solvents have been published which show a variation in the sign of the 280 m$\mu$ region but otherwise retain the same qualitative features of the spectra illustrated in Fig. 12. Specifically, the sign of the CD in the 280 m$\mu$ region is as follows for poly-L-tyrosine in these solvents: negative in trimethylphosphate (Quadrifoglio et al., 1970; Damle, 1970; Engel et al., 1971), positive in aqueous solution at pH = 10.6 and $\mu = 0.1$ (Friedman and Ts'o, 1971), and positive near 280 m$\mu$ but negative at slightly longer wavelengths in propanediol containing 3% (v/v) trimethyl phosphate (Engel et al., 1971).

roughly 198 m$\mu$, which is of negative rotational strength and corresponds to absorption polarized parallel to the chains; another band, at somewhat shorter wavelengths ($\sim$195 m$\mu$), which is of net positive rotational strength and corresponds to absorption polarized in the direction of the hydrogen bonds between chains; and, finally, another band at still shorter wavelengths ($\sim$175 m$\mu$) which is of net negative rotational strength and corresponds to absorption polarized perpendicular to the plane of the sheet. Assuming the two long-wavelength bands together constitute only one band, the splitting being relatively small, Pysh's computations lead to a net exciton rotational strength of +0.027 DBM for the 195–198 m$\mu$ region. The nonexciton contribution to the $\pi^0$–$\pi^-$ transition in this region was found to be two to four times greater than this. Combining Pysh's computed exciton and nonexciton rotational strengths for the 195–198 m$\mu$ region then, one finds a value which is roughly less than an order of magnitude smaller than experimentally determined values, which range from +0.175 to +0.343 (as reported in the works cited above). Pysh's basic prediction that the near-UV CD spectrum of the antiparallel $\beta$-structure is constituted by three bands (one of $n_1$–$\pi^-$ origin and two of $\pi^0$–$\pi^-$ origin) is consistent with the findings of Timasheff et al. (1967), who report three extrema in the CD spectrum of poly-L-lysine in sodium dodecyl sulfate; however, the sign (positive) of the shortest wavelength extremum here (at approximately 190 m$\mu$) is opposite to the predicted sign of Pysh's shortest wavelength band.

Also applying first-order perturbation theory to this problem, Zubkov and Vol'kenshtein (1967) computed a value of $-0.0176$ DBM for the rotational strength of the $n_1$–$\pi^-$ transition of a 21 × 21 (21 chains of 21 residues each) $\beta$-sheet, in essential agreement with the Pysh result. In a subsequent investigation (Zubkov and Vol'kenshtein, 1970), the relative importance of the nonexciton and exciton contributions of the $\pi^0$–$\pi^-$ transition was carefully evaluated. The exciton system of rotational strengths was grouped so as to constitute two CD bands, one band, at 194 m$\mu$, being positive and the other, at 179 m$\mu$, being negative. The signs, net rotational strengths, and positions of these exciton bands are in basic agreement with Pysh's findings (assuming the merger of the two long-wavelength exciton bands in Pysh's calculation). With respect to the nonexciton contribution of the $\pi^0$–$\pi^-$ transition, Zubkov and Vol'kenshtein conclude, as did Pysh, that this contribution is as important as the exciton contribution in determining the $\pi^0$–$\pi^-$ optical activity of the antiparallel $\beta$-structure. (Note that a different conclusion was reached concerning the relative importance of the exciton and nonexciton terms in the case of the $\alpha$-helix, as discussed in Section II,G,1.)

Whereas Pysh did not consider the effects of distributing the nonexciton contribution over the $\pi^0$–$\pi^-$ exciton system (see footnote 24) Zubkov and Vol'kenshtein specifically did. It was found that the amplitudes of the exciton bands are greatly enhanced by nonexciton contributions and, in addition, that the shortest wavelength $\pi^0$–$\pi^-$ band (which is negative when only exciton effects are taken into account) becomes positive. This contrasts Pysh's prediction of a negative rotational strength for this band and it also brings the theory into closer agreement with the $\beta$-CD spectrum reported by Timasheff et al. (1967) which exhibits a small short-wavelength maxima near 190 m$\mu$ below a larger maxima near 197 m$\mu$. (See the preceding paragraph.) In passing the reader must be forewarned that Zubkov and Vol'kenshtein's results should be regarded with some caution since it is unclear whether or not they corrected for the fact that the antiparallel $\beta$-structure coordinates they report using (i.e., those of Marsh et al., 1955) describe D- rather than L-amino acids. The virtual agreement with Pysh's results seems to render this possibility unlikely however.

Utilizing the matrix formalism and his other modifications introduced earlier in the theoretical treatment of the $\alpha$-helix, Woody (1969) systematically calculated $n_1$–$\pi^-$ and $\pi^0$–$\pi^-$ rotational strengths for various $\beta$-structures consisting of 1, 2, 4, or 6 chains of 2 to 24 residues in length. These results are now summarized. Considering two proposed sets of coordinates for the antiparallel sheet (Fraser and MacRae, 1962; Arnott et al., 1967), Woody calculated values for $R_{n\pi}$ which were found to be negative for all sheet sizes except one (6 × 8), ranging from +0.005 to −0.058 DBM; generally, $R_{n\pi}$ became less negative as the sheet widened (more chains) and either remained fairly constant or became more negative as the sheet lengthed (longer chains). In agreement with Pysh, the $\pi^0$–$\pi^-$ exciton rotational strengths were found clustered into three distinct regions according to the polarization of the absorption intensity. For sheets comprised of more than eight residues, the rotational strengths of these three bands varied as follows: between +0.1035 and +0.1597 DBM for the long-wavelength band, between +0.0005 and +0.1245 DBM for the intermediate-wavelength band, and between −0.0831 and −0.244 DBM for the short-wavelength band. Therefore, for large sheets, Woody's conclusions concerning the rotational strengths (as well as the relative positions and polarizations) of the three $\pi^0$–$\pi^-$ exciton bands are essentially in agreement with conclusions reached earlier by Pysh (1966), except that Woody finds positive rather than negative rotational strength for the long-wavelength band; thus, Woody's calculations are in better agreement with the experimental CD spectrum reported by Timasheff et al. (1967), as just described, since here two adjacent positive CD bands are predicted. For sheets of eight or fewer residues, the rotational strengths (and the positions) of the bands were found to vary over a much greater range of values there being in some instances even a change in the sign of the intermediate exciton band. While Woody's calculation appears to provide a good qualitative account of the optical activity of the antiparallel $\beta$-sheet and is basically consistent with earlier calculations, it should be borne in mind, in view of conclusions reached in the foregoing calculations, that the nonexciton effects of transitions other than the $n_1$–$\pi^-$ transition were not taken into account here.

A few other interesting observations were made by Woody. The exciton band with the most absorption intensity—the intermediate band polarized along the hydrogen bonds of the chains—was found to shift to the red as the width of the sheet increased; this predicted trend appears to be experimentally supported in that the absorption maximum of low molecular weight $\beta$-poly-L-serine is near 187 m$\mu$ (Quadrifoglio and Urry, 1968b) whereas the maximum of high molecular weight poly-L-lysine is near 194 m$\mu$ (Rosenheck and Doty, 1961). Another interesting observation was that the CD of single, extended, 2- to 24-residue polypeptide chains with $\beta$-structure dihedral angles, qualitatively resembles the CD of larger multichain $\beta$-structures.

Many of the qualitative results concerning the optical activity of antiparallel $\beta$-structures reported in the preceding calculations have been supported in a number of recent calculations. Urry (1968c), applying his semiempirical approach described earlier, found values for $R_{n\pi}$ which ranged from $-0.168$ DBM (for a $2 \times 10$ sheet) to $-0.56$ DBM (for an infinite two-chain sheet); on the whole, in Urry's particular approach to this problem there is a tendency to overestimate the rotational strength. Applying the matrix formalism in much the same way as Woody (1969), Madison (1969) investigated the $\pi^0$–$\pi^-$, $n_1$–$\pi^-$ optical activity of $\beta$-structures of various sizes and arrived at many of the same conclusions as Pysh (1966) and Woody (1969). In addition, Madison concludes that the electric–magnetic coupling mechanism is the dominant factor in generating the optical activity of this structure and that small deviations from the perfect $\beta$-structure, in which the hydrogen bonded atoms are in the symmetry plane of the amide, could lead to significant changes in the optical activity. Pysh (1970a) has also applied the matrix formalism to the problem of calculating two- as well as three-state CD curves for $2 \times 10$ antiparallel $\beta$-sheets. The latter curve is illustrated in Fig. 13 along with the experimental CD spectrum of high molecular weight poly-L-lysine (Townend et al., 1966). The qualitative agreement between the two is probably as good as can be expected for the following reasons: the overall shape of the calculated curve is very sensitive to the choice of band halfwidths, band centers, and the exact coordinates and dimensions of the sheet; the experimental curves on the other hand show a dependence on the solvent, on the $C_\alpha$-side chain (Iizuka and Yang, 1966; Stevens et al., 1968; Timasheff, 1970), as well as on the size of the individual polypeptide chains (Li and Spector, 1969).

Rotational strengths and CD curves for the parallel $\beta$-structure have also been theoretically calculated in several of the foregoing papers (Pysh, 1966, 1970a; Woody, 1969; Madison, 1969; Zubkov and Vol'kenshtein, 1970) and these results are now briefly summarized. The exciton splitting of the $\pi^0-\pi^-$ transition is found to give rise to essentially two bands, one of absorption intensity polarized along the direction of the chains and one, at shorter wavelengths, of absorption intensity polarized slightly out of the plane of the sheet, perpendicular to the chains. There is basic disagreement between these works as to the predicted signs of these CD bands primarily because Pysh (1966) and, it appears, Zubkov and Vol'kenshtein (1970) both inadvertently used coordinates for poly-D- rather than poly-L-amino acids. [In the former work, the coordinates of Pauling and Corey (1951) were used whereas in the latter work it is reported that the slightly refined coordinates of Pauling and Corey (1953) were used, these also being for D-amino acids.] If one reverses the signs of the CD bands calculated in these two papers in order to relate the results to poly-L-amino acids, one finds that the agreement among all of these works is greatly improved. However, Pysh's (1966) results now predict positive rotational strength for the $n_1-\pi^-$ transition, whereas the calculations by Woody, Madison, and subsequently, Pysh (1970a) predict negative rotational strength. It appears likely that the $n_1-\pi^-$ is in fact negative not only because the latter three calculations concur on this point but also because, as argued by Woody (1969), a chain in the parallel conformation is not very different from a chain in the antiparallel conformation which is predicted to exhibit negative $n_1-\pi^-$ rotational strength. In summary then, these calculations indicate that the CD of the parallel $\beta$-pleated sheet should be very similar to the long-wavelength portion of the CD of the antiparallel $\beta$-sheet, there being a long-wavelength negative band adjacent to a shorter-wavelength positive band of slightly greater size.

There are unfortunately no experimental results to date with which to compare the calculated CD of the parallel $\beta$-structure since there are no known model systems in this configuration. Obviously, because of the similarity of the CD of the antiparallel and parallel conformations, problems could arise in distinguishing these too. Woody (1969) has suggested certain tests whereby they could be distinguished:

In the antiparallel form, the strongest absorption band may be either red- or blue-shifted (relative to the monomer $\pi^0-\pi^-$ transition wavelength), depending on the sheet width, but it is associated with a small positive CD band, while in the parallel $\beta$-structure, the major component in absorption is always blue-shifted and is associated with a large negative CD band.

Woody also points out that in comparing the difference in wavelength between the maximum CD and maximum absorption

. . . narrow structures of both parallel and antiparallel types have similar values for this difference, as would be expected for their near identity in the limit of a single-strand structure. As the structure becomes wider, however, this difference approaches a value of about 5 m$\mu$ for the antiparallel structure . . . while for parallel $\beta$-structures, the limiting value appears to be about 13 m$\mu$.

## 4. *Poly*-L-*proline*

Extending his earlier investigation (Pysh, 1966) concerning the $\beta$-structure, Pysh (1967) calculated rotational strengths and dipole strengths for two known forms of poly-L-proline: the right-handed helical form I (Traub and Shmueli, 1963) where the prolyl residues are in the *cis* configuration, and the left-handed helical form II (Cowan and McGavin, 1955; Sasisekharan, 1959) where the prolyl residues are in the *trans* configuration. Pysh made the assumption, which may (Nielsen and Schellman, 1967; Schellman and Nielsen, 1967a) or may not (Madison, 1969) be a good one, that the transition moment of the $\pi^0-\pi^-$ transition of a tertiary amide is the same as it is for the primary amide myristamide (cf. Peterson and Simpson, 1957). He found that for both structures there is substantial exciton splitting which is comparable to that of the $\alpha$-helix. Pysh predicted that in both cases the splitting would lead to a short-wavelength negative CD band, associated with absorption intensity polarized perpendicular to the helix axis; and a long-wavelength positive CD band, associated with absorption intensity polarized parallel to the helix axis. For both structures, it was also predicted that the $n_1-\pi^-$ rotational strength will be negative and relatively weak being roughly an order of magnitude smaller than it is found to be for the $\alpha$-helix.

In a subsequent investigation, Pysh (1970a) arrived at a similar set of conclusions, this time using the matrix formalism. However, in contrast to the earlier calculation, the origin of the $\pi^0-\pi^-$ transition was placed at 200 m$\mu$ rather than at 206 m$\mu$ because the latter value here is most likely too high, as pointed out by Rosenheck *et al.* (1969). In the case of poly-L-proline I, Pysh concluded that the majority, 70%, of the absorption intensity resides in the low energy, parallel exciton component located near 209 m$\mu$ (see Fig. 14). This conclusion is in strong

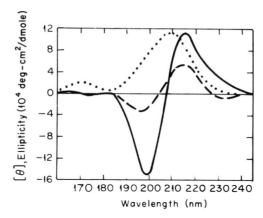

Fig. 14. The three-state solution (including the $n_1$–$\pi^-$, $\pi^0$–$\pi^-$ and $n'$–$\pi^-$ peptide transitions) for the CD spectrum (solid curve) of poly-L-proline I. Also shown is the calculated oscillator strength (dotted line), in arbitrary units, and the experimental CD spectrum (dashed line) of poly-L-proline I in trifluoroethanol for 13–18 minutes as determined by Bovey and Hood (1967). (Pysh, 1970a.)

agreement with experiment: Bovey and Hood (1967) observe an absorption maximum for this structure near 210 m$\mu$; and Rosenheck et al. (1969), through linear dichroism measurements, detect a major absorption band centered at 211 m$\mu$ polarized parallel to the proline helix axis. Pysh also concluded that the optical activity of form I, like that of the $\alpha$-helix, will be mainly attributable to the exciton rotational strength; this is not too surprising in view of the fact that poly-L-proline I is a "tight" helix having a residue translation distance, 1.85 Å, nearly as small as that of an $\alpha$-helix, 1.50 Å.

Together these findings predict a CD for poly-L-proline I which has large positive ellipticity at long wavelengths, near the absorption maximum, adjacent to a shorter-wavelength region of large negative ellipticity, as illustrated in Fig. 14. These expectations are borne out in the experimental absorption and CD (see Fig. 14) curves of Bovey and Hood (1967) as well as in those of Timasheff et al. (1967). Although there is excellent qualitative agreement between the experimental and calculated CD curves depicted in Fig. 14, there appears to be rather poor quantitative agreement. However, it should be noted that an error was made here in reproducing the spectrum of Bovey and Hood; for the positive band there should have been a maximum ellipticity nearly twice that indicated and, thus, the experimental and calculated positive bands are brought into closer quantitative agreement. It will be observed in Fig. 14 that the calculated CD curve does not reproduce the small negative band near 236 m$\mu$, which is also present in the Timasheff *et al.* CD spectrum. Yet Pysh calculates a negative rotational strength for the $n_1$–$\pi^-$ band which should be the longest wavelength band. This band is relatively small and Pysh offers several equally good explanations for why this band may not stand in the calculated curve. [It should be noted here that much experimental evidence (Schellman and Nielsen, 1967b; Madison and Schellman, 1970b) indicates that the transition wavelength of the $n_1$–$\pi^-$ transition is extremely sensitive to solvent and therefore the exact conformation of the poly-L-proline. Thus it is impossible to make a sure assignment for this transition in any calculation.]

For poly-L-proline II, theory and experiment are largely at odds. Pysh (1970a) predicts that 84% of the absorption intensity should reside in a perpendicularly polarized band near 200 m$\mu$. Experimentally, Bovey and Hood (1967) observe an absorption maximum for this structure at 204 m$\mu$. But Rosenheck *et al.* (1969) resolve this absorption intensity into a parallel-polarized band centered near 209 m$\mu$ and a somewhat smaller perpendicularly polarized band centered at 191 m$\mu$. Furthermore, Pysh predicts a strong negative CD band (of perpendicularly polarized absorption) near 190 m$\mu$; an equally intense positive CD band (of parallel-polarized absorption) near 206 m$\mu$; and a weak negative $n_1$–$\pi^-$ CD band. In contrast, the experimental spectra for this structure (Carver *et al.*, 1966; Bovey and Hood, 1967; Timasheff *et al.*, 1967) exhibit broad, deep, negative ellipticity centered near 207 m$\mu$ in the region of parallel-polarized absorption and very shallow, positive ellipticity centered at 221 m$\mu$, the general region of the spectrum in which one anticipates finding the $n_1$–$\pi^-$ transition. In addition to several *ad hoc* reasons, Pysh (1970a) suggests that one possible explanation for these rather serious discrepancies is that poly-L-proline II may in reality exist in a relatively large distribution of slightly different conformations in contrast to the poly-L-proline I which appears to be confined to a steep potential energy well. Madison (1969) and Madison and Schellman (1970a,c) have also extensively examined the optical activity of the two forms of poly-L-proline. Many of the results of their calculations concerning long poly-L-proline I and II chains qualitatively confirm Pysh's findings. In addition, it is shown that the qualitative features of the calculated optical activity are nearly independent of the number of residues in the chain (from 2 to 20 residues). Also, it was shown (1) that the optical activity of poly-L-proline II varies markedly with relatively small variations in both the conformation and the assumed direction of the transition moment of the $\pi^0$–$\pi^-$ transition and (2) that much better agreement with experiment than previously reported by Pysh (1970a) could be achieved by varying these parameters within fairly reasonable limits.

The particularly interesting feature of poly-L-proline II from a biological standpoint is its structural similarity to the polypeptide chains of collagen which, as determined from X-ray studies (Rich and Crick, 1961; Ramachandran, 1963), consists of three extended left-hand helical chains intertwined to form a right-hand superhelix. It is of interest then to compare the CD spectra of collagen (Timasheff et al., 1967; Madison, 1969) and poly-L-proline II. Indeed, one finds close qualitative similarity between the two; both show a shallow long-wavelength maximum (at 220 m$\mu$) followed by a deep, shorter, wavelength minimum (at 200 m$\mu$). Pysh (1967) considered the optical activity of a collagen model composed of three poly-L-proline chains, seventeen residues each, where every third residue is replaced by glycine. A small positive value for the $n_1$–$\pi^-$ rotational strength was calculated, this appearing to be in agreement with experiment; Pysh's calculation also indicates substantial $\pi^0$–$\pi^-$ splitting, although no rotational strengths were reported here for this region. Subsequently, Madison (1969) calculated rotational strengths for three collagen structures proposed by Rich and Crick (1961), Ramachandran and Sasisekharan (1968), and Traub et al. (1969). Using the model (Gly-Pro-Gly)$_8$ it was found that qualitative features of the CD are independent of the association of the chains; that is, the same results obtained for either separate chains or chains intertwined in the triple helix. As similarly found for poly-L-proline II, the theoretically predicted optical activity of the collagen structures, which derive from X-ray diffraction studies on fibers, differs significantly from experiment. However, just as with poly-L-proline II, introducing small changes (within reasonable limits) in the chain structure and the transition moment directions brings the theoretical CD curve into much better agreement. It is concluded from these results that the structures of collagen and poly-L-proline II as proposed by X-ray studies should be reexamined since significantly modified structures are found to give better qualitative results in the optical activity calculations. It is pointed out in connection with this conclusion that the exact structures of these molecules cannot be determined from the X-ray diffraction patterns (there being too little information) and that models must therefore be constructed, the merits of which are judged on how well they reproduce the X-ray pattern. This is to say, then, that the possibility remains that other model structures may produce essentially the same X-ray pattern but more reasonable, calculated CD curves.

## 5. Random Coil

As compared with other protein structures discussed in previous sections, the theoretical efforts to account for the optical activity of polypeptides which lack long-range order have been less successful. Part of the reason for this no doubt relates in an important way to the uncertainties which now surround both the experimental and theoretical characterization of the unordered state(s) of polypeptides. As far as experiment is concerned here, distinctly different CD spectra have been reported for unordered homogeneous polypeptides in solution as compared with both unordered homogeneous polypeptides in films and unordered proteins themselves (see Fig. 16). As far as the theory is concerned, it remains to be established whether or not the statistical methods utilized in the following calculations are indeed appropriate for determining optical activity parameters of one or many aperiodic structures; moreover, it is unclear whether or not these methods can also account for the similarities and differences between the different types of experimental spectra noted above.

In an attempt to calculate $\pi^0-\pi^-$ and $n_1-\pi^-$ rotational strengths for the random coil, Zubkov et al. (1969, 1971) applied the Monte-Carlo method of statistical analysis. This method can be summarized as follows: The ($\phi$, $\psi$) steric map of the L-alanyl residue (Ramachandran et al., 1963) was subdivided into several regions, the greatly simplifying and tenuous assumption being made that the dipeptide conformation energy for any $\phi-\psi$ pair in a given region is uniformly the same. Next, the conformation of a decameric or pentameric chain was randomly generated restricting, at the same time, the possible $\phi$ and $\psi$ angles to only one or two regions. Finally, rotational strengths were calculated and then the process was repeated until the mean results of the trials converged to yield what was assumed to be the optical activity of the random coil. Convergence was found to follow from 400 randomly generated pentamers or 100 decamers, and the results for the pentamers and decamers were practically identical.

For the various regions and combinations of regions of the steric map, $\pi^0-\pi^-$ CD curves were calculated including both exciton and nonexciton contributions; the $n_1-\pi^-$ bands were not included here because of the indeterminacy of the wavelength of this transition. When the $\phi-\psi$ pairs were restricted to a region which is largely forbidden energetically, the CD curve was satisfyingly not found to resemble experimental results. Basically, two types of curves were found when the regions were energetically allowed: type I exhibited a short-wavelength minimum ($[\theta]_{min} = -1.3$ to $-4.3 \times 10^4$) adjacent to a longer wavelength maximum of lesser amplitude ($[\theta]_{max} = +0.3$ to $+1.8 \times 10^4$); whereas type II exhibited a single minimum ($[\theta]_{min} = -1.7$ to $-2.0 \times 10^4$). Therefore, as far as their shapes are concerned, type I qualitatively resembles curve I in Fig. 16 and type II resembles curve II. In some instances, one can imagine even better agreement between theory and experiment if the $n_1-\pi^-$ band is also included here: being negative, it would either tend to reduce the calculated amplitude of the long-wavelength band of the type I curve or it might even produce a negative long-wavelength shoulder (as seen in curve II of Fig. 16) in some of the type II curves calculated. In spite of the fact that the calculated shapes and the experimental curves are qualitatively similar for the most part, a serious and as yet unexplained discrepancy is found in the wavelength positions of the extrema; the calculated curves as a whole are blue-shifted a great distance (10–15 m$\mu$) relative to the experimental curves.

In a similar fashion, Aebersold and Pysh (1970) calculated CD and absorption curves for random statistical ensembles of pentamers, octamers, and decamers of poly-L-alanine. Fundamentally, the following distinguish this calculation from the preceding one: the steric map was divided into different shaped and different sized regions; a distinct potential energy (as a fraction of $\phi$ and $\psi$) was calculated for each chain conformation and this was used to weight each calculated spectrum according to a Boltzmann distribution; the effects of modifying the potential function were also tested; and finally, exciton, but not nonexciton, rotational strengths were evaluated. As far as the exciton results are concerned, they are in qualitative agreement with the findings of Zubkov et al. above. In contrast, however, all the theoretical CD curves qualitatively resembled only curve I in Fig. 16 with the amplitudes of the extrema too low by one or two orders of magnitude. In some instances—depending on the form of the potential, the boundaries placed on $\phi$ and $\psi$, and the oligomer length—the positions of the calculated extrema paired well with corresponding experimental extrema; however, it should be pointed out that this may be more the result of using a somewhat large wavelength (198 m$\mu$) for the amide $\pi^0-\pi^-$ transition. As one primary criterion for testing the success of their calculation, it was checked to see if their parameters reproduced the large red-shift of the negative CD extremum (near 198 m$\mu$ experimentally) relative to the absorption maximum (near 191 m$\mu$). In some cases the anticipated shift emerged, in a qualitative sense; the largest wavelength difference, however, was only 3 m$\mu$.

Aebersold and Pysh are cautious in interpreting their results, emphasizing that errors could easily arise from several sources—e.g., the potential function, omission of nonexciton effects, or the way in which the conformational space is considered. Although the calculation in some cases at least appears to render qualitative agreement with experiment, it is concluded that the results really indicate either that the presently available potential functions are correct and the existing theory of optical activity is not, or that just the reverse obtains.

In another calculation concerning the random polypeptide conformation, Tonelli (1969) assumed that its optical activity could be estimated by averaging: (1) the nonexciton rotational strengths of the central residue of a tripeptide; and (2) the exciton rotational strengths of a dipeptide. Averaging was carried out according to Boltzmann statistics and all conformations (angles) of the tripeptide and dipeptide were included. The results of this calculation are contrary to those found by Zubkov et al. and Aebersold and Pysh, discussed above: in opposition to the conclusion reached in the former work, Tonelli calculates a large positive value for $R_{n\pi}$ and a positive nonexciton contribution for the $\pi^0$–$\pi^-$ transition; in contrast to both works, Tonelli calculates a very low value for the exciton rotational strength. As Aebersold and Pysh point out in their calculation, at least pentamers were needed for convergence, suggesting that Tonelli was led to underestimate considerably the exciton splitting and rotational strength of the $\pi^0$–$\pi^-$ transition because he only considered nearest-neighbor interactions.

## 6. Model Compounds

Schellman and his colleagues have undertaken an interesting series of experimental and theoretical investigations concerning the optical activity of several protein-related model compounds. The compounds studied include variously substituted linear and cyclic dipeptides (Schellman and Nielsen, 1967b; Bayley et al., 1969; Nielsen and Schellman, 1971); o-, m-, and p-tyrosine (Hooker and Schellman, 1970); and several L-proline derivatives (Madison, 1969; Madison and Schellman, 1970a,b,c). The theoretical optical activity of these compounds was systematically studied by varying their individual conformational angles and then calculating the rotational strengths of their lowest-lying transitions—the $n_1$–$\pi^-$, $\pi^0$–$\pi^-$ transitions in the case of the peptide- and amide-containing compounds; the aromatic $^1L_b$, $^1L_a$, $^1B_b$, and $^1B_a$ $\pi$–$\pi^*$ transitions and the carboxyl $\pi$–$\pi^*$ transition in the case of the tyrosine compounds. The most favorable regions of conformation space were also screened by energy calculations and the signs and magnitudes of the calculated rotational strengths of these regions were compared with experimental spectra obtained under various ranges of solvent and temperature conditions. Although the lack of rigidity of most of these compounds prevents a definitive assessment of the conformations of these compounds, the calculations are useful in providing plausible explanations for the experimental results. In addition, the calculations provide information as to the relative importance of each mechanism in generating optical activity as a function of conformation.

One particularly interesting aspect of the findings in these detailed investigations is their relationship to the optical activity of more complex polypeptide structures. Being cautious to emphasize the dangers arising in comparing the optical properties of small and large molecules, Schellman and his colleagues (cf. Bayley *et al.*, 1969; Madison, 1969; Madison and Schellman, 1970c) compare their calculated rotational strengths and CD curves of the appropriate model compounds with experimentally determined parameters for the $\alpha$-helix, poly-L-proline I and II, and collagen. In each case the magnitude of $R_{n\pi}$ is found to agree qualitatively with experiment, this being not too surprising since the optical activity of this transition tends to be dominated by short-range nearest-neighbor interactions. In contrast, the optical activity of the $\pi^0-\pi^-$ transition was generally not reproduced, the main reason being that longer-range interactions are in effect here. However, it was found that the calculated and the experimentally determined $\pi^0-\pi^-$ optical activity of the *cis* conformers of $N$-acetyl-L-proline-$N,N$-dimethyl amide did, in fact, closely resemble that of poly-L-proline I, its subunits also being in the *cis* configuration. The same was true for the experimental and calculated CD of this model compound in the *trans* configuration as compared, respectively, with the experimental and calculated CD of poly-L-proline II, whose subunits are also in the *trans* configuration. In the latter case, it was shown that a large disparity existed between the calculated and experimental CD for the model compound. This difference mimicked that one observed for poly-L-proline II as discussed in Section II,G,4. Therefore, Madison (1969) and Madison and Schellman (1970a) used this model compound, varying its conformation and transition moments, as a guide to probing the sources of this problem for poly-L-proline II.

In the course of investigating the low-temperature CD spectra of several cyclic D- and L-dipeptides, Strickland *et al.* (1970a) attempted to determine the approximate geometry of $C$-L-Tyr-L-Tyr diketopiperazine by carrying out a simple calculation involving the exciton splitting of the two phenolic $^1L_b$ transitions. Using Tinoco's (1963) expressions to calculate the exciton splitting and rotational strengths of this dimer, it was argued that the most likely conformation, which is also consistent with NMR data (as cited therein), is one in which the two phenolic groups are folded over the diketopiperazine ring so as to give the molecule a U-shaped appearance. Although this conclusion reached by Strickland *et al.* might appear to be in error, at first sight, because Tinoco's (1963) equation (our Eq. 64) has incorrect signs, Strickland has informed us in a recent private communication that he used Tinoco's (1963) expression in terms of spherical polar coordinates which has correct signs (see footnote 20).

Recently, the optical activity of the decapeptide antibiotic gramicidin S has been theoretically investigated by Pysh (1970b) and, independently, by Bayley (1971). This compound, being cyclic, is restricted to a relatively small portion of conformation space and it therefore affords a unique opportunity for the investigation of the optical activity of a peptide structure which is of intermediate complexity between the simple mono- and dipeptide compounds heretofore discussed and larger polypeptides and proteins. Several different structures have been proposed for gramicidin S on the basis of spectral measurements, X-ray diffraction analyses, and energy calculations. Pysh calculated the CD for three such structures, one proposed by Momany *et al.* (1969) and two proposed by Scott *et al.* (1967). Transforming the CD into ORD curves, using the Kronig-Kramers transforms, it was found by comparison of the ORD $b_0$ parameters (Moffitt, 1956b) that only the former structure gave a negative value for $b_0$, as is found experimentally. Although this result is not alone sufficient to determine the actual conformation of gramacidin S or, more likely, the equilibrium distribution of conformations, Pysh argues that optical activity calculations such as this are useful for screening potential structures. In a more extensive treatment of this problem, Bayley calculated $n_1-\pi^-$ and $\pi^0-\pi^-$ rotational strengths for the proposed gramicidin S structures (to that time). It was concluded that only one gave the appropriate strong negative value for $R_{n\pi}$ observed experimentally; this structure is the one proposed by Hodgkin and Oughton (1957) and Schwyzer (1958), and is also the one in best agreement with NMR data.

## 7. *Proteins*

Madison (1969) has calculated the optical activity of various segments of the backbone of lysozyme in either $\alpha$-helical, $\beta$-pleated sheet, or random-coil conformations, as determined by the X-ray diffraction studies of Blake *et al.* (1967). The calculated CD spectra for the four irregular helical segments of lysozyme were compared with the calculated CD spectra for perfectly ordered helices of equal and longer lengths. The results for the irregular and ideal helices were found to be in qualitative agreement with one another in spite of the large structural variation in the residues of the irregular helices. Surprisingly, the spectra calculated for three of the four helical segments of lysozyme agreed better with the calculated spectra of the longer idealized helices than did the comparably sized, shorter idealized helices; however, Madison points out that this may be a chance result rather than one of more general implications. Comparing the calculated CD of the $\beta$-structure in lysozyme with that for a perfectly ordered $\beta$-sheet of identical size, it was found that these differed qualitatively. Moreover, the calculated CD spectra of the various "unordered" regions did not resemble either type of experimental spectrum for the random coil polypeptide (see Section II,G,5; Fig. 16). On the basis of these findings and those discussed earlier in connection with the length dependence of the optical activity of the $\alpha$-helix (Section II,G,1) and the conformation sensitivity of the $\beta$-structure optical activity (Section II,G,3), it is concluded by Madison that there may be no good polypeptide models which exhibit CD comparable to that exhibited by the unique structures in proteins. (See, however, Section III for empirical approaches.)

Utilizing the refined three-dimensional coordinates for sperm whale myoglobin (Kendrew, 1962; Watson, 1966) and horse heart oxyhemoglobin (Perutz et al., 1968) as determined from X-ray studies, Hsu and Woody (1971) probed in depth the optical activity of the heme chromophores of these proteins. Although the heme group itself is planar and therefore not inherently optically active, when bound in the relatively rigid structures of these two proteins its allowed $\pi$–$\pi^*$ transitions develop Cotton effects. Considering the possible interactions between the heme and its environment in myoglobin and in hemoglobin, Hsu and Woody ruled out the following as having the dominant effects in generating the heme optical activity in both these proteins: (1) Because there is presently no evidence from X-ray studies that the heme becomes nonplanar upon binding, the possibility of the heme becoming inherently optically active was considered unlikely. (2) Furthermore, because the heme itself has very high ($D_{4h}$) effective symmetry (see argument in Schellman, 1966) and because the distribution of charges and partial charges which surround the heme in the protein are a priori expected to have largely cancelling effects, the one-electron mechanism through perturbation was judged of only secondary importance. Specifically considered in this vein was the mixing of the magnetically allowed $d$-$d$ transitions of the iron with the $\pi$–$\pi^*$ transitions of the porphyrin ring; however, this was also ruled out because (1) the CD spectra of hemoglobin and complexes of globin with iron-free porphyrin are qualitatively very similar (Ruckpaul et al., 1970) and (2) reduction of the iron and ligand changes produce only small alterations in the heme CD. Hence by the process of elimination, Hsu and Woody concluded that the dominant source of heme optical activity here most likely stems from the electric–electric (coupled-oscillator mechanism) and electric–magnetic coupling of the heme $\pi$–$\pi^*$ transitions with peptide and side-chain aromatic transitions of nearby globin residues.

Calculating rotational strengths, Hsu and Woody found that there is no one interaction, peptide or side chain, fully accounting for the heme optical activity. In the case of hemoglobin, in fact, the heme of one chain is found to interact substantially not only with the aromatics of the same chain but also with those of neighboring chains. The calculated rotational strength (+0.1 DBM) of the Soret band of horse heart oxyhemoglobin compares favorably with the experimentally determined value (+0.23 DBM) for human oxyhemoglobin, the assumption being, of course, that only minor structural differences exist between these two variants. For myoglobin, the calculated rotational strength (+0.3 DBM) for the Soret band agrees reasonably well with the range of experimental values (+0.5 ± 0.05 DBM) determined for metmyoglobin (Beychok, 1967; Willick et al., 1969). Calculated rotational strengths for the N, L, and Q bands of both proteins were also found to agree qualitatively with experiment. Much poorer agreement was achieved, however, for the computed Soret rotational strengths of the separated $\alpha$- and $\beta$-chains of hemoglobin; it could not be determined whether the difficulty here reflects an error in the calculation or simply a conformational change in the subunits when separate from the tetramer.

Because of the near degeneracy of the two transitions constituting the Soret band (these being polarized perpendicular to one another and in the plane of the heme), Hsu and Woody found that the shape, but not the net rotational strength, of this band was sensitive to the directions of the transition moments. Screening possible orientations for these two transition moments, they concluded that if the moment axes are oriented toward the bridging methine carbons away from the pyrrole nitrogen of the porphyrin ring then the qualitative features of the Soret CD of both myoglobin and hemoglobin can be reproduced. In myoglobin the Soret CD appears as a single positive band, whereas in hemoglobin it appears as a strong positive band adjacent to a shorter-wavelength weak negative band. It is also predicted in connection with these findings that one could possibly observe dramatic changes in the Soret CD accompanying the binding of ligands not as a result of a conformational change in the protein but as a consequence of changing the directions of polarization of these two Soret transitions.

In summary, the general results of Hsu's and Woody's calculation appear to be very satisfactory, particularly in view of the complexity of the systems considered here. Of the very long list of potential sources of error in this calculation perhaps the most serious relates to the reliability of the transition monopole charge distributions used for the heme, tryptophan, and histidine chromophores. Because of the scarcity of experimental information concerning the transition directions and moments of these groups, Hsu and Woody were forced to determine the charge distributions solely on theoretical grounds using available MO wavefunctions. Until the spectral properties of these chromophores are subjected to further experimental investigation, the conclusions reached here must be viewed with some reservations. [During the final stages of preparation of this manuscript two very interesting papers appeared in which the optical activity spectra of a few more proteins are theoretically examined. In addition to summarizing many of their previous findings (as discussed in several of the preceding sections here), Madison and Schellman (1972) calculate molar ellipticities for various backbone segments of $\alpha$-chymotrypsin, lysozyme, ribonuclease S and myoglobin. The segments studied include the nonperiodic regions as well as the $\alpha$- and $\beta$-structures of these proteins as determined by X-ray diffraction studies. In the other paper, Strickland (1972) has theoretically analyzed the near-UV tyrosyl CD bands of RNase S and he utilized the results of his calculation to examine the structural differences between RNase S and RNase A, the latter being structurally less well-defined and showing altered near-UV optical activity.]

## III. Secondary Structure of Proteins

### A. CD Spectra of Polypeptides in $\alpha$-Helical, $\beta$-Pleated Sheet and Aperiodic Conformations

The CD spectra of poly-L-lysine in the three conformational states so far characterized for this polypeptide are shown in Fig. 15 (Greenfield and Fasman, 1969). The $\beta$-form shown is that of a solution brought to pH 11.1, heated at 52°C for 15 min, then cooled to 22°C, at a concentration of 0.01%. The helical form is also produced at pH 11.1, in an unheated solution, and the aperiodic form is at pH 5.7 (Greenfield and Fasman, 1969). Sarkar and Doty (1966) and Li and Spector (1969) have demonstrated that the $\beta$-form of poly-L-lysine produced at neutral pH with sodium dodecyl sulfate exhibits a different spectrum than that produced in water at high pH, the main difference being a reduced ellipticity at 218 nm in SDS.

The spectrum of the $\alpha$-helix is very similar to that of poly-$\alpha$-L-glutamic acid or of a helical copolymer containing glutamyl, lysyl, and alanyl residues (Holzwarth and Doty, 1965); a number of other helical

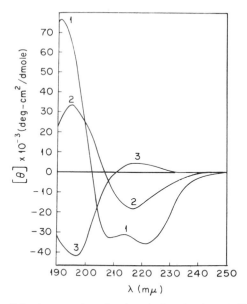

FIG. 15. Circular dichroism spectra of poly-L-lysine, in the α-helical (1), β (2), and random (3) conformations. Taken from Greenfield and Fasman (1969).

polypeptides give very similar spectra (Beychok, 1968) with some variations in the ellipticity values at the negative maxima of 222–223 nm and 207–209 nm. In their review article, Gratzer and Cowburn (1969) note an overall variability of about 10% in ellipticity values at 222–223 nm from various laboratories using various polypeptides with side chains that do not absorb at that wavelength. At 208–209 nm, the overall variability is closer to 25%. At the positive maximum near 191 nm, the variation is about 30%.

Uncertainty in values for the β-structures are greater for several reasons. In addition to the distinct spectra of poly-L-lysine in water at pH 11 and in SDS at neutral pH, there is a solvent dependence not noted with the α-helical polypeptides (Fasman and Potter, 1967; Quadrifoglio and Urry, 1968a; Stevens et al., 1968). Furthermore, the earlier demonstration by ORD of at least two different types of spectra exhibited by antiparallel, pleated sheets (Fasman and Potter, 1967) has been confirmed by CD spectra. The CD spectrum of poly-S-carboxymethyl-L-cysteine differs qualitatively from that of poly-L-lysine, with a negative band at 227 nm ($\theta = -5.9 \times 10^3$) and a positive band at 198 nm ($\theta = 14.9 \times 10^3$) (Stevens et al., 1968).

The CD spectrum of a polypeptide in a parallel β-pleated sheet conformation has so far not been reported (see, however, Section II,G,3, for theoretical calculations).

The random-coil CD spectrum presents a somewhat different kind of problem with respect to estimates of the secondary structure of proteins. Briefly stated, the main difficulty is that the CD spectra of denatured proteins are substantially different from those of poly-L-lysine (or poly-L-glutamate) in the random-coil forms. This, of course, raises the significant quandary of the appropriate rotatory parameters for aperiodic, unordered backbone regions of proteins. Moreover, even if the denatured protein CD spectra were not different from the spectra of random-coil polypeptides, we would still not safely conclude that the unique conformations of aperiodic segments of backbone in the native protein generate the same CD curve as the denatured protein.

These matters have not been overlooked by investigators in this field. That extensively denatured proteins do not, generally, show a positive band between 210 and 220 nm has been frequently noted (Beychok, 1967; Tiffany and Krimm, 1969; Dearborn and Wetlaufer, 1970; Fasman et al., 1970; Jirgensons, 1970; Pflumm and Beychok, 1969a; Gratzer and Cowburn, 1969). An exception is phosvitin which near neutral pH in aqueous solutions shows a positive band at 216–219 nm (Timasheff et al., 1967; Perlman and Grizzuti, 1971). Tiffany and Krimm (1969) demonstrated that the positive band exhibited by Na-PGA in water or dilute salt is abolished in 3 $M$ guanidine hydrochloride with a resultant negative shoulder, instead, at 218–220 nm. Dearborn and Wetlaufer (1970) reported, however, that poly-L-lysine in 6 $M$ guanidine hydrochloride at pH 6 exhibited the same CD spectrum as in 0.01 $M$ NaCl. Fasman et al. (1970), in a study of the CD spectra of films of Na-PGA and of poly-L-lysine showed that the polymers in the random conformation differed notably from the CD of the corresponding polymers in solution. In films, the polymers exhibited negative bands at 200–205 nm and negative shoulders at 215–230 nm. Furthermore, these spectra were quite similar to those of denatured proteins (Fig. 16).

There is some disagreement as to the conformation of the homopolypeptides in strong salt solutions, guanidine hydrochloride, and a number of other solvents, but there appears to be no disagreement that the CD spectra of random-coil or unordered homopolypeptides differ in two important regards from spectra of denatured proteins. In the latter (1) there is no positive ellipticity peak at 216–218 nm and (2) the negative ellipticity peak near 200 nm is less intense and may be red-shifted in some cases by as much as 4 nm, though it frequently occurs at 198 nm as it does in the homopolypeptides.

In terms of the quantitative analysis of secondary structure of dis-

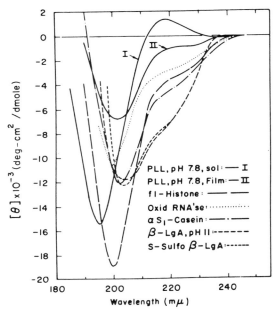

Fig. 16. Circular dichroism of unordered poly-L-lysine and of several proteins in unordered conformations. Taken from Fasman et al. (1970). Curve I appears to have been drawn to half-scale in the original figure.

solved proteins, this raises the important question of how to estimate contributions of aperiodic segments (neither helical or pleated sheet) to the CD spectra of proteins. To this difficulty must be added the uncertain contribution made by side-chain transitions to the far-UV CD spectra of proteins (Beychok, 1966).

## B. Estimates of Protein Secondary Structure

Greenfield and Fasman (1969) showed that the CD curves of poly-L-lysine in the $\beta$- and random-coil forms are isodichroic at 208 nm, with an average ellipticity of $-4000$ deg-cm$^2$/dmole. Thus, they suggested estimating the fraction of residues in $\alpha$-helical conformation using the observed residue ellipticity at 208 nm, which for the poly-L-lysine $\alpha$-helix is $-32,600 \pm 4000$ deg-cm$^2$/dmole

$$\% \ \alpha\text{-helix} = \frac{-[\theta]_{208\,\text{nm}} - 4000}{33,000 - 4000} \times 100 \tag{78}$$

To find the proportions of $\beta$- and random-coil structures, the experimental curve is compared to a family of computed curves generated by different percentages of $\beta$- and random-coil structures at the given $\alpha$-helix value (method I). More refined estimates are achieved by digital computation which minimizes the difference between experimental curves and computed mixtures after initial estimates by method I (method II). Comparison of these results with the structures obtained by X-ray analysis of several proteins is given in Table I, which is taken directly from the paper by Greenfield and Fasman (1969).

The secondary structures of myoglobin, ribonuclease A, and lysozyme are adequately estimated. For carboxypeptidase A, the α-helix content is underestimated, the β-structure overestimated. [It should be noted that carboxypeptidase A contains parallel pleated sheet sections as well as antiparallel β-structure (Reeke et al., 1967b).] The content of α-helix is overestimated for chymotrypsin and chymotrypsinogen. Method I, which is much simpler, gives almost equally satisfying results; the table also shows that use of the β-parameters for the structure found in SDS

TABLE I
STRUCTURAL CONTENT (%) OF PROTEINS AS ESTIMATED BY X-RAY DIFFRACTION AND BY CIRCULAR DICHROISM USING METHOD II[a]

| Protein | X-Ray structure | | | Circular dichroism calculated structure | | | | | |
| --- | --- | --- | --- | --- | --- | --- | --- | --- | --- |
|  |  |  |  | α | | β | | Random coil | |
|  | α | β | Random coil | b | c | b | c | b | c |
| Myoglobin | 65–72, 77[d] | 0 | 32–23 | 68.3 | 68.2 | 4.7 | 7.9 | 27.0 | 23.9 |
| Lysozyme | 28–42[d] | 10 | 62–48 | 28.5 | 29.9 | 11.1 | 9.3 | 60.4 | 61.0 |
| RNase | 6–18[d] | 36 | 58–46 | 9.3 | 12.0 | 32.6 | 43.4 | 58.1 | 44.5 |
| RNase S | 15[e] | 31 | 54 |  |  |  |  |  |  |
| Carboxypeptidase A | 23–30[d] | 18 | 59–52 | 13.0 | 15.9 | 30.6 | 39.9 | 56.4 | 44.2 |
| α-Chymotrypsin[f] | 3[g] |  |  | 11.8 | 13.4 | 22.8 | 31.9 | 65.5 | 54.8 |
| Chymotrypsinogen |  |  |  | 13.8 | 17.3 | 25.2 | 28.9 | 60.9 | 53.8 |

[a] The best fit to the experimental circular dichroic curve was found by minimizing the variance between the experimental circular dichroic curve and a linear combination of circular dichroic curves for the α-helix, β-structure, and random coil from 208 to 240 nm. Taken from Greenfield and Fasman (1969), where original references to X-ray structure analysis may be found.

[b] Values used in calculations are for poly-L-lysine in $H_2O$.

[c] Values used in calculations are the same as in footnote b with data for β-poly-L-lysine in $H_2O$.

[d] Lower value represents true regular α-helix; upper value, total helix including $3_{10}$- and distorted helices.

[e] Helical type not distinguished by authors.

[f] α-Tosylchymotrypsin used for X-ray work.

[g] Only one short-chain section α-helix included; isolated helical turns and β-structure not reported, although they may be present.

does not greatly alter the estimates by method II. Sigler *et al.* (1968) did not explicitly give the $\beta$-structure content of $\alpha$-tosylchymotrypsin; recently, Saxena and Wetlaufer (1971) have interpreted the X-ray structure presented by Birktoft *et al.* (1970) to be comprised of 22% $\beta$-structure, which is the lower value of two calculated by Greenfield and Fasman in Table I.

There are, of course, two other isodichroic points, one at approximately 204 nm for $\alpha$ and random coil, the second at about 198 nm for $\alpha$ and $\beta$. Myer (1969) has suggested these, also, for estimation of secondary structure. From Fig. 15, it is clear that both of these are somewhat less appealing. The ellipticities at both points are large relative to that of the third structure. At both wavelengths, the $\alpha$-helix curve is changing steeply. Furthermore, measurements at 198 nm are not always possible and, when possible, are much less precise than at 208 nm. Sonenberg and Beychok (1971) did, however, utilize an isodichroic point at 204 nm in estimating $\beta$-structure in human and bovine growth hormones and found very good agreement with values determined using digital computation.

A rather different approach is to use the crystal structure values and attempt to construct the experimental CD curve with minor adjustments in the characteristic CD parameters of the model compounds. This offers the possibility of calibrating the model compound parameters for more confident use in unknown structures after many fits to known structures have been accomplished. Pflumm and Beychok (1969a) used this approach with pancreatic RNase A and achieved reasonable success. Their fit required, however, that the aperiodic segments of RNase A have an intensity at 198 nm only $\frac{2}{3}$ as great as that of the random coil form of poly-L-lysine (see, however, Sec. III,A). In addition, a positive band due to (presumed) aromatic contributions centered near 226 nm was included. The intensity of this band, 1500 deg-cm$^2$/dmole at the maximum, was suggested to be compatible with expected contributions of aromatics. The presence of the positive band was furthermore demanded by residual positive ellipticity in the interval 230–240 nm which could not otherwise be accounted for unless the parameters of the $\alpha$-helix were substantially modified (Schellman and Lowe, 1968).

Saxena and Wetlaufer (1971) have recently proposed a different basis for interpreting the far-UV CD spectra of proteins, utilizing the known structures of RNase, lysozyme, and myoglobin and the experimental CD curves of these proteins. For each of the three protein spectra at a given wavelength,[26] the ellipticity is taken to be the same as the value for each of the three structures (a, $\beta$, and "remainder") multiplied by its known

---

[26] The values used for ribonuclease appear to be at variance with those published by others (cf. Pflumm and Beychok, 1969a).

fraction in the protein. This generates three simultaneous equations at each wavelength which can be solved at any wavelength, for the value of the ellipticity of $\alpha$, $\beta$, and "remainder" (aperiodic structures) in the native protein. The spectra generated in this way are shown in Fig. 17. The spectrum of the $\alpha$-helix in the three proteins is seen to be quite similar to that of poly-L-lysine. That of the $\beta$-structure is slightly different, the negative band being shifted to 220 nm and more intense than the poly-L-lysine spectrum. The positive band occurs at 195 nm as in the model compound. A small positive band occurs at around 235 nm which is not seen in poly-L-lysine.

The most striking difference is in the "remainder" curve which shows a positive band at 222 nm four times as intense as is observed for the homopolypeptides. The negative band, at 192–193 nm, is about two thirds as intense as in poly-L-lysine.

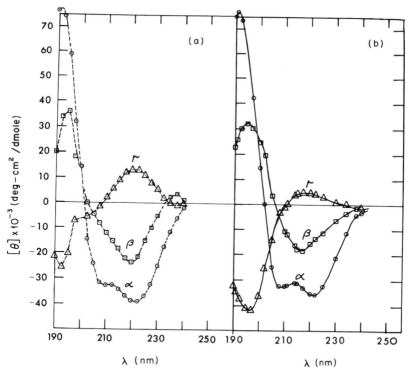

FIG. 17. (a) CD spectra of three polypeptide structural modes computed from X-ray diffraction structural data and experimental CD spectra of lysozyme, myoglobin, and ribonuclease. (b) CD spectra of three polypeptide structural modes obtained experimentally for three structural forms of poly-L-lysine by Greenfield and Fasman (1969). Taken from Saxena and Wetlaufer (1971).

With these spectra, Saxena and Wetlaufer then compute secondary structures of carboxypeptidase A, $\alpha$-chymotrypsin, and chymotrypsinogen. The spectra are computed first by trial and then best visual fit to the experimental CD curves. Their results are shown in Table II taken directly from their paper. For carboxypeptidase, the agreement is excellent. For chymotrypsin, the $\alpha$-helix is overestimated somewhat more than by the Greenfield and Fasman method. For chymotrypsinogen, both methods overestimate $\alpha$-helix somewhat. The $\beta$-fit is very good in the three examples using both methods.

Side-chain contributions are not explicitly taken into account in this procedure. They are somehow weighted into the three computed spectra and it is surprising that three proteins (six, considering the fits) of quite different secondary structure generate the same spectra, particularly so for the very large positive "remainder" band. This method should be calibrated against many more proteins as the structures become available. It may be possible, as any discrepancies arise, to treat the side chains and disulfide with an empirical factor. It is important not to overlook very anomalous spectra. Avidin and strepavidin, for example, give spectra which are clearly dominated by aromatic residues (Green and Melamed, 1966) and which will not be fitted by secondary structure parameters only. Almost simultaneously, Chen and Yang (1971) developed a procedure quite similar to that of Saxena and Wetlaufer and have examined several spectra not included by Saxena and Wetlaufer. Their computed

TABLE II
STRUCTURAL CONTENT (%) OF THREE PROTEINS AS ESTIMATED BY X-RAY DIFFRACTION AND BY TWO INDEPENDENT CIRCULAR DICHROISM APPROACHES[a]

| Protein | $\alpha$-Helix | $\beta$ | Remainder |
|---|---|---|---|
| Carboxypeptidase | | | |
| X-ray diffraction[b] | 23–30 | 18 | 59–52 |
| CD (this work) | 26 | 18 | 56 |
| CD (polylysine)[c] | 13–16 | 31–40 | 56–44 |
| $\alpha$-Chymotrypsin | | | |
| X-ray diffraction[d] | 8 | 22 | 70 |
| CD (this work) | 20 | 20 | 60 |
| CD (polylysine)[c] | 12–13 | 23–32 | 65–55 |
| Chymotrypsinogen | | | |
| CD (this work) | 15 | 29 | 56 |
| CD (polylysine)[c] | 14–17 | 25–29 | 61–54 |

[a] Taken from Saxena and Wetlaufer (1971).
[b] Reeke et al. (1967a).
[c] Greenfield and Fasman (1969).
[d] These estimates are based on Saxena and Wetlaufer's interpretation of Fig. 2 and its accompanying text in the paper of Birktoft et al. (1970).

CD parameters are qualitatively similar to those of Saxena and Wetlaufer but differ quantitatively in several significant respects.[27]

In Section V of this review, secondary structure estimates of several individual proteins will be discussed. We may note here, however, that there are a number of proteins for which high-resolution crystal structures are now available but for which critical CD analyses, of the kind just discussed, have not yet been presented. In some instances, preliminary ORD and CD data seemed to disagree with the subsequently solved structures. These should now be subjected to careful analysis to see whether the discrepancies persist and whether the procedures of Greenfield and Fasman (1969), Pflumm and Beychok (1969a), Saxena and Wetlaufer (1971), and Chen and Yang (1971) quantitatively account for any discrepancies. Cytochrome $c$, for example, on the basis of ORD results, was estimated to be comprised of about 25% $\alpha$-helix (Urry and Doty, 1965; Ulmer, 1965). Preliminary crystallographic data (Dickerson et al., 1967) suggested no $\alpha$-helix whatever in the molecule. From the CD spectra of Myer (1968a,b), we estimate the residue ellipticity of horse heart ferricytochrome $c$ at 208 nm to be $-8900$ deg-cm$^2$/dmole. According to Eq. (78) this would amount to 17% $\alpha$-helix. A recent high resolution X-ray structure of the protein (Dickerson et al., 1971; Dickerson, 1972) shows that there is, indeed, about 20% of $\alpha$-helical structure. The CD estimate is thus close enough to warrant a Greenfield-Fasman computer analysis and to test the Saxena-Wetlaufer approach.

There are a number of other proteins for which high-resolution X-ray structures are available but for which only incomplete ORD and/or CD analyses have been attempted, including lactate dehydrogenase (Adams et al., 1970), subtilisin BPN (Wright et al., 1969), carbonic anhydrase C (Friborg et al., 1967), and papain (Dreuth et al., 1968). Insulin (Adams et al., 1969), elastase (Shotton and Hartley, 1970; Shotton and Watson, 1970), and staphylococcal nuclease (Cotton, 1969) are discussed in detail in Section V.

## IV. Side-Chain Optical Activity in Model Compounds and Proteins

A. TYROSYL RESIDUES

### 1. Near Ultraviolet

Early studies on the free amino acid L-tyrosine in acid or neutral pH revealed a positive band centered near 275 nm, with maximum ellipticity

---

[27] An important conclusion reached by Chen and Yang is that $\alpha$-helix, but not $\beta$-structure, is best estimated in CD spectra from residue ellipticity values at the 222–224 nm extremum, in accordance with the theoretical analysis of Madison (1969).

of about 1300 deg-cm$^2$/dmole (Beychok and Fasman, 1964; Legrand and Viennet, 1965). The ethyl ester also shows a positive band, but $N$-acetyl-L-tyrosine ethyl ester and $N$-acetyl-L-tyrosineamide show negative bands in aqueous solution with intensities of about $-600$ deg-cm$^2$/dmole (Beychok, 1966; Shiraki, 1969). Ionization of the phenolic hydroxyl shifts these bands to longer wavelength; in all cases, the bands are then positive with maximum intensities close to 1000 deg-cm$^2$/dmole. The signs and intensities of the CD bands of the un-ionized tyrosine derivatives are markedly solvent dependent (Horwitz et al., 1970). With recent instrumentation, it becomes apparent that, at least with $N$-acetyl-L-tyrosineamide and $N$-acetyl-L-tyrosine ethyl ester, the main band has a vibronic shoulder about 6 nm toward the red from the maximum. The shoulder can be resolved at low temperature and is frequently noted in protein spectra (see Section V,D), even at room temperature (Shiraki, 1969; Horwitz et al., 1970; Menendez and Herskovitz, 1970).

Incorporation of tyrosine into peptides and proteins modifies somewhat the near-UV CD characteristics and adds conformational dependence to the bands. Poly-L-tyrosine, itself, at pH 11.2, in helical and random-coil conformations, is shown in Fig. 18. The degree of ionization is not precisely known in either form, but the polymer in the helical conformation is probably mainly un-ionized (Beychok and Fasman, 1964). The helical form shows a positive band near 270 nm with a residue el-

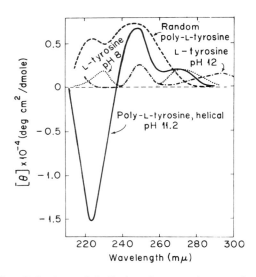

FIG. 18. Circular dichroism of helical poly-L-tyrosine, random poly-L-tyrosine (both at pH 11.2), and L-tyrosine at the pH values indicated. Taken from Beychok and Fasman (1964).

lipticity of about 2000 deg-cm²/dmole. The random-coil form of the polypeptide shows no CD band in this region. Friedman and Ts'o (1971) have presented similar results at pH 11.2 but note that a helical form at pH 10.6 (with less ionization), while exhibiting positive dichroism near 280, reveals no distinct band. A significant result here is the sharp conformation dependence and the fact that the band near 270 nm is positive, in contrast to that of $N$-acetyl-L-tyrosineamide in aqueous neutral solution.

It is perhaps worthwhile to note, in anticipation of protein results presented later, that tyrosyl optical activity in the near UV need not result from coupling of tyrosine moments with those of other molecules since the free amino acids generate bands of intensity comparable to those found in tyrosine-containing polypeptides.

Schechter et al. (1971) have synthesized an interesting series of oligo- and polypeptides of known tyrosine composition and sequence. The peptides possessed the structure (L-Tyr-L-Ala-L-Glu)$_n$ with $n = 1, 2, 3, 4, 7, 9$, and $13$. A high molecular weight polymer with $n$ approximately 200 was also characterized. At pH 7.4, the tripeptide ($n = 1$) gives a positive band at 270 nm with a mean residue ellipticity of 115 (on a tyrosine or molar basis, the value of $[\theta]$ would be 345)'. As $n$ increases, the band diminishes gradually (becomes more negative). The polymer which is $\alpha$-helical shows a very small positive peak at 287 ($[\theta] = 40$) and a larger negative band at 273 ($[\theta] = -360$). On a tyrosine basis, $[\theta]$ would be $-1080$.

These results may be compared with those achieved by Ziegler and Bush (1971), who investigated cyclic hexapeptides with one and two side chains, the remainder of the residues being glycyl. $c$-Gly$_5$-L-Tyr exhibits a negative band at 277 nm with $[\theta]_{max}$ on a tyrosine (molar) basis of $-600$. A linear hexapeptide, H-Gly$_2$-Tyr-Gly$_3$-OH shows a negative band at 280 nm, with $[\theta]_{max} = -300$. $c$-Gly$_2$-L-Tyr-Gly$_2$-L-His shows a negative band at 277 nm with $[\theta]_{max} = -720$ (tyrosine basis).

From these various results, we have only rough guidelines as to what kind of CD behavior will be exhibited by tyrosyl residue in proteins. Un-ionized tyrosyl residues may ordinarily be expected to contribute bands in the near-UV CD spectrum centered near 274–280 nm with shoulders at 282 to 287 nm. The bands may be positive or negative; both are encountered in proteins. Pancreatic ribonuclease and insulin, for example, display negative tyrosyl CD bands (see Section V); egg white lysozyme (Halper et al., 1971; Ikeda and Hamaguchi, 1969) and T$_1$ ribonuclease exhibit positive bands. Many other examples of each kind occur. The sign, intensity, and position of ellipticity for a particular residue is influenced by environment and, perhaps, by other factors

(Horwitz et al., 1970). For a residue that is not coupled with other aromatic residues, the intensity should fall between 350 and 1000. Coupling interactions may alter this range considerably.

An interesting example is provided by tyrosine B26 in beef insulin. The residue occurs in a C-terminal octapeptide which is liberated on tryptic digestion (Menendez and Herskovitz, 1970). The CD of the separated peptide appears to be identical (or very similar) to its contribution in the intact protein. The tyrosyl band is negative with maximum ellipticity at about 275 nm of each −500 (tyrosine basis).

In attempting to identify tyrosine contributions to the near-UV CD spectra, investigators have occasionally made use of the reaction of tyrosine residues with $N$-acetyl imidazole (see, for example, Ettinger and Timasheff, 1971a). The products of this reaction are (mainly) $O$-acetylated tyrosine residues which absorb at 263 nm with diminished extinction at 280 nm ($\Delta\epsilon$ = 1160 per mole) (Riordan et al., 1965). It may be noted here that the CD spectrum of $N,O$-diacetyl tyrosineamide shows two bands, at 269 and 278 nm, corresponding exactly in position to the bands of $N$-acetyltyrosineamide, but with the ellipticity at 269 reduced by about 40% (R. Sella and S. Beychok, unpublished results).

## 2. Far Ultraviolet

The far-ultraviolet CD bands of $N$-acetyltyrosineamide at pH 7 and 12 and of $N,O$-diacetyltyrosineamide at pH 7.3 are shown in Fig. 19. For $N$-acetyltyrosineamide, the results coincide closely with those of Shiraki (1969). The spectrum, moreover, of the ethyl ester is very similar to that of the amide (Beychok, 1966; Menendez and Herskovitz, 1970). Shiraki's spectrum shows another positive band at 200 nm, with maximum ellipticity ($[\theta]_{200}$ = 52,000), or slightly more than three times as great as the band at 226 nm. The main effect of $O$-acetylation is to blue-shift the band 5 nm with accompanying slight diminution in maximum intensity and slight broadening of the band, so that the rotational strengths are comparable. Ionization of the phenolic hydroxyl brings about the expected red-shift without significant change in maximum ellipticity. It has been noted before (Beychok, 1966) that the red-shifted band is at 237–238 nm, rather than at 245 nm, which is the position of the absorption band of ionized residues.

It is difficult to state what may be expected for tyrosyl CD bands in proteins at, or near, 226 nm because of the peptide ellipticity bands. In poly-L-tyrosine, a complex set of electronic and magnetic interactions make it unlikely that one could seek additivity (see Section II,G, and Fig. 18). If the peptide contribution in the helix is −36,000 deg-cm$^2$/dmole, this would give a value of 21,000 deg-cm$^2$/dmole (of tyrosyl resi-

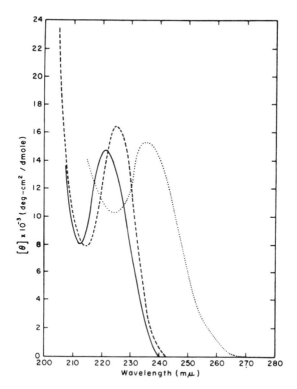

FIG. 19. CD spectra of $N$-acetyl-L-tyrosineamide and $N,O$-diacetyl-L-tyrosineamide. (– – –) $N$-Acetyl-L-tyrosineamide, pH 7; (· · ·) $N$-acetyl-L-tyrosineamide, pH 12; (——) $N,O$-diacetyl-L-tyrosineamide, pH 7.3. Spectra were recorded on a Cary model 60 spectropolarimeter. Taken from Pflumm and Beychok (1969a).

dues), not very different from the value for $N$-acetyl-L-tyrosineamide. Schechter et al. (1971), in their study of oligo- and polypeptides of the structure (Tyr-Ala-Glu)$_n$ find about 12,000 deg-cm²/dmole on a tyrosine basis for $n = 1$ (tripeptide). However, with $n = 200$, the helical polymer gives $[\theta] = 8500$ deg-cm²/dmole on a residue basis and this would indicate a *possible* tyrosyl contribution of 85,000 per tyrosine residue, which is five times the amino acid value.

In no case that we are aware of has it so far been possible to isolate the far-UV tyrosyl band in a protein. Some generalities, however, may be emerging. So far, all assignments in model compounds suggest that the 226 nm L-tyrosyl band is always positive, irrespective of conformation and irrespective of the sign of the longer-wavelength bands. Various oligopeptides containing a single tyrosyl residue, such as oxytocin, exhibit molar ellipticities in the range 15–25,000 deg-cm²/dmole (Beychok and Breslow, 1968). It is possible that all tyrosyl residues make a molar contribution of at least this magnitude in native proteins. In the ionized state, tyrosyl residues appear to generate bands of this magnitude between 240 and 250 nm in alkali-denatured proteins even when all near-UV Cotton effects have disappeared (see, for example, Pflumm and Beychok, 1969b). The conformation dependence of the 226 nm band (or 240 nm band in ionized states) thus appears less sharp than that of the

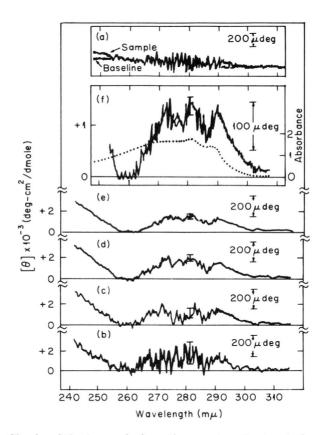

FIG. 20. Circular dichroism and absorption spectra of L-tryptophan in H₂O at pH 7.0. Concentration, 7.42 mg per 100 ml; 1 cm path length; temperature, 25°C. (a) Original recording from the instrument at sensitivity of 0.002°/full scale, gain setting of 4, and scanning speed of 2 nm/min; (b) readout from the computer after a single scan; (c) readout after four scans; (d) readout after eight scans; (e) readout after 16 scans; (f) 5-fold y-axis expansion of (e). (———) Actual CD tracings; (– – –) electrical mean of the computer output; (· · ·) absorption spectrum. Taken from Myer and MacDonald (1967).

near-UV bands. The upper limit of intensity, however, cannot presently be set with any confidence.

## B. Tryptophanyl Residues

### 1. Near Ultraviolet

Figure 20, from the paper by Myer and MacDonald (1967), shows the near-UV CD spectrum of L-tryptophan in water. The spectrum is very similar to that of $N$-acetyltryptophanamide in water (Shiraki, 1969), except that the intensities of the bands of the latter are somewhat lower and the amide shows a fourth band (negative, $[\theta] = -670$) at 266 nm sharply overlapped with a positive band at 270.5. It is now evident (see Section V) that, in proteins, tryptophan ellipticity is signaled by a well-resolved peak, which may be negative or positive at 290–292 nm. The shorter-wavelength bands may be obscured or overlapped with tyrosine and disulfide bands, but the 292 nm band is ordinarily present and at least partly resolved.

Figure 21 shows the spectrum of poly-L-tryptophan in ethylene glycol monomethyl ether, where it is $\alpha$-helical, and spectra of copolymers of

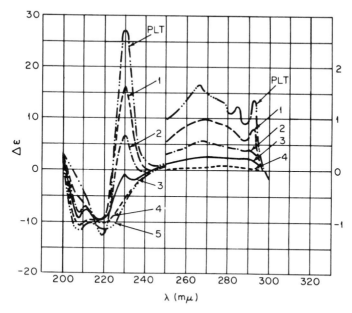

Fig. 21. CD spectra of poly-L-tryptophan (PLT) and copolymers of L-Trp and $\alpha$-ethyl-L-glutamate. The numbers on the curves refer to mole fraction of Trp in the initial monomer mixture: (1) 0.84, (2) 0.68, (3) 0.50, (4) 0.32, (5) 0.16. Taken from Peggion et al. (1968).

L-tryptophan with γ-ethyl-L-glutamate, the mole fractions of Trp varying from 0.84 to 0.32. The longest-wavelength peak is quite evident in three of the samples. On a tryptophan basis, the ellipticity is 4300 deg-cm$^2$/dmole for poly-L-tryptophan, 2985 deg-cm$^2$/dmole for the copolymer of mole fraction 0.84, and 1940 deg-cm$^2$/dmole for the copolymer of mole function 0.68. The copolymer appears to be α-helical at all ratios of tryptophan: ethyl glutamate is the solvent used. There is no example of a protein in which it has been established that a single tryptophan residue, not electronically coupled to other aromatic residues, generates ellipticity equal to that in poly-L-tryptophan. However, in cases where several tryptophan residues are present, it has so far not been possible to assign bands unequivocally to particular tryptophan residues. In a number of cases cited in Section V, the ellipticities would achieve values in the range 3000–5000 if *all* the CD at 292 nm were attributable to a single residue, and there is a smaller number of instances in which the average tryptophan residue contribution is as much as 3000–4000, suggesting individual contributions substantially larger than average.

Staphylococcal nuclease contains a single tryptophan residue in a total chain of 149. The unresolved band near 292 has an intensity of 1340 deg-cm$^2$/dmole on a Trp basis. This positive band is overlapped on the short-wavelength side by a larger negative band, so this represents a minimum contribution, though it is not clear whether the band maximum is being reduced by the overlap (Omenn et al., 1969).

In Section II, there is a discussion of the known UV transitions of aromatic residues. Much of the analysis depends on the work of Strickland and his associates (1969, 1970) and earlier work by Edelhoch et al. (1968). In connection with the analysis of protein and polypeptide near-UV spectra, it should be noted that the greatest part of the near-UV rotational strengths of tryptophan-containing compounds resides in a broad band centered near 270 nm, rather than in one or two bands near 292 and 285 nm. The analyses by Strickland and co-workers attributes the 270 nm activity to the $^1L_a$ transition and that near 290 nm to the $^1L_b$ transition. The $^1L_b$ transition has two vibronic components which may usually be identified as a couplet. The longer-wavelength band of the couplet occurs near 292 (this is the band discussed in detail above) and its counterpart band is ordinarily at about 7 nm shorter, near 285 nm. There are $^1L_a$ vibronic components at still longer wavelengths. Increasingly in proteins a very small band is noted at 300–305 nm (see, for example, Fasman et al., 1966).

Very recently, Halper et al. (1971) have found, in human lysozyme, a band at 313 nm, well removed from any other band so far discerned in proteins (see Section V,C), but the band has not yet definitely been assigned.

## 2. Far Ultraviolet

Figure 22, from the paper by Shiraki (1969), shows the far-UV CD spectrum of N-acetyl-L-tryptophanamide in water. The positive band at 227 nm is virtually coincident with that of N-acetyl-L-tyrosineamide (the ellipticities in the figure are given as *in vacuo* values; comparison with uncorrected ellipticity values may be made by multiplying the numbers by 1⅓). Shiraki points out that amide contributions at 225 nm are less than 10% of those of the aromatic ring chromophores, as judged from values for N-acetyl-L-alanineamide and glycyl-L-leucineamide.

N-acetyltryptophanamide differs notably from N-acetyl-L-tyrosineamide in that the ellipticity becomes negative just below 220 nm and reaches a shoulder at 212 nm and a negative maximum at 196 nm, whereas tyrosine is everywhere positive over this wavelength interval.

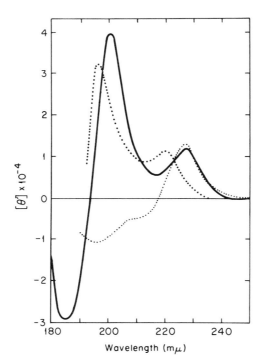

FIG. 22. Far-ultraviolet CD of N-acetyl-L-tyrosineamide (ActyrA), N-acetyl-L-phenylalanineamide (AcPheA), and N-acetyl-L-tryptophanamide (ActrpA) in water. (———) AcTyrA. $1.346 \times 10^{-3} M$; cell length, 1 mm (above 204 nm), 0.1 mm (below 210 nm), 0.1 mm (below 210 nm). (– – –) AcPheA, $1.4 \times 10^{-3} M$; cell length, 1 mm (above 205 nm), 0.1 mm (below 230 nm). (· · ·) AcTrpA, $3.27 \times 10^{-5} M$; cell length, 1 mm. Taken from Shiraki (1969).

The free amino acid tryptophan exhibits a very similar spectrum in the far UV, differing only slightly in intensity (Myer and MacDonald, 1967).

As in the case of tyrosyl residues, there is evidence that, at least occasionally, tyrotophanyl residues are contributing substantially more to regions of the far-UV CD spectrum than is expected from the model compounds. Human carbonic anhydrase C exhibits a value of approximately $-3000$ deg-cm$^2$/dmole at 222 nm (Beychok et al., 1966). From the recently found crystal structure, it is known that 20% of the residues are in $\alpha$-helical segments and about 35% in pleated-sheet conformation (Liljas et al., 1972). This should generate a value of $-10,000$ to $-12,000$ deg-cm$^2$/dmole at 222 nm. To account for the discrepancy, if altogether due to the 27 aromatics, would require that each of the aromatic residues contributes bands of intensity four or five times greater than the acetylated amino acid amides. The same kind of result was observed by Green and Melamed (1966) with avidin and strepavidin, in the former of which the ellipticity is actually positive at 222 nm; in these cases, however, the secondary structure has not been elucidated in the crystalline state by X-ray diffraction analysis.

Adler et al. (1972) have shown that helical copolymers of $N^5$-(2-hydroxyethyl)-L-glutamine with various amounts of randomly incorporated L-tryptophan generate spectra which are sharply dependent on the mole percent of tryptophan. At 14.8%, the residue ellipticities of the $\alpha$-helix at 222 nm are reduced by about 13,000 deg-cm$^2$/dmole or about 80,000 per mole of tryptophan, which is not very far from the value for poly-L-tryptophan (Peggion et al., 1968).

These very large contributions are, however, extreme rather than normal, in proteins. Were this not so, the generally good fits of far-UV CD spectra to known structure, as exemplified by the results in Table I and of the analysis of Saxena and Wetlaufer (1971) or Chen and Yang (1971), would not be obtained. It is more likely that small discrepancies in the fits, by any of these methods, arise in part from aromatic contributions and that such discrepancies will probably be better estimated by contributions of the order of magnitudes of the amino acid values. We shall return to this question again.

## C. Phenylalanyl Residues

### 1. Near Ultraviolet

The near-ultraviolet CD spectra of $N$-acetylphenylalanineamide is shown in Fig. 23, where it is compared to the other two aromatic amino acids. The characteristic fine structure spectrum is seen. Phenylalanine

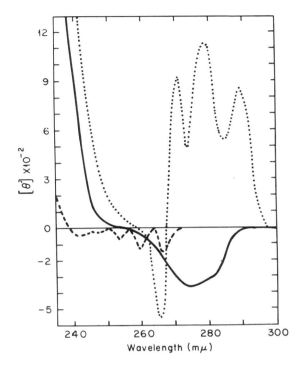

Fig. 23. Near-ultraviolet CD spectra of AcTyrA (———) $1.346 \times 10^{-3} M$; AcPheA (– – –) $1.4 \times 10^{-2} M$; and AcTrpA (· · ·) $3.27 \times 10^{-4} M$ in water. Taken from Shiraki (1969).

optical activity in the near-UV region is frequently overlooked because of its very low intensity relative to that of Trp and Tyr residues. The fine structure is, however, sometimes seen imposed on the smoother envelopes of the other two aromatics and, when observed, is probably diagnostic.

## 2. Far Ultraviolet

While the near-ultraviolet transitions of phenylalanine are only of very low rotational strengths, the far-ultraviolet bands are of strength quite comparable to those of tryptophan and tyrosine; this fact is often overlooked. Figure 22 shows the spectrum. The 220 nm band is at a wavelength 7 nm shorter than that of the other two aromatics, but the intensity of the maximum is 90% of those and the widths of the three bands are comparable. As in $N$-acetyltyrosineamide, and in contrast to $N$-acetyltryptophanamide, a positive, shorter wavelength band, very intense, occurs in this case at 196 nm.

In estimating aromatic contributions in proteins in the far UV, it is necessary to include phenylalanyl residues. It may be pointed out that, in water, the three spectra coincide at 222 nm with a molecular ellipticity of about 12,000 deg-cm²/dmole. It would seem to us worthwhile, as a preliminary procedure, to assign all aromatic residues a positive contribution of that magnitude when evaluating features of secondary structure from CD values at 222–224 nm. More precise analysis should take account of the difference in position of the phenylalanine band, especially for proteins of high phenylalanine content, as well as the negative shoulder and band of tryptophan at shorter wavelengths.

It remains to be seen how often uncharacteristically enhanced rotational strengths are encountered in proteins, as is observed in carbonic anhydrase and avidin.

D. Cystinyl Residues (Disulfide)

When they occur in proteins, cystine residues may generate CD bands in the near and far UV because of certain special features of the side-chain chromophore, the disulfide bond. The UV absorption spectrum and the CD spectrum are both interesting in this connection and are shown for cystine itself, in acid, in Fig. 24. Cystine is the only non-

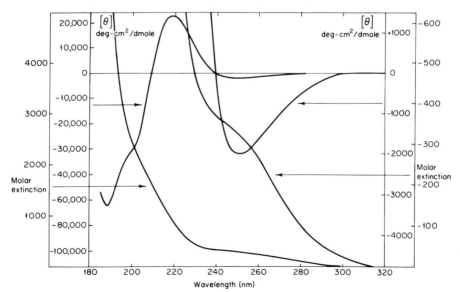

Fig. 24. UV absorption and CD spectra of L-cystine in 1 $M$ HClO₄. Plotted for curve resolution. Taken from Kahn (1972).

aromatic side chain with transitions in the near-UV region. Much of the relation between these transitions and the conformation of the disulfide is the result of the fact that the disulfide normally occurs in a skewed conformation, so that disulfides may occur as right-handed or left-handed nonsuperimposable screws (Beychok, 1965, 1966). Early studies of the optical activity of disulfides sought to establish the geometric basis of CD spectra (Beychok, 1965; Coleman and Blout, 1968). The longest wavelength absorption band in aliphatic disulfides occurs near 250 nm. The position of this band depends, predominantly, on the dihedral angle between the planes defined by $R_1$—S—S— and —S—S—$R_2$. In open-chain disulfides, this angle is between 80 and 110°, most frequently near 90°. In small rings, or other constraining situations, the angle may be considerably smaller, and the longest wavelength band then occurs at longer wavelengths (up to ~350 nm). The UV absorption thus provides a way to examine this aspect of conformation. Since the extinction coefficient is weak, disulfide bands are ordinarily not readily observed in proteins. The corresponding CD spectral bands can be very pronounced. The $C_2$ symmetry of such chiral chromophores confers inherent dissymmetry and the optical activity associated with opposite chirality must be oppositely signed unless the optical activity induced by external perturbations is great enough to overcome the oppositely signed inherent activity. This was a matter of some controversy in early studies; the controversy was settled by several experimental techniques and confirmed theoretically, as well. CD measurements on crystals of opposite handedness (Kahn and Beychok, 1968; Imanishi and Isemura, 1969; Ito and Takagi, 1970) showed that corresponding transitions were oppositely signed, throughout the accessible UV. Carmack and Neubert (1967) achieved comparable results on constrained ring disulfides of known chirality, as have Ludescher and Schwyzer (1971). The signs of the longer-wavelength transitions, in particular, are dependent on screw sense. There is an empirical and theoretical basis for the commonly held notion that the longest-wavelength transition in an inherently dissymmetric chromophore is positive if the chromophore is right-handed, and negative if it is left-handed. (The rigorous basis for rules of this kind for coupled dipoles is given in Section II.) If the rule obtains, then determination of the sign of the longest wavelength CD band reveals the chiral sense of a particular disulfide. It turns out that the right-hand rule for chiral molecules is true in disulfides and other chromophores of $C_2$ symmetry only in certain ranges of dihedral angles (Linderberg and Michl, 1970; Kahn, 1972). For the dihedral angles when it is true, it is necessary to identify the longest wavelength transition.

Beychok and Breslow (1968) showed in oxytocin and derivatives of oxytocin that the 250 nm prominent disulfide band is accompanied, in several situations, by another at or near 270–280 nm but of opposite sign. This posed a significant quandary which has recently been resolved by Kahn (1972) in an exhaustive experimental and theoretical study of cystine, oxytocin, homocystine, and oxidized glutathione. For details, the reader is referred to Kahn (1972). The main results may be summarized as follows: The near-ultraviolet band usually found near 250 nm (for dihedral angles close to 90°) in the absorption and CD of open-chain disulfides in solution is comprised of two electronic transitions. Their maximum dichroic intensities are ±2000 deg-cm²/dmole or less. In the solid state, these two transitions are well resolved; the longer-wavelength one has a molecular ellipticity of ±6000 to ±8000 deg-cm²/dmole. In crystals, only a single rotamer occurs, whereas in solution, two conformers equilibrate. One of these is preferred and is largely responsible for the overall appearance of the solution spectra.

The separation of the first two transitions is a measure of the deviation of the disulfide dihedral angle from a conformation of minimum local energy. In this conformation, near 100°, the transitions are degenerate and the separation is zero. For a mixture of conformers, or for a single conformer, the average of the first (longest) two wavelengths is governed by the —C—S—S— bond angle.

Table III, from Kahn (1972), summarizes band positions, band separations, and preferred screw senses for homocystine, cystine, oxidized glutathione, and two oxytocin derivatives.

In a discussion of neurophysin II (Breslow, 1970) later, it will be shown that a negative band, unequivocally assigned to a disulfide transition, occurs at 280 nm with an (unresolved) rotational strength greater than the positive band accompanying it at 248 nm. In addition, the resolution of disulfide contributions in ribonuclease performed by Horwitz et al. (1969), in conjunction with findings cited above, is compatible with the expected contributions, near 270 nm, of disulfides in single conformer states.

## V. Selected Proteins

### A. Hemoglobin and Myoglobin

The optical activity of hemoglobin and myoglobin are interesting for historical reasons and because both proteins are being subjected to continuous further investigation. (This is especially true of the abnormal hemoglobins.) The ORD spectrum of myoglobin was actually the first published which penetrated the $n_1$–$\pi^0$ Cotton effect (Beychok and Blout,

TABLE III

SUMMARY OF REVISED SEPARATIONS, AVERAGES, BOND AND DIHEDRAL ANGLES, FAVORED SCREW SENSES, AND MECHANISMS UNDERLYING SCREW SENSE ASSIGNMENT[a]

| B trans | A trans | Separations | Dihedral angles (°) | Averages | Bond angles (°) | Screw sense | Mechanism |
|---|---|---|---|---|---|---|---|
| *Homocystine* | | | | | | | |
| 275 | 229 | 46 | 60 | 252 | 104 | L | Strong differential perturbation |
| *2-Ileu-oxytocin* | | | | | | | |
| 289.5 | 247.5 | 42 | 65 | 268.5 | 95–100 | L | Strong differential perturbation |
| *Deamino-2-ileu-oxytocin* | | | | | | | |
| 255 | 231 | 24 | 85 | 243 | 104 | R | Strong differential perturbation |
| *Cystine* | | | | | | | |
| 256 | 239.5 | 16.5 | 90 | 247.5 | 104 | R → L | Some perturbation |
| *GSSG* | | | | | | | |
| 261 | 254.5 | 6.5 | 100 | 257.5 | 104 | L | Inherent asymmetry |

[a] Taken from Kahn (1972).

1961; Urnes et al., 1961). Moreover, the CD spectrum of myoglobin was also the first of any protein and is in the historic paper by Holzwarth et al. (1962). These papers not only established the sense of helix in the polypeptide models used as references until then, which was essential for theoretical and experimental progress in this field, but removed much uneasiness about possibilities of significant difference in backbone structure of proteins in the crystalline and dissolved states.

The main features of the CD spectra of these proteins have been reviewed many times and we shall accordingly not do so. Rather, we will focus on aspects which require further attention and which have not been discussed extensively in either original papers or reviews.

Figure 25 shows the far-UV CD spectrum of horse methemoglobin. This spectrum is very similar to that of metmyoglobin.

In reviewing the literature we found that the average value for sperm whale metmyoglobin tends to be about 500 deg-cm$^2$/dmole more negative

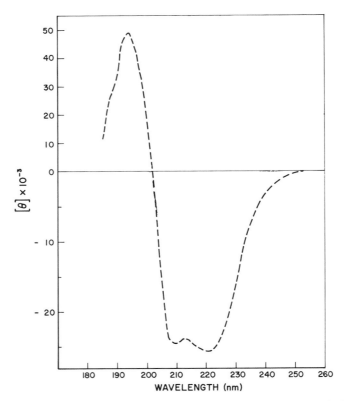

FIG. 25. Far-UV CD spectrum of horse methemoglobin at pH 7. Taken from results of Waks et al. (1971).

than that of horse hemoglobin (met-, cyanomet-, oxy-, and carbonmonoxy-) at 222.5 nm, 1000 deg-cm$^2$/dmole more negative at 207–208 nm, and 200–300 deg-cm$^2$/dmole more positive at 192 nm. These differences are barely greater than instrument uncertainty and error in concentration determination, but they are persistent. Human hemoglobin values (same derivatives) show more variation in the literature, but probably differ somewhat more from horse hemoglobin than the latter differs from myoglobin (Y. K. Yip and M. Waks, unpublished observations).

At 222.5 nm, the separated $\alpha$-subunits are 2–4% more negative and the $\beta$-chains about 2–4% more positive than hemoglobin (Beychok et al., 1967).

The contribution of heme itself, while dominant elsewhere, is apparently negligible in this spectral region, since different ligands generate great differences in heme Cotton effects at other wavelengths in the spectrum but none at 222.5 nm (Beychok, 1964). [The apparent exception of deoxyhemoglobin is not due to heme transitions at or near 222.5 nm, as shown by Frankel (1970).]

An excellent fit to the myoglobin spectrum was achieved by Greenfield and Fasman (1969) using poly-L-lysine reference values. (Their method II; see Table I and accompanying text in Section III,B.) The computed fit yielded 68% $\alpha$, 5% $\beta$, and 27% random. The small $\beta$-content should not be dismissed, out of hand, as inadequacy of fit. Madison and Schellman (1972; and see Section II,G,3) have demonstrated that the generation of characteristic $\beta$-structure optical activity does not require an actual pleated-sheet structure; a single chain with appropriate $\phi,\psi$ angles is all that is needed; moreover, the main features of this are already seen in a dimer unit of such a chain. There are a number of interhelical and terminal helical residues ($N$-termini of helical segments) in the $\beta$-space of the $\phi,\psi$ map for myoglobin (Dickerson and Geis, 1969), and it is not yet possible to indicate the nature of coupling mechanisms between these residues and their immediate neighbors which are, in general, outside the $\beta$-space.

In their fit, Greenfield and Fasman did not explicitly include possible side-chain contributions. In myoglobin (and hemoglobin), the aromatics make up about 8% of the molecule. The data from the model compounds (Section IV) would yield just under (positive) 1000 deg-cm$^2$/dmole at 222.5 nm.

A final and important aspect of the far-ultraviolet spectra is the great similarity between myoglobin and hemoglobin itself. While it is true that the secondary structures of the molecules are notably alike, they differ (Kendrew et al., 1960; Perutz et al., 1968) in ways that have significant implications with respect to analysis of the CD. The residue C-helix in myoglobin is an $\alpha$-helix, but in both chains of hemoglobin, the residues are in a 3.0$_{10}$-helix. Moreover, in myoglobin, the 20-residue long helix, E, has a 7° kink but is an otherwise regular $\alpha$-helix. In the $\alpha$-chain of hemoglobin it is irregular throughout, linear hydrogen bonding being rare in that segment; in the $\beta$-chain, only the last three residues form an irregular helix. Altogether as many as 26–28 residues (about 18%) which are $\alpha$-helical in myoglobin are substantially deviant from a regular $\alpha$-helix in the $\alpha$-chain, but this has little, if any, effect on the position and intensity of the spectrum.

In the near UV, an interesting aspect of the hemoglobin spectra is the intense band due to heme near 260 nm, which differs greatly between $\alpha$- and $\beta$-chains and is the mean of the values of the separated chains (Beychok et al., 1967). This band is of significant diagnostic value in hemoglobins which do not contain a full complement of heme, since the intensity reveals whether the bound heme is associated with α- or β-chains.

Except for the likelihood that the difference in intensity of the band in α- and β-chains arises from immediate environmental differences, nothing has been adduced as to the mechanism generating this induced optical activity and the specific reasons for the twofold difference in intensity at the peak. This difference is maintained with different ligands, as well as in deoxyhemoglobin, although the ligand has a pronounced effect on the band in both α- and β-chains. Since additivity is observed when comparing the sum of the bands to the intensity in hemoglobin, the mechanism is not sensitive to the cooperative phenomena in hemoglobin.

The Soret Cotton effect is somewhat more intense than the 260 nm band; in Section II,G,7, a discussion is presented of attempts to account for the mechanism generating the Soret ellipticity.

## B. Ribonuclease

The far- and near-UV CD spectra of pancreatic ribonuclease (RNase) and a number of its derivatives have been extensively studied and have yielded much valuable information and numerous insights. A considered and thoughtful review of the near-UV CD spectrum of RNase (and a number of other proteins) was presented by Timasheff (1970) as part of a larger review of physical probes of enzyme structures in solution. We shall only add a few observations stemming from recent work.

Figure 26 shows the far-UV spectra of RNase A, RNase S, and RNase S-protein (Richards and Vithayathil, 1959). The CD spectrum of RNase A has been computed (Pflumm and Beychok, 1969a) from reference values, assuming 11.5% α-helix and 33.5% β-structure. At the extremes of the spectrum, the fit requires two special assumptions; to obtain good agreement at very short wavelengths, the aperiodic segments are assumed to generate a negative band at 198 nm only two thirds as intense as the reference polypeptides (see Section III,A); to achieve satisfactory superposition on the long-wavelength side (>230 nm) requires inclusion of a positive band, presumed to arise from aromatics. The intensity of this band is 1500 deg-cm$^2$/dmole at 226 nm. The nine tyrosine + phenylalanine residues would give rise to about 900 deg-cm$^2$/dmole if the amino acid values were used.

Greenfield and Fasman (1969) achieved a fit with secondary structure parameters quite close to those of Pflumm and Beychok. They did not include an aromatic contribution nor did they attempt to fit above 230 nm. At the short-wavelength end, Greenfield and Fasman had to

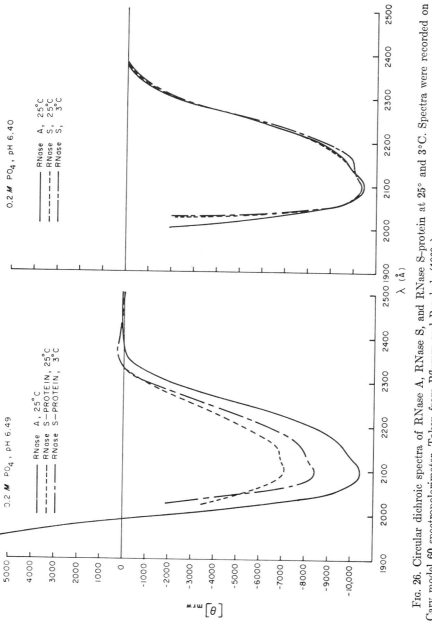

FIG. 26. Circular dichroic spectra of RNase A, RNase S, and RNase S-protein at 25° and 3°C. Spectra were recorded on Cary model 60 spectropolarimeter. Taken from Pflumm and Beychok (1969a).

make no special assumptions about the random coil parameters, which is an improvement over the Pflumm-Beychok resolution. In the main, the far-UV CD is thus reasonably well accounted for, using polypeptide reference values. The shortness and irregularities of the helical segments appear to pose no special problems except, possibly, below 200 nm where, indeed, helix length becomes a severe complication (Madison and Schellman, 1972), as discussed in the sections on theory and calculation.

The near-UV spectrum at a number of pH values is shown in Fig. 27. While there has been little disagreement as to the experimentally measured spectra, interpretations regarding the negative band at 270 nm have been in conflict (Simpson and Vallee, 1966; Simmons and Glazer, 1967; Beavan and Gratzer, 1968). The small positive band near 240, at neutral pH, almost certainly is due in part or altogether to the long-wavelength side of the far-UV aromatic bands, as discussed above (Beychok, 1966; Simons and Blout, 1968; Pflumm and Beychok, 1969a). The band, however small, is probably too intense if the aromatic ellipticity bands are centered at 226 and 220 (tyrosine and phenylalanine, respectively) and have rotational strengths comparable to the model compounds. It is not possible at present to decide between the alternatives of enhanced band strengths or long-wavelength shifts of some of the bands.

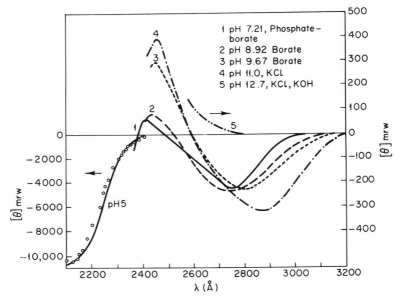

FIG. 27. CD spectra of RNase A at several pH values. Taken from Pflumm and Beychok (1969a).

Moreover, a disulfide contribution to the band has not been ruled out. As noted above, Timasheff (1970) has reviewed the various aspects of this interesting feature of the spectrum in depth.

A rather startling finding was that the intensity of this band is more than doubled on decreasing the temperature from 25° to 4°C, with little change elsewhere in the spectrum (Pflumm and Beychok, 1969a). The origins of the subtle conformational change that this signifies have not been examined and would seem to represent a worthwhile area for future investigation.

The large negative band near 270 nm is very complex. Horwitz et al. (1970) have presented a distinctive and persuasive analysis of that band, which exploits their investigations of model compounds and other proteins at very low temperature. Their results and analysis for RNase A are seen in Fig. 28. At 77°K, vibronic components of the tyrosine transitions are sharpened and resolved, whereas the disulfide continues to present a smooth envelope. Moreover, the location of the 0—0 transition of tyrosine, which is well resolved, depends on exposure to solvent. Using curve fitting procedures in conjunction with the resolved absorption spectrum at 77°K (see Fig. 6), the authors arrive at the resolution shown in Fig. 28. The three exposed tyrosine residues generate 35–45% of the 270 nm band, one buried residue contributes 15–20%,

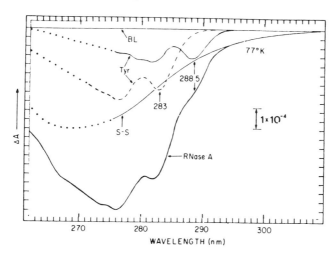

Fig. 28. CD spectrum of 3.6 m$M$ RNase A in water–glycerol (1:1, v/v) with 25 m$M$ sodium phosphate at 77°K. S—S designates disulfide. Dotted lines represent regions of extrapolation. The exposed tyrosyl CD spectrum in this figure has $\Delta A_{277}/\Delta A_{238} = 1.2$; cell path length, 0.20 mm. Additional shoulders are present at 261 and 255 nm (not shown in figure). Division indicated by the double-headed arrow is $1 \times 10^{-4} \Delta A$. From Horwitz et al. (1970).

and the remaining 40–50% of the total band area arises from disulfide. These findings are in general accord with findings due to chemical modification and pH change.

The proposed disulfide contribution may be considered now in light of studies of model compounds cited in Section IV,D. There are four cystinyl residues in pancreatic ribonuclease. One of these is right handed, two are definitely left handed, and the fourth is probably left handed (Wyckoff et al., 1970). If we assume that the ellipticity of the right-handed disulfide cancels a left-handed one, then the disulfide band arises from the two left-handed disulfides. With the dihedral angles near 100°, the sign of the band (negative) accords with the screw sense. Moreover, if half the intensity of the band is presumed to arise from disulfides, then on a molar (rather than residue) basis, the disulfide ellipticity is about $-12,500$ deg-cm$^2$/dmole. Each of the two would thus be generating about $-6250$ deg-cm$^2$/dmole, which is in the range of values generated by cystine in the crystalline state and calculated from theory for the longest wavelength band (Kahn, 1972). The resolution of Horwitz et al. (1970) is thus entirely compatible with expected values for nonequilibrating (rotational equilibration) disulfides.

C. Lysozymes

Greenfield and Fasman (1969; see Table I) computed the far-UV spectrum of hen egg white lysozyme using 28.5% $\alpha$-helix, 11.1% $\beta$, and 60.4% random coil. The $\beta$- and aperiodic region contents are in excellent agreement with the X-ray results of Blake et al. (Blake et al., 1965, 1967; Phillips, 1966, 1967). The value for $\alpha$ agrees perfectly with the content of regular $\alpha$-helix. Inclusion of $3_{10}$- and distorted helices increases the content of helix to 42%, however. As pointed out earlier, the hemoglobin and myoglobin CD spectra appear relatively indifferent to distortion of helix; $3_{10}$-segments, which are common toward the C-termini of the helical segments and which comprise all of C helix in hemoglobin, seemingly generate far-UV spectra (at least between 200 and 230 nm) very much like those of the $\alpha$-helix. In distorted helices, the exact geometry is probably important. Clearly, as those cases illustrated, this question requires further elucidation before quantitative resolutions can be accepted in proteins of unknown crystalline conformation.

Again, as in the previous cases, aromatic bands have not been included. At 222 nm, the minimum aromatic contribution should be $+1200$ deg-cm$^2$/dmole on a mean residue basis, which is a significant value relative to the observed residue ellipticity of $-9050$. At shorter wavelengths, however, the tryptophan and {tyrosine + phenylalanine} may oppose each other, leading to lesser significance of the aromatic contribution. In the near UV (see below, this section), a complex set of tryptophan interactions dominate the spectrum and we must therefore anticipate a possibly complex set of ellipticity bands due to aromatic coupling.

Whatever the actual complications, they are in no sense severe, since the secondary structure analysis is very good, indeed. The three-dimensional structure of the protein is well-enough known, though, that, together with a recently improved theory, it may be possible now to compute expected aromatic contributions in the spectral range 200–240 nm.

The near-UV spectrum of hen egg white lysozyme at pH 5.8 is shown as the dashed line in Fig. 29. Very similar spectra have been frequently reported and compared to spectra resulting from binding of inhibitors, pH variation, chemical modification, and solvent variation (Beychok, 1965; Glazer and Simmons, 1965, 1966; Ikeda et al., 1967; Ikeda and Hamaguchi, 1969; Cowburn et al., 1970; Teichberg et al., 1970). Glazer and Simmons (1966) first noted that the spectrum is altered when the simple inhibitor $N$-acetyl-D-glucosamine (NAG) is bound, implicating aromatic residues, particularly tryptophan residues, of the active site in the near-UV CD spectrum.

The two longest wavelength bands are due to tryptophan. At 292–294 nm, the maximum intensity, on a tryptophan basis, is 1075 deg-cm$^2$/dmole, which is in good agreement with the value found for $N$-acetyl-L-tryptophanamide (Fig. 23). However, this assumes that all tryptophan residues contribute equally. If this is not so, then one or more residues must be contributing more than the amino acid value. Teichberg et al. (1970) have studied a derivative in which Trp 108, in the active site, is selectively oxidized by reaction with iodine. This derivative exhibits a sharply altered near-UV spectrum, all the ellipticity between 260 and 300 nm now being negative. Teichberg et al. consider Trp 108 as predominantly responsible for the band at 294 nm and furthermore responsible for most of the observed enhancement of the near-UV positive bands when the competitive inhibitor tri-$N$-acetyl-D-glucosamine (tri-NAG) is complexed. However, since the optical activity of the oxindole derivative is not characterized, and because three of the Trp residues in the active site, residues 62, 63, and 108, are close enough to interact with one another via dipole coupling, the predominance of Trp 108 may be only apparent. There is little doubt, though, that Trp 108 is contributing to the tryptophan band system. Tyrosine contributions to the spectrum are considerably smaller and have been analyzed by Ikeda and Hamaguchi (1969).

Figure 29 shows, also, the near-UV spectra of a closely related pro-

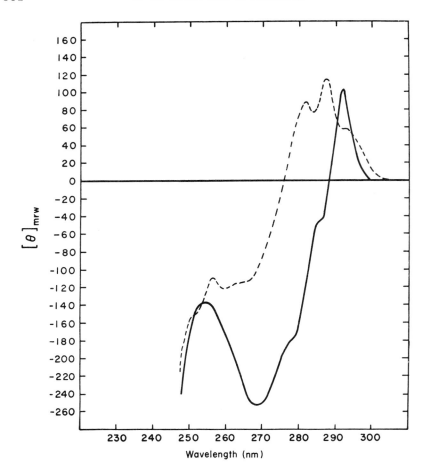

Fig. 29. Circular dichroism spectra of human and hen egg-white lysozyme at pH 5.8, 25°C. (———) Human lysozyme; (– – –) hen egg-white lysozyme. Each spectrum is an average of measurements on six different solutions, each measurement performed in duplicate. The concentration range included is 0.4–0.8 gm/liter. Cell path length, 1.0 cm. Acetate buffer, ionic strength 0.10. Taken from Halper et al. (1971).

tein, human lysozyme, isolated from the urine of individuals with monocytic or monomyelocytic leukemia. There is extensive sequence homology between human and egg white lysozyme, 78 of the 130 positions being equal. The numbers of aromatics, however, differ somewhat in the two, there being five Trp, six Tyr, and two Phe in the human enzyme and six Trp, three Tyr, and three Phe in the hen egg white lysozyme (Jolles and Jolles, 1972; Canfield et al., 1971). The two spectra appear very different on visual inspection, but curve resolution reveals very considerable similarity (Halper et al., 1971). For example, each of the three longest wavelength bands (at neutral pH) in the hen egg white lysozyme is paralleled by bands of the same size, position, and comparable intensity in the human enzyme. The negative shoulders are due to these bands set on a much more intense negative band centered near 270 nm. The latter band is mainly due to an intense $^1L_a$ transition associated with tryptophan residues. The total tryptophan intensity in the human enzyme is greater than that in the hen egg white proteins, notwithstanding the smaller number of tryptophan residues. Halper et al. (1971) suggest that this may arise from cancellation of tryptophan optical activity in hen egg lysozyme due to dipole coupling interactions. Cowburn et al. (1972) have recently attempted a theoretical treatment of possible coupling terms in the near UV of these two proteins and a number of closely related α-lactalbumins. Their results show that coupling in hen egg white lysozyme can indeed generate the complex spectrum observed and that the known residue substitutions in the several homologous proteins may explain the variability in spectra. A significant problem is uncertainty in the two tryptophan transition moment directions ($^1L_a$, $^1L_b$). Thus, their results are tentative, but the approach is promising.

Figure 30 shows the pH dependence of the near-UV CD spectrum of the human enzyme. The relatively minor variations between 270 and 292 nm in the pH interval demonstrate the weak contribution of tyrosine, or at least of the four tyrosine residues which have reversibly titrated at pH 11.4 (Latovitzki et al., 1971).

The striking feature of the spectrum, and the reason for its inclusion here, is the pH-dependent band which appears at 313 nm. This band has no parallel in any protein without a bound prosthetic group. It is the longest wavelength intrinsic band due (presumably) to a side-chain chromophore. Ikeda et al. (1972) have recently shown that the neutral and acid spectrum has a very weak band (the mean residue molecular ellipticity is 10°) at 305 nm. In other proteins, for example chymotrypsin, bands have been observed at wavelengths as long as 303–305 nm (Fasman et al., 1966). Such bands are assumed to arise from a rotationally allowed long-wavelength vibronic component of the $^1L_a$ indole transition (Strickland et al., 1969).

It may be concluded that a special aromatic interaction, unique to this protein, occurs. The band is modified by tri-NAG binding, but its pH dependence is unchanged (Mulvey, Gualtieri, and Beychok, 1973). Chemical modification studies may reveal the specific residues generating the band.

In the far UV, the CD spectra are very similar (Halper et al., 1971;

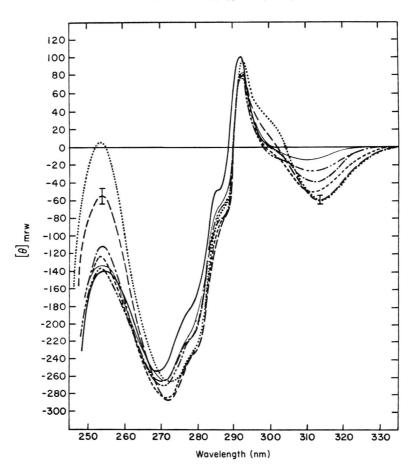

Fig. 30. Circular dichroism spectra of human lysozyme at several pH values between 5.8 and 11.4; ionic strength 0.10, 25°C. (———), pH 5.8, acetate buffer. (——), pH 8.89, glycine–KOH. (- - - -), pH 9.15, glycine–KOH. (- - -), pH 9.60, glycine–KOH. (- - - -), pH 9.96, glycine–KOH. (- - -), pH 10.83, glycine–KOH. (· · ·), pH 11.40, KOH–KCl. Taken from Halper et al. (1971).

Cowburn et al., 1972), suggesting that human lysozyme is similar in secondary structure to hen egg white lysozyme, in accord with the results of the X-ray diffraction studies (Blake and Swan, 1971).

D. INSULIN

Figure 31 shows the near-UV CD of insulin. When treated with trypsin, insulin is hydrolyzed at one or two peptide bonds near the C-terminus

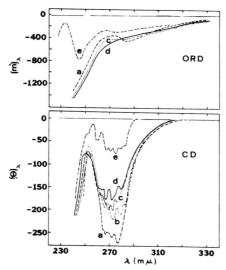

FIG. 31. The optical rotatory dispersion (ORD) and circular dichroism (CD) of insulin and its trypsin-modified derivatives in the aromatic absorption region. Curve (a), zinc insulin, pH 7.5, 0.1 $M$ Cl$^-$, 0.01 $M$ phosphate; curve (b), zinc insulin, pH 2.2, 0.1 $M$ Cl$^-$; curve (c), zinc-free insulin, pH 6.9, 0.1 $M$ Cl$^-$, 0.01 $M$ phosphate; curve (d), desoctapeptide insulin, pH 6.8, 0.1 $M$ Cl$^-$, 0.01 $M$ phosphate; curve (e), B-chain heptapeptide, 0.1 $M$ phosphate buffer, pH 7.0. Protein concentration 0.021–0.053%. Taken from Menendez and Herskovitz (1970).

of the B-chain, resulting in either a heptapeptide or an octapeptide, and the remaining protein, known as desoctapeptide insulin (DOI). The DOI spectra (curve d in the figure) and that of the heptapeptide (curve e) are also shown.

Curve (a) is that of Zn-insulin (hexamer) near neutral pH; curve (b) is Zn-insulin (dimer) in acid. Curve (c) is Zn-free insulin near neutral pH (dimer). The intensities of the peak for the unmodified insulins are very similar to those reported by other groups (Beychok, 1965; Mercola et al., 1967; Morris et al., 1968; Ettinger and Timasheff, 1971a). The fine structure is particularly detailed in this figure; a shoulder near 282–283 nm is ordinarily observed.

The difference between curves (a) and (b) represents the effect of the reduction of molecular weight from the hexamer to the dimer. It was first reported by Mercola et al. (1967) and is an important finding, representing a rare case (at present) of a CD change specifically resulting from disruption of contact interfaces among identical subunits. The intensity of the 270 nm band is strictly dependent on molecular weight, any number of procedures leading to dimer formation resulting in an average intensity of $-190$ to $-210$ deg-cm$^2$/dmole; the hexamer, however formed, exhibits a maximum of $-260$ to $-280$ deg-cm$^2$/dmole. The effects of Zn binding, pH variation, concentration variation, etc., are all consistent in this connection (C. Menendez-Botet and S. Beychok, unpublished observations).

The heptapeptide (residues B23–B29) which generates curve (e) contains two phenylalanine residues, which are responsible for the fine structure on the short-wavelength side of the small band, and one tyrosine residue, which is responsible for most or all of the intensity of curve (e). From the figure, one may judge that curves (e) and (d) are approximately additive to yield curve (c), since all the results are presented on a mean residue basis, and thus the tyrosine of the free heptapeptide is not significantly different in environment or interactions from the point of view of the asymmetry generating the optical activity. Indeed, the magnitude of its ellipticity, as described in Section IV,A, is close to that of $N$-acetyl-L-tyrosineamide. Thus, residue B26 is responsible for about 3.5% of the near-UV band at 270 nm (on a molar basis, the heptapeptide ellipticity is approximately 350 and that of insulin about 10,000).

The near-UV CD spectra of insulin have also been thoroughly examined by Ettinger and Timasheff (1971a). Figure 32 shows spectra of insulin and insulin reacted with $N$-acetylimidazole at pH 7.5, in borate buffer. The dotted curve in the figure represents incomplete reaction of tyrosine residues, while the dashed curve represents complete reaction (4 equiv/mole of insulin monomer). The results suggest a significant, possibly dominant, contribution of tyrosine residues to the spectrum. It remains difficult, however, to sort out the tyrosine contributions from the disulfide, since disulfides very close to tyrosine residues may be affected by tyrosine modification. The disulfide contribution in the native and modified insulin spectra may be inferred by the persistent ellipticity between 300 and 310 nm, a region in which tyrosyl and $O$-acetyltyrosyl residues should not contribute. Ettinger and Timasheff (1971a) suggest that Tyr residues A14 and A19 are probably significant sources of intensity in the 270 nm band on the basis of these results, the effects of temperature on the CD spectra, and considerations based on known hydrogen bonding of these residues in the crystalline state (Adams *et al.*, 1969).

The far-ultraviolet CD spectra of insulin have been recently analyzed by Menendez and Herskovitz (1970) and Ettinger and Timasheff (1971a,b). Menendez and Herskovitz have fitted their data with 30% $\alpha$-helix, 15% $\beta$, and 55% random, which is in accord with the X-ray data (Adams *et al.*, 1969). Ettinger and Timasheff (1971a) point out

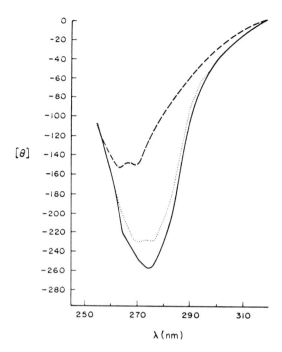

Fig. 32. Near-ultraviolet circular dichroism spectra of insulin in pH 7.5 borate buffer, subsequent to reaction with $N$-acetylimidazole: (———) unreacted insulin; insulin treated with a five-times excess of reagent (calculated with respect to four tyrosine equivalents per mole of insulin); the product contained 1.3 acetylated tyrosines (· · ·); insulin treated with a 120-times excess of reagent, the product containing 4.0 acetylated tyrosines (- - -). Taken from Ettinger and Timasheff (1971a).

that the far-UV spectra have two anomalous features: a positive extremum at 196 nm and a high magnitude at 209 relative to 222 nm. In a detailed consideration of the origins of these peculiarities, Ettinger and Timasheff concluded that the most likely explanation is an unusual $\alpha$-helical contribution.

The far-UV CD spectra of proinsulin were measured and analyzed by Frank and Veros (1968), who found that these spectra indicated no $\alpha$-helix in the connecting peptide.

E. IMMUNOGLOBULINS

The immunoglobulins are considered as a group here because the far-UV CD spectra are very similar for many different individual immunoglobulins. In the near UV, there are distinctive features to be found in purified myeloma immunoglobulins and in several specific antibodies, but little analysis of distinctive features has been presented until now (see, however, Cathou et al., 1968).

Figure 33, from Litman et al. (1970), illustrates the main features of the far-UV CD spectra of 7 S IgG-immunoglobulins and fragments derived from these. The prominent aspect is a negative band at 217 nm, observed in all immunoglobulins (see also, for example, Cathou et al., 1968; Ikeda et al., 1968; Ross and Jirgensons, 1968; Ghose and Jirgensons, 1971; Bjork et al., 1971). As Fig. 33 reveals, the 217 nm band occurs also in light chains and in Fab and Fc fragments, which have been reduced and alkylated. An exception in Fig. 33 is identified as component II, which is the C-terminal fragment of the heavy chains (120 residues). Bjork et al. (1971) have also demonstrated the presence of the 217 nm band in reduced and alkylated C- and N-termini (constant and variable regions) of Bence-Jones proteins (light chains).

In addition to the main band at 217 nm, the intact molecule frequently shows a shoulder at 225–230 nm and another between 235 and 240 nm. In Fab, the latter is sometimes a band rather than a shoulder. Cathou et al. (1968) reported a small positive band near 230 nm.

The characteristic 217 nm band is weak, with a maximum amplitude

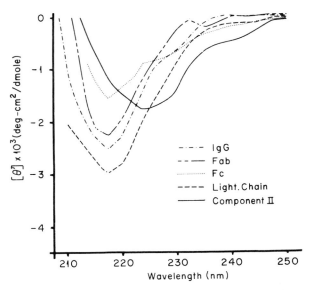

FIG. 33. Circular dichroism spectra of IgG and its enzymically and chemically derived subunits. Light chains determined in 0.02 M sodium acetate–0.15 M NaCl (pH 5.5). Spectra of other preparations determined in 0.02 M sodium phosphate, pH 7.3, 0.15 M NaCl. Taken from Litman et al. (1970).

close to $-3000$ deg-cm$^2$/dmole in all IgG molecules. Most authors agree that the spectrum indicates little or no $\alpha$-helix in the molecule and its fragments. The 217 nm band is usually ascribed to $\beta$-structure, which seems reasonable, though no actual fits have been attempted. Experimental difficulties generally preclude measurements below 200 nm. The spectra tend to zero between 206 and 209 nm; Ross and Jirgensons (1968) reported a positive maximum at 200 nm of intensity slightly weaker than that at 217 nm.

The obvious problem with interpretation of these spectra is that it would take about 50% of $\beta$-structure (poly-L-lysine-type spectrum) to overcome the remaining 50% of aperiodic structure which would generate an intense negative band at 198–200 nm. However, the intensity at 217 nm is far too weak for the required 50% $\beta$.

While a quantitative interpretation of secondary structure is awaited, there is much empirical value in examining these spectra from a different point of view. An important recent concept in immunoglobulin conformation is Edelman's domain theory (Edelman and Gall, 1969). According to this proposal, domains of conformation within the chains are established within intrachain disulfide loops. These domains are considered to be more or less conformationally independent. Moreover, the folding of each is presumed to be largely uninfluenced by the presence or absence of other domains. Additivity studies of the far-UV CD spectra of fragments and of separated chains are obviously relevant in this connection.

Experiments of Bjork and Tanford (1971) are revealing for this purpose. Figure 34 shows the characteristic spectrum of rabbit IgG-immunoglobulin as the solid line. The dashed curve is the spectrum obtained after separation of light and heavy chains and subsequent recombination following renaturation of the separated chains. The dotted lines represent spectra calculated from the sums of the spectra of separated heavy and light chains. When light chains are renatured after reduction and acidification, a mixture of monomers and dimers is formed, which can be separated on Sephadex G-75. The recombination curve Fig. 34a is for recombination of heavy-chain dimer with light-chain monomer and in Fig. 34b recombination with light-chain dimer. These are indistinguishable from one another. Both are virtually identical with the native IgG spectra and measurably different from the sum (dotted) curves.

Taken together with hydrodynamic and immunochemical studies of the separated renatured chains and the recombined immunoglobulins, these results led Bjork and Tanford to a number of significant conclusions: chain separation of nonspecific rabbit IgG is accompanied by a conformational change that is reversible within the limits detectable by optical activity measurements, provided uncombined chains are removed; the

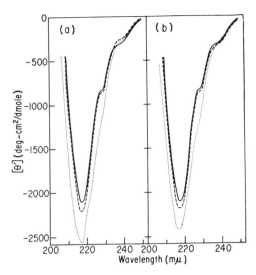

Fig. 34. Circular dichroism spectra of native IgG (solid line) and IgG, reconstituted at pH 5.5 (broken line) from heavy-chain dimer and (a) light-chain monomer, (b) light-chain dimer. The solvent was 0.1 M NaCl (pH 5.5) and protein concentrations were 0.3–0.55 mg/ml. The dotted lines represent the curves calculated for equimolar mixtures of heavy-chain dimer and either light-chain monomer or dimer. Taken from Bjork and Tanford (1971).

internal folding of the chains of recombined rabbit IgG is identical with that of the chains of native IgG; finally, there is no evidence that recombination of chains which were originally paired with other partners leads to conformational alteration detectable by optical and hydrodynamic methods. Stevenson and Dorrington (1970), however, observed different ORD spectra in reconstituted and native IgG. This may be due to failure to remove all uncombined chains, according to Bjork and Tanford.

The question of additivity of domains within light chains has been investigated by Bjork et al. (1971). They measured ORD and CD spectra of the variable and constant halves of a Bence-Jones protein (achieved by enzymic digestion of the protein, which was of λ type). Both fragments give the typical 217 nm band. In other respects the spectra were quite different, but calculated spectra of an equimolar mixture were only very slightly, if at all, different from the intact light chain. This finding according to Bjork et al. suggests that the two domains occur as independently folded regions.

A very different and highly interesting application of optical activity measurements to immunoglobulin conformation and reactivity is illustrated in Fig. 35, taken from the work of Glaser and Singer (1971). These authors studied the induced circular dichroism of DNP-lysine and TNP-lysine when bound to various antibodies. In the uppermost curve of Fig. 35 is shown a TNP Cotton effect(s) when the hapten is bound to a myeloma protein (MOPC-315) which has a high binding affinity for nitrophenyl ligands ($K = 2 \times 10^{-7}$ $M^{-1}$ for DNP-lysine). The middle and lower curves are extrinsic Cotton effects produced by binding of TNP lysine to mouse anti-TNP and anti-DNP antibodies, respectively. The latter two antibody preparations are specific but represent heterogeneous pools of proteins whereas the MOPC-315 is a homogeneous protein.

It is clear from the figure that the same chromophore (which shows no measurable circular dichroism at these wavelengths, albeit covalently attached to an optically active amino acid) generates quite differently shaped curves on binding. Here, the induced circular dichroism reflects

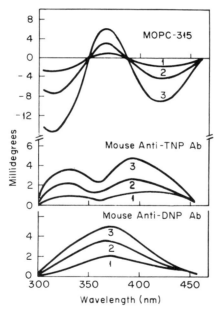

FIG. 35. The circular dichroism spectra produced upon binding of TNP lysine to (top) MOPC-315; (middle) mouse anti-TNP antibodies; and (bottom) mouse anti-DNP antibodies. The protein concentration in each case was 28 $\mu M$. The curves labeled (1), (2), and (3) were obtained at different concentrations of added TNP lysine: 7, 20, and 47 $\mu M$, respectively, for the top set; 10, 20, and 50 $\mu M$ for the middle set; and 10, 20, and 40 $\mu M$ for the bottom set. At the highest hapten concentration in each set, the binding sites were saturated, since further addition of hapten did not affect the spectra. The spectra for the mixtures containing MOPC-315 were obtained in a cell of 1.0 cm path length; the others were obtained in a 2.0 cm cell. Taken from Glaser and Singer (1971).

the mutual interaction of the residues in the antibody site with TNP lysine. (Possibly, coupling of the extrinsic chromophore transitions in this wavelength interval with aromatic residues in the binding site is the main source of the induced activity.)

Glaser and Singer demonstrate that the CD spectra obtained when different proportions of the available sites on MOPC-315 are occupied are consistent with the existence of a single type of binding site. At low concentrations of hapten, when virtually all hapten is bound, the CD bands are proportional to hapten concentration; at higher concentrations, a plateau was reached. Moreover, the presence of isodichroic points indicated only a single type of binding site.

With the heterogeneous antibody preparation, the initially added hapten is bound to antibody of highest affinity and, at higher concentrations, to sites of decreasing affinity. The shapes of the spectra change somewhat and the CD spectra are not proportional to bound hapten concentration.

The method clearly offers an additional means of obtaining binding constants with pure protein if the number of sites per molecule is assumed; moreover, chemical modification and labeling procedures along with sequence studies may allow identification of the residues responsible for the induced optical activity. In any event, the sensitivity is good in this case and should provide a useful additional probe of antibody binding site structure and reactivity.

## F. Elastase

Elastase is a pancreatic hydrolase with much similarity in sequence and structure to trypsin. The complete sequence (Shotton and Watson, 1970) and three-dimensional structure (Shotton and Hartley, 1970) have been elucidated. The protein is included in our review here because of an excellent and systematic investigation by Visser and Blout (1971) of the effects of SDS on its structure, which included a comparative study of many widely different proteins.

Figure 36, taken from Visser and Blout (1971), shows the CD spectra of native elastase, elastase in SDS, and elastase after prolonged dialysis. Native elastase is a protein with only very little $\alpha$-helix (about 5%) and the CD spectrum is consistent with this. In SDS, the spectrum is sharply transformed to one reflecting appreciable helix content. This phenomenon, that is, increased helicity in SDS, has been previously observed and is displayed by many proteins. (For a review of the effects of SDS on protein conformation, see Tanford, 1968.)

By far the most dramatic result is obtained after dialysis to remove

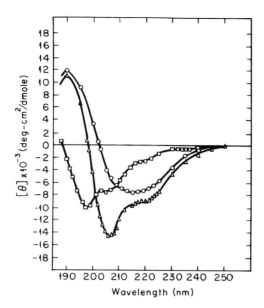

Fig. 36. The effect of sodium dodecyl sulfate treatment and subsequent dialysis on the circular dichroism spectrum of elastase. Solid sodium dodecyl sulfate was added to 0.025% elastase and then circular dichroism spectra were obtained after 1 hr. Similar results were obtained with 0.2% and 2% sodium dodecyl sulfate: elastase (□—□) plus sodium dodecyl sulfate (△—△) and subsequent dialysis (○—○). Taken from Visser and Blout (1971).

SDS. The resulting CD spectrum is not that of native elastase, but rather a remarkably typical $\beta$-spectrum (poly-L-lysine-type). In view of the well-known problems of making quantitative estimates of $\beta$-content, the authors properly qualify their own estimate, but suggest that 50% is a likely value. Visser and Blout extended these SDS studies to a large number of quite different proteins, in each case examining the appropriate IR spectra for confirmation of the secondary structural features. In all cases, the general pattern was the same. Additional helix was not induced in hemoglobin and myoglobin; ribonuclease was an exception to the $\beta$-structure produced after dialysis. Visser and Blout discuss the exceptions as well as the implications of the general findings in great detail, and the reader is referred to the original paper.

The fact that dialysis leads to a structure notably different from the native in almost all cases must be viewed in light of the probability that not all SDS is removed during even exhaustive dialysis. The results nonetheless have quite striking implications for protein structure studies, considering especially the widespread use of SDS. Attention is called here to the extremely low concentrations of SDS which are capable of eliciting this family of phenomena.

### G. Lactate Dehydrogenase

The crystal structure of lactate dehydrogenase has been determined (Adams et al., 1969). The molecule contains approximately 30% of residues in helical conformation and 12–15% in $\beta$-conformation. Chen and Yang (1971) utilized the protein as one of six in their analysis of the ORD and CD parameters to be used in empirical calculation of protein secondary structure. At least from the point of view of $b_0$, $m_{233}$ and $\theta_{222}$, then, the protein generates typical spectra.

The CD spectrum of native dogfish $M_4$ lactate dehydrogenase is shown in Fig. 37, taken from Levi and Kaplan (1971). The notable aspects of the figure are contained in the comparison of this spectrum with those of lactate dehydrogenase which had been dissociated and inactivated by lithium chloride, and then reassociated and reactivated in the presence or absence of the coenzyme DPNH.

Recovery of activity, measured by enzyme assay, was studied along with reassociation measured by sedimentation analysis, sulfhydryl reactivity, fluorescence, immunochemical properties, and ORD and CD spectra. The CD spectrum of untreated enzyme is seen to differ from the reactivated enzymes, which in turn differ from one another depending on whether reduced DPN is present or not during the reactivation. All of the physical and chemical techniques concur in demonstrating a difference in conformation between native and reactivated enzyme.

DPNH was found to have a significant effect on the rate of recovery of the enzymic activity but not on the overall recovery. As compared to the native enzyme, the reassociated enzyme gave no significant differences in $K_m$ values for pyruvate, lactate, DPN+, and DPNH. However, small differences in specific activity were observed when various coenzyme analogs were employed.

These very slight differences in enzymic activity are considerably amplified in other kinds of measurements, including those mentioned above as well as stability to heat and urea denaturation.

As revealed in Fig. 37, lactate dehydrogenase reactivated in the presence of reduced DPN showed structural characteristics closer to the native enzyme than dehydrogenase which was reassociated in the absence of the reduced coenzyme. However, subsequent incubation of the latter reassociated dehydrogenase with DPNH did not convert the reactivated enzyme to a structure similar to the native one.

The differences in CD between these three forms are actually quite

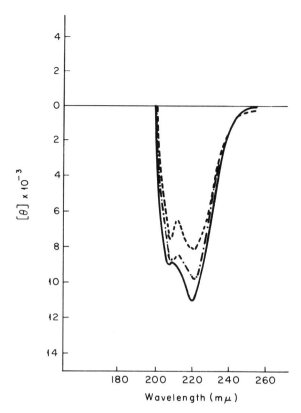

Fig. 37. Circular dichroism spectra of native and reactivated dogfish $M_4$ lactate dehydrogenase. Samples were in $0.1\,M$ phosphate buffer, pH 7.0. Untreated sample (——); sample reactivated in the presence of DPNH (-·-·-); sample reactivated in the absence of DPNH (- - -). Enzyme concentration, $5.0 \times 10^{-7}$. Taken from Levi and Kaplan (1971).

large. It would be difficult, though not impossible, to explain these changes as solely arising from tertiary folding without secondary structure changes.

Levi and Kaplan finally concluded that the presence of DPNH plays a role, at least *in vitro*, not only in enhancing the rate of folding but in altering the final conformation achieved. The fact that DPNH added subsequent to refolding has no effect, suggests the presence of barriers, the nature of which require additional investigation.

Supporting evidence for these findings also comes from the work of Teipel and Koshland (1971), who denatured a series of different enzymes, including lactate dehydrogenase, and found instances of recovery of altered structure after reversal of denaturation. These authors also studied the influence of coenzymes, inhibitors, and substrates on rates of recovery as well as conformation and activity of the final states achieved.

## H. Staphylococcal Nuclease

In Section IV,B,1, brief mention was made of tryptophanyl CD in the near UV of staphylococcal nuclease. In the work referred to, Omenn *et al.* (1969) also examined tyrosyl CD and performed a very informative series of CD measurements on a complex of the enzyme with an inhibitor, deoxythymidine 3′,5′-diphosphate (pdTp). Omenn *et al.* (1969) showed that the tyrosyl contribution to a negative band at 277 nm is $-1600$ deg-cm$^2$/dmole averaged over the seven tyrosyl residues. Several of these are known to be in the active site. When the inhibitor pdTp is added, the near-UV spectrum changes profoundly, a main positive peak appearing near 280 nm with longer-wavelength fine structure. The changes are large enough to allow a binding constant to be measured. The inhibitor itself, however, is optically active, generating a large positive CD band at 273 nm. To meet this difficulty, Omenn *et al.* computed difference spectra at a number of pdTp concentrations. The inhibitor constant obtained was $K_I \sim 10^{-6} M$, in good agreement with values obtained by other methods.

The binding of the inhibitor leaves the UV bands largely unaffected, indicating no major conformational alterations.

$Ca^{2+}$ ion is known to be necessary to the interaction with inhibitor. The ion itself causes no change in the intrinsic CD spectrum of the protein or of the inhibitor. In the absence of added $Ca^{2+}$, however, there were detectable CD changes. Since EDTA reversed these effects, it was concluded that the reagents contained sufficient calcium to bring about binding.

In the far UV, staphylococcal nuclease exhibits a negative maximum at 220 nm. This value and measurements of the residue rotation, $m'_{233}$, have led to estimates of helicity of about 15% (Taniuchi and Anfinsen, 1968). The ellipticity value is $-9950$ deg-cm$^2$/dmole at 220 nm. The Chen-Yang procedure, or estimates based on poly-L-lysine reference values, would give a helix content of closer to 25%. This range brackets the X-ray value of slightly over 20% (Arnone *et al.*, 1969, 1971).

The far-UV ellipticity measurements have proved of great utility in a series of studies of complementation of overlapping fragments of staphylococcal nuclease carried out by Anfinsen and his associates (Taniuchi and Anfinsen, 1969, 1971). In these investigations, it was discovered that a limited tryptic digestion of staphylococcal nuclease in the presence of pdTp and $Ca^{2+}$ leads to the production of large, separable fragments of the enzyme. These can be reassociated or each can be recombined with other fragments produced by CNBr reaction or proteolytic digestion of modified enzyme. The fragments themselves are devoid of enzymic activity whereas different noncovalent complexes of the fragments possess varying activities. For example, the negative CD between 215 and 240 nm of the fragment nuclease-(1–126) suggests a disordered structure, which is uninfluenced by pdTp and $Ca^{2+}$; similarly, nuclease-(99–149) and nuclease T-(49–149) are featureless. A mixture of nuclease-(1–126) and nuclease-(99–149), in contrast, yields a CD spectrum differing from the sum of the individual components and one which is further changed when pdTp and $Ca^{2+}$ are added. The helix content of this complex was estimated by Taniuchi and Anfinsen (1971) to be about 7%, or approximately one-half that of nuclease itself.

Using fluorescence, NMR, and a variety of other techniques, Anfinsen and his associates, in a highly systematic series of investigations, have concluded that essentially the entire amino acid sequence is required to furnish the minimum information—at least in the case of this enzyme—necessary to determine the stable and functional structure of the protein. Moreover, stable structures closely resembling the native enzyme can be formed in several ways from flexible and disordered fragments when the minimum informational requirement is fulfilled.

## I. Neurophysins

The neurophysins are selected for review here because they exhibit the most striking and intense near-UV transitions unequivocally assigned to disulfides. This class of proteins will doubtless serve as models for disulfide optical activity in the years to come. The neurophysins are a group of closely related proteins found in the hypothalamus and posterior pituitary glands of mammals. They occur as noncovalent complexes with the hormones oxytocin and vasopressin. The amino acid analyses of each of the major bovine neurophysins have been reported (Rauch et al., 1969). It has been demonstrated that these proteins bind 1 mole of either oxytocin or vasopressin per mole of monomer (Breslow and Abrash, 1966). The neurophysins have a high disulfide bond content—six to seven disulfide bonds per 10,000 gm—and a low content of aromatic amino acids, i.e., no tryptophanyl, one tyrosyl, and three phenylalanyl residues per mole. Breslow (1970) suggested that the neurophysins are thus unique systems for the observation of disulfide optical activity.

Figure 38, taken from Breslow (1970), shows the near-UV CD spectra of neurophysin II at pH 6.2 and 11.3, as well as after reaction with dithi-

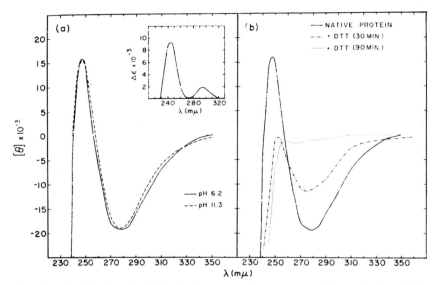

Fig. 38. (a) CD (circular dichroism) spectra of neurophysin II at pH 6.2 and 11.3. The inset shows the difference in the absorption spectrum of neurophysin II between pH 11.3 and pH 6.2. (b) CD spectrum of neurophysin II alone and in the presence of 0.02 $M$ dithiothreitol, at pH 6.2, at the time intervals noted. Taken from Breslow (1970).

othreitol for 30 and 90 min. The inset in Fig. 38a shows the tyrosine difference spectrum between pH 6.2 and 11.3; calculation indicates that this represents more than two-thirds ionization of the single tyrosyl residue. The CD spectra at pH 6.2 and 11.3 are, however, virtually, identical. The spectrum thus arises solely from disulfide transitions, a situation unique, at this time, in proteins and found only in several derivatives of oxytocin and in still simpler disulfide-containing peptides. Reduction of the protein, with DTT (Fig. 38b), abolishes both bands, at apparently unequal rates. Breslow reports that preliminary curve resolution indicates little overlap between the positive and negative bands, the shorter-wavelength band probably being centered at 245 nm and the longer band at 277 nm.

The average negative ellipticity at 277 nm for each of the disulfides is almost 3000 deg-cm$^2$/dmole. It was shown in Section IV,D that in the solid state (in agreement also with calculation) a single conformer with appropriate geometry can generate 6000–8000 deg-cm$^2$/dmole in the longest wavelength transition. The value observed with neurophysin is thus in no sense anomalous, but obviously large-scale cancellation of opposite screw conformers is not occurring. A very crude approximation would suggest that three fourths of the disulfide bonds have the same (probably left) screw sense. The situation may thus be reminiscent of the band in ribonuclease, described in Section V,B. However, in ribonuclease there is no corresponding positive band and the entire spectrum is obscured by tyrosine transitions in the near and far UV. The same may be said of insulin.

The interactions of neurophysin with oxytocin, vasopressin, and a peptide, S-methyl-Cys-Tyr-Phe-amide, were also studied. Binding of oxytocin is accompanied by increased negative ellipticity above 290 nm with appearance of a small negative band at 291 nm; there is diminished negative dichroism between 288 and 260 nm and increased positive ellipticity in the 248 nm band. Qualitatively similar changes occur when neurophysin I is bound to oxytocin, when lysine vasopressin is bound to neurophysin II, and when the peptide is bound to neurophysin II.

Breslow points out that it is the disulfide of the protein, not of oxytocin, which is responsible for the CD changes above 290 nm, since peptide binding gives a comparable result. Changes between 275 and 288 nm may be due to neurophysin disulfide or to an altered contribution of tyrosyl residues of either protein or hormone. Tyrosine is probably involved here because the tyrosine ring of the hormone is known to participate in the binding (Breslow and Abrash, 1966).

## J. Concanavalin A

Concanavalin A, a plant lectin isolated from jack beans, is currently under active investigation because of its importance in studies of normal and transformed cell surfaces. The protein is mitogenic and possesses hemagglutinating activity. It is known to bind $\alpha$-D-gluco- and $\alpha$-D-mannopyranosides (free, or as exposed groupings on cell surface polysaccharides, etc.) and it precipitates with mannans and dextrans (Goldstein et al., 1965). The protein has a subunit molecular weight of 27,000, but exists near neutral pH as a tetramer of molecular weight 106,000 (Pflumm et al., 1971).

The far- and near-UV CD spectra of concanavalin A at several pH values are shown in Fig. 39a and b, respectively. Near neutral pH, there are two sharp and one diffuse positive maxima center near 291, 283, and 260 nm. The entire near-UV envelope is somewhat intensified when $\alpha$-methyl-D-mannoside is bound (Pflumm et al., 1971). The far UV is unchanged when the sugar is bound, indicating tertiary changes or interactions predominantly involving tyrosyl and tryptophanyl residues in the binding site.

Raising the pH has a profound effect on the entire UV CD spectrum, as shown in Fig. 39. The positive ellipticity at wavelengths longer than

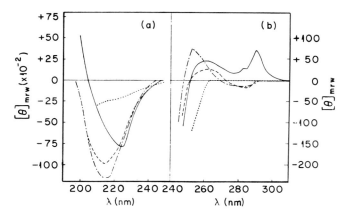

FIG. 39. Far ultraviolet (a) and near ultraviolet (b) circular dichroism spectra of concanavalin A at several pH values between 7.4 and 10.6. (———) Phosphate-buffered 0.9% NaCl solution, pH 8.25; (– – –) 0.05 M carbonate buffer, pH 9.0; (–··–) 0.1 M KCl, pH 10.6; (– – – –) pH 10.6 solution back-titrated to pH 7.4. The spectra were taken at room temperature. Taken from Pflumm et al. (1971).

260 is abolished and a band appears instead near 250 nm. Note that back titration does not restore the original spectrum. Pflumm et al. (1971) conclude that an irreversible transformation occurs between pH 8 and 9. The parallel changes in the far UV demonstrate the deep-seated changes which are involved.

The far-UV spectrum near neutral pH is very similar to that reported also by Kay (1970). There are a number of noteworthy features which deserve attention. Most striking is that a negative maximum occurs at 222.5–225 nm, $\theta = -7500$. However, the shape is clearly not that generated by $\alpha$-helix but is actually more reminiscent of a $\beta$-structure. Pflumm et al. (1971) point out that the characteristics of the band resemble those of the $\beta$-structure of poly-$S$-carboxymethyl-L-cysteine, first reported by Stevens et al. (1968). As mentioned earlier in this article, this polypeptide forms an antiparallel pleated sheet but its CD spectrum is quite different from that of poly-L-lysine, except for the characteristic $\beta$-biphasic shape in this spectral region. Still more striking, in Fig. 39, is the change wrought in the irreversible transformation. At higher pH, the band shifts to the characteristic poly-L-lysine $\beta$-position, 217 nm, with increase in intensity. It will be very interesting, in this regard, to see whether the crystal structure, when available, reveals any unusual features in the $\beta$-regions of this protein which might help to explain the highly anomalous CD spectrum prior to transformation. The presence of $\alpha$-helical segments appears, from this work, highly unlikely.

## VI. Concluding Remarks

We have, for the most part, omitted the more complex systems now under investigation by various spectral techniques, including CD. An excellent review of membrane CD and its implications for structure and function studies of cell membranes is available (Singer, 1971). Recent nucleohistone studies will doubtless soon lead to further investigations of chromatin (Slayter et al., 1972) and related nucleic acid–protein complexes. It is our belief that the basic theoretical and experimental systems described herein will provide a sufficient foundation for the analysis of these and other extraordinarily complex biological systems.

The theoretical advances in CD related to the secondary structure of proteins have been astounding in the last few years. The same is true for the analysis of near-UV CD and details of tertiary structure, but in this case calculations and comparison with experiment are only just beginning, although the general theory is virtually complete. The main problems are the extreme sensitivity of CD to certain geometric arrangements of side-chain chromophores, the still uncertain transition moment directions, etc. As the wavefunctions improve, so too will the calculations. While these developments are awaited CD remains an exceptionally valuable, sometimes unique, tool in monitoring slight as well as profound conformational alterations in proteins, whether simple or complex.

## Appendix: Electromagnetic Units

Because there are several commonly used systems of electromagnetic units (see the C. R. C. *Handbook of Physics and Chemistry*, 48th ed.), there is occasional confusion as to which units should be used. Fortunately, when dealing with the electromagnetic properties of molecules, the Gaussian system is almost always used. In this system there are three basic units (the permeability and electric constant of free space are both set equal to one): mass, $m$ (in grams); length, $l$ (in cm); and time, $t$ (in sec). Loosely referred to as the cgs system, the Gaussian system is, in fact, a partial combination of two other cgs systems; specifically, electrically related parameters are reported in terms of electrostatic units (esu) and magnetically related parameters are reported in terms of electromagnetic units (emu). Note that with this convention the electric field strength and the magnetic field strength of an electromagnetic wave are always equal in magnitude.

Table IV gives the Gaussian units for several parameters often en-

TABLE IV

Gaussian Electromagnetic Units[a]

| Symbol | Units | Physical concept | Gaussian units and nomenclature | MKS conversion factor[b] | |
|---|---|---|---|---|---|
| $\mathbf{E}$ | $[m^{1/2}l^{-1/2}t^{-1}]$ | Electric field strength | 1 cgs = 1 esu = 1 statvolt/cm | $\equiv$ (c) $\times 10^{-6}$ volt/m | Upper limit for light sources other than lasers, $|\mathbf{E}| < 0.003$ cgs |
| $\mathbf{H}$ | $[m^{1/2}l^{-1/2}t^{-1}]$ | Magnetic field strength | 1 cgs = 1 emu = 1 gauss | $\equiv (1/4\pi) \times 10^3$ amp/m | Upper limit for light sources other than lasers, $|\mathbf{H}| < 0.003$ cgs |
| $q$ | $[m^{1/2}l^{3/2}t^{-1}]$ | Charge | 1 cgs = 1 esu = 1 statcoul | $\equiv [1/(c)] \times 10$ coul | Charge on the electron is $(-e) = -4.8029 \times 10^{-10}$ esu |
| $\mu$ | $[m^{1/2}l^{5/2}t^{-1}]$ | Electric dipole moment | 1 cgs = 1 esu-cm = 1 statcoul-cm | $\equiv [1/(c)] \times 10^{-1}$ coul-m | 1 Debye unit = 1 D = $10^{-18}$ esu-cm |
| $\mathbf{m}$ | $[m^{1/2}l^{5/2}t^{-1}]$ | Magnetic dipole moment | 1 cgs = 1 emu = erg/gauss = gauss-cm$^3$ | $\equiv 10^{-3}$ amp-m$^2$ | 1 Bohr magneton = $0.9273 \times 10^{-20}$ erg/gauss |
| $D$ | $[m l^5 t^{-2}]$ | Electric dipole strength | 1 cgs = (1 statcoul-cm)$^2$ | | |
| $R$ | $[m l^5 t^{-2}]$ | Rotational strength | 1 cgs = 1 esu-cm-erg/gauss | | 1 Debye-Bohr magneton = 1 DBM = $0.9273 \times 10^{-38}$ cgs |
| $\alpha$ | $[l^3]$ | Polarizability | 1 cgs = 1 cm$^3$ | | |
| $\beta$ | $[l^4]$ | Rotatory parameter | 1 cgs = 1 cm$^4$ | | |

[a] Adapted from CRC *Handbook of Physics and Chemistry*, 48th ed.
[b] (c) denotes the value of the vacuum speed of light in cm/sec without the units; i.e., (c) $\approx 3 \times 10^{10}$.

countered in discussions concerning CD. As an indication of the magnitude of these parameters, the corresponding MKS units and the factors for converting between the two systems are also included in the table. In the far right column is a list of other quantities and unit conventions which are of use.

#### Acknowledgments

We wish to thank Professor Gerald D. Fasman for making available to us a review article prior to its publication. We are indebted to Prof. I. Tinoco, Jr., for reviewing parts of our rederivation of his general expression and for comments on several of the terms in his and our equations. We also wish to express our gratitude to Prof. C. R. Cantor, Drs. P. C. Kahn, M. McCann, M. Waks, Y. K. Yip, C. Menendez-Botet, M. C. Pflumm, R. Mulvey and Mr. J. Harding, for many valuable discussions, and to S. R. Sears for her excellent editorial assistance.

Previously unpublished work presented in this article was supported in part by grants from the National Institutes of Health and the National Science Foundation. D. W. S. is a predoctoral Trainee of the National Institutes of Health.

#### References

Adams, M. J., Ford, G. C., Koekoek, R., Lentz, P. J., Jr., McPherson, A., Jr., Rossmann, M. G., Smiley, I. E., Schevitz, R. W., and Wonacott, A. J. (1970), *Nature (London)* **227**, 1098.
Adams, M. J., Blundell, T. C., Dodson, E. J., Vijayan, M., Baker, E. N., Harding, M. M., Hodgkin, D. C., Rimmer, B., and Sheats, S. (1969). *Nature (London)* **224**, 491.
Adler, A. J., Greenfield, N., and Fasman, G. D. (1972). *In* "Methods in Enzymology," Vol. 24. Academic Press, New York (in press).
Aebersold, D., and Pysh, E. S. (1970). *J. Chem. Phys.* **53**, 2156.
Ando, T. (1968). *Progr. Theor. Phys.* **40**, 471.
Ando, T., and Nakano, H. (1966). *Progr. Theor. Phys.* **35**, 1163.
Arnone, A., Bier, C. J., Cotton, F. A., Hazen, E. E., Jr., Richardson, D. C., and Richardson, J. S. (1969). *Proc. Nat. Acad. Sci. U. S.* **64**, 420.
Arnott, S., Dover, S. D., and Elliott, A. (1967). *J. Mol. Biol.* **30**, 201.
Atkins, P. W., and Woolley, R. (1970). *Proc. Roy. Soc., Ser. A* **314**, 251.
Barnes, D. G., and Rhodes, W. (1968). *J. Chem. Phys.* **48**, 817.
Barnes, E. E., and Simpson, W. T. (1963). *J. Chem. Phys.* **39**, 670.
Basch, H., Robin, M. B., and Kuebler, N. A. (1967). *J. Chem. Phys.* **47**, 1201.
Basch, H., Robin, M. B., and Kuebler, N. A. (1968). *J. Chem. Phys.* **49**, 5007.
Bayley, P. M. (1971). *Biochem. J.* **125**, 90p.
Bayley, P. M., Nielsen, E. B., and Schellman, J. A. (1969). *J. Phys. Chem.* **73**, 228.
Beavan, G. H., and Gratzer, W. B. (1968). *Biochim. Biophys. Acta* **168**, 456.
Beychok, S. (1964). *Biopolymers* **2**, 575.
Beychok, S. (1965). *Proc. Nat. Acad. Sci. U. S.* **53**, 999.
Beychok, S. (1966). *Science* **154**, 1288.
Beychok, S. (1967). *In* "Poly-α-Amino Acids" (G. D. Fasman, ed.), pp. 293–337. Dekker, New York.
Beychok, S. (1968). *Annu. Rev. Biochem.* **37**, 437.
Beychok, S., and Blout, E. R. (1961). *J. Mol. Biol.* **3**, 769.

Beychok, S., and Breslow, E. (1968). *J. Biol. Chem.* **243**, 151.
Beychok, S., and Fasman, G. D. (1964). *Biochemistry* **3**, 1675.
Beychok, S., Armstrong, J. McD., Lindblow, C., and Edsall, J. T. (1966). *J. Biol. Chem.* **241**, 5150.
Beychok, S., Tyuman, I., Benesch, R. E., and Benesch, R. (1967). *J. Biol. Chem.* **242**, 2560.
Birktoft, J. J., Blow, D. M., Henderson, R., and Steitz, T. A. (1970). *Phil. Trans. Roy. Soc. London, Ser. B* **257**, 67.
Bjork, I., and Tanford, C. (1971). *Biochemistry* **10**, 1289.
Bjork, I., Karlsson, F. A., and Berggård, I. (1971). *Proc. Nat. Acad. Sci. U. S.* **68**, 1707.
Blake, C. C. F., and Swan, I. D. A. (1971). *Nature (London), New Biol.* **232**, 12.
Blake, C. C. F., Koenig, D. F., Mair, G. A., North, A. C. T., Phillips, D. C., and Sarma, V. R. (1965). *Nature (London)* **206**, 757.
Blake, C. C. F., Mair, G. A., North, A. C. T., Phillips, D. C., and Sarma, V. R. (1967). *Proc. Roy. Soc., Ser. B* **314**, 365.
Blum, L., and Frisch, H. L. (1970). *J. Chem. Phys.* **52**, 4379.
Blum, L., and Frisch, H. L. (1971). *J. Chem. Phys.* **54**, 4140.
Bohm, D. (1951). *In* "Quantum Theory," p. 427. Prentice-Hall, Englewood Cliffs, New Jersey.
Bouman, T. D., and Moscowitz, A. (1968). *J. Chem. Phys.* **48**, 3115.
Bovey, F. A., and Hood, F. P. (1967). *Biopolymers* **5**, 325.
Bradley, D. F., Tinoco, I., Jr., and Woody, R. W. (1963). *Biopolymers* **1**, 239.
Breslow, E. (1970). *Proc. Nat. Acad. Sci. U. S.* **67**, 493.
Breslow, E., and Abrash, L. (1966). *Proc. Nat. Acad. Sci. U. S.* **56**, 640.
Breslow, E., Beychok, S., Hardman, K., and Gurd, F. R. N. (1965). *J. Biol. Chem.* **240**, 304.
Briat, B., and Djerassi, C. (1968). *Nature (London)* **217**, 918.
Bush, C. A. (1970). *J. Chem. Phys.* **53**, 3522.
Bush, C. A., and Brahms, J. (1967). *J. Chem. Phys.* **46**, 79.
Caldwell, D. J., and Eyring, H. (1963). *Rev. Mod. Phys.* **35**, 577.
Canfield, R. E., Kammerman, S., Sobel, J. H., and Morgan, F. J. (1971). *Nature (London), New Biol.* **232**, 16.
Carmack, M., and Neubert, L. A. (1967). *J. Amer. Chem. Soc.* **89**, 7134.
Carver, J. P., Schechter, E., and Blout, E. R. (1966). *J. Amer. Chem. Soc.* **88**, 2550.
Cassim, J. Y., and Yang, J. T. (1969). *Biochemistry* **8**, 1947.
Cassim, J. Y., and Yang, J. T. (1970). *Biopolymers* **9**, 1475.
Cathou, R. E., Kulczycki, A., Jr., and Haber, E. (1968). *Biochemistry* **7**, 3958.
Chen, A. K., and Woody, R. W. (1971). *J. Amer. Chem. Soc.* **93**, 29.
Chen, Y.-H., and Yang, J. T. (1971). *Biochem. Biophys. Res. Commun.* **44**, 1285.
Chirgadze, Y. N., Venyminov, S. Y., and Lobachev, V. M. (1971). *Biopolymers* **10**, 809.
Chiu, Y. N. (1969). *J. Chem. Phys.* **50**, 5336.
Chiu, Y. N. (1970). *J. Chem. Phys.* **52**, 1042.
Coleman, D. L., and Blout, E. R. (1968). *J. Amer. Chem. Soc.* **90**, 2405.
Condon, E. U. (1937). *Rev. Mod. Phys.* **9**, 432.
Condon, E. U., Altar, W., and Eyring, H. (1937). *J. Chem. Phys.* **5**, 753.
Cowan, P. M., and McGavin, S. (1955). *Nature (London)* **176**, 501.
Cowburn, D. A., Bradbury, E. M., Crane-Robinson, C., and Gratzer, W. B. (1970). *Eur. J. Biochem.* **14**, 83.
Cowburn, D. A., Brew, K., and Gratzer, W. B. (1972). *Biochemistry* **11**, 1228.

Craig, D. P., and Wamsley, S. H. (1968). "Excitons in Molecular Crystals; Theory and Application." Benjamin, New York.
Damle, V. (1970). *Biopolymers* **9**, 937.
Davydov, A. S. (1948). *J. Exp. Theor. Phys.* **18**, 210.
Davydov, A. S. (1951). *J. Exp. Theor. Phys.* **21**, 673.
Davydov, A. S. (1962). *In* "Theory of Molecular Excitons" (M. Kasha and M. Oppenheimer, eds.), McGraw, New York.
Dearborn, D. G., and Wetlaufer, D. B. (1970). *Biochem. Biophys. Res. Commun.* **39**, 314.
DeTar, D. F. (1969). *Anal. Chem.* **41**, 1406.
Deutsche, C. W. (1969). *Naturwissenschaften* **56**, 495.
Deutsche, C. W. (1970). *J. Chem. Phys.* **53**, 3650.
Deutsche, C. W., and Moscowitz, A. (1968). *J. Chem. Phys.* **49**, 3257.
Deutsche, C. W., and Moscowitz, A. (1970). *J. Chem. Phys.* **53**, 2630.
Deutsche, C. W., Lightner, D. A., Woody, R. W., and Moscowitz, A. (1969). *Annu. Rev. Phys. Chem.* **20**, 407.
Dickerson, R. E. (1972). *Sci. Am.* **226**, 58.
Dickerson, R. E., and Geis, I. (1969). "The Structure and Action of Proteins." Harper, New York.
Dickerson, R. E., Kopka, M. L., Weinzierl, J. E., Varnum, J. C., Eisenberg, D., and Margoliash, E. (1967). *Abstr. 154th Nat. Meet. Amer. Chem. Soc., Chicago*.
Dickerson, R. E., Takano, T., Eisenberg, D., Kallai, D. B., Samson, L., Cooper, A., and Margoliash, E. (1971). *J. Biol. Chem.* **246**, 1511.
Disch, R. L., and Sverdlik, D. I. (1969). *Anal. Chem.* **41**, 82.
Djerassi, C. (1960). "Optical Rotatory Dispersion." McGraw-Hill, New York.
Donovan, J. W. (1969). *In* "Physical Principles and Techniques of Protein Chemistry" (S. J. Leach, ed.), Part A, p. 101. Academic Press, New York.
Drenth, J., Jansonius, J. N., Koekoek, R., Swen, H. M., and Wolthers, B. G. (1968). *Nature (London)* **218**, 929.
Edelhoch, H., Lippoldt, R. E., and Wilchek, M. (1968). *J. Biol. Chem.* **243**, 4799.
Edelman, G. M., and Gall, W. E. (1969). *Annu. Rev. Biochem.* **38**, 415.
Emeis, C. A., Oosterhoff, L. J., and de Vries, G. (1967). *Proc. Roy. Soc., Ser. A* **297**, 54.
Engel, J., Liehl, E., and Sorg, C. (1971). *Eur. J. Biochem.* **21**, 22.
Ettinger, M. J., and Timasheff, S. N. (1971a). *Biochemistry* **10**, 824.
Ettinger, M. J., and Timasheff, S. N. (1971b). *Biochemistry* **10**, 831.
Fasman, G. D., and Potter, J. (1967). *Biochem. Biophys. Res. Commun.* **27**, 209.
Fasman, G. D., Bodenheimer, E., and Lindblow, C. (1964). *Biochemistry* **3**, 1665.
Fasman, G. D., Foster, R. J., and Beychok, S. (1966). *J. Mol. Biol.* **19**, 240.
Fasman, G. D., Hoving, H., and Timasheff, S. N. (1970). *Biochemistry* **9**, 3316.
Feynman, R. P., Leighton, R. B., and Sands, M. (1965). "The Feynman Lectures on Physics," Vol. III. Addison-Wesley, Reading, Massachusetts.
Frank, B. H., and Veros, A. J. (1968). *Biochem. Biophys. Res. Commun.* **32**, 155.
Frankel, R. F. (1970). Ph.D. Dissertation, Harvard University, Cambridge, Massachusetts.
Fraser, R. O. B., and MacRae, T. P. (1962). *J. Mol. Biol.* **5**, 457.
Frenkel, J. (1936). *Phys. Z. Sowjetunion* **9**, 158.
Fretto, L., and Strickland, E. H. (1971a). *Biochim. Biophys. Acta* **235**, 473.
Fretto, L., and Strickland, E. H. (1971b). *Biochim. Biophys. Acta* **235**, 489.
Fridborg, K., Kannan, K. K., Liljas, A., Lundin, J., Strandberg, B., Tilander, B., and Wirén, G. (1967). *J. Mol. Biol.* **25**, 505.

Friedman, S., and Ts'o, P. O. P. (1971). *Biochem. Biophys. Res. Commun.* **42**, 510.
Ghose, A. C., and Jirgensons, B. (1971). *Biochim. Biophys. Acta* **261**, 14.
Glaser, M., and Singer, S. J. (1971). *Proc. Nat. Acad. Sci. U. S.* **68**, 2477.
Glazer, A. N., and Simmons, N. S. (1965). *J. Amer. Chem. Soc.* **87**, 3991.
Glazer, A. N., and Simmons, N. S. (1966). *J. Amer. Chem. Soc.* **88**, 2335.
Gō, N. (1965). *J. Chem. Phys.* **43**, 1275.
Gō, N. (1966). *J. Phys. Soc. Jap.* **21**, 1579.
Gō, N. (1967). *J. Phys. Soc. Jap.* **23**, 88.
Goldstein, I. J., Hollerman, C. E., and Smith, E. E. (1965). *Biochemistry* **4**, 876.
Gould, R. R., and Hoffmann, R. (1970). *J. Amer. Chem. Soc.* **92**, 1813.
Gratzer, W. B., and Cowburn, D. A. (1969). *Nature (London)* **222**, 426.
Gratzer, W. B., Holzwarth, G. M., and Doty, P. (1961). *Proc. Nat. Acad. Sci. U. S.* **47**, 1785.
Green, N. M., and Melamed, M. (1966). *Biochem. J.* **100**, 614.
Greenfield, N., and Fasman, G. D. (1969). *Biochemistry* **8**, 4108.
Halper, J. P., Latovitzki, N., Bernstein, H., and Beychok, S. (1971). *Proc. Nat. Acad. Sci. U. S.* **68**, 517.
Hameka, H. F. (1964a). *J. Chem. Phys.* **41**, 3612.
Hameka, H. F. (1964b). *Ann. Phys. (Leipzig)* [7] **26**, 122.
Hameka, H. F. (1965). "Advanced Quantum Chemistry," p. 235. Addison-Wesley, Reading, Massachusetts.
Harnung, S. E., Ong, E. C., and Weigang, O. E. (1971). *J. Chem. Phys.* **55**, 5711.
Harris, R. A. (1965). *J. Chem. Phys.* **43**, 959.
Harris, R. A. (1969). *J. Chem. Phys.* **50**, 3947.
Hodgkin, D. C., and Oughton, B. M. (1957). *Biochem. J.* **65**, 752.
Hohn, E. G., and Weigang, O. E. (1968). *J. Chem. Phys.* **48**, 1127.
Holzwarth, G., and Doty, P. (1965). *J. Amer. Chem. Soc.* **87**, 218.
Holzwarth, G., Gratzer, W. B., and Doty, P. (1962). *J. Amer. Chem. Soc.* **84**, 3194.
Hooker, T. M., and Schellman, J. A. (1970). *Biopolymers*, **9**, 1319.
Horwitz, J., and Heller, J. (1971). *Biochemistry* **10**, 1402.
Horwitz, J., and Strickland, E. H. (1971). *J. Biol. Chem.* **246**, 3749.
Horwitz, J., Strickland, E. H., and Kay, E. (1968). *Anal. Biochem.* **23**, 363.
Horwitz, J., Strickland, E. H., and Billups, C. (1969). *J. Amer. Chem. Soc.* **91**, 184.
Horwitz, J., Strickland, E. H., and Billups, C. (1970). *J. Amer. Chem. Soc.* **92**, 2119.
Hsu, M-C., and Woody, R. W. (1971). *J. Amer. Chem. Soc.* **93**, 3515.
Hutchinson, D. A. (1968). *Can. J. Chem.* **46**, 599.
Iizuka, E., and Yang, J. T. (1966). *Proc. Nat. Acad. Sci. U. S.* **55**, 1175.
Ikeda, K., and Hamaguchi, K. (1969). *J. Biochem. (Tokyo)* **66**, 513.
Ikeda, K., Hamaguchi, K., Imanishi, M., and Amano, T. (1967). *J. Biochem. (Tokyo)* **62**, 315.
Ikeda, K., Hamaguchi, K., and Migita, S. (1968). *J. Biochem. (Tokyo)* **63**, 654.
Ikeda, K., Hamaguchi, K., Miwa, S., and Nishina, T. (1972). *J. Biochem.* **71**, 371.
Imanishi, A., and Isemura, T. (1969). *J. Biochem. (Tokyo)* **65**, 309.
Ito, N., and Takagi, T. (1970). *Biochim. Biophys. Acta* **221**, 430.
Jauch, J. M., and Rohrlich, F. (1955). "The Theory of Photons and Electrons." Addison-Wesley, Reading, Massachusetts.
Jirgensons, B. (1970). *Biochim. Biophys. Acta* **200**, 9.
Johnson, W. C., and Tinoco, I., Jr. (1969). *Biopolymers* **7**, 727.
Jolles, J., and Jolles, P. (1972). *FEBS Lett.* **22**, 31.
Kahn, P. C. (1972). Ph.D. Dissertation, Columbia University, New York.

Kahn, P. C., and Beychok, S. (1968). *J. Amer. Chem. Soc.* **90**, 4168.
Kartha, G., Bello, J., and Harker, D. (1967). *Nature (London)* **213**, 862.
Kauzmann, W., and Eyring, H. (1941). *J. Chem. Phys.* **9**, 41.
Kauzmann, W., Walter, J. E., and Eyring, H. (1940). *Chem. Rev.* **26**, 339.
Kay, C. M. (1970). *FEBS Lett.* **9**, 78.
Kendrew, J. C. (1962). *Brookhaven Symp. Biol.* **15**, 216.
Kendrew, J. C., Dickerson, R. E., Strandberg, B. E., Hart, R. G., Davies, D. R., Phillips, D. C., and Shore, V. C. (1960). *Nature (London)* **185**, 422.
Kirkwood, J. G. (1937). *J. Chem. Phys.* **5**, 479.
Krueger, W. C., and Pschigoda, L. M. (1971). *Anal. Chem.* **43**, 675.
Latovitzki, N., Halper, J. P., and Beychok, S. (1971). *J. Biol. Chem.* **246**, 1457.
Lawetz, V., and Hutchinson, D. A. (1969). *Can. J. Chem.* **47**, 579.
Legrand, M., and Viennet, R. (1965). *Bull. Soc. Chim. Fr.* p. 679.
Levi, A. S., and Kaplan, N. O. (1971). *J. Biol. Chem.* **246**, 6409.
Li, L. K., and Spector, A. (1969). *J. Amer. Chem. Soc.* **91**, 220.
Liljas, A., Kannan, K. K., Bergsten, P-C., Waara, I., Fridborg, K., Strandberg, B., Carlbom, U., Jarup, L., Lövgren, S., and Petef, M. (1972). *Nature (London) New Biol.* **235**, 131.
Lin, S. H. (1971). *J. Chem. Phys.* **55**, 3546.
Linderberg, J., and Michl, J. (1970). *J. Amer. Chem. Soc.* **92**, 2619.
Litman, B. J., and Schellman, J. A. (1965). *J. Phys. Chem.* **69**, 978.
Litman, G. W., Good, R. A., Frommel, D., and Rosenberg, A. (1970). *Proc. Nat. Acad. Sci. U. S.* **67**, 1085.
Loxsom, F. M. (1969a). *Int. J. Quantum Chem.* **35**, 147.
Loxsom, F. M. (1969b). *J. Chem. Phys.* **51**, 4899.
Loxsom, F. M. (1970). *Phys. Rev. B* **1**, 858.
Loxsom, F. M., Tterlikkis, L., and Rhodes, W. (1971). *Biopolymers* **10**, 2405.
Ludescher, U., and Schwyzer, R. (1971). *Helv. Chim. Acta* **54**, 1637.
McLachlan, A. D., and Ball, M. A. (1964). *Mol. Phys.* **8**, 581.
McMaster, W. M. (1954). *Amer. J. Phys.* **22**, 351.
Madison, V., and Schellman, J. (1970a). *Biopolymers* **9**, 65.
Madison, V., and Schellman, J. (1970b). *Biopolymers* **9**, 511.
Madison, V., and Schellman, J. (1970c). *Biopolymers* **9**, 569.
Madison, V., and Schellman, J. (1972). *Biopolymers* **11**, 1041.
Madison, V. S. (1969). Ph.D. Dissertation, University of Oregon, Eugene.
Markussen, J. (1971). *Int. J. Protein Res.* **3**, 201.
Marsh, R. E., Corey, R. B., and Pauling, L. (1955). *Biochim. Biophys. Acta* **16**, 1.
Mayers, D. F., and Urry, D. W. (1971). *Tetrahedron Lett.* No. 1 p. 9.
Menendez, C. J., and Herskovitz, T. T. (1970). *Arch. Biochem. Biophys.* **140**, 286.
Mercola, D. A., Morris, J. W. S., Arguilla, E. R., and Bramer, W. W. (1967). *Biochim. Biophys. Acta* **133**, 224.
Moffitt, W. (1956a). *J. Chem. Phys.* **25**, 467.
Moffitt, W. (1956b). *Proc. Nat. Acad. Sci. U. S.* **42**, 736.
Moffitt, W., and Moscowitz, A. (1959). *J. Chem. Phys.* **30**, 648.
Moffitt, W., Fitts, D. D., and Kirkwood, J. G. (1957). *Proc. Nat. Acad. Sci. U. S.* **43**, 723.
Moffitt, W., Woodward, R. B., Moscowitz, A., Klyne, W., and Djerassi, C. (1961). *J. Amer. Chem. Soc.* **88**, 3937.
Momany, F. A., Vanderkooi, G., Tuttle, R. W., and Scheraga, H. A. (1969). *Biochemistry* **8**, 744.

Morris, J. W. S., Mercola, D. A., and Arguilla, E. R. (1968). *Biochim. Biophys. Acta* **160**, 145.
Moscowitz, A. (1957). Ph.D. Dissertation, Harvard University, Cambridge, Massachusetts.
Moscowitz, A. (1960a). In "Optical Rotatory Dispersion" (C. Djerassi, ed.), p. 269. McGraw-Hill, New York.
Moscowitz, A. (1960b). *Rev. Mod. Phys.* **32**, 440.
Moscowitz, A. (1962). *Advan. Chem. Phys.* **4**, 113.
Moscowitz, A. (1965). In "Modern Quantum Chemistry" (O. Sinanoglu, ed.), Part I, p. 31. Academic Press, New York.
Moscowitz, A., Wellman, K. M., and Djerassi, C. (1963a). *J. Amer. Chem. Soc.* **85**, 3515.
Moscowitz, A., Wellman, K. M., and Djerassi, C. (1963b). *Proc. Nat. Acad. Sci. U. S.* **50**, 799.
Mulliken, R. S. (1939). *J. Chem. Phys.* **7**, 14.
Mulvey, R. S., Gualtieri, R. J., and Beychok, S. (1973). In "Lysozyme Conference, Proceedings" (E. F. Osserman, R. E. Canfield, and S. Beychok, eds.), Academic Press, New York.
Myer, Y. P. (1968a). *Biochim. Biophys. Acta* **154**, 84.
Myer, Y. P. (1968b). *Biochemistry* **7**, 765.
Myer, Y. P. (1969). *Polym. Repr., Amer. Chem. Soc., Div. Polym. Chem.* **10**, No. 1, 307.
Myer, Y. P., and MacDonald, L. H. (1967). *J. Amer. Chem. Soc.* **89**, 7142.
Nakano, H., and Kimura, H. (1969). *J. Phys. Soc. Jap.* **27**, 519.
Nielsen, E. B., and Schellman, J. A. (1967). *J. Phys. Chem.* **71**, 2297.
Nielsen, E. B., and Schellman, J. A. (1971). *Biopolymers* **10**, 1559.
Omenn, G. S., Cuatrecasas, P., and Anfinsen, C. B. (1969). *Proc. Nat. Acad. Sci. U. S.* **64**, 923.
Ooi, T., Scott, R. A., Vanderkooi, G., and Scheraga, H. A. (1967). *J. Chem. Phys.* **46**, 4410.
Pao, Y., Longworth, R., and Kornegay, R. L. (1965). *Biopolymers* **3**, 519.
Pauling, L., and Corey, R. B. (1951). *Proc. Nat. Acad. Sci. U. S.* **37**, 729.
Pauling, L., and Corey, R. B. (1953). *Proc. Nat. Acad. Sci. U. S.* **39**, 253.
Pauling, L., Corey, R. B., and Branson, H. R. (1951). *Proc. Nat. Acad. Sci. U. S.* **37**, 205.
Peggion, E., Cosani, A., Verdini, A. S., Del Pra, A., and Mammi, M. (1968). *Biopolymers* **6**, 1477.
Perlmann, G. E., and Grizzuti, K. (1971). *Biochemistry* **10**, 258.
Perutz, M. F., Muirhead, H., Cox, J. M., and Goaman, L. C. G. (1968). *Nature (London)* **219**, 131.
Peterson, D. L., and Simpson, W. T. (1957). *J. Amer. Chem. Soc.* **79**, 2375.
Pflumm, M. N., and Beychok, S. (1969a). *J. Biol. Chem.* **244**, 3973.
Pflumm, M. N., and Beychok, S. (1969b). *J. Biol. Chem.* **244**, 3982.
Pflumm, M. N., Wang, J. L., and Edelman, G. M. (1971). *J. Biol. Chem.* **246**, 4369.
Phillips, D. C. (1966). *Sci. Amer.* **215**, 78.
Phillips, D. C. (1967). *Proc. Nat. Acad. Sci. U. S.* **57**, 484.
Philpott, M. R. (1972). *J. Chem. Phys.* **56**, 683.
Power, E. A., and Shail, R. (1959). *Proc. Cambridge Phil. Soc.* **55**, 87.
Pysh, E. S. (1966). *Proc. Nat. Acad. Sci. U. S.* **56**, 825.

Pysh, E. S. (1967). *J. Mol. Biol.* **23**, 587.
Pysh, E. S. (1970a). *J. Chem. Phys.* **52**, 4723.
Pysh, E. S. (1970b). *Science* **167**, 290.
Quadrifoglio, F., and Urry, D. W. (1968a). *J. Amer. Chem. Soc.* **90**, 2755.
Quadrifoglio, F., and Urry, D. W. (1968b). *J. Amer. Chem. Soc.* **90**, 2760.
Quadrifoglio, F., Ius, A., and Crescenzi, V. (1970). *Makromol. Chem.* **136**, 241.
Ramachandran, G. N. (1963). *In* "Aspects of Protein Structure" (G. N. Ramachandran, ed.), p. 39. Academic Press, New York.
Ramachandran, G. N., and Sasisekharan, V. (1968). *Advan. Protein Chem.* **23**, 283.
Ramachandran, G. N., Ramakrishnan, C., and Sasisekharan, V. (1963). *J. Mol. Biol.* **7**, 95.
Rauch, R., Hollenberg, M. D., and Hope, D. B. (1969). *Biochem. J.* **115**, 473.
Reeke, G. N., Hartsuck, J. A., Ludwig, M. L., Quiocho, F. A., Steitz, T. A., and Lipscomb, W. N. (1967a). *Proc. Nat. Acad. Sci. U. S.* **58**, 981.
Reeke, G. N., Hartsuck, J. A., Ludwig, M. L., Quiocho, F. A., Steitz, T. A., and Lipscomb, W. N. (1967b). *Proc. Nat. Acad. Sci. U. S.* **58**, 2220.
Rhodes, W. (1970a). *In* "Spectroscopic Approaches to Biomolecular Conformation" (D. W. Urry, ed.), p. 123. Amer. Med. Ass., Chicago, Illinois.
Rhodes, W. (1970b). *J. Chem. Phys.* **53**, 3650.
Rich, A., and Crick, F. H. C. (1961). *J. Mol. Biol.* **3**, 483.
Richards, F. M., and Vithayathil, P. J. (1959). *J. Biol. Chem.* **234**, 1459.
Riordan, J. F., Wacker, W. E. C., and Vallee, B. L. (1965). *Biochemistry* **4**, 1758.
Rosenfeld, V. L. (1928). *Z. Phys.* **52**, 161.
Rosenheck, K., and Doty, P. (1961). *Proc. Nat. Acad. Sci. U. S.* **47**, 1775.
Rosenheck, K., Miller, H., and Zakaria, A. (1969). *Biopolymers* **7**, 614.
Ross, D. L., and Jirgensons, B. (1968). *J. Biol. Chem.* **243**, 2829.
Ruckpaul, K., Rein, H., and Jung, F. (1970). *Naturwissenshaften* **57**, 131.
Sarkar, P., and Doty, P. (1966). *Proc. Nat. Acad. Sci. U. S.* **55**, 981.
Sasisekharan, V. (1959). *Acta Crystallogr.* **12**, 897.
Saxena, V. P., and Wetlaufer, D. B. (1971). *Proc. Nat. Acad. Sci. U. S.* **68**, 969.
Schechter, B., Schechter, I., Ramachandran, J., Conway-Jacobs, A., and Sela, M. (1971). *Eur. J. Biochem.* **20**, 301.
Schellman, J. A. (1966). *J. Chem. Phys.* **44**, 55.
Schellman, J. A. (1968). *Accounts Chem. Res.* **1**, 144.
Schellman, J. A., and Lowe, M. J. (1968). *J. Amer. Chem. Soc.* **90**, 1070.
Schellman, J. A., and Nielsen, E. B. (1967a). *J. Phys. Chem.* **71**, 3914.
Schellman, J. A., and Nielsen, E. B. (1967b). *In* "Conformation of Biopolymers" (G. N. Ramachandran, ed.), Vol. 1, p. 109. Academic Press, New York.
Schellman, J. A., and Oriel, P. (1962). *J. Chem. Phys.* **37**, 2114.
Schwyzer, R. (1958). *In* "Ciba Foundation Symposium: Amino Acids and Peptides with Anti-Metabolic Activity" (K. Elliot, and G. E. W. Wolstenholme, eds.). p. 171. J. and A. Churchill Ltd., London.
Scott, R. A., Vanderkooi, G., Tuttle, R. W., Shames, P. M., and Scheraga, H. A. (1967). *Proc. Nat. Acad. Sci. U. S.* **58**, 2204.
Shiraki, M. (1969). *Sci. Pap. Coll. Gen. Educ., Univ. Tokyo* **19**, 151.
Shiraki, M., and Imahori, K. (1966). *Sci. Pap. Coll. Gen. Educ., Univ. Tokyo* **16**, 215.
Shotton, D. M., and Hartley, B. S. (1970). *Nature (London)* **225**, 802.
Shotton, D. M., and Watson, H. C. (1970). *Phil. Trans. Roy. Soc. London, Ser. B*

Sigler, P. B., Blow, D. M., Matthews, B. W., and Henderson, R. (1968). *J. Mol. Biol.* **35**, 143.
Simmons, N. S., and Glazer, A. N. (1967). *J. Amer. Chem. Soc.* **89**, 5040.
Simons, E. R., and Blout, E. R. (1968). *J. Biol. Chem.* **243**, 218.
Simpson, R. J., and Vallee, B. L. (1966). *Biochemistry* **5**, 2531.
Singer, S. J. (1971). In "Membrane Structure and Function" (L. I. Rothfield, ed.), p. 146. Academic Press, New York.
Slayter, H. S., Shih, T. Y., Adler, A. J., and Fasman, G. D. (1972). *Biochemistry* **11**, 3044.
Sonenberg, M., and Beychok, S. (1971). *Biochim. Biophys. Acta* **229**, 88.
Stephen, M. J. (1958). *Proc. Cambridge Phil. Soc.* **54**, 81.
Stevens, L., Townend, R., Timasheff, S. N., Fasman, G. D., and Potter, J. (1968). *Biochemistry* **10**, 3717.
Stevenson, G. T., and Dorrington, K. J. (1970). *Biochem. J.* **118**, 703.
Straus, J. H., Gordon, A. S., and Wallach, D. F. H. (1969). *Eur. J. Biochem.* **11**, 201.
Strickland, E. H. (1972). *Biochemistry* **11**, 3465.
Strickland, E. H., Horwitz, J., and Billups, C. (1969). *Biochemistry* **8**, 3205.
Strickland, E. H., Wilchek, M., Horwitz, J., and Billups, C. (1970). *J. Biol. Chem.* **245**, 4168.
Strickland, E. H., Horwitz, J., Kay, E., Shannon, L. M., Wilchek, M., and Billups, C. (1971) *Biochem.* **10**, 2631.
Strickland, E. H., Wilchek, M., Horwitz, J., and Billups, C. (1972a). *J. Biol. Chem.* **247**, 572.
Strickland, E. H., Billups, C., and Kay, E. (1972b). *Biochemistry* **11**, 3657.
Taniuchi, H., and Anfinsen, C. B. (1969). *J. Biol. Chem.* **244**, 3864.
Taniuchi, H., and Anfinsen, C. B. (1971). *J. Biol. Chem.* **246**, 2291.
Tanford, C. (1968). *Advan. Protein Chem.* **23**, 121.
Teichberg, V. I., Kay, C. M., and Sharon, N. (1970). *Eur. J. Biochem.* **16**, 55.
Teipei, F. W., and Koshland, D. E., Jr. (1971). *Biochemistry* **10**, 798.
Tiffany, M. L., and Krimm, S. (1969). *Biopolymers* **8**, 347.
Timasheff, S. N. (1970). In "The Enzymes" (P. D. Boyer, ed.), 3rd ed., Vol. 2, p. 371. Academic Press, New York.
Timasheff, S. N., and Gorbunoff, M. J. (1967). *Annu. Rev. Biochem.* **36**, 13.
Timasheff, S. N., Susi, H., Townend, R., Stevens, L., Gorbunoff, M. J., and Kumosinski, T. F. (1967). In "Conformation of Biopolymers" (G. N. Ramachandran, ed.), Vol. 1, p. 173. Academic Press, New York.
Tinoco, I., Jr. (1962). *Advan. Chem. Phys.* **4**, 113.
Tinoco, I., Jr. (1963). *Radiat. Res.* **20**, 133.
Tinoco, I., Jr. (1964). *J. Amer. Chem. Soc.* **86**, 297.
Tinoco, I., Jr. (1965). In "Molecular Biophysics" (B. Pullman and M. Weissbluth, eds.), p. 274. Academic Press, New York.
Tinoco, I., Jr., and Bush, C. A. (1964). *Biopolym. Symp.* **1**, 235.
Tinoco, I., Jr., and Cantor, C. R. (1970). *Methods Biochem. Anal.* **18**, 81.
Tinoco, I., Jr., Halpern, A., and Simpson, W. T. (1962). In "Polyamino Acids, Polypeptides, and Proteins" (M. A. Stahmann, ed.), p. 183. Univ. of Wisconsin Press, Madison.
Tinoco, I., Jr., Woody, R. W., and Bradley, D. F. (1963). *J. Chem. Phys.* **38**, 1317.
Tonelli, A. E. (1969). *Macromolecules* **2**, 635.
Townend, R., Kumosinski, T. F., Timasheff, S. N., Fasman, G. D., and Davidson, B. (1966). *Biochem. Biophys. Res. Commun.* **23**, 163.

# AUTHOR INDEX

Chance, B., 350, *354*
Chang, H. C., 275, *299*
Changeux, J. P., 237, *242*
Chantrenne, H. K., 31, *72*
Charlwood, P. A., 10, 47, *72*
Chen, A. K., 414, *440*, 515, 516, *586*
Chen, Y. H., 385, 401, 403, *440*, 540, 541, 550, 576, *586*
Chirgadze, Y. N., 454, *586*
Chiu, Y. N., 454, 486, *586*
Chonacky, N. J., 174, *239*
Christiansen, J. A., 125, *137*
Claesson, S., 125, *137*
Clark, A. M., 39, *72*
Cleeman, J. C., 177, *239*
Coates, J. H., 14, *72*
Cohen, C., 422, *440*
Cohen, G., 14, 48, *72*, 167, *240*
Cohn, E. J., 25, *72*
Cohn, M., 247, 248, 249, 250, 251, 253, 266, 274, 276, 277, 279, 281, 282, 283, 284, 286, 287, 288, 289, 290, 291, 293, 294, *298*, *299*, *300*
Coleman, D. L., 416, *440*, 553, *586*
Coll, H., 130, *139*
Condon, E. U., 367, 368, *440*, 450, 458, 459, 460, 461, 466, 471, 473, 476, 483, 487, *586*
Connick, R. E., 266, 280, *299*, *300*
Connolly, T. N., 274, *299*
Conrad, H., 174, 192, 193, *240*
Conway-Jacobs, A., *443*, 543, 545, *591*
Coon, M. J., 290, *299*
Cooper, L., 352, *353*
Corey, R. B., 474, 515, 519, 521, *589*, *590*
Corwin, A. H., 17, 33, 34, *72*
Cosani, A., 415, *440*, *442*, 547, 550, *590*
Cottam, G. L., 290, *299*
Cotton, F. A., 578, *585*
Courtney, R. C., 274, *298*
Cowan, P. M., 422, 423, *441*, 522, *586*
Cowburn, D. A., 417, *441*, 476, 535, 563, 565, 566, *586*, *587*, *588*
Cox, J. M., 192, *242*, 531, 557, *590*
Crabbé, P., 361, *441*
Craig, D. P., 475, 496, *587*
Craig, L. C., 133, 134, *137*
Cramer, F., 209, 238, *242*
Crane-Robinson, C., 247, *298* 563, *586*
Crescenzi, V., 515, 517, *591*

Crick, F. H. C., 474, 525, *591*, *593*
Cuatrecasas, P., 548, 578, *590*
Cullis, A. F., 192, *240*
Cunningham, B. A., 199, 204, *240*
Cunningham, L., 289, *299*

## D

Damaschun, G., 157, 158, 172, 177, 178, *240*
Damle, V. N., 415, *441* 517, *587*
Danchin, A., 276, 277, *298*, *299*
Davidson, B., 392, 399, 400, 401, 428, *441*
Davies, D. R., 188, 204, *241*, *243*, 557, *589*
Davies, M., 121, 127, 131, *138*, *139*
Davis, H. F., 133, *138*
Davydov, A. S., 380, *441*, 474, *587*
Dayhoff, M. O., 39, *72*, 74
De, P. K., 393, *443*
Dearborn, D. G., 535, *587*
Debye, P., 155, 157, 166, 196, *240*
de Clerk, K., 326, *353*
Del Pra, A., 415, *442*, 547, 550, *590*
Dennis, U. E., 317, 350, *354*
Deranleau, D. A., 276, 277, *299*
DeRosier, D. J., 50, *72*
DeTar, D. F., 457, *587*
Deutsche, C. W., 454, 473, 486, 487, 494, 506, 514, *587*
DeVoe, J. R., 349, *353*
deVries, G., 370, *441*, 457, *587*
Dickerson, R. E., 188, *241*, 541, 557, *587*, *589*
Diefenbach, H., 287, 289, *298*, *300*
Diesselhorst, H., 6, *74*
Dimbat, M., 122, 132, 135, *137*
Disch, R. L., 449, *587*
Djerassi, C., 361, 438, *441*, 449, 468, 493, 494, 506, *586*, *587*, *589*, *590*
Dodson, E. J., 541, 568, 576, *585*
Donnan, F. G., 79, *138*
Donovan, J. W., 466, *587*
Dorrington, K. J., *592*
Doty, P., 373, 374, 375, 376, 384, 386, 387, 388, 389, 390, 392, 394, 397, 399, 400, 402, 411, 412, 421, 427, 428, 434, *440*, *441*, *442*, *443*, *444*, 447, 451, 455, 465, 475, 507, 508, 511, 520, 522, 523, 524, 531, 533, 541, 555, *588*, *591*, *593*
Dover, S. D., 410, *440*, 519, *585*
Dowden, B. F., 326, 349, *354*

## AUTHOR INDEX

Downie, A. R., 388, 392, 427, 428, *440*, *441*
Draper, J. C., 180, *239*
Drenth, J., 541, *587*
Duffield, J. J., 430, *440*
Duke, B. J., 304, 351, *353*
DuMond, J. W. M., 180, *240*
Dunstone, J. R., 120, *138*
Durchschlag, H., 234, 236, 237, *240*

### E

Easterday, R. L., 275, *299*
Echols, G. H., 210, *240*
Edelhoch, H., 548, *587*
Edelman, G. M., 168, 183, 187, 199, 200, 201, 202, 203, 204, *240*, *243*, 571, 581, 582, *587*, *590*
Edelstein, S. J., 44, 45, *72*, *74*
Ederer, D., 317, 350, *353*
Edmond, E., 120, *138*
Edsall, J. T., 25, *72*, 94, 119, 120, *138*, 412, 413, 417, *442*, 550, *586*
Ehrenberg, W., 172, *240*
Eigen, M., 235, *241*
Eisenberg, D., 541, *587*
Eisenberg, H., 14, 15, 48, 58, 65, *72*, *74*, 97, 137, *137*, 167, 214, *240*
Eisinger, J., 249, 291, *299*
Elliot, A., 392, 410, 427, 428, *440*, 519, *585*
Emeis, C. A., 370, *441*, 457, *587*
Engel, J., 517, *587*
Engelborghs, Y., 229, *243*
Enoksson, B., 125, *138*
Epand, R. F., 384, 408, *441*
Ettinger, M. J., 544, 567, 568, 569, *587*
Eyring, H., 22, *73*, 368, *440*, 450, 466, 471, 472, 473, 476, 483, 506, 511, *586*, *589*

### F

Fabry, M. E., 293, 294, *299*
Fahey, P. F., 43, 61, *72*
Farquhar, S., 120, *138*
Fasman, G. D., 361, 373, 392, 399, 400, 401, 402, 403, 412, 414, 415, 416, 417, 422, 423, 424, 426, 428, *440*, *441*, *442*, *443*, *444*, 507, 508, 515, 517, 520, 521, 533, 534, 535, 536, 537, 539, 540, 541, 542, 548, 550, 557, 558, 562, 565, 582, 583, *585*, *586*, *587*, *592*
Federov, B. A., 182, *240*
Feigin, L. A., 182, *243*

Feinstein, A., 204, *240*
Fenn, M. D., 27, *72*
Ferris, T. G., 23, 24, 25, 27, 30, 49, *73*
Feughelman, M., 296, *299*
Feynman, R. P., 474, *587*
Fiat, D., 266, *299*
Filmer, D., 237, *241*
Filmer, D. L., 158, 237, *240*, 241
Fischbach, F. A., 215, 218, 224, 225, *240*
Fisher, H. F., 70, *72*
Fite, W., 293, *300*
Fitts, D., 380, 395, *442*
Fitts, D. D., *441*, 472, 475, 487, 508, *589*
Flory, P. J., 107, *137*, *138*
Ford, G. C., 541, *585*
Foster, R. J., 548, 565, *587*
Fournet, G., 146, 158, 180, *240*
Frank, B. H., 569, *587*
Frankel, R. F., 557, *587*
Franks, A., 172, *240*
Franks, F., 294, *300*
Fraser, R. D. B., 312, 314, 323, 325, 326, 331, 349, 350, 351, *353*, *354*, 392, 410, 427, 428, *440*, *441*, 519, *587*
Freer, S. T., 188, 194, *242*
Frenkel, J., 474, *587*
Fresco, J. R., 350, *354*
Fresnel, A., 364, *441*
Fretto, L., 468, *587*
Fridborg, K., 541, *587*, *589*
Friedman, S., 517, 543, *588*
Frisch, H. L., 486, *586*
Frommel, D., 570, *589*
Fuoss, R. M., 125, 128, *138*

### G

Gaber, B., 293, 294, *299*
Gage, F. W., 18, *73*
Gall, W. E., 168, 183, 187, 199, 200, 201, 202, 203, 204, *240*, *243*, 571, *587*
Garfinkel, D., 350, *354*
Garfinkel, L., 350, *354*
Gates, V., 193, *243*
Gee, A., 101, 102, 109, 114, *139*
Geffcken, W., 39, *72*
Geil, P. H., 215, 220, *239*
Geis, I., 557, *587*
George, K. P., 430, *440*
Gerber, B. R., 23, 24, 25, 30, 32, *73*
Gerold, V., 181, *240*
Gersonde, K., 421, *441*

Ghose, A. C., 570, *588*
Gibbs, T. C., 304, 326, 329, 349, 351, *353, 354*
Giddings, J. C., 326, *354*
Gill, S. J., 22, *73*
Gillespie, L. J., 41, *72*
Gladney, H. M., 326, 349, *354*
Glaser, M., 572, 573, *588*
Glatter, O., 156, 168, 181, 229, *240, 243*
Glazer, A. N., 417, *441*, 563, *588*
Glogovsky, R. L., 22, *73*
Gō, N., 486, 487, *588*
Goaman, L. C. G., 531, 557, *590*
Godschalk, W., 18, 70, *73*
Godwin, R. W., 194, 195, *242*
Goldstein, I. J., 581, *588*
Good, N. E., 274, *299*
Good, R. A., 570, *589*
Goodman, M., 407, 415, 427, *441*
Goodrich, R., 42, *73*
Gorbunoff, M. J., 396, 426, *443*, 517, 518, 519, 523, 524, 525, 535, *592*
Gordon, A. S., *592*
Gordon, J. A., 55, *73*
Gosney, I., 27, *72*
Gottlieb, P. D., 199, 204, *240*
Gough, G., 287, 288, *300*
Gould, R. R., 459, *588*
Gratzer, W. B., 394, 397, 417, 422, *441*, 447, 451, 455, 475, 476, 535, 565, 566, *586, 587, 588*
Green, D. W., 205, *240*
Green, N. M., 202, 204, *243*, 417, *441*, 540, 550, *588*
Green, S. B., 350, *354*
Greenfield, N., 401, *441*, 533, 534, 536, 537, 539, 540, 541, 550, 557, 558, 562, *585, 588*
Grizzuti, K., 535, *590*
Gross, J., 423, 425, 426, *440*
Groves, M. L., 37, *74*
Grunberg-Managa, M., *298*
Grushka, E., 326, *354*
Gualtieri, R. J., 565, *590*
Gucker, F. T., Jr., 18, *73*
Güntelberg, A. V., 14, *73*, 79, 97, 100, 108, 110, 125, 126, *138*
Guidotti, G., 107, 112, 113, 114, 115, 116, 131, 135, 136, *138*
Guinier, A., 146, 149, 158, 172, 180, *240*
Gurd, F. R. N., 507, *586*

Guschlbauer, W., 350, *354*
Gutfreund, H., 117, *138*
Gutknecht, W. F., 326, 351, *354*
Gutowsky, H. S., 250, *298*

## H

Haager, O., 155, 168, 183, 187, 200, 201, 202, 203, *241, 243*
Haber, E., 570, *586*
Hach, K. M., 158, *243*
Hagen, A., 230, *241*
Hall, S. R., 351, *353*
Halper, J. P., 418, *441*, 543, 548, 564, 565, 566, *588, 589*
Halpern, A., 475, 507, *592*
Haly, A. R., 296, 298, *299*
Hamaguchi, K., 393, 417, 418, *441, 442*, 543, 563, 565, 570, *588*
Hameka, H. F., 454, 459, 486, *588*
Hamilton, W. C., 305, 310, 312, 316, 317, 352, 353, *354*
Hanby, W. E., 392, 427, 428, *440*
Hansen, A. T., 127, *138*
Hanson, A. W., 468, 562, *593*
Harding, M. M., 541, 568, 576, *585*
Hardman, K., 507, *586*
Hardman, K. D., 468, *593*
Harker, D., 468, *589*
Harnung, S. E., 469, *588*
Harrington, W. F., 61, *73*, 423, 426, *441*
Harris, R. A., 459, 486, *588*
Harrison, P. M., 215, 218, 224, 225, *240, 241*
Harry, J. B., 111, 130, 131, 133, *138*
Hart, M., 168, 172, *239*
Hart, R. G., 188, *241*, 557, *589*
Hartley, B. S., 541, 574, *591*
Hartsuck, J. A., 537, 540, *591*
Haschemeyer, R. H., 351, *354*
Haselkorn, R., 50, *72*
Hashizume, H., 430, *441*
Hawes, R. C., 430, *440*
Hazelwood, C. F., 294, *299*
Hazen, E. E., Jr., 578, *585*
Heath, R. L., 350, *354*
Hecht, H. G., 349, *354*
Heikens, D., 178, *241*
Heine, S., 181, *241*
Heller, J., 468, *588*
Henderson, R., 538, 542, *586, 592*

Hendricks, R. W., *241*
Herbst, M., 163, 168, 183, 184, 185, 187, 200, 201, 202, 203, 211, 212, 213, 214, 215, 216, 217, 230, 231, 232, 233, 234, 235, 238, *242*, *243*
Hermans, J., Jr., 429, *441*
Hermans, P. H., 178, *241*
Herrmann, R., *243*
Herskovitz, T. T., 542, 544, 567, 568, *589*
Hewitt, L. F., 79, *138*
Hight, R., 149, 181, *243*
Hill, R. L., 197, *239*
Hill, T. L., 117, *138*
Hill, W. E., 207, 208, *241*
Hindeleh, A. M., 351, *354*
Hipp, N. J., 37, *74*
Hodgins, M. G., 39, 42, *72*, *73*
Hodgkin, D. C., 530, 541, 568, 576, *585*, *588*
Hoffmann, R., 459, *588*
Hofschneider, P., 230, *241*
Hohn, E. G., 478, 493, 496, 503, 505, *588*
Hohn, T., *243*
Holeysovosoka, H., 284, *299*
Hollenberg, M. D., 579, *591*
Hollenberg, P. F., 290, *299*
Hollerman, C. E., 581, *588*
Holter, H., 126, *138*
Holzwarth, G., 395, *442*, 447, 451, 455, 465, 475, 507, 508, 511, 533, *588*
Holzwarth, G. M., 394, 397, *441*
Hood, F. P., *440*, 522, 524, *586*
Hooker, T. M., 497, 522, *588*
Hooper, P. B., 430, *440*
Hooton, B. R., 289, *299*
Hope, D. B., 579, *591*
Horwitz, J., 419, *441*, 467, 468, 470, 525, 542, 544, 548, 554, 561, 562, 565, *588*, *592*
Hossfeld, F., 174, 181, *241*, *243*
Hoving, H., 535, 536, *587*
Hoy, T. G., 225, *240*
Hsu, M. C., 421, *441*, 531, *588*
Huber, C. P., 351, *353*
Hulbert, C. W., 39, *72*
Hunter, M. J., 21, 38, 47, 60, *73*
Hutchinson, D. A., 486, *588*, *589*
Hyman, A., 196, *241*

I

Ibers, J. A., 349, *354*
Ifft, J. B., 70, *73*
Iizuka, E., 384, 393, 399, 400, 412, 428, *442*, 517, 521, *588*
Ikeda, K., 417, 418, *442*, 543, 563, 565, 570, *588*
Ikkai, T., 32, *73*
Imahori, K., 376, 379, 382, 386, 387, 389, 390, 391, 392, 393, 399, 414, 428, 430, *441*, *442*, *443*, *444*, 515, 516, 555, *591*, *593*
Imanishi, M., 417, 418, *442*, 553, 563, *588*
Inagami, T., 468, *593*
Incardona, N. L., 219, *242*
Isemura, T., 553, *588*
Ito, N., 553, *588*
Ius, A., 515, 517, *591*
Izawa, S., 274, *299*

J

Jacobsen, C. F., 24, 25, 31, *74*
Jacobsson, G., 125, *137*, *138*
Jaenicke, R., 45, *73*, 430, *442*
Jagodzinsky, H., 172, *241*
James, E., 287, *299*
Jansonius, J. N., 541, *587*
Jarup, L., *589*
Jauch, J. M., 487, *588*
Jeffrey, P. D., 106, 131, *138*
Jensen, C. E., 125, *137*
Jirgensons, B., 362, 374, 402, 412, *442*, 535, 570, 571, *588*, *591*
Johansen, G., 23, *73*
Johansson, J., 172, *241*
Johnson, C., 395, *442*
Johnson, D. J., 351, *354*
Johnson, F. H., 22, *73*
Johnson, L. N., 468, *593*
Johnson, P., 122, *137*
Johnson, W. C., 476, *588*
Johnston, T. S., 349, *354*
Jolles, J., 564, *588*
Jolles, P., 564, *588*
Jones, R. N., 304, 312, 323, 350, *355*
Josephs, R., 61, *73*
Jullander, I., 125, *138*
Jung, F., 531, *591*

## K

Kaesberg, P., 146, 158, 168, 174, 215, 219, 220, *239*, *240*, *242*, *243*
Kahn, P. C., 449, 552, 553, 554, 555, 556, 562, *588*, *589*, *593*
Kahovec, L., 165, 177, 181, *241*
Kammerman, S., 564, *586*
Kannan, K. K., 541, *587*, *589*
Kaper, J. M., 50, *73*
Kaplan, N. O., 576, 577, *589*
Karlsson, F. A., 570, 572, *586*
Kartha, G., 468, *589*
Karush, F., *243*
Kasarda, D., 32, *74*
Kasper, S., 351, *354*
Katchalski, E., 403, 416, 423, *442*, *444*
Katz, L., 178, *241*
Katz, S., 23, 24, 25, 27, 30, 49, *73*
Kauzmann, W., 23, 24, 25, 26, 49, 61, *72*, *73*, *74*, 367, *442*, 450, 466, 471, 472, *589*
Kawahara, K., 18, *73*
Kay, C. M., 47, *73*, 418, *442*, 532, 563, 582, *589*, *592*, *593*
Kay, E., 414, 420, *443*, 468, *588*, *592*
Kayne, F. J., 291, *299*
Keech, D. B., 275, 276, *298*
Kegeles, G., 61, *73*
Keller, W. D., 329, 351, *354*
Kellerman, M., 188, 194, *242*
Kelly, M. J., 42, *73*
Kendrew, J. C., 188, *241*, 531, *589*
Kendrew, J. G., 557, *589*
Kent, P., 181, *241*
Kerrigan, F. J., 349, *354*
Kimura, H., 486, *590*
King, T. P., 133, *137*
Kirdani, R. Y., 349, *354*, *355*
Kirkwood, J. G., 367, 380, 395, *441*, *442*, 466, 471, 472, 473, 474, 475, 476, 483, 487, 508, *589*
Kirschner, K., 234, 235, 236, 237, *240*, *241*
Kirste, R. G., 164, 165, 191, 209, 210, *241*, *243*
Klemperer, E., 386, 428, *441*
Kley, G., 172, *240*
Kliman, H. L., 22, 61, *73*
Klimanek, P., 158, *241*
Klotz, I. M., 4, *73*, 80, *138*
Klyne, W., 493, 494, 506, *589*

Knox, J. R., 468, 562, *593*
Koekoek, R., 541, *585*, *587*
Koenig, D. F., 188, 195, 197, *239*, 393, *440*, 562, *586*
Koenig, S. H., 293, 294, *299*
Konigsberg, W., 133, 134, *137*
Konishi, E., 391, *444*
Kopka, M. L., 541, *587*
Kornegay, R. L., 515, *590*
Koshland, D. E., Jr., 237, *241*, 577, *592*
Kowalik, J., 304, 305, 308, 310, 313, 352, *354*
Kowalsky, A., 247, *299*
Kratky, C., 176, *241*
Kratky, O., 43, *73*, *74*, 144, 146, 147, 149, 159, 163, 164, 165, 166, 167, 168, 169, 170, 174, 176, 177, 178, 179, 181, 182, 183, 184, 185, 186, 187, 199, 200, 201, 202, 203, 209, 218, 220, 222, 223, 226, 227, 228, 229, 230, 231, 232, 233, 234, 235, 236, 237, 238, *239*, *240*, *241*, *242*, *243*
Krausz, L. M., 26, *73*
Kraut, J., 188, 194, *242*, 541, *593*
Krigbaum, W. R., 194, 195, 196, *242*
Krimm, S., 409, *442*, *443*, 535, *592*
Krivacic, J., 23, 24, 30, *73*, 193, *242*, 436, *444*
Krueger, W. C., 457, *589*
Kruis, A., 39, *72*
Kuby, S. A., 286, *299*
Kuebler, N. A., 465, *585*
Kügler, F. R., 195, 196, *242*
Kuhn, W., 367, *442*
Kulczychi, A., Jr., 570, *586*
Kumosinski, T. F., 396, 426, *443*, 507, 508, 517, 518, 519, 520, 523, 524, *592*
Kupke, D. W., 39, 42, 43, 50, 61, *72*, *73*, *75*, 79, 81, 92, 105, 106, 119, 121, 122, 123, 124, 133, 135, *138*
Kurtz, J., 423, *442*

## L

Lake, J. A., 181, *242*
Lamb, A. B., 39, *73*
Lampe, J., 421, *442*
Landsberg, M., 414, 415, *441*
Lane, M. D., 275, *299*
Langbein, G., 41, *72*
Lanz, H., 22, 25, 36, 45, *74*

Lapanje, S., 107, 120, 131, *138*
Latovitzki, N., 418, *441*, 543, 548, 564, 565, 566, *588*, *589*
Lauffer, M. A., 27, 45, 67, 71, *73*, *74*, 100, 117, 131, *137*, *138*
Laurent, T. C., 121, *138*
Law, A. D., 350, *353*
Lawetz, V., 486, *589*
Leaver, I. H., 349, 350, *354*
Lee, B., 468, 562, *593*
Lee, C. S., 287, 288, *300*
Lee, R. E., 39, *73*
Lee, S., 350, *354*
Legrand, M., 515, 542, *589*, *593*
Leigh, J. S., Jr., 248, 249, 281, 286, 289, 290, *298*, *299*, *300*
Leighton, R. B., 474, *587*
Lely, J. A., 172, *242*
Lentz, P. J., Jr., 541, *585*
Leonard, B. R., 149, 170, *242*
Leopold, H., 43, *73*, *74*, 166, 176, *242*, *243*
Levenberg, K., 311, 312, *354*
Levi, A. S., 576, 577, *589*
Levy, E. J., 326, *354*
Lewin, S. Z., 33, 35, 36, 37, *72*
Lewis, G. N., 18, *74*
Li, L. K., 521, 533, *589*
Licht, A., *243*
Liehl, E., 517, *587*
Lightner, D. A., 473, 494, 506, 514, *587*
Liljas, A., 541, *587*, *589*
Lin, S. H., 469, *589*
Lindblow, C., 414, 415, 417, *441*, 515, 550, *586*, *587*
Linderberg, J., 553, *589*
Linderstrøm-Lang, K. U., 14, 22, 24, 25, 31, 36, 37, 45, 61, *72*, *73*, *74*, 79, 97, 100, 108, 110, 117, 125, 126, *138*, 374, *442*
Lindley, H., *442*
Lippoldt, R. E., 548, *587*
Lipscomb, W. N., 537, 540, *591*
Litjens, E. C., 50, *73*
Litman, B. J., 375, *442*, 494, *589*
Litman, G. W., 570, *589*
Littlewood, A. B., 326, 329, 349, *353*, *354*
Lobachev, V. M., 454, *586*
Lövgren, S., *589*
Longtin, B., 38, *74*
Longworth, R., 515, *590*
Lonsdale, K., 351, *354*

Lontie, R., 226, 229, *242*, *243*
Lotz, W. E., Jr., 39, *72*
Lowe, M. J., 407, *443*, 538, *591*
Lowry, T. M., 375, *442*
Loxsom, F. M., 475, 486, 487, 514, *589*
Ludescher, U., 553, *589*
Ludwig, M. L., 537, 540, *591*
Luenberger, D. G., 317, 350, *354*
Lui, N. S. T., 289, *299*
Lundberg, R. D., 386, *441*, *442*
Lundin, J., 541, *587*
Lusebrink, T. R., 329, 351, *354*
Luz, Z., 250, *299*
Luzzati, V., 164, 168, 172, 180, 205, 206, 207, 209, *242*, *243*
Lynch, L. J., 295, 296, 297, 298, *299*
Lynen, F., 163, 184, 185, 229, 230, 231, 232, 233, 234, 235, 238, *242*

## M

McCabe, W. J., 398, *444*
McConnell, H. M., 248, 285, *299*
McCubbin, W. D., 418, *442*
McDonald, C. C., 247, *299*
MacDonald, L. H., 468, 546, 547, 550, *590*
McFarland, B. G., 248, 285, *299*
McGavin, S., 422, 423, *441*, 522, *586*
MacInnes, D. A., 39, *72*, *74*
McKeekin, T. L., 37, *74*
McLachlan, A. D., 486, 487, *589*
McLaughlin, A., 289, *300*
McMaster, W. M., 487, *589*
McMullen, D., 350, *354*
McPherson, A., Jr., 541, *585*
MacRae, T. P., 350, 351, *354*, 410, *441*, 519, *587*
Macurdy, L. B., 17, 34, *74*
Madison, V., 351, *354*, 369, 394, 396, 406, 408, 409, 410, 411, 424, 426, *442*, 524, 528, 529, 533, 557, 560, *589*
Madison, V. S., 508, 511, 513, 520, 521, 522, 524, 525, 528, 529, 530, 541, *589*
Magar, M. E., 351, *354*
Mair, G. A., 188, 195, 197, *239*, 393, *440*, 530, 562, *586*
Malmon, A. G., 149, 158, 197, *242*
Mammi, M., 415, *442*, 547, 550, *590*
Mandel, R., 395, *442*
Margoliash, E., 541, *587*
Margulies, S., 321, *354*

Mark, J. E., 409, *442*
Markussen, J., *589*
Marquardt, D. W., 310, 311, 312, 350, *354*
Marrack, J., 79, *138*
Marsden, K. H., 287, 288, 295, 296, 297, *299, 300*
Marsh, R. E., 519, *589*
Marshall, S. W., 331, 351, *354*
Martell, A. E., 274, *298*
Martin, A. J., 326, *354*
Martinez, H. M., 385, 401, 403, *440*
Maruyama, H., 275, *299*
Matthews, B. W., *592*
Mayer, A., 174, 192, 193, *240*
Mayers, D. F., 494, *589*
Mazzarella, L., 192, *242*
Mazzone, H. M., 219, *242*
Mead, D. J., 125, 128, *138*
Meiboom, S., 250, *299*
Meiron, J., 308, 311, *355*
Melamed, M. D., 417, *441*, 540, 550, *588*
Menendez, C. J., 542, 544, 567, 568, *589*
Mercola, D. A., 567, *589, 590*
Mescaanti, L., 400, *443*
Metcalfe, J. C., 247, *299*
Michl, J., 553, *589*
Migchelsen, C., 298, *299*
Migita, S., 565, *588*
Mildvan, A. S., 247, 248, 251, 274, 275, 276, 277, 279, 282, 284, 286, 290, 292, *299*
Miles, D. W., 368, *442*
Milevskaya, I. S., 396, 409, *444*, 526, *593*
Miller, H., 522, 523, 524, *591*
Miller, I., 133, *137*
Miller, R. S., 275, *299*
Miller, W. G., 107, *138*
Mittlebach, P., 149, 150, 151, 152, 153, 158, 160, 161, 162, *242*
Miwa, S., 565, *588*
Moffitt, W., 380, 389, 395, *442*, 456, 460, 466, 469, 472, 474, 475, 476, 483, 487, 493, 494, 506, 508, 530, *589*
Momany, F. A., 530, *589*
Momoki, K., 351, *355*
Monod, J., 237, *242*
Montague, R., Jr., 39, *72*
Moore, L. D., 79, *139*
Morales, M., 22, *74*
Morgan, F. J., 564, *586*
Morgan, L. O., 281, 291, *298*

Moring-Claesson, I., 168, 170, 182, 199, 203, 226, 227, 228, *243*
Morris, J. W. S., 567, *589, 590*
Morrison, J. F., 286, 287, *299*, 350, *355*
Moscowitz, A., 351, *355*, 369, *442*, 450, 454, 456, 459, 460, 461, 462, 466, 468, 469, 473, 493, 494, 501, 506, 514, *586, 587, 589, 590*
Moser, C. E., 18, *73*
Müller, J. J., 172, 177, *240*
Muirhead, H., 192, *240, 242*, 531, 557, *590*
Mulliken, R. S., 460, *590*
Mulvey, 565, *590*
Myer, Y. P., 468, 538, 541, 546, 547, 550, *590*
Myers, D. V., 412, 413, 417, *442*
Myers, M. N., 326, *354*

## N

Nachmansohn, D., 22, *75*
Nagy, B., 407, *442*
Nakano, H., 486, *585, 590*
Nelson, C. A., 202, *242*
Nelson, J. A., 331, 351, *354*
Nemethy, G., 237, *241*
Neubert, L. A., 553, *586*
Nicholls, B. L., 294, *299*
Nicolaieff, A., 164, 209, *242*
Nielsen, E. B., 369, 407, *440*, 463, 483, 484, 494, 501, 503, 522, 524, 528, 529, 530, *585, 590, 591*
Nishina, T., 565, *588*
Noda, L., 287, 288, *300*
Noelken, M. E., 202, *242*
Noguchi, H., 22, 23, 24, 25, 27, 30, 32, *73, 74*
Noltmann, E. A., 286, *299*
Norland, K., 416, *442*
North, A. C. T., 188, 192, 195, 197, *239, 240*, 393, *440*, 530, 562, *586*

## O

Oberdorfer, R., 199, *241*
Oesterhelt, D., 163, 184, 185, 230, 231, 232, 233, 234, 235, 238, *242*
Ogston, A. G., 117, 120, 121, 131, *138, 139*
Ohnishi, S., 248, *299*
Oikawa, K., 418, *442*
Oliveira, R. J., 22, 61, *72*
Omenn, G. S., 548, 578, *590*

Ong, E. C., 469, *588*
Ooi, T., 32, *73*, 515, *590*
Oosterhoff, L. J., 370, *441*, 457, *587*
O'Reilly, D. E., 250, *299*
Oriel, P. J., 368, 394, *443*, 477, 494, 503, 504, 506, 507, 510, 511, 514, *591*
Osborne, M. R., 304, 305, 308, 310, 313, 352, *353*, *354*
Oster, G., 149, *242*
O'Sullivan, W. J., 253, 275, 276, 277, 279, 283, 284, 286, 287, 288, 289, *298*, *299*, *300*
Otteson, M., 107, *138*
Oughton, B. M., 530, *588*
Overbeek, J. T., 94, *138*

## P

Paglini, S., 130, *138*
Paletta, B., 199, *241*
Pao, Y., 515, *590*
Papoušek, D., 312, 350, *355*
Parry, D. A. D., 350, 351, *354*
Partridge, S. M., 133, *138*
Pasteur, L., 362, *442*
Pauling, L., 474, 515, 519, 521, *589*, *590*
Peacocke, A. R., 291, *300*
Pearson, J. E., 281, *300*
Pedersen, K. O., 48, *74*
Peggion, E., 415, *440*, *442*, 547, 550, *590*
Pekar, A. H., 116, *137*
Pepinsky, R., 351, *355*
Peret, R., 179, *242*
Perlmann, G. E., 39, *72*, 535, *590*
Perone, S. P., 326, 351, *354*
Perutz, M. F., 188, 192, *240*, *242*, 430, *442*, 531, 557, *590*
Petef, M., *589*
Peterson, D. L., 502, 522, *590*
Pflumm, M. N., 535, 538, 541, 545, 546, 558, 559, 560, 561, 581, 582, *590*
Phelps, F. P., *72*
Phelps, R. A., 117, *138*
Phillip, H. J., 124, 134, *138*
Phillips, D. C., 188, 195, 197, *239*, 393, *440*, 530, 557, 562, *586*, *589*, *590*
Phillips, W. D., 247, *299*
Philpott, M. R., 486, 487, *590*
Pigliacampi, J., 114, *139*
Pilz, I., 14, *74*, 156, 163, 168, 170, 178, 179, 182, 183, 184, 185, 186, 187, 198, 199, 200, 201, 202, 203, 204, 209, 211, 212, 213, 214, 215, 216, 217, 226, 227, 228, 229, 230, 231, 232, 233, 234, 235, 238, *241*, *242*, *243*
Pitha, J., 304, 312, 323, 350, *355*
Plato, F., 6, 35, 41, *74*
Plíva, J., 312, 350, *355*
Polissar, M. J., 22, *73*
Poole, C. P., Jr., 250, *299*
Porod, G., 149, 158, 159, 163, 164, 165, 167, 169, 177, 181, 199, 232, *241*, *242*, *243*
Porteus, J. O., 182, *243*
Portzehl, H., 120, 125, 135, *138*, *139*
Potter, J., 402, 412, *441*, *443*, 521, 534, 582, *587*, *592*
Power, E. A., 486, *590*
Preston, B. N., 121, 131, *139*
Pring, M., 350, *354*
Priore, R. L., 349, *354*, *355*
Pschigoda, L. M., 457, *589*
Puchwein, G., 168, 183, 187, 200, 201, 202, 203, 234, 236, 237, *240*, *243*
Pürschel, H. V., 157, 158, *240*
Purcell, E. M., 262, *298*
Pysh, E. S., 396, 408, 410, 424, *440*, *442*, 465, 483, 488, 513, 517, 519, 520, 521, 522, 523, 524, 525, 527, 530, *585*, *590*, *591*

## Q

Quadrifoglio, F., 411, *443*, 507, 508, 509, 510, 515, 517, 520, 534, *591*
Quiocho, F. A., 537, 540, *591*

## R

Rainford, P., 22, 32, *74*
Ramachandran, G. N., 525, 526, *591*
Ramachandran, J., *443*, 543, 545, *591*
Ramakrishnan, C., 526, *591*
Ramsay, G. C., 349, *354*
Randall, M., 18, 38, *74*
Rasper, J., 24, 25, 26, *73*, *74*
Rauch, R., 579, *591*
Ray, D. R., 39, *74*
Record, B. R., 125, *137*
Redfield, A. G., 293, *298*, *300*
Reed, G. H., 266, 276, 277, 279, 281, 283, 284, 286, 287, 288, *300*
Reeke, G. N., 537, 540, *591*
Reiff, T. R., 127, *139*

Rein, H., 531, *591*
Rein, N., 421, *442*
Reinhardt, W. P., 351, *355*
Reithel, F. J., 42, *73*, 106, 131, *137*, *139*
Reisler, E., 14, 48, *74*
Reuben, J., 251, 253, 266, 277, 281, 282, 291, 293, 294, *300*
Rhodes, L., 61, *73*
Rhodes, W., 422, *441*, 465, 486, 487, 514, *585*, *589*, *591*
Rich, A., 525, *591*
Richards, E. G., 350, *354*
Richards, F. M., 468, 558, 562, *591*, *593*
Richards, M., 468, *593*
Richards, R. E., 291, *300*
Richards, T. W., 38, *74*
Richardson, D. C., 578, *585*
Richardson, J. S., 578, *585*
Riley, D. P., 149, *242*
Rimmer, B., 541, 568, 576, *585*
Riordan, J. F., 544, *590*
Ritland, H. N., 174, *243*
Roberts, S. M., 349, *355*
Robertson, J. M., 351, *355*
Robin, M. B., 465, *585*
Roess, L. C., 149, 181, *243*
Rohns, G., 223, *239*
Rohrlich, F., 487, *588*
Rolfson, F. B., 130, *139*
Rollett, J. S., 352, *355*
Rollett, R. S., *442*
Ronish, E. W., 409, *443*
Roos, B., 350, *355*
Roppert, J., 181, *241*
Rosen, I. G., 407, *441*
Rosenberg, A., 570, *589*
Rosenfeld, V. L., 366, *443*, 459, 471, 476, 483, 487, *591*
Rosenheck, K., 410, 411, *443*, 475, 520, 522, 523, 524, *591*
Ross, D. L., 570, 571, *591*
Ross, P. A., 173, *243*
Rossmann, M. G., 192, *240*, 541, *585*
Rowe, A. J., 204, *240*
Rowe, D. S., 109, 127, 135, *139*
Ruckpaul, K., 531, *591*
Ruland, W., 179, *242*
Rupley, J. A., 23, 24, 30, *73*, 193, *242*, *243*
Rupprecht, A., 298, *299*
Rutishauser, 199, 204, *240*

## S

Sage, H. J., 416, *443*
Sands, M., 474, *587*
Sarkar, P., 533, *591*
Sarker, P. K., 392, 399, 400, 402, 428, *443*
Sarma, V. R., 188, 195, 197, 204, *239*, *243*, 393, *440*, 530, 562, *586*
Sasa, T., 287, *300*
Sasisekharan, V., 522, 525, 526, *591*
Sato, H., 351, *355*
Saxena, V. P., *443*, 538, 539, 540, 541, 550, *591*
Scatchard, G., 41, 70, *72*, *74*, 79, 91, 92, 94, 96, 97, 99, 101, 102, 103, 109, 110, 114, 117, 118, 135, *139*
Schachman, H. K., 34, 36, 44, 67, 71, *72*, *74*
Schechter, B., *443*, 543, 545, *591*
Schechter, E., 371, 376, 377, 378, 379, 395, 396, 399, 400, 405, 424, 426, *440*, *443*
Schechter, I., *443*, 457, 461, 463, 465, 507, 508, 511, 524, 543, 545, *586*, *591*
Scheel, K., 6, *74*
Schelen, W., 421, *442*
Schellman, C., 361, 371, *443*
Schellman, C. G., 412, *443*
Schellman, J. A., 351, *354*, 361, 368, 369, 371, 374, 394, 406, 407, 408, 409, 411, 412, 424, 426, *440*, *442*, *443*, 463, 472, 477, 478, 483, 484, 488, 489, 490, 491, 492, 493, 494, 496, 497, 501, 503, 504, 506, 507, 510, 511, 514, 522, 524, 528, 529, 531, 533, 538, 557, 560, *585*, *589*, *590*, *591*
Schelten, J., 181, *243*
Scheraga, H. A., 384, 406, 408, 429, *441*, *444*, 503, 504, 510, 511, 512, 515, 530, *589*, *590*, *591*, *593*
Schettler, P. D., 326, *354*
Schevitz, R. W., 541, *585*
Schillinger, W. E., 293, 294, *299*
Schlimme, E., 209, 238, *242*
Schmadebeck, R. L., 329, 350, *355*
Schmidt, P. W., 149, 181, *243*
Schmier, I., 397, *400*
Schmitt, E. E., 427, *441*
Schmitz, P. J., 178, 199, *241*, *242*
Schneider, R., 192, *240*
Schonbaum, G. R., 532, *593*
Schubert, D., 218, 220, 222, *243*
Schulz, G. V., 209, 210, *241*

Schumb, W. C., 41, *72*
Schuster, I., 234, 236, 237, *240*
Schwaiger, S., 192, *240*
Schwartz, M. K., 350, *355*
Schwyzer, R., 530, 553, *589*, *591*
Scott, R. A., 515, 530, *590*, *591*
Sederholm, C. H., 329, 351, *354*
Segal, D. M., 525, *593*
Sela, M., *243*, 423, 426, *441*, *443*, 543, 545, *591*
Sekora, A., 186, 187, 199, *241*, *242*
Senter, J. P., 40, 42, 60, *74*
Shaffer, L., 178, *243*
Shail, R., 486, *590*
Shames, P. M., 530, *591*
Shannon, L. M., 414, 420, *443*, 468, *592*
Sharon, N., 563, *592*
Sheard, B., 247, 291, *300*
Sheats, S., 541, 568, 576, *585*
Shechter, E., 351, *353*
Shehedrin, B. M., 182, *243*
Shen, A. L., 117, 118, *139*
Shih, T. Y., 583, *592*
Shipley, J. W., 38, *74*
Shiraki, M., 399, 414, 430, *441*, *443*, 515, 516, 544, 547, 549, 551, *591*
Shmueli, U., 422, *443*, 522, 525, *593*
Shore, V. C., 557, *589*
Shotton, D. M., 541, 574, *591*
Shull, C. G., 149, 181, *243*
Shulman, R. G., 249, 291, *299*
Sick, H., 421, *441*
Sigler, P. B., *592*
Silpananta, P., 121, *138*
Silverton, E. W., 204, *243*
Simmons, N. S., 397, 412, 417, *440*, *441*, *443*, 563, *588*
Simons, E. R., 560, *592*
Simpson, R. J., 560, *592*
Simpson, R. T., 418, *443*
Simpson, W. T., 475, 502, 507, 522, *585*, *590*, *592*
Singer, S. J., 572, 573, 583, *588*, *592*
Singh, R. M. M., 274, *299*
Skala, Z., 169, 174, 181, *241*
Slayter, H. S., 583, *592*
Smiley, I. E., 541, *585*
Smith, E. E., 581, *588*
Snatzke, G., 361, *443*
Snyder, C. F., *72*
Sobel, J. H., 564, *586*

Sober, H. A., 403, *444*
Solomon, I., 266, *300*
Sommer, B., 410, *443*
Sonenberg, M., 538, *592*
Sørenson, S. P. L., 99, 127, *139*
Sorg, C., 517, *587*
Soucek, D. A., 116, *137*
Spanner, D. C., 80, *139*
Sparks, C. J., 172, *243*
Speakman, J. C., 351, *355*
Spector, A., 521, 533, *589*
Squire, P. G., 351, *355*
Srere, P. A., 251, *300*
Stabinger, H., 43, *73*, *74*, 166, *242*, *243*
Staverman, A. J., 124, *139*
Steinberg, D., 352, *353*
Steiner, H., 223, *239*
Steiner, R. F., 111, 116, 130, 131, 133, *138*, *139*
Steitz, T. A., 537, 538, 540, 542, *586*, *591*
Stejskal, E. O., 262, *300*
Stephen, M. J., 486, 487, *592*
Stevens, C. L., 27, 45, *74*
Stevens, L., 396, 412, 426, *443*, 517, 518, 519, 521, 523, 524, 525, 534, 535, 582, *592*
Stevenson, G. T., *592*
Stokes, A. R., 350, *354*
Stone, H., 331, 350, *355*
Stott, V., *74*
Stracher, A., 133, *137*
Strandberg, B. E., 188, *241*, 541, 557, *587*, *589*
Straus, J. H., *592*
Streyer, L., *440*
Strickland, E. H., 414, 419, 420, *441*, *443*, 467, 468, 470, 525, 542, 544, 548, 554, 561, 562, 565, *587*, *588*, *592*
Stross, F. H., 122, 132, 135, *137*
Stryer, L., 420, *443*
Strzelecka-Golaszewska, H., 407, *442*
Stuhrmann, H. B., 164, 165, 191, 209, 210, *241*, *243*
Suelter, C. H., 291, *299*
Sund, H., 211, 212, 213, 214, 215, 216, 217, *242*, *243*
Susi, H., 396, 426, *443*, 517, 518, 519, 523, 524, 525, 535, *592*
Suzuki, E., 312, 314, 323, 325, 326, 331, 349, 350, 351, *353*, *354*, *355*
Svedberg, T., 48, *74*, 125, *138*

## AUTHOR INDEX

Sverdlik, D. I., 449, *587*
Swalen, J. D., 326, 349, *354*
Swan, I. D. A., 566, *586*
Swen, H. M., 541, *587*
Swift, T. J., 280, *300*
Swinehart, D. F., 42, *73*
Syneček, V., 182, *243*
Szymanski, B. M., 249, 291, *299*

### T

Taggart, V. G., 393, *443*
Tait, M. J., 294, *300*
Takagi, T., 553, *588*
Tanford, C., 4, 18, *73*, *74*, 79, 92, 94, 95, 97, 107, 119, 120, 131, *138*, *139*, 202, *242*, 393, *443*, 571, 572, 574, *586*, *592*
Tang, L. H., 116, *137*
Taniuchi, H., 578, 579, *592*
Tanner, J. E., 262, *300*
Taylor, E. W., 385, *440*
Taylor, J. S., 248, 287, 289, *298*, *300*
Tefft, R. F., 41, *72*
Teichberg, V. I., 563, *592*
Teipei, F. W., 577, *592*
Teller, D. C., 351, *355*
TenEyck, L. F., 61, *74*
Terbojevich, M., 415, *440*
Terry, W. D., 204, *243*
Thiery, J. M., 370, *443*
Thiesen, M., 6, *74*
Thomas, H. P., 174, 192, 193, *240*, *243*
Thomas, J. O., 45, *74*
Thompson, J. D., 207, 208, *241*
Thygesen, J. E., 23, *73*
Tiffany, M. L., 409, *443*, 535, *592*
Tilander, B., 541, *587*
Timasheff, S. N., 205, 206, 207, *243*, 396, 400, 412, 426, *443*, 507, 508, 517, 518, 519, 520, 521, 523, 524, 525, 534, 535, 536, 544, 558, 561, 567, 568, 569, 582, *587*, *592*
Tinoco, I., Jr., 351, *355*, 361, 369, 384, 394, 395, 399, 403, 406, 426, *442*, *443*, *444*, 449, 450, 452, 459, 461, 462, 463, 465, 466, 475, 476, 477, 478, 479, 495, 496, 501, 502, 503, 506, 507, 508, 509, 511, 512, 513, 514, 517, 525, 530, *586*, *593*
Tinoco, I. J., 476, *588*
Todd, A., *443*
Tonelli, A. E., 528, *592*
Tonioli, C., 415, *441*

Tooney, N., 392, 399, 400, 428, *441*
Townend, R., 396, 400, 412, 426, *443*, 507, 508, 517, 518, 519, 520, 521, 523, 524, 525, 534, 535, 582, *592*
Traub, W., 422, *443*, 522, 525, *593*
Trombka, J. I., 329, 350, *355*
Tsernoglou, D., 468, 562, *593*
Ts'o, P. O. P., 517, 543, *588*
Tsuboi, M., 391, *444*
Tterlikkis, L., 486, 487, 514, *589*
Tunnicliff, D. D., 349, *355*
Tuttle, R. W., 530, *589*, *591*
Tyuman, I., 556, 557, *586*

### U

Uhr, M. L., 286, *299*
Ulmer, D. D., 362, 420, *443*, 541, *593*
Ulrich, D. V., 39, *75*
Urnes, P., 384, 386, 387, 388, 412, 434, *444*, 555, *593*
Urry, D. W., 361, 362, 367, 368, 369, 370, 371, 410, 411, 420, 436, *442*, *443*, *444*, 466, 494, 506, 507, 508, 509, 510, 511, 512, 517, 520, 534, 541, *589*, *591*, *593*

### V

Valentine, R. C., 202, 204, *243*
Vallee, B. L., 362, 418, 420, *443*, *444*, 544, 560, *591*, *592*
Vanaman, T. C., 197, *239*
van Bruggen, E. F. J., 229, *243*
van Deinse, A., 186, 187, *242*
Vandendriessche, L., 31, *72*, *75*
vander Haar, F., 209, 238, *242*
Vanderkooi, G., 515, 530, *589*, *590*, *591*, *593*
van Ryssel, T. W., 172, *242*
Varnum, J. C., 541, *587*
Vaughan, P. A., 196, *241*
Velluz, L., *593*
Venyminov, S. Y., 454, *586*
Verdini, A. S., 415, *440*, *442*, 547, 550, *590*
Veros, A. J., 569, *587*
Viennet, R., 515, 542, *589*
Vijayan, M., 541, 568, 576, *585*
Vink, H., 133, *139*
Vinograd, J., 70, *73*
Virden, R., 279, 289, *300*
Visser, L., 574, 575, *593*
Vithayathil, P. J., 558, *591*
Vogel, H., 174, 192, 193, *240*

Voigt, B., 235, *241*
Voigt, W., 362, *444*
Volkenstein, M. V., 396, 409, 410, *444*, 512, 518, 521, 526, *593*
Volkova, L. A., 182, *240*
Vonk, C. G., 181, *243*
von Nordstrand, R. A., 158, *243*
Voronin, L. A., 182, *240*
Vournakis, J. N., 406, *444*, 503, 504, 510, 511, 512, *593*

## W

Waara, I., *589*
Wacker, W. E. C., 362, 420, *444*, 544, *591*
Wada, A., 373, 391, 428, *441*, *444*
Wagenbreth, H., 6, *75*
Wagner, R. H., 79, 94, 104, 122, 134, *139*
Waks, W., 556, *593*
Walker, L. R., 349, *355*
Wallach, D. F. H., *592*
Walter, J. E., 450, 466, 471, 472, *589*
Wamsley, S. H., 475, 496, *587*
Wang, J. L., 581, 582, *590*
Ward, R. L., 251, *300*
Warren, J. R., 55, *73*
Warwicker, J. O., 352, *355*
Watson, H. C., 531, 541, 574, *591*, *593*
Watson, J. D., 474, *593*
Wawra, H., 177, 186, 187, 223, *239*, *241*, *242*
Waxdal, M. J., 199, 204, *240*
Webb, M. B., 146, 168, *239*
Weber, H. H., 22, *75*, 120, 125, 135, *139*
Weeks, J., 101, 102, 109, 114, *139*
Weidinger, A., 178, *241*
Weigang, O. E., 466, 469, 470, 478, 493, 496, 503, 505, *588*, *593*
Weiner, H., 248, 292, *299*, *300*
Weinzierl, J. E., 541, *587*
Wellman, K. M., 468, *590*
Wells, J. D., 117, 120, *138*
Westerberg, A. W., 349, *355*
Westort, C., 22, 61, *72*
Wetlaufer, D. B., *443*, 535, 538, 539, 540, 541, 550, *587*, *591*
White, R. A., 215, *243*
Wilchek, M., 468, 525, 548, *587*, *592*
Wilenzick, R. M., 331, 351, *354*
Wilkinson, D. H., 349, *355*
Willick, G. E., 532, *593*
Windle, J. J., 296, *300*

Winget, G. D., 274, *299*
Winter, W., 274, *299*
Wirén, G., 541, *587*
Witters, R., 226, *242*
Witz, J., 164, 205, 206, 207, 209, *242*, *243*
Woessner, D. E., 250, *298*
Wohlleben, K., 172, *241*
Woldbye, F., 365, *444*
Wollmer, A., 421, *441*
Wolthers, B. G., 541, *587*
Wonacott, A. J., 351, *353*, 541, *585*
Woodward, R. B., 493, 494, 506, *589*
Woody, R. W., 384, 394, 395, 396, 399, 403, 406, 410, 411, 414, 421, 426, *440*, *441*, *443*, *444*, 466, 473, 477, 494, 502, 503, 506, 507, 509, 510, 511, 512, 513, 514, 515, 516, 517, 519, 520, 521, 531, *586*, *587*, *588*, *592*, *593*
Woolley, R., 486, *585*
Worthington, C. R., 170, *243*
Wright, C. S., 541, *593*
Wright, H. T., 188, 194, *242*
Wright, M., 215, 220, *239*
Wu, Y. V., 117, 118, *139*
Wyckoff, H. W., 468, 562, *593*
Wyld, G. E. A., 349, *355*
Wyman, J., 237, *242*

## Y

Yahara, I., 392, *442*, *444*
Yamaoka, K. K., 374, *444*
Yan, J. F., 406, *444*, 503, 504, 510, 511, 512, 515, *593*
Yang, J. T., 22, 27, *74*, 361, 370, 371, 373, 374, 375, 376, 383, 384, 385, 389, 390, 393, 395, 398, 399, 400, 401, 403, 412, 427, 428, 430, 438, *440*, *441*, *442*, *444*, 457, 507, 508, 513, 517, 521, 540, 541, 550, 576, *586*, *588*
Yaron, A., 403, *444*
Yiengst, M., 127, *139*
Yonath, A., 525, *593*
Yphantis, D. A., 427, *441*

## Z

Zakaria, A., 522, 523, 524, *591*
Ziegler, S., 543, *593*
Zipper, P., 218, 220, 222, *243*
Zosa, M., 135, *139*
Zubkov, V. A., 396, 409, 410, *444*, 512, 518, 521, 526, *593*

# SUBJECT INDEX

## A

Absolute intensity
  determination of, 177–179
  molecular weight determination and, of X-ray scattering, 165
Absorption spectra, vibronic fine structure at low temperature, 469–471
$N$-Acetyl-L-phenylalanineamide, far-ultraviolet circular dichroism spectrum of, 549
$N$-Acetyl-L-proline-$N,N$-dimethylamide, optical activity of, 529
$N$-Acetyl-L-tryptophanamide, far-ultraviolet circular dichroism spectrum of, 549
$N$-Acetyl-L-tyrosineamide, far ultraviolet CD spectrum of, 544, 545, 549
Activation analysis, least squares in data analysis, 349
Activity, relative, in osmosis, 81
Activity coefficient, 84, 88
Albumin
  bovine serum, see Serum albumin
  chloride ion binding to, 70
Alcohol dehydrogenase, proton relaxation rate enhancement by, 291–293
Aldolase, osmotic pressure and subunit evaluation, 106
Allosteric effect
  in conformational transitions, 430
  small-angle X-ray scattering and, 234–238
Amino acid analyzer, least squares in data analysis, 303, 342–344
Anisotropy, effect on small-angle X-ray scattering curve, 147, 149
Anisotropy constant, 461, 470
Apoferritin, protein shell, small-angle X-ray scattering by, 222–226
Apotransferrin, proton relaxation rate, 294
Apparent specific volume, see Volume
Association (of proteins)
  and osmotic pressure, 92, 98, 108, 114
  self-, density changes and, 60
  and volume change of proteins, 12
Association constants
  from osmotic pressures, 110
  from osmotic pressures of nonideal systems, 111
Association equilibria, osmometry $vs$ sedimentation, 112
Asymmetry pressure, in osmometry, 134
ATP, hydrolysis of, volume changes in, 32
Avidin, spectra and structure of, 540
Axial ratio, osmotic pressure and, 119, 120

## B

Bacteriophages, protein shells, small-angle X-ray scattering, 216–218, 220–222
Beer-Lambert law, 450
$\gamma$-Benzyl-L-glutamate, Moffitt parameter for oligomers of, 391, 392
Binding, see also Interactions
  chloride ion, to albumin, 70
  of creatine kinase to substrates, and proton relaxation rate, 285
  ion, to proteins, 117
  of paramagnetic species to protein, 251–253
  preferential, effect on osmotic pressures, 13
  selective, of ions, 92, 96, 99, 117
  sites, determination of, 293
  volume changes and, 49
Bovine serum albumin, see Serum albumin
Bovine visual pigment, low-temperature near-ultraviolet spectra, 468
BSA, see Serum albumin
Buoyancy, partial specific volume and, 45, 48
Buoyancy corrections, in density measurements, 34

## C

Cameras, small-angle X-ray scattering, 170–177, 238
$d$-Camphor 10-sulfonate, optical rotatory

dispersion and circular dichroism curves, 370
Carbonic anhydrase B, rotatory dispersion curves for, 413
Carbonic anhydrase C, optical activity and structure of, 541
Carboxypeptidase A
  low-temperature near-ultraviolet spectra, 468
  secondary structure of, 537, 540
Cauchy bands, in data analysis, 322, 324, 328
Cellophane membranes, for osmometry, 133
Cellulose acetate membranes, for osmometers, 131, 133
Chain length, effect on optical activity of $\alpha$-helix, 509
Charge on proteins, effect on osmotic pressure, 93
Chelating agents, as impurities in proton relaxation rate measurements, 253
Chemical potential
  definition, 85, 86
  free energy, 5, 51
  standard, in osmosis, 88
Chloride ion, binding to albumin, 70
Chromatin, optical activity and structure of, 583
Chromatography, least squares in data analysis, 326, 349
Chymotrypsin, small-angle X-ray scattering by, 193–195
$\alpha$-Chymotrypsin
  molar ellipticity of, 533
  secondary structure of, 537, 540
Chymotrypsinogen
  optical rotatory dispersion and helical content, 387
  secondary structure of, 537, 540
  small-angle X-ray scattering by, 193–195
Chymotrypsinogen A, low-temperature near-ultraviolet spectra of, 468
Circular birefringence, 450–452
Circular dichroism, 445–593
  bands, and optical rotatory dispersion bands, interrelationships of, 369
  conformation and, of proteins, 583
  definition, 450
  least squares in data analysis, 351

optical rotatory dispersion and, 454–457
  relative advantages of, 436–439
phenomenological description of, 364–366
quantum mechanical treatment of, 473
rotational strength and, 460
solvent and temperature effects on, 466–469
Circular dichroism spectra
  low-temperature, 468
  of polypeptides in $\alpha$-helical, $\beta$-pleated sheet and aperiodic conformations, 533–536
  of poly-L-tyrosine, 514–517
  random-coil, and secondary structure of proteins, 535
  resolution of, 460–466
  vibronic fine structure at low temperature, 469–471
Collagen
  optical activity of, 525, 529
  optical rotatory dispersion and structure of, 421–426
Collagen–water system, proton relaxation rate and, 298
Collodion membranes, for osmometers, 133
Component, definition of, in osmosis, 82
Computer programs, small-angle X-ray scattering data and, 238
Concanavalin A, ultraviolet and circular dichroism spectra of, 581, 582
Concentration
  measurements, accuracy of, 20
  protein, determination for osmometry, 136
Concentration units, in osmosis, 90
Condon–Altar–Eyring theory, of optical activity, 472
Conformation
  helical, 377
  at low temperature, 468
  optical activity and, 528, 529
  osmotic pressure and, 120
  of poly-L-tyrosine, 515–517
  of protein main chain and optical rotatory dispersion, 357–444
  random-coil, 377
  rotational strength and, 458
  transitions, and density measurements, 60

## SUBJECT INDEX

ultraviolet spectra and, 553–555
volume changes and, 26, 27, 32
β-Conformation
   Moffitt equation and, 390–394
   optical activity and, 410–412
   of polypeptides, optical activity of, 517–522
Conformation changes
   of antibodies upon reaction with antigen, 237
   denaturation and, and small-angle X-ray scattering, 209, 210
   helix-to-coil and coil to β-structure, 427–430
   Cotton effect, 365, 366
   near-ultraviolet, 417
   optical rotatory dispersion curves and, 455
   peptide, 394–412
   side-chain chromophores and, 412–414
Countercurrent distribution, least squares in data analysis, 349
Creatine kinase
   binding to substrates, proton relaxation rate and, 285–289
   effect on proton relaxation rate, 249
   spin-labeled, 289
   ternary complex with $Mn^{2+}$, proton relaxation rate enhancement by, 283
Crystallization, volume changes during, 30
Crystal structures, of ribonuclease A and S, 468
L-Cystine, ultraviolet and circular dichroism spectra of, 552, 554, 555
Cystinyl residues (disulfide), optical activities of, 552–554
Cytochrome c, spectra and structure of, 541

## D

Data analysis, *see also* Parameter optimization
   analytical representation, asymmetric band shapes, 325
      Cauchy bands, 322, 324, 328
      derivative functions, 326
      four-parameter symmetrical band shape, 325
      Gaussian bands, 321, 327–329
      sum and product bands, 323
      baseline functions, 331
      computational procedure, 331–352
      digital standard shape functions, 329
      least squares in, 301–355
         application of iterative nonlinear, 349–353
Degree of hydration, estimation of, 60
Degree of interaction, estimation of, 60
Dehydrogenases, *see also* Lactate dehydrogenase
   glutamate, cross-section X-ray scattering curve, 211–217
   proton relaxation rates for, 291–293
Denaturation
   acid, of myoglobin, and conformation changes, 209
   of bovine serum albumin, 391
      and conformational changes, 209
   conformation changes and, 209, 210
Densimetry
   hydrostatic weighing method, 43
   magnetic, 39–43
   resonance frequency method, 43
Density
   composition and, 16–21
   definition, 6
   measurement, accuracy of, 17, 32
      applications, 45–62
      buoyancy corrections in, 34
      by density-gradient columns, 36, 37
      and dry weight, 7, 8, 21
      by hydrostatic weighing method, 43
      by isopycnic temperature method, 37–39
      by magnetic densimeter, 39–43
      for osmotic molecular weight experiments, 103
      and reaction kinetics, 61
      and relaxation phenomena, 62
   partial specific volume derivation from, 62–66
   pressure effects on, 61
   of pure water, 6, 32
   tables, 41
   temperature dependence, 60
   and volume change measurements, 1–75
Density-composition tables, use of, 17, 18, 34, 53, 56
Density-gradient columns, 36
Deoxyhemoglobin, *see* Hemoglobin

Deprotonation, of proteins, volume change and, 26
Dextran, osmotic pressure of mixture with serum, 109
Dialysis, equilibration, 55
Dilatometers
 capillary, 22–25
 in solid-phase studies, 30
Dilatometry, volume change measurements, 22–25
Dipeptides
 cyclic and linear, optical activities of, 528, 529
 cyclic D- and L-, low-temperature circular dichroism spectra of, 529
Dipole moment, electric transition, 459
Dipole moment operators, electric and magnetic, for polymers, 476
Disulfides, optical activities of, 553
DNA, hydrolysis, volume changes in, 31
Donnan effects
 on osmotic pressure, 13, 14, 92–102, 109, 118
 of plasma extender, 109
 pH and, 96
Drude equation
 one-term, for optical rotatory dispersion, 372–375, 380
 two-term, and Moffitt equation, 389
 for optical rotatory dispersion, 375–380, 381
Dry weight, density measurements and, 7, 8, 21

## E

Elastase, circular dichroism spectrum and structure of, 541, 574–576
Electromagnetic units, Gaussian, 583–585
Electron density
 of proteins, 166
 radial, distribution, by Fourier transformation, 214–226
Electronic spectra, least squares in data analysis, 350
Electron spin resonance, least squares in data analysis, 349
Ellipsoids, *see also* Shape
 X-ray scattering curves for, 145, 149–152, 190, 191
Ellipticity
 chain length effect on, 511
 definition, 452–454
 molecular, reduced, 454
 protein, 533
 quantum mechanical expression for, 473
 reduced molar, 366
Energy, partial molal free, 5, 85
Enolase, effect on proton relaxation rate, 249
Enzyme kinetics, least squares in data analysis, 350
Enzymes, proton relaxation rates for, 249, 251–253, 286–294
Equilibrium
 osmotic, 12–15, 48, 51
 thermodynamic, 80
α-Ethyl-L-glutamate, copolymer with L-tryptophan, circular dichroism spectrum of, 547, 548
Extinction coefficient
 determination of, 56
 partial molar, 454
 protein, 21
 values (variable) of, 47

## F

Ferricytochrome c, spectra and structure of, 541
Ferritin, protein shell, small-angle X-ray scattering by, 222–226
Flagellin
 polymerization kinetics and volume changes, 32
 volume changes during polymerization, 30
Flotation temperature method, *see* Isopycnic temperature method
Fourier transformation, radial electron density distribution by, 214–226
Free energy, partial molal, definition, 85

## G

G-actin, volume change on transformation to F-actin, 32
Gamma-ray and pulse height spectra, least squares in data analysis, 350
Gaussian bands, in data analysis, 321, 327–329
Gaussian curve, small-angle X-ray scattering curves and, 146
Gaussian functions

and circular dichroism spectra, 461, 462, 466
of electromagnetic units, 583–585
Gauss-Newton method, of parameter optimization, 305, 309, 312
Gel phases, osmotic pressure and, 120
Gibbs-Donnan effect, *see* Donnan effect
Gibbs free energy, 5, 84
γ-Globulin, osmotic pressure of mixture with serum albumin, 109
L-Glutamate, poly-, volume changes on titration of, and benzyl ester, 27
Glutamate dehydrogenase, cross-section X-ray scattering curves, 211–217
L-Glutamic acid, copolymer with L-tyrosine, optical rotatory dispersion and helical content, 387
Glutathione, oxidized, ultraviolet spectra and conformations of, 554, 555
Glyceraldehyde 3-phosphate dehydrogenase (GPDH)
  allosteric behavior of yeast, and small-angle X-ray scattering, 234–238
  conformational changes in, 430
Gramicidin S, optical activity and structure of, 530
Guinier plot, in small-angle X-ray scattering, 147, 184, 186

## H

α-Helix
  optical rotatory dispersion, 377, 380, 394, 397
    and circular dichroism curves for, 406–408
  rotational strength of, 475, 506–514
  screw sense of, 426
Helix to random-coil transition, volume changes in, 27
Hemocyanin
  small-angle X-ray scattering by, 170, 182, 226–229
  subunits, small-angle X-ray scattering curves of, 229
Hemoglobin
  conformational changes in, 430
  deoxy- and oxy-, osmotic pressures of mixtures, 115, 116
  deoxy-, small-angle X-ray scattering by, 192, 193

dissociation, osmometry and, 107, 112–116, 135
hybridization of, 112
liganded and unliganded, 114, 135
optical activity of, 554–558
optical rotatory dispersion and helical content, 385, 387
osmotic pressure data for, 135, 136
oxy-, optical activity of, 531, 532
  small-angle X-ray scattering by, 192, 193
rotational strength of, 532
volume changes during crystallization, 30
Henry's law, 83
Histone
  lysine-rich, volume change on titration, 26
  optical rotatory dispersion and helical content, 387
Homocystine, ultraviolet CD spectra and conformation of, 554, 555
Homopolypeptides, circular dichroism and conformation of, 535
Hyaluronic acid, osmotic pressure of, 120
Hybridization
  of hemoglobin, 112–114
  of protein subunits, 112–116
Hydration, interaction parameter and, 60
  of proteins and nonideality, 99
Hydration value
  for proteins, and partial specific volume, 45
  of simple proteins, 70
Hydrolysis, kinetics of, of proteins, 31
Hydrostatic weighing, density measurement by, 43

## I

Ideal solution, in osmosis, 88, 91
Immunoglobulins
  small-angle X-ray scattering curves, 197–204
  ultraviolet and circular dichroism spectra of, 569–574
  X-ray scattering curves for, effect of concentration on, 183
Impurities, effect on osmotic pressure estimation of molecular weights, 104
Infrared spectra, least squares in data analysis, 350

Insulin
  optical rotatory dispersion and helical content, 387
  ultraviolet and circular dichroism spectra of, 541, 566–569
Intensive property, in osmosis, 84
Interactions
  metal ion–protein, proton relaxation rate and, 282
  parameter, and hydration, 60
    for turnip yellow mosaic virus, 58, 59
    preferential parameter, and density, 48, 50–60
    and exclusion of diffusible components, 66–70
  preferential salt, 99
  protein–protein, 108
  proton–proton dipolar, 294
  thermodynamic parameter, 50, 54, 57–60
Ion binding, see Binding
Isopotential specific volume, see Volume
Isopycnic temperature method, for measuring density, 37–39

## K

Keratin fibers, proton relaxation rate, and absorption of water, 295–298
Kirkwood theory, of optical activity, 472–475
Kronig-Kramers transforms, and optical rotatory dispersion, 370

## L

$\alpha$-Lactalbumin, bovine, small-angle X-ray scattering by, 195–197
Lactate dehydrogenase
  circular dichroism spectra and structure of, 541, 576–578
  Moffitt equation for, 385
$\alpha$-Lactoglobulin, small-angle X-ray scattering by, 205–207
$\beta$-Lactoglobulin
  optical rotatory dispersion and helical content, 387
  $\beta$-structure and optical rotatory dispersion, 400
Lamellar particles, X-ray scattering curves for, 163

Least squares
  damped, in parameter optimization, 311, 314
  in data analysis, 301–355
  parameter optimization, applications, 349–353
    convergence of nonlinear, 309–316
    and rotational strength, 508
  symbols, 301, 302
Lorentz factor, and solvent effects on optical activity, 466
Lowry's plot, 375
Lysine-rich histone, volume change on titration, 26
Lysozyme
  $\alpha$-helix content, 408
  Moffitt parameter for, 393
  molar ellipticity of, 533
  optical activity and structure of, 530
  optical rotatory dispersion by, 378
    and helical content, 385, 387
  secondary structure of, 537
  small-angle X-ray scattering by, 195–197
  ultraviolet and CD spectra of, 548, 562–566

## M

Macroion, definition, 93
Macromolecules
  characteristics of hollow bodies, 226–234
  homogeneity of, for small-angle X-ray scattering, 185–188
Manganese ion, binding in enzymic reactions, 248, 251
Mass spectra, least squares in data analysis, 350
Membrane phase, in osmometry, 132
Membranes
  circular dichroism and structure of, 583
  for osmometers, 124, 131–135
  for osmometry of large proteins, 134
  of small proteins, 133
  semipermeable, 81
Metal ions, paramagnetic, interaction with enzymes, 248
Methemoglobin, optical activity of, 555, 556
Metmyoglobin
  conformation changes on acid denaturation of, 209, 210

SUBJECT INDEX 615

optical activity of, 555, 557
radii of gyration, 189
rotational strength of, 532
Mixtures, osmotic pressures of, 109, 114–116
Moffitt equation
and $\beta$-conformation, 390–394
and Drude two-term equation, 389
for proteins containing $\alpha$-helix and unordered structures, 380–390
Moffitt, Fitts and Kirkwood theory, of optical activity, 475
Moffitt-Yang plot, 381–385
Molecular ellipticity, reduced, 454
Molecular rotation, 454
Molecular weight
and mass per unit length, 165
number-average value, 90
osmotic, at infinite dilution, 135
osmotic pressure and, 79
of ovalbumin and plakalbumin, 108
of proteins, osmotic pressure and, 102–106
specific volume and, 48
Mole fraction, in osmosis, 88
Monochromatization, Bragg's reflection or total reflection, 172–174
Mössbauer spectra, least squares in data analysis, 351
Mucopolysaccharides, osmotic pressures and, 120
Myoglobin
acid denaturation of, and conformation changes, 209
$\alpha$-helix content, 408
molar ellipticity of, 533
optical activity of, 531, 532, 554–558
optical rotatory dispersion and helical content, 385, 387
secondary structure of, 537
small-angle X-ray scattering, 188–192
Myosin, osmotic pressure of, 120

## N

Neurophysins ultraviolet and circular dichroism spectra of, 554, 579–581
Nomenclature, angular, in X-ray scattering, 146
Nonideality
and hydration of proteins, 99

and protein nonidentity, 100
solution, in osmosis, 91
Nuclear magnetic resonance spectroscopy
least squares in data analysis, 351
pulsed, apparatus for, 256, 267, 268
techniques, 245–300
relaxation, phenomenological description of, 257
theory of, 253–267
relaxation time, measurement of spin-spin by spin echo method, 260–262
transverse or spin–spin, 257
symbols, 246
Nutation, definition, 255

## O

Oligopeptides, stepwise synthesis of, 427
Optical activity
basic relations for optically active molecules, 362–371
chain length effect on, 509
Condon-Altar-Eyring theory, 472
conformation and, 528, 529
Cotton effects, 394–412
coupled-oscillator mechanism, 472, 495–501
electric–magnetic coupling mechanism, 472, 501–506
of helical polymers, 507
of $\alpha$-helix, 506–514
Kirkwood theory, 472–475
magnetic electric coupling, 369
matrix formalism, 483–488, 511, 513
Moffitt, Fitts and Kirkwood theory, 475
one-electron mechanism, 368, 472, 489–495
phenomenological description of, 364–366, 449–454
quantum mechanical theory of, 366–369, 471–488
of random coils, 526–528
Rosenfeld's theory, 471, 472
rotational strength and, 457–460
side-chain, 412–414
and helical content, 414
symmetry rules and origins of, 488–506
theory and applications of, 449–533
Tinoco's theory, 369, 476, 506
Optical rotatory dispersion (ORD)
circular dichroism and, 454–457

interrelationships of, 369
relative advantages of, 436–439
Drude equations for, 372–380
history and scope, 360–362
instrumental factors, 434
instrumentation, 430–434
least squares in data analysis, 351
and main chain conformation of proteins, 357–444
phenomenological description of, 365
of poly-L-tyrosine, 514–517
rotational strength and, 459
scanning speed, 435
sensitivity, 434
solution and cells for, 435
turbidity and orientation of sample, 435
visible, 371–394

Osmometers
dynamic, 121–125
electronic, 127–131
Fuoss-Mead, 125
interferometric, 125
Mechrolab, 129
Melabs, 127, 128
membranes, 124, 131–135
Rolfson-Coll, 130
static, 121–123
ultrasensitive, 125–127

Osmometry
asymmetry pressure in, 134
and molecular weight determinations, 105
of ovalbumin and plakalbumin, 108

Osmosis
definition of, 82
diagramatic representation of, 81

Osmotic equilibrium, 12–15, 48, 51

Osmotic pressure, 77–139
apparent specific volume and, 119
applications, 102–121
axial ratio and, 119, 120
conformation and, 120
data treatment, 135
definition of, 82
description of, 80
diagramatic representation of, 81
effect of impurities on molecular weights, 104
of temperature and pressure on, 117
excluded volume and, 118, 121
gel phases and, 120
and hybridization, 112
and limited proteolysis, 107
of mixtures, 109, 114–116
molecular weight and, 79, 102–106
and protein association, 110
and protein mixtures, 109
of random-coil polymers, 107
reduced, 91
and salt binding, 92
and selective ion binding, 96, 99
shape and, 118–120
and subunit evaluation, 106, 108
symbols, 77, 78
techniques, 121–137
theory, 87–102

Ovalbumin
Moffitt parameter for, 391
optical rotatory dispersion and helical content, 387
osmotic comparison with plakalbumin, 108, 125

Oxyhemoglobin, see Hemoglobin
Oxytocin, optical activities of, and derivatives, 553–555

## P

Papain
Moffitt equation for, 385
optical activity and structure of, 541

Parameter optimization
computational procedure, flow chart, 331–334
computer program, input and output, 334–348
constrained, 307–309
Gauss-Newton method, 305, 309, 312
iteration process, damped least squares, 311
measures of goodness of fit, 310
scaled parameter adjustments, 312–314
temporary constraints, 314
iterative refinement, 307
least squares in, 302–304
statistical aspects, choice of weighting function, 316
errors in parameter estimates, 317
significance tests, 317–321
theory of, by least squares, 304–309

Partial molal free energy, definition, 5, 85
Partial molal volume, see Volume

# SUBJECT INDEX

Partial specific volume, *see* Volume
Permeability coefficient, in osmometry, 124
Peroxidase A1, horse radish, low-temperature near-ultraviolet spectra, 468
Phase, definition of, 80
Phenylalanine derivatives, low-temperature near-ultraviolet spectra of, 468
Phenylalanyl residues, circular dichroism spectra of, 550–552
Phosphoenolpyruvate carboxykinase, proton relaxation rate enhancement by, 275, 276
Plakalbumin, osmotic comparison with ovalbumin, 108, 125
Plasma expander, serum protein–dextran mixture as, 109
Pleated sheet, *see* $\beta$-Conformation
Polarography, least squares in data analysis, 326, 351
Poly-$O$-acetyl-L-serine, Moffitt parameter for, 391, 392
Poly-L-alanine
  random pentamers, octamers and decamers, optical activities of, 527
  rotational strength of $\alpha$-helical, 507, 509, 510
Poly-$\gamma$-benzyl-L-glutamate, optical rotatory dispersion data, 376, 381, 385
Poly-$O$-benzyl-L-serine, Moffitt parameter for, 391, 392
Poly-$S$-carbobenzoxymethyl-L-cysteine, Moffitt parameter for, 391
Poly-$S$-carboxymethyl-L-cysteine, circular dichroism spectrum of, 534
Poly-L-glutamate, volume changes on titration of, and benzyl ester, 27
Poly-L-glutamic acid
  optical rotatory dispersion data, 374, 376, 379, 381–385
  optical rotatory dispersion of helical and random coil forms, 397–399
Poly-L-isoleucine, $\beta$-structure and Cotton effects, 399
Poly-L-lysine
  circular dichroism spectrum of, 518, 520, 533–536
  copolymer with tyrosine, optical rotatory dispersion curves and Cotton effects, 402
  far-ultraviolet optical rotatory dispersion in $\alpha$-helical, antiparallel $\beta$- and random coil conformations, 401
  Moffitt parameter for, 391, 392
  $\beta$-structure and Cotton effects, 399–405
Polymerization
  kinetics of, of flagellin, 30, 32
  volume changes during, 30, 31
Polymers
  multielectronic large, quantum mechanical treatment of optical properties of, 473–488
  optical activity, Tinoco theory, 369
  osmotic pressures of synthetic, 132
Poly-$\gamma$-methyl-L-glutamate, absorption spectra and circular dichroism of $\alpha$-helical, 464
Polypeptides
  circular dichroism spectra of $\alpha$-helical, $\beta$-pleated sheet and aperiodic conformations, 533–536
  helix-to-coil and coil-to-$\beta$-structure, 428–430
  optical activity of $\beta$-structures, 517
  rotational strength of $\alpha$-helical, 508
Poly-L-proline
  optical activity of, 522–525
  optical rotatory dispersion curves for, 422–426
Poly-L-proline I, 422–426, 529
Poly-L-proline II, 422–426, 529
$\beta$-Poly-L-serine, optical activity of, 520
Poly-L-tryptophan, circular dichroism spectra of, and of copolymer with $\alpha$-ethyl-L-glutamate, 547, 548
Poly-L-tyrosine
  circular dichroism of, 542
  conformations of, 515–517
  optical activity of, 514–517
Potential energy, in osmosis, 82
Pressure
  effect on density, 61
  on osmotic pressure, 117
Prisms, small-angle X-ray scattering curves for, 149, 151, 152
L-Proline, derivatives, optical activities of, 528
Proteins
  associations, effect on osmotic pressure, 92, 98, 108, 114
  charge on, effect on osmotic pressure, 93
  circular dichroism and structure of, 583

circular dichroism spectra at low temperature, and conformation, 468
concentration, determination for osmometry, 136
  measurements of, 20–21
conformation and osmotic pressures of, 120
conformation changes in, and volume changes, 26, 27, 32
$\beta$-conformation Cotton effects, 399
crystallization, volume changes during, 30
density and volume change measurements of solutions, 3
dry-weight analysis, 21
far-ultraviolet optical rotatory dispersion of helical and random coil forms, 397–399
hydration value and partial specific volume, 45
hydration values for simple, 70
hydrolysis of, kinetics of, and volume changes, 31
internal solvation of, 167
ion binding to, 117
ionization and volume changes, 25
membranes for osmometry of large, 134
  for osmometry of small, 133
mixtures, osmotic pressure of, 109, 114
modifications and proton relaxation rates, 284
molecular weight and osmotic pressure, 102–106
optical rotatory dispersion and main chain conformation of, 357–444
partial specific volume of, 8–10
protonation and deprotonation and volume changes, 26
secondary structure, estimates of, 535–541
  from optical rotatory dispersion curves, 405
self-association, density changes and, 60
  effect on osmotic pressure, 92
  volume changes and, 12
shape and X-ray scattering curves for, 148–158
side-chain optical activity in, 541–554
small-angle X-ray scattering by, 143
solutions, concentration effect on X-ray scattering curves, 182

subunit evaluation by osmotic pressure, 106, 108
tobacco mosaic virus, polymerization and volume changes in, 27
Protein–water solution, partial specific volume, 8
Proteolysis, and osmotic pressure, 107
Protonation, of protein carboxyl groups and volume change, 26
Proton relaxation rate
  and absorption of water by protein, 294–298
  definition, 250
  effect of buffer on, 274
    of concentration of $Mn^{2+}$ on, 274
    of paramagnetic species, 265–267
    of pH on, 274
    of purity of sample on, 275
    of substrate on, 275
    of temperature on, 275, 279
  enhancement of, 248–250
  data analysis, 276–279
  frequency (field) dependence of, 281, 293
  information available from, 251–253
  limitations of measurements of, 253
  manganous ion–bovine serum albumin complex enhancement of, 282
  manganous ion–creatine kinase complex, proton relaxation rate enhancement by, 283
  measurement of longitudinal, and apparatus, 269–271
    of spin–spin, 271–273
  principles, 259–262
  for spin-labeled compounds, 285
Pycnometers, 33, 43
Pycnometry, 33–36
Pyruvate kinase, proton relaxation rate enhancement by, 290, 291

## Q

Quantum mechanical theory, of optical activity, 471–488

## R

Radial electron density distribution
  by Fourier transformation, 214–226
  of particles with spherical symmetry, 158
Radicals, stabilized free, and electron para-

magnetic resonance spectra, 248
Radius of gyration
  definition, 147
  of fatty acid synthetase, 185
  scattering curve form and shape of particle, 147, 148
Random coil
  helix transition to, volume changes in, 27
  optical activity and, 408–410, 526–528, 535
Random-coil conformation, 377
Random coil polymer chains, second virial coefficients of, 107
Raoult's law, 83
Reaction kinetics, density measurements and, 61
Refractive index, in volume change measurements, 22
Relaxation, see also Proton relaxation time
  density measurements and, 62
Resolution, of circular dichroism spectra, 460–466
Resonance frequency, density measurement and, 43
Ribonuclease
  Cotton effect and optical rotatory dispersion and circular dichroism curves for, 407
  exclusion parameters and, 71
  optical rotatory dispersion and helical content, 379, 385, 387
Ribonuclease A
  absorption spectra, 467, 468
  low-temperature near-ultraviolet spectra of, 468
  secondary structure of, 537
  structure of, 533
Ribonuclease S
  absorption spectra, 468
  low-temperature near-ultraviolet spectra of, 468
  molar ellipticity of, 533
Ribonucleases, circular dichroism spectra, 558–562
Ribosomes, small-angle X-ray scattering by, 207, 208
RNase, see Ribonuclease
Rods, see also Shape
  X-ray scattering curves for, 159–162

Rosenfeld equation, and optical rotatory dispersion, 369, 371
Rosenfeld theory, of optical activity, 471, 472
Rotational strength
  definition, 366, 367
  exciton and nonexciton, 508, 512
  of $\alpha$-helix, 506
  isolated-subunit formalism, 477
  optical activity and, 457–460
  static-field formalism, 477, 480
  total electronic, 469, 470

## S

Scatchard components
  definition, 96
  and ion binding to proteins, 117
Sedimentation data, partial specific volume and, 44, 48
Sedimentation equilibrium
  least squares in data analysis, 351
  and osmotic equilibria, 112
Sephadex, for osmotic pressure measurement, 117, 120
Serum, osmotic pressure of mixture with dextran, 109
Serum albumin,
  denaturation and conformation changes, 209
  and Moffitt parameter, 391, 393
  and volume changes, 27
  manganous ion binding and proton relaxation rate, 282
  optical rotatory dispersion and helical content, 379, 387
  optical rotatory dispersion data, 375
  osmotic pressure of mixture with $\gamma$-globulin, 109
  with oxyhemoglobin, 115, 116
  proton relaxation rate and protein modification, 284
Shape
  osmotic pressure and, 118–120
  small-angle X-ray scattering curve and, 145, 148–158
Side chains, Cotton effects due to, in proteins, 414–421
Silk fibroin
  Moffitt parameter for, 391
  $\beta$-structure and Cotton effects, 399, 400

Solvation
  effect on circular dichroism, 467
  internal, of proteins, 167
Solvents
  effect on circular dichroism, 466–469
  on optical rotatory dispersion, 378, 379, 385–387
Specific gravity, density and, 34
Specific inner surface, definition, 165
Specific viscosity, see Viscosity
Specific volume, see Volume
Spectropolarimeters, 430–434
Spheres, see also Shape
  radial density distribution in, 158
  small-angle X-ray scattering curves for hollow, 149, 154
Squash mosaic virus, protein shell, X-ray scattering curves, 219, 220
Standard chemical potential, see Chemical potential
Staphylococcal nuclease, ultraviolet and circular dichroism spectra of, 541, 548, 578
Stereochemistry, rotational strength and, 458
Strepavidin, spectra and structure of, 540
Subtilisin BPN, optical activity and structure of, 541
Sucrose, absolute densities of solutions, 41
Symbols
  glossary, circular dichroism, 446, 447
  for density and volume change measurements, 1–75
  for least squares in data analysis, 301, 302
  for nuclear magnetic resonance spectroscopy, 246
  for optical rotatory dispersion, 358–360
  osmotic pressure, 77, 78
  for small-angle X-ray scattering, 141, 142
Synthetase, fatty acid, small-angle X-ray scattering by, 184, 229–234

T

Temperature
  effect on circular dichroism, 466–469
  on density, 60
  on osmotic pressure, 117

Tinoco's theory, of optical activity, 369, 476, 506
Titration, least squares in data analysis, 351
Tobacco mosaic virus protein, polymerization and volume changes, 27
$\alpha$-Tosylchymotrypsin, secondary structure of, 538
Transferrin, proton relaxation rate, 294
Tropomyosin, optical rotatory dispersion and helical content, 379, 385, 387
L-Tryptophan
  circular dichroism and absorption spectra, 546, 547
  derivatives, low-temperature near-ultraviolet spectra of, 468
Tryptophanyl residues, optical activities of, 547–550
Turnip yellow mosaic virus (TYMV)
  exclusion parameters and, 71
  interaction parameter for, 58, 59
  partial specific volumes of, 50
Tyrosine
  copolymer with L-glutamic acid, optical rotatory dispersion and helical content, 387
  with poly-L-lysine, optical rotatory curves and Cotton effects, 402
  derivatives, low-temperature near-ultraviolet spectra, 468
  $o$-, $m$- and $p$-, optical activities of, 528
L-Tyrosineamide, $N,O$-diacetyl-, circular dichroism spectrum, 544, 545
Tyrosyl residues, optical activity of, 541–547

U

Ultraviolet spectra
  conformations and, 553–555
  far, of tryptophan residues, 549, 550
  of tyrosyl residues, 544–547
  near, of tryptophan residues, 547, 548
  of tyrosine residues, 541–544

V

van't Hoff limiting law
  deviations from, 92, 94, 108, 118, 135
  osmotic pressure, 89, 102
Vibronic fine structure, of circular dichroism and absorption spectra, 469–471

Virial coefficient
  osmotic, 91
  second, and apparent specific volume, 104
    ion effects, 94
    osmotic pressures of oxyhemoglobin and deoxyhemoglobin mixtures, 115
    and random coil polymers, 107
    thermodynamic expression for, 97
  third, osmotic, 135
Viruses protein shells of spherical, and radial electron density distribution, 215–226
Viscosity, specific, rate of change and volume change, 32
Volume
  apparent and partial specific, differences, 11
  apparent isopotential specific, 14
  apparent specific, changes in, 27
  and osmotic pressure, 119
  of proteins, 10–12
  and second virial coefficient, 104
  change, applications, 25–32
    and chemical changes, 5, 25
    on conformational transitions, 26
    during crystallization, 30
    and density measurements, 1–75
    direct measurement, and dilatometers, 22–25
    kinetics of, 31
    on mixing, 28–30, 45
    during polymerization, 30, 31
    on protonation and deprotonation, 26
    in solid-phase studies, 30
  contraction on ion–water dipole interaction, 25
  excluded, effect on osmotic pressure, 118, 121
  isopotential specific, 12–16
  partial molal, definition, 4, 88
  partial specific, in binding studies, 49
    from buoyancy data, 45, 48
    definition and evaluation, 3, 4, 8–10
    derivation from density, 62–66
    determination of changes in, 49
    purity of proteins and, 47
    reported values for, 46
    and sedimentation data, 44, 48

  and tail end of X-ray scattering curve, 163–165

## W

Water
  density of pure, 6, 32
  formation from hydrogen and hydroxyl ions, volume change, 26
Weighing, accuracy of, by density method, 17
Wool, osmotic pressure and subunit evaluations of, 106

## X

X-ray diffraction, least squares in data analysis, 351
X-ray scattering
  small-angle, 141–243
    allosteric effects and conformation changes by, 234–238
    applications of, 188–238
    calculation of, 149
    cameras for, 170–177, 238
    and changes in conformation on denaturation, 209, 210
    collimation (smearing) effects, 179–182
    curves for ellipsoids, 145, 149–152
    curves, for lamellar particles, 163
    curves for rod-like particles, 159–163
    curves, volume and tail end of, 163–165
    determination of cross-section of elongated particles, 211
    experimental technique, 168–188
    fundamentals, 143–146
    of hollow bodies, 149, 153
    prospects, 238, 239
    shape and size effect on, 148
    special requirements for, 169
    subsidiary maxima of, 150–153
    symbols, 141, 142
    theory of particulate, 143–168
    and X-ray crystal study, 188

## Y

Yang plot, 375, *see also* Moffitt-Yang plot

# *Molecular Biology*

An International Series of Monographs and Textbooks

Editors

BERNARD HORECKER

*Department of Molecular Biology*
*Albert Einstein College of Medicine*
*Yeshiva University*
*Bronx, New York*

NATHAN O. KAPLAN

*Department of Chemistry*
*University of California*
*At San Diego*
*La Jolla, California*

JULIUS MARMUR

*Department of Biochemistry*
*Albert Einstein College of Medicine*
*Yeshiva University*
*Bronx, New York*

HAROLD A. SCHERAGA

*Department of Chemistry*
*Cornell University*
*Ithaca, New York*

HAROLD A. SCHERAGA. Protein Structure. 1961

STUART A. RICE AND MITSURU NAGASAWA. Polyelectrolyte Solutions: A Theoretical Introduction, *with a contribution by Herbert Morawetz*. 1961

SIDNEY UDENFRIEND. Fluorescence Assay in Biology and Medicine. Volume I—1962. Volume II—1969

J. HERBERT TAYLOR (Editor). Molecular Genetics. Part I—1963. Part II—1967

ARTHUR VEIS. The Macromolecular Chemistry of Gelatin. 1964

M. JOLY. A Physico-chemical Approach to the Denaturation of Proteins. 1965

SYDNEY J. LEACH (Editor). Physical Principles and Techniques of Protein Chemistry. Part A—1969. Part B—1970. Part C—1973

KENDRIC C. SMITH AND PHILIP C. HANAWALT. Molecular Photobiology: Inactivation and Recovery. 1969

RONALD BENTLEY. Molecular Asymmetry in Biology. Volume I—1969. Volume II—1970

JACINTO STEINHARDT AND JACQUELINE A. REYNOLDS. Multiple Equilibria in Protein. 1969

DOUGLAS POLAND AND HAROLD A. SCHERAGA. Theory of Helix-Coil Transitions in Biopolymers. 1970

JOHN R. CANN. Interacting Macromolecules: The Theory and Practice of Their Electrophoresis, Ultracentrifugation, and Chromatography. 1970

# Molecular Biology

*An International Series of Monographs and Textbooks*

WALTER W. WAINIO. The Mammalian Mitochondrial Respiratory Chain. 1970

LAWRENCE I. ROTHFIELD (Editor). Structure and Function of Biological Membranes. 1971

ALAN G. WALTON AND JOHN BLACKWELL. Biopolymers. 1973

WALTER LOVENBERG (Editor). Iron-Sulfur Proteins. Volume I, Biological Properties — 1973. Volume II, Molecular Properties — 1973

A. J. HOPFINGER. Conformational Properties of Macromolecules. 1973

R. D. B. FRASER AND T. P. MACRAE. Conformation in Fibrous Proteins. 1973

In preparation

OSAMU HAYAISHI (Editor). Molecular Mechanisms of Oxygen Activation